Bowes and Church's

Food Values of Portions Commonly Used

Bowes and Church's

Food Values of Portions Commonly Used

15th edition   revised by

Jean A. T. Pennington, Ph.D., R.D.

HarperPerennial

A Division of HarperCollins*Publishers*

The author has attempted to provide readers with the most
accurate food composition data available as of the date of
manuscript submission. However, the field of food
composition is a dynamic one. Because of changes in product
formulation, changes in product size, development of new
products, discontinuation of products, sampling concerns,
and advances in analytical methodologies, the information in
these tables should be used as reasonable approximations.
Individuals who are on restricted diets for medical purposes
may need to contact food manufacturers for more specific
information.

Library of Congress Cataloging-in-Publication Data

Bowes, Anna dePlanter.
   [Food values of portions commonly used]
   Bowes and Church's food values of portions commonly
used. — 15th ed. / revised by Jean A.T. Pennington.
     p.   cm.
   Includes index.
   ISBN 0-06-055157-7 : $ ISBN 0-06-273156-4
(pbk.) : $
   1. Food—Composition—Tables.   2. Nutrition—Tables.
I. Church, Helen Nichols.   II. Pennington, Jean A.
Thompson.   III. Title.   IV. Title: Food values of portions
commonly used.
TX551.B64   1989
641.1—dc19   89-1715
                           CIP

93 94 95 96 97 CWI 10 9 8 7 6 5 4 3 2 1

    94 95 96 97 CWI 10 9 8 7   (pbk.)

Dedication

This edition of *Food Values of Portions Commonly Used* is dedicated to **Helen Nichols Church**, who was coauthor with her husband, Charles F. Church, M.D., on the ninth through twelfth editions (1963 to 1975) of this book.

Helen N. Church earned her bachelor's degree in home economics from the University of Illinois. She served as the clinical dietitian at Cornell University Medical College Clinic and the New York Hospital Clinic. She also directed field work in nutrition for Columbia University graduate students, served as dietitian-in-charge for the first Howard Johnson Restaurant (in Quincy, Massachusetts), worked with agricultural experimental stations in the northeastern United States, and provided dietetic counseling to numerous patients. Her research in nutrition has concerned methods to measure dietary intake, reliability of dietary data obtained by interviewers, and the nutritional and dietary status of industrial workers.

The contribution of Helen N. Church to the field of nutrition and dietetics, and particularly to *Food Values of Portions Commonly Used*, is gratefully acknowledged.

Preface

The tradition of *Bowes and Church's Food Values of Portions Commonly Used* now spans more than 50 years. The purpose of this edition remains the same as that of the first and subsequent editions—"to supply authoritative data on the nutritional values of foods in a form for quick and easy reference."

The nutrient values in this edition have been revised according to the most recent data available from the food industry, USDA food composition publications, and the scientific literature. There are two major changes from the previous edition:

- Values for dietary fiber, which were previously listed in a supplementary table, are now included in the main table. Values for crude fiber, which were previously listed in the main table, have been deleted.

- Following the listing of sources used to compile data for this edition (page 287), there is an extended listing of recent literature references that provides additional data on the composition of foods and some information about methods used to analyze foods for various substances.

I would like to express my gratitude to the many companies and trade associations who responded to my request for food composition data for this edition. I gratefully acknowledge the reviews of this book provided by Kathryn Fleming, Ph.D., Carolyn Miles, Ph.D., and Suzanne Murphy, Ph.D., and I thank Patricia Cleary and the staff at J. B. Lippincott for their editorial assistance and support.

Jean A. T. Pennington, Ph.D., R.D.

June 1987

Preface to the First Edition

The purpose of this book is to supply authoritative data on the nutritional values of foods in a form for quick and easy reference.

In teaching nutrition to students of medicine, dentistry, dental hygiene and public health nursing, food values based on common measures or portions frequently served have been found most useful. This basis of calculation is particularly well suited to the practical study of comparative food values, as well as to the approximate analysis of diets from records of daily food intake. For calculations of diets from weighed portions the actual weight of each food is given in grams or ounces.

Anna dePlanter Bowes
Charles F. Church

November 1937

Contents

Supplementary Tables

Explanatory Notes

In the main table, entitled "Nutrient Content of Foods," foods are grouped on the basis of food type with concerns for common usage. To conserve space, each food is listed only once, although some foods may be applicable to several sections. Please use the index if you cannot locate an item.

For most sections in the main table, each food has two lines of nutrient data. The heading at the top of the page indicates the nutrients and units of measurement for the numerical values listed in the two lines. For two sections—"Infant Formulas" and "Special Dietary Formulas, Commercial and Hospital"—there are four lines of nutrient data per food because of the complete nutrient profiles that were available for these formulated products.

The abbreviated heading names and other abbreviations and symbols used throughout the text are listed on pages xxi–xxii.

Each food is identified by name, description, brand name (if applicable), and serving portion. Foods are presented primarily in their table-ready form, although ingredient items (e.g., flour, baking soda, herbs) are also listed. Where available, brand names are used to help identify products such as ready-to-eat breakfast cereals, candy bars, and entrees.

The serving portion for most foods is listed in both household units and grams. Serving portions are those suggested by the food manufacturer or USDA food composition tables, or are practical sizes adopted by the author.

Supplementary tables are used to present information on nutrients or substances where there was not enough data to warrant a listing in the main table. Information provided in the supplementary tables is also listed by food groups. Most of the foods in the supplementary tables are also listed in the main table and the serving portions are usually consistent.

The data in the tables presented in this book have been taken directly from the sources indicated on pages 288–297. Values from various sources have not been averaged, and values for multi-ingredient foods have not been calculated from ingredients.

Precautionary Notes

For those unfamiliar with the use of food composition tables, please note the following:

- These tables should be used as approximate guides to the nutrient content of foods. Persons on special diets for various disease conditions may require more specific nutrient composition data from food manufacturers.

- Blank spaces denote lack of information. Do not assume that missing values are zeros.

- The nutrient values presented here are mean values; some of these values may have large standard deviations and wide ranges. Causes for nutrient variations in foods include soil type, season, geography, genetics, diet, processing, method of preparation, sampling scheme, and method of analysis. Because of nutrient variation and the fact that the data are collected from various sources (the food industry, USDA publications, and the scientific literature), apparent inconsistencies may occur. For example, portion sizes and gram weights of similar foods may vary among sources, and nutrient values may vary according to analytical methodology.

- The values presented here may not be representative of the entire food supply. Representativeness depends upon the sampling scheme and the number of samples collected and analyzed.

- The mineral content of water varies from one location to another. The mineral content of beverages made by addition of water to powders or frozen concentrates and of foods cooked in water (for example, rice, oatmeal, pasta, vegetables) may vary depending on the mineral content of the water used. Likewise, the mineral content of commercial beverages (for example, beer, carbonated sodas, juice drinks) depends on the mineral content of the water in the area where the beverages are bottled. Individuals on medically restricted diets may need to obtain information on the mineral content of their home tap water and on the mineral content of the specific beverages they consume.

- Information presented in these tables may not be the same as that provided on food labels. There are several reasons for such differences.
  - Portions sizes may differ. The household portions and gram weights presented here may differ from those used on the label.
  - Nutrient values on food labels are rounded and adjusted to be in compliance with government labeling regulations. Values presented here are generally mean values that have not been adjusted.
  - The product may have been reformulated with different ingredients or different proportions of ingredients.

Food and Nutrition Board, National Academy of Sciences—National Research Council

Recommended Daily Dietary Allowances,[a] Revised 1980

Designed for the Maintenance of Good Nutrition of Practically All Healthy People in the U.S.A.

	Age (years)	Weight (kg)	(lb)	Height (cm)	(in)	Protein (g)	Fat-Soluble Vitamins			Water-Soluble Vitamins							Minerals					
							Vitamin A (μg RE)[b]	Vitamin D (μg)[c]	Vitamin E (mg α TE)[d]	Vitamin C (mg)	Thiamin (mg)	Riboflavin (mg)	Niacin (mg NE)[e]	Vitamin B_6 (mg)	Folacin[f] (μg)	Vitamin B_{12} (μg)	Calcium (mg)	Phosphorus (mg)	Magnesium (mg)	Iron (mg)	Zinc (mg)	Iodine (μg)
Infants	0.0–0.5	6	13	60	24	kg × 2.2	420	10	3	35	0.3	0.4	6	0.3	30	0.5[g]	360	240	50	10	3	40
	0.5–1.0	9	20	71	28	kg × 2.0	400	10	4	35	0.5	0.6	8	0.6	45	1.5	540	360	70	15	5	50
Children	1–3	13	29	90	35	23	400	10	5	45	0.7	0.8	9	0.9	100	2.0	800	800	150	15	10	70
	4–6	20	44	112	44	30	500	10	6	45	0.9	1.0	11	1.3	200	2.5	800	800	200	10	10	90
	7–10	28	62	132	52	34	700	10	7	45	1.2	1.4	16	1.6	300	3.0	800	800	250	10	10	120
Males	11–14	45	99	157	62	45	1000	10	8	50	1.4	1.6	18	1.8	400	3.0	1200	1200	350	18	15	150
	15–18	66	145	176	69	56	1000	10	10	60	1.4	1.7	18	2.0	400	3.0	1200	1200	400	18	15	150
	19–22	70	154	177	70	56	1000	7.5	10	60	1.5	1.7	19	2.2	400	3.0	800	800	350	10	15	150
	23–50	70	154	178	70	56	1000	5	10	60	1.4	1.6	18	2.2	400	3.0	800	800	350	10	15	150
	51+	70	154	178	70	56	1000	5	10	60	1.2	1.4	16	2.2	400	3.0	800	800	350	10	15	150
Females	11–14	46	101	157	62	46	800	10	8	50	1.1	1.3	15	1.8	400	3.0	1200	1200	300	18	15	150
	15–18	55	120	163	64	46	800	10	8	60	1.1	1.3	14	2.0	400	3.0	1200	1200	300	18	15	150
	19–22	55	120	163	64	44	800	7.5	8	60	1.1	1.3	14	2.0	400	3.0	800	800	300	18	15	150
	23–50	55	120	163	64	44	800	5	8	60	1.0	1.2	13	2.0	400	3.0	800	800	300	18	15	150
	51+	55	120	163	64	44	800	5	8	60	1.0	1.2	13	2.0	400	3.0	800	800	300	10	15	150
Pregnant						+30	+200	+5	+2	+20	+0.4	+0.3	+2	+0.6	+400	+1.0	+400	+400	+150	[h]	+5	+25
Lactating						+20	+400	+5	+3	+40	+0.5	+0.5	+5	+0.5	+100	+1.0	+400	+400	+150	[h]	+10	+50

[a] The allowances are intended to provide for individual variations among most normal persons as they live in the United States under usual environmental stresses. Diets should be based on a variety of common foods in order to provide other nutrients for which human requirements have been less well defined.

[b] Retinol equivalents. 1 Retinol equivalent = 1 μg retinol or 6 μg β carotene.

[c] As cholecalciferol. 10 μg cholecalciferol = 400 IU vitamin D.

[d] α-tocopherol equivalents. 1 mg d-α-tocopherol = 1 α TE

[e] 1 NE (niacin equivalent) is equal to 1 mg of niacin or 60 mg of dietary tryptophan.

[f] The folacin allowances refer to dietary sources as determined by *Lactobacillus casei* assay after treatment with enzymes (conjugases) to make polyglutamyl forms of the vitamin available to the test organism.

[g] The RDA for Vitamin B_{12} in infants is based on average concentration of the vitamin in human milk. The allowances after weaning are based on energy intake (as recommended by the American Academy of Pediatrics) and consideration of other factors such as intestinal absorption.

[h] The increased requirement during pregnancy cannot be met by the iron content of habitual American diets nor by the existing iron stores of many women; therefore the use of 30–60 mg of supplemental iron is recommended. Iron needs during lactation are not substantially different from those of nonpregnant women, but continued supplementation of the mother for 2 to 3 months after parturition is advisable in order to replenish stores depleted by pregnancy.

Reproduced from National Academy of Sciences: Recommended Dietary Allowances, 9th rev. ed., Washington, D.C., 1980.

Food and Nutrition Board, National Academy of Sciences—National Research Council

Estimated Safe and Adequate Daily Dietary Intakes of Selected Vitamins and Minerals[a]

	Age (years)	Vitamins			Trace Elements[b]							Electrolytes		
		Vitamin K (µg)	Biotin (µg)	Pantothenic Acid (mg)	Copper (mg)	Manganese (mg)	Fluoride (mg)	Chromium (mg)	Selenium (mg)	Molybdenum (mg)		Sodium (mg)	Potassium (mg)	Chloride (mg)
Infants	0–0.5	12	35	2	0.5–0.7	0.5–0.7	0.1–0.5	0.01–0.04	0.01–0.04	0.03–0.06		115–350	350–925	275–700
	0.5–1	10–20	50	3	0.7–1.0	0.7–1.0	0.2–1.0	0.02–0.06	0.02–0.06	0.04–0.08		250–750	425–1275	400–1200
Children	1–3	15–30	65	3	1.0–1.5	1.0–1.5	0.5–1.5	0.02–0.08	0.02–0.08	0.05–0.1		325–975	550–1650	500–1500
and	4–6	20–40	85	3–4	1.5–2.0	1.5–2.0	1.0–2.5	0.03–0.12	0.03–0.12	0.06–0.15		450–1350	775–2325	700–2100
Adolescents	7–10	30–60	120	4–5	2.0–2.5	2.0–3.0	1.5–2.5	0.05–0.2	0.05–0.2	0.10–0.3		600–1800	1000–3000	925–2775
	11+	50–100	100–200	4–7	2.0–3.0	2.5–5.0	1.5–2.5	0.05–0.2	0.05–0.2	0.15–0.5		900–2700	1525–4575	1400–4200
Adults		70–140	100–200	4–7	2.0–3.0	2.5–5.0	1.5–4.0	0.05–0.2	0.05–0.2	0.15–0.5		1100–3300	1875–5625	1700–5100

[a] Because there is less information on which to base allowances, these figures are not given in the main table of the RDA and are provided here in the form of ranges of recommended intakes.

[b] Since the toxic levels for many trace elements may be only several times usual intakes, the upper levels for the trace elements given in this table should not be habitually exceeded.

Reproduced from National Academy of Sciences: Recommended Dietary Allowances, 9th rev. ed., Washington, D.C., 1980.

Mean Heights and Weights and Recommended Energy Intake[a]

	Age (years)	Weight		Height		Energy Needs (with range)	
		(kg)	(lb)	(cm)	(in)	(kcal)	(MJ)
Infants	0.0–0.5	6	13	60	24	kg × 115 (95–145)	kg × .48
	0.5–1.0	9	20	71	28	kg × 105 (80–135)	kg × .44
Children	1–3	13	29	90	35	1300 (900–1800)	5.5
	4–6	20	44	112	44	1700 (1300–2300)	7.1
	7–10	28	62	132	52	2400 (1650–3300)	10.1
Males	11–14	45	99	157	62	2700 (2000–3700)	11.3
	15–18	66	145	176	69	2800 (2100–3900)	11.8
	19–22	70	154	177	70	2900 (2500–3300)	12.2
	23–50	70	154	178	70	2700 (2300–3100)	11.3
	51–75	70	154	178	70	2400 (2000–2800)	10.1
	76+	70	154	178	70	2050 (1650–2450)	8.6
Females	11–14	46	101	157	62	2200 (1500–3000)	9.2
	15–18	55	120	163	64	2100 (1200–3000)	8.8
	19–22	55	120	163	64	2100 (1700–2500)	8.8
	23–50	55	120	163	64	2000 (1600–2400)	8.4
	51–75	55	120	163	64	1800 (1400–2200)	7.6
	76+	55	120	163	64	1600 (1200–2000)	6.7
Pregnant						+300	
Lactating						+500	

[a] The data in this table have been assembled from the observed median heights and weights of children together with desirable weights for adults for the mean heights of men (70 inches) and women (64 inches) between the ages of 18 and 34 years as surveyed in the U. S. population (HEW/NCHS data).

The energy allowances for the young adults are for men and women doing light work. The allowances for the two older groups represent mean energy needs over these age spans, allowing for a 2% decrease in basal (resting) metabolic rate per decade and a reduction in activity of 200 kcal/day for men and women between 51 and 75 years, 500 kcal for men over 75 years and 400 kcal for women over 75 years. The customary range of daily energy output is shown in parentheses for adults, and is based on a variation in energy needs of ±400 kcal at any one age, emphasizing the wide range of energy intakes appropriate for any group of people.

Energy allowances for children through age 18 are based on median energy intakes of children of these ages followed in longitudinal growth studies. The values in parentheses are 10th and 90th percentiles of energy intake, to indicate the range of energy consumption among children of these ages.

Reproduced from National Academy of Sciences: Recommended Dietary Allowances, 9th rev. ed., Washington, D.C., 1980.

United States Recommended Daily Allowances (U.S. RDA)[a]

	Unit	Infants (birth–12 mo.)	Children under 4 yrs.	Adults and children 4 or more yrs.	Pregnant or lactating women
Protein[b]	g	25	28	65	—
Protein[c]	g	18	20	45	—
Vitamin A	IU	1500	2500	5000	8000
Vitamin D	IU	400	400	400	400
Vitamin E	IU	5	10	30	30
Vitamin C	mg	35	40	60	60
Folic Acid	mg	0.1	0.2	0.4	0.8
Thiamin (B_1)	mg	0.5	0.7	1.5	1.7
Riboflavin (B_2)	mg	0.6	0.8	1.7	2.0
Niacin	mg	8	9	20	20
Vitamin B_6	mg	0.4	0.7	2.0	2.5
Vitamin B_{12}	mcg	2	3	6	8
Biotin	mg	0.05	0.15	0.3	0.3
Pantothenic Acid	mg	3	5	10	10
Calcium	g	0.6	0.8	1.0	1.3
Phosphorus	g	0.5	0.8	1.0	1.3
Iodine	mcg	45	70	150	150
Iron	mg	15	10	18	18
Magnesium	mg	70	200	400	450
Copper	mg	0.6	1.0	2.0	2.0
Zinc	mg	5	8	15	15

[a] The U.S. RDAs are nutrient standards set by the Food and Drug Administration in 1973 using the Recommended Dietary Allowances of the National Academy of Sciences, National Research Council. The U.S. RDAs are established for four age–sex groups. Generally, the highest values in the RDA table were selected for use within each U.S. RDA category. The nutritional information on food labels is expressed as percent of the U.S. RDA.

[b] Protein efficiency ratio less than casein.
[c] Protein efficiency ratio greater than or equal to casein.

Reproduced from the FDA consumer memo, "Nutrition Labels and U.S. RDA." 81-2146, 1981.

Abbreviations and Symbols

am	American		iso	isoleucine
amt	amount		IU	international unit
ap	as purchased		jce	juice
arg	arginine		jr	junior food
avg	average		K	potassium
bbq	barbeque		kcal	calorie(s)
bio	biotin		lb	pound
blank space	insufficient information for nutrient values		leu	leucine
			lys	lysine
Ca	calcium		marb	marbled
caff	caffeine		marg	margarine
cal	calorie(s)		mcg	microgram(s)
calif	California		mct	medium-chain triglycerides
chln	choline		med	medium
cho	carbohydrate		met	methionine
choc	chocolate		mg	milligram(s)
chol	cholesterol		Mg	magnesium
cinn	cinnamon		Mn	manganese
ckd	cooked		micro ckd	microwave cooked
Cl	chloride		Mo	molybdenum
cnd*	canned		Na	sodium
combo	combination		nfdm	nonfat dry milk solids
conc	concentrate		nia	niacin
cond	condensed		orig	original
Cr	chromium		oz	ounce
crm	cream		P	phosphorus
Cu	copper		pant	pantothenic acid
cys	cystine		phe	phenylalanine
dfibr	dietary fiber		pkg	package
enr	enriched		pkt	packet
Fe	iron		prep	prepared
fl oz	fluid ounce		pro	protein
fol	folate		pufa	polyunsaturated fatty acids
frzn*	frozen		rd	round
g	gram(s)		RE	retinol equivalents
his	histidine		recon	reconstituted
H$_2$O	water		reg	regular
hp	heaping		rib	riboflavin
I	iodine		rte	ready-to-eat
imit	imitation		rts	ready-to-serve
inos	myo-inositol		sce	sauce
inst	instant		Se	selenium
			sec	second

* *Canned (cnd)* and *frozen (frzn)* refer to commercially canned and frozen foods.

(*continued*)

sep	separable		vit B-1	vitamin B-1 (thiamine)
sfa	saturated fatty acids		vit B-2	vitamin B-2 (riboflavin)
std	standard		vit B-6	vitamin B-6 (pyridoxine)
str	strained baby food		vit B-12	vitamin B-12 (cobalamin)
sub	substitute		vit C	vitamin C (ascorbic acid)
t	teaspoon		vit D	vitamin D
T	tablespoon		vit E	vitamin E (tocopherol)
thi	thiamin		vit K	vitamin K
thr	threonine		vol	volume
tr	trace		whpd	whipped
try	tryptophan		wt	weight
tyr	tyrosine		w/	with
unenr	unenriched		w/o	without
val	valine		Zn	zinc
van	vanilla		&	and
veg	vegetable		0	zero/none
vit A	vitamin A		/	per/or

Conversion Tables

Volume Measures

$$1 \text{ t} = \frac{1}{3}\text{ T} = \frac{1}{6}\text{ fl oz} = 4.9 \text{ ml}$$
$$3 \text{ t} = 1\text{ T} = \frac{1}{2}\text{ fl oz} = 14.8 \text{ ml}$$
$$2 \text{ T} = \frac{1}{8}\text{ cup} = 1 \text{ fl oz} = 29.6 \text{ ml}$$
$$4 \text{ T} = \frac{1}{4}\text{ cup} = 2 \text{ fl oz} = 59.1 \text{ ml}$$
$$5\frac{1}{3} \text{ T} = \frac{1}{3}\text{ cup} = 2\frac{2}{3}\text{ fl oz} = 78.9 \text{ ml}$$
$$8 \text{ T} = \frac{1}{2}\text{ cup} = 4 \text{ fl oz} = 118.3 \text{ ml}$$
$$10\frac{2}{3} \text{ T} = \frac{2}{3}\text{ cup} = 5\frac{1}{3}\text{ fl oz} = 157.7 \text{ ml}$$
$$12 \text{ T} = \frac{3}{4}\text{ cup} = 6 \text{ fl oz} = 177.4 \text{ ml}$$
$$14 \text{ T} = \frac{7}{8}\text{ cup} = 7 \text{ fl oz} = 207.0 \text{ ml}$$
$$16 \text{ T} = 1\text{ cup} = 8 \text{ fl oz} = 236.6 \text{ ml}$$

$$1 \text{ ml} = .034 \text{ fl oz} = 1 \text{ cc} = .001 \text{ liter}$$
$$1 \text{ liter} = 34 \text{ fl oz} = 1000 \text{ ml}$$

$$1 \text{ pint} = 2 \text{ cups} = .473 \text{ liter} = 473 \text{ ml}$$
$$1 \text{ quart} = 2 \text{ pt} = .946 \text{ liter} = 946 \text{ ml}$$
$$1 \text{ gallon} = 4 \text{ quarts} = 3.785 \text{ liter} = 3785 \text{ ml}$$
$$1 \text{ liter} = 1.057 \text{ quarts} = 0.264 \text{ gallon} = 1000 \text{ ml}$$

Weight Measures

$$1 \text{ g} = .035 \text{ oz} = .001 \text{ kg} = 1000 \text{ mg} = 1,000,000 \text{ mcg}$$
$$1 \text{ mg} = .001 \text{ g} = 1000 \text{ mcg}$$
$$1 \text{ oz} = 28.35 \text{ g} \text{ (often rounded to 28 g)}$$
$$1 \text{ lb} = 16 \text{ oz} = 453.59 \text{ g} = .454 \text{ kg}$$
$$1 \text{ kg} = 2.21 \text{ lb} = 1000 \text{ g}$$
$$.1 \text{ kg} = 100 \text{ g} = 3.52 \text{ oz}$$

Heat Measures

$$1 \text{ kilojoule} = 0.239 \text{ kilocalories}$$
$$1 \text{ kilocalorie} = 4.184 \text{ kilojoules}$$

The relationship between volume measures and weight measures is variable depending on the food. For example, note the weights of 1 cup of the following foods:

Food	Weight of 1 level cup
almonds, chopped	127 g
cottage cheese	233 g
orange juice	244 g
peas, green, canned	172 g
pickle relish	243 g
chicken noodle soup	230 g
milk, whole	244 g
mayonnaise	221 g
shredded wheat	35 g
puffed wheat	12 g
peanut butter	251 g

The volume weight of water is a commonly used reference point for other food measures.

$$1 \text{ T water} = 15 \text{ g} = 15 \text{ cc}$$
$$1 \text{ cup water} = 237 \text{ g}$$
$$1 \text{ fl oz water} = 29.54 \text{ g} \text{ (often rounded to 30 g)}$$
$$1 \text{ cc water} = 1 \text{ g} = 1 \text{ ml}$$
$$1 \text{ liter water} = 1 \text{ kg} = 1000 \text{ g}$$
$$1 \text{ quart water} = 946 \text{ g} = .946 \text{ kg}$$

Nutrient Content of Foods

	KCAL	H₂O (g)	FAT (g)	PUFA (g)	CHOL (mg)	A (RE)	C (mg)	B-2 (mg)	B-6 (mg)	FOL (mcg)	Na (mg)	Ca (mg)	Mg (mg)	Zn (mg)	Mn (mg)
	WT (g)	PRO (g)	CHO (g)	SFA (g)	DFIB (g)	A (IU)	B-1 (mg)	NIA (mg)	B-12 (mcg)	PANT (mg)	K (mg)	P (mg)	Fe (mg)	Cu (mg)	

1. BEVERAGES
1.1. ALCOHOLIC BEVERAGES
1.1.1. ALES, BEERS & MALT LIQUORS

	KCAL/WT	H₂O/PRO	FAT/CHO	PUFA/SFA	CHOL/DFIB	A(RE)/A(IU)	C/B-1	B-2/NIA	B-6/B-12	FOL/PANT	Na/K	Ca/P	Mg/Fe	Zn/Cu	Mn
ale, Blatz Cream	155		0.0	0.0	0						14	17			
12 fl oz	360	1.1	11.1	0.0							77		.02		
beer	146	328.8	0.0	0.0	0	0	0	.09	.18	21	19	18	23	.06	.043
12 fl oz	356	0.9	13.2	0.0		0	.02	1.6	.06	.21	89	44	.11	.032	
Anheuser-Busch	153	327.0	0.0	0.0	0		0	.14	.16		9	16	22	.01	.040
12 fl oz	355	1.5	13.2	0.0	0.0		.01	1.4		.41	142	57	.01	.014	
Black Label	136		0.0	0.0	0						17	15			
12 fl oz	360	1.2	11.2	0.0							81		.02		
Blatz	142		0.0	0.0	0						14	16			
12 fl oz	360	1.1	11.8	0.0							82		.02		
Coors Premium	141							.09			15	12			
12 fl oz	360	1.0	11.6				.04	1.4			83				
Heileman's Old Style	146		0.0	0.0	0						10	17			
12 fl oz	360	1.1	11.5	0.0							81		.02		
Heileman's Special Export	155		0.0	0.0	0						14	17			
12 fl oz	360	1.1	11.1	0.0							77		.02		
Heileman's Special Export Dark	155		0.0	0.0	0						14	17			
12 fl oz	360	1.1	11.1	0.0							77		.02		
Herman Joseph's	157										14	9			
12 fl oz	360	1.3	13.5								124				
Killian's	172										15	10			
12 fl oz	360	1.4	14.9								116				
Rainier	142		0.0	0.0	0						13	23			
12 fl oz	360	1.2	11.6	0.0							61		.02		
Schmidt	142		0.0	0.0	0						15	29			
12 fl oz	360	1.1	11.8	0.0							61		.02		
Stroh's American Lager	145	323.0	0.0	0.0	0	0	0	.07	.18		25	22	25	.03	
12 fl oz	355	9.9	13.4	0.0	0.0	0	.01	1.8		.17	64	48	.00	.000	
beer, light	100	337.0	0.0	0.0	0	0	0	.11	.12	15	10	18	17	.11	.057
12 fl oz	354	0.7	4.8	0.0		0	.03	1.4	.02	.13	64	43	.12	.085	
Anheuser-Busch	117	334.0	0.0	0.0	0		0	.10	.16		9	15	23	.01	.040
12 fl oz	355	1.1	9.2	0.0	0.0		.00	0.1		.29	128	46	.01	.014	
Blatz	96		0.0	0.0	0						15	17			
12 fl oz	360	0.9	3.2	0.0							61		.02		
Coors	103							.09			18	13			
12 fl oz	360	0.7	4.9				.04	0.1			67				
Heileman's Old Style	99		0.0	0.0	0						7	17			
12 fl oz	360	0.9	6.4	0.0							54		.02		
Heileman's Special Export	115		0.0	0.0	0						7	17			
12 fl oz	360	1.0	7.5	0.0							68		.02		
beer, LA (low alcohol)															
Anheuser-Busch	111	330.0	0.0	0.0	0		0	.09			9	18	18	.00	.030
12 fl oz	355	0.7	15.6	0.0	0.0		.04	0.2	.00	.00	174	39	.01	.014	
Blatz	73		0.0	0.0	0						8	16			
12 fl oz	360	0.9	7.0	0.0							45		.02		
Heileman's Old Style	73		0.0	0.0	0						8	16			
12 fl oz	360	0.7	7.0	0.0							45		.02		
malt liquor															
Blatz Old Fashioned Private Stock	149		0.0	0.0	0						14	17			
12 fl oz	360	1.1	11.5	0.0							80		.02		
Colt 45	156		0.0	0.0	0						14	17			
12 fl oz	360	1.2	11.0	0.0							93		.02		

1.1.2. COCKTAILS & COCKTAIL MIXES

	KCAL / WT (g)	H_2O (g) / PRO (g)	FAT (g) / CHO (g)	PUFA (g) / SFA (g)	CHOL (mg) / DFIB (g)	Vitamins: A (RE) / A (IU)	C (mg) / B-1 (mg)	B-2 (mg) / NIA (mg)	B-6 (mg) / B-12 (mcg)	FOL (mcg) / PANT (mg)	Minerals: Na (mg) / K (mg)	Ca (mg) / P (mg)	Mg (mg) / Fe (mg)	Zn (mg) / Cu (mg)	Mn (mg)
bloody mary (tomato jce, vodka & lemon	116	127.3	0.1	0.0	0	51	20	.03	.11	20	332	10	11	.14	.073
jce)—*5 fl oz cocktail*	148	0.8	4.8	0.0		508	.05	0.6	.00	.24	216	21	.55	.102	
bourbon & soda	105	100.8	0.0	0.0	0	0	0	.00	.00	0	16	4	1	.09	
4 fl oz cocktail	116	0.0	0.0	0.0		0	.00	0.0	.00	.00	2	2			
daiquiri															
cnd		154.4									83	1			
6.8 fl oz can	207										22		.02		
rum, lime jce & sugar	111	41.9	0.0	0.0	0	0	1	.00	.00	1	3	2	1	.04	
2 fl oz cocktail	60	0.0	4.1	0.0		2	.01	0.0	.00	.01	13	4	.09	.026	
gin & tonic (tonic water, gin & lime jce)	171	193.0	0.0	0.0	0	0	1	.00	.01	1	10	4	2		
7.5 fl oz cocktail	225	0.0	15.8	0.0		2	.01	0.0	.00	.01	12	2			
manhattan (whiskey & vermouth)	128	37.7	0.0	0.0	0		0	.00	.00	0	2	1	1	.03	.024
2 fl oz cocktail	57	0.0	1.8	0.0			.01	0.1	0	.01	15	4	.05	.016	
martini (gin & vermouth)	156	47.3	0.0	0.0	0		0	.00	.00	0	2	1	1	.01	
2.5 fl oz cocktail	70	0.0	0.2	0.0			.00	0.0	.00	.00	13	2	.06	.004	
pina colada															
cnd	525	121.8	16.9	0.3							158	1			
6.8 fl oz can	222	1.3	61.3	14.6							184		.07		
pineapple jce, rum, sugar & coconut	262	91.7	2.6	0.5	0	0	7	.02		14	9	11		.19	
cream—*4.5 fl oz cocktail*	141	0.6	39.9	1.2		3	.04	0.2	.00		100	10	.31	.116	
screwdriver (orange jce & vodka)	174	178.6	0.1	0.0	0	13	67	.03	.08	75	2	16	17	.09	.023
7 fl oz cocktail	213	1.2	18.4	0.0		133	.14	0.3	.00	.27	325	29	.17	.079	
tequila sunrise															
cnd		166.2									119	1			
6.8 fl oz can	211										22		.04		
orange jce, tequila, lime jce &	189	137.2	0.2	0.0	0	17	33	.03	.09		7	10	12	.11	
grenadine—*5.5 fl oz cocktail*	172	0.6	14.7	0.0		166	.07	0.3	.00		178	17	.47	.072	
tom collins (club soda, gin, lemon jce &	121	202.9	0.0	0.0	0	0	4	.00	.01	2	39	10	3	.17	
sugar)—*7.5 fl oz cocktail*	222	0.1	3.0	0.0		2	.01	0.0	.00	.01	18	1			
whiskey sour															
cnd		160.7									91	1			
6.8 fl oz cocktail	209										22		.02		
lemon jce, whiskey & sugar	123	69.3	0.1	0.0	0	1	11	.01	.02	5	10	5	4	.05	.015
3 fl oz cocktail	90	0.2	5.0	0.0		7	.19	0.1	.00	.04	48	6	.07	.027	
prep from bottled mix	158	76.9	0.0	0.0	0	1	2	.01	.00	0	66	1	1	.06	.006
2 fl oz mix & 1.5 fl oz whiskey	106	0.0	13.8	0.0	0.1	14	.01	0.0	.00	.01	19	6	.08	.010	
prep from powdered mix—*17 g pkt w/*	169	71.2	0.0	0.0	0	1	1	.00	.00	0	48	47	4	.05	.006
1.5 fl oz water & 1.5 fl oz whiskey	103	0.1	16.4	0.0	0.1	5	.00	0.0	.00	.01	4	5	.08	.034	
whiskey sour mix															
bottled[a]	55	50.5	0.0	0.0	0	1	2	.01	.00	0	66	1	1	.05	.000
2 fl oz cocktail	65	0.0	13.8	0.0	0.1	14	.01	0.0	.00	.01	18	4	.07	.000	
powder[a]	64	0.1	0.0	0.0	0	1	1	.00	.00	0	46	45	3	.02	.000
1 pkt	17	0.1	16.2	0.0	0.1	5	.00	0.0		.01	3	2	.07	.022	

1.1.3. DISTILLED SPIRITS

	KCAL / WT (g)	H_2O (g) / PRO (g)	FAT (g) / CHO (g)	PUFA (g) / SFA (g)	CHOL (mg) / DFIB (g)	Vitamins: A (RE) / A (IU)	C (mg) / B-1 (mg)	B-2 (mg) / NIA (mg)	B-6 (mg) / B-12 (mcg)	FOL (mcg) / PANT (mg)	Minerals: Na (mg) / K (mg)	Ca (mg) / P (mg)	Mg (mg) / Fe (mg)	Zn (mg) / Cu (mg)	Mn (mg)
gin, 90 proof	110	26.1	0.0	0.0	0	0	0	.00	.00	0	1	0	0	.00	
1.5 fl oz jigger	42	0.0	0.0	0.0		0	.00	0.0	.00	.00	0	0	.00	.002	
gin/rum/vodka/whiskey, 94 proof	116	25.3	0.0	0.0	0	0	0	.00	.00	0	0	0	0	.02	.008
1.5 fl oz jigger	42	0.0	0.0	0.0		0	.00	0.0	.00	.00	1	2	.02	.009	
gin/rum/vodka/whiskey, 100 proof	124	24.2	0.0	0.0	0	0	0	.00	.00	0	0	0	0	.02	.008
1.5 fl oz jigger	42	0.0	0.0	0.0		0	.00	0.0	.00	.00	1	2	.02	.009	

[a] Contains no alcohol.

	KCAL / WT (g)	H₂O (g) / PRO (g)	FAT (g) / CHO (g)	PUFA (g) / SFA (g)	CHOL (mg) / DFIB (g)	A (RE) / A (IU)	C (mg) / B-1 (mg)	B-2 (mg) / NIA (mg)	B-6 (mg) / B-12 (mcg)	FOL (mcg) / PANT (mg)	Na (mg) / K (mg)	Ca (mg) / P (mg)	Mg (mg) / Fe (mg)	Zn (mg) / Cu (mg)	Mn (mg)
rum, 80 proof	97	28.0	0.0	0.0	0	0	0	.00	.00	0	0	0	0	.03	
1.5 fl oz jigger	42	0.0	0.0	0.0		0	.00	0.0	.00	.00	1	2	.05	.021	
vodka, 80 proof	97	28.0	0.0	0.0	0	0	0	.00	.00	0	0	0	0	.00	.000
1.5 fl oz jigger	42	0.0	0.0	0.0		0	.00	0.0	.00	.00	0	2	.00	.004	
whiskey, 86 proof	105	26.8	0.0	0.0	0	0	0	.00	.00	0	0	0	0	.02	.006
1.5 fl oz jigger	42	0.0	0.1	0.0		0	.00	0.0	.00	.00	1	2	.01	.009	

1.1.4. LIQUEURS

	KCAL / WT (g)	H₂O (g) / PRO (g)	FAT (g) / CHO (g)	PUFA (g) / SFA (g)	CHOL (mg) / DFIB (g)	A (RE) / A (IU)	C (mg) / B-1 (mg)	B-2 (mg) / NIA (mg)	B-6 (mg) / B-12 (mcg)	FOL (mcg) / PANT (mg)	Na (mg) / K (mg)	Ca (mg) / P (mg)	Mg (mg) / Fe (mg)	Zn (mg) / Cu (mg)	Mn (mg)
coffee, 53 proof	174	16.1	0.1	0.1	0	0	0	.01		0	4	1	1	.01	
1.5 fl oz	52	0.0	24.4	0.1		0	.00	0.1		.00	15	3	.03	.021	
coffee, 63 proof	160	21.5	0.1	0.1	0	0	0	.01		0	4	1	1	.01	
1.5 fl oz	52	0.0	16.7	0.1		0	.00	0.1		.00	15	3	.03	.021	
coffee w/ cream, 34 proof	154	21.9	7.4	0.3			0	.03		0	43	7	1	.08	
1.5 fl oz	47	1.3	9.8	4.5			.00	0.0		.04	15	23	.06	.019	
creme de menthe	186	14.2	0.1	0.1	0			.00	.00	0	3	0	0		.020
1.5 fl oz	50	0.0	20.8	0.0			.00	0.0	.00	.00	0	0	.04	.040	

1.1.5. WINES & WINE BEVERAGES

	KCAL / WT (g)	H₂O (g) / PRO (g)	FAT (g) / CHO (g)	PUFA (g) / SFA (g)	CHOL (mg) / DFIB (g)	A (RE) / A (IU)	C (mg) / B-1 (mg)	B-2 (mg) / NIA (mg)	B-6 (mg) / B-12 (mcg)	FOL (mcg) / PANT (mg)	Na (mg) / K (mg)	Ca (mg) / P (mg)	Mg (mg) / Fe (mg)	Zn (mg) / Cu (mg)	Mn (mg)
sparkling cooler, citrus, La Croix	215		0.0	0.0	0						10				
12 fl oz	360	0.5	30.0	0.0										.07	
sparkling cooler, strawberry, La Croix	215		0.0	0.0	0						10				
12 fl oz	360	0.5	30.0	0.0										.07	
wine, dessert, dry	74	42.7	0.0	0.0	0		0	.01	.00	0	5	5	5	.04	.070
2 fl oz	59	0.1	2.4	0.0			.01	0.1	.00	.02	54	6	.14	.027	
wine, dessert, sweet	90	42.7	0.0	0.0	0		0	.01	.00	0	5	5	5	.04	.070
2 fl oz	59	0.1	7.0	0.0			.01	0.1	.00	.02	54	6	.14	.027	
wine, table, all types	72	91.6	0.0	0.0	0	0	0	.02	.03	1	8	9	10	.07	.149
3.5 fl oz	103	0.2	1.4	0.0		0	.00	0.1	.01	.03	91	14	.42	.014	
wine, table, red	74	91.1	0.0	0.0	0	0	0	.03	.04	2	6	8	13	.10	.615
3.5 fl oz	103	0.2	1.8	0.0		0	.01	0.1	.01	.04	115	14	.44	.021	
wine, table, rose	73	91.6	0.0	0.0	0		0	.02	.03	1	5	9	10	.06	.108
3.5 fl oz	103	0.2	1.5	0.0			.00	0.1	.01	.03	102	15	.39	.054	
wine, table, white	70	92.3	0.0	0.0	0	0	0	.01	.01	0	5	9	11	.07	.473
3.5 fl oz	103	0.1	0.8	0.0		0	.00	0.1	.00	.02	82	14	.33	.022	

1.2. CARBONATED BEVERAGES

	KCAL / WT (g)	H₂O (g) / PRO (g)	FAT (g) / CHO (g)	PUFA (g) / SFA (g)	CHOL (mg) / DFIB (g)	A (RE) / A (IU)	C (mg) / B-1 (mg)	B-2 (mg) / NIA (mg)	B-6 (mg) / B-12 (mcg)	FOL (mcg) / PANT (mg)	Na (mg) / K (mg)	Ca (mg) / P (mg)	Mg (mg) / Fe (mg)	Zn (mg) / Cu (mg)	Mn (mg)
apple Slice	196	308.4	0.0	0.0							5	0			
12 fl oz	360	0.0	48.0	0.0	0.0						67	0	.00		
cherry coke, Coca-Cola	154		0.0								14				
12 fl oz	370	0.0	40.0								0	54			
cherry cola Slice	164	313.2	0.0	0.0							5	0			
12 fl oz	360	0.0	43.2	0.0	0.0						67	0	.00		
cherry RC	171		0.0								1				
12 fl oz	360	0.0	42.8								12	47			
Coca-Cola	155		0.0								6				
12 fl oz	370	0.0	40.0								0	54			
Coca-Cola Classic	144		0.0								14				
12 fl oz	369	0.0	38.0								0	60			
cola	151	330.8	0.1		0	0	0	.00	.00	0	14	9	3	.05	.130
12 fl oz	370	0.1	38.5			0	.00	0.0	.00	.00	4	46	.13	.041	
cola, RC	171		0.0								1				
12 fl oz	360	0.0	42.8								12	47			
cola, RC 100, caffeine-free	171		0.0								1				
12 fl oz	360	0.0	42.8								12	47			
cream soda	191	321.5	0.0	0.0	0	0	0	.00	.00	0	43	19	3	.24	
12 fl oz	371	0.0	49.3	0.0		0	.00	0.0	.00	.00	4	0	.19	.030	
Dr. Nehi	163		0.0								26				
12 fl oz	360	0.0	40.8								1	46			

	KCAL / WT (g)	H₂O (g) / PRO (g)	FAT (g) / CHO (g)	PUFA (g) / SFA (g)	CHOL (mg) / DFIB (g)	A (RE) / A (IU)	C (mg) / B-1 (mg)	B-2 (mg) / NIA (mg)	B-6 (mg) / B-12 (mcg)	FOL (mcg) / PANT (mg)	Na (mg) / K (mg)	Ca (mg) / P (mg)	Mg (mg) / Fe (mg)	Zn (mg) / Cu (mg)	Mn (mg)
fruit punch, Nehi	200		0.0								16				
12 fl oz	360	0.0	49.9								0	0			
ginger ale	124	333.9	0.0	0.0	0	0	0	.00	.00	0	25	12	3	.18	
12 fl oz	366	0.1	31.9	0.0		0	.00	0.0	.00	.00	5	1	.66	.066	
ginger ale, Fanta	126		0.0								28				
12 fl oz	367	0.0	32.0								0	0			
ginger ale, Nehi	152		0.0								1				
12 fl oz	360	0.0	37.9								0	0			
grape, Fanta	172		0.0								14				
12 fl oz	372	0.0	44.0								0	0			
grape, Nehi	192		0.0								15				
12 fl oz	360	0.0	47.9								0	0[a]			
grape soda	161	330.3	0.0	0.0	0	0	0	.00	.00	0	57	12	4	.26	
12 fl oz	372	0.0	41.7	0.0			.00		.00	.00	3	0	.31	.082	
Kick	195		0.0								49				
12 fl oz	360	0.0	48.7								0	0			
lemon-lime soda	149	329.4	0.0	0.0	0	0	0	.00	.00	0	41	9	2	.18	.048
12 fl oz	368	0.0	38.4	0.0		0	.00	0.1	.00	.00	4	1	.25	.044	
mandarin orange Slice	193	307.2	0.0	0.0							22	0			
12 fl oz	360	0.0	50.4	0.0	0.0						85	0	.00		
Mello Yello	174		0.0				0				24				
12 fl oz	372	0.0	44.0								8	0			
Mr. Pibb	142		0.0								22				
12 fl oz	369	0.0	38.0								0	42			
Mountain Dew	179	326.4	0.0	0.0							31				
12 fl oz	360	0.0	44.4	0.0	0.0						10	0			
orange, Fanta	176		0.0								14				
12 fl oz	372	0.0	46.0								0	0			
orange, Nehi	209		0.0								22				
12 fl oz	360	0.0	52.1								0	60			
orange soda	177	325.9	0.0	0.0	0	0	0	.00	.00	0	46	19	4	.38	
12 fl oz	372	0.0	45.8	0.0		0	.00		.00	.00	9	4	.23	.056	
peach, Nehi	203		0.0								33				
12 fl oz	360	0.0	50.8								0	0			
pepper type soda	151	328.9	0.4		0	0	0	.00	.00	0	38	12	1	.15	
12 fl oz	368	0.0	38.2			0	.00	0.0	.00	.00	2	41	.14	.022	
Pepsi Cola	160	328.8	0.0	0.0							2	0			
12 fl oz	360	0.0	39.6	0.0	0.0							55	.00		
Pepsi Free	160	328.8	0.0	0.0							2	0			
12 fl oz	360	0.0	39.6	0.0	0.0						13	55	.00		
quinine water, Nehi	142		0.0								0				
12 fl oz	360	0.0	35.5								0	0			
root beer	152	330.4	0.0	0.0	0	0	0	.00	.00	0	49	19	4	.26	
12 fl oz	370	0.1	39.2	0.0		0	.00	0.0	.00	.00	3	2	.18	.026	
root beer, Fanta	156		0.0								20				
12 fl oz	370	0.0	40.0								0	0			
root beer, Nehi	193		0.0								18				
12 fl oz	360	0.0	48.0								0	0			
root beer, Ramblin'	176		0.0								20				
12 fl oz	373	0.0	46.0								0	0			
Slice	152	328.8	0.0	0.0							11	0			
12 fl oz	360	0.0	39.6	0.0	0.0						100	0	.00		
Sprite	142		0.0								46				
12 fl oz	368	0.0	36.0								0	0			
strawberry, Nehi	192		0.0								14[b]				
12 fl oz	360	0.0	47.9								0	0			
tonic water/quinine water	125	333.6	0.0	0.0	0	0	0	.00	.00	0	15	5	1		
12 fl oz	366	0.0	32.2	0.0		0	.00	0.0	.00	.00	1	0			
Upper 10	169		0.0								40				
12 fl oz	360	0.0	42.2								0	0			
Upper 10, salt-free	173		0.0								0				
12 fl oz	360	0.0	43.2								25	0			

[a] For cnd soda; if in bottles P is 73 mg.

[b] For cnd soda; if in bottles Na is 0 mg & P is 94 mg.

	KCAL / WT (g)	H₂O (g) / PRO (g)	FAT (g) / CHO (g)	PUFA (g) / SFA (g)	CHOL (mg) / DFIB (g)	A (RE) / A (IU)	C (mg) / B-1 (mg)	B-2 (mg) / NIA (mg)	B-6 (mg) / B-12 (mcg)	FOL (mcg) / PANT (mg)	Na (mg) / K (mg)	Ca (mg) / P (mg)	Mg (mg) / Fe (mg)	Zn (mg) / Cu (mg)	Mn (mg)

1.3. CARBONATED BEVERAGES, LOW CALORIE

	KCAL / WT	H₂O / PRO	FAT / CHO	PUFA / SFA	CHOL / DFIB	A(RE) / A(IU)	C / B-1	B-2 / NIA	B-6 / B-12	FOL / PANT	Na / K	Ca / P	Mg / Fe	Zn / Cu	Mn
club soda	0	354.8	0.0	0.0	0	0	0	.00	.00	0	75	17	4	.36	
12 fl oz	355	0.0	0.0	0.0		0	.00	0.0	.00	.00	6	0			
diet apple Slice[a]	20	349.2	0.0	0.0							5	0			
12 fl oz	360	0.0	4.8	0.0	0.0						67	0	.00		
diet cherry coke, Coca-Cola	1		0.0								8				
12 fl oz	354	0.0	0.3								19	28			
diet cherry cola Slice[b]	20	349.2	0.0	0.0							5	0			
12 fl oz	360	0.0	4.8	0.0	0.0						67	0	.00		
diet cherry RC	2		0.0								1				
12 fl oz	360	0.0	0.4								60	41			
diet coke, Coca-Cola	1		0.0								8				
12 fl oz	354	0.0	0.3								18	24			
diet cola, aspartame sweetened	2	354.3	0.0	0.0	0	0	0	.08	.00	0	21[c]	12	4	.28	
12 fl oz	355	0.2	0.3	0.0		0	.02	0.0	.00	.00	0	30	.11		
diet grape, Nehi	3		0.0								0				
12 fl oz	360	0.0	0.8								44	0			
diet lemon-lime, Nehi	5		0.0								0				
12 fl oz	360	0.0	1.1								83	0			
diet mandarin orange Slice[d]	19	349.2	0.0	0.0							22	0			
12 fl oz	360	0.0	4.2	0.0	0.0						85	0	.00		
diet orange, Nehi	2		0.0								0				
12 fl oz	360	0.0	0.6								63	0			
diet Pepsi[e]	1	354.0	0.0	0.0							2	0			
12 fl oz	360	0.0	0.2	0.0	0.0						30	29	.00		
diet Pepsi Free[e]	1	354.0	0.0	0.0							2	0			
12 fl oz	360	0.0	0.2	0.0	0.0						100	50	.00		
diet RC	1		0.0								1				
12 fl oz	360	0.0	0.4								60	41			
diet RC, caffeine-free	1		0.0								1				
12 fl oz	360	0.0	0.4								60	41			
diet RC 100	2		0.0								1				
12 fl oz	360	0.0	0.4								60	41			
diet Rite cola, salt-free	2		0.0								1				
12 fl oz	360	0.0	0.4								60	41			
diet Slice[f]	26	348.0	0.0	0.0							11	0			
12 fl oz	360	0.0	6.0	0.0	0.0						100	0	.00		
diet soda, sodium saccharin sweetened	2	354.3	0.0	0.0	0	0	0	.00	.00	0	57	14	3	.18	.060
12 fl oz	355	0.1	0.3	0.0		0	.00	0.0	.00	.00	7	38[g]	.14	.089	
diet Sprite	4		0.0								0				
12 fl oz	354	0.0	0.0								56	0			
diet strawberry, Nehi	3		0.0								0				
12 fl oz	360	0.0	0.7								44	0			
diet Upper 10	5		0.0								0				
12 fl oz	360	0.0	1.1								83	0			
Fresca	4		0.0				0				0				
12 fl oz	354	0.0	0.3								76	0			
Pepsi Light[e]	1	354.0	0.0	0.0							2	0			
12 fl oz	360	0.0	0.1	0.0	0.0						46	26	.00		
Tab	1		0.0								8				
12 fl oz	354	0.0	0.3								18	46			

[a] Contains 170.0 mg aspartame.
[b] Contains 168.5 mg aspartame.
[c] Cola w/aspartame & sodium saccharine contains 32 mg Na.
[d] Contains 187.1 mg aspartame.
[e] Contains 177.0 mg aspartame.
[f] Contains 136.6 mg aspartame.
[g] For cola & pepper types: other flavors have 0 mg P.

	KCAL / WT (g)	H₂O (g) / PRO (g)	FAT (g) / CHO (g)	PUFA (g) / SFA (g)	CHOL (mg) / DFIB (g)	A (RE) / A (IU)	C (mg) / B-1 (mg)	B-2 (mg) / NIA (mg)	B-6 (mg) / B-12 (mcg)	FOL (mcg) / PANT (mg)	Na (mg) / K (mg)	Ca (mg) / P (mg)	Mg (mg) / Fe (mg)	Zn (mg) / Cu (mg)	Mn (mg)

1.4. CEREAL GRAIN BEVERAGES

	KCAL/WT	H₂O/PRO	FAT/CHO	PUFA/SFA	CHOL/DFIB	A(RE)/A(IU)	C/B-1	B-2/NIA	B-6/B-12	FOL/PANT	Na/K	Ca/P	Mg/Fe	Zn/Cu	Mn
powder	9	0.1	0.1	0.0	0			.00			2	1	6	.01	.025
1 t	2.3	0.1	1.9	0.0				0.4	.00		42	13	.11	.005	
prep from powder															
w/ water	9	177.7	0.1	0.0	0			.00			7	5	7	.06	.027
6 fl oz water & 1 t powder	180	0.1	1.9	0.0				0.4	.00		43	13	.12	.016	
w/ water, Postum	11	178.0	0.0	0.0	0		0	.01	.02	5	3	8	10	.25	
6 fl oz	181	0.2	2.6	0.0		0	.02	0.7	.00	.05	97	20	.20	.055	
w/ water, Postum, coffee flavored	11	177.0	0.0	0.0	0		0	.01	.01	5	3	8	10	.25	
6 fl oz	180	0.2	2.6	0.0		0	.02	0.7	.00	.05	96	20	.20	.055	
w/ whole milk	121	160.9	6.2	0.3	25	57	2	.30	.08	9	91	219	30	.71	.030
6 fl oz milk & 1 t powder	185	6.1	10.4	3.8		230	.07	0.5	.65	.57	319	184	.20	.022	

1.5. COFFEE

	KCAL/WT	H₂O/PRO	FAT/CHO	PUFA/SFA	CHOL/DFIB	A(RE)/A(IU)	C/B-1	B-2/NIA	B-6/B-12	FOL/PANT	Na/K	Ca/P	Mg/Fe	Zn/Cu	Mn
brewed	4	175.7	0.0	0.0	0		0	.00	.00	0	4	3	10	.03	.048
6 fl oz	177	0.1	0.8	0.0			.00	0.4	.00		96	2	.72	.012	
inst powder	4	0.1	0.0	0.0	0	0	0	.00	.00	0	1	3	6	.01	.031
1 rd t	1.8	0.2	0.7	0.0		0	.00	0.5	.00	.00	64	5	.08	.003	
cappuccino flavor, sugar sweetened	62	0.2	2.1	0.0		0	0	.01	.00	0	98	4	7	.04	.028
2 rd t	14	0.4	10.7	1.8	0.0	0	.02	0.3	.00	.01	118	26	.14	.016	
decaffeinated	4	0.1	0.0	0.0	0	0	0	.02	.00	0	0	3	6	.00	.022
1 rd t	1.8	0.2	0.8	0.0		0	.00	0.5	.00	.00	63	5	.07	.001	
french flavor, sugar sweetened	57	0.3	3.4	0.1								4			
2 rd t	11.5	0.5	6.6	2.9				0.7			136	41			
mocha flavor, sugar sweetened	51	0.2	1.9	0.0		0	0	.00	.00	0	31	4	8	.11	.051
2 rd t	11.5	0.5	8.4	1.6	0.0	0	.00	0.3	.00	.01	119	29	.23	.050	
w/ chicory	6	0.1	0.0	0.0	0		0	.39			5	2	4	.01	.022
1 rd t	1.8	0.2	1.3	0.0			.01		.00		61	5	.09	.001	
prep from inst powder	4	177.2	0.0	0.0	0	0	0	.00	.00	0	6	6	8	.05	.032
6 fl oz water & 1 rd t powder	179	0.2	0.7	0.0		0	.00	0.5	.00	.00	64	6	.09	.013	
amaretto, General Foods	51	177.8	2.4	0.0	0		0	.03	.00	0	23	4	10	.02	
6 fl oz water & 11.5 g powder[a]	189	0.2	6.8	2.0		0	.00	0.8	.00	.00	230	57	.14	.029	
amaretto, sugar-free, General Foods	36	177.5	2.6	0.0	0		0	.03	.00	0	21	4	9	.01	
6 fl oz water & 7.7 g powder[a]	185	0.3	3.0	2.1		0	.00	0.7	.00	.00	253	35	.06	.020	
cappuccino flavor, sugar sweetened	62	177.9	2.1	0.0	0	0	0	.01	.00	0	104	7	9	.08	.029
6 fl oz water & 2 rd t powder	192	0.4	10.7	1.8		0	.02	0.3	.00	.01	119	26	.15	.027	
decaffeinated	4	177.2	0.0	0.0	0	0	0	.03	.00	0	6	6	7	.05	.023
6 fl oz water & 1 rd t powder	179	0.2	0.8	0.0		0	.00	0.5	.00	.00	63	5	.08	.013	
francais, General Foods	55	177.7	3.2	0.1	0		0	.02	.00	0	27	3	11	.01	
6 fl oz water & 11.5 g powder[a]	189	0.3	6.0	2.6		0	.00	0.2	.00	.00	278	73	.13	.029	
francais, sugar-free, General Foods	36	177.5	2.5	0.0	0		0	.03	.00	0	22	4	9	.01	
6 fl oz water & 7.7 g powder[a]	185	0.3	3.1	2.1		0	.00	0.7	.00	.00	260	34	.06	.020	
french flavor, sugar sweetened	57	177.7	3.4	0.1								8	2	.05	.000
6 fl oz water & 2 rd t powder	189	0.5	6.6	2.9				0.7			137	41	.01	.011	
irish creme, General Foods	55	177.5	2.4	0.0	0		0	.03	.00		19	8	10	.04	
6 fl oz water & 12.8 g powder[a]	190	0.3	8.1	2.0		0	.00	0.6	.00	.00	243	52	.18	.039	
irish creme, sugar-free, General Foods	32	178.1	2.2	0.0	0		0	.02	.00	0	14	4	9	.03	
6 fl oz water & 7.1 g powder[a]	185	0.3	3.0	1.8		0	.00	0.5	.00	.00	234	31	.09	.027	
irish mocha mint, General Foods	52	177.7	2.2	0.0	0		0	.02	.00		22	5	14	.00	
6 fl oz water & 11.5 g powder[a]	189	0.2	7.6	1.8		0	.00	0.1	.00	.00	248	65	.36	.015	
irish mocha mint, sugar-free, General Foods—6 fl oz water & 6.4 g powder[a]	27	177.8	1.9	0.0	0		0	.02	.00	0	18	4	11	.11	
	184	0.5	2.8	1.5		1	.00	0.2	.00	.01	211	33	.30	.066	

[a] 2 rd t.

						Vitamins					Minerals				
	KCAL	H₂O (g)	FAT (g)	PUFA (g)	CHOL (mg)	A (RE)	C (mg)	B-2 (mg)	B-6 (mg)	FOL (mcg)	Na (mg)	Ca (mg)	Mg (mg)	Zn (mg)	Mn (mg)
	WT (g)	PRO (g)	CHO (g)	SFA (g)	DFIB (g)	A (IU)	B-1 (mg)	NIA (mg)	B-12 (mcg)	PANT (mg)	K (mg)	P (mg)	Fe (mg)	Cu (mg)	
mocha flavor, sugar sweetened	51	176.7	1.9	0.0		0	0	.00	.00	0	36	7	9	.15	.053
6 fl oz water & 2 rd t powder	188	0.5	8.4	1.6		0	.00	0.3	.00	.01	119	29	.24	.060	
orange cappuccino, General Foods	59	177.3	2.0	0.0	0		0	.03	.00	0	106	4	12	.02	
6 fl oz water & 14 g powder [a]	191	0.2	9.7	1.7		0	.00	0.3	.00	.00	125	51	.24	.032	
orange cappuccino, sugar-free, General	29	177.6	1.9	0.0	0		0	.03	.00	0	62	4	10	.01	
Foods—6 fl oz water & 6.7 g powder [a]	184	0.2	2.8	1.6		0	.00	0.6	.00	.00	128	30	.15	.022	
suisse mocha, General Foods	53	177.7	2.5	0.0	0		0	.02	.00		23	4	11	.01	
6 fl oz water & 11.5 g powder [a]	189	0.2	7.5	1.7		0	.00	0.2	.00	.00	153	57	.02	.019	
suisse mocha, sugar-free, General Foods	30	177.8	2.1	0.0	0		0	.02	.00	0	20	3	9	.05	
6 fl oz water & 6.4 g powder [a]	184	0.4	2.8	1.7		1	.00	0.3	.00	.00	151	31	.19	.038	
vienna, General Foods	59	177.2	2.0	0.0	0		0	.00	.00	0	105	4	11	.01	
6 fl oz water & 14 g powder [a]	191	0.2	9.9	1.7		0	.00	0.4	.00	.00	112	32	.04	.027	
vienna, sugar-free, General Foods	30	177.6	2.0	0.0	0		0	.00	.00	0	95	0	9	.01	
6 fl oz water & 6.7 g powder [a]	184	0.2	2.7	1.7		0	.00	0.7	.00	.00	122	21	.00	.020	
w/ chicory	6	177.2	0.0	0.0	0			.01			10	6	6	.05	.023
6 fl oz water & 1 rd t powder	179	0.2	1.3	0.0			.00	0.4	.00		61	5	.09	.013	

1.6. FRUIT JUICE DRINKS & FRUIT FLAVORED BEVERAGES [b]

	KCAL	H₂O (g)	FAT (g)	PUFA (g)	CHOL (mg)	A (RE)	C (mg)	B-2 (mg)	B-6 (mg)	FOL (mcg)	Na (mg)	Ca (mg)	Mg (mg)	Zn (mg)	Mn (mg)
	WT (g)	PRO (g)	CHO (g)	SFA (g)	DFIB (g)	A (IU)	B-1 (mg)	NIA (mg)	B-12 (mcg)	PANT (mg)	K (mg)	P (mg)	Fe (mg)	Cu (mg)	
appleberry jce works, Campbell's	96		0.0				0	.00			39	28			
6 fl oz	182	0.3	23.8			0	.01	0.1			215		1.30		
apple, Hawaiian Punch	90		0.0				60[c]	.00			13	1			
6 fl oz	185	0.0	22.0			1	.00	0.0			21	2			
apple jce works, Campbell's	97		0.1				5	.00			30	19			
6 fl oz	182	0.3	23.7			0	.00	0.3			125		.90		
berry blend, from powder, Crystal Light	3	236.6	0.0	0.0	0		6	.00	.00	0	0		17	.00	
8 fl oz	238	0.1	0.2	0.0	0.0	0	.00	0.0	.00	.00	46	0			
caribbean cooler, from powder, Crystal	3	236.7	0.0	0.0	0		6	.00	.00	0	1		16	.00	
Light—8 fl oz	238	0.1	0.3	0.0	0.0	0	.00	0.0	.00	.00	46				
cherry, Hawaiian Punch	90		0.0				60[c]	.00			17	5			
6 fl oz	185	0.0	23.0			5	.00	0.0			26	3			
cherry jce works, Campbell's	96		0.1				7	.01			18	20			
6 fl oz	182	0.2	23.6			0	.01	0.4			118		.90		
citrus blend, from powder, Crystal Light	3	236.5	0.0	0.0	0		6	.00	.00	0	0		21	.00	
8 fl oz	238	0.1	0.2	0.0	0.0	0	.00	0.0	.00	.00	46	0			
citrus fruit jce drink, from frozn conc	114	217.9	0.0	0.0	0	10	67[c]		.06	5	7	21	16	.13	.181
8 fl oz	248	0.8	28.4	0.0	0.2	103	.04	0.4	.00	.33	277	25	2.79	.079	
cranapple drink, Ocean Spray	130		0.0				60				4	13	4	.07	
6 fl oz		0.0	32.0								51	5	.10	.014	
cranberry jce cocktail															
bottled	108	162.4	0.1		0	1	67[c]	.02	.04	1	4	7	4	.14	.367
6 fl oz	190	0.0	27.4			7	.02	0.1	.00	.11	34	4	.28	.034	
from frozn conc	102	160.6	0.0		0	2	18[c]	.02	.02	0	6	9	4	.07	.077
6 fl oz	187	0.0	26.2			18	.01	0.0	.00	.26	27	3	.17	.019	
low cal, bottled [d]	33	169.5	0.0		0		57[c]				6	16	3	.04	
6 fl oz	178	0.0	8.5						.00		39	1	.07		
cranberry-apple jce drink, bottled	123	152.4	0.0		0		59[c]	.04			4	13	3	.08	
6 fl oz	184	0.1	31.5				.01	0.1	.00		50	5	.11	.013	
cranberry-apricot jce drink, bottled	118	153.6	0.0		0		0	.02			4	17	6	.07	
6 fl oz	184	0.3	29.9				.01	0.2	.00		113	10	.28	.028	
cranberry-grape jce drink, bottled	103	157.5	0.2	0.1	0		59[c]	.03			5	15	6	.07	
6 fl oz	184	0.3	25.8	0.1			.02	0.2	.00		44	7	.02	.013	
crangrape drink, Ocean Spray	110		0.0				60	.03			7	7	4		
6 fl oz		0.0	26.0				.02	0.2			37	7	.25	.022	
cranraspberry drink, Ocean Spray	110		0.0				60				6				
6 fl oz		0.0	27.0								35				
fruit juicy red, Hawaiian Punch	90		0.0				60[c]	.00			17	3			
6 fl oz	185	0.0	22.0			10	.00	0.0			30	2			
fruit punch, from powder, Crystal Light	3	236.8	0.0	0.0	0		6	.00	.00	0	0	33		.00	
8 fl oz	238	0.1	0.1	0.0	0.0	0	.00	0.0	.00	.00	45	15			

[a] 2 rd t.
[b] Fruit & veg juices are listed in section 14.
[c] Vit C added.
[d] Sweetened w/ calcium saccharin & corn sweeteners.

	KCAL	H₂O (g)	FAT (g)	PUFA (g)	CHOL (mg)	A (RE)	C (mg)	B-2 (mg)	B-6 (mg)	FOL (mcg)	Na (mg)	Ca (mg)	Mg (mg)	Zn (mg)	Mn (mg)
	WT (g)	PRO (g)	CHO (g)	SFA (g)	DFIB (g)	A (IU)	B-1 (mg)	NIA (mg)	B-12 (mcg)	PANT (mg)	K (mg)	P (mg)	Fe (mg)	Cu (mg)	
fruit punch drink															
cnd	87	163.7	0.0	0.0	0	3	55ᵃ	.04	.00	2	41	14	4	.23	.372
6 fl oz	186	0.1	22.1	0.0	0.2	26	.04	0.0	.00	.03	47	2	.38	.095	
from frzn conc	113	217.8	0.0	0.0	0	3	108ᵃ	.03	.02	2	11	9	6	.09	.252
8 fl oz	247	0.1	28.8	0.2	0.0	27	.03	0.1	.00	.02	31	2	.22	.074	
from powder	97	236.8	0.0	0.0	0	0	31ᵃ	.01	.00	0	38	41	3	.09	.008
2 rd t in 8 fl oz water	262	0.0	24.8	0.0	0.0	1	.00	0.0	.00	.00	2	52	.14	.047	
fruit punch jce drink, from frozn conc	123	216.5	0.5	0.1	0	1	14	.16	.03	0	12	18	9	.54	.149
8 fl oz	248	0.2	30.4	0.1	0.2	15	.00	0.1	.00	.07	191	0	.57	.060	
fruit punch, low sugar, Hawaiian Punch	30		0.0				60ᵃ	.00			20	5			
6 fl oz	185	0.0	8.0			32	.00	0.0			30	2			
grape drink, cnd	84	166.3	0.0	0.0	0	0	64ᵃ	.01	.01	1	12			.21	
6 fl oz	188	0.0	21.6	0.0		2	.01	0.0	.00	.01	10		.31	.023	
grape, from powder, Crystal Light	3	236.8	0.0	0.0	0		6	.00	.00	0	0		14	.00	
8 fl oz	238	0.1	0.3	0.0	0.0	0	.00	0.0	.00	.00	0	0			
grape, Hawaiian Punch	90		0.0				60ᵃ	.00			13	7			
6 fl oz	185	0.0	23.0			15	.00	0.1			30	7			
grape jce drink, cnd	94	163.6	0.0			0	30ᵃ	.02	.04	2	2	6			
6 fl oz	188	0.2	24.2			4	.02	0.2	.00	.02	66	8	.19		
grape jce works, Campbell's	98		0.2				15	.02			15	14			
6 fl oz	182	0.4	23.5			0	.00	0.3			71		.70		
island fruit cocktail, Hawaiian Punch	90		0.0				60ᵃ	.00			19	2			
6 fl oz	185	0.0	22.0			19	.00	0.0			29	2			
kool-aid, from powder, all flavors	98	220.2	0.0	0.0	0		6	.00	.00	0	8	15		.00	
8 fl oz	246	0.0	25.1	0.0	0.0	0	.00	0.0	.00	.00	1	8			
kool-aid, from sugar sweetened powder, all flavors—8 fl oz	80	220.0	0.0	0.0	0		6	.00	.00	0	8	26		.00	
	241	0.0	20.4	0.0	0.0	0	.00	0.0	.00	.00	1	9			
kool-aid kooler, from powder, all flavors	127	227.5	0.0	0.0	0		60	.01	.02	3	2	4	3	.03	
8.5 fl oz	262	0.2	33.2	0.0	0.0	5	.01	0.1	.00	.03	51	4	.11	.018	
kool-aid, sugar-free, from powder, all flavors—8 fl oz	3	236.8	0.0	0.0	0		7	.00	.00	0	8	27		.00	
	238	0.1	0.3	0.0	0.0	0	.00	0.0	.00	.00	4	26			
lemonade															
from frzn conc	100	221.6	0.1	0.0	0	5	10ᵇ	.05	.02	6	8	8	5	.09	.012
6 fl oz	248	0.1	26.0	0.0		53	.02	0.0	.00	.03	38	5	.41	.045	
from powder	102	236.7	0.0	0.0	0	0	9		.01	4	13	71	3	.10	
8 fl oz	240	0.0	26.9	0.0		0	.01	0.0	.00	.02	33	34	.15		
from powder, aspartame sweetened	5	236.5	0.0	0.0	0	0	6ᵃ	.00	.00		7	51	3	.06	.002
8 fl oz	238	0.1	1.3			0	.00	0.0	.00	.00	1	23	.10	.017	
from powder, Country Time	82	220.0	0.0	0.0	0		9	.00	.00	0	21			.00	
8 fl oz	242	0.0	20.5	SFA	DFIB	A	.00	NIA	B-12	PANT	12	0		.000	
from powder, Country Time sugar-free	5	237.0	0.0	0.0	0		9	.00	.00	0	0	33		.00	
8 fl oz	239	0.1	0.2	0.0	0.0	0	.00	0.0	.00	.00	28	15		.000	
from powder, Crystal Light	5	237.0	0.0	0.0	0		6	.00	.00	0	0		25	.00	
8 fl oz	239	0.1	0.2	0.0	0.0	0	.00	0.0	.00	.00	62				
lemonade flavored drink, from powder	113	236.9	0.1	0.0	0	0	34ᵃ	.00	.00	0	19	29	3	.07	
2 T powder in 8 fl oz water	266	0.0	28.8	0.0		0	.00	0.0	.00	.00	1	3	.04	.027	
lemonade, pink															
from powder, Country Time	82	220.1	0.0	0.0	0		9	.00	.00	0	21			.00	
8 fl oz	242	0.0	20.4	0.0	0.0	0	.00	0.0	.00	.00	12	0			
from powder, Country Time sugar-free	5	237.0	0.0	0.0	0		9	.00	.00	0	0	33			
8 fl oz	239	0.1	0.2	0.0	0.0	0	.00	0.0	.00	.00	28	15			
lemon lime															
from powder, Country Time	82	220.1	0.0	0.0	0		9	.00	.00	0	20			.00	
8 fl oz	242	0.0	20.5	0.0	0.0	0	.00	0.0	.00	.00	7	0			
from powder, Country Time sugar-free	5	237.0	0.0	0.0	0		9	.00	.00	0	0	33		.00	
8 fl oz	239	0.1	0.2	0.0	0.0	0	.00	0.0	.00	.00	17	15			
from powder, Crystal Light	4	236.3	0.0	0.0	0		6	.00	.00	0	0		11	.00	
8 fl oz	238	0.1	0.1	0.0	0.0	0	.00	0.0	.00	.00	6	0			

ᵃ Vit C added.　　　　ᵇ Range is 1 to 18 mg.

	KCAL	H₂O (g)	FAT (g)	PUFA (g)	CHOL (mg)	A (RE)	C (mg)	B-2 (mg)	B-6 (mg)	FOL (mcg)	Na (mg)	Ca (mg)	Mg (mg)	Zn (mg)	Mn (mg)
	WT (g)	PRO (g)	CHO (g)	SFA (g)	DFIB (g)	A (IU)	B-1 (mg)	NIA (mg)	B-12 (mcg)	PANT (mg)	K (mg)	P (mg)	Fe (mg)	Cu (mg)	
limeade, from frozn conc	102	219.6	0.1	0.0	0		7	.01			6	7	2	.05	.000
8 fl oz	247	0.1	27.1	0.0			.01	0.1			33	3	.06	.012	
orange & apricot jce drink, cnd	128	216.8	0.3	0.1	0	145	50ᵃ	.03				13			
8 fl oz	250	0.8	31.8	0.0		1450	.05	0.5	.00		201	20	.25		
orange breakfast drink, from frozn conc	84	165.7	0.0	0.0	0	1	104ᵇ	2.0	.13ᵇ	61ᵇ	18	221ᵇ	20	.10	
6 fl oz	188	0.2	21.3	0.0	0.2	11	.20ᵇ	0.5ᵇ	.00	.36ᵇ	254	63	.15	.182	
orange drink, cnd	94	161.6	0.0	0.0	0	3	64ᵃ	.01	.02		31	12	3	.16	.028
6 fl oz	186	0.0	24.0	0.0		33	.01	0.1	.00	.03	33	3	.53	.006	
orange drink, from powder, Crystal Light	4	236.2	0.0	0.0	0		6	.00	.00	0	0	57		.00	
8 fl oz	238	0.1	0.3	0.0	0.0	0	.00	0.0	.00	.00	46	27			
orange flavor breakfast drink															
from frzn conc	91	162.2	0.3	0.1	0	0	129ᵃ	.07	.00		17	61	2	.07	
6 fl oz	186	0.0	22.7	0.0		0	.22ᶜ	0.0	.00	.01	231	40	.15		
prep from powder	86	163.8	0.0	0.0	0	413ᵈ	91ᵃ	.03	.00	107ᵉ	9	46	2	.07	.011
3 rd t in 6 oz water	186	0.0	21.9	0.0		1376ᵈ	.00	0.0	.00	.00	37	29	.15	.028	
orange flavor gelatin drink, prep from powder—1 pkt in 4 fl oz water	67	118.9	0.2	0.1			50ᵃ				32	1		.03	.000
	136	6.2	10.5	0.0							3		.01	.007	
orange, Hawaiian Punch	100		0.0				60ᵃ	.00			19	6			
6 fl oz	185	0.0	24.0			37	.00	0.0			37	5			
orange jce works, Campbell's	90		0.3				20	.02			56	15			
6 fl oz	182	0.5	21.4			23	.01	0.5			185		1.10		
paradise punch, from powder, Crystal Light—8 fl oz	3	236.9	0.0	0.0	0		6	.00	.00	0	0	13		.00	
	238	0.1	0.1	0.0	0.0	0	.00	0.0	.00	.00	45	0			
pineapple & grapefruit jce drink, cnd	117	219.7	0.2	0.1	0	9	115ᵃ	.04	.11	26	34	18	15	.15	1.033
8 fl oz	250	0.6	29.0	0.0		88	.08	0.7	.00	.13	154	14	.77	.113	
pineapple & pink grapefruit jce drink, cnd, Dole—6 fl oz	101		0.1				60ᵃ	.02			0	9			
	188	0.4	25.4			20	.04	0.2			116	9	.40		
pineapple & orange jce drink, cnd	125	217.0	0.0		0	133ᶠ	56ᵃ	.05	.12	27	9	13	14	.14	.903
8 fl oz	250	3.1	29.4			1328ᶠ	.08	0.5	.00	.14	116	10	.67	.103	
pink grapefruit jce cocktail, Ocean Spray	80		0.0				100				15	14	8	.11	
6 fl oz		0.0	20.0								125	16	.24	.060	
strawberry jce works, Campbell's	94		0.0				2	.00			53	56			
6 fl oz	182	0.7	22.8			0	.00	0.1			289		1.60		
tang, orange, from powder	88	161.5	0.0	0.0	0		60	.17	.20	80	2	20		.00	
6 fl oz	185	0.0	21.9	0.0	0.0	501	.00	2.0	.00	.00	45	12	1.80		
tang, orange, sugar-free, from powder	5	177.9	0.0	0.0	0		60	.00	.00	80	1	20		.00	
6 fl oz	180	0.1	0.4	0.0	0.0	501	.00	0.0	.00	.00	51	21	.00		
thirst quencher, bottled	60	225.3	0.1	0.0	0	0	0	.00	.00	0	96	0	1	.05	.000
8 fl oz	241	0.0	15.2	0.0		0	.01	0.0	.00	.00	26	22	.12	.048	
tropical fruit, Hawaiian Punch	90		0.0				60ᵃ	.00			8	3			
6 fl oz	185	0.0	22.0			14	.00	0.1			31	2			
very berry, Hawaiian Punch	90		0.0				60ᵃ	.00			22	6			
6 fl oz	185	0.0	22.0			9	.00	0.0			28	3			
wild fruit, Hawaiian Punch	90		0.0				60ᵃ	.00			19	2			
6 fl oz	185	0.0	23.0			14	.00	0.0			26	2			

1.7. MALT BEVERAGES, NONALCOHOLIC

	KCAL	H₂O (g)	FAT (g)	PUFA (g)	CHOL (mg)	A (RE)	C (mg)	B-2 (mg)	B-6 (mg)	FOL (mcg)	Na (mg)	Ca (mg)	Mg (mg)	Zn (mg)	Mn (mg)
malt beverage	32	353.2	0.0	0.0	0							25	31	.03	.047
12 fl oz	360	1.1	5.1	0.0								110	.05	.040	
malt beverage, Kingsbury	60		0.0	0.0	0						20	25			
12 fl oz	360	0.6	12.8	0.0							50		.02		

ᵃ Vit C added.
ᵇ Added nutrients.
ᶜ Thiamin added.
ᵈ Vit A added.
ᵉ Folacin added.
ᶠ Carotene added for color.

	KCAL	H₂O (g)	FAT (g)	PUFA (g)	CHOL (mg)	A (RE)	C (mg)	B-2 (mg)	B-6 (mg)	FOL (mcg)	Na (mg)	Ca (mg)	Mg (mg)	Zn (mg)	Mn (mg)
	WT (g)	PRO (g)	CHO (g)	SFA (g)	DFIB (g)	A (IU)	B-1 (mg)	NIA (mg)	B-12 (mcg)	PANT (mg)	K (mg)	P (mg)	Fe (mg)	Cu (mg)	

1.8. TEA, HOT/ICED

tea, black

brewed 3 min	2	177.4	0.0	0.0	0	0	0	.03	.00	9	5	0	5	.04	
6 fl oz	178	0.0	0.4	0.0		0	.00	0.0	.00		66	1	.04	.018	
inst powder	2	0.0	0.0	0.0	0	0	0	.01	.01	1	1	0	3	.02	.518
1 t	0.7	0.1	0.4	0.0		0	.00	0.1	.00	.03	46	3	.03	.006	
w/ lemon flavor	4	0.1	0.0	0.0	0	0	0	.02			7	0	2	.02	.429
1 rd t	1.4	0.1	1.1	0.0		0	.00	0.1	.00		48	1	.01	.005	
w/ sugar & lemon flavor	87	0.1	0.1	0.0	0	0	0	.05		10		1	3	.02	.673
3 rd t	23	0.1	22.1	0.0		0	.00	0.1	.00		49	3	.04	.006	
w/ sodium saccharin & lemon flavor	5	0.1	0.0	0.0	0	0	0	.01		5	17	0	2	.01	.486
2 t	1.6	0.1	1.3	0.0		0	.00	0.1	.00		41	2	.14	.002	
prep from inst powder	2	236.2	0.0	0.0	0	0	0	.01	.01	1	8	5	5	.08	.519
1 t powder in 8 fl oz water	237	0.1	0.4	0.0		0	.00	0.1	.00	.03	47	3	.04	.019	
berry, Crystal Light	3	236.8	0.0	0.0	0		6	.00	.00	0	1		8	.00	
8 fl oz	238	0.1	0.5	0.0	0.0	0	.00	0.0	.00	.00	15	0			
citrus, Crystal Light	3	236.8	0.0	0.0	0		6	.00	.00	0	1		8	.00	
8 fl oz	238	0.1	0.5	0.0	0.0	0	.00	0.0	.00	.00	15	0			
Crystal Light	3	236.9	0.0	0.0	0		6	.00	.00	0	1		10	.00	
8 fl oz	238	0.1	0.4	0.0	0.0	0	.00	0.0	.00	.00	15	0			
tropical, Crystal Light	3	236.8	0.0	0.0	0		6	.00	.00	0	1		8	.00	
8 fl oz	238	0.1	0.5	0.0	0.0	0	.00	0.0	.00	.00	15	0			
w/ lemon, Crystal Light	2	237.0	0.0	0.0	0		6	.00	.00	0	0		8	.00	
8 fl oz	238	0.1	0.3	0.0	0.0	0	.00	0.0	.00	.00	15	0			
w/ lemon flavor	4	236.5	0.0	0.0	0	0	0	.02			14	5	4	.08	.431
1 rd t powder in 8 fl oz water	238	0.1	1.1	0.0		0	.00	0.1	.00		49	2	.02	.019	
w/ sugar & lemon flavor	87	236.3	0.1	0.0	0	0	0	.05		10		6	5	.08	.673
3 rd t powder in 8 fl oz water	259	0.1	22.1	0.0		0	.00	0.1	.00		50	3	.05	.021	
w/ sodium saccharin & lemon flavor	5	236.3	0.0	0.0	0	0	0	.01		5	24	5	4	.07	.488
2 t powder in 8 fl oz water	238	0.1	1.3	0.0		0	.00	0.1	.00		41	2	.15	.017	
tea, herb, brewed	1	177.5	0.0	0.0	0	0	0	.01	.00	1	2	4	2	.06	.078
6 fl oz	178	0.1	0.3	0.0		0ᵃ	.02	0.0	.00	.02	15	0	.14	.027	

1.9. WATER

bottled, Perrier	0	191.9	0.0	0.0	0	0	0	.00	.00	0	3	26	1	.00	.000
6.5 fl oz bottled	192	0.0	0.0	0.0		0	.00	0.0	.00	.00	0	0	.00	.000	
bottled, Poland Spring	0	237.0	0.0	0.0	0	0	0	.00	.00	0	1	3	2		
8 fl oz	237	0.0	0.0	0.0		0	.00	0.0	.00	.00	0		.01		
mineral, La Croix	0		0.0	0.0	0						1	20			
12 fl oz	360	0.0	0.0	0.0.							1		.01		
municipal	0	236.9	0.0	0.0	0	0	0	.00	.00	0	7	5	2	.06	.002
8 fl oz	237	0.0	0.0	0.0		0	.00	0.0	.00	.00	1	0	.01	.014	

2. CANDY

almond joy, Peter Paul	136	2.6	2.3		1			.04			58	20			
1 oz	28	1.5	16.4	0.4			.01	0.2					.44		
baby ruth	260		12.0								120				
2 oz bar	57	4.0	36.0								120				
bonkers fruit candyᵇ	20		0.0								0				
1 piece		0.0	5.0								0				

ᵃ Chamomile tea contains 36 IU of vitamin A.

ᵇ Grape, orange, strawberry or watermelon.

	KCAL	H₂O (g)	FAT (g)	PUFA (g)	CHOL (mg)	A (RE)	C (mg)	B-2 (mg)	B-6 (mg)	FOL (mcg)	Na (mg)	Ca (mg)	Mg (mg)	Zn (mg)	Mn (mg)
	WT (g)	PRO (g)	CHO (g)	SFA (g)	DFIB (g)	A (IU)	B-1 (mg)	NIA (mg)	B-12 (mcg)	PANT (mg)	K (mg)	P (mg)	Fe (mg)	Cu (mg)	
bridge mix	140		6.0								15				
14 pieces (1 oz)	28	1.0	20.0								55				
butterfinger	260		12.0								100				
2 oz bar	57	4.0	38.0								120				
butterscotch candy	113	0.4	1.0				0	.00			19	5			
1 oz	28	0.0	26.9	0.5		40	.00	0.0			1	2	.40		
candied															
apricots	94	3.4	0.1												
1 oz	28	0.2	24.2												
cherries	96	3.4	0.1												
1 oz (8 cherries)	28	0.1	24.6												
citron	89	5.0	0.1								82	24			
1 oz	28	0.1	22.7								34	7	.20		
figs	84	5.9	0.1												
1 oz	28	1.0	20.6												
ginger root	96	3.4	0.1												
1 oz	28	0.1	24.7												
grapefruit peel	90	4.9	0.1												
1 oz	28	0.1	22.9												
lemon peel	90	4.9	0.1												
1 oz	28	0.1	22.9												
orange peel	90	4.9	0.1												
1 oz	28	0.1	22.9												
pear	85	5.9	0.2												
1 oz	28	0.4	21.3												
pineapple	90	5.0	0.1												
1 oz	28	0.2	22.7												
sweet potato	143	51.0	2.8				9	.03			36	31			
1 piece	85	1.1	29.1			5360	.05	0.3			162	37	.80		
candy corn	182	3.8	1.0				0	.00			106	7			
1/4 cup	50	0.1	44.8	0.3		0	.00	0.0			2	3	.55		
caramello, Cadbury	141	1.2	6.7		6			.07			54	58			
1 oz	28	1.8	18.4		0.1	31	.02	0.0					.28		
caramels	113	2.1	2.9				0	.05			64	42			
1 oz	28	1.1	21.7	1.6		0	.01	0.1			54	35	.40		
caramels w/ nuts	121	2.0	4.6				0	.05			58	40			
1 oz	28	1.3	20.0	1.6		10	.03	0.1			66	39	.40		
charleston chew, choc/strawberry/van	240		6.0								80				
2 oz	57	2.0	44.0								60				
cherry candy, Y & S Bites	100		1.0								85				
1 oz	28	1.0	23.0												
chocolate															
bittersweet	135	0.5	11.3				0	.05			1	16			
1 oz	28	2.2	13.3	6.3		10	.01	0.3			174	81	1.40		
german sweet, Bakers	141	0.2	9.4	0.3	0		0	.02	.00	1	1	7	27	.35	
1 oz square	28	1.0	16.9	5.6		6	.00	0.1	.00	.02	78	36	.64	.201	
semi-sweet	144	0.3	10.1				0	.02			1	9			
1 oz	28	1.2	16.2	5.7		10	.00	0.1			92	43	.70		
semi-sweet, Bakers	134	0.3	9.1	0.4	0		0	.03	.00	1	1	11	40	.53	
1 oz square	28	1.5	16.3	5.4		8	.01	0.2	.00	.03	115	56	.94	.294	
special dark, Hershey	221		12.4					.10			4	8	47	.62	.328
1.45 oz bar	41	2.5	24.8			8	.01	0.3			139	66	.86	.328	
sweet	150	0.3	10.0				0	.04			9	27			
1 oz	28	1.2	16.4	5.6		0	.01	0.1			76	40	.40		
choc chips															
Bakers	196	0.4	8.8	0.3	0		0	.09	.03	2	26	65	35	.46	
1/4 cup	43	1.7	31.3	6.9		4	.03	0.1	.08	.21	222	81	.87	.204	
german sweet, Bakers	203	0.3	12.3	0.4	0		0	.04	.01	2	1	13	50	.64	
1/4 cup	43	1.8	27.7	7.2		10	.01	0.3	.00	.03	144	69	1.18	.465	

	KCAL	H₂O (g)	FAT (g)	PUFA (g)	CHOL (mg)	Vitamins A (RE)	C (mg)	B-2 (mg)	B-6 (mg)	FOL (mcg)	Minerals Na (mg)	Ca (mg)	Mg (mg)	Zn (mg)	Mn (mg)
	WT (g)	PRO (g)	CHO (g)	SFA (g)	DFIB (g)	A (IU)	B-1 (mg)	NIA (mg)	B-12 (mcg)	PANT (mg)	K (mg)	P (mg)	Fe (mg)	Cu (mg)	
semi-sweet, Bakers	202	0.2	11.8	0.4	0		0	.04	.01	1	1	13	47	.60	
1/4 cup	43	1.7	28.4	6.9		9	.01	0.2	.00	.03	136	65	1.11	.439	
semi-sweet, Hershey	228		12.5					.04			4	13	47	.73	.344
1/4 cup (1.5 oz)	43	2.4	26.1			26	.02	0.1			129	45	1.29	.301	
choc coated															
almonds	161	0.6	12.4				0	.15			17	58			
1 oz (6-8 almonds)	28	3.5	11.2	2.1		0	.03	0.5			155	97	.80		
choc fudge	122	1.7	4.5				0	.04			65	29			
1 oz	28	1.1	20.7	1.5		0	.01	0.1			55	31	.40		
choc fudge w/ nuts	128	1.7	5.9				0	.04			58	29			
1 oz	28	1.4	19.1	1.5		0	.02	0.1			62	39	.40		
coconut	124	1.8	5.0				0	.02			56	14			
1 oz	28	0.8	20.4	2.9		0	.01	0.1			47	22	.30		
mint	45	0.6	1.2				0	.01			20	6			
1 small	11	0.2	8.9	0.4		0	.00	0.0			10	6	.10		
fondant	116	1.6	3.0				0	.02			52	16			
1 oz	28	0.5	23.0	0.9		0	.01	0.0			26	15	.30		
fudge, caramel & peanuts	123	2.3	5.1				0	.06			58	51			
1 oz	28	2.2	18.2	1.4		0	.05	0.5			85	53	.40		
nougat & caramel	118	2.2	3.9				0	.05			49	36			
1 oz	28	1.1	20.6	1.2		10	.02	0.1			60	35	.50		
peanuts	159	0.3	11.7				0	.05			17	33			
1 oz (8-16 peanuts)	28	4.6	11.1	3.0		0	.10	2.1			143	84	.40		
peanuts, Nabisco	160		9.0								15				
14 pieces (1 oz)	28	4.0	14.0								120				
raisins	120	1.3	4.8				0	.06			18	43			
1 oz	28	1.5	20.0	2.7		40	.02	0.1			171	49	.70		
raisins, Nabisco	130		5.0								20				
29 pieces (1 oz)	28	1.0	21.0								150				
van creams	123	2.1	4.8				0	.02			52	36			
1 oz	28	1.1	19.9	1.4		0	.01	0.0			50	31	.20		
choc covered cherries, Welch's Cortina	175		4.5								20				
2 pieces	43	1.0	32.0								35				
choc disks, sugar-coated	132	0.3	5.6				0	.06			20	38			
1 oz	28	1.5	20.6	3.1		30	.02	0.1			71	40	.40		
choc flavored roll	112	1.6	2.3				0	.02			56	19			
1 oz	28	0.6	23.4	0.7		0	.01	0.0			35	34	.50		
choc fudgies, Kraft	33	0.6	1.0	0.5	0		0	.02	.00	5	19	8	1		
1 piece	8	0.2	6.2	0.1		0	.00	0.0	.02	.04	20	7	.05		
choc kisses, Hershey	222		12.5					.14			33	83	27	.53	.123
9 pieces (1.5 oz)	41	3.4	23.7			25	.03	0.1			168	127	.53	.164	
chuckles candy[a]	105		0.3								12				
1 oz	28	0.0	24.1								0				
cough drops															
Beech-Nut	10		0.0								0				
1 piece		0.0	3.0								0				
honey & wild cherry, Pine Brothers	130		0.0								0				
13 pieces	28	0.0	31.0								75				
fudge bar, Nabisco	170		5.0								40				
2 pieces	39	1.0	29.0								50				
fudge, choc	113	2.3	3.5				0	.03			54	22			
1 oz	28	0.8	21.3	1.2		0	.01	0.1			42	24	.30		
fudge, choc w/ nuts	121	2.2	4.9				0	.03			48	22			
1 oz	28	1.1	19.6	1.2		0	.01	0.1			50	32	.30		
fudge squares w/ nuts, Nabisco	130		5.0								25				
2 pieces (1 oz)	28	1.0	21.0								30				
fudge, van	113	2.8	3.1				0	.04			59	32			
1 oz	28	0.9	21.2	0.8		0	.01	0.0			36	24	.10		
fudge, van w/ nuts	120	2.6	4.6				0	.04			53	31			
1 oz	28	1.2	19.5	0.8		0	.01	0.0			32	32	.20		

[a] Includes fruit jellies & jells, jelly candy, jujubes, licorice jellies & jelly eggs, marshmallow eggs, nougat centers & eggs, orange slices, spearmint leaves, drops, strings, berries & softies.

						Vitamins					Minerals				
	KCAL	H₂O (g)	FAT (g)	PUFA (g)	CHOL (mg)	A (RE)	C (mg)	B-2 (mg)	B-6 (mg)	FOL (mcg)	Na (mg)	Ca (mg)	Mg (mg)	Zn (mg)	Mn (mg)
	WT (g)	PRO (g)	CHO (g)	SFA (g)	DFIB (g)	A (IU)	B-1 (mg)	NIA (mg)	B-12 (mcg)	PANT (mg)	K (mg)	P (mg)	Fe (mg)	Cu (mg)	
good stuff bar	250		14.0								90				
1.79 oz bar	51	4.0	29.0								150				
gum drops	98	3.3	0.2				0	.00			10	2			
1 oz	28	0.0	24.8			0	.00	0.0			1	0	.10		
hard candy	109	0.4	0.3				0	.00			9	6			
1 oz	28	0.0	27.6			0	.00	0.0			1	2	.50		
jelly beans	104	1.8	0.1				0	.00			3	3			
1 oz	28	0.0	26.4			0	.00	0.0			0	1	.30		
junior mints	120		3.0								10				
12 pieces (1 oz)	28	1.0	24.0								25				
kit kat wafer, Hershey	244		13.2					.13			53	101	21	.46	.138
1.625 oz bar	46	3.1	28.5			32	.02	0.2			138	83	.41	.092	
krackel bar, Hershey	249		13.6					.17			73	95	28	.52	.141
1.65 oz bar	47	3.1	28.6			33	.02	0.3			160	113	.33	.188	
life savers, all flavors[a]	9.1		0.0								1				
1 piece		0.0	3.0								0				
life savers flavor pop, all flavors	40		0.0								0				
1 pop		0.0	10.0								0				
life savers sours	10		0.0								0				
1 piece		0.0	3.0								0				
lollipop, life savers[b]	45		0.0								10				
1 pop		0.0	11.0								0				
maple sugar	99	2.2	0.0	0.0							4	41			
1 oz piece	28	0.0	25.5	0.0							69	3	.40		
marshmallow	19	1.0	0.0				0	.00			2	1			
1 large	6	0.1	4.8			0	.00	0.0			0	0	.10		
marshmallows, miniature	147	8.0	0.0				0	.00			18	8			
1 cup, not packed	46	0.9	37.0			0	.00	0.0			3	3	.70		
milk choc	147	0.3	9.2				0	.10			27	65			
1 oz	28	2.2	16.1	5.1		80	.02	0.1			109	65	.30		
milk choc, Cadbury	151	0.2	8.2	1.0	7			.09			46	70	16	.06	
1 oz	28	2.2	17.2	4.9	0.1	71	.02	0.2			123	70	.78	.090	
milk choc bar, Hershey	254		14.5					.17			35	94	31	.61	.141
1.65 oz bar	47	3.8	27.1			28	.04	0.2			193	156	.52	.188	
milk choc chips	148		7.8					.06			34	43	15	.25	.084
1 oz	28	1.6	17.8			17	.01	0.2			90	76	.31	.084	
milk choc stars, Nabisco	160		8.0								35				
13 pieces (1 oz)	28	2.0	19.0								85				
milk choc w/ almonds	151	0.4	10.1				0	.12			23	65			
1 oz	28	2.6	14.5	4.5		70	.02	0.2			125	77	.50		
milk choc w/ almonds, Cadbury	153	0.3	8.9		6			.12			41	74			
1 oz	28	2.9	15.4		0.3	65	.02	0.2					.44		
milk choc bar w/ almonds, Hershey	246		15.4					.13			55	116	40	.70	.264
1.55 oz bar	44	5.0	21.7			40	.01	0.2			189	132	.75	.220	
milk choc w/ fruit & nuts, Cadbury	148	0.6	8.0		6			.09			39	67			
1 oz	28	2.4	16.7		0.2	40	.02	0.5					.68		
milk choc w/ hazel nuts, Cadbury	153	0.3	8.7		4			.10			44	70			
1 oz	28	2.6	16.1		0.3		.02	0.2					.39		
milk choc w/ peanuts	154	0.3	10.8				0	.07			19	49			
1 oz	28	4.0	12.6	4.5		50	.07	1.4			138	83	.40		
mints	100	2.1	0.6				0	.00			58	4			
1/4 cup	28	0.0	24.7			0	.00	0.0			1	2	.30		
mints, Breath Savers, sugar-free[c]	8		0.0								0				
1 piece		0.0	2.0								0				
mounds, Peter Paul	135	2.8	7.0		0			.01			53	5			
1 oz	28	1.1	16.8		0.4		.00	0.1					.56		
Mr. Goodbar, Hershey	296		19.4					.14			18	58	49	.94	.416
1.85 oz bar	52	7.2	23.7			21	.03	2.4			234	146	.62	.260	
naturally nut 'n fruit bars, all flavors,	143		7.0								79				
Planters[d]—*1 oz*	28	2.5	17.0								100				
nips, coffee/carmel/licorice, Pearson's	120		3.0								70				
1 oz	28	1.0	23.0								57				

[a] Values are averages for 17 flavors.
[b] Values are for assorted, carnival & swirled flavors.
[c] Values are for 4 flavors.
[d] Values are averages for almond apricot, almond pineapple, peanut raisin & walnut apple.

	KCAL / WT (g)	H₂O (g) / PRO (g)	FAT (g) / CHO (g)	PUFA (g) / SFA (g)	CHOL (mg) / DFIB (g)	A (RE) / A (IU)	C (mg) / B-1 (mg)	B-2 (mg) / NIA (mg)	B-6 (mg) / B-12 (mcg)	FOL (mcg) / PANT (mg)	Na (mg) / K (mg)	Ca (mg) / P (mg)	Mg (mg) / Fe (mg)	Zn (mg) / Cu (mg)	Mn (mg)
parfait, choc/coffioca/mint/peanut butter,	120		3.0								70				
Pearson's—*1 oz*	28	1.0	23.0								69				
peanut, Planters	140		9.0								70				
1 oz	28	4.0	13.0								130				
peanut bar, Planters	240		14.0								110				
1.6 oz bar	45	8.0	21.0								130				
peanut brittle	119	0.6	2.9				0	.01			9	10			
1 oz	28	1.6	23.0	0.6		0	.05	1.0			43	27	.70		
peanut butter chips, Reese's	228		12.8					.09			108	47	47	.86	.602
1/4 cup (1.5 oz)	43	8.9	19.3			9	.02	3.5			217	133	.73	.172	
peanut butter cups, Reese's	281		16.7					.11			148	40	43	.71	.408
2 pieces (1.8 oz)	51	6.4	25.9			10	.02	2.0			204	122	.56	.204	
peppermint patty, Nabisco	110		0.0								10				
2 patties (1 oz)	28	1.0	25.0								25				
peppermint patty, York	120	2.3	3.2	0				.01			10	4			
1 oz	28	0.5	22.2		0.1		.00	0.0					.30		
pom pom caramels	200		6.0								140				
2 oz box	57	2.0	30.0								90				
reese's pieces, Hershey	270		11.4					.13			83	73	50	.83	.330
1.95 oz pkg	55	8.0	33.5				.03	3.1			215	132	.83	.165	
rolo caramels in milk choc, Hershey	264		12.0					.14			94	74	17	.39	.110
9 pieces (1.93 oz)	55	2.8	37.1			33	.03	0.0			143	88	.28	.055	
skor toffee bar, Hershey	220		13.8					.14			92	45			
1.4 oz bar	40	1.9	22.6				.01	.04					.16		
strawberry twizzlers	100		1.0								95				
1 oz	28	1.0	23.0												
sugar babies milk caramels	180		2.0								85				
1 pkg	46	1.0	40.0								50				
sugar-coated almonds	129	0.6	5.3				0	.08			6	28			
1 oz	28	2.2	19.9	0.4		0	.01	0.3			72	47	.50		
sugar daddy milk caramel pop	150		1.0								85				
1 pop	39	1.0	33.0								50				
sugar mama milk caramel candy	90		3.0								30				
3/4 oz piece	21	0.0	17.0								30				
thin mints, Cortina	120		2.0								10				
3 pieces	28	1.0	23.0								20				
whatchamacallit bar, Hershey	270		14.9					.12			122	69	33	.71	.306
1.8 oz bar	51	5.7	28.8				.03	1.7			179	102	.41	.153	

3. CEREALS, COOKED OR TO-BE-COOKED

	KCAL / WT (g)	H₂O (g) / PRO (g)	FAT (g) / CHO (g)	PUFA (g) / SFA (g)	CHOL (mg) / DFIB (g)	A (RE) / A (IU)	C (mg) / B-1 (mg)	B-2 (mg) / NIA (mg)	B-6 (mg) / B-12 (mcg)	FOL (mcg) / PANT (mg)	Na (mg) / K (mg)	Ca (mg) / P (mg)	Mg (mg) / Fe (mg)	Zn (mg) / Cu (mg)	Mn (mg)
bulgur (parboiled wheat), cnd	227	75.6	0.9				0	.04			809	27			
1 cup	135	8.4	47.3			0	.07	3.2			117	270	1.80		
corn grits, inst, white hominy, enr	82	116.4	0.2					.09	.05	1	344	7	5	.09	
1 pkt prep	137	2.1	17.8			0	.17	1.3	.00	.43	29	17	1.02	.014	
Quaker	79	1.9	0.1					.07	.05	2	385	7	5	.00	.000
1 pkt (4/5 cup ckd)	23	1.9	17.7			0	.12	1.0	.00	.00	28	17	.81	.010	
w/ cheddar cheese flavor, Quaker	104	1.9	1.0					.07	.07	10	497	14	10	.00	.000
1 oz pkt	28	2.2	21.6			0	.12	1.0	.00	.00	40	38	.81	.020	
w/ cheese flavor	107	115.6	0.9					.13			481	14			
1 pkt prep	142	2.8	21.3			0	.17	1.5			40		1.16		
w/ imit bacon bits	104	114.2	0.5					.12	.05	8	531	6	9	.18	
1 pkt prep	141	3.0	21.5			0	.20	1.6	.00	.06	60	28	1.29	.034	
w/ imit bacon bits, Quaker	101	2.0	0.4					.07	.05	2	544	7	9	.00	.000
1 oz pkt	28	2.7	21.6			0	.12	1.0	.00	.00	62	28	.81	.030	
w/ imit ham bits	103	114.2	0.4					.15	.06	10	657	7	9	.17	
1 pkt prep	141	2.9	21.4			0	.25	1.7	.00	.74	52	33	1.51	.027	

	KCAL	H₂O (g)	FAT (g)	PUFA (g)	CHOL (mg)	A (RE)	C (mg)	B-2 (mg)	B-6 (mg)	FOL (mcg)	Na (mg)	Ca (mg)	Mg (mg)	Zn (mg)	Mn (mg)
	WT (g)	PRO (g)	CHO (g)	SFA (g)	DFIB (g)	A (IU)	B-1 (mg)	NIA (mg)	B-12 (mcg)	PANT (mg)	K (mg)	P (mg)	Fe (mg)	Cu (mg)	
w/ imit ham bits, Quaker	99	2.1	0.3					.07	.06	1	665	7	9	.00	.000
1 oz pkt	28	2.7	21.3			0	.12	1.0	.00	1.00	56	33	.81	.030	
corn grits, quick, yellow hominy, enr,	101	3.0	0.2					.07	.05	3	1	1	6	.00	.000
Quaker—*3 T (1 oz)*	28	2.4	22.4			161	.12	1.0	.00	.00	45	20	.81	.010	
corn grits, reg/quick															
enr, ckd	146	206.5	0.5					.15	.06	1	0ᵃ	1	11	.17	.041
1 cup	242	3.5	31.4	0.6		ᵇ	.24	2.0			54	29	1.55	.029	
unenr, ckd	146	206.4	0.5					.02	.06	1	0ᵃ	1	11	.17	.041
1 cup	242	3.5	31.4	0.6		ᵇ	.05	0.5			54	29	.48	.029	
white hominy, enr, Quaker/Aunt	101	3.0	0.2					.07	.03	3	1	1	6	.00	.000
Jemima—*3 T (1 oz)*	28	2.4	22.4			0	.12	1.0	.00	.00	39	20	.81	.010	
cream of rice, ckd	95	160.1	0.1					.00	.05	6	1ᶜ	6	6	.29	.264
3/4 cup	183	1.6	21.1				.10	0.8			37	32	.30	.062	
cream of wheat, inst															
ckd	115	152.8	0.4					.10		8	5ᵈ	44	11	.31	
3/4 cup	181	3.3	23.8				.20	1.3			36	32	9.00	.069	
mix & eat	102	116.4	0.3			375		.30	.50	100	241	20	8	.24	
1 pkt prep	142	2.8	21.4			1250	.40	5.0		.13	38	20	8.10	.041	
mix & eat, flavoredᵉ	132	117.2	0.4			375		.20	.50	100	241	40	9	.23	
1 pkt prep	150	2.5	28.9			1250	.40	5.0			55	20	8.10	.057	
cream of wheat, quick, ckd	96	155.1	0.4					.10		7	104ᶠ	38	9	.26	
3/4 cup	179	2.7	20.0				.10	1.1		.13	35	75	7.60	.050	
cream of wheat, reg, ckd	100	163.8	0.4					.10		7	2ᵍ	38	7	.24	
3/4 cup	188	2.9	20.8				.10	1.1		.14	33	31	7.70	.056	
farina															
ckd, enr	87	153.8	0.1					.09	.02	4	1ʰ	3	3	.12	.00
3/4 cup	175	2.5	18.5			0	.14	1.0	.00	.10	22	21	.88	.019	
ckd, unenr	87	153.8	0.1					.02	.02	4	1ʰ	3	3	.12	.00
3/4 cup	175	2.5	18.5			0	.02	0.1	.00	.098	22	21	.04	.019	
dry, quick, creamy wheat, Quaker	101	3.1	0.2					.08	.02	19	1	6	6	.00	.000
2 1/2 T (1 oz)	28	3.1	21.7			0	.12	1.0	.00	.00	31	26	.81	.060	
maltex, ckd	135	151.5	0.8					.07	.06	16	7ⁱ	14	43	1.40	
3/4 cup	187	4.3	29.7				.19	1.8		.17	200	134	1.35	.252	
malt-o-meal, plain/choc, ckd	92	157.6	0.2			0	0	.20	.01	4	2ʲ	4		.13	
3/4 cup	180	2.6	19.4				.30	4.4		.104	35	18	7.10	.020	
maypo, ckd	128	148.9	1.8			527	21	.60	.70	7	6ᵏ	94	38	1.12	
3/4 cup	180	4.4	23.9			1754	.50	7.0	2.10	.25	158	186	6.30	.119	
oatmeal, inst	104	151.8	1.7		0	455		.29	.74	150	286	163	35		
1 pkt prep	177	4.4	18.1			1514	.53	5.5	.00	.35	100	133	6.32		
Quaker	105	2.7	1.7					.17	.40	80	281	100	35	1.00	1.000
1 pkt (3/4 cup ckd)	28	4.5	18.1			1000	.30	3.0	.00	.00	101	128	8.10	.130	
w/ apples & cinn	135	115.9	1.6			435		.28	.70	137	222	158	31		
1 pkt prep	149	3.9	26.3			1450	.48	5.1	.00	.35	107	117	6.07		
w/ apples & cinn, Quaker	134	2.5	1.6					.17	.40	80	181	100	31	1.00	1.000
1 pkt (3/4 cup ckd)	35	3.9	26.0			1000	.30	3.0	.00	.00	108	110	4.50	.120	
w/ bran & raisins	158	155.7	1.9			479		.63	.76	155	247	173	57	1.35	1.587
1 pkt prep	195	4.9	30.4			1596	.56	8.1	.00	.45	236	206	7.61	.284	
w/ bran & raisins, Quaker	153	3.5	1.7					.17	.40	80	240	100	71	1.00	2.000
1 pkt (3/4 cup ckd)	43	5.2	29.2			1000	.30	3.0	.00	.00	227	156	4.50	.220	

ᵃ Na is 540 mg if salt is added according to label directions.
ᵇ White corn grits contain only a trace of vitamin A; yellow corn grits contain 145 IU.
ᶜ Na is 317 mg if salt is added according to label directions.
ᵈ Na is 273 mg if salt is added according to label directions.
ᵉ Apple w/cinn, banana & spice, or maple & brown sugar.
ᶠ Na is 347 mg if salt is added according to label directions.
ᵍ Na is 252 mg if salt is added according to label directions.
ʰ Na is 576 mg if salt is added according to label directions.
ⁱ Na is 142 mg if salt is added according to label directions.
ʲ Na is 243 mg if salt is added according to label directions.
ᵏ Na is 194 mg if salt is added according to label directions.

	KCAL	H₂O (g)	FAT (g)	PUFA (g)	CHOL (mg)	A (RE)	C (mg)	B-2 (mg)	B-6 (mg)	FOL (mcg)	Na (mg)	Ca (mg)	Mg (mg)	Zn (mg)	Mn (mg)
	WT (g)	PRO (g)	CHO (g)	SFA (g)	DFIB (g)	A (IU)	B-1 (mg)	NIA (mg)	B-12 (mcg)	PANT (mg)	K (mg)	P (mg)	Fe (mg)	Cu (mg)	
w/ cinn & spice	177	118.0	1.9			475		.34	.77	153	280	172	51	.97	1.476
• 1 pkt prep	161	4.8	35.1			1581	.56	5.7	.00	.38	104	146	6.65	.119	
w/ cinn & spice, Quaker	176	2.9	1.8					.17	.40	80	258	100	40	1.00	1.000
1 pkt (3/4 cup ckd)	46	5.0	34.8			1000	.30	3.0	.00	.00	106	143	8.10	.130	
w/ honey & graham, Quaker	136	2.1	1.7					.17	.40	80	224	100	35	1.00	1.000
1 pkt (3/4 cup ckd)	35	3.6	26.6			1000	.30	3.0	.00	.00	87	113	4.50	.120	
w/ maple & brown sugar	163	115.3	1.9			451		.32	.74	145	280	162	85		
1 pkt prep	155	4.6	31.9			1503	.53	5.4	.00	.32	102	143	6.35		
w/ maple & brown sugar, Quaker	163	2.7	1.9					.17	.40	80	228	100	85	1.00	1.000
1 pkt (3/4 cup ckd)	43	4.6	31.9			1000	.30	3.0	.00	.00	98	136	8.10	.200	
w/ peaches & cream, Quaker	136	2.0	2.0					.17	.40	80	134	100	35	1.00	1.000
1 pkt (3/4 cup ckd)	35	3.5	26.0			1000	.30	3.0	.00	.00	130	60	4.50	.110	
w/ raisins & spice	161	118.4	1.8			440		.36	.75	150	225	165	57		
1 pkt prep	158	4.3	31.8			1466	.51	5.5	.00	.37	150	133	6.58		
w/ raisins & spice, Quaker	159	3.4	1.8					.17	.40	80	217	100	57	1.00	1.000
1 pkt (3/4 cup ckd)	43	4.5	31.4			1000	.30	3.0	.00	.00	155	123	4.50	.160	
w/ raisins, dates & walnuts, Quaker	150	2.5	3.6					.17	.40	80	155	100	39	1.00	1.000
1 pkt (3/4 cup)	37	4.0	25.4			1000	.30	3.0	.00	.00	128	128	4.50	.160	
w/ strawberries & cream, Quaker	136	2.0	1.9					.17	.40	80	172	100	34	1.00	1.000
1 pkt (3/4 cup)	35	3.4	26.3			1000	.30	3.0	.00	.00	158	121	4.50	.100	
oatmeal, quick/old fashioned, Quaker	109	2.6	1.9					.03	.06	29	1	14	40	1.00	1.000
1/3 cup dry (2/3 cup ckd)	28	4.6	18.4			0	.19	0.2	.00	.00	105	128	1.08	.090	
oatmeal, quick/reg, ckd	108	149.3	1.8	0.7	0			.04	.04	7	1ᵃ	15	42	.86	1.024
3/4 cup (from 1/3 cup or 1 oz dry)	175	4.5	18.9	0.3	1.6	29	.19	0.2	.00	.35	99	133	1.19	.096	
oatmeal, quick/reg, Ralston Purina	100		1.8				1	.06	.06	14	8	15	41	.95	
1 oz dry	28	4.6	18.0		3.4	8	.23	0.3	.00	.26	103	132	1.25	.147	
ralston, ckd	100	163.5	0.6				1	.13	.09	13	3ᵇ	10	45	1.06	
3/4 cup	190	4.2	21.2		3.2	35	.15	1.5	.08	.25	115	111	1.24	.150	
roman meal, ckd	111	149.7	0.7					.09	.09	18	2ᶜ	22	82	1.34	
3/4 cup	181	4.9	24.8					.18	2.3		.28	227	162	1.59	.241
roman meal w/ oats, ckd	127	146.8	1.5					.16	.28		7ᵈ	21	56		
3/4 cup	180	5.4	25.6					.23	2.4		.19		176	1.05	.112
wheatena, ckd	101	155.4	0.8					.04	.04	13	4ᵉ	8	37	1.26	1.496
3/4 cup	182	3.7	21.5					.02	1.0		.08	140	109	1.02	.095
wheat harts, ckd, Ralston Purina	110		1.0								0ᶠ				
3/4 cup (from 1 oz dry)		4.0	21.0								110				
whole wheat hot natural cereal, ckd,	113	152.1	0.7					.09	.05	19	1ᵍ	13	40	.87	1.061
Quaker—3/4 cup	182	3.7	24.9			0	.13	1.6	.00	.30	129	126	1.13	.151	

4. CEREALS, READY-TO-EAT

	KCAL	H₂O (g)	FAT (g)	PUFA (g)	CHOL (mg)	A (RE)	C (mg)	B-2 (mg)	B-6 (mg)	FOL (mcg)	Na (mg)	Ca (mg)	Mg (mg)	Zn (mg)	Mn (mg)
	WT (g)	PRO (g)	CHO (g)	SFA (g)	DFIB (g)	A (IU)	B-1 (mg)	NIA (mg)	B-12 (mcg)	PANT (mg)	K (mg)	P (mg)	Fe (mg)	Cu (mg)	
almond delight, Ralston Purina	110		1.6				15		.50	100	199	24		1.50	
3/4 cup (1 oz)	28	2.1	23.0		1.5		.38	5.0	1.50				1.80		
alpha-bits, Post	110	0.3	0.7	0.2	0	371	0	.42	.49	99	178	9	22	1.48	
1 cup (1 oz)	28	2.3	24.1	0.2	0.6	1235	.37	4.9	1.48	.15	59	57	2.67	.086	
apple jacks, Kellogg's	110	0.7	0.1			375	15	.40	.50	100	125	3	6	3.70	
1 cup (1 oz)	28	1.5	25.7		0.2	1250	.40	5.0		.10	23	30	4.50	.099	
body buddies, brown sugar & honey,	110		1.0								290				
General Mills—1 cup (1 oz)	28	2.0	24.0								50				
body buddies, natural fruit flavor, General	110		1.0								280				
Mills—1 cup (1 oz)	28	2.0	24.0								40				
booberry, General Mills	110		1.0								210				
1 cup (1 oz)	28	1.0	24.0								45				

ᵃ Na is 280 mg if salt is added according to label directions.
ᵇ Na is 357 mg if salt is added according to label directions.
ᶜ Na is 148 mg if salt is added according to label directions.
ᵈ Na is 405 mg if salt is added according to label directions.
ᵉ Na is 433 mg if salt is added according to label directions.
ᶠ Na is 190 mg if salt is added according to label directions.
ᵍ Na is 424 mg if salt is added according to label directions.

	KCAL	H_2O (g)	FAT (g)	PUFA (g)	CHOL (mg)	Vitamins A (RE)	C (mg)	B-2 (mg)	B-6 (mg)	FOL (mcg)	Minerals Na (mg)	Ca (mg)	Mg (mg)	Zn (mg)	Mn (mg)
	WT (g)	PRO (g)	CHO (g)	SFA (g)	DFIB (g)	A (IU)	B-1 (mg)	NIA (mg)	B-12 (mcg)	PANT (mg)	K (mg)	P (mg)	Fe (mg)	Cu (mg)	
bran															
Kellogg's all-bran	71	0.9	0.5			375	15	.40	.50	100	320	23	106	3.70	
1/3 cup (1 oz)	28	4.0	21.1		8.5	1250	.40	5.0		.49	350	264	4.50	.324	
Loma Linda	60	1.0	1.0		0			.08	.08		115	31		.95	
1/3 cup (1 oz)	28	4.0	19.0				.11	0.8	.00	.25	220		1.70		
100%	76	0.8	1.4	0.8			27	.80	.90		196	20	134	2.46	
1/2 cup (1 oz)	28	3.5	20.7	0.3	8.4		.70	9.0	2.70	.55	354	344	3.49	.447	
bran buds, Kellogg's	73	0.8	0.7			375	15	.40	.50	100	174	19	90	3.70	
1/3 cup (1 oz)	28	3.9	21.6		7.9	1250	.40	5.0		.55	474	246	4.50	.300	
bran chex, Ralston Purina	90	0.7	0.7			6	15	.16	.50	100	299	17	73	3.75	
2/3 cup (1 oz)	28	2.9	23.0		6.1	62	.38	5.0	1.50	.29	203	290	4.50	.209	
bran flakes															
Kellogg's	93	0.9	0.5			375		.40	.50	100	264	14	52	3.70	
3/4 cup (1 oz)	28	3.6	22.2		4.0	1250	.40	5.0	1.50		180	139	8.10	.210	
Post	87	0.9	0.4	0.2	0	371	0	.42	.49	99	227	16	68	1.48	
2/3 cup (1 oz)	28	3.0	22.2	0.1	5.6	1235	.37	4.9	1.48		199	166	8.00	.240	
Ralston Purina	92	0.7	0.4			375	15	.40	.50	100	264	22	68	1.18	
3/4 cup (1 oz)	28	3.3	22.6		3.5	1250	.40	5.0	1.50	.12	184	158	4.50	.207	
bran muffin crisp, General Mills	130		1.0								250				
2/3 cup (1.2 oz)	35	3.0	30.0		4.0						190				
cap'n crunch, Quaker	121	0.7	2.6	0.4		0		.26	.50	100	185	5	8	2.00	.000
3/4 cup (1 oz)	28	1.4	22.9	1.7	0.3	0	.37	5.0	1.00	2.00	37	20	4.50	.020	
cap'n crunch's															
choco crunch, Quaker	116	0.6	1.7					.26	.50	100	172	5	8	2.00	.000
3/4 cup (1 oz)	28	1.5	23.8			0	.37	5.0	1.00	2.00	45	20	4.50	.040	
crunchberries, Quaker	120	0.7	2.6	0.4				.26	.50	100	166	9	8	2.00	.000
3/4 cup (1 oz)	28	1.4	22.9	1.6	0.3	0	.37	5.0	1.00	2.00	40	20	4.50	.010	
peanut butter, Quaker	127	0.5	3.8	0.8		0		.26	.50	100	210	6	8	2.00	.000
3/4 cup (1 oz)	28	2.3	20.9	1.5	0.3	0	.37	5.0	1.00	2.00	54	20	4.50	.050	
cheerios, General Mills	111	1.4	1.8	0.8		375	15	.42	.50	6	290	48	39	.79	.763
1 1/4 cups (1 oz)	28	4.3	19.6	0.3	2.0	1250	.37	5.0	1.50	.29	105	134	4.45	.143	
cinnamon toast crunch, General Mills	120		3.0								220				
3/4 cup (1 oz)	28	1.0	23.0								40				
circus fun, General Mills	110		1.0								160				
1 cup (1 oz)	28	1.0	24.0								25				
cocoa krispies, Kellogg's	110	0.7	0.4			375	15	.40	.50	100	217	5	9	1.50	
3/4 cup (1 oz)	28	1.5	25.2		0.1	1250	.40	5.0			42	37	1.80	.090	
cocoa pebbles, Post	112	0.6	1.1	0.0	0	371	0	.42	.49	99	158	5	11	1.48	
7/8 cup (1 oz)	28	1.3	24.5	1.0	0.1	1235	.37	4.9	1.48		41	22	1.78	.056	
cocoa puffs, General Mills	110		1.0								200				
1 cup (1 oz)	28	1.0	25.0								45				
cookie crisp, choc chip, Ralston Purina	110	0.5	1.0			375	15	.26	.50	100	190	9	8	2.25	
1 cup (1 oz)	28	1.5	25.0		1.0	1250	.38	5.0	1.50	2.00	42	30	4.50	.005	
cookie crisp, vanilla wafer, Ralston Purina	110	0.5	1.1					.26	.50	100	203	8	7	2.25	
1 cup (1 oz)	28	1.5	25.0		1.0		.38	5.0	1.50	2.00	29	47	4.50	.047	
corn bran, Quaker	109	0.4	0.9			0		.26	.50	100	244	23	8	3.00	.000
2/3 cup (1 oz)	28	2.2	23.3		5.4	0	.37	5.0	1.00	2.00	62	20	8.10	.060	
corn chex, Ralston Purina	110	0.5	0.2				15	.10	.50	100	293	2	3	.12	.024
1 cup (1 oz)	28	2.0	25.0		1.1		.38	5.0	1.50	.03	23	10	1.80	.023	
corn flakes															
country, General Mills	110		1.0								310				
1 cup (1 oz)	28	2.0	25.0								30				
Kellogg's	110	0.7	0.1			375	15	.40	.50	100	351	1	3	.08	.023
1 1/4 cups (1 oz)	28	2.3	24.4		0.3	1250	.40	5.0		.05	26	18	1.80	.019	
low sodium	113	0.9	0.1						.05		3	12	3	.08	
1 cup (1 oz)	28	2.2	25.2				.00	0.1			21	14	.63	.025	
Post Toasties, Post	108	1.0	0.3	0.1	0	371	0	.42	.49	99	277	2	3	.27	
1 cup (1 oz)	28	1.9	24.0	0.2	0.4	1235	.37	4.9	1.48		31	11	.36	.023	

	KCAL	H₂O (g)	FAT (g)	PUFA (g)	CHOL (mg)	A (RE)	C (mg)	B-2 (mg)	B-6 (mg)	FOL (mcg)	Na (mg)	Ca (mg)	Mg (mg)	Zn (mg)	Mn (mg)
	WT (g)	PRO (g)	CHO (g)	SFA (g)	DFIB (g)	A (IU)	B-1 (mg)	NIA (mg)	B-12 (mcg)	PANT (mg)	K (mg)	P (mg)	Fe (mg)	Cu (mg)	
Ralston Purina	110	0.8	0.1				15		.50	100	274	1	3		.05
1 cup (1 oz)	28	2.0	25.0		1.4	1250	.38	5.0	1.50	.06	25	10	1.00	.023	
corn total, General Mills	110		1.0								310				
1 cup (1 oz)	28	2.0	24.0								30				
count chocula, General Mills	110		1.0								210				
1 cup (1 oz)	28	2.0	24.0								65				
cracklin bran, Kellogg's	108	1.0	4.1			375	15	.40	.50	100	230	19	55	1.50	
1/3 cup (1 oz)	28	2.6	19.4		4.3	1250	.40	5.0		.32	168	114	1.80	.159	
crisp rice, low Na	114	0.9	0.1					.05			3	19	11	.43	
1 cup (1 oz)	28	1.6	25.8				.00	0.4			22	29	.87	.069	
crispy oatmeal & raisin chex, Ralston Purina—*3/4 cup (1.3 oz)*	140		0.5				4	.32	.50	100	168	12	27	3.75	
	38	3.0	31.0		1.7		.38	5.0	1.50	.08	111	85		.124	
crispy rice	112	0.7	0.1				1	.00	.05	3	208	5	12	.47	
1 cup (1 oz)	28	1.8	25.1		0.4	1250	.10	2.0	.08	.11	27	31	.70	.062	
crispy wheats & raisins, General Mills	99	2.0	0.5			375		.40	.50		180	47	23	.34	
3/4 cup (1 oz)	28	2.0	23.1		1.3	1250	.40	5.0	1.50		100	77	4.50	.085	
C.W. Post	126	0.7	4.4	0.4		375		.40	.50	100	49	14	20	.48	
1/4 cup (1 oz)	28	2.6	20.3	3.3	0.7	1250	.40	5.0	1.50		58	66	4.50	.110	
C.W. Post w/ raisins	123	1.1	4.0	0.4		375		.40	.50	100	44	14	20	.45	
1/4 cup (1 oz)	28	2.4	20.3	3.0	0.5	1250	.40	5.0	1.50		72	64	4.50	.109	
dairy crisp, Pet	120		4.0								140				
1/4 cup (1 oz)	28	3.0	19.0												
fiber one, General Mills[a]	60		1.0								230				
1/2 cup (1 oz)	28	4.0	21.0		12.0						340				
fortified oat flakes, Post	105	0.8	0.5	0.2	0	371	0	.42	.49	99	247	33	41	1.02	
2/3 cup (1 oz)	28	6.4	19.9	0.1	0.8	1235	.37	4.9	1.48		144	95	8.00	.170	
frankenberry, General Mills	110		1.0								210				
1 cup (1 oz)	28	1.0	24.0								45				
froot loops, Kellogg's	111	0.7	0.5			375	15	.40	.50	100	145	3	7	3.70	
1 cup (1 oz)	28	1.7	25.0		0.2	1250	.40	5.0		.10	26	24	4.50	.060	
frosted mini-wheats, Kellogg's[b]	102	1.4	0.3			375	15	.40	.50	100	8	9	23	1.50	
4 biscuits (1 oz)	28	2.9	23.4		2.1	1250	.40	5.0			97	74	1.80	.130	
frosted rice krinkles	109	0.5	0.1			375		.40	.50	100	179	4	8	1.50	
7/8 cup (1 oz)	28	1.4	25.8			1250	.40	5.0	1.50		14	20	1.80	.047	
fruit & fiber, Post															
harvest medley	91	1.3	1.0	0.3	0		0	.42	.49	99	192	18	61	1.48	
1/2 cup (1 oz)	28	2.8	21.7	0.1	4.2	1235	.37	4.9	1.48	.06	201	147	4.44	.222	
mountain trail	91	1.4	1.2	0.3	0		2	.42	.49	99	177	16	59	1.48	
1/2 cup (1 oz)	28	2.8	21.4	0.5	4.2	1235	.37	4.9	1.48	.06	200	141	4.44	.226	
tropical fruit	93	1.1	1.4	0.3	0		0	.42	.49	99	178	15	59	1.48	
1/2 cup (1 oz)	28	2.6	21.7	0.8	4.3	1235	.37	4.9	1.48		207	136	4.45	.224	
w/ dates, raisins & walnuts	89	1.5	0.8	0.5	0		0	.42	.49	99	178	15	57	1.48	
1/2 cup (1 oz)	28	2.7	21.9	0.1	4.2	1235	.37	4.9	1.48		197	138	4.44	.214	
fruity pebbles, Post	112	0.8	1.0	0.0	0	371	0	.42	.49	99	144	4	7	1.48	
7/8 cup (1 oz)	28	1.1	24.6	1.0	0.2	1235	.37	4.9	1.48		21	17	1.78	.039	
golden grahams, General Mills	109	0.6	1.1	0.2		375	15	.42	.50		280	17	12	.25	
3/4 cup (1 oz)	28	1.6	24.1	0.8	0.5	1250	.37	5.0	1.50	.12	55	41	4.45	.226	
graham crackos, Kellogg's	102	0.9	0.2			375	15	.40	.50	100	185	13	24	1.50	
3/4 cup (1 oz)	28	2.1	24.5		1.7	1250	.40	5.0			102	62	1.80	.074	
granola															
hearty, Post	127	0.5	4.1	0.4	0		0	.42	.49	99	76	11	23	.71	
1/4 cup (1 oz)	28	2.4	20.7	3.0	0.9	1235	.37	4.9	1.48		62	60	4.44	.116	
hearty w/ raisins, Post	123	0.8	3.8	0.4	0		0	.42	.49	99	79	11	22	.66	
1/4 cup (1 oz)	28	2.2	20.8	2.7	0.9	1235	.37	4.9	1.48		74	57	4.44	.113	
homemade	138	0.9	7.7	4.0			0	.07	.10	23	3	18	33	1.04	
1/4 cup (1 oz)	28	3.5	15.6	1.4		10	.17	0.5		.17	142	115	1.13	.162	
Kretschmer Sun Country															
w/ almonds	130		5.0					.04			10	23			
1/4 cup (1 oz)	28	3.0	19.0	1.0			.10				85		.85		

[a] Contains aspartame.　　　　[b] Sugar frosted, brown sugar cinnamon or apple flavored.

	KCAL / WT (g)	H₂O (g) / PRO (g)	FAT (g) / CHO (g)	PUFA (g) / SFA (g)	CHOL (mg) / DFIB (g)	A (RE) / A (IU)	C (mg) / B-1 (mg)	B-2 (mg) / NIA (mg)	B-6 (mg) / B-12 (mcg)	FOL (mcg) / PANT (mg)	Na (mg) / K (mg)	Ca (mg) / P (mg)	Mg (mg) / Fe (mg)	Zn (mg) / Cu (mg)	Mn (mg)
w/ raisins	130		5.0					.04			10	23			
1/4 cup (1 oz)	28	3.0	20.0				.10					85	.85		
w/ raisins & dates	130		4.0					.04			10	23			
1/4 cup (1 oz)	28	3.0	20.0				.10					85	.85		
Nature Valley[a]	126	1.1	4.9	0.7				.05		21	58	18	29	.55	
1/3 cup (1 oz)	28	2.9	18.9	3.3	1.0		.10	0.2		.23	98	89	.95	.092	
grape-nut flakes, Post	104	0.9	0.8	0.2	0	371	0	.42	.49	99	158	10	30	1.19	
7/8 cup (1 oz)	28	2.8	22.7	0.4	1.9	1235	.37	4.9	1.48	.29	94	82	8.00	.128	
grape-nuts, Post	104	1.0	0.1	0.1	0	371	0	.42	.49	99	188	11	27	1.19	
1/4 cup (1 oz)	28	3.1	23.1	0.0	1.8	1235	.37	4.9	1.48	.27	85	71	2.67	.118	
grape-nuts w/ raisins, Post	98	2.3	0.3	0.1	0		0	.42	.49	99	138	11	22	.64	
1/4 cup (1 oz)	28	2.7	22.1	0.1	1.7	1235	.37	4.9	1.48		97	64	.71	.103	
halfsies, Quaker	113	0.7	1.1					.26	.50	100	243	12	8	2.00	.000
3/4 cup (1 oz)	28	1.8	24.0			0	.37	5.0	1.00	2.00	37	29	8.10	.050	
heartland natural	123	1.2	4.4					.04		16	72	19	36	.75	
1/4 cup (1 oz)	28	2.9	19.4		1.3		.09	0.4			95	103	1.07		
w/ coconut	125	0.9	4.6					.04		15	57	18	37	.74	
1/4 cup (1 oz)	28	3.0	19.2		1.4		.09	0.5			104	103	1.45		
w/ raisins	120	1.4	4.0					.04		11	58	17	36	.73	
1/4 cup (1 oz)	28	2.8	19.6		1.3		.08	0.4			107	97	1.04		
honey & nut corn flakes, Kellogg's	113	1.1	1.5			375	15	.40	.50	100	225	3	6	.11	
3/4 cup (1 oz)	28	1.8	23.3		0.3	1250	.40	5.0			36	13	1.80	.050	
honey bran	97	0.7	0.6			375	15	.40	.50	19	164	13	37	.73	
7/8 cup (1 oz)	28	2.5	23.2		3.1	1250	.40	5.0	1.50	.16	122	107	4.50	.135	
honey buc wheat crisp, General Mills	110		1.0								260				
3/4 cup (1 oz)	28	2.0	24.0								50				
honeycomb, Post	110	0.4	0.3			371	0	.42	.49	99	158	4	10	1.48	
1 1/3 cups (1 oz)	28	1.4	25.4		0.4	1235	.37	4.9	1.48	.11	32	23	2.67	.057	
honey nut cheerios, General Mills	107	0.9	0.7	0.3		375	15	.42	.50		250	20	33	.74	
3/4 cup (1 oz)	28	3.1	22.8	0.1		1250	.37	5.0	1.50		95	105	4.45	.198	
ice cream cones, choc chip, General Mills—3/4 cup (1 oz)	110		2.0								190				
	28	1.0	23.0								40				
ice cream cones, vanilla, General Mills	110		2.0								190				
3/4 cup (1 oz)	28	1.0	23.0								45				
kaboom, General Mills	110		1.0								290				
1 cup (1 oz)	28	2.0	23.0								65				
king vitamin, Quaker	112	0.7	1.1					.68	.60	160	250	6	14	.00	.000
1 1/4 cups (1 oz)	28	2.2	23.4			1500	.60	8.0	2.00	.00	49	40	8.10	.040	
kix, General Mills	110	0.9	0.7	0.3		375	15	.42	.50		290	35	12	.25	
1 1/2 cups (1 oz)	28	2.5	23.4	0.2	0.4	1250	.37	5.0	1.50		45	39	8.10	.047	
life, cinn, Quaker	110	1.4	1.6			0		.42	.06	21	150	60	47	1.00	1.400
2/3 cup (1 oz)	28	5.2	18.8		0.9	0	.37	5.0	.00	.00	175	100	8.10	.160	
life, Quaker	111	1.3	1.8			0		.42	.04	21	150	60	12	1.00	.000
2/3 cup (1 oz)	28	5.2	18.6		0.9	0	.37	5.0	.00	.00	172	100	8.10	.170	
life w/ raisins, Quaker	105	2.4	1.8					.42	.04	18	161	60	45	1.00	1.000
2/3 cup (1 oz)	28	5.5	16.8			0	.37	5.0	.00	.00	232	100	8.10	.210	
lucky charms, General Mills	110	0.8	1.1	0.4		375	15	.40	.50		180	32	24	.50	
1 cup (1 oz)	28	2.6	23.2	0.2	0.6	1250	.40	5.0	1.50	.18	75	78	4.50	.098	
most, Kellogg's	95	1.3	0.3			1502	60	1.70	2.00	400	150	43	56	1.50	
2/3 cup (1 oz)	28	4.0	21.6		3.5	5000	1.50	20.0	6.00		185	197	18.00	.159	
Mr. T, Quaker	121	0.6	2.6					.34	.60	8	189	9	10	.00	.000
1 cup (1 oz)	28	1.3	23.1			0	.30	3.0	1.00	.00	37	24	2.70	.010	
nutri-grain, Kellogg's															
barley	106	0.8	0.2			375	15	.40	.50	100	192	8	22	3.70	
3/4 cup (1 oz)	28	3.1	23.5		1.7	1250	.40	5.0	1.50		75	87	1.00	.170	
corn	108	0.8	0.7			375	15	.40	.50	100	187	1	18	3.70	
2/3 cup (1 oz)	28	2.3	23.9		1.8	1250	.40	5.0	1.50		66	81	.60	.082	
rye	102	0.9	0.2			375	15	.40	.50	100	193	6	22	3.70	
3/4 cup (1 oz)	28	2.5	24.0		1.8	1250	.40	5.0	1.50		51	74	.80	.097	
wheat	102	0.9	0.3			375	15	.40	.50	100	193	8	22	3.70	
3/4 cup (1 oz)	28	2.5	24.0		1.8	1250	.40	5.0	1.50		77	106	.80	.153	

[a] Cinn & raisin, coconut & honey, fruit & nut or toasted oat.

	KCAL / WT (g)	H₂O (g) / PRO (g)	FAT (g) / CHO (g)	PUFA (g) / SFA (g)	CHOL (mg) / DFIB (g)	A (RE) / A (IU)	C (mg) / B-1 (mg)	B-2 (mg) / NIA (mg)	B-6 (mg) / B-12 (mcg)	FOL (mcg) / PANT (mg)	Na (mg) / K (mg)	Ca (mg) / P (mg)	Mg (mg) / Fe (mg)	Zn (mg) / Cu (mg)	Mn (mg)
pac man, General Mills	110		1.0								200				
1 cup (1 oz)	28	1.0	25.0								30				
product 19, Kellogg's	108	0.9	0.2			1502	60	1.70	2.00	400	325	3	11	.43	
3/4 cup (1 oz)	28	2.8	23.5		0.3	5000	1.50	20.0	6.00	.18	44	40	18.00	.079	
puffed rice	57	0.4	0.1					.01	.01	3	0	1	3	.15	.213
1 cup (1/2 oz)	14	0.9	12.8		0.1		.02	0.4		.05	16	14	.15	.024	
puffed rice, Quaker	55	0.4	0.1					.02	.01	3	1	3	6	.00	.000
1 cup (1/2 oz)	14	0.9	12.7			0	.10	1.2	.00	.00	16	26	.41	.020	
puffed wheat	52	0.4	0.2					.03	.02	5	1	4	21	.33	
1 cup (1/2 oz)	14	2.1	11.3		0.5		.03	1.5		.07	49	50	.67	.058	
puffed wheat, Quaker	54	0.4	0.2					.10	.01	5	1	3	13	.00	.000
1 cup (1/2 oz)	14	2.2	10.8			0	.08	2.0	.00	.00	50	48	.67	.060	
Quaker 100% natural	136	0.5	5.6			0		.07	.05	14	11	38	34	1.00	1.000
1/4 cup (1 oz)	28	3.7	17.8		1.0	0	.09	0.5	.00	.00	150	99	.85	.120	
w/ apples & cinn	133	0.5	5.3	0.4				.07	.03	5	15	40	29	1.00	1.000
1/4 cup (1 oz)	28	3.2	18.2	4.2	1.3	0	.11	0.4	.00	.00	141	104	.89	.140	
w/ raisins & dates	132	0.8	5.0	0.4				.03	.04	12	11	36	32	1.00	1.000
1/4 cup (1 oz)	28	3.5	18.2	3.5	1.1	0	.08	0.4	.00	.00	156	99	.86	.120	
quisp, Quaker	121	0.6	2.6	0.3		0		.34	.60	8	189	9	10	.00	.000
1 cup (1 oz)	28	1.3	23.1	1.4	0.4	0	.30	3.0	1.00	.00	37	24	2.70	.010	
raisin bran															
Kellogg's	115	3.1	0.7			375		.40	.50	100	269	13	48	3.80	
3/4 cup (1.3 oz)	37	4.0	27.9		4.0	1250	.40	5.0	1.50		192	137	4.50	.214	
Post	86	2.0	0.4	0.1	0	371	0	.42	.49	99	178	15	52	1.48	
1/2 cup (1 oz)	28	2.5	22.0	0.0	3.7	1235	.37	4.9	1.48		200	129	4.44	.199	
Ralston Purina	120	2.8	0.2			375	1	.40	.50	100	328	18	57	1.13	
3/4 cup (1.3 oz)	38	3.0	31.4		4.8	1250	.40	5.0	1.50	.28	194	167	4.50	.206	
raisin nut bran, General Mills	110		3.0								140				
1/2 cup (1 oz)	28	3.0	20.0		3.0						150				
raisins, rice & rye, Kellogg's	124	2.2	0.1			375	0	.40	.50	100	280	8	16	3.80	
1 cup (1.3 oz)	37	2.1	31.5			1250	.40	5.0	1.50		115	40	4.50	.130	
rice chex, Ralston Purina	110	0.7	0.3				15	.04	.50	100	252	4	7	.32	
1 1/8 cups (1 oz)	28	1.7	25.0		0.5	17	.38	5.0	1.50	.10	32	28	1.80	.068	
rice krispies, Kellogg's	112	0.7	0.2			375	15	.40	.50	100	340	4	10	.48	.281
1 cup (1 oz)	28	1.9	24.8		0.1	1250	.40	5.0		.20	30	34	1.80	.071	
rice krispies, frosted, Kellogg's	109	0.7	0.1			375	15	.40	.50	100	240	1	5	.31	
1 cup (1 oz)	28	1.3	25.7		0.1	1250	.40	5.0		.13	21	27	1.80	.071	
rice toasties, Post	108	1.1	0.1	0.1	0	0					238	4	12	.38	
3/4 cup (1 oz)	28	2.0	23.9	0.0	0.4		.30	4.0			34	30	.83	.070	
rocky road, General Mills	120		2.0								110				
2/3 cup (1 oz)	28	2.0	23.0								50				
ruskets biscuits, Loma Linda	110	2.0	1.0		0			.43			85	14		.59	
2 biscuits	30	4.0	23.0				.31	3.0			100		3.10		
seven grain cereal, crunchy, Loma Linda	110	1.0	2.0		0			.04	.03		90	13		.36	
1/2 cup (1 oz)	28	2.0	21.0				.01	0.7	.00	.15	85		.42		
seven grain cereal, Loma Linda	110	1.0	1.0		0			.04	.06		75	17		.56	
1 cup (1 oz)	28	5.0	20.0				.03	1.1	.00	.34	140		.56		
shredded wheat	102	1.5	0.6					.08	.07	14	3	11	37	.93	
1 oz	28	3.1	22.6		2.6		.07	1.5		.24	102	100	1.20	.188	
	83	1.5	0.3					.06	.06	12	0	10	40	.59	.725
1 biscuit	24	2.6	18.8		2.2		.07	1.1		.19	77	86	.74	.118	
Quaker	104	1.8	0.4					.08	.05	19	1	11	32	1.00	1.000
1 oz	28	3.1	22.0			0	.07	1.3	.00	.00	107	116	.86	.140	
shredded wheat 'n bran, Nabisco	110		1.0								0				
1 oz	28	3.0	23.0								135				
s'mores crunch, General Mills	120		2.0								250				
3/4 cup (1 oz)	28	1.0	24.0								230				

	KCAL	H₂O (g)	FAT (g)	PUFA (g)	CHOL (mg)	Vitamins					Minerals				
						A (RE)	C (mg)	B-2 (mg)	B-6 (mg)	FOL (mcg)	Na (mg)	Ca (mg)	Mg (mg)	Zn (mg)	Mn (mg)
	WT (g)	PRO (g)	CHO (g)	SFA (g)	DFIB (g)	A (IU)	B-1 (mg)	NIA (mg)	B-12 (mcg)	PANT (mg)	K (mg)	P (mg)	Fe (mg)	Cu (mg)	
special K, Kellogg's	111	0.6	0.1			375	15	.40	.50	100	265	8	16	3.70	
1 1/3 cups (1 oz)	28	5.6	21.3		0.2	1250	.40	5.0		.15	49	55	4.50	.128	
sugar corn pops, Kellogg's	108	1.0	0.1			375	15	.40	.50	100	103	1	2	1.50	
1 cup (1 oz)	28	1.4	25.6		0.2	1250	.40	5.0		.06	17	28	1.80	.060	
sugar frosted flakes															
Kellogg's	108	0.7	0.1			375	15	.40	.50	100	230	1	2	.04	
3/4 cup (1 oz)	28	1.4	25.7		0.3	1250	.40	5.0			18	21	1.80	.060	
Ralston Purina	111	0.4	0.4			375	15	.40	.50	2	184	3	2	.61	
3/4 cup (1 oz)	28	1.5	25.5		0.4	1250	.40	5.0	1.50	.01	18	7	.70	.019	
sugar smacks, Kellogg's	106	0.9	0.5			375	15	.40	.50	100	75	3	13	.28	
3/4 cup (1 oz)	28	2.0	24.7		0.4	1250	.40	5.0		.13	42	31	1.80	.079	
sugar sparkled flakes, Post	108	0.5	0.1	0.0	0	371	0	.42	.49	99	170	3	3	.05	
3/4 cup (1 oz)	28	1.2	25.7	0.0	0.3	1235	.37	4.9	1.48		27	7	.36	.044	
sun flakes, Ralston Purina															
corn & rice	110		0.9				15	.13	.50	100	237	9	12	.19	
1 cup (1 oz)	28		24.0		1.0	1250	.38	5.0	1.50	.34	31	25	1.80		
wheat & rice	135		3.4				15		.50	100	182				
1 cup (1 oz)	28	2.3	24.0			1250	.38	5.0	1.50	.48	70		1.80		
super golden crisp, Post	104	0.4	0.2	0.1	0		0	.42	.49	99	44	5	19	1.48	
7/8 cup (1 oz)	28	2.0	25.2	0.0	0.4	1235	.37	4.9	1.48		48	47	2.67	.089	
super sugar crisp	106	0.4	0.3			375		.40	.50	100	25	6	17	1.50	
7/8 cup (1 oz)	28	1.8	25.6		0.4	1250	.40	5.0	1.50	.10	105	51	1.80	.077	
tasteeos, Ralston Purina	111	0.6	0.8			375	15	.40	.50	11	216	13	31	.81	
1 1/4 cups (1 oz)	28	3.6	22.4		1.0	1250	.40	5.0	1.50	.14	84	113	4.50	.167	
team flakes	111	1.1	0.5			375	15	.40	.50		175	4	13	.39	
1 cup (1 oz)	28	1.8	24.3		0.3	1250	.40	5.0	1.50	.21	48	44	1.74	.146	
toasted wheat & raisins, Nabisco	100		1.0								0				
1 oz	28	2.0	23.0								110				
total, General Mills	100	1.1	0.6	0.3		1502	60	1.68	2.00	400	280	48	32	.67	.425
1 cup (1 oz)	28	2.8	22.3	0.1	2.0	5000	1.48	19.8	6.20	.20	110	110	18.00	.122	
trix, General Mills	109		0.4			375	15	.42	.50		170	6	6	.13	
1 cup (1 oz)	28	1.5	25.2		0.1	1250	.37	5.0	1.50	.08	25	19	4.45	.045	
waffelos, Ralston Purina	115	0.7	1.2			375	15	.40	.50	3	118	8	6	.23	
1 cup (1 oz)	28	1.0	24.5		0.3	1250	.40	5.0	1.50	.05	25	231	4.50	.032	
wheat & raisin chex, Ralston Purina	130		0.3					.58	.50	100	198	25	34	.88	
3/4 cup (1.3 oz)	37	3.4	31.0		3.4		.38	5.0	1.50	.12	150	112	5.40	.156	
wheat chex, Ralston Purina	100	0.7	0.7				15	.10	.50	100	200	11	33	.74	
2/3 cup (1 oz)	28	2.9	23.0		3.7		.38	5.0	1.50	.12	107	109	4.50	.135	
wheaties, General Mills	99	1.3	0.5	0.2		375	15	.42	.50	9	270	43	31	.63	.425
1 cup (1 oz)	28	2.7	22.6	0.1	2.0	1250	.37	5.0	1.50	.21	110	98	4.45	.131	

5. CHEESE & CHEESE PRODUCTS

5.1. CHEESE

	KCAL	H₂O (g)	FAT (g)	PUFA (g)	CHOL (mg)	A (RE)	C (mg)	B-2 (mg)	B-6 (mg)	FOL (mcg)	Na (mg)	Ca (mg)	Mg (mg)	Zn (mg)	Mn (mg)
	WT (g)	PRO (g)	CHO (g)	SFA (g)	DFIB (g)	A (IU)	B-1 (mg)	NIA (mg)	B-12 (mcg)	PANT (mg)	K (mg)	P (mg)	Fe (mg)	Cu (mg)	
american, processed	106	11.1	8.9	0.3	27	82	0	.10	.02	2	406	124	6	.85	
1 oz	28	6.3	0.5	5.6	0.0	343	.01	0.0	.20	.14	46	211	.11		
blue	100	12.0	8.2	0.2	21	65	0	.11	.05	10	396	150	7	.75	
1 oz	28	6.1	0.7	5.3	0.0	204	.01	0.3	.35	.49	73	110	.09		
brick	105	11.7	8.4	0.2	27	86	0	.10	.02	6	159	191	7	.74	
1 oz	28	6.6	0.8	5.3	0.0	307	.00	0.0	.36	.08	38	128	.07		
brie	95	13.7	7.9		28		0	.15	.07	18	178	52			
1 oz	28	5.9	0.1		0.0	189	.02	0.1	.47	.20	43	53	.14		
camembert	85	14.7	6.9	0.2	20	71	0	.14	.06	18	239	110	6	.68	
1 oz	28	5.6	0.1	4.3	0.0	262	.01	0.2	.37	.39	53	98	.09		
caraway	107	11.1	8.3			82	0	.13			196	191	6		
1 oz	28	7.1	0.9		0.0	299	.01	0.1	.08	.05		139			
cheddar	114	10.4	9.4	0.3	30	86	0	.11	.02	5	176	204	8	.88	
1 oz	28	7.1	0.4	6.0	0.0	300	.01	0.0	.23	.12	28	145	.19		

	KCAL / WT (g)	H₂O (g) / PRO (g)	FAT (g) / CHO (g)	PUFA (g) / SFA (g)	CHOL (mg) / DFIB (g)	A (RE) / A (IU)	C (mg) / B-1 (mg)	B-2 (mg) / NIA (mg)	B-6 (mg) / B-12 (mcg)	FOL (mcg) / PANT (mg)	Na (mg) / K (mg)	Ca (mg) / P (mg)	Mg (mg) / Fe (mg)	Zn (mg) / Cu (mg)	Mn (mg)
cheddar	403	36.8	33.1	0.9	105	303	0	.38	.07	18	620	721	28	3.11	
3.5 oz	100	24.9	1.3	21.1	0.0	1059	.03	0.1	.83	.41	98	512	.68		
cheddar, grated	455	41.5	37.5	1.1	119	342	0	.42	.08	21	701	815	31	3.51	
1 cup, not packed	113	28.1	1.5	23.8	0.0	1197	.03	0.1	.94	.47	111	579	.77		
cheddar, med, Kraft	114	10.5	9.3	0.3	28		0	.11	.02	6	174	208	7	1.01	.010
1 oz	28	7.1	0.5	5.4		366	.01	0.1	.23	.11	20	143	.07	.004	
cheshire	110	10.7	8.7		29	69	0	.08			198	182	6		
1 oz	28	6.6	1.4		0.0	279	.01				27	131	.06		
colby	112	10.8	9.1	0.3	27	78	0	.11	.02		171	194	7	.87	
1 oz	28	6.7	0.7	5.7	0.0	293	.00	0.0	.23	.06	36	129	.22		
cottage cheese, creamed	29	22.3	1.3	0.0	4	14	0	.00	.02	4	114	17	2	.20	
1 rd T	28	3.5	0.8	0.8	0.0	48	.00	0.0	.18	.06	24	37	.07		
cottage cheese, creamed	117	89.2	5.1	0.2	17	54	0	.18	.08	14	457	68	6	.42	
4 oz	113	14.1	3.0	3.2	0.0	184	.02	0.1	.70	.24	95	149	.16		
cottage cheese, creamed	217	165.8	9.5	0.3	31	101	0	.34	.14	26	850	126	11	.78	
1 cup, not packed	210	26.2	5.6	6.0	0.0	342	.04	0.3	1.31	.45	177	277	.29		
cottage cheese, creamed w/ fruit	279	163.0	7.7	0.2	25		0	.29	.12	22	915	108	9	.66	
1 cup	226	22.4	30.1	4.9		139	.04	0.2	1.12	.38	151	236	.25		
cottage cheese, dry curd	123	115.7	0.6	0.0	10	12	0	.21	.12	21	19	46	6	.68	
1 cup, not packed	145	25.0	2.7	0.4	0.0	44	.04	0.2	1.20	.24	47	151	.33		
cottage cheese, low fat, 1% fat	164	186.4	2.3	0.0	10	25	0	.37	.15	28	918	138	12	.80	
1 cup	226	28.0	6.2	1.5	0.0	84	.05	0.3	1.43	.49	183	302	.32		
cottage cheese, low fat, 2% fat	203	179.2	4.4	0.1	19	45	0	.42	.17	30	918	155	14	.95	
1 cup	226	31.1	8.2	2.8	0.0	158	.05	0.3	1.61	.55	217	340	.36		
cream cheese	99	15.2	9.9	0.4	31	124	0	.06	.01	0	84	23	2	.15	
1 oz (2T)	28	2.1	0.8	6.2	0.0	405	.01	0.0	.12	.08	34	30	.34		
cream cheese, light, Philadelphia Brand	62	18.3	4.7	0.1	16		0	.07	.01	5	160	38	3	.21	.005
1 oz	28	2.9	1.8	2.8		168	.01	0.0	.09	.16	52	45	.04	.008	
cream cheese, Philadelphia Brand	98	15.5	9.5	0.3	28		0	.04	.01	5	86	20	2	.17	.010
1 oz	28	2.3	0.7	5.4		349	.01	0.1	.06	.10	25	27	.02	.009	
cream cheese w/ herb & garlic, Cremerie	103	14.4	9.9	0.3	29		0	.05	.01	4	216	29	3	.17	
1 oz	28	0.9	2.5	5.5		373	.01	0.1	.07	.10	37	30	.10	.015	
edam	101	11.8	7.9	0.2	25	72	0	.11	.02	5	274	207	8	1.06	
1 oz	28	7.1	0.4	5.0	0.0	260	.01	0.0	.44	.08	53	152	.12		
feta	75	15.7	6.0	0.2	25		0				316	140	5	.82	
1 oz	28	4.0	1.2	4.2	0.0						18	96	.18		
fontina	110	10.8	8.8	0.5	33		0	.06				156	4	.99	
1 oz	28	7.3	0.4	5.4	0.0	333	.01	0.0					.06		
gjetost	132	3.8	8.4	0.3							170	113			
1 oz	28	2.7	12.1	5.3	0.0			0.2				126			
gouda	101	11.8	7.8	0.2	32	49	0	.10	.02	6	232	198	8	1.11	
1 oz	28	7.1	0.6	5.0	0.0	183	.01	0.0		.10	34	155	.07		
gruyere	117	9.4	9.2	2.4	31		0	.08	.02	3	95	287			
1 oz	28	8.5	0.1	5.4	0.0	346	.02	0.0	.45	.16	23	172			
havarti, Casino	121	10.4	10.6	0.3	34		0	.09	.01	11	144	176	5	.85	.005
1 oz	28	6.1	0.3	6.5		361	.01	0.0	.16	.08	13	114	.04	.015	
limburger	93	13.7	7.7	0.1	26		0	.14	.02	16	227	141	6	.60	
1 oz	28	5.7	0.1	4.8	0.0	363	.02	0.1	.30	.33	36	111	.04		
monterey	106	11.6	8.6				0	.11			152	212	8	.85	
1 oz	28	6.9	0.2		0.0	269					23	126	.20		
mozzarella	80	15.4	6.1	0.2	22	68	0	.07	.02	2	106	147	5	.63	
1 oz	28	5.5	0.6	3.7	0.0	225	.00	0.0	.19	.02	19	105	.05		
mozzarella, low moisture	90	13.7	7.0	0.2	25	78	0	.08	.02	2	118	163	6	.70	
1 oz	28	6.1	0.7	4.4	0.0	256	.01	0.0	.21	.02	21	117	.06		
mozzarella, part skim	72	15.3	4.5	0.1	16	50	0	.09	.02	2	132	183	7	.78	
1 oz	28	6.9	0.8	2.9	0.0	166	.01	0.0	.23	.02	24	131	.06		
mozzarella, part skim, low moisture	79	13.8	4.9	0.1	15	54	0	.10	.02	3	150	207	7	.89	
1 oz	28	7.8	0.9	3.1	0.0	178	.01	0.0	.26	.03	27	149	.07		
muenster	104	11.8	8.5	0.2	27	90	0	.09	.02	3	178	203	8	.80	
1 oz	28	6.6	0.3	5.4	0.0	318	.00	0.0	.42	.05	38	133	.12		
neufchatel	74	17.6	6.6	0.2	22	75	0	.06	.01	3	113	21	2	.15	
1 oz	28	2.8	0.8	4.2	0.0	321	.00	0.0	.08	.16	32	39	.08		

	KCAL	H₂O (g)	FAT (g)	PUFA (g)	CHOL (mg)	Vitamins A (RE)	C (mg)	B-2 (mg)	B-6 (mg)	FOL (mcg)	Minerals Na (mg)	Ca (mg)	Mg (mg)	Zn (mg)	Mn (mg)
	WT (g)	PRO (g)	CHO (g)	SFA (g)	DFIB (g)	A (IU)	B-1 (mg)	NIA (mg)	B-12 (mcg)	PANT (mg)	K (mg)	P (mg)	Fe (mg)	Cu (mg)	
parmesan, grated	23	0.9	1.5	0.0	4		0	.02	.01	0	93	69	3	.16	
1 T	5	2.1	0.2	1.0	0.0	35	.00	0.0		.03	5	40	.05		
parmesan, hard	111	8.3	7.3	0.2	19		0	.09	.03	2	454	336	12	.78	
1 oz	28	10.1	0.9	4.7	0.0	171	.01	0.1		.13	26	197	.23		
pimento, processed	106	11.1	8.8	0.3	27			.10	.02	2	405	174	6	.84	
1 oz	28	6.3	0.5	5.6		358	.01	0.0	.20	.14	46	211	.12		
port du salut	100	12.9	8.0	0.2	35		0	.07	.02	5	151	184			
1 oz	28	6.7	0.2	4.7	.0.0	378		0.0	.43	.06		102			
provolone	100	11.6	7.6	0.2	20	75	0	.09	.02	3	248	214	8	.92	
1 oz	28	7.3	0.6	4.8	0.0	231	.01	0.0	.42	.14	39	141	.15		
ricotta, part skim	171	92.3	9.8	0.3	38	140	0	.23	.03		155	337	18	1.66	
1/2 cup	124	14.1	6.4	6.1	0.0	536	.03	0.1	.36		155	226	.55		
ricotta, whole milk	216	88.9	16.1	0.5	63	166	0	.24	.05		104	257	14	1.44	
1/2 cup	124	14.0	3.8	10.3	0.0	608	.02	0.1	.42		130	196	.47		
romano	110	8.8	7.6		29		0	.11		2	340	302			
1 oz	28	9.0	1.0		0.0	162		0.0				215			
roquefort, sheep's milk	105	11.2	8.7	0.4	26		0	.17	.04	14	513	188	8	.59	
1 oz	28	6.1	0.6	5.5	0.0	297	.01	0.2	.18	.49	26	111	.16		
swiss	107	10.6	7.8	0.3	26	72	0	.10	.02	2	74	272	10	1.11	
1 oz	28	8.1	1.0	5.0	0.0	240	.01	0.0	.48	.12	31	171	.05		
swiss, processed	95	12.0	7.1	0.1	15	65	0	.08	.01		388	219	8	1.02	
1 oz	28	7.0	0.6	2.9	0.0	229	.00	0.0	.35	.07	61	216	.17		
tilsit, whole milk	96	12.2	7.4	0.2	29		0	.10			213	198	4	.99	
1 oz	28	6.9	0.5	4.8	0.0	296	.02	0.1	.60	.10	18	142	.06		

5.2. CHEESE PRODUCTS

	KCAL	H₂O (g)	FAT (g)	PUFA (g)	CHOL (mg)	A (RE)	C (mg)	B-2 (mg)	B-6 (mg)	FOL (mcg)	Na (mg)	Ca (mg)	Mg (mg)	Zn (mg)	Mn (mg)
	WT (g)	PRO (g)	CHO (g)	SFA (g)	DFIB (g)	A (IU)	B-1 (mg)	NIA (mg)	B-12 (mcg)	PANT (mg)	K (mg)	P (mg)	Fe (mg)	Cu (mg)	
cheese food															
american	93	12.2	7.0	0.2	18		0	.13			337	163	9	.85	
1 oz	28	5.6	2.1	4.4	0.0	259	.01	0.0	.32	.16	79	130	.24		
american, cold pack	94	12.2	6.9	0.2	18		0	.13	.04	2	274	141	8	.85	
1 oz	28	5.6	2.4	4.4	0.0	200	.01	0.0	.36	.28	103	113	.24		
american, grated, Kraft	131	1.1	7.1	0.2	18		0	.31	.07	10	790	266	20	.76	
1 oz	28	7.8	9.0	4.1		260	.06	0.1	.91	.90	310	312	.20	.030	
hot pepper, Land O'Lakes	92		6.9	0.3	18		0	.13			357	169			
1 oz	28	5.6	1.9	4.3		126	.01	0.1						.18	
jalapeno, Land O'Lakes	90		7.0	1.0	20						360				
1 oz	28	6.0	2.0	4.0											
la chedda, Land O'Lakes	90		7.0	1.0							335				
1 oz	28	6.0	2.0	4.0							80				
onion, Land O'Lakes	90		7.0	1.0	15						330				
1 oz	28	6.0	2.0	4.0											
pepperoni, Land O'Lakes	90		7.0	1.0	20						395				
1 oz	28	6.0	1.0	4.0							65				
salami, Land O'Lakes	100		8.0	1.0	20						400				
1 oz	28	5.0	2.0	5.0											
sharp cheddar w/ wine, cold pack, Land O'Lakes—1 oz	93		6.9	0.2	18		0	.12			270	139			
1 oz	28	5.5	2.3	4.3		197	.01	0.0			102	112	.24		
swiss	92	12.4	6.8		23		0	.11			440	205	8	1.01	
1 oz	28	6.2	1.3		0.0	243	.00	0.0	.65	.14	81	149	.17		
cheese nuggets															
cheddar, breaded, frzn, Banquet	414		30.0				0	.01			1015	328			
3 oz	85	13.0	24.0			300	.02	0.3			46		.50		
mozzarella, breaded, frzn, Banquet	288		16.0				0	.16			750	355			
3 oz	85	15.0	21.0			310	.02	0.3			81		.50		
cheese products															
Light n' Lively, sharp cheddar	72	14.0	4.1	0.1	12		0	.12	.03	5	404	192	11	.92	
1 oz	28	5.9	2.9	2.4		205	.02	0.1	.22	.21	64	141	.08	.008	

	KCAL	H₂O (g)	FAT (g)	PUFA (g)	CHOL (mg)	A (RE)	C (mg)	B-2 (mg)	B-6 (mg)	FOL (mcg)	Na (mg)	Ca (mg)	Mg (mg)	Zn (mg)	Mn (mg)
	WT (g)	PRO (g)	CHO (g)	SFA (g)	DFIB (g)	A (IU)	B-1 (mg)	NIA (mg)	B-12 (mcg)	PANT (mg)	K (mg)	P (mg)	Fe (mg)	Cu (mg)	
Light n' Lively, singles	73	13.9	4.3	0.1	15		0	.12	.03	6	411	190	9	.88	.010
1 oz	28	6.4	2.1	2.5		196	.02	0.1	.33	.19	75	157	.08	.014	
Light n' Lively, swiss	71	13.9	3.7	0.1	12		0	.11	.03	3	366	214	11	.93	
1 oz	28	6.3	3.1	2.0		100	.02	0.1	.32	.21	69	150	.09	.032	
cheese sce															
cnd, Campbell's	60		4.1				0	.04			288	63			
2 oz	57	2.3	3.5			161	.00	0.1			23		.20		
from dry mix	158	108.0	8.5	0.8	26		1	.28			783	285	23	.48	
1/2 cup	139	8.0	11.6	4.6	0.0		.07	0.1			277	218	.13		
Land O'Lakes	41		2.9		6		0	.05			205	47			
1 oz	28	1.5	2.2			57	.01	0.0			29	100	.05		
aged cheddar	50		3.7		8		0	.05			232	69			
1 oz	28	2.3	1.8			84	.01	0.0			29	124	.06		
la chedda cheddar	37		2.5		4		0	.04			181	39			
1 oz	28	1.2	2.4			45	.01	0.0			28	80	.04		
cheese spread															
american	47	7.7	3.4	0.1	9		0	.03	.01	1	218	91	5	.16	
1 T	16	2.7	1.4	2.2	0.0	127	.00	0.0	.06	.11	39	115	.05		
american	82	13.5	6.0	0.2	16		0	.12	.03	2	381	159	8	.73	
1 oz	28	4.7	2.5	3.8	0.0	223	.01	0.0	.11	.19	69	202	.09		
Cheez Whiz, Kraft	77	14.5	5.7	0.1	16		0	.11	.02	4	370	147	8	.69	
1 oz	28	4.6	1.8	3.1		193	.02	0.1	.18	.16	52	255	.06	.019	
Easy Cheese, all flavors[a]	80		6.0								345				
1 oz	28	4.0	2.0								73				
Land O'Lakes, golden velvet	80		6.0	1.0	15						380				
1 oz	28	5.0	2.0	4.0							70				
Velveeta sharp	84	13.0	6.1	0.2	21		0	.13	.02	6	454	154	8	.61	.010
1 oz	28	5.2	2.2	3.6		343	.02	0.0	.33	.19	88	292	.06	.015	

6. CHIPS, PRETZELS, POPCORN & SIMILAR SNACK FOODS

	KCAL	H₂O	FAT	PUFA	CHOL	A (RE)	C	B-2	B-6	FOL	Na	Ca	Mg	Zn	Mn
bugles, General Mills	150		8.0								290				
1 oz	28	2.0	18.0								20				
bugles, nacho cheese, General Mills	160		9.0								270				
1 oz	28	2.0	17.0								35				
cheese balls/curls, Planters	160		11.0								280				
1 oz	28	2.0	14.0								40				
cheese 'n crunch, Nabisco	160		11.0								190				
41 pieces (1 oz)	28	2.0	15.0								55				
cheese straws	272	13.0	17.9				0	.10			433	155			
10 pieces	60	6.7	20.7	6.4[b]		230	.01	0.2			38	124	.40		
chex party mix	130		4.7				6	.10	.30	60	320		16	.30	
2/3 cup (1 oz)	28	3.0	19.0				.30	4.0	.90			40	1.80	.080	
nacho	120		4.3				6	.10	.30	60	430		16	.30	
2/3 cup (1 oz)	28	2.9	18.0				.30	4.0	.90			40	1.80	.080	
sweet & nutty	130		4.1				9	.07	.30	60	240		8	.30	
3/4 cup (1 oz)	28	3.1	20.0				.30	4.0	.90			40	1.80	.080	
chipsters light 'n crispy potato snacks, Nabisco—57 pieces (1 oz)	120		5.0								580				
	28	1.0	19.0								180				
corn chips	153	0.3	8.8				0	0.3			218	2	22	.43	
1 oz	28	1.7	16.6			65	.05	0.0	.00			31	.039		
corn diggers, Nabisco	150		8.0								260				
36 pieces (1 oz)	28	2.0	17.0								25				

[a] Values are averages for american, cheddar, cheddar 'n chives, cheese 'n bacon, nacho & sharp cheddar.

[b] If made w/ veg shortening; 7.9 g if made w/ lard.

	KCAL / WT (g)	H₂O (g) / PRO (g)	FAT (g) / CHO (g)	PUFA (g) / SFA (g)	CHOL (mg) / DFIB (g)	A (RE) / A (IU)	C (mg) / B-1 (mg)	B-2 (mg) / NIA (mg)	B-6 (mg) / B-12 (mcg)	FOL (mcg) / PANT (mg)	Na (mg) / K (mg)	Ca (mg) / P (mg)	Mg (mg) / Fe (mg)	Zn (mg) / Cu (mg)	Mn (mg)
doo dads, all flavors, Nabisco[a]	140		6.0								393				
1/2 cup (1 oz)	28	3.0	18.0								80				
peanut butter boppers, General Mills[b]	166		10.2								115				
1 oz bar	28	4.0	14.4								108				
pizza crunchies, Planters	160		10.0								160				
1 oz	28	2.0	15.0								45				
popcorn, popped	23	0.2	0.3				1	.01			0	1			
1 cup	6	0.8	4.6	0.0				0.1				17	.20		
fat & salt added	41	0.3	2.0				0	.01			175	1			
1 cup	9	0.9	5.3	0.9[c]				0.2				19	.20		
Pillsbury microwave, frzn															
butter flavor	192	0.9	11.5				3	.03			273	9			
3 cups	37	2.8	18.9			51	.04	0.6			91	69	.63		
original flavor	192	0.9	11.5				2	.03			301	9			
3 cups	37	2.8	18.9			51	.05	0.6			91	69	.63		
salt free	139	0.8	5.9				2	.03			5	2			
3 cups	30	2.8	18.9			51	.03	0.6			90	66	.60		
pop secret, natural/butter flavor, General Mills—*4 cups*	230		13.5								430				
	50	4.0	27.5								95				
sugar-coated	134	1.4	1.2				0	.02			0	2			
1 cup	35	2.1	29.9	0.4				0.4				47	.50		
potato chips	148	0.7	10.1	5.2	0	0	12	.01	.14	13	133[d]	7	17	.30	.125
1 oz	28	1.8	14.7	2.6		0	.04	1.2	.00	.11	369	43	.34	.057	
from dried potatoes	164	0.5	13.1	0.9	0		2	.02			216	6	16	.17	
1 oz	28	1.6	12.4	4.0			.04	0.9			312	40	.43	.028	
Pringle's	167	0.5	13.0		0		2	.02			215	7	16	.17	
1 oz	28	1.6	10.8				.04	0.9			312	40	.43	.028	
cheez-ums	167	0.5	13.0		2		2	.03			240	31			
1 oz	28	1.9	10.5				.05	0.7					.45		
light style	147	0.6	8.2		0		4	.02			152	8	16	.20	
1 oz	28	1.9	16.5				.05	1.0			301	48	.45	.028	
rippled	167	0.6	11.9		0		2	.02			249	9	18	.20	
1 oz	28	1.7	13.3				.04	0.9			286	48	.34	.057	
sour cream 'n onion	167	0.6	13.0		0		3	.03			146	18			
1 oz	28	1.8	10.8				.05	0.7					.40		
potato sticks	148	0.6	9.8	5.0	0	0	13	.03	.09	11	71	5	18	.28	.119
1 oz	28	1.9	15.2	2.5		0	.03	1.4	.00	.11	351	49	.65	.090	
pretzels	111	0.7	1.0				0	.07	.01	4	451	7	7	.30	
1 oz	28	2.6	22.4			0	.09	1.2	.00	.08	28	25	.55	.041	
pretzels, all shapes, Mister Salty[e]	110		1.2								543				
1 oz	28	3.0	21.5								37				
sour cream & onion puffs, Planters	160		10.0								300				
1 oz	28	1.0	16.0								45				
taro chips	110	0.6	5.9	0.8	0	0		.01			85	10	19		
10 chips	23	0.5	15.5	1.8		0	.01	0.0	.00		189	30	.31		
tortilla chips, reg/nacho, Planters	150		8.0								155				
1 oz	28	2.0	18.0								63				

7. CREAMS & CREAM SUBSTITUTES

creamers															
liquid															
coffee rich, non-dairy creamer	22	10.4	1.6	0.5	0						7	0	0	.00	
1/2 oz	14	0.0	2.1	0.3							5	5	.02	.000	

[a] Values are averages for cheddar 'n bacon, cheddar 'n herb, original & zesty cheese.
[b] Values are averages for cookie crunch, fudge chip, fudge graham, honey crisp & peanut crunch.
[c] If butter; 1.4 g if coconut oil.
[d] If no salt is added, Na is 2 mg.
[e] Values are averages for 15 shapes.

	KCAL	H₂O (g)	FAT (g)	PUFA (g)	CHOL (mg)	A (RE)	C (mg)	B-2 (mg)	B-6 (mg)	FOL (mcg)	Na (mg)	Ca (mg)	Mg (mg)	Zn (mg)	Mn (mg)
	WT (g)	PRO (g)	CHO (g)	SFA (g)	DFIB (g)	A (IU)	B-1 (mg)	NIA (mg)	B-12 (mcg)	PANT (mg)	K (mg)	P (mg)	Fe (mg)	Cu (mg)	
poly rich, non-dairy creamer	22	10.5	1.4	0.7	0						5	0	0	.00	
1/2 oz	14	0.0	2.1	0.3							11	5	.00	.003	
liquid/frzn[a]	20	11.6	1.5	0.0	0	1	0	.00	.00	0	12	1	0	.00	
1/2 fl oz	15	0.2	1.7	0.3	0.0	13	.00	0.0	.00	.00	29	10	.00		
liquid/frzn[b]	20	11.6	1.5	0.0	0	1	0	.00	.00	0	12	1	0	.00	
1/2 fl oz	15	0.2	1.7	1.4	0.0	13	.00	0.0	.00	.00	29	10	.00		
powdered[b]	11	0.0	0.7	0.0	0	0	0	.00	.00	0	4	0	0	.00	
1 t	2	0.1	1.1	0.7	0.0	4	.00	0.00	.00	.00	16	8	.00		
powdered, coffee-mate, Carnation	10		1.0	0.0	0						2	1			
1 t	2	1.0	1.0	0.7	0.0						18	1			
half & half cream	20	12.1	1.7	0.0	6	16	0	.02	.01	0	6	16	2	.08	
1 T	15	0.4	0.6	1.1	0.0	65	.01	0.0	.05	.04	19	14	.01		
light (coffee/table) cream	29	11.1	2.9	0.1	10	27	0	.02	.01	0	6	14	1	.04	
1 T	15	0.4	0.6	1.8	0.0	108	.01	0.0	.03	.04	18	12	.01		
medium (25% fat) cream	37	10.3	3.8	0.1	13	35	0	.02	.01	0	6	14	1	.04	
1 T	15	0.4	0.5	2.3	0.0	141	.00	0.0	.03	.04	17	11	.01		
sour cream															
cultured	26	8.5	2.5	0.0	5	23	0	.02	.00	1	6	14	1	.03	
1 T	12	0.4	0.5	1.6	0.0	95	.00	0.0	.04	.04	17	10	.01		
half & half, cultured	20	12.0	1.8	0.0	6	17	0	.02	.00	2	6	16	2	.08	
1 T	15	0.4	0.6	1.1	0.0	68	.01	0.0	.05	.05	19	14	.01		
imitation	59	20.2	5.5	0.0	0	0	0	.00	.00	0	29	1			
1 oz	28	0.7	1.9	5.0	0.0	0	.00	0.0	.00	.00	46	13			
sour cream dip, flavored, Land O'Lakes	70		5.0	1.0	10						315				
1.7 oz	48	2.0	4.0	3.0											
whipped cream, pressurized	8	1.8	0.7	0.0	2	6	0	.00	.00		4	3	0	.01	
1 T	3	0.1	0.4	0.4	0.0	27	.00	0.0	.01	.01	4	3	.00		
whipping cream															
heavy, fluid	52	8.7	5.6	0.2	21	63	0	.02	.00	1	6	10	1	.03	
1 T	15	0.3	0.4	3.5	0.0	220	.00	0.0	.03	.04	11	9	.00		
light, fluid	44	9.5	4.6	0.1	17	44	0	.02	.00	1	5	10	1	.04	
1 T	15	0.3	0.4	2.9	0.0	169	.00	0.0	.03	.04	15	9	.00		
whipped topping															
from mix, Dream Whip	9	3.3	0.4	0.0	1		0	.00	.00	0	3	5	1	.02	
1 T	5	0.2	1.0	0.3		10	.00	0.0	.01	.01	6	4	.00	.001	
from mix, w/ nutrasweet, D-Zerta	7	3.9	0.6	0.0	0		0	.01	.00	0	7	3	1	.01	
1 T (from 1.112 g dry mix)	5	0.1	0.3	0.5		9	.00	0.0	.01	.02	8	5	.00	.000	
from mix, prep w/ whole milk	8	2.7	0.5	0.0	0	2	0	.01	.00	0	3	4	0	.01	
1 T	4	0.1	0.7	0.4	DFIB	14	.00	0.0	.01	.01	6	3	.00		
frzn	13	2.0	1.0	0.0	0	3	0	.00	.00	0	1	0	0	.00	
1 T	4	0.1	0.9	0.9	0.0	34	.00	0.0	.00	.00	1	0	.00		
frzn, Cool Whip extra creamy, dairy	16	2.4	1.2	0.0	0		0	.00	.00	0	4	2	0	.01	
1 T	5	0.1	1.2	1.2		38	.00	0.0	.00	.00	2	3	.00	.000	
frzn, Cool Whip non-dairy	11	2.1	0.8	0.0	0		0	.00			1	0	0	.00	
1 T	4	0.1	1.0	0.8		27	.00		.00		0	0	.00		
liquid, non-dairy topping, Richwhip[c]	20	4.2	1.6								4	0	0	.00	
1/4 oz	7		1.2	1.3							0	0	.00	.000	
pressurized	11	2.4	0.9	0.0	0	2	0	.00	.00	0	2	0	0	.00	
1 T	4	0.0	0.6	0.8	0.0	19	.00	0.0	.00	.00	1	1	.00		
pressurized, non-dairy topping, Richwhip—*1/4 oz*	20	4.2	1.7								3	0	0	.00	
	7	0.0	1.1	1.4							1	0	.00	.000	
prewhipped, non-dairy topping, Richwhip—*1 T*	12	2.0	0.9		0			.00			1	0	0	.00	
	4	0.1	1.0	0.9			.00	0.0			1	2	.01	.003	

[a] Contains hydrogenated veg oil & soy protein; veg oils are usually soybean, cottonseed, safflower, or blends thereof.

[b] Contains lauric acid oils and Na caseinate; lauric oils include modified coconut oil, hydrogenated coconut oil, and/or palm kernel oil.

[c] 1 fl oz if whipped.

8. DESSERTS
8.1. BROWNIES & OTHER BAR COOKIES

Each food item occupies two rows. Column headers are paired (top value / bottom value):

Food / Portion	KCAL / WT (g)	H₂O / PRO (g)	FAT / CHO (g)	PUFA / SFA (g)	CHOL (mg) / DFIB (g)	A (RE) / A (IU)	C / B-1 (mg)	B-2 / NIA (mg)	B-6 (mg) / B-12 (mcg)	FOL (mcg) / PANT (mg)	Na / K (mg)	Ca / P (mg)	Mg / Fe (mg)	Zn / Cu (mg)	Mn (mg)
brownie															
black forest, from mix, Pillsbury	160		6.0								100				
2" square	65	2.0	25.0								65				
double fudge, from mix, Pillsbury	160		6.0								105				
2" square	60	2.0	24.0								60				
from mix	86	2.1	4.0				0	.02			33	9			
1 brownie	20	1.0	12.6	0.8		20	.03	0.1			34	27	.40		
frzn, Am Hosp Co	216		2.1		50			.06			180	10			
2.25 oz	64	5.8	43.5			50	.04	1.4			24		1.20		
fudge, from mix, Pillsbury	150		6.0								100				
2" square	35	2.0	21.0								35				
fudge 'n nut, Almost Home	160		7.0								75				
1 brownie	35	1.0	23.0								50				
fudge, refrig dough, Pillsbury	139	5.4	5.3				0	.05			112	12			
1 bar	34	1.3	21.6			0	.04	0.8			60	20	.85		
Rainbo Snacks	230		12.0												
1 brownie	51	2.0	30.0												
rocky road fudge, from mix, Pillsbury	170		7.0								100				
2" square		2.0	24.0								50				
w/ nuts, homemade	97	2.0	6.3				0	.02			50	8			
1 brownie (3x1x7/8")	20	1.3	10.2	1.4		40	.04	0.1			38	30	.40		
w/ nuts & choc icing	81	2.6	3.5				0	.05	.03	2	47	10	8	.29	
1 brownie	20	0.9	12.7			59	.06	0.3	.04	.10	40	21	.48	.091	
w/ nuts & choc icing, frzn	103	3.1	5.0				0	.02			49	10			
1 brownie (1 1/2x 1 3/4x7/8")	25	1.2	14.9	1.7		50	.02	0.1			44	31	.40		
w/ walnuts, from mix, PIllsbury	150		8.0								90				
2" square	40	2.0	19.0								40				
brownie mix, Duncan Hines	120	1.0	2.8	0				.04			98	6	12	.17	
1/24 pkg (1 brownie)	28	1.4	22.1				.04	0.5				22	1.15	.056	
choc chip oatmeal fudge jumbles, from mix, Pillsbury—1 bar	100		4.0								60				
		1.0	14.0								30				
coconut bar	45	0.3	2.2				0	.01			13	7			
1 bar (3x1 1/4x1/4")	9	0.6	5.8	0.8		14	.00	0.0			21	11	.13		
peanut butter oatmeal fudge jumbles, from mix, Pillsbury—1 bar	100		4.0								55				
		1.0	14.0								55				

8.2. CAKES

Food / Portion	KCAL / WT (g)	H₂O / PRO (g)	FAT / CHO (g)	PUFA / SFA (g)	CHOL (mg) / DFIB (g)	A (RE) / A (IU)	C / B-1 (mg)	B-2 / NIA (mg)	B-6 (mg) / B-12 (mcg)	FOL (mcg) / PANT (mg)	Na / K (mg)	Ca / P (mg)	Mg / Fe (mg)	Zn / Cu (mg)	Mn (mg)
amaretto bundt, frzn, Am Hosp Co	289		13.2		48			.04			288	28			
2.5 oz	71	7.3	32.6			52	.08	0.1			92		.19		
angel food															
cake mix, Duncan Hines	131	1.0	0.2	0				.14			119	20	3	.02	
1/12 pkg	34	3.4	28.9				.00	0.1			85	32	.05	.003	
homemade	161	18.9	0.1				0	.14	.01	5	161	7	7	.12	
1 piece	60	4.8	35.7			0	.07	0.7	.02	.14	59	17	.51	.023	
from mix	126	20.0	0.2				0	.11	.01	4	269	44	4	.05	
1/12 cake	53	3.2	28.6			0	.03	0.1	.00	.12	70	91	.23	.032	
apple crumb, frzn, Am Hosp Co	214		8.7					.11			213	49			
2 oz	57	2.5	31.2			51	.06	0.4			68		.60		
applesce spice, from mix, Pillsbury	250		11.0								300				
1/12 cake		3.0	36.0								80				
apple spice, frzn, Am Hosp Co	316		12.9		9		2	.11			472	35			
3 oz	85	3.6	46.0			910	.05	1.7			60		.93		
banana, from mix, Pillsbury	250		11.0								290				
1/12 cake		3.0	36.0								80				

	KCAL	H_2O (g)	FAT (g)	PUFA (g)	CHOL (mg)	A (RE)	C (mg)	B-2 (mg)	B-6 (mg)	FOL (mcg)	Na (mg)	Ca (mg)	Mg (mg)	Zn (mg)	Mn (mg)
	WT (g)	PRO (g)	CHO (g)	SFA (g)	DFIB (g)	A (IU)	B-1 (mg)	NIA (mg)	B-12 (mcg)	PANT (mg)	K (mg)	P (mg)	Fe (mg)	Cu (mg)	
banana, frzn, Am Hosp Co	308		13.7		39		0	.12			306	71			
3 oz	85	6.3	40.3			280	.01	0.8			103		.70		
blueberry crumb, frzn, Am Hosp Co	210		7.7					.10			200	35			
2 oz	57	2.8	32.3			46	.05	0.7			47		.60		
boston cream pie, homemade[a]	311	35.5	9.7				0	.11			192	69			
1/8 cake	103	5.2	51.4	3.0[b]		220	.03	0.2			92	104	.50		
bundt ring, from mix, Pillsbury[c]	248		9.8								295				
1/16 cake		2.8	37.4									73			
butter recipe, from mix, Pillsbury	260		12.0								370				
1/12 cake		3.0	34.0									35			
caramel w/ caramel icing, homemade	398	21.9	15.5				0	.07			265	88			
1/12 cake	105	3.9	62.1	4.7		210	.02	0.1			67	100	1.60		
carrot cake mix, Duncan Hines Deluxe	187	1.0	4.0		0			.06			253	62	3	.09	
1/12 pkg	44	1.6	36.1			1175	.08	0.7			48	110	.62	.022	
carrot, frzn, Am Hosp Co	326		17.1		24		0	.10			310	28			
3 oz	85	5.6	36.8			442	.00	0.9			122		1.10		
carrot 'n spice, from mix, Pillsbury	260		11.0								330				
1/12 cake		3.0	36.0								85				
cheesecake	257	39.0	16.3				4	.11	.05	15	189	48	9	.36	
1 piece	85	4.6	24.3			216	.03	0.4	.42	.49	83	75	.41	.051	
cheesecake, from mix															
Jell-O[d]	278	43.0	12.9	1.0	28		0	.21	.03	6	350	142	18	.29	
1/8 cheesecake	99	4.9	36.3	8.0		357	.10	0.5	.18	.22	199	187	.39	.038	
Royal[d]	225		9.0								370				
1/8 cheesecake		5.0	31.0								210				
cheesecake, lite, from mix, Royal[e]	210		10.0								380				
1/8 cheesecake		5.0	23.0								250				
cheese crumb, frzn, Am Hosp Co	202		9.3					.08			191	27			
2 oz	57	4.3	25.5			124	.02	0.2			40		.60		
cherry crumb, frzn, Am Hosp Co	193		7.3					.10			168	37			
2 oz	57	3.1	29.0			159	.05	0.6			65		.70		
choc chip cake mix, Duncan Hines Deluxe	189	1.2	4.4		0			.08			249	57	6	.13	
1/12 pkg	44	1.8	35.6				.11	1.0				119	.92	.040	
choc chip crumb, frzn, Am Hosp Co	260		13.8								231	30			
2 oz	57	3.7	30.3			89	.03	0.5			127		.80		
choc chip, from mix, Pillsbury	270		14.0								290				
1/12 cake		3.0	33.0								60				
choc chip pudding, frzn, Am Hosp Co	355		17.8		9		0	.18			336	33			
3 oz	85	3.5	45.1			1	.00	0.3			108	68			
choc, frzn, Am Hosp Co	326		13.7		27		0	.21			293	35			
3 oz	85	7.5	43.7			201	.02	1.3			234		4.20		
choc mint, from mix, Pillsbury	250		12.0								370				
1/12 cake		3.0	32.0								85				
choc mousse, frzn, Nutra Choice	150		6.0		5						273	65			
2.2 oz serving	62	2.8	21.0										.79		
coffee cake															
apple cinn, from mix, Pillsbury	240		7.0								160				
1/8 cake		3.0	40.0								90				
from mix	232	21.6	6.9				0	.12			310	44			
1/6 cake	72	4.5	37.7	2.0		120	.13	1.0			78	125	1.20		
mix, Aunt Jemima	162	1.1	4.4					.04	.01	28	225	58	13	.00	.000
1.3 oz dry mix	37	1.8	28.7			8	.12	1.0	.00	.00	32	94	.73	.080	
cottage pudding															
homemade	186	14.4	6.1				0	.09			161	49			
1/8 cake	54	3.5	29.3	1.7		80	.08	0.6			48	62	.80		
w/ choc sce, homemade	235	20.6	6.5				0	.10			172	53			
1/8 cake & 1 T sce	74	3.9	42.0	2.0		70	.09	0.7			104	81	1.00		
w/ strawberry sce, homemade	204	25.6	6.2				8	.11			163	51			
1/8 cake & 1 T sce	70	3.6	33.9	1.7		80	.08	0.8			65	65	.80		

[a] Layer cake w/ custard filling & powdered sugar topping.
[b] Made w/ veg shortening; 4.5 g if made w/ butter.
[c] Values are averages for boston cream, choc macaroon, choc mousse, lemon blueberry, pineapple cream, pound, tunnel of fudge & tunnel of lemon.
[d] Prepared w/ vitamin D fortified whole milk.
[e] Prepared w/ vitamin D fortified 2% low fat milk.

	KCAL	H₂O (g)	FAT (g)	PUFA (g)	CHOL (mg)	A (RE)	C (mg)	B-2 (mg)	B-6 (mg)	FOL (mcg)	Na (mg)	Ca (mg)	Mg (mg)	Zn (mg)	Mn (mg)
	WT (g)	PRO (g)	CHO (g)	SFA (g)	DFIB (g)	A (IU)	B-1 (mg)	NIA (mg)	B-12 (mcg)	PANT (mg)	K (mg)	P (mg)	Fe (mg)	Cu (mg)	
dark choc, from mix, Pillsbury	250		12.0								380				
1/12 cake		3.0	32.0								65				
date nut roll, Dromedary	80		2.0								160				
1/2" slice		1.0	13.0								60				
dessert cup, Hostess	60		0.0		9		0				120				
1 dessert cup	21	1.0	14.0			0									
devil's food															
cake mix, Duncan Hines Deluxe	189	0.6	4.1		0			.07			363	66	21	.35	
1/12 pkg	44	2.5	35.6				.07	0.7			145	119	1.98	.176	
from mix	312	21.7	11.3				0	.07			241	54			
1/12 cake	92	4.0	53.6	4.4		140	.03	0.3			120	97	.70		
from mix, Pillsbury	270		14.0								370				
1/12 cake		4.0	32.0								130				
homemade	227	13.8	11.3				0	.11	.02	6	160	68	23	.48	
1 piece	60	3.4	30.4			53	.08	0.7	.13	.24	95	76	1.03	.142	
pudding type, from mix	191	21.7	9.3				0	.08	.03	6	341	54	16	.41	
1 piece	60	2.9	24.8			47	.06	0.8	.26	.12	129	128	1.21	.101	
w/ choc icing, frzn	323	17.9	15.0				0	.07			357	46			
1/6 cake	85	3.7	47.3	7.7		370	.02	0.2			101	78	.70		
w/ choc icing, homemade	233	11.6	10.8				0	.08	.02	4	108	50	21	.40	
1 piece	60	2.6	34.2			39	.05	0.5	.10	.17	87	62	.83	.140	
w/ whipped cream filling & choc icing,	315	25.2	18.6				0	.07			162	68			
frzn—*1/6 cake*	85	3.0	37.2	6.2		230	.02	0.2			96	104	.50		
french crumb, frzn, Am Hosp Co	229		8.5					.14			173	26			
2 oz	57	3.4	34.6			35	.07	0.8			58		.80		
fruitcake															
dark, homemade	163	7.8	6.6				0	.06			68	31			
1 piece	43	2.1	25.7	1.4		50	.06	0.3			213	49	1.10		
light, homemade	167	8.0	7.1				0	.05			83	29			
1 piece	43	2.6	24.7	1.6		30	.04	0.3			100	49	.70		
fudge cake mix, Duncan Hines Butter	185	1.6	4.2		0			.12			242	22	17	.26	
Recipe—*1/12 pkg*	44	2.5	34.8				.09	1.0			176	53	1.14	.088	
fudge marble, from mix, Pillsbury	270		12.0								300				
1/12 cake		4.0	36.0								55				
german choc, from mix, Pillsbury	250		11.0								340				
1/12 cake		3.0	36.0								60				
gingerbread															
from mix	174	23.3	4.3				0	.06			192	57			
1/9 cake	63	2.0	32.2	1.1		0	.02	0.5			173	63	1.00		
from mix, Dromedary	100		2.0								190				
2" square		1.0	19.0								90				
from mix, Pillsbury	190		4.0								310				
3" square		2.0	36.0								160				
homemade	267	17.1	12.9				0	.11	.05	6	99	48	15	.32	
1 piece	70	3.0	35.0			28	.13	1.1	.07	.27	237	41	2.11	.134	
golden cake mix, Duncan Hines Butter	185	1.5	3.9		0			.08			158	35	2	.09	
Recipe—*1/12 pkg*	44	1.6	36.5				.09	0.8			23	57	.53	.018	
honey spice, from mix	363	23.4	11.1				0	.09			252	73			
1/12 cake	103	4.2	62.7	3.3		160	.02	0.2			84	199	.80		
lemon delight, frzn, Am Hosp Co	271		15.1		18			.05			50	38			
3 oz	85	3.0	34.1			5	.02	0.2			113		.30		
lemon, from mix, Pillsbury	220		9.0								260				
1/12 cake		3.0	31.0								30				
marble streusel w/ icing, from mix	224	19.0	9.8				0	.08	.03	6	278	34	9	.26	
1 piece	65	2.5	32.7			38	.07	0.6	.13	.22	71	129	.88	.061	
mocha, from mix, Pillsbury	250		12.0								380				
1/12 cake		3.0	33.0								75				
oats 'n brown sugar, from mix, Pillsbury	260		12.0								310				
1/12 cake		3.0	35.0								80				

	KCAL	H₂O (g)	FAT (g)	PUFA (g)	CHOL (mg)	A (RE)	C (mg)	B-2 (mg)	B-6 (mg)	FOL (mcg)	Na (mg)	Ca (mg)	Mg (mg)	Zn (mg)	Mn (mg)
	WT (g)	PRO (g)	CHO (g)	SFA (g)	DFIB (g)	A (IU)	B-1 (mg)	NIA (mg)	B-12 (mcg)	PANT (mg)	K (mg)	P (mg)	Fe (mg)	Cu (mg)	
pineapple upside-down, homemade	236	25.0	9.1				4	.08	.05	9	179	54	13	.41	
1 piece	75	2.5	37.4			291	.12	0.8	.06	.25	128	47	1.19	.090	
pound															
from mix, Dromedary	150		6.0								340				
1/2" slice		2.0	21.0								65				
frzn, Am Hosp Co	256		14.8		8		2	.22			290	34			
2.25 oz	64	3.2	27.5			0	.11	1.1			61		1.10		
modified, homemade	119	5.6	5.4				0	.03			52	12			
1 piece	29	1.9	15.9	1.4		80	.01	0.1			23	30	.20		
old fashioned, homemade	142	5.2	8.9				0	.03			33	6			
1 piece	30	1.7	14.1	2.3		80	.01	0.1			18	24	.20		
snackin' cake mix[a]	192		5.8								252				
1/9 cake		2.0	32.8								89				
sponge, homemade	188	21.6	3.1				0	.13	.04	15	164	25	7	.80	
1 piece	66	4.8	35.7			125	.09	0.7	.33	.53	59	65	1.11	.033	
strawberry															
delight mousse, frzn, Nutra Choice	135		4.1		4						179	54			
2.2 oz serving	62	2.0	15.8										.56		
from mix, Pillsbury	260		11.0								300				
1/12 cake		3.0	37.0								85				
stir 'n frost cake mix w/ frosting[b]	237		6.6								246				
1/6 pkg		2.0	42.0								103				
streusel swirl, from mix, Pillsbury[c]	263		11.0								235				
1/16 cake		3.0	38.0								44				
white															
cake mix, Duncan Hines Deluxe	189	1.2	4.0	0				.09			251	75	3	.13	
1/12 pkg	44	2.1	36.1				.12	1.0				145	.70	.022	
from mix	333	20.0	10.2				0	.08			216	94			
1/12 cake	95	3.7	59.7	3.7		60	.02	0.2			110	170	.50		
from mix, Pillsbury	240		10.0								290				
1/12 cake		3.0	35.0								75				
homemade	285	19.4	11.6				0	.16	.02	5	346	87	11	.27	
1 piece	78	4.6	40.9			29	.12	1.1	.08	.22	80	66	.84	.031	
pudding type, from mix	219	26.6	9.8				0	.08	.01	4	293	34	4	.11	
1 piece	71	2.5	31.2			0	.06	0.7	.04	.21	34	115	.46	.026	
w/ choc icing, homemade	298	15.8	12.1				0	.12	.02	4	219	62	15	.29	
1 piece	78	3.3	45.8			23	.08	0.7	.06	.15	84	58	.75	.086	
yellow															
cake mix, Duncan Hines Deluxe	189	0.6	3.6	0				.09			271	53	3	.09	
1/12 pkg	44	1.9	37.0				.12	0.9				110	.66	.022	
from mix	310	23.6	10.4				0	.07			209	84			
1/12 cake	92	3.8	53.0	3.7		130	.02	0.2			100	167	.60		
from mix, Pillsbury	250		11.0								290				
1/12 cake		3.0	34.0								30				
homemade	283	17.9	12.4				0	.15	.03	9	329	89	11	.41	
1 piece	75	4.3	38.9			74	.13	1.1	.19	.35	72	80	1.02	.033	
pudding type, from mix	231	23.6	10.8				0	.11	.02	7	299	43	5	.41	
1 piece	69	3.1	30.5			56	.12	1.0	.18	.23	39	139	.87	.066	
w/ choc icing, homemade	292	14.6	12.3				0	.11	.02	6	208	62	14	.37	
1 piece	75	3.2	43.8			52	.08	0.7	.13	.24	79	66	.86	.085	

8.3. CAKES, CUPCAKES & SNACK CAKES

	KCAL	H₂O (g)	FAT (g)	PUFA (g)	CHOL (mg)	A (RE)	C (mg)	B-2 (mg)	B-6 (mg)	FOL (mcg)	Na (mg)	Ca (mg)	Mg (mg)	Zn (mg)	Mn (mg)
	WT (g)	PRO (g)	CHO (g)	SFA (g)	DFIB (g)	A (IU)	B-1 (mg)	NIA (mg)	B-12 (mcg)	PANT (mg)	K (mg)	P (mg)	Fe (mg)	Cu (mg)	
choco bliss, Hostess	200		7.0	0			0				170				
1 piece	50	2.0	34.0			0	.00								
choco diles, Hostess	240		11.0		22		0				280				
1 piece	57	2.0	35.0			0		0.0							
cinn nut rolls, Rainbo Snacks	330		11.0												
2 rolls	85	5.0	52.0												

[a] Values are averages for 5 flavors.
[b] Values are averages for 7 flavors.
[c] Values are averages for cinn, dutch apple, lemon & pecan brown sugar.

	KCAL / WT (g)	H₂O (g) / PRO (g)	FAT (g) / CHO (g)	PUFA (g) / SFA (g)	CHOL (mg) / DFIB (g)	A (RE) / A (IU)	C (mg) / B-1 (mg)	B-2 (mg) / NIA (mg)	B-6 (mg) / B-12 (mcg)	FOL (mcg) / PANT (mg)	Na (mg) / K (mg)	Ca (mg) / P (mg)	Mg (mg) / Fe (mg)	Zn (mg) / Cu (mg)	Mn (mg)
crumb cake, Hostess	130		4.0		10	0					95				
1 piece	35	1.0	22.0			0									
cupcakes															
choc, Hostess	170		6.0		3	0					250				
1 cupcake	50	2.0	29.0			0									
choc, Rainbo Snacks	350		10.0												
2 cupcakes	68	4.0	62.0												
orange, Hostess	150		5.0		13	0					175				
1 cupcake	43	1.0	28.0			0									
w/ choc icing, from mix	172	10.7	6.0				0	.05			161	62			
1 cupcake	48	2.2	28.4			80	.02	0.1			56	95	.40		
w/ choc icing, homemade	173	10.1	6.5				0	.04			108	30			
1 cupcake	47	2.0	27.9			80	.01	0.1			54	49	.30		
w/ white icing, homemade	172	9.7	5.5				0	.03			107	24			
1 cupcake	47	1.6	29.8			90	.01	0.0			29	35	.10		
ding dongs/big wheels, Hostess	170		9.0		6	0					130				
1 piece	38	1.0	21.0			0	.00								
filled twins, Rainbo Snacks	300		7.0												
2 twins	85	3.0	56.0												
fruitcake, supreme, Rainbo Snacks	110		5.0												
1 oz	28	1.0	16.0												
fruit loaf, Hostess	400		9.0		7	0					520				
1 piece	142	4.0	77.0			0									
ho hos, Hostess	120		6.0		13	0					90				
1 piece	28	1.0	17.0			0	.00	0.0							
honey buns, glazed, Hostess	450		27.0		24	0					650				
1 piece	106	5.0	49.0			0									
honey buns, Rainbo Snacks	380		23.0												
1 piece	85	4.0	41.0												
hoot 'n toots, Rainbo Snacks	450		16.0												
2 pieces	92	3.0	73.0												
hostess O's, Hostess	240		11.0		14	0					265				
1 piece	64	3.0	33.0			0									
lil' angles, Hostess	90		2.0		2	0					95				
1 piece	28	1.0	14.0			0		0.0							
sno balls, Hostess	150		4.0		2	0					170				
1 piece	43	1.0	28.0			0									
suzy Q's, banana, Hostess	240		9.0		21	0					195				
1 piece	64	2.0	38.0			0									
suzy Q's, choc, Hostess	240		10.0		16	0					300				
1 piece	64	2.0	37.0			0									
tiger tail, Hostess	210		6.0		25	0					240				
1 piece	64	2.0	38.0			0									
twinkies, Hostess	160		5.0		20	0					150				
1 piece	43	1.0	26.0			0									

8.4. COOKIES (COMMERCIAL UNLESS OTHERWISE INDICATED)

	KCAL / WT (g)	H₂O (g) / PRO (g)	FAT (g) / CHO (g)	PUFA (g) / SFA (g)	CHOL (mg) / DFIB (g)	A (RE) / A (IU)	C (mg) / B-1 (mg)	B-2 (mg) / NIA (mg)	B-6 (mg) / B-12 (mcg)	FOL (mcg) / PANT (mg)	Na (mg) / K (mg)	Ca (mg) / P (mg)	Mg (mg) / Fe (mg)	Zn (mg) / Cu (mg)	Mn (mg)
animal crackers	112	0.8	2.4				0	.03			79	14			
10 pieces	26	1.7	20.8	0.6		30	.01	0.1			25	30	.10		
animal crackers, Barnum's	130		4.0								120				
11 cookies	28	2.0	21.0								30				
apple/blueberry/cherry fruit sticks,	70		2.0								73				
Almost Home—*1 cookie*	19	0.0	14.0								23				
apple newtons	147		2.7								60				
2 cookies	37	1.3	28.0								47				
applesce raisin, iced, Almost Home	140		8.0								70				
2 cookies	28	2.0	17.0								60				
arrowroot biscuit, National	130		4.0								80				
6 cookies	28	2.0	21.0								25				
blueberry newtons	147		2.6								107				
2 cookies	38	1.3	28.0								33				

	KCAL / WT (g)	H₂O (g) / PRO (g)	FAT (g) / CHO (g)	PUFA (g) / SFA (g)	CHOL (mg) / DFIB (g)	A (RE) / A (IU)	C (mg) / B-1 (mg)	B-2 (mg) / NIA (mg)	B-6 (mg) / B-12 (mcg)	FOL (mcg) / PANT (mg)	Na (mg) / K (mg)	Ca (mg) / P (mg)	Mg (mg) / Fe (mg)	Zn (mg) / Cu (mg)	Mn (mg)
brown edge wafers, Nabisco	140		6.0								80				
5 cookies	28	1.0	20.0								30				
butter cookies	229	2.3	8.5				0	.03			209	63			
10 cookies	50	3.1	35.5			330	.02	0.2			30	47	.30		
butter flavored, Nabisco	130		5.0								140				
6 cookies	28	2.0	20.0								35				
cherry newtons	147		2.7								107				
2 cookies	38	1.3	26.7								40				
choc cakes, Mallomars	130		6.0								35				
2 cookies	28	1.0	18.0								55				
choc chip	99	0.6	4.4				0	.01			84	8			
2 cookies	21	1.1	14.6			26	.01	0.1			28	24	.38		
Almost Home	130		5.0								100				
2 cookies	28	1.0	20.0								35				
Chips Ahoy	130		6.0								110				
2 cookies	28	1.0	18.0								25				
Chips Ahoy Chewy	140		7.0								95				
3 cookies	28	2.0	18.0								35				
Chips 'n More original/coconut	150		7.5								83				
2 cookies	28	1.5	18.0								40				
cookie mix, Duncan Hines	144	0.9	7.3	0				.06			81	11	10	.17	
1/18 pkg (2 cookies)	28	1.3	18.2				.04	0.5			50	21	.50	.084	
from refrig dough	47	0.5	2.3				0	.02	.00	1	36	3	2	.05	
1 cookie	10	0.5	6.6			6	.01	0.2		.02	13	7	.22	.011	
frzn, Am Hosp Co	250		14.0		18		0	.03			194	20			
1.75 oz	50	2.0	30.0			329	.04	0.2			61		.90		
homemade	46	0.3	2.7				0	.02	.00	1	21	3	4	.04	
1 cookie	10	0.5	6.4			4	.02	0.1	.01	.03	21	8	.25	.032	
refrig dough, Continental	110		5.0								75				
2 cookies	25	1.0	16.0								25				
refrig dough, Pillsbury	208	6.5	9.7				0	.17			151	9			
3 cookies	47	1.8	28.4			0	.05	0.6			74	30	.93		
w/ pecans refrig dough	110		5.0								95				
2 cookies	25	1.0	15.0								25				
choc chip macaroons, frzn, Am Hosp Co	327		16.1		9		4	.01			218	33			
2.5 oz	71	4.8	40.6				.00	1.1			315				
choc chip snaps, Nabisco	130		4.0								100				
6 cookies	28	2.0	21.0								30				
choc choc chip, frzn, Am Hosp Co	240		14.0		18		0	.04			144	21			
1.75 oz	50	2.0	29.0			330	.04	0.2			125		1.00		
choc coated graham crackers	124	0.5	6.2				0	.08			106	30			
2 cookies	26	1.4	17.6	1.8		20	.02	0.4			84	54	.60		
Nabisco	150		7.0								70				
3 cookies	28	2.0	19.0								80				
choc middles, Nabisco	150		8.0								65				
2 cookies	28	2.0	18.0								50				
choc peanut bar, Ideal	150		8.0								130				
2 cookies	28	2.0	17.0								80				
choc sandwich	49	0.2	2.1					.02	.00	0	63	3	5	.09	
1 cookie	10	0.5	7.1				.02	0.2	.00		22	10	.36	.040	
Oreo	140		6.0								170				
3 cookies	28	1.0	20.0								50				
Oreo double stuff	140		7.0								120				
2 cookies	28	1.0	19.0								35				
Oreo mint creme	140		6.0								160				
2 cookies	28	1.0	20.0								50				

	KCAL	H₂O (g)	FAT (g)	PUFA (g)	CHOL (mg)	A (RE)	C (mg)	B-2 (mg)	B-6 (mg)	FOL (mcg)	Na (mg)	Ca (mg)	Mg (mg)	Zn (mg)	Mn (mg)
	WT (g)	PRO (g)	CHO (g)	SFA (g)	DFIB (g)	A (IU)	B-1 (mg)	NIA (mg)	B-12 (mcg)	PANT (mg)	K (mg)	P (mg)	Fe (mg)	Cu (mg)	
choc snaps, Nabisco	130		4.0								140				
7 cookies	28	2.0	21.0								50				
choc/van double dip creme sandwich, I Scream—2 cookies	150		7.0								70				
	28	1.0	20.0								37				
choc/van sandwich, Giggles	140		6.0								60				
2 cookies	28	1.0	17.0								17				
choc wafers, Famous	130		4.0								200				
5 cookies	28	2.0	21.0								75				
cinn raisin, Almost Home	140		7.0								95				
2 cookies	28	3.0	16.0								85				
coconut macaroon soft cakes, Nabisco	190		9.0								65				
1 cookie	38	1.0	23.0								95				
cone for ice cream	45	1.1	0.3			0		.03			28	19			
1 cone	12	1.2	9.3	0.1		0	.01	0.1			29	24	.05		
Comet Cups	20		0.0								5				
1 cone		0.0	4.0								5				
Comet Cups, choc flavored	25		0.0								5				
1 cone		0.0	5.0								10				
Comet Cups, sugar	40		0.0								35				
1 cone		1.0	9.0								15				
creme sandwich, Baronet/Cameo	140		5.5								80				
3 cookies	28	1.0	20.0								35				
devil's food cakes, Nabisco	140		1.0								90				
1 cookie	38	1.0	30.0								50				
double choc cookie mix, Duncan Hines	135	2.0	6.2		0			.06			77	4	11	.16	
1/18 pkg (2 cookies)	27	1.4	18.4				.05	0.6			84	22	.78	.081	
dutch apple fruit sticks, iced, Almost Home—1 cookie	70		1.0								40				
	19	0.0	14.0								40				
fig bars	53	1.7	1.0				0	.02	.02	1	45	10	4	.09	
1 bar	14	0.5	10.6		0.2	16	.02	0.2	.00	.05	41	8	.34	.041	
fig newtons	100		2.0								100				
2 cookies	28	1.0	20.0								60				
fudge, caramel & peanut bar, Heyday	140		8.0								45				
1 cookie	28	2.0	15.0								80				
fudge choc chip, Almost Home	130		5.0								130				
2 cookies	28	1.0	20.0								60				
fudge choc chip, Chips 'n More	140		6.0								90				
3 cookies	28	2.0	19.0								45				
fudge choc chip raisin, Almost Home	130		5.0								85				
2 cookies	28	1.0	18.0								60				
fudge choc sandwich, Gaiety	150		7.0								120				
3 cookies	28	2.0	19.0								50				
fudge 'n choc/van creme sandwich, Almost Home—1 cookie	140		6.0								115				
	32	1.5	20.0								38				
fudge striped shortbread, Nabisco	150		7.0								110				
3 cookies	28	1.0	19.0								30				
gingersnaps	59	0.4	1.2				0	.01			80	10			
2 cookies	14	0.8	11.2	0.3		10	.01	0.1			65	7	.32		
homemade	34	0.2	1.6				0	.01	.00	1	20	3	1	.03	
1 cookie	7	0.3	4.7			3	.01	0.1	.01	.02	14	4	.16	.009	
Nabisco	120		3.0								200				
4 cookies	28	2.0	22.0								100				
golden sugar cookie mix, Duncan Hines	118	0.8	5.0		0			.05			66	4	1	.02	
1/18 pkg (2 cookies)	24	1.2	16.8				.06	0.6			10	10	.26	.000	
graham, Bugs Bunny	120		4.0								130				
9 cookies	28	2.0	20.0								60				
imported danish, Nabisco	150		8.0								70				
5 cookies	28	1.0	18.0								25				

	KCAL	H$_2$O (g)	FAT (g)	PUFA (g)	CHOL (mg)	A (RE)	C (mg)	B-2 (mg)	B-6 (mg)	FOL (mcg)	Na (mg)	Ca (mg)	Mg (mg)	Zn (mg)	Mn (mg)
	WT (g)	PRO (g)	CHO (g)	SFA (g)	DFIB (g)	A (IU)	B-1 (mg)	NIA (mg)	B-12 (mcg)	PANT (mg)	K (mg)	P (mg)	Fe (mg)	Cu (mg)	
ladyfingers	79	4.2	1.7				0	.03			15	9			
2 ladyfingers	22	1.7	14.2	0.5		145	.01	0.0			15	36	.35		
macaroons	181	1.7	8.8				0	.06			13	10			
2 cookies	38	2.0	25.1			0	.02	0.2			176	32	.30		
macaroons, frzn, Am Hosp Co	390		22.0		15		0	.08			225	63			
3 oz	85	4.0	48.0			55	.03	0.2			235		1.20		
marshmallow[a]	147	3.5	4.7				0	.02			75	8			
2 cookies	36	1.4	26.0			95	.00	0.0			33	20	.20		
marshmallow puffs, Nabisco	120		4.0								55				
1 cookie	28	1.0	20.0								50				
marshmallow sandwich, Nabisco	120		3.0								80				
4 cookies	28	1.0	22.0								20				
marshmallow twirls cakes, Nabisco	130		5.0								55				
1 cookie	28	1.0	19.0								55				
molasses	137	1.3	3.4				0	.02			125	17			
1 cookie	33	2.0	24.7	0.9		30	.01	0.2			45	27	.70		
molasses, Pantry	130		4.0								130				
2 cookies	28	2.0	21.0								120				
mystic mint sandwich	150		8.0								95				
2 cookies	28	1.0	19.0								70				
nilla wafers, Nabisco	130		4.0								95				
7 cookies	28	1.0	21.0								30				
nutter butter peanut butter sandwich, Nabisco—*2 cookies*	140		6.0								100				
	28	3.0	18.0								55				
nutter butter peanut creme patties, Nabisco—*4 cookies*	150		8.0								95				
	28	3.0	17.0								70				
oatmeal, Bakers Bonus	130		5.0								90				
2 cookies	28	2.0	20.0								35				
oatmeal creme sandwich, Almost Home	140		5.0								150				
1 cookie	32	2.0	21.0								60				
oatmeal, homemade	62	0.4	2.6				0	.02	.01	2	45	11	5	.13	
1 cookie	13	0.8	8.9			5	.03	0.2	.01	.07	31	18	.41	.014	
oatmeal raisin	235	1.5	8.0				0	.04			84	11			
4 cookies	52	3.2	38.2	2.1		30	.06	0.3			192	53	1.50		
cookie mix, Duncan Hines	134	1.3	6.2		0			.04			63	20	13	.22	
1/18 pkg (2 cookies)	28	1.9	18.2				.07	0.4			70	42	.50	.056	
from refrig dough	61	0.6	2.6				0	.02	.01	2	37	4	4	.08	
1 cookie	13	0.7	8.9			10	.02	0.2		.06	23	14	.28	.014	
frzn, Am Hosp Co	191		7.8		13		1	.00			160	22			
1.6 oz	45	3.0	26.4			222	.02	0.4			80		1.20		
iced/uniced, Almost Home	130		5.0								90				
2 cookies	28	2.0	19.5								55				
refrig dough, Pillsbury	200	7.7	8.4				0	.18			198	11			
3 cookies	48	3.1	28.1			0	.11	1.6			80	52	1.01		
party grahams, Nabisco	140		7.0								100				
3 cookies	28	1.0	19.0								50				
peanut butter															
Almost Home	140		7.0								95				
2 cookies	28	3.0	16.0								85				
cookie mix, Duncan Hines	135	0.8	7.3		0			.04			111	9	11	.16	
1/18 pkg (2 cookies)	26	2.5	15.1				.03	1.1			60	25	.31	.052	
from refrig dough	50	0.5	2.6				0	.02	.00	2	57	12	4	.08	
1 cookie	10	0.8	5.9			15	.02	0.4		.01	19	24	.19	.016	
refrig dough, Continental	110		5.0								135				
2 cookies	25	2.0	15.0								40				
refrig dough, Pillsbury	199	7.2	8.5				0	.17			193	11			
3 cookies	47	3.1	27.4			0	.12	1.5			71	53	1.08		
peanut butter choc chip, frzn, Am Hosp Co—*1.75 oz*	231		13.0		35		0	.03			58				
	50	4.4	25.7			64	.07	1.1			104				

[a] Choc coated cookie w/ marshmallow.

	KCAL	H₂O (g)	FAT (g)	PUFA (g)	CHOL (mg)	A (RE)	C (mg)	B-2 (mg)	B-6 (mg)	FOL (mcg)	Na (mg)	Ca (mg)	Mg (mg)	Zn (mg)	Mn (mg)
	WT (g)	PRO (g)	CHO (g)	SFA (g)	DFIB (g)	A (IU)	B-1 (mg)	NIA (mg)	B-12 (mcg)	PANT (mg)	K (mg)	P (mg)	Fe (mg)	Cu (mg)	
peanut butter creme sandwich, Almost Home—*1 cookie*	140		6.0								120				
	32	2.0	20.0								60				
peanut butter fudge, Almost Home	140		7.0								90				
2 cookies	28	3.0	16.0								85				
pecan shortbread, Nabisco	150		9.0								80				
2 cookies	28	2.0	16.0								20				
pinwheels, choc & marshmallow cakes	130		5.0								35				
1 cookie	28	1.0	20.0								40				
raisin nut, frzn, Am Hosp Co	270		11.0		20		0	.04			213	19			
2 oz	58	4.0	41.0			354	.04	0.3			145		1.10		
sandwich type															
choc/van	198	0.9	9.0				0	.02			193	10			
4 cookies	40	1.9	27.7	2.4		0	.02	0.2			15	96	.30		
w/ peanut butter filling	116	0.6	4.7				0	.01			42	10			
2 cookies	25	2.4	16.4			50	.01	0.7			43	28	.20		
shortbread	75	0.5	3.5				0	.01			9	11			
2 cookies	15	1.1	9.8	0.9		6	.01	0.1			10	23	.08		
homemade	42	0.2	2.3				0	.02	.00	1	36	2	1	.04	
1 cookie	8	0.5	4.9			85	.02	0.2	.00	.02	5	9	.15	.005	
Lorna Doone	140		7.0								135				
4 cookies	28	2.0	18.0								30				
social tea biscuits, Nabisco	130		4.0								105				
6 cookies	28	2.0	21.0								30				
sugar															
from refrig dough	35	0.3	1.7				0	.01	.00	1	38	7	1	.02	
1 cookie	7	0.3	4.6			5	.01	0.2		.02	5	13	.14	.003	
homemade	71	1.3	2.7				0	.02			51	12			
2 cookies	16	1.0	10.9	0.7		18	.02	0.2			12	16	.22		
refrig dough, Continental	110		6.0								85				
2 cookies	25	1.0	15.0								15				
refrig dough, Pillsbury	201	7.4	8.1				0	.18			192	4			
3 cookies	48	1.8	30.2			0	.12	1.1			72	21	.98		
sugar rings, Bakers Bonus	130		5.0								100				
2 cookies	28	2.0	20.0								25				
sugar wafers	92	0.3	3.7				0	.01			36	7			
2 cookies	19	0.9	13.9	0.9		26	.00	0.0			11	15	.06		
Biscos	150		7.0								35				
8 cookies	28	1.0	20.0								15				
w/ peanut butter filling	66	0.3	2.7				0	.01			24	6			
2 cookies	14	1.4	9.4			28	.01	0.4			25	16	.12		
super heroes, Nabisco	135		5.0								120				
11 cookies	28	2.0	20.0								50				
van flavored creme sandwiches, Cookie Break—*3 cookies*	140		6.0								95				
	28	1.0	20.0								20				
van wafers	92	0.6	3.2				0	.01			50	8			
5 cookies	20	1.1	14.9			25	.00	0.0			15	12	.10		
waffle cremes, Biscos	150		7.0								30				
3 cookies	28	1.0	20.0								5				

	KCAL / WT (g)	H₂O (g) / PRO (g)	FAT (g) / CHO (g)	PUFA (g) / SFA (g)	CHOL (mg) / DFIB (g)	A (RE) / A (IU)	C (mg) / B-1 (mg)	B-2 (mg) / NIA (mg)	B-6 (mg) / B-12 (mcg)	FOL (mcg) / PANT (mg)	Na (mg) / K (mg)	Ca (mg) / P (mg)	Mg (mg) / Fe (mg)	Zn (mg) / Cu (mg)	Mn (mg)

8.5. CUSTARDS

	KCAL / WT (g)	H₂O / PRO	FAT / CHO	PUFA / SFA	CHOL / DFIB	A(RE) / A(IU)	C / B-1	B-2 / NIA	B-6 / B-12	FOL / PANT	Na / K	Ca / P	Mg / Fe	Zn / Cu	Mn
baked, homemade	305	204.6	14.6				1	.50			209	297			
1 cup	265	14.3	29.4	6.8		9.30	.11	0.3			387	310	1.10		

8.6. DOUGHNUTS

	KCAL / WT (g)	H₂O / PRO	FAT / CHO	PUFA / SFA	CHOL / DFIB	A(RE) / A(IU)	C / B-1	B-2 / NIA	B-6 / B-12	FOL / PANT	Na / K	Ca / P	Mg / Fe	Zn / Cu	Mn
cake	105	5.3	5.8				0	.05	.01	2	139	11	6	.13	
1 doughnut	25	1.3	12.2			14	.06	0.4		.10	27	55	.37	.033	
cake, Hostess	110		7.0		7		0				135				
1 doughnut	28	1.0	12.0			0									
choc coated donette, Hostess	60		3.0		4		0				50	0			
1 piece	13	1.0	6.0			0		0.0					.00		
choc coated, Hostess	130		8.0		4		0				150				
1 doughnut	28	1.0	14.0			0									
cinn, Hostess	110		6.0		6		0				140				
1 doughnut	28	1.0	15.0			0									
cinn apple, Earth Grains	310		17.0		25						270				
1 doughnut	71	3.0	35.0								75				
devil's food, Earth Grains	330		21.0		20						320				
1 doughnut	74	3.0	33.0								125				
glazed old fashioned, Earth Grains	310		18.0		20						270				
1 doughnut	71	3.0	34.0								65				
glazed old fashioned, Hostess	230		12.0		11		0				200				
1 doughnut	57	2.0	30.0			0									
krunch, Hostess	110		4.0		4		0				130				
1 doughnut	28	1.0	16.0			0									
old fashioned, Hostess	180		10.0		9		0				220				
1 doughnut	43	2.0	22.0			0									
powdered sugar, Hostess	110		5.0		6		0				140				
1 doughnut	28	1.0	15.0			0									
powdered sugar donette, Hostess	40		2.0		2		0	.00			40	0			
1 piece	9	0.0	5.0			0		0.0					.00		
powdered sugar old fashioned, Earth Grains—*1 doughnut*	290		19.0		20						260				
	61	3.0	27.0								60				
sugar gem, Rainbo Snacks	310		16.0												
6 doughnuts	71	3.0	38.0												
yeast	176	11.9	11.3				0	.07			99	16			
1 doughnut	42	2.7	16.0	2.8		30	.07	0.6			34	32	.60		

8.7. FROZEN DESSERTS

	KCAL / WT (g)	H₂O / PRO	FAT / CHO	PUFA / SFA	CHOL / DFIB	A(RE) / A(IU)	C / B-1	B-2 / NIA	B-6 / B-12	FOL / PANT	Na / K	Ca / P	Mg / Fe	Zn / Cu	Mn
bon bon ice cream nuggets, Carnation															
bavarian mint	163		12.1				0	.06	.01	2	28	39	6	.18	.019
5 nuggets	48	1.3	13.0			137	.01	0.0	.11	.11	58	36	.17	.051	
choc	179		12.9				0	.08	.02	2	50	44	7	.19	.019
5 nuggets	49	1.6	15.4			160	.02	0.1	.12	.14	78	49	.31	.050	
choc peanut butter	192		14.7				0	.09	.02	2	53	55	7	.22	.011
5 nuggets	49	2.4	13.8			142	.03	0.5	.16	.15	101	62	.27	.044	
van	167		12.1				0	.06	.01	1	28	39	6	.18	.019
5 nuggets	48	1.3	13.0			137	.01	0.0	.11	.11	58	36	.17	.051	
dreamy tofu, Giant Foods[a]	135		4.8		0						46				
1/2 cup		1.8	21.5												
frozen dessert, Weight Watchers															
choc mint treat	60		1.0				1	.11			50	90			
1.75 fl oz		3.0	12.0			12	.04	0.2			125	72	.10		
choc treat	100		1.0				1	.17			75	135			
2.75 fl oz		4.0	18.0			17	.06	0.4			190	112	.20		

[a] Values are averages for 4 flavors, choc, mint choc swirl, peanut butter swirl & wildberry swirl.

								Vitamins				Minerals			
	KCAL	**H₂O (g)**	**FAT (g)**	**PUFA (g)**	**CHOL (mg)**	**A (RE)**	**C (mg)**	**B-2 (mg)**	**B-6 (mg)**	**FOL (mcg)**	**Na (mg)**	**Ca (mg)**	**Mg (mg)**	**Zn (mg)**	**Mn (mg)**
	WT (g)	**PRO (g)**	**CHO (g)**	**SFA (g)**	**DFIB (g)**	**A (IU)**	**B-1 (mg)**	**NIA (mg)**	**B-12 (mcg)**	**PANT (mg)**	**K (mg)**	**P (mg)**	**Fe (mg)**	**Cu (mg)**	
cookies 'n cream	120		1.0				2	.18			105	151			
4.6 fl oz		4.0	22.0			19	.03	0.3				98	.40		
double fudge	60		1.0				0	.10			45	82			
1.75 fl oz		3.0	12.0			10	.05	0.2			140	69	.10		
dutch choc	100		1.0				1	.16			75	135			
4.4 fl oz		4.0	18.0			17	.07	0.4				111	.20		
golden van	100		1.0				1	.17			75	142			
4.6 fl oz		4.0	18.0			21	.05	0.1				109	.10		
neapolitan	100		1.0				2	.17			75	140			
4.6 fl oz		4.0	18.0			20	.06	0.2				110	.10		
orange-van treat	60		1.0				1	.11			50	92			
1.75 fl oz		3.0	12.0			14	.03	0.1			130	71	.00		
van sandwich bar	130		2.0				0	.14			170	98			
2.75 fl oz		3.0	26.0			10	.10	0.7				54	.60		
frozen yogurt, Honey Hill Farms[a]	118		3.5												
4.5 fl oz	100	3.5	18.7												
fruit & cream bar, Dole															
blueberry	90	44.9	1.4		5		2	.06			20	31			
1 bar	68	1.1	19.4			21	.01	0.1			57		.17		
peach	90	45.0	1.4		5		16	.07			19	27			
1 bar	68	1.0	19.4			226	.01	0.2			67		.07		
strawberry	90	45.0	1.4		5		6	.06			22	32			
1 bar	68	1.1	19.3			11	.01	0.1			73		.17		
fruit ice															
boysenberry, Haagen-Dazs	90		0.0								20				
1/2 cup		0.0	22.0												
cassis, Haagen-Dazs	130		0.0								10				
1/2 cup		0.0	32.0												
lemon, Haagen-Dazs	140		0.0								20				
1/2 cup		0.0	34.0												
lime	247	129.1	0.0				2	.00			0	0			
1 cup	193	0.8	62.9			0	.00	0.0			6	0	.00		
orange, Haagen-Dazs	140		0.0								20				
1/2 cup		0.0	34.0												
raspberry, Haagen-Dazs	100		0.0								20				
1/2 cup		1.0	24.0												
fruit juice bar															
all flavors, Jell-O[b]	42	40.7	0.0	0.0	0		5	.01	.01	3	4	3	2	.03	
1 bar	52	0.6	10.1	0.0		15	.00	0.1	.00	.02	35	3	.10	.012	
banana, Dole	80	53.9	0.0				0	.02			13	2	3		
1 bar	75	0.4	19.7			61	.00	0.2			119	9	.11		
dark sweet cherry, Dole	70	59.0	0.0				4				13	0			
1 bar	75	0.0	17.0			0					55				
gelatin pops, all flavors, Jell-O[c]	31	36.1	0.0	0.0	0		0	.00	.00	0	20	0	0	.00	
1 pop	44	0.5	6.9	0.0		0	.00	0.0	.00	.00	1	0	.00	.000	
grape/raspberry/strawberry, Weight Watchers—*1.7 oz bar*	35		0.0				d	.02			10	10			
		0.0	9.0			2	.00	0.0					.10		
mandarin orange, Dole	70	55.4	0.1				20	.02			14	2	2		
1 bar	75	0.4	18.2			113	.05	0.1			87	10	.11		
orange, Dole	70	53.5	0.0				18	.02			6	6	6		
1 bar	75	0.3	16.2			190	.01	0.1			70	9	.33		
pina colada, Dole	90	60.0	3.0				4				2	0			
1 bar	79	0.0	16.0			0	.06				48		.04		
pineapple, Dole	70	52.4	0.0				5	.02			4	9	8		
1 bar	75	0.3	17.1			17	.00	0.1			51	5	.28		
raspberry, Dole	70	57.7	0.0				2	.02			14	1	1		
1 bar	75	0.2	16.1			11	.01	0.1			20	3	.28		
strawberry, Dole	70	53.3	0.0				3	.02			6	6	4		
1 bar	75	0.2	16.3			6	.00	0.1			64	8	.20		

[a] Values are averages for 26 flavors.
[b] Values are averages for 8 flavors.
[c] Values are averages for 12 flavors.
[d] 6 mg for grape, 7 mg for raspberry, and 27 mg for strawberry.

	KCAL / WT (g)	H₂O (g) / PRO (g)	FAT (g) / CHO (g)	PUFA (g) / SFA (g)	CHOL (mg) / DFIB (g)	A (RE) / A (IU)	C (mg) / B-1 (mg)	B-2 (mg) / NIA (mg)	B-6 (mg) / B-12 (mcg)	FOL (mcg) / PANT (mg)	Na (mg) / K (mg)	Ca (mg) / P (mg)	Mg (mg) / Fe (mg)	Zn (mg) / Cu (mg)	Mn (mg)
ice bean frzn dessert[a]	137		7.8		0						68				
4 fl oz		2.6	15.0												
ice cream															
butter pecan, Haagen-Dazs	310		24.0								100				
1/2 cup		5.0	19.0												
carob, Haagen-Dazs	260		17.0								100				
1/2 cup		6.0	20.0												
cherry van, Haagen-Dazs	260		17.0								45				
1/2 cup		4.0	22.0												
choc, Haagen-Dazs	280		17.0								120				
1/2 cup		6.0	25.0												
choc chip, Haagen-Dazs	310		18.0								180				
1/2 cup		7.0	25.0												
choc swiss almond, Haagen-Dazs	250		17.0								70				
1/2 cup		3.0	21.0												
coffee, Haagen-Dazs	270		17.0								90				
1/2 cup		6.0	24.0												
cookies 'n cream, Haagen-Dazs	270		17.0								90				
1/2 cup		5.0	24.0												
elberta peach, Haagen-Dazs	250		16.0								100				
1/2 cup		4.0	27.0												
french van, soft serve	377	103.4	22.5	1.0	153	199	1	.45	.10	9	153	236	25	1.99	
1 cup	173	7.0	38.3	13.5	0.0	794	.08	0.2	1.00	1.07	338	199	.43		
honey van, Haagen-Dazs	270		17.0								100				
1/2 cup		5.0	24.0												
macadamia nut, Haagen-Dazs	260		19.0								40				
1/2 cup		4.0	20.0												
maple walnut, Haagen-Dazs	290		19.0								70				
1/2 cup		6.0	23.0												
mocha chip, Haagen-Dazs	270		18.0								70				
1/2 cup		5.0	22.0												
oreo cookies 'n cream, choc/mint/van	140		8.0								100				
3 fl oz		2.0	16.0								110				
pralines 'n cream, Haagen-Dazs	260		16.0								110				
1/2 cup		4.0	26.0												
rum raisin, Haagen-Dazs	260		16.0								220				
1/2 cup		4.0	25.0												
strawberry, Haagen-Dazs	270		16.0								80				
1/2 cup		5.0	25.0												
van, Haagen-Dazs	270		17.0								60				
1/2 cup		5.0	24.0												
van, reg (10% fat)	269	80.9	14.3	0.5	59	133	1	.33	.06	3	116	176	18	1.41	
1 cup	133	4.8	31.7	8.9	0.0	543	.05	0.1	.63	.65	257	134	.12		
van, rich (16% fat)	349	87.1	23.7	0.9	88	219	1	.28	.05	2	108	151	16	1.21	
1 cup	148	4.1	32.0	14.7	0.0	897	.04	.12	.54	.56	221	115	.10		
van chip, Haagen-Dazs	280		17.0								50				
1/2 cup		5.0	25.0												
van swiss almond, Haagen-Dazs	340		24.0								80				
1/2 cup		7.0	28.0												
ice cream bar															
choc coated w/ dark choc, Haagen-Dazs—*1 bar*	360		25.0								50				
		5.0	35.0								270				
cookies 'n cream on a stick	220		15.0								100				
1 bar		3.0	19.0								120				
oreo cookies 'n cream sandwich	240		11.0								300				
1 sandwich		4.0	31.0								175				
van coated w/ dark choc, Haagen-Dazs	330		23.0								60				
1 bar		4.0	28.0								200				
van coated w/ milk choc, Haagen-Dazs	330		23.0								60				
1 bar		4.0	28.0								200				

[a] Avg of 12 flavors, almond espresso, carob, choc, honey van, mocha, peanut butter carob chip, strawberry, toasted almond fudge, tofu almond fudge, tofu strawberry, tofu van & wildberry.

	KCAL	H₂O (g)	FAT (g)	PUFA (g)	CHOL (mg)	A (RE)	C (mg)	B-2 (mg)	B-6 (mg)	FOL (mcg)	Na (mg)	Ca (mg)	Mg (mg)	Zn (mg)	Mn (mg)
	WT (g)	PRO (g)	CHO (g)	SFA (g)	DFIB (g)	A (IU)	B-1 (mg)	NIA (mg)	B-12 (mcg)	PANT (mg)	K (mg)	P (mg)	Fe (mg)	Cu (mg)	

ice milk

	KCAL	H₂O	FAT	PUFA	CHOL	A(RE)	C	B-2	B-6	FOL	Na	Ca	Mg	Zn	Mn
van	184	89.9	5.6	0.2	18	52	1	.35	.09	3	105	176	19	.55	
1 cup	131	5.2	29.0	3.5	0.0	214	.08	0.1	.88	.66	256	129	.18		
van, soft serve	223	121.9	4.6	0.2	13	44	1	.54	.13	5	162	274	29	.86	
1 cup	175	8.0	38.4	2.9	0.0	175	.12	0.2	1.37	1.03	412	202	.28		

le tofu, frzn desserts, Brightsong Light Foods

	KCAL	H₂O	FAT	PUFA	CHOL	A(RE)	C	B-2	B-6	FOL	Na	Ca	Mg	Zn	Mn
choc	155	70.4	7.7								33	12			
3.5 oz	100	1.4	20.0										.20		
strawberry	163	69.5	8.5								24	22			
3.5 oz	100	0.9	20.0										.20		
van	155	70.6	7.8								33	12			
3.5 oz	100	0.8	20.6										.20		

mellorine, Fresh 'n Frosty

	KCAL	H₂O	FAT	PUFA	CHOL	A(RE)	C	B-2	B-6	FOL	Na	Ca	Mg	Zn	Mn
choc	372	127.8	13.6		6			.49			142	214	41	.76	
12 fl oz	206	9.0	53.2			639	.10	0.3			395	195	1.44	.210	
strawberry	376	126.4	13.2		6			.51			166	247	28	.68	
12 fl oz	205	9.2	54.8			620	.12	0.4			349	212	.41	.100	
van	368	127.6	12.8		6			.51			166	247	26	.66	
12 fl oz	206	9.4	53.9			602	.12	0.3			351	214	.31	.120	

mocha mix frzn dessert[a]

	KCAL	H₂O	FAT	PUFA	CHOL	A(RE)	C	B-2	B-6	FOL	Na	Ca	Mg	Zn	Mn
	144		8.0	3.1	0						80				
1/2 cup		0.6	17.5	2.1											

pudding pops, Jell-O

	KCAL	H₂O	FAT	PUFA	CHOL	A(RE)	C	B-2	B-6	FOL	Na	Ca	Mg	Zn	Mn
choc	75	29.9	1.9	0.0	2		0	.09	.01	1	80	69	10	.18	
1 pop	47	2.0	12.7	1.8		51	.02	0.1	.11	.17	91	54	.28	.037	
choc caramel swirl	73	30.4	1.9	0.0	1		0	.09	.01	1	63	64	7	.14	
1 pop	47	1.9	12.3	1.8		67	.02	0.1	.17	.17	74	49	.16	.019	
choc fudge	69	30.9	2.1	0.0	1		0	.08	.02	1	78	66	10	.18	
1 pop	47	1.9	11.3	2.0		53	.01	0.1	.29	.16	113	53	.22	.046	
choc van swirl	73	30.4	1.9	0.0	1		0	.09	.01	1	63	64	7	.14	
1 pop	47	1.9	12.4	1.8		66	.02	0.1	.17	.17	74	49	.16	.021	
choc w/ choc chips	79	29.9	2.7	0.0	1		0	.08	.02	1	75	64	12	.20	
1 pop	48	1.9	12.8	2.3		51	.01	0.1	.28	.15	127	55	.27	.068	
choc w/ choc coating	127	24.5	7.4	0.5	1		0	.08	.01	1	75	63	12	.22	
1 pop	49	2.0	14.4	5.4		54	.02	0.1	.26	.16	124	56	.26	.064	
double choc swirl	70	30.9	2.1	0.0	1		0	.08	.02	1	78	66	9	.16	
1 pop	47	1.9	11.3	2.0		53	.01	0.1	.29	.16	104	52	.18	.038	
milk choc	70	30.9	2.1	0.0	1		0	.08	.02	1	78	67	8	.15	
1 pop	47	1.9	11.4	2.0		54	.01	0.0	.29	.16	95	51	.14	.030	
van	71	31.0	2.0	0.0	1		0	.09	.02	1	46	60	5	.10	
1 pop	47	1.7	12.1	1.9		81	.02	0.0	.22	.17	57	44	.04	.004	
van w/ choc chips	79	30.5	2.7	0.0	1		0	.08	.01	1	50	63	8	.15	
1 pop	48	1.8	12.4	2.3		52	.01	0.1	.28	.15	88	48	.13	.036	
van w/ choc coating	127	25.0	7.4	0.5	1		0	.08	.01	1	52	62	8	.17	
1 pop	49	1.8	14.1	5.5		56	.02	0.0	.26	.16	89	50	.13	.036	

sherbet

	KCAL	H₂O	FAT	PUFA	CHOL	A(RE)	C	B-2	B-6	FOL	Na	Ca	Mg	Zn	Mn
fruit flavors, Land O'Lakes	127		1.7	0.1	4		1	.06			24	44			
1/4 cup	97	1.2	27.3	1.0		66					174	34			
orange	270	127.5	3.8	0.1	14	39	4	.09	.03	14	88	103	15	1.33	
1 cup	193	2.2	58.7	2.4		185	.03	0.1	.16	.06	198	74	.31		

sorbet, Dole

	KCAL	H₂O	FAT	PUFA	CHOL	A(RE)	C	B-2	B-6	FOL	Na	Ca	Mg	Zn	Mn
mandarin orange	110	75.3	0.1				25	.05			9				
1/2 cup	104	0.5	28.0			312	.02				138		.38		
peach	120	73.7	0.1				13	.05			11	3	3		
1/2 cup	104	0.6	29.6			314	.01	0.3			61	3	.57		
pineapple	120	75.0	0.1				12	.06			11	9	8		
1/2 cup	104	0.5	28.0			12	.05	0.1			79	5	.47		

[a] Avg of 8 flavors, choc chip, dutch choc, mocha, almond fudge & neapolitan.

	KCAL	H₂O (g)	FAT (g)	PUFA (g)	CHOL (mg)	A (RE)	C (mg)	B-2 (mg)	B-6 (mg)	FOL (mcg)	Na (mg)	Ca (mg)	Mg (mg)	Zn (mg)	Mn (mg)
	WT (g)	PRO (g)	CHO (g)	SFA (g)	DFIB (g)	A (IU)	B-1 (mg)	NIA (mg)	B-12 (mcg)	PANT (mg)	K (mg)	P (mg)	Fe (mg)	Cu (mg)	
raspberry	110	75.7	0.0				1	.04			12				
1/2 cup	104	0.4	27.9								37		.43		
strawberry	110	75.3	0.1				4	.04			11	8	3		
1/2 cup	104	0.5	27.9			6	.02	0.2			62	6	.52		
sorbet & cream, Haagen-Dazs															
orange & van	210		9.0								45				
1/2 cup		3.0	29.0												
raspberry & van	180		9.0								40				
1/2 cup		3.0	24.0												
tofulite, Barricini Foods[a]	165		8.5		0						70				
4 fl oz		1.5	23.5												
tofutti, Tofu Time															
hard pack[b]	220		13.3		0						105				
4 fl oz		2.8	21.3								29	47			
soft serve[c]	158		8.0		0						64				
4 fl oz		1.3	19.8								28	24			

8.8. FRUIT BETTYS/TURNOVERS

	KCAL	H₂O	FAT	PUFA	CHOL	A (RE)	C	B-2	B-6	FOL	Na	Ca	Mg	Zn	Mn
apple brown betty, homemade	325	138.7	7.5				2	.09			329	39			
1 cup	215	3.4	63.9	3.2		220	.13	0.9			215	47	1.30		
turnover, apple, Pillsbury	170	23.4	7.8				0	.06			320	6			
1 turnover	57	2.1	23.1			0	.09	0.9			32	123	.74		
turnover, blueberry, Pillsbury	166	23.9	7.8				0	.06			316	6			
1 turnover	57	2.0	21.4			5	.09	0.9			23	123	.57		
turnover, cherry, Pillsbury	173	22.2	7.8				5	.07			314	6			
1 turnover	57	2.1	23.4			52	.09	0.9			32	124	.80		

8.9. GELATIN DESSERTS, ALL FLAVORS, FROM MIX

	KCAL	H₂O	FAT	PUFA	CHOL	A (RE)	C	B-2	B-6	FOL	Na	Ca	Mg	Zn	Mn
Jell-O[d]	81	118.9	0.0	0.0	0		0	.00	.00	0	54	0	0	.00	
1/2 cup	140	1.6	18.8	0.0		0	.00	0.0	.00	.00	0	30	.02	.004	
Royal[e]	80		0.0								95				
1/2 cup		2.0	19.0								0				
sugar-free, D-Zerta[f]	8	118.7	0.0	0.0	0		0	.00	.00	0	4			.00	
1/2 cup	121	1.6	0.1	0.0	0.0	0	.00	0.0	.00	.00	50	0			
sugar-free, Featherweight	10		0.0				9								
1/2 cup		2.0	1.0												
sugar-free, Jell-O[g]	8	118.6	0.0	0.0	0		0	.00	.00	0	55			.00	
1/2 cup	121	1.4	0.2	0.0		0	.00	0.0	.00	.00	0	33			
sugar-free, Royal[h]	12		0.0								71				
1/2 cup		2.0	0.0								0				
w/ bananas & grapes, from mix	161	196.3	0.2				7				82				
1 cup	240	3.1	39.4												

---

[a] Values are averages for banana pecan, choc, rum raisin, strawberry, van & wildberry.

[b] Values are averages for choc supreme, maple walnut, van almond bark & wildberry supreme.

[c] Values are averages for banana pecan, choc, maple walnut, peanut butter, van & wildberry.

[d] Values are averages for 16 flavors.

[e] Values are averages for 13 flavors.

[f] Values are averages for cherry, lemon, lime, orange, raspberry & strawberry.

[g] Values are averages for 11 flavors.

[h] Values are averages for 5 flavors.

	KCAL / WT (g)	H₂O (g) / PRO (g)	FAT (g) / CHO (g)	PUFA (g) / SFA (g)	CHOL (mg) / DFIB (g)	A (RE) / A (IU)	C (mg) / B-1 (mg)	B-2 (mg) / NIA (mg)	B-6 (mg) / B-12 (mcg)	FOL (mcg) / PANT (mg)	Na (mg) / K (mg)	Ca (mg) / P (mg)	Mg (mg) / Fe (mg)	Zn (mg) / Cu (mg)	Mn (mg)

8.10. GRANOLA DESSERTS

all flavors

	KCAL / WT (g)	H₂O / PRO	FAT / CHO	PUFA / SFA	CHOL / DFIB	A-RE / A-IU	C / B-1	B-2 / NIA	B-6 / B-12	FOL / PANT	Na / K	Ca / P	Mg / Fe	Zn / Cu	Mn
dandy bar, Nature Valley[a]	163		7.3								108				
1 bar	31	2.3	22.5								105				
granola bar	109	1.0	4.2					.03			67	14			
1 bar	24	2.3	16.0				.07		.00	.13	78	66	.76		
granola bar, Nature Valley[b]	120		5.1								73				
1 bar	24	2.1	16.1								66				
caramel grand slam, New Trail	245		11.8					.14			117	58			
1.8 oz bar	51	4.2	31.0				.07	0.2					.59		
caramel nut dipps, Quaker	148	1.4	6.3					.06	.00	10	63	30	16	.00	.000
1.1 oz bar	31	1.9	21.0			0	.02	0.4	.00	.00	87	51	.56	.120	
caramel nut, Smores	120		3.9					.10	.04	8	81	14	15	.50	
1 oz bar	28	2.0	21.0		0.7	31	.04	0.7		.10	53	47	1.16	.082	
choc chip, chewy, Quaker	129	1.4	4.7					.03	.03	6	79	26	22	1.00	.000
1 oz bar	28	2.2	19.4			0	.07	0.3	.00	.00	99	65	.74	.110	
choc chip dipps, Quaker	138	0.9	6.4					.06	.00	7	65	29	18	.00	.000
1 oz bar	29	1.8	18.4			0	.02	0.2	.00	.00	90	56	.77	.100	
choc chip, New Trail	185		8.9					.06			89	19			
1.3 oz bar	37	2.7	23.5				.04	0.4					.54		
choc chip, Smores	120		2.9					.11	.03	9	60	11	16	.27	
1 oz bar	28	1.5	22.0		1.0	36	.06	0.5		.06	73	38	1.17	.085	
choc chip raisin, Smores	120		3.9					.07			60				
1 oz bar	28	2.3	21.0				.03						.10		
choc fudge, Smores	130		4.1					.07			65				
1 oz bar	28	2.1	22.0				.03						.10		
choc, graham & marshmallow, chewy, Quaker—*1 oz bar*	126	1.7	4.4					.05	.01	6	93	25	20	.00	.000
	28	1.9	19.6			0	.04	0.3	.00	.00	79	57	.79	.080	
chunky nut & raisin, chewy, Quaker	133	1.6	6.1					.03	.03	8	81	24	26	1.00	.000
1 oz bar	28	2.4	17.2			0	.05	0.8	.00	.00	118	68	.64	.110	
cocoa, choc covered, New Trail	200		11.6					.06			59	43			
1.3 oz bar	37	2.1	21.4				.02	0.1					.40		
honey & oats, chewy, Quaker	125	1.7	4.4					.03	.02	7	93	28	21	1.00	1.000
1 oz bar	28	2.4	19.0			0	.08	0.3	.00	.00	97	65	.76	.080	
honey & oats dipps, Quaker	137	1.1	6.1					.07	.00	5	66	30	16	.00	.000
1 oz bar	28	1.7	18.8			0	.03	0.2	.00	.00	82	51	.62	.080	
honey graham, choc covered, New Trail	200		11.7					.04			52	26			
1.3 oz bar	37	1.9	21.4				.04	0.1					.36		
mint choc chip dipps, Quaker	140	0.8	6.4					.06	.00	4	56	30	18	.00	.000
1 oz bar	28	1.8	18.8			0	.03	0.2	.00	.00	86	49	.70	.120	
peanut butter & choc chip, chewy, Quaker	131	1.6	5.5					.03	.03	9	92	23	25	1.00	.000
1 oz bar	28	3.0	17.4			0	.04	0.8	.00	.00	108	74	.49	.110	
peanut butter & choc chip, New Trail	185		9.4					.11			85	19			
1.3 oz bar	37	3.9	20.9				.03	0.5					.63		
peanut butter & choc chip, Smores	130		4.4					.09	.04	9	69	11	18	.45	
1 oz bar	28	2.2	20.0		0.9	30	.07	0.7		.07	76	49	1.58	.088	
peanut butter, chewy, Quaker	130	1.4	5.0					.03	.03	9	106	26	24	1.00	.000
1 oz bar	28	3.2	18.0			0	.07	0.9	.00	.00	109	71	.62	.190	
peanut butter dipps, Quaker	141	1.0	7.0					.06	.00	7	89	23	19	.00	.000
1 oz bar	28	2.5	17.0			0	.02	0.6	.00	.00	87	54	.62	.090	
peanut butter, New Trail	185		8.8					.07			89	28			
1.3 oz bar	37	4.6	21.3				.06	1.0					.67		
peanut butter, Smores	120		3.9					.10	.04	9	65	14	18	.30	
1 oz bar	28	2.1	20.0		0.7	31	.04	0.8		.10	64	50	1.44	.068	
peanut, choc covered, New Trail	196		11.2					.06			78	40			
1.3 oz bar	37	3.4	20.2				.04	0.1					.34		
raisin & almond dipps, Quaker	139	1.0	6.4					.07	.00	5	84	26	16	.00	.000
1 oz bar	28	1.7	18.6			0	.02	0.2	.00	.00	81	49	.75	.080	

[a] Values are averages for choc almond, dark choc, milk choc & peanut butter. [b] Values are averages for almond, choc chip, cinnamon, coconut, oats 'n honey, peanut & peanut butter.

	KCAL	H₂O (g)	FAT (g)	PUFA (g)	CHOL (mg)	Vitamins A (RE)	C (mg)	B-2 (mg)	B-6 (mg)	FOL (mcg)	Minerals Na (mg)	Ca (mg)	Mg (mg)	Zn (mg)	Mn (mg)
	WT (g)	PRO (g)	CHO (g)	SFA (g)	DFIB (g)	A (IU)	B-1 (mg)	NIA (mg)	B-12 (mcg)	PANT (mg)	K (mg)	P (mg)	Fe (mg)	Cu (mg)	
raisin & cinn, chewy, Quaker	130	1.7	5.1					.03	.03	6	83	28		.00	.000
1 oz bar	28	2.3	18.6			0	.08	0.3	.00	.00	104	62	.71	.080	
rocky road dipps, Quaker	140	0.9	6.5					.05	.00	5	60	30	17	.00	.000
1 oz bar	28	2.2	18.2			0	.02	0.5	.00	.00	87	50	.57	.100	

8.11. PASTRIES (DANISH PASTRIES/FILLED PASTRIES/SWEET ROLLS/TOASTER PASTRIES)

	KCAL	H₂O	FAT	PUFA	CHOL	A (RE)	C	B-2	B-6	FOL	Na	Ca	Mg	Zn	Mn
	WT	PRO	CHO	SFA	DFIB	A (IU)	B-1	NIA	B-12	PANT	K	P	Fe	Cu	
bear claw, Earth Grains	250		15.0												
2 oz claw	58	4.0	27.0												
charlotte russe w/ whipped cream filling, homemade[a]—*1 serving*	326	51.9	16.6				0	.11			49	52			
	114	6.7	38.2	8.3		840	.03	0.1			73	104	.80		
cream puff w/ custard filling	303	75.8	18.1				0	.22			108	105			
1 cream puff	130	8.5	26.7	5.6		460	.05	0.1			157	148	.90		
cream puff, bavarian, Rich's	146	11.7	7.6		35			.06			66	9	8	.24	
1 cream puff	38	2.1	17.0				.03	0.3			34	17	.62	.010	
danish pastry	161	11.2	8.8				0	.10			161	45	6	.35	
1 piece	42	2.6	18.8			45	.10	0.9			39	43	.78		
apple cinn, frzn, Am Hosp Co	455		24.2		37		4	.03			522	71			
4 oz	113	10.9	48.5			183	.05	3.3			157		2.90		
apple, Earth Grains	230		13.0												
2 oz danish	58	3.0	26.0												
apple, Hostess	360		20.0		19		0				410				
1 piece	99	4.0	43.0			0									
apple w/ icing, refrig dough, Pillsbury—*1 roll*	240	25.9	11.1				0	.10			260	16			
	74	2.2	31.8			6		1.1			47	44	.98		
butterhorn, Hostess	330		18.0		8		0				520				
1 piece	81	5.0	39.0			0									
caramel w/ nuts, refrig dough, Pillsbury	311	17.9	16.4				0	.12			488	18			
2 rolls	78	3.9	38.2			0	.19	1.6			58	211	1.18		
cherry, Earth Grains	230		13.0												
2 oz danish	56	4.0	26.0												
cinn, Earth Grains	220		10.0												
2 oz danish	58	4.0	29.0												
cinn, refrig dough	125		4.0								360				
1 roll	39	2.0	21.0								35				
cinn raisin, frzn, Am Hosp Co	419		17.8		29		4	.03			591	95			
4 oz	113	9.2	55.4			430	.05	2.5			199		2.60		
cinn raisin w/ icing, refrig dough, Pillsbury—*2 rolls*	291	20.0	13.8				0	.12			442	16			
	77	3.1	38.9			6	.17	1.4			54	190	.95		
orange, refrig dough	130		5.0								320				
1 roll	39	2.0	18.0								25				
orange w/ icing, refrig dough, Pillsbury	289	20.8	13.9				2	.12			477	10			
2 rolls	77	3.2	37.7			8	.18	1.5			33	200	1.00		
raspberry, Hostess	270		7.0		20		0				295				
1 piece	81	5.0	46.0			0									
raisin, refrig dough	135		5.0								290				
1 roll	39	2.0	21.0								50				
eclair, choc, Rich's	205	19.3	9.7		35			.08			113	10	10	.26	
1 eclair	57	2.3	27.3				.03	0.4			47	33	.84	.120	
eclair w/ custard filling & choc icing	239	56.2	13.6				0	.16			82	80			
1 eclair	100	6.2	23.2	4.4		340	.04	0.1			122	112	.70		
raisin bun	179	20.8	1.9				0	.07			250	49			
1 bun	65	4.5	36.7			0	.04	0.5			159	59	.91		
sweet roll	154	10.5	6.8				0	.13	.04	13	170	11	8	.20	
1 roll	42	2.6	21.4			0	.16	0.9		.16	47	30	.76	.020	
sweet roll, refrig dough															
cinn	110		3.0								300				
1 roll	34	2.0	18.0								35				

[a] 4 ladyfingers w/ 1/3 cup whipped cream filling.

	KCAL	H₂O (g)	FAT (g)	PUFA (g)	CHOL (mg)	A (RE)	C (mg)	B-2 (mg)	B-6 (mg)	FOL (mcg)	Na (mg)	Ca (mg)	Mg (mg)	Zn (mg)	Mn (mg)
	WT (g)	PRO (g)	CHO (g)	SFA (g)	DFIB (g)	A (IU)	B-1 (mg)	NIA (mg)	B-12 (mcg)	PANT (mg)	K (mg)	P (mg)	Fe (mg)	Cu (mg)	
cinn, butter flavored	130		7.0								320				
1 roll	34	2.0	15.0								30				
cinn, frzn, Am Hosp Co	520		14.5		12		7	.03			695	100			
4.75 oz	135	12.5	84.0			438	.13	4.5			200		2.70		
cinn w/ icing, Hungry Jack	291	22.1	14.2				0	.13			580	13			
2 rolls	79	3.3	37.9			5	.18	1.5			32	234	.95		
cinn w/ icing, Pillsbury	230	19.4	9.4				0	.13			517	13			
2 rolls	67	2.9	33.6			3	.17	1.5			40	215	1.27		
cinn w/ icing, Pillsbury Best Quick	211	13.7	9.3				0	.11			266	13			
1 roll	57	2.4	29.4			13		1.1			29	82	1.08		
toaster pastry	195	6.4	5.7					.17	.19	40	229	96	9	.29	
1 pastry	50	1.9	35.2			482	.16	2.1	.00	.11	84	96	2.00	.075	
toaster strudel, Pillsbury															
apple spice	187	14.8	8.6				0	.11			190	8			
1 pastry	54	2.8	28.1			11	.16	1.4			38	69	.97		
blueberry	194	12.3	7.6				0	.11			201	6			
1 pastry	54	2.2	27.9			1	.11	1.1			39	45	.70		
cherry	194	15.4	8.6				0	.11			201	8			
1 pastry	54	2.8	26.2			12	.16	1.4			41	71	1.03		
cinnamon	193	17.3	8.4				0	.11			198	5			
1 pastry	54	2.2	25.9			0	.13	1.1			36	45	.70		
raspberry	187	17.8	7.5				0	.11			201	6			
1 pastry	54	2.2	26.5			2	.13	1.1			41	46	.70		
strawberry	199	12.2	7.5				1	.11			202	6			
1 pastry	54	2.2	27.1			1	.13	1.1			41	46	.70		
toastettes, all flavors, Nabisco[a]	200		5.1								193				
1 pastry		2.0	35.9								45				

8.12. PIES & PIE CRUSTS

	KCAL	H₂O (g)	FAT (g)	PUFA (g)	CHOL (mg)	A (RE)	C (mg)	B-2 (mg)	B-6 (mg)	FOL (mcg)	Na (mg)	Ca (mg)	Mg (mg)	Zn (mg)	Mn (mg)
	WT (g)	PRO (g)	CHO (g)	SFA (g)	DFIB (g)	A (IU)	B-1 (mg)	NIA (mg)	B-12 (mcg)	PANT (mg)	K (mg)	P (mg)	Fe (mg)	Cu (mg)	
apple															
frzn, baked	231	43.5	9.1				1	.01			195	8			
1/6 pie	92	1.8	36.5	2.3		10	.01	0.2			66	19	.20		
frzn, Banquet	253		11.0				3	.03			282	12			
1/6 of 20 oz pie	94	2.0	37.0			15	.02	0.3			49		1.00		
homemade	282	60.0	11.9				0	.09	.03	6	181	11	9	.20	
1/8 pie	118	2.4	43.0			22	.13	1.1	.00	.17	100	27	1.06	.063	
banana cream															
from mix, Jell-O[b]	233	46.9	11.9	0.7	28		0	.13	.03	6	267	67	11	.31	
1/8 pie	92	3.3	28.8	7.3		375	.09	0.7	.20	.24	104	154	.43	.039	
from mix, Banquet	177		10.0				1	.06			146	32			
1/6 of 14 oz pie	66	2.0	21.0			7	.02	0.2			94		1.00		
banana custard, homemade	252	62.0	10.6				1	.15			221	75			
1/8 pie	114	5.1	35.0	3.4		290	.05	0.3			231	93	.60		
blackberry															
frzn, Banquet	268		11.0				5	.03			342	23			
1/6 of 20 oz pie	94	3.0	40.0			49	.02	0.3			65		1.00		
homemade	287	60.2	13.0				5	.02			316	22			
1/8 pie	118	3.1	40.6	3.2		110	.02	0.4			118	31	.60		
blueberry															
frzn, Banquet	266		11.0				3	.03			342	17			
1/6 of 20 oz pie	94	3.0	40.0			25	.03	0.3			45		1.00		
homemade	286	60.2	12.7				4	.02			316	13			
1/8 pie	118	2.8	41.2	3.2		40	.02	0.4			77	27	.70		
butterscotch, homemade	304	51.4	12.5				0	.11			244	86			
1/8 pie	114	5.0	43.7	4.5		300	.03	0.2			108	92	1.00		

[a] Values are averages for apple, blueberry, brown sugar, cinn, cherry, fudge & strawberry, frosted & unfrosted.

[b] Prepared w/ vitamin D fortified whole milk.

	KCAL	H₂O (g)	FAT (g)	PUFA (g)	CHOL (mg)	A (RE)	C (mg)	B-2 (mg)	B-6 (mg)	FOL (mcg)	Na (mg)	Ca (mg)	Mg (mg)	Zn (mg)	Mn (mg)
	WT (g)	PRO (g)	CHO (g)	SFA (g)	DFIB (g)	A (IU)	B-1 (mg)	NIA (mg)	B-12 (mcg)	PANT (mg)	K (mg)	P (mg)	Fe (mg)	Cu (mg)	
cherry															
frzn, baked	282	39.4	11.7				2	.02			222	12			
1/6 pie	97	2.1	42.9	2.9		280	.02	0.2			79	23	.20		
frzn, Banquet	252		11.0				2	.04			258	13			
1/6 of 20 oz pie	94	3.0	36.0			185	.03	0.3			72		1.00		
homemade	308	55.0	13.3				0	.02			359	17			
1/8 pie	118	3.1	45.3	3.5		520	.02	0.6			124	30	.40		
choc chiffon	266	26.7	12.4				0	.08			204	19			
1/8 pie	81	5.5	35.4	4.2		250	.02	0.2			89	79	1.00		
choc cream															
frzn, Banquet	185		10.0				0	.05			106	38			
1/6 of 14 oz pie	66	2.0	24.0			7	.02	0.2			86		1.00		
homemade	301	56.4	17.3				0	.19	.05	10	311	96	29	.75	
1/8 pie	114	5.2	33.6			301	.11	0.8	.42	.55	162	124	1.23	.136	
choc mint, from mix, Royal[a]	260		15.0								280				
1/8 pie		5.0	25.0								260				
choc mousse, from mix, Jell-O[a]	249	47.2	14.6	0.5	29		0	.14	.03	3	437	74	30	.57	
1/8 pie	95	3.6	27.5	8.3		392	.05	0.6	.20	.19	270	220	1.03	.193	
coconut cream															
from mix, Jell-O[a]	260	46.8	16.5	0.7	29		1	.10	.04	4	309	67	16	.36	
1/8 pie	94	2.7	26.6	11.1		381	.03	0.1	.20	.24	132	159	.37	.078	
frzn, Banquet	187		11.0				0	.05			113	30			
1/6 of 14 oz pie	66	2.0	22.0			1	.01	0.2			77		1.00		
coconut custard															
frzn, baked	249	51.2	12.0				0	.16			252	95			
1/6 pie	100	6.0	29.5	5.0		160	.04	0.2			172	115	.60		
homemade	268	63.2	14.3				0	.22			282	107			
1/8 pie	114	6.8	28.4	5.7		260	.07	0.3			186	132	.80		
custard, homemade	249	66.2	12.7				0	.18			327	109			
1/8 pie	114	7.0	26.7	4.3		260	.06	0.3			156	129	.70		
lemon chiffon	254	28.8	10.2				2	.06			211	19			
1/8 pie	81	5.7	35.5	2.7		140	.02	0.2			66	67	.70		
lemon cream, frzn, Banquet	173		9.0				2	.04			111	30			
1/6 of 14 oz pie	66	2.0	23.0			2	.02	0.2			70		1.00		
lemon meringue, homemade	350	66.4	13.1				4	.14	.03	13	260	18	8	.39	
1/6 pie	140	4.5	55.1	2.7		195	.11	0.8	.22	0.4	62	56	1.05	.032	
mince, homemade	320	50.7	13.6				1	.05			529	33			
1/8 pie	118	3.0	48.6	3.6		0	.08	0.5			210	45	1.20		
mincemeat, frzn, Banquet	258		11.0				1	.03			364	19			
1/6 of 20 oz pie	94	3.0	38.0			2	.03	0.3			110		1.00		
peach															
frzn, Banquet	244		11.0				14	.03			275	11			
1/6 of 20 oz pie	94	3.0	35.0			232	.02	0.5			69		1.00		
homemade	301	56.1	12.6				4	.05			316	12			
1/8 pie	118	3.0	45.1	3.1		860	.02	0.8			176	34	.60		
pecan, homemade	431	20.1	23.6				0	.07			228	48			
1/8 pie	103	5.3	52.8	3.3		160	.16	0.3			127	106	2.90		
pie crust															
from mix	742	29.9	46.5				0	.05		1301	65				
1/2 yield from 10 oz pkg	160	10.2	70.4	11.5		0	.05	0.8			89	136	.65		
from sticks/mix, Pillsbury	270		16.8				0	.15			416	10			
1/6 of 2 crust pie		4.2	24.8			0	.26	1.9			77	36	1.02		
homemade, enr, baked	900	26.8	60.1				0	.25[b]		1100	25				
1 pie shell	180	11.0	78.8	14.9		0	.36[b]	3.2[b]			89	90	3.10[b]		
mix, Flako	244	4.1	14.4					.09	.01	10	314	48	9	.00	.000
1.7 oz dry mix	48	3.4	25.0			0	.17	1.3	.00	.00	35	55	1.12	.060	

[a] Prepared w/ vitamin D fortified whole milk.

[b] Unenr pie shell contains .05 mg thiamin, .05 mg riboflavin, 0.9 mg niacin & .90 mg iron.

	KCAL / WT (g)	H₂O (g) / PRO (g)	FAT (g) / CHO (g)	PUFA (g) / SFA (g)	CHOL (mg) / DFIB (g)	A (RE) / A (IU)	C (mg) / B-1 (mg)	B-2 (mg) / NIA (mg)	B-6 (mg) / B-12 (mcg)	FOL (mcg) / PANT (mg)	Na (mg) / K (mg)	Ca (mg) / P (mg)	Mg (mg) / Fe (mg)	Zn (mg) / Cu (mg)	Mn (mg)
ready-to-eat, Pillsbury	241	12.4	14.9								214				
1/8 of 2 crust pie	54	1.7	23.8								30				
pineapple, homemade	299	56.6	12.6				1	.02			320	15			
1/8 pie	118	2.6	45.0	3.1		20	.05	0.5			85	25	.60		
pineapple chiffon, homemade	233	33.3	9.8				1	.07			207	19			
1/8 pie	81	5.3	31.7	2.6		280	.03	0.3			79	62	.70		
pineapple custard	251	61.9	9.9				1	.10			212	57			
1/8 pie	114	4.6	36.6	3.0		210	.05	0.5			111	74	.50		
pumpkin															
frzn, Banquet	197		8.0				2	.07			341	53			
1/6 of 20 oz pie	94	3.0	29.0			1938	.04	0.4			138		1.00		
homemade	241	67.5	12.8				0	.11			244	58			
1/8 pie	114	4.6	27.9	4.5		2810	.03	0.6			182	79	.60		
raisin, homemade	319	50.2	12.6				1	.04			336	21			
1/8 pie	118	3.1	50.7	3.1		10	.04	0.4			227	47	1.10		
rhubarb, homemade	299	55.9	12.6				4	.05			319	76			
1/8 pie	118	3.0	45.1	3.1		60	.02	0.4			188	31	.80		
strawberry cream, frzn, Banquet	168		9.0				6	.05			112	30			
1/6 of 14 oz pie	66	2.0	22.0			4	.02	0.2			80		1.00		
strawberry, homemade	184	54.3	7.3				23	.04			180	15			
1/8 pie	93	1.8	28.7	1.8		40	.02	0.4			112	23	.70		
sweet potato, homemade	243	67.6	12.9				5	.14			249	79			
1/8 pie	114	5.1	27.0	4.6		2730	.06	0.3			186	96	.60		

8.13. PIES, SNACK (FRIED)

	KCAL / WT (g)	H₂O (g) / PRO (g)	FAT (g) / CHO (g)	PUFA (g) / SFA (g)	CHOL (mg) / DFIB (g)	A (RE) / A (IU)	C (mg) / B-1 (mg)	B-2 (mg) / NIA (mg)	B-6 (mg) / B-12 (mcg)	FOL (mcg) / PANT (mg)	Na (mg) / K (mg)	Ca (mg) / P (mg)	Mg (mg) / Fe (mg)	Zn (mg) / Cu (mg)	Mn (mg)
apple, fried	258	40.6	13.1				0	.11			288	9	10		.24
1 snack pie	90	2.7	32.6				.18	1.2			74	68	1.25	.045	
apple, Hostess	390		20.0		18						540				
1 snack pie	128	5.0	45.0			0									
berry, Hostess	390		20.0		18										
1 snack pie	128	3.0	48.0			0									
blueberry, Hostess	390		20.0		18						450				
1 snack pie	128	3.0	49.0			0									
cherry, Hostess	420		20.0		18						530				
1 snack pie	128	5.0	55.0			0									
fried, Rainbo Snacks	410		19.0												
1 snack pie		5.0	57.0												
lemon, Hostess	400		22.0		30		0				470				
1 snack pie	128	3.0	53.0			0									
peach, Hostess	400		20.0		18						445				
1 snack pie	128	4.0	53.0			0									
pudding, Hostess	470		17.0		18		0				360				
1 snack pie	142	4.0	75.0			0									
strawberry, Hostess	340		14.0		13						400				
1 snack pie	128	3.0	56.0			0									

8.14. PUDDINGS

	KCAL / WT (g)	H₂O (g) / PRO (g)	FAT (g) / CHO (g)	PUFA (g) / SFA (g)	CHOL (mg) / DFIB (g)	A (RE) / A (IU)	C (mg) / B-1 (mg)	B-2 (mg) / NIA (mg)	B-6 (mg) / B-12 (mcg)	FOL (mcg) / PANT (mg)	Na (mg) / K (mg)	Ca (mg) / P (mg)	Mg (mg) / Fe (mg)	Zn (mg) / Cu (mg)	Mn (mg)
all flavors															
from inst mix, Royal[a]	181		4.4								375				
1/2 cup		3.7	31.5								208				
from reg mix, Royal[a]	162		3.7								160				
1/2 cup		3.6	27.7								191				
sugar-free, from inst mix, Royal[b]	103		2.3								473				
1/2 cup		4.3	16.3								243				
apple tapioca, homemade	293	175.0	0.3				0	.00			128	8			
1 cup	250	0.5	73.5			30	.00	0.0			65	10	.50		

[a] Values are averages for 11 flavors; pudding is prepared w/ vitamin D fortified whole milk.

[b] Values are averages for 3 flavors; pudding is prepared w/ vitamin D fortified 2% low fat milk.

	KCAL / WT (g)	H₂O (g) / PRO (g)	FAT (g) / CHO (g)	PUFA (g) / SFA (g)	CHOL (mg) / DFIB (g)	A (RE) / A (IU)	C (mg) / B-1 (mg)	B-2 (mg) / NIA (mg)	B-6 (mg) / B-12 (mcg)	FOL (mcg) / PANT (mg)	Na (mg) / K (mg)	Ca (mg) / P (mg)	Mg (mg) / Fe (mg)	Zn (mg) / Cu (mg)	Mn (mg)
banana cream															
from inst mix, Jell-O[a]	165	108.3	4.3	0.2	17		1	.20	.05	6	443	147	17	.47	
1/2 cup	147	4.0	28.4	2.6		154	.05	0.1	.44	.39	187	314	.09	.017	
sugar-free, from inst mix, Jell-O[b]	88	108.9	2.4	0.1	9		1	.20	.05	6	426	150	17	.47	
1/2 cup	130	4.2	12.6	1.5		250	.05	0.1	.44	.39	192	302	.08	.013	
bread, homemade	496	155.3	16.2				3	.50			533	289			
1 cup	265	14.8	75.3	7.7		800	.16	1.3			570	302	2.90		
butter pecan, from inst mix, Jell-O[a]	170	107.8	4.8	0.3	17		1	.21	.05	7	443	150	19	.48	
1/2 cup	147	4.2	28.1	2.6		154	.05	0.1	.44	.39	195	319	.14	.027	
butterscotch															
from inst mix, Jell-O[a]	164	108.3	4.3	0.2	17		1	.20	.05	6	481	147	17	.47	
1/2 cup	147	4.0	28.3	2.6		154	.05	0.1	.44	.39	187	340	.09	.017	
Rich's	133	58.6	5.9					.09			128	26	5	.14	
3 oz	85	1.7	18.3				.02				85	47	.20	.020	
sugar-free, from inst mix, D-Zerta[c]	68	112.3	0.2	0.0	2		1	.17	.05	6	66	154	14	.49	
1/2 cup	130	4.3	12.1	0.1		251	.04	0.1	.46	.40	237	124	.05	.016	
sugar-free, from inst mix, Jell-O[b]	90	109.5	2.4	0.1	9		1	.20	.05	6	426	150	17	.48	
1/2 cup	131	4.2	12.8	1.5		251	.05	0.1	.45	.39	193	303	.08	.010	
choc															
from inst mix	325	178.6	6.5				2	.39			322	374			
1 cup	260	7.8	63.4	3.6		340	.08	0.3			335	237	1.30		
from inst mix, Jell-O[a]	176	107.5	4.5	0.2	17		1	.21	.06	6	514	149	27	.61	
1/2 cup	150	4.4	31.0	2.7		155	.05	0.1	.44	.39	297	379	.43	.096	
from non-inst mix	322	182.0	7.8				2	.39			335	265			
1 cup	260	8.8	59.3			340	.05	0.3			354	247	.80		
homemade	385	171.1	12.2				1	.36			146	250			
1 cup	260	8.1	66.8	6.7		390	.05	0.3			445	255	1.30		
Rich's	141	57.4	7.1					.08			136	22	13	.25	
3 oz	85	1.7	18.2				.02	0.2			111	53	.64	.030	
sugar-free, from inst mix, D-Zerta[c]	65	112.3	0.4	0.0	2		1	.18	.05	6	68	159	23	.60	
1/2 cup	130	4.6	11.3	0.3		252	.05	0.1	.46	.41	324	136	.31	.076	
sugar-free, from inst mix, Featherweight—1/2 cup	80		0.5		2		1	.17	.05	10	100	190	20	.50	
	134	5.0	15.0			240	.05	0.2	.46	.42	440	194	1.85	.010	
sugar-free, from inst mix, Jell-O[b]	96	109.4	2.7	0.1	9		1	.21	.06	6	414	153	25	.59	
1/2 cup	133	4.5	14.1	1.6		252	.05	0.1	.45	.40	281	303	.34	.074	
choc fudge															
from inst mix, Jell-O[a]	174	107.5	4.6	0.2	17		1	.21	.06	6	479	150	31	.67	
1/2 cup	150	4.6	30.7	2.7		156	.05	0.2	.44	.39	356	387	.58	.130	
sugar-free, from inst mix, Jell-O[b]	100	109.7	3.0	0.1	9		1	.22	.06	6	366	155	35	.72	
1/2 cup	135	4.9	15.2	1.7		254	.05	0.2	.45	.41	404	318	.66	.147	
choc fudge mousse, from inst mix, Jell-O[a]	139	54.2	5.6	0.2	9		1	.17	.05	5	84	119	36	.72	
1/2 cup	86	5.2	19.6	4.0		82	.04	0.2	.39	.36	369	137	.81	.184	
choc mousse, from inst mix, Jell-O[a]	141	54.2	5.4	0.1	9		1	.17	.05	5	83	119	31	.66	
1/2 cup	86	4.9	20.2	3.9		81	.04	0.2	.39	.35	233	129	.58	.142	
choc tapioca, from inst mix, Jell-O[a]	169	108.5	4.6	0.2	17		1	.21	.06	6	169	150	31	.67	
1/2 cup	147	4.7	27.6	2.9		157	.05	0.2	.44	.39	294	143	.49	.123	
coconut cream, from inst mix, Jell-O[a]	178	107.8	6.4	0.2	17		1	.20	.05	6	356	148	20	.46	
1/2 cup	147	4.3	26.7	4.5		154	.05	0.1	.44	.38	209	293	.22	.038	
french van, from inst mix, Jell-O[a]	165	108.3	4.3	0.2	17		1	.20	.05	6	439	149	17	.47	
1/2 cup	147	4.0	28.4	2.6		154	.05	0.1	.44	.39	187	315	.09	.017	
golden egg custard, from inst mix, Jell-O[a]	160	107.5	5.4	0.3	73		1	.31	.09	11	197	199	26	.69	
1/2 cup	143	5.2	23.1	3.0		230	.08	0.2	.69	.80	283	173	.34	.027	
lemon, from inst mix, Jell-O[a]	168	107.7	4.3	0.2	17		1	.20	.05	6	396	147	17	.46	
1/2 cup	147	4.0	29.1	2.6		154	.05	0.1	.44	.38	187	314	.10	.019	
milk choc, from inst mix, Jell-O[a]	178	107.5	4.7	0.2	17		1	.21	.06	6	506	154	26	.60	
1/2 cup	150	4.5	30.9	2.8		161	.05	0.1	.45	.40	281	381	.39	.090	
pineapple cream, from inst mix, Jell-O[a]	165	108.3	4.3	0.2	17		1	.20	.05	6	395	147	17	.47	
1/2 cup	147	4.0	28.6	2.6		154	.05	0.1	.44	.39	187	314	.09	.017	

[a] Prepared w/ vitamin D fortified whole milk.
[b] Prepared w/ vitamin A & D fortified 2% low fat milk.
[c] Prepared w/ vitamin A & D fortified skim milk.

	KCAL	H₂O (g)	FAT (g)	PUFA (g)	CHOL (mg)	Vitamins A (RE)	C (mg)	B-2 (mg)	B-6 (mg)	FOL (mcg)	Minerals Na (mg)	Ca (mg)	Mg (mg)	Zn (mg)	Mn (mg)
	WT (g)	PRO (g)	CHO (g)	SFA (g)	DFIB (g)	A (IU)	B-1 (mg)	NIA (mg)	B-12 (mcg)	PANT (mg)	K (mg)	P (mg)	Fe (mg)	Cu (mg)	
pistachio															
from inst mix, Jell-O[a]	170	107.8	4.8	0.3	17		1	.21	.05	7	442	149	19	.46	
1/2 cup	147	4.2	28.2	2.6		154	.05	0.1	.44	.39	195	319	.15	.027	
sugar-free, from inst mix, Jell-O[b]	95	108.9	3.0	0.2	9		1	.21	.05	7	427	152	20	.48	
1/2 cup	131	4.3	12.8	1.5		250	.05	0.1	.44	.39	200	307	.13	.021	
prune whip, homemade	203	74.4	0.3				3	.18			213	29			
1 cup	130	5.7	48.0			600	.03	0.7			377	43	1.70		
rennin dessert															
caramel/fruit/van, from mix	238	199.3	9.0				3	.40			115	293			
1 cup	250	8.0	32.0	5.0		380	.08	0.3			320	230	.00		
choc, from mix	260	198.6	9.7				3	.38			133	311			
1 cup	255	8.7	36.0	5.3		360	.08	0.3			319	245	.00		
homemade	227	206.8	8.9				3	.38			209	283			
1 cup	255	7.9	29.6	4.9		360	.08	0.3			321	212	.00		
rice, from inst mix, Jell-O[a]	175	108.9	4.1	0.2	17		1	.20	.05	6	158	150	20	.55	
1/2 cup	149	4.8	30.0	2.6		154	.11	0.6	.44	.41	187	125	.54	.016	
rice w/ raisins, homemade	387	174.4	8.2				0	.37			188	260			
1 cup	265	9.5	70.8	4.5		290	.08	0.5			469	249	1.10		
tapioca cream, homemade	221	118.5	8.4				2	.30			257	173			
1 cup	165	8.3	28.2	3.9		480	.07	0.2			223	180	.70		
vanilla															
(blancmange), homemade	283	193.8	9.9				2	.41			166	298			
1 cup	255	8.9	40.5	5.5		410	.08	0.3			352	232	.00		
from inst mix, Jell-O[a]	168	107.7	4.3	0.2	17		1	.20	.05	6	441	147	17	.48	
1/2 cup	147	4.0	29.0	2.6		154	.05	0.1	.44	.38	187	313	.10	.019	
Rich's	129	58.5	5.9					.08			162	26	5	.17	
3 oz	85	1.6	18.4				.02	0.0			85	48	.17	.030	
sugar-free, from inst mix, D-Zerta	69	112.4	0.2	0.0	2		1	.17	.05	6	66	154	14	.49	
1/2 cup	130	4.3	12.1	0.1		251	.04	0.1	.46	.40	205	124	.05	.016	
sugar-free, from inst mix, Jell-O[b]	90	109.5	2.4	0.1	9		1	.20	.05	6	425	150	17	.48	
1/2 cup	131	4.2	12.9	1.5		251	.05	0.1	.45	.39	193	304	.08	.010	
vanilla/butterscotch. sugar-free, from inst mix, Featherweight—*1/2 cup*	80		0.3		2		1	.17	.05	10	100	190	14	.49	
	134	4.0	16.0			240	.05	0.1	.46	.40	330	170	1.60	.010	
vanilla tapioca, from inst mix, Jell-O[a]	160	108.3	4.1	0.2	17		1	.20	.05	6	170	147	17	.48	
1/2 cup	145	4.0	27.4	2.6		154	.05	0.1	.44	.38	186	115	.09	.055	

8.15. SAUCES, SYRUPS & TOPPINGS FOR DESSERTS

	KCAL	H₂O (g)	FAT (g)	PUFA (g)	CHOL (mg)	A (RE)	C (mg)	B-2 (mg)	B-6 (mg)	FOL (mcg)	Na (mg)	Ca (mg)	Mg (mg)	Zn (mg)	Mn (mg)
	WT (g)	PRO (g)	CHO (g)	SFA (g)	DFIB (g)	A (IU)	B-1 (mg)	NIA (mg)	B-12 (mcg)	PANT (mg)	K (mg)	P (mg)	Fe (mg)	Cu (mg)	
cake & cookies decorator, Pillsbury															
all flavors except choc	71	2.5	2.0				0	.00			0	0			
1 T	18	0.0	13.7			0	.00	0.0			5	0	.00		
choc	61	2.5	1.8				0	.00			0	0			
1 T	16	0.1	11.4			0	.00	0.0			27	11	.13		
choc syrup	92	11.9	0.8				0	.03			20	6			
2 T	38	0.9	23.5	0.4		0	.01	0.2			106	35	.60		
choc flavored syrup, Hershey	73		0.4					.01			18	4	20	.22	.112
2 T (1 oz)	28	0.9	16.4			6	.01	0.1			48	36	.50	.140	
choc fudge topping, Hershey	95		3.8					.06			36	28	13	.22	.084
2 T (1 oz)	28	1.3	14.3			25	.01	0.1			60	48	.34	.084	
fudge topping	124	9.5	5.1				0	.08			33	48			
2 T	38	1.9	20.3	2.9		60	.02	0.2			107	60	.50		
fudge topping, Kraft	65	6.1	2.5	0.5	0		0	.02	.00		41	15	1	.10	
1 T	20	0.4	10.3	0.2		1	.00	0.0	.03	.02	20	19	.03	.042	
icing/frosting, from mix															
choc fudge	293	11.9	11.1				0	.03			121	13			
1/4 cup	78	1.7	51.9	2.7		210	.00	0.1			49	51	.78		
coconut almond, Pillsbury	160		10.0								85				
amt for 1/12 cake		1.0	16.0								85				

[a] Prepared w/ vitamin D fortified whole milk.   [b] Prepared w/ vitamin A & D fortified 2% low fat milk.

	KCAL / WT (g)	H₂O (g) / PRO (g)	FAT (g) / CHO (g)	PUFA (g) / SFA (g)	CHOL (mg) / DFIB (g)	A (RE) / A (IU)	C (mg) / B-1 (mg)	B-2 (mg) / NIA (mg)	B-6 (mg) / B-12 (mcg)	FOL (mcg) / PANT (mg)	Na (mg) / K (mg)	Ca (mg) / P (mg)	Mg (mg) / Fe (mg)	Zn (mg) / Cu (mg)	Mn (mg)
coconut pecan, Pillsbury	150		7.0								105				
amt for 1/12 cake		1.0	20.0								60				
fluffy white, Pillsbury	60		0.0					0	.01		63	1			
amt for 1/12 cake		0.3	14.6			0	.00	0.0			4	1	.00		
icing/frosting, homemade															
choc	259	9.9	9.6					0	.07		42	41			
1/4 cup	69	2.2	46.4	5.3		145	.01	0.1			134	76	.83		
white	300	8.9	5.2					0	.01		39	12			
1/4 cup	80	0.4	65.1	2.9		215	.00	0.0			16	10	.00		
icing/frosting mix															
choc, Duncan Hines	160	7.0	6.6		7				.02		82	6			
1/12 pkg	39	0.5	24.6				.01	0.1					.90		
dark dutch fudge, Duncan Hines	156	7.4	6.6		7				.01		94	7			
1/12 pkg	39	0.6	23.8				.00	0.1					.78		
milk choc, Duncan Hines	156	7.0	6.6		7				.02		82	7			
1/12 pkg	39	0.7	24.2					0.1					.66		
van, Duncan Hines	158	7.0	6.6		7				.01		84	4			
1/12 pkg	39	0.1	25.0				.00	0.0					.16		
icing/frosting, ready-to-spread															
all flavors, creamy deluxe, General Mills[a] *1/12 tub*	164		7.3								89				
	33	1.0	23.9								53				
caramel pecan, Pillsbury	159	9.5	8.7				0		.00		70	2			
amt for 1/12 cake	38	0.2	21.3			3	.02	0.0			35	6	.06		
choc chip, Pillsbury	151	5.0	5.2				0		.00		67	1			
amt for 1/12 cake	37	0.1	26.3			1	.00	0.0			19	5	.08		
choc fudge, Pillsbury	150	6.8	6.5				0		.00		82	1			
amt for 1/12 cake	38	0.5	23.7			0	.00	0.1			93	39	.46		
choc mint, Pillsbury	148	7.6	6.5				0		.00		83	1			
amt for 1/12 cake	38	0.5	22.8			0	.00	0.1			106	16	.91		
coconut almond, Pillsbury	151	7.4	9.1				0		.02		59	7			
amt for 1/12 cake	35	0.6	17.1			0	.01	0.1			59	17	.23		
coconut pecan, Pillsbury	160	7.8	10.0				0		.00		63	5			
amt for 1/12 cake	37	0.5	17.8			3	.02	0.0			57	14	.19		
cream cheese, Pillsbury	160	5.3	6.2				0		.00		117	1			
amt for 1/12 cake	38	0.0	26.2			0	.00	0.0			10	1	.02		
double dutch, Pillsbury	141	8.1	6.3				0		.00		45	1			
amt for 1/12 cake	37	0.7	21.1			0	.00	0.1			134	21	1.18		
lemon, Pillsbury	160	5.3	6.2				0		.00		79	0			
amt for 1/12 cake	38	0.0	26.2			0	.00	0.0			10	1	.02		
milk choc, Pillsbury	149	6.7	6.2				0		.00		59	1			
amt for 1/12 cake	37	0.4	22.9			0	.00	0.1			77	14	.38		
mocha, Pillsbury	149	5.7	6.1				0		.00		58	1			
amt for 1/12 cake	37	0.2	23.7			0	.00	0.1			63	9	.48		
sour cream van, Pillsbury	158	4.4	5.9				1		.00		76	0			
amt for 1/12 cake	37	0.0	26.3			31	.00	0.0			17	5	.00		
strawberry, Pillsbury	160	5.3	6.1				0		.00		78	0			
amt for 1/12 cake	38	0.0	26.2			0	.00	0.0			10	1	.02		
van, Pillsbury	160	5.3	6.1				0		.00		74	0			
amt for 1/12 cake	38	0.0	26.2			0	.00	0.0			10	0	.02		
marshmallow creme, Kraft	90	4.9	0.0	0.0	0		0	.01	.00	5	17	9	0		
1 oz	28	0.3	23.0	0.0		0	.00	0.0	.00	.00	9	3	.00		

[a] Values are averages for 20 flavors.

	KCAL / WT(g)	H₂O(g) / PRO(g)	FAT(g) / CHO(g)	PUFA(g) / SFA(g)	CHOL(mg) / DFIB(g)	A(RE) / A(IU)	C(mg) / B-1(mg)	B-2(mg) / NIA(mg)	B-6(mg) / B-12(mcg)	FOL(mcg) / PANT(mg)	Na(mg) / K(mg)	Ca(mg) / P(mg)	Mg(mg) / Fe(mg)	Zn(mg) / Cu(mg)	Mn(mg)

9. EGGS, EGG DISHES & EGG SUBSTITUTES
9.1. EGGS, CHICKEN

Food	KCAL / WT	H₂O / PRO	FAT / CHO	PUFA / SFA	CHOL / DFIB	A(RE) / A(IU)	C / B-1	B-2 / NIA	B-6 / B-12	FOL / PANT	Na / K	Ca / P	Mg / Fe	Zn / Cu	Mn
boiled, hard/soft	79	37.3	5.6	0.7	274	78	0	.14	.06	24	69	28	6	.72	
1 large	50	6.1	0.6	1.7	0.0	260	.04	0.0	.66	.86	65	90	1.04		
fried	83	33.1	6.4	0.7	246	83	0	.13	.05	22	144	26	5	.64	
1 large	46	5.4	0.5	2.4	0.0	286	.03	0.0	.58	.76	58	80	.92		
omelet, plain	95	48.8	7.1	0.7	248	89	0	.16	.06	22	155	47	8	.70	
1 large egg	64	6.0	1.4	2.8	0.0	311	.04	0.0	.64	.82	85	97	.93		
poached	79	37.1	5.6	0.7	273	78	0	.13	.05	24	146	28	6	.72	
1 large	50	6.0	0.6	1.7	0.0	259	.04	0.0	.62	.86	65	90	1.04		
scrambled w/ milk & fat	95	48.8	7.1	0.7	248	89	0	.16	.06	22	155	47	8	.70	
1 large egg	64	6.0	1.4	2.8	0.0	311	.04	0.0	.64	.82	85	97	.93		
white, fresh/frzn	16	29.1	0.0	0.0	0	0	0	.09	.00	5	50	4	3	.01	.002
white of 1 large egg	33	3.4	0.4	0.0	0.0	0	.00	0.0	.02	.08	45	4	.01	.009	
whole, dried, stabilized (glucose reduced)	31	0.1	2.2	0.3	101	31	0	.06	.02	10	27	11	2	.28	
1 T	5	2.4	0.1	0.7	0.0	102	.02	0.0	.53	.34	26	86	.41		
whole, fresh/frzn	79	37.3	5.6	0.7	274	78	0	.15	.06	32	69	28	6	.72	.021
1 large	50	6.1	0.6	1.7	0.0	260	.04	0.0	.77	.86	65	90	1.04	.033	
yolk, fresh	63	8.3	5.6	0.7	272	94	0	.07	.05	26	8	26	3	.58	.019
yolk of 1 large egg	17	2.8	0.0	1.7	0.0	313	.04	0.0	.65	.75	15	86	.95	.024	

9.2. EGGS, OTHER

Food	KCAL / WT	H₂O / PRO	FAT / CHO	PUFA / SFA	CHOL / DFIB	A(RE) / A(IU)	C / B-1	B-2 / NIA	B-6 / B-12	FOL / PANT	Na / K	Ca / P	Mg / Fe	Zn / Cu	Mn
duck, whole	130	49.6	9.6	0.9	619		0	.28	.18	56	102	45	12	.99	
1 egg	70	9.0	1.0	2.6	0.0	930	.11	0.1	3.78		156	154	2.70		
goose, whole	276	101.4	19.1	2.4			0								
1 egg	144	20.0	1.9	5.2	0.0										
quail, whole	14	6.7	1.0	0.1	76		0	.07	.01			6			
1 egg	9	1.2	0.0	0.3	0.0	27	.01	0.0				20	.33		
turkey, whole	135	57.3	9.4	1.3	737		0	.37				78			
1 egg	79	10.8	0.9	2.9	0.0		.09	0.0				134	3.24		

9.3. EGG (CHICKEN) DISHES

Food	KCAL / WT	H₂O / PRO	FAT / CHO	PUFA / SFA	CHOL / DFIB	A(RE) / A(IU)	C / B-1	B-2 / NIA	B-6 / B-12	FOL / PANT	Na / K	Ca / P	Mg / Fe	Zn / Cu	Mn
omelette															
cheddar cheese, frzn, Am Hosp Co	313		27.1		328		0	.00			398	245			
4 oz	113	15.9	1.3				.07				128		1.58		
mushroom & cheese, frzn, Am Hosp Co—3 oz	152		11.6		245		0	.16			138	59			
Co—3 oz	85	9.8	1.9			334	.04	0.3			101		.90		
mushroom, cheese & onion, frzn, Am Hosp Co—4 oz	252		20.2		373		1	.29			310	145			
	113	15.3	2.3			653	.07				160		1.60		
plain, frzn, Am Hosp Co	199		17.7		298		0	.03			222	41			
3 oz	85	8.9	0.9			767	.07	0.1			100		1.39		
spanish, frzn, Am Hosp Co[a]	199		16.3		365		6	.23			197	47			
4 oz	113	10.3	2.8			671	.07	0.3			164		1.60		
western, frzn, Am Hosp Co[b]	207		16.7		366		9	.24			281	46			
4 oz	113	11.1	3.0			611	.09	0.4			164		1.60		
quiche															
bacon & onion, Pour-a-Quiche	230		18.0		240						385				
4.3 oz	123	13.0	6.0								170				
florentine, frzn, Am Hosp Co	410		19.6		113						652				
7 oz	198	28.1	30.2								448				
ham, Pour-a-Quiche	230		17.0		235						360				
4.3 oz	123	13.0	4.0								165				
lorraine, frzn, Am Hosp Co	531		19.9		210						1313				
7 oz	198	24.8	63.3								421				
spinach & onion, Pour-a-Quiche	220		16.0		230						365				
4.3 oz	123	12.0	6.0								210				

[a] Contains tomatoes, onions, mushrooms, green peppers & celery.　　[b] Contains onions, mushrooms, ham, red & green peppers & tomatoes.

	KCAL / WT (g)	H₂O (g) / PRO (g)	FAT (g) / CHO (g)	PUFA (g) / SFA (g)	CHOL (mg) / DFIB (g)	A (RE) / A (IU)	C (mg) / B-1 (mg)	B-2 (mg) / NIA (mg)	B-6 (mg) / B-12 (mcg)	FOL (mcg) / PANT (mg)	Na (mg) / K (mg)	Ca (mg) / P (mg)	Mg (mg) / Fe (mg)	Zn (mg) / Cu (mg)	Mn (mg)
three cheeses, Pour-a-Quiche	236		18.7		250		1	.30			326	289			
4.3 oz	123	12.9	4.2			696	.06	0.1			154	253	.96		
scrambled egg croquette, frzn, Am Hosp Co—3 oz	190		13.2		289		3	.14			272	110			
	85	8.3	9.6			819	.23	1.3			113		1.88		
scrambled egg w/ bacon pastry wrap, frzn, Am Hosp Co—4.5 oz	277		19.8		247		11	.23			363	41			
	128	7.9	16.7			369	.17				119		1.99		

souffle

	KCAL / WT (g)	H₂O (g) / PRO (g)	FAT (g) / CHO (g)	PUFA (g) / SFA (g)	CHOL (mg) / DFIB (g)	A (RE) / A (IU)	C (mg) / B-1 (mg)	B-2 (mg) / NIA (mg)	B-6 (mg) / B-12 (mcg)	FOL (mcg) / PANT (mg)	Na (mg) / K (mg)	Ca (mg) / P (mg)	Mg (mg) / Fe (mg)	Zn (mg) / Cu (mg)	Mn (mg)
cheese, homemade	207	61.8	16.2				0	.23			346	191			
1 cup	95	9.4	5.9	8.2		760	.05	0.2			115	185	1.00		
spinach[a]	218	100.6	18.4	3.1	184	675	3	.31	.12	62	763	230	37	1.29	
1 cup	136	11.0	2.8	7.1		3461	.09	0.5	.68		202	231	1.34	.120	

9.4. EGG SUBSTITUTES

by type

	KCAL / WT (g)	H₂O (g) / PRO (g)	FAT (g) / CHO (g)	PUFA (g) / SFA (g)	CHOL (mg) / DFIB (g)	A (RE) / A (IU)	C (mg) / B-1 (mg)	B-2 (mg) / NIA (mg)	B-6 (mg) / B-12 (mcg)	FOL (mcg) / PANT (mg)	Na (mg) / K (mg)	Ca (mg) / P (mg)	Mg (mg) / Fe (mg)	Zn (mg) / Cu (mg)	Mn (mg)
frzn[b]	96	43.9	6.7	3.7	1	81		.23	.08		120	44		.59	
1/4 cup	60	6.8	1.9	1.2	0.0	810	.07			1.00	128	43	1.19		
liquid[c]	40	38.9	1.6	0.8	0	102	0	.14			33	25		.61	
1.5 fl oz	47	5.6	0.3	0.3	0.0	1015	.05	.05	.14	1.27	155	57	.99		
powdered[d]	44	0.4	1.3	0.2	57		0	.17			79	32			
0.35 oz	10	5.5	2.2	0.4	0.0	122	.02	0.0			74	47	.31		

by brand

	KCAL / WT (g)	H₂O (g) / PRO (g)	FAT (g) / CHO (g)	PUFA (g) / SFA (g)	CHOL (mg) / DFIB (g)	A (RE) / A (IU)	C (mg) / B-1 (mg)	B-2 (mg) / NIA (mg)	B-6 (mg) / B-12 (mcg)	FOL (mcg) / PANT (mg)	Na (mg) / K (mg)	Ca (mg) / P (mg)	Mg (mg) / Fe (mg)	Zn (mg) / Cu (mg)	Mn (mg)
country morning, Land O'Lakes	173		12.1		594		0	.52			180	52			
1/2 cup	121	14.6	1.3			1138	.06				133	197	2.07		
egg beaters, Fleischmann's	25		0.0		0						80				
1/4 cup		5.0	1.0								85				
egg beaters w/ cheez, Fleischmann's	130		6.0		5						440				
1/2 cup		14.0	3.0								110				
egg vantage lite, frzn, Am Hosp Co[e]	25		0.0		0						100				
1 serving	50	6.0	0.0												
scrambled, Land O'Lakes	143		9.1		466		0	.38			173	77			
1/2 cup	121	12.1	3.0			1071	.06				150	190	1.67		
scramblers, Morningstar	58	44.5	2.4	1.5	1			.63	.15		94	46		.77	
1/4 cup	57	6.5	2.7	0.3		1144	.33		2.39	1.84	98		1.28		

10. ENTREES & MEALS
10.1. BOX MIX ENTREES

	KCAL / WT (g)	H₂O (g) / PRO (g)	FAT (g) / CHO (g)	PUFA (g) / SFA (g)	CHOL (mg) / DFIB (g)	A (RE) / A (IU)	C (mg) / B-1 (mg)	B-2 (mg) / NIA (mg)	B-6 (mg) / B-12 (mcg)	FOL (mcg) / PANT (mg)	Na (mg) / K (mg)	Ca (mg) / P (mg)	Mg (mg) / Fe (mg)	Zn (mg) / Cu (mg)	Mn (mg)
chicken helper main dishes[f]—1/5 pkg w/	530		27.8								1166				
chicken & other ingredients		32.6	36.8								498				
chili con carne, concentrate, Oscar Mayer	78	14.5	5.8		14		1	.08	.06	3	426	6	8	.65	
1 oz	28	3.9	2.5				.02	1.1	1.05	.12	98	37	.81	.060	
burgers, ckd, made w/ burger stuffin, Oscar Mayer															
bacon & cheese	265	44.4	19.8		71						588	75	18	4.05	
1 burger	88	21.3	0.5								246	276	1.95	.230	
mexican style	233	52.3	18.4		53						701	90	18	2.94	
1 burger	90	15.6	1.1								198	210	2.24	.350	
mushroom	224	50.6	17.3		65						440	75	14	3.19	
1 burger	87	16.1	1.2								192	217	1.62	.230	
onion	229	48.0	16.9		88						452	81	17	4.10	
1 burger	86	16.9	2.4								185	236	1.75	.160	
pizza	244	51.0	18.0		62						615	109	16	3.32	
1 burger	92	18.2	2.3								263	205	2.28	.220	
hamburger helper main dishes[g]	335		14.9								1027				
1/5 pkg w/ 1/5 lb ground beef		19.4	30.8								436				

[a] Contains whole milk, spinach, egg white, cheddar cheese, egg yolk, butter, flour, salt & pepper.
[b] Contains egg white, corn oil & nfdm.
[c] Contains egg white, hydrogenated soybean oil & soy protein.
[d] Contains egg white solids, whole egg solids, sweet whey solids, nfdm & soy protein.
[e] Contains egg white, flavors & colors.
[f] Values are averages for 5 main dishes.
[g] Values are averages for 15 main dishes.

	KCAL	H₂O (g)	FAT (g)	PUFA (g)	CHOL (mg)	A (RE)	C (mg)	B-2 (mg)	B-6 (mg)	FOL (mcg)	Na (mg)	Ca (mg)	Mg (mg)	Zn (mg)	Mn (mg)
	WT (g)	PRO (g)	CHO (g)	SFA (g)	DFIB (g)	A (IU)	B-1 (mg)	NIA (mg)	B-12 (mcg)	PANT (mg)	K (mg)	P (mg)	Fe (mg)	Cu (mg)	
macaroni & cheese deluxe dinner, prep,	255	90.3	7.5	0.6	18		0	.24	.05	12	652[a]	123	29	1.27	.300
Kraft—*3/4 cup*	147	10.8	36.3	4.2		388	.26	1.7	.23	.35	94	345	1.94	.116	
pasta shells & cheese, prep, Velvetta	261	93.9	9.8	0.7	25		0	.26	.05	22	858[b]	186	30	1.46	.330
3/4 cup	150	12.1	31.8	5.4		372	.23	1.8	.39	.36	117	434	1.91	.135	
pizza, cheese, from Contadina Pizzeria Kit															
thick crust	295	56.0	3.8	0.7	0		12	.38	.00	0	820	63	1	.01	
1/4 pizza	128	8.8	56.0	0.2		782	.69	4.8	.02	.04	322	163	3.88		
thin crust	209	53.0	3.0	0.7	0		12	.26	.00	0	650	44	1	.01	
1/4 pizza	104	6.2	39.2	0.2		782	.48	3.4	.01	.03	285	114	2.87		
tuna helper main dishes[c]	300		13.8								943				
1/5 pkg w/ 1.3 oz tuna		14.8	29.0								253				

10.2. CANNED ENTREES

	KCAL	H₂O (g)	FAT (g)	PUFA (g)	CHOL (mg)	A (RE)	C (mg)	B-2 (mg)	B-6 (mg)	FOL (mcg)	Na (mg)	Ca (mg)	Mg (mg)	Zn (mg)	Mn (mg)
	WT (g)	PRO (g)	CHO (g)	SFA (g)	DFIB (g)	A (IU)	B-1 (mg)	NIA (mg)	B-12 (mcg)	PANT (mg)	K (mg)	P (mg)	Fe (mg)	Cu (mg)	
beans, baked															
w/ bacon, Special Recipe	316		12.6				5	.12			1024	120			
6.9 oz	195	10.5	40.2			194	.12	0.6			311		4.20		
w/ beef	321	189.7	9.2	0.5	59	57	5	.12	.24		1264	119	67	3.19	1.596
1 cup	266	17.0	45.0	4.5		565	.14	2.5	.67		851	215	4.26	.798	
w/ beef in bbq sce, Special Recipe	262		5.7				8	.13			814	85			
6.9 oz	195	15.7	36.9			707	.12	1.6			493		3.70		
w/ chicken in bbq sce, Special Recipe	254		5.4				8	.11			803	90			
6.9 oz	195	14.7	36.6			761	.12	1.4			535		3.20		
w/ franks	366	178.2	16.9	2.1	15	40	6	.14	.12	77	1105	123	71	4.79	1.079
1 cup	257	17.3	39.6	6.0	5.9	395	.15	2.3	.87	.36	604	267	4.45	.547	
w/ franks, Campbell's	352		13.8				5	.11			1154	122			
8 oz	227	14.6	42.5			293	.08	2.1			463		3.90		
w/ franks in tomato sce, Campbell's	346		13.5				5	.11			1134	119			
7.9 oz	223	14.4	41.8			288	.08	2.1			455				
w/ franks, Van Camp's beanee weenee	326	173.1	15.4					.17			990	95			
1 cup	235	15.2	31.7			294	.13	2.0			565		4.27		
w/ pork	268	180.8	3.9	0.5	17	45	5	.10	.16	92	1048	133	85	3.69	.913
1 cup	253	13.1	50.6	1.5	6.6	451	.13	1.1	.06	.25	781	274	4.31	.544	
beef & veg stew	194	202.1	7.6				7	.12			1007	29			
1 cup	245	14.2	17.4			2380	.07	2.5			426	110	2.20		
Bounty	144		3.4				6	.10			987	21			
7.6 oz	216	12.1	16.2			3936	.03	2.4			287		2.20		
Campbell's	144		3.5				9	.11			962	23			
7.5 oz	213	12.0	16.2			4233	.02	2.1			337		2.10		
Wolf Brand	179	173.8	7.5					.30	.17	13	1043	34	26	2.00	.000
scant cup	213	9.6	18.3			8090	.02	3.5	2.00	.00	417	136	1.89	.110	
beef ravioli, Franco-American	223		4.7				5	.18			1092	35			
7.5 oz	213	9.0	36.1			1692	.15	2.8			324		2.30		
beef raviolio's, Franco-American	243		7.3				8	.18			900	38			
7.5 oz	213	10.2	34.2			467	.25	3.3			436		2.40		
beef sirloin w/ onions, Prego Plus	150		6.2				12	.05			402	48			
4 oz	113	4.5	19.1			830	.05	0.8			285		1.60		
chicken a la king, Swanson	182		11.7				0	.15			657	45			
5.3 oz	149	10.9	8.3			39	.04	2.8			177		.90		
chicken & dumplings, Swanson	220		11.2				0	.12			963	25			
7.5 oz	213	11.4	18.4			674	.03	2.7			151		1.10		
chicken & veg stew															
Bounty	166		7.0				5	.09			1055	29			
7.5 oz	213	10.5	15.1			7317	.03	3.4			315		1.20		
Swanson	164		7.1				6	.09			962	27			
7.6 oz	216	9.7	15.4			6236	.04	3.2			289		1.00		

[a] 585 mg Na if prepared w/ unsalted cooking water.
[b] 717 mg Na if prepared w/ unsalted cooking water.
[c] Values are averages for 6 main dishes.

	KCAL	H₂O (g)	FAT (g)	PUFA (g)	CHOL (mg)		A (RE)	C (mg)	B-2 (mg)	B-6 (mg)	FOL (mcg)		Na (mg)	Ca (mg)	Mg (mg)	Zn (mg)	Mn (mg)
	WT (g)	PRO (g)	CHO (g)	SFA (g)	DFIB (g)		A (IU)	B-1 (mg)	NIA (mg)	B-12 (mcg)	PANT (mg)		K (mg)	P (mg)	Fe (mg)	Cu (mg)	
chicken w/ veg & pasta, Prego	120		2.4					5	.07				965	50			
9.49 oz	269	10.8	13.7				2467	.05	2.9				195		1.30		
chilee weenee, Van Camp's	309	178.1	15.7						.20				1057	68			
1 cup	236	14.4	27.6				2228	.11	2.1				527		4.63		
chili-mac																	
Bounty	232		10.2					4	.14				1250	74			
7.8 oz	220	10.2	25.0				374	.11	2.3				306		2.70		
Wolf Brand	317	154.2	19.9						.53	.13	40		854	28	28	2.00	.000
scant cup	213	11.5	22.9				2450	.08	3.5	2.00	.00		337	136	1.87	.170	
chili w/ beans	286	192.6	14.0	0.9	43		86	4	.27	.34			1330	119	115	5.10	
1 cup	255	14.6	30.4	6.0			860	.12	0.9	.03	3.62		932	393	8.75		
Bounty	292		13.4					4	.11				1101	82			
7.8 oz	220	16.2	26.7				1290	.08	1.8				477		3.40		
Bounty, hot	287		12.5					3	.10				1189	88			
7.8 oz	220	16.3	27.3				1356	.08	1.7				477		3.50		
Van Camp's	352	176.3	23.2						.22				1215	63			
1 cup	235	14.9	20.9				3428	.09	2.4				537		4.83		
Wolf Brand	345	161.6	22.0						.82	.25	32		1013	59	57	2.00	.000
1 cup	227	15.0	21.8				3940	.09	2.7	.00	.00		633	252	2.54	.230	
Wolf Brand, extra spicy	324	151.5	20.6						.77	.23	30		926	55	53	2.00	.000
scant cup	213	14.1	20.6				3690	.08	2.5	.00	.00		593	236	2.38	.220	
chili w/o beans																	
Van Camp's	412	169.6	33.5						.28				1499	46			
1 cup	231	15.4	12.1				4934	.06	3.2				370		3.48		
Wolf Brand	387	157.5	26.6						1.27	.10	25		1042	64	30	2.00	.000
1 cup	227	20.7	16.2				4770	.11	3.9	1.00	.00		717	146	3.86	.130	
Wolf Brand, extra spicy	363	147.7	24.9						1.19	.09	23		962	60	28	2.00	.000
scant cup	213	19.4	15.3				4470	.10	3.6	1.00	.00		672	137	3.62	.120	
cowpeas, common w/ pork	199	186.2	3.8	0.6	17		0	1	.12				840	41	104	2.49	.941
1 cup	240	6.6	39.7	1.5			0	.15	1.0	.03			427	231	3.41	.408	
dumplings & chicken, Bounty	200		9.2					0	.10				1047	26			
7.5 oz	213	9.5	19.7				2708	.03	2.6				131		1.10		
linguini w/ white clam sce, Prego	391		10.5					3	.20				753	45			
11 oz	312	17.8	56.4				0	.60	2.5				107		6.20		
macaroni & cheese	228	192.5	9.6					0	.24				730	199			
1 cup	240	9.4	25.7	4.2			260	.12	1.0				139	182	1.00		
Franco-American	168		5.4					0	.21				876	93			
7.4 oz	209	6.4	23.5				567	.25	2.2				103		1.50		
macaroni w/ meat sce, Hearty	192		4.8					5	.18				814	97			
7.5 oz	213	9.8	27.4				270	.21	3.3				390		2.30		
noodles & chicken, Bounty	204		11.7					1	.15				1020	61			
7.5 oz	213	7.9	16.8				728	.09	2.3				145		1.50		
noodle weenee, Van Camp's	245	187.2	8.5						.20				1245	47			
1 cup	238	9.3	32.9				672	.12	2.1				289		5.98		
pasta twists w/ meat sce, Hearty	208		6.1					5	.20				873	43			
7.5 oz	213	9.6	28.7				481	.14	3.3				458		2.10		
pasta w/ chicken cacciatore sce, Prego	359		9.1					6	.22				933	57			
12.2 oz	347	26.1	43.0				1140	.49	6.4				577		4.60		
pizzo's, Franco-American	163		0.6					7	.19				1048	22			
7.5 oz	213	5.0	34.3				778	.18	2.4				352		1.60		
potatoes & beef in gravy, Bounty	142		0.9					9	.09				822	26			
7.5 oz	213	10.4	23.0				0	.04	2.0				362		1.60		
ravioli w/ meat sce																	
Franco-American	284		10.8					7	.21				995	75			
7.5 oz	213	11.3	35.4				1091	.23	3.4				495		1.80		

	KCAL / WT (g)	H₂O (g) / PRO (g)	FAT (g) / CHO (g)	PUFA (g) / SFA (g)	CHOL (mg) / DFIB (g)	A (RE) / A (IU)	C (mg) / B-1 (mg)	B-2 (mg) / NIA (mg)	B-6 (mg) / B-12 (mcg)	FOL (mcg) / PANT (mg)	Na (mg) / K (mg)	Ca (mg) / P (mg)	Mg (mg) / Fe (mg)	Zn (mg) / Cu (mg)	Mn (mg)
Hearty	284		10.8				7	.21			995	75			
7.5 oz	213	11.3	35.4			1091	.23	3.4			495		1.80		
sausage, italian & green peppers, Prego Plus—*4 oz*	164		8.2				19	.05			478	48			
	113	3.4	19.1			1058	.07	1.0			362		1.30		
skettee weenee, Van Camp's	243	186.6	7.4					.20			1128	44			
1 cup	238	9.4	34.7			644	.11	2.2			253		6.31		
spaghetti 'n beef, Franco-American	213		8.2				5	.20			1075	32			
7.5 oz	213	9.0	25.8			819	.20	3.3			395		2.30		
spaghettio's, Franco-American															
w/ franks	210		8.4				4	.18			974	32			
7.4 oz	209	7.6	26.1			519	.20	3.2			302		2.20		
w/ meatballs	204		7.5				4	.18			946	29			
7.4 oz	209	9.1	24.9			524	.19	3.1			318		2.20		
w/ tomato & cheese sce	165		1.3				0	.16			912	29			
7.5 oz	213	4.9	33.5			707	.23	2.4			219		1.50		
spaghetti w/ meatballs & tomato sce	258	195.0	10.3				5	.18			1220	53			
1 cup	250	12.3	28.5	2.2		1000	.15	2.3			245	113	3.30		
Franco-American	215		7.4				4	.20			834	29			
7.4 oz	209	9.6	27.5			567	.18	3.2			338		2.30		
Prego	450		13.4				11	.31			919	90			
13 oz	369	20.2	62.1			1779	.87	7.6			686		9.30		
spaghetti w/ meat sce															
Franco-American	211		8.1				5	.21			1101	28			
7.5 oz	213	8.5	26.2			928	.22	3.5			391		2.20		
Hearty	176		4.8				5	.17			797	37			
7.4 oz	209	9.5	23.5			389	.20	3.2			397		2.10		
Prego	390		8.3				11	.38			870	50			
12.5 oz	354	21.1	57.7			1376	.92	8.0			641		6.60		
spaghetti w/ tomato sce & cheese	190	200.3	1.5				10	.28			955	40			
1 cup	250	5.5	38.5			930	.35	4.5			303	88	2.80		
Franco-American	180		1.3				0	.19			958	28			
7.8 oz	220	5.4	36.9			768	.28	2.8			227		1.60		
spaghetti w/ tomato sce, Franco-American	171		1.2				0	.18			910	27			
7.4 oz	209	5.1	35.0			730	.27	2.6			215		1.60		
spanish rice, Van Camp's	150	188.2	2.7					.10			1358	35			
1 cup	227	3.1	28.2			1228	.05	1.2			260		2.92		
tamales, Wolf Brand	328	150.4	24.5					.49	.26	23	1181	68	43	2.00	.000
scant cup	213	8.3	24.9			2150	.08	1.9	.00	.00	279	141	2.19	.110	
tamales w/ sce, Van Camp's	293	153.8	16.2					.16			1132	36			
1 cup	227	8.3	28.6			1906	.05	2.1			236		2.71		
UFO's, Franco-American	185		2.3				8	.24			781	38			
7.5 oz	213	5.8	35.1			697	.22	3.6			341		2.30		
UFO's w/ meteors, Franco-American	234		8.5				7	.24			784	28			
7.5 oz	213	9.2	30.0			637	.17	3.6			361		2.40		
veal & mushrooms, Prego Plus	143		5.0				14	.05			395	45			
4 oz	113	5.1	19.5			856	.04	0.8			338		1.30		

10.3. FROZEN ENTREES (1-2 ITEMS/ENTREE)

	KCAL / WT (g)	H₂O (g) / PRO (g)	FAT (g) / CHO (g)	PUFA (g) / SFA (g)	CHOL (mg) / DFIB (g)	A (RE) / A (IU)	C (mg) / B-1 (mg)	B-2 (mg) / NIA (mg)	B-6 (mg) / B-12 (mcg)	FOL (mcg) / PANT (mg)	Na (mg) / K (mg)	Ca (mg) / P (mg)	Mg (mg) / Fe (mg)	Zn (mg) / Cu (mg)	Mn (mg)
beef & macaroni, Am Hosp Co	191		6.3		35		12	.19			770	23			
6 oz entree	170	16.6	16.9			792	.17	4.3			521		3.00		
beef & noodles w/ sce, Am Hosp Co	186		7.2		35		1	.23			72	12			
6 oz entree	170	19.8	9.6			78	.11	6.0			109		3.60		
beef & peppers w/ sce, Am Hosp Co	155		3.0		32		46	.17			614	16			
6 oz entree	170	19.8	11.4			816	.08	3.4			362		2.40		

	KCAL / WT (g)	H₂O (g) / PRO (g)	FAT (g) / CHO (g)	PUFA (g) / SFA (g)	CHOL (mg) / DFIB (g)	A (RE) / A (IU)	C (mg) / B-1 (mg)	B-2 (mg) / NIA (mg)	B-6 (mg) / B-12 (mcg)	FOL (mcg) / PANT (mg)	Na (mg) / K (mg)	Ca (mg) / P (mg)	Mg (mg) / Fe (mg)	Zn (mg) / Cu (mg)	Mn (mg)
beef & veg szechwan style, Van de Kamp's—*11 oz entree*	370		15.0								940				
	312	20.0	35.0												
beef brisket bbq pastry wrap, Am Hosp Co—*4.5 oz entree*	301		18.0		24		0	.16			347				
	128	12.8	22.0			194	.14	3.0			326		2.41		
beef burgundy, Am Hosp Co *6 oz entree*	131		4.3		45		1	.15			279	15			
	170	20.4	2.6			86	.05	3.3			297		2.88		
beef burgundy w/ parsley noodles, frzn, Light & Elegant—*9 oz entree*	230		4.0		55		1	.23			1240	14			
	255	23.0	25.0			906	.15	1.7			200	172	3.10		
beef, chipped & creamed, Banquet *4 oz entree*	90		2.0		25		0	.18			818	107			
	113	9.0	10.0			15	.06	0.9			206		1.10		
beef julienne w/ rice & peppers, frzn, Light & Elegant—*8.5 oz entree*	260		7.0				4	.10			990	27			
	241	21.0	27.0			1781	.09	3.5			240	152	4.90		
beef marsala w/ noodles, Prego *11.3 oz*	384		15.0				3	.24			849	37			
	319	27.5	34.7			467	.18	4.7			187		4.00		
beef patties, charbroiled w/ mushroom gravy, Banquet—*1/6 of 32 oz pkg*	210		15.0				1	.12			850	38			
	151	10.0	8.0			57	.10	1.1			170		1.50		
beef pot roast w/ sce & mixed veg, Am Hosp Co—*6 oz entree*	139		4.1		43		5	.17			341	19			
	170	20.0	5.7			906	.09	4.0			313		2.64		
beef ravioli, Am Hosp Co *5 oz entree*	185		5.7		59		6	.16			446	83			
	142	9.2	24.4			895	.08	1.2			348		1.15		
beef short ribs w/ bbq sce, Am Hosp Co *5.6 oz entree*	345		20.9		59		5	.16			702	19			
	159	26.0	13.3			716	.09	4.6			266		3.50		

beef, sliced

	KCAL / WT (g)	H₂O (g) / PRO (g)	FAT (g) / CHO (g)	PUFA (g) / SFA (g)	CHOL (mg) / DFIB (g)	A (RE) / A (IU)	C (mg) / B-1 (mg)	B-2 (mg) / NIA (mg)	B-6 (mg) / B-12 (mcg)	FOL (mcg) / PANT (mg)	Na (mg) / K (mg)	Ca (mg) / P (mg)	Mg (mg) / Fe (mg)	Zn (mg) / Cu (mg)	Mn (mg)
& veg, frzn, Banquet *10 oz entree*	300		9.0		100		1	.27			770	78			
	284	38.0	17.0			520	.45	4.1			382		4.30		
w/ bbq sce, Am Hosp Co *5 oz*	225		8.4		40		0	.13			378	53			
	142	18.9	18.5			592	.04	3.7			590		2.60		
w/ bbq sce, Banquet *4 oz entree*	90		2.0		40		0	.09			789	16			
	113	9.0	10.0			339	.05	1.1			137		1.50		
w/ gravy, Banquet *4 oz entree*	90		3.0		40		0	.10			426	10			
	113	12.0	4.0			29	.03	1.2			81		2.00		
w/ gravy, Banquet *1/4 of 32 oz pkg*	160		6.0				2	.21			900	17			
	227	19.0	9.0			73	.07	2.2			150		2.50		
w/ mustard & tomato sce, Am Hosp Co—*6 oz entree*	240		10.9		48		8	.15			678	38			
	170	17.5	17.9			828	.07	3.1			378		2.82		
beef stew, Am Hosp Co *6 oz entree*	150		6.0		45		8	.11			846	26			
	170	12.0	12.0			1530	.07	2.7			222		1.86		
beef stew, Banquet *1/4 of 32 oz pkg*	254		13.0				7	.10			977	23			
	227	12.0	21.0			1862	.08	2.7			308		2.00		
beef stroganoff, Am Hosp Co *6 oz*	185		9.1		44		1	.22			350	20			
	170	20.5	5.3			189	.06				277		3.20		
beef stroganoff w/ parsley noodles, Light & Elegant—*9 oz entree*	260		6.0		65		2	.26			790	45			
	255	24.0	27.0			944	.20	2.1			230	79	3.00		
beef tamale pie, Am Hosp Co *6 oz entree*	178		6.6		37		24				429	154			
	170	12.6	16.8			915		2.4			446		2.10		
beef teriyaki, Am Hosp Co *6 oz entree*	154		4.4		18		20	.15			816	46			
	170	12.3	16.1			1290	.07	1.7			382		2.20		
beef teriyaki w/ rice & pea pods, Light & Elegant—*8 oz entree*	240		3.0		45		2	.10			625	30			
	227	18.0	37.0			122	.40	2.5			215	152	5.60		
beef wellington, Am Hosp Co *6.5 oz entree*	245		15.5		113						427				
	184	24.2	2.1								641				

burrito

	KCAL / WT (g)	H₂O (g) / PRO (g)	FAT (g) / CHO (g)	PUFA (g) / SFA (g)	CHOL (mg) / DFIB (g)	A (RE) / A (IU)	C (mg) / B-1 (mg)	B-2 (mg) / NIA (mg)	B-6 (mg) / B-12 (mcg)	FOL (mcg) / PANT (mg)	Na (mg) / K (mg)	Ca (mg) / P (mg)	Mg (mg) / Fe (mg)	Zn (mg) / Cu (mg)	Mn (mg)
beef, red hot, El Charrito *5 oz burrito*	340		17.0		20						780				
	142	10.0	39.0								260				
crispy fried, Van de Kamp's *6 oz entree*	365		15.0								825				
	170	10.0	40.0												
grande, El Charrito *6 oz burrito*	430		16.0		25						820				
	170	17.0	53.0								540				
green chili, El Charrito *5 oz burrito*	370		16.0		20						650				
	142	12.0	51.0								380				

	KCAL	H₂O (g)	FAT (g)	PUFA (g)	CHOL (mg)	A (RE)	C (mg)	B-2 (mg)	B-6 (mg)	FOL (mcg)	Na (mg)	Ca (mg)	Mg (mg)	Zn (mg)	Mn (mg)
	WT (g)	PRO (g)	CHO (g)	SFA (g)	DFIB (g)	A (IU)	B-1 (mg)	NIA (mg)	B-12 (mcg)	PANT (mg)	K (mg)	P (mg)	Fe (mg)	Cu (mg)	
green chili grande, El Charrito	410		14.0		20						1030				
6 oz entree	170	16.0	56.0								510				
jalapeno grande, El Charrito	410		15.0		25						1060				
6 oz burrito	170	16.0	53.0								560				
red chili, El Charrito	380		18.0		20						620				
5 oz burrito	142	12.0	40.0								380				
red chili grande, El Charrito	410		15.0		25						830				
6 oz burrito	170	16.0	53.0								570				
red hot, El Charrito	374		18.0		20						650				
5 oz burrito	142	12.0	41.0								410				
sirloin, grande, Van de Kamp's	440		15.0								1120				
11 oz entree	312	25.0	45.0												
cabbage stuffed w/ rice, grd beef & tom	172		6.1		43		39	.16			479	43			
sce, Am Hosp Co—7.4 oz entree	210	10.9	18.3			1113	.11	2.6			447		2.56		
cabbage stuffed w/ rice, tom, corn & crm	211		6.1		105		20				930	73			
sce, Am Hosp Co—7.1 oz entree	201	11.3	27.6			2570	.13	3.4			585		1.60		
cannelloni florentine (veal, spinach &	362		21.6		89		18	.48			689	120			
beef), Am Hosp Co—8 oz entree	227	16.6	25.3			3360	.62	6.5			272		4.20		
cheese blintz, Am Hosp Co	432		25.6		436				2.00		246	336			
8 oz entree	227	19.2	31.2			1568	.33	3.9			312		4.80		
chicken a la kiev, Am Hosp Co	400		27.6		140						395				
6 oz entree	170	31.2	6.6								501				
chicken a la king															
Am Hosp Co	150		4.3		42		11				404	46			
6 oz entree	170	15.5	12.4			468	.06	4.3			251		1.02		
Banquet	110		5.0		40		0	.14			551	64			
4 oz entree	113	8.0	9.0			72	.06	1.0			137		.60		
chicken a la reine pastry wrap, Am Hosp	266		17.2		27		1	.15			387	19			
Co—4.5 oz entree	128	9.8	17.9			213	.14	3.4			160		1.43		
chicken almond, cantonese w/ rice, Van	440		15.0								1220				
de Kamp's—11 oz entree	312	25.0	50.0												
chicken & biscuits, Am Hosp Co	265		12.5		46		0	.22			962	84			
6.6 oz entree	187	15.2	22.4			349	.12	3.8			250		1.30		
chicken & dumplings	430		25.0				0	.14			928	18			
1/4 of 32 oz pkg	227	18.0	31.0			77	.06	3.5			133		.90		
chicken & noodles w/ sce, Am Hosp Co	169		6.8		41		6	.30			456	55			
6 oz entree	170	15.0	12.0			348	.31	4.3			144		1.50		
chicken breast, Am Hosp Co															
hawaiian (w/ pineapple stuffing)	231		2.8		107						303				
8 oz entree	227	35.6	16.0								731				
jamaica (w/ chili sce, green peppers,	287		5.6		84		20	.25			1018	126			
raisins & almonds)—7.5 oz entree	213	24.7	34.5				.15	9.5			1022		2.90		
mandarin (w/ stuffing & orange sce)	253		10.7		87		27	.23			270	37			
8 oz entree	227	26.0	13.1			460	.13	9.3			476		2.20		
new englander (w/ stuffing & crm sce)	335		14.4		101						330				
7 oz entree	198	35.5	16.0								500				
pilgrim (w/ cranberry stuffing & crm	243		5.4		84						185				
sce)—7 oz entree	198	30.6	18.1								448				
w/ cornbread stuffing & gravy	360		20.9		46		1	.22			630	33			
8 oz entree	227	28.4	14.6			712	.09	9.7			387		1.84		
w/ wild rice stuffing	318		16.5		107						189				
8 oz entree	227	35.9	6.5								672				
chicken cacciatore, Banquet	260		5.0		40		1	.21			510	42			
10 oz entree	284	17.0	35.0			678	.19	2.3			191		3.00		
chicken cordon blue, Am Hosp Co	272		10.3		108		17				327				
6.5 oz entree	184	25.7	19.2								411				
chicken, french, Banquet	190		4.0		35		1	.16			850	52			
10 oz entree	284	12.0	25.0			1084	.14	2.2			455		1.90		
chicken fricassee, Am Hosp Co	152		6.4		41		118	.12			152	34			
6 oz entree	170	16.6	7.1			301	.07	4.7			250		1.20		

Item	KCAL / WT (g)	H₂O (g) / PRO (g)	FAT (g) / CHO (g)	PUFA (g) / SFA (g)	CHOL (mg) / DFIB (g)	A (RE) / A (IU)	C (mg) / B-1 (mg)	B-2 (mg) / NIA (mg)	B-6 (mg) / B-12 (mcg)	FOL (mcg) / PANT (mg)	Na (mg) / K (mg)	Ca (mg) / P (mg)	Mg (mg) / Fe (mg)	Zn (mg) / Cu (mg)	Mn (mg)
chicken, glazed w/ veg rice, Light & Elegant—8 oz entree	240		4.0		75		1	.16			660	15			
	227	29.0	23.0			156	.15	7.9			300	348	1.90		
chicken in bbq sce w/ corn & pecan rice, Light & Elegant—8 oz entree	300		6.0				5	.15			900	22			
	227	26.0	35.0			943	.19	6.9			450	896	1.50		
chicken in cheese sce w/ rice & broccoli, Light & Elegant—8.75 oz entree	293		11.0		50		11	.21			800	204			
	248	19.0	30.0			377	.12	1.9			170		1.60		
chicken kiev, Am Hosp Co 6.5 oz entree	240		9.5		126		11	.14			442				
	184	23.6	14.9				.00				451				
chicken leg, Am Hosp Co															
hawaiian (w/ pineapple stuffing) 6 oz entree	226		7.9		62		4	.27			277	28			
	170	24.6	14.0			114	.14	6.8			273		1.61		
normandy (w/ grain-based onion & apple stuffing)—6 oz entree	222		7.5		61		1	.28			309	35			
	170	24.7	14.0			66	.13	6.7			279		1.60		
saint laurent (w/ ham, broccoli & cheese stuffing)—6 oz entree	222		11.8		73		10	.32			309	115			
	170	27.4	1.6			549	.14	6.8			304		1.60		
w/ apple bread, almond & raisin stuffing—6 oz entree	326		17.9		92		1	.25			266	254			
	170	27.9	13.4			73	.10	13.3			297		2.00		
w/ cranberry & crouton stuffing 6 oz entree	254		11.2		61		2	.26			247	25			
	170	24.2	14.3			72	.10	6.7			244		1.60		
chicken parmigiana w/ parsley noodles, Light & Elegant—8 oz entree	260		6.0				7	.20			680	73			
	227	28.0	23.0			640	.33	6.6			320		2.80		
chicken piccata w/ rice, Prego 11 oz	342		18.5				6	.14			1148	96			
	312	21.0	22.9			893	.18	8.0			135		2.90		
chicken princess, Am Hosp Co[a] 6 oz entree	110		4.2		34		3	.08			433	12			
	170	12.0	9.0			53	.04	2.8			250		1.20		
chicken salad, Am Hosp Co 1 oz entree	45		2.3		8		1	.02			78	5			
	28	4.2	1.8			15	.01	1.4			67		3.40		
chicken stew, Am Hosp Co 6 oz entree	126		2.4		28		3	.08			107	18			
	170	16.2	10.2			530	.05	2.2			306		4.60		
chicken tortilla casserole, Am Hosp Co 6 oz entree	215		10.8		20		14	.09			415	269			
	170	14.4	14.4			524	.08	5.9			209		.66		
chicken wellington, Am Hosp Co[b] 7.5 oz entree	386		20.8		79		1	.24			534	21			
	213	26.3	23.3			148	.15	8.4			331		2.11		
chicken w/ bbq sce, Am Hosp Co 5 oz entree	175		4.5		51		8	.20			132	12			
	142	27.5	6.0			581	.07	10.4			498		1.70		
chicken w/ broccoli, Light & Elegant 9.5 oz entree	290		11.0				1	.21			805	204			
	270	19.0	30.0			377	.11	1.8			180	240	1.60		
chicken w/ crm sce & veg, Am Hosp Co[c] 6 oz entree	179		8.3		44		52	.17			410	47			
	170	15.4	10.5			666	.08				329		.90		
chicken w/ veloute sce 5 oz entree	165		4.0		55		0	.44			64	41			
	142	27.0	5.5			79	.06	11.6			338		1.30		
chili & cheese pastry wrap, Am Hosp Co 4.5 oz entree	339		25.3		15		10	.12			548	67			
	128	7.1	20.7			545	.15	2.0			233		2.09		
chili relleno casserole, Am Hosp Co 6 oz entree	377		27.5		124		101	.26			679	483			
	170	18.9	13.4			2118	.13	1.0			208		1.38		
chili w/ meat sce, Am Hosp Co 6 oz entree	176		6.0		28		8				1140	16			
	170	9.6	21.0			472		3.1			429				
chow mein															
beef, Am Hosp Co 6 oz entree	83		3.3		15		17	.08			658	22			
	170	8.0	5.2			251	.07	1.1			100		1.08		
beef mandarin, Van de Kamp's 11 oz entree	310		10.0								1700				
	312	20.0	40.0												
chicken mandarin, Van de Kamp's 11 oz entree	340		10.0								1180				
	312	20.0	40.0												
vegetable, Am Hosp Co 6 oz entree	56		1.0				38	.11			597	24			
	170	2.1	9.5			358	.09	0.6			213		.90		
coquille saint-jacques, Am Hosp Co[d] 4 oz entree	126		4.9		25						115				
	113	10.1	10.2								199				
crepes a la reine, Am Hosp Co[e] 4.5 oz entree	130		5.7		106						144				
	128	13.5	7.0								295				

[a] Chicken w/ oyster sce, bamboo shoots, mushrooms, snow peas, water chestnuts & scallions.
[b] Breast meat wrapped in pastry dough w/ madeira wine sce.
[c] Vegetables include mashed potatoes, mushrooms & red peppers.
[d] Scallops, potatoes & mushrooms w/ sce.
[e] Contains chicken & cream sce.

	KCAL / WT (g)	H₂O (g) / PRO (g)	FAT (g) / CHO (g)	PUFA (g) / SFA (g)	CHOL (mg) / DFIB (g)	A (RE) / A (IU)	C (mg) / B-1 (mg)	B-2 (mg) / NIA (mg)	B-6 (mg) / B-12 (mcg)	FOL (mcg) / PANT (mg)	Na (mg) / K (mg)	Ca (mg) / P (mg)	Mg (mg) / Fe (mg)	Zn (mg) / Cu (mg)	Mn (mg)
crepes de la mer, Am Hosp Co[a]	187		9.3								436				
5 oz entree	142	13.3	14.2								128				
eggplant parmigiana, Am Hosp Co	223		12.0		5		19	.19			982	218			
6.4 oz entree	181	10.8	17.9			1436	.15	1.7			462		2.16		
egg rolls, cantonese, Van de Kamp's	280		5.0								550				
5.25 oz	149	10.0	40.0												
enchilada casserole, Am Hosp Co[b]	270		15.9		31		9	.16			847	97			
6 oz entree	170	15.5	16.3			1302	.10	2.8			463		1.86		
enchiladas															
beef & cheese, El Charrito	880		42.0		70						2520				
6 enchiladas (16.25 oz entree)	461	23.0	96.0								880				
beef, Banquet	264		8.0				0	.13			1477	52			
1/4 of 32 oz pkg	227	11.0	37.0			179	.19	1.9			92		1.00		
beef, El Charrito	560		31.0		55						1550				
3 enchiladas (11 oz entree)	312	17.0	53.0								490				
beef, El Charrito	880		49.0		75						2130				
6 enchiladas (16.25 oz entree)	461	27.0	85.0								770				
beef, shredded, Van de Kamp's	180		10.0								930				
5.5 oz entree	156	10.0	15.0												
beef, Van de Kamp's	250		15.0								1200				
7.5 oz entree	213	10.0	20.0												
beef, Van de Kamp's	340		15.0								1480				
4 enchiladas (8.5 oz)	241	15.0	30.0												
cheese, El Charrito	470		20.0		30						1720				
3 enchiladas (11 oz entree)	312	12.0	64.0								540				
cheese, El Charrito	780		30.0		45						2680				
6 enchiladas (16.25 oz entree)	461	19.0	109.0								880				
cheese ranchero, Van de Kamp's	250		15.0								540				
5.5 oz entree	156	10.0	20.0												
cheese, Van de Kamp's	270		15.0								965				
7.5 oz entree	213	10.0	25.0												
cheese, Van de Kamp's	370		20.0								1175				
4 enchiladas (8.5 oz)	241	15.0	30.0												
chicken, El Charrito	440		13.0		60						1550				
3 enchiladas (11 oz entree)	312	23.0	59.0								630				
chicken suiza, Van de Kamp's	220		10.0								590				
5.5 oz entree	156	10.0	20.0												
chicken, Van de Kamp's	250		10.0								1105				
7.5 oz entree	213	15.0	25.0												
fettuccini & chicken w/ sce, Am Hosp Co	122		4.3		18		2	.08			182	40			
4 oz entree	113	7.2	13.6			358	.08	1.8			126		.88		
fettuccini primavera, Am Hosp Co[c]	305		14.2		45		10	.15			1410	251			
6 oz entree	170	11.6	32.3			1588	.19	1.5			165		1.43		
fish, batter-dipped & chips, Van de	440		25.0								640				
Kamp's—*7 oz entree*	198	20.0	35.0												
french bread pizza															
cheese, Am Hosp Co	417		18.5		38		8	.40			941	316			
5.75 oz entree	163	17.8	44.9			635	.47	3.6			238		3.10		
cheese, Pillsbury Microwave	339	83.1	13.5				6	.31			574	281			
1 piece	155	14.1	40.3			1319	.31	3.6			256	214	2.95		
pepperoni, Am Hosp Co	425		20.1		59		7	.45			1165	189			
5.6 oz entree	159	16.1	44.8			574	.62	5.4			275		4.40		
pepperoni, Pillsbury Microwave	410	88.9	20.2				5	.51			1163	245			
1 piece	171	16.6	40.5			607	.63	6.5			316	207	4.79		
sausage, Pillsbury Microwave	409	95.6	19.1				6	.34			1018	239			
1 piece	177	18.1	40.0			598	.46	4.4			329	228	3.19		
sausage & pepperoni, Pillsbury	429	97.3	21.9				11	.31			1149	246			
Microwave—*1 piece*	185	18.1	41.1			670	.37	4.6			346	228	3.15		
green pepper steak, Banquet	310		5.0		50		1	.19			1125	28			
10 oz entree	284	21.0	44.0			307	.21	2.7			221		3.30		

[a] Contains shrimp, crab, langostinos & scallops w/ cream sce.
[b] Contains ground beef, tortillas, jack cheese & sce.
[c] Contains peas, broccoli, carrots & mushrooms.

	KCAL / WT (g)	H₂O (g) / PRO (g)	FAT (g) / CHO (g)	PUFA (g) / SFA (g)	CHOL (mg) / DFIB (g)	A (RE) / A (IU)	C (mg) / B-1 (mg)	B-2 (mg) / NIA (mg)	B-6 (mg) / B-12 (mcg)	FOL (mcg) / PANT (mg)	Na (mg) / K (mg)	Ca (mg) / P (mg)	Mg (mg) / Fe (mg)	Zn (mg) / Cu (mg)	Mn (mg)
green pepper, stuffed, Am Hosp Co															
w/ beef, rice & sce	227		10.2		64		74	.16			465	27			
7.5 oz entree	213	15.5	18.2			325	.09	2.6			330		2.11		
w/ ground beef, rice & tomatoes	172		6.1		54		26	.16			188	28			
7.5 oz entree	213	13.9	15.4			980	.14	6.1			590		2.80		
ham & potato casserole, Am Hosp Co	176		8.6			47	12	.12			811	76			
6 oz entree	170	9.8	14.8			189	.25	1.7			391		1.40		
italian sausage & peppers, Am Hosp Co	327		29.4		59		361	.13			665	14			
6 oz entree	170	7.2	9.6			934	.28	1.9			155		1.80		
lasagna															
Am Hosp Co	244		10.8		21		19	.30			816	167			
6 oz entree	170	15.6	21.0			1308	.53	4.4			271		2.80		
beef & mushroom, Van de Kamp's	430		25.0								970				
11 oz entree	312	25.0	30.0												
florentine, Light & Elegant	280		5.0		25		23	.38			980	280			
11.25 oz entree	319	24.0	35.0			4080	.39	1.8			720	387	3.60		
italian sausage, Van de Kamp's	440		25.0								1190				
11 oz entree	312	25.0	35.0												
primavera (spinach), Am Hosp Co	178		9.6		30		14	.10			378	126			
6 oz entree	170	7.4	15.5			3558	.11	2.3			245		1.38		
w/ meat sce, Am Hosp Co	256		11.4		35		7	.17			496	142			
6 oz entree	170	16.8	21.6			1416	.19	2.9			475		2.22		
w/ meat sce, Banquet	330		11.0				1	.30			1000	244			
1/4 of 32 oz pkg	227	19.0	39.0			1291	.50	3.2			340		2.40		
lumache, Am Hosp Co															
w/ italian sausage & tomato sce	138		8.7		35		0	.04			531	51			
3 oz entree	85	6.3	8.1			83	.05	0.4			201		.90		
w/ ricotta cheese & tomato sce	139		7.3		19		1	.08			313	121			
3.58 oz entree	101	8.0	10.4			491	.07	5.3			169		.58		
macaroni & cheese															
Am Hosp Co	196		7.1		17		4	.24			358	269			
6 oz entree	170	11.0	21.8			88	.18	0.9			73		.69		
Banquet	344		17.0				0	.13			930	210			
8 oz entree	227	11.0	36.0			605	.05	0.7			135		1.00		
Banquet	230		8.0				0	.19			790	153			
1/4 of 32 oz pkg	227	9.0	31.0			127	.26	0.9			40		1.40		
w/ bread crumb topping, Light &	300		9.0		5		0	.43			1010	328			
Elegant—*9 oz entree*	255	16.0	38.0			302	.34	1.5			216	334	2.00		
manicotti, Am Hosp Co															
vegetarian w/ tomato sce	192		6.7		31		0	.16			403	257			
5.6 oz entree	159	8.9	24.0			470	.12	0.1			158		.78		
w/ ricotta cheese & tomato sce	229		13.5		56		2	.04			421	18			
5 oz entree	142	13.0	12.5			163	.04	0.3			225		.10		
meatballs															
swedish w/ sce, Am Hosp Co	220		13.2		42		3	.18			528	36			
6 oz entree	170	12.0	13.2			204	.16	7.2			294		1.20		
w/ tomato sce, Am Hosp Co	186		8.4		58		0	.17			942	38			
6 oz entree	170	15.6	12.6			2214	.14	4.2			198		3.00		
meatloaf															
Am Hosp Co	147		7.2		58		4	.12			465	11			
3 oz entree	85	17.1	3.4			90	.06	3.4			285		2.10		
Banquet	240		15.0		45		19	.15			827	55			
5 oz entree	142	12.0	12.0			303	.14	1.6			385		2.50		
w/ brown gravy, Am Hosp Co	169		7.8		58		5	.13			742	11			
5 oz entree	142	18.0	6.8			90	.07	5.7			304		2.10		

	KCAL	H₂O (g)	FAT (g)	PUFA (g)	CHOL (mg)	A (RE)	C (mg)	B-2 (mg)	B-6 (mg)	FOL (mcg)	Na (mg)	Ca (mg)	Mg (mg)	Zn (mg)	Mn (mg)
	WT (g)	PRO (g)	CHO (g)	SFA (g)	DFIB (g)	A (IU)	B-1 (mg)	NIA (mg)	B-12 (mcg)	PANT (mg)	K (mg)	P (mg)	Fe (mg)	Cu (mg)	
w/ tomato sce, Am Hosp Co	182		8.1		58		16	.14			623	12			
5 oz entree	142	17.9	9.3			722	.10	4.0			458				
mostaccioli w/ meat sce, Banquet	240		4.0				1	.23			1180	35			
1/4 of 32 oz pkg	227	12.0	38.0			1092	.43	2.8			330		3.60		
noodles & beef, Banquet	283		18.0				1	.12			877	18			
1/4 of 32 oz pkg	227	10.0	21.0			100	.19	2.3			99		1.50		
osso bucco (veal shank w/ veg & sce), Am	321		17.2		119		5	.39			673	32			
Hosp Co—*11.7 oz entree*	332	32.8	8.8			1102	.30	10.4			754		4.90		
pasta shells & sce, Banquet	310		8.0		35		0	.28			950	178			
10 oz entree	284	17.0	41.0			687	.37	2.9			335		3.50		
pizza, bacon, Totino's Party	271	49.4	12.7				4	.21			721	114			
1/3 pizza	103	7.7	28.8			467	.21	2.3			173	97	1.34		
pizza, canadian style bacon															
Big Slice Lite	61	15.7	2.3		1		1	.04			74	74			
1 oz slice	28	2.9	6.9			144	.03	1.5					.50		
Celeste	541	112.0	26.0					.73	.29	141	1593	491	62	4.00	.000
7.75 oz pizza	220	27.3	49.6			1553	.37	2.9	2.00	.00	444	684	1.89	.260	
Celeste	329	69.8	16.9					.44	.19	100	976	301	35	3.00	.000
1/4 of 19 oz pizza	135	16.6	27.7			900	.20	2.0	1.00	.00	290	418	1.21	.160	
Totino's Classic	320	71.5	11.9				6	.86			817	212			
1/4 pizza	143	15.9	37.2			183	1.00	3.3			225	232	2.29		
Totino's Party	230	55.2	8.4				3	.21			670	112			
1/3 pizza	103	10.1	28.3			383	.21	2.1			169	62	1.24		
pizza, cheese															
Celeste	497	84.4	24.5					.44	.15	35	828	375	48	4.00	.000
6.5 oz pizza	184	21.2	48.0			927	.15	1.2	2.00	.00	342	386	1.36	.240	
Celeste	317	62.1	16.6					.39	.10	14	673	204	30	3.00	.000
1/4 of 18 oz pizza	126	14.2	27.8			779	.08	1.1	1.00	.00	268	266	1.00	.140	
Fox Deluxe	170	46.5	6.1				2	.17			504	117			
1/3 pizza	83	6.6	21.6			337	.17	1.8			106	159	1.16		
Pillsbury Heat 'n Eat Microwave	269	54.2	10.4				5	.22			667	241			
4.1 oz	116	12.8	31.3			537	.16	1.7			202	204	1.16		
Pillsbury Microwave	480		19.0								1190				
7.1 oz	201	21.0	56.0								360				
Totino's Classic	351	58.0	14.8				6	.26			811	282			
1/4 pizza	138	15.2	40.0			722	.18	2.5			273	239	1.66		
Totino's Extra	251	44.7	12.2				3	.19			462	229			
1/4 pizza	94	11.3	24.4			451	.12	1.3			185	182	1.03		
Totino's Microwave	254	51.8	9.4				5	.21			638	194			
3.9 oz	111	11.1	31.1			501	.16	1.9			199	171	1.22		
Totino's Party	255	47.5	11.8				3	.17			635	151			
1/3 pizza	97	8.9	27.2			458	.15	1.6			131	190	1.07		
Totino's Slices	170	35.9	6.8				2	.13			346	125			
1 slice	71	7.2	19.9			258	.11	1.1			87	110	.78		
pizza, combination															
Pillsbury Microwave	673	114.8	36.2				18	.43			1550	390			
9 oz	255	27.0	59.4			1025	.43	4.6			500	390	3.06		
Totino's Classic	461	68.9	25.4				12	.30			1033	271			
1/4 pizza	164	18.0	39.4			758	.28	3.1			342	262	2.13		
pizza, deluxe															
Celeste	582	123.8	31.8					.82	.26	80	1308	332	61	5.00	.000
8.25 oz pizza	234	22.7	51.2			1226	.30	3.5	2.00	2.00	498	480	2.57	.350	
Celeste	378	86.3	22.1					.55	.17	57	953	267	38	3.00	.000
1/4 of 22 oz pizza	158	15.5	29.3			950	.20	2.7	2.00	2.00	352	357	1.85	.220	
pizza, ground beef															
Am Hosp Co	300		12.9		22		8	.18			320	118			
5.66 oz entree	160	19.7	26.3			536	.16	5.1			671		2.52		

	KCAL / WT (g)	H₂O (g) / PRO (g)	FAT (g) / CHO (g)	PUFA (g) / SFA (g)	CHOL (mg) / DFIB (g)	A (RE) / A (IU)	C (mg) / B-1 (mg)	B-2 (mg) / NIA (mg)	B-6 (mg) / B-12 (mcg)	FOL (mcg) / PANT (mg)	Na (mg) / K (mg)	Ca (mg) / P (mg)	Mg (mg) / Fe (mg)	Zn (mg) / Cu (mg)	Mn (mg)
Fox Deluxe	180	45.7	6.2				2	.17			457	69			
1/3 pizza	83	8.1	22.4			221	.17	2.2			144	66	1.66		
Totino's Party	279	51.9	13.3				3	.21			650	111			
1/3 pizza	106	11.7	27.9			413	.15	2.1			193	86	1.70		
pizza, mexican style, Totino's Party	239	38.5	13.9				6	.18			590	113			
1/3 pizza	89	7.6	21.0			348	.18	1.9			163	82	1.78		
pizza, nacho, Totino's Party	230	47.6	12.5				7	.18			480	191			
1/3 pizza	92	8.6	20.2			362	.14	1.5			98	121	1.29		
pizza, pepperoni															
Big Slice Lite	67	14.7	2.3		1		1	.04			57	74			
1 oz slice	28	3.1	7.0			144	.03	1.5					.57		
Celeste	546	86.1	29.6					.71	.21	97	1417	336	53	4.00	.000
6.75 oz pizza	191	20.2	49.7			1316	.25	2.6	2.00	.00	416	443	2.04	.290	
Celeste	368	65.4	21.3					.57	.18	101	1061	186	39	4.00	.000
1/4 of 19 oz pizza	135	15.0	29.2			1343	.14	1.6	1.00	.00	284	386	1.42	.160	
Fox Deluxe	169	46.9	5.8				2	.17			505	67			
1/3 pizza	83	6.6	22.4			232	.17	2.2			116	62	1.41		
Pillsbury Heat 'n Eat Microwave	346	58.1	18.2				5	.29			1070	217			
4.6 oz	130	13.4	31.2			673	.20	2.5			226	228	1.69		
Pillsbury Microwave	595	110.9	28.0				18	.43			1571	390			
8.5 oz	241	26.8	59.5			901	.39	4.6			458	381	2.89		
Totino's Classic	409	63.2	19.9				7	.27			1104	245			
1/4 pizza	152	17.2	40.3			711	.26	3.2			333	234	1.82		
Totino's Extra	260	40.0	13.7				4	.17			686	159			
1/4 pizza	94	10.2	24.1			507	.15	1.9			115	175	1.50		
Totino's Microwave	292	50.2	14.1				5	.23			1267	130			
4 oz	113	9.9	30.5			571	.23	2.5			191	163	1.70		
Totino's Party	259	53.0	12.4				3	.15			739	114			
1/3 pizza	105	9.7	28.0			302	.17	2.0			156	162	1.26		
Totino's Slices	190	36.7	9.2				3	.15			526	84			
1 slice	74	6.7	20.0			312	.15	1.5			87	106	1.11		
pizza, sausage															
Am Hosp Co	342		20.8		40		26	.24			698	146			
6 oz entree	170	10.6	28.1			1155	.26	2.2			170		2.20		
Big Slice Lite	65	15.0	2.5		1		1	.04			74	80			
1 oz slice	28	3.1	7.2			113	.03	1.5					.60		
Celeste	571	103.7	31.7					.77	.23	109	1374	371	60	4.00	.000
7.5 oz pizza	213	22.6	48.8			1359	.30	2.9	2.00	.00	456	505	2.28	.340	
Celeste	376	70.6	21.7					.51	.17	89	988	322	40	3.00	.000
1/4 of 20 oz pizza	142	15.6	29.7			1048	.18	2.0	1.00	.00	311	385	1.52	.180	
Fox Deluxe	180	42.9	6.5				2	.17			532	71			
1/3 pizza	85	6.5	23.0			234	.26	2.1			129	65	1.36		
Pillsbury Heat 'n Eat Microwave	360	53.0	19.7				5	.27			1127	219			
4.8 oz	136	13.3	31.3			681	.26	2.3			262	241	1.90		
Pillsbury Microwave	647	112.1	34.2				21	.40			1421	387			
8.75 oz	248	25.3	59.0			1019	.42	4.2			476	377	2.98		
Totino's Classic	439	68.0	23.5				7	.26			974	240			
1/4 pizza	162	16.2	40.5			757	.29	2.9			366	238	2.11		
Totino's Extra	279	34.1	15.5				4	.20			767	141			
1/4 pizza	98	9.8	24.3			472	.20	2.0			176	166	1.57		
Totino's Microwave	301	47.6	14.9				5	.24			925	132			
4.2 oz	119	10.1	31.2			534	.24	2.3			242	162	1.67		
Totino's Party	270	46.2	13.1				3	.21			772	114			
1/3 pizza	105	9.8	28.4			391	.21	1.9			174	85	1.37		
Totino's Slices	200	34.3	10.1				3	.16			558	86			
1 slice	78	6.9	20.3			282	.16	1.4			121	107	1.09		
pizza, sausage & mushroom															
Celeste	592	127.7	32.3					.87	.34	82	1347	362	60	5.00	.000
8.5 oz pizza	241	23.9	51.3			1200	.34	3.9	2.00	2.00	549	504	2.41	.340	

	KCAL	H₂O (g)	FAT (g)	PUFA (g)	CHOL (mg)	A (RE)	C (mg)	B-2 (mg)	B-6 (mg)	FOL (mcg)	Na (mg)	Ca (mg)	Mg (mg)	Zn (mg)	Mn (mg)
	WT (g)	PRO (g)	CHO (g)	SFA (g)	DFIB (g)	A (IU)	B-1 (mg)	NIA (mg)	B-12 (mcg)	PANT (mg)	K (mg)	P (mg)	Fe (mg)	Cu (mg)	
Celeste	387	85.2	22.4					.51	.19	102	1033	310	45	3.00	.000
1/4 of 25 oz pizza	177	16.9	29.4			1065	.24	2.9	2.00	.00	361	407	1.61	.220	
pizza, sausage & pepperoni															
Fox Deluxe	170	42.1	6.2				2	.20			505	67			
1/3 pizza	81	6.3	21.9			223	.22	2.0			120	62	1.30		
Pillsbury Heat 'n Eat Microwave	378	55.0	21.3				5	.28			1202	218			
4.9 oz	139	13.9	31.3			678	.25	2.6			270	243	1.95		
Totino's Extra	289	37.4	16.2				4	.20			793	160			
1/4 pizza	101	10.8	24.7			503	.20	2.0			199	183	1.62		
Totino's Microwave	307	48.9	15.5				5	.23			1091	140			
4.2 oz	119	10.5	30.9			562	.24	2.4			227	171	1.74		
Totino's Party	278	50.1	13.2				4	.22			791	125			
1/3 pizza	109	10.5	29.4			421	.19	2.0			172	88	1.42		
Totino's Slices	200	35.9	9.9					.16			645	92			
1 slice	78	7.0	20.3			307	.15	1.5			108	112	1.17		
pizza, suprema															
Celeste	678	130.8	39.3					.92	.28	120	1693	454	64	5.00	.000
9 oz pizza	255	26.5	54.0			1943	.33	3.7	3.00	3.00	528	602	2.73	.360	
Celeste	381	87.9	24.1					.68	.20	67	1043	295	41	3.00	.000
1/4 of 23 oz pizza	163	16.8	29.2			1167	.20	2.2	2.00	.00	342	408	1.78	.230	
pizza, vegetable, Totino's Party	219	61.6	7.9				10	.22			631	118			
1/3 pizza	108	7.9	29.2			591	.14	1.8			158	72	1.40		
pork chops w/ bbq sce, Am Hosp Co	293		13.0		80		6	.26			499	56			
7.2 oz entree	204	21.6	22.3			529	1.0	5.1			725		3.30		
pork chops w/ brown gravy, Am Hosp Co	252		16.2				1	.20			367	9			
6.5 oz entree	184	22.1	4.4			33	.43	4.3			339		2.50		
pork, ground															
w/ cheese sce & veg, Am Hosp Co[a]	217		13.1		48		10	.19			365	167			
6 oz entree	170	16.0	8.8			2184	.32	2.3			286		2.10		
w/ brown sce, apple stuffing & mixed veg, Am Hosp Co—*6 oz entree*	227		13.9		48		1	.16			612	17			
	170	15.5	10.2			698	.36	3.1			318		.96		
pork rib shaped patties, char-broiled w/ bbq sce, Am Hosp Co—*6.7 oz entree*	467		20.3		191		9	.17			1506	63			
	190	19.5	51.6			697	.35	4.6			766		3.01		
pot pie															
beef, Banquet	449		24.0				1	.05			1292	23			
8 oz pie	227	14.0	44.0			655	.18	0.9			113		2.00		
chicken, Banquet	450		24.0				1	.07			966	38			
8 oz pie	227	15.0	44.0			669	.21	2.0			109		1.90		
tuna, Banquet	395		18.0				0	.06			565	58			
8 oz pie	227	16.0	43.0			170	.19	3.5			162		.70		
turkey, Banquet	437		23.0				0	.07			1263	45			
8 oz pie	227	13.0	46.0			134	.20	2.2			159		1.60		
rigatoni w/ ham & peas, Banquet	280		8.0		45		1	.31			1215	110			
10 oz entree	284	16.0	36.0			91	.67	2.6			212		2.40		
salisbury steak															
Banquet	230		18.0		35		0	.12			766	49			
5 oz entree	142	10.0	7.0			29	.09	1.1			180		1.90		
champignon, Am Hosp Co	194		9.9		111		0	.11			253	20			
5.5 oz entree	156	17.6	8.7			80	.06	3.6			253		3.30		
w/ gravy, Banquet	210		15.0				1	.12			850	38			
1/6 of 32 oz pkg	151	10.0	8.0			57	.10	1.1			170		1.50		
w/ onion sce, Am Hosp Co	280		19.3		68		1	.13			1165	7			
6.7 oz entree	190	16.7	9.9			105	.07	2.4			386		1.90		
scrod fillet, Seafood Elites															
w/ broccoli & mozzarella	140		7.0		66						180				
5 oz entree	142	19.0	1.0												

[a] Vegetables include butternut squash, green beans & red peppers.

	KCAL	H₂O (g)	FAT (g)	PUFA (g)	CHOL (mg)	Vitamins — A (RE)	C (mg)	B-2 (mg)	B-6 (mg)	FOL (mcg)	Minerals — Na (mg)	Ca (mg)	Mg (mg)	Zn (mg)	Mn (mg)
	WT (g)	PRO (g)	CHO (g)	SFA (g)	DFIB (g)	A (IU)	B-1 (mg)	NIA (mg)	B-12 (mcg)	PANT (mg)	K (mg)	P (mg)	Fe (mg)	Cu (mg)	
w/ lemon & wild rice	150		5.0		49						320				
5 oz entree	142	20.0	7.0												
w/ spinach & cheddar	140		6.0		63						225				
5 oz entree	142	21.0	1.0												
seafood (cod & surimi) w/ rice, spinach & crm sce, Am Hosp Co—6 oz entree	198		9.1		37		5	.08			356	75			
	170	16.9	12.3			2512	.08	1.2			282		.90		
seafood natural herbs, Armour Classic Lites—11.5 oz entree	250		6.0		20						1240				
	326	12.0	38.0								340				
seafood (scallops) creole, Am Hosp Co	137		3.8		21		38	.10			372	223			
6 oz entree	170	10.4	15.4			1220	.11	1.8			537		1.56		
seafood (shrimp, cod & surimi) & pasta w/ sce, Am Hosp Co—6 oz entree	180		8.4		56		1	.08			316	26			
	170	10.6	15.6			287	.09	1.3			140		.78		
shrimp creole w/ rice & peppers, Light & Elegant—10 oz entree	218		2.0		120		1	.09			1050	68			
	283	14.0	36.0			720	.17	1.2			220	225	1.80		
sirloin tips supreme, Banquet	160		1.0		40		1	.20			1030	49			
10 oz entree	284	13.0	24.0			2200	.14	2.2			365		3.20		
sole fillet, Seafood Elites															
w/ broccoli & mozzarella	150		7.0		53						205				
5 oz entree	142	18.0	2.0												
w/ lemon & wild rice	150		7.0		47						325				
5 oz entree	142	16.0	8.0												
w/ spinach & cheddar	130		6.0		58						200				
5 oz entree	142	16.0	2.0												
spaghetti w/ meat sce															
Am Hosp Co	173		4.5		26		14	.30			654	30			
6 oz entree	170	14.4	18.6			702	.57	5.8			612		3.40		
Banquet	270		8.0				14	.12			1242	29			
8 oz entree	227	14.0	35.0			898	.12	3.2			421		3.00		
Light & Elegant	290		8.0		35		10	.15			700	100			
10.25 oz entree	290	16.0	40.0			787	.25	3.4			794	252	6.00		
sweet & sour pork, Am Hosp Co	222		9.7		38		44	.08			801	19			
6 oz entree	170	12.6	21.0			660	.09	2.6			300		1.33		
sweet & sour pork w/ rice, Van de Kamp's—11 oz entree	430		15.0								790				
	312	15.0	65.0												
swiss steak w/ brown sce, Am Hosp Co	224		11.8		60		5				738	5			
5.7 oz entree	162	25.8	3.7			0	.06	6.1			298		1.60		
taquitos, beef, shredded w/ guacamole, Van de Kamp's—8 oz entree	490		25.0								990				
	227	15.0	45.0												
tortellini besciamella (ham), Am Hosp Co	276		15.8		94		2	.16			650	149			
6 oz entree	170	13.4	19.9			485	.23	2.0			200		1.62		
tostada, beef supreme, Van de Kamp's	530		30.0								900				
8.5 oz entree	241	25.0	40.0												
tuna & cheese pastry wrap, Am Hosp Co	287		18.8		14		10	.12			387	74			
4.5 oz entree	128	12.3	17.0			386	.14	4.5			68		1.63		
tuna & noodles, Am Hosp Co	135		3.1		14		1	.04			518	23			
6 oz entree	170	13.8	12.9			77	.10	3.8			140		1.30		
turkey breast, sliced w/ apple stuffing & gravy, Am Hosp Co—6 oz entree	339		14.8		55		2	.14			333	33			
	170	23.1	28.5			302	.16	7.2			304		1.32		
turkey, carrots & peas w/ supreme sce, Am Hosp Co—6 oz entree	132		3.6		33		6	.15			80	57			
	170	17.8	7.8			4122	.11	3.8			178		7.80		
turkey, sliced w/ gravy	95	120.8	3.7	0.7				.18	.14		786	20	12	.99	
5 oz entree	142	8.4	6.6	2.2		59	.03	2.6				114	1.31		
Banquet	110		5.0		45		0	.09			586	19			
5 oz entree	142	8.6	7.0			8	.04	1.7			83		.70		
Banquet	160		8.0				2	.14			1010	31			
1/4 of 32 oz pkg	227	14.0	8.0			45	.01	2.9			150		1.00		
w/ cranberry dressing, Am Hosp Co	283		17.0		83		2	.15			441	30			
6 oz entree	170	18.7	13.8			846	.06	3.9			224		1.20		

	KCAL / WT (g)	H₂O / PRO (g)	FAT / CHO (g)	PUFA / SFA (g)	CHOL / DFIB (mg/g)	A (RE) / A (IU)	C / B-1 (mg)	B-2 / NIA (mg)	B-6 / B-12 (mg/mcg)	FOL / PANT (mcg/mg)	Na / K (mg)	Ca / P (mg)	Mg / Fe (mg)	Zn / Cu (mg)	Mn (mg)
w/ white & wild rice stuffing, Light & Elegant—*8 oz entree*	230		5.0		55		1	.14			1020	18			
	227	20.0	25.0			857	.12	4.6			280	121	1.00		
turkey tetrazzini															
Am Hosp Co	228		9.0		23		0	.17			66	250			
6 oz entree	170	20.8	15.9			238	.06	3.5			196		.66		
Banquet	270		4.0		25		0	.32			1025	138			
10 oz entree	284	19.0	38.0			286	.20	3.3			247		2.60		
veal & beef patties, Am Hosp Co	136		6.3		41		2	.15			189	14			
3 oz entree	85	14.4	5.4			69	.18	6.9			206		.09		
veal & peppers w/ tomato sce, Am Hosp Co—*6 oz entree*	106		1.4		31		62	.66			569	22			
	170	15.0	8.4			2148	.06	4.1			365				
veal blanquette (veal, onion & mushrooms w/ sce), Am Hosp Co—*6 oz entree*	148		5.6		61		3	.18			419	37			
	170	18.6	5.7			606	.09	3.5			467		1.50		
veal cordon blue, Am Hosp Co	242		9.0		96						416				
5.5 oz entree	156	33.0	7.3								634				
veal florentine, Am Hosp Co[a]	208		11.1		60		6	.13			315	106			
6 oz entree	170	15.5	11.6			1767	.13	3.9			418		2.80		
veal italiano, Am Hosp Co[b]	195		9.1		47		11	.21			267	75			
5.83 oz entree	165	17.8	10.5			718	.24	9.7			472		.61		
veal marsala, Am Hosp Co	164		9.1		52		2	.17			366	14			
6 oz entree	170	13.0	7.7			226	.11	4.5			288		2.22		
veal parmigiana															
Am Hosp Co	279		17.9		67		2	.15			858	133			
5 oz entree	142	17.0	11.8			122	.11	2.0			429		1.67		
Banquet	282		18.0				5	.10			961	92			
1/5 of 32 oz pkg	181	15.0	21.0			378	.06	2.3			213		1.40		
breaded, Banquet	230		11.0		35		0	.12			842	66			
5 oz entree	142	10.0	20.0			478	.13	1.6			237		1.70		
veal stew, Am Hosp Co[c]	108		1.8		39		5	.22			108	24			
6 oz entree	170	15.6	7.2			4002	.19	6.6			156		3.00		

10.4. FROZEN MEALS (USUALLY 3 COURSE MEALS)

	KCAL / WT (g)	H₂O / PRO (g)	FAT / CHO (g)	PUFA / SFA (g)	CHOL / DFIB (mg/g)	A (RE) / A (IU)	C / B-1 (mg)	B-2 / NIA (mg)	B-6 / B-12 (mg/mcg)	FOL / PANT (mcg/mg)	Na / K (mg)	Ca / P (mg)	Mg / Fe (mg)	Zn / Cu (mg)	Mn (mg)
beans & franks, Banquet	500		19.0				20	.23			1377	153			
10.25 oz meal	291	19.0	64.0			4741	.33	3.1			685		5.00		
beef, Banquet Extra Helping	864		46.0				13	.39			1731	97			
16 oz meal	454	40.0	72.0			4864	.24	8.2			904		5.00		
beef burgundy, Armour Dinner Classics	330		15.0		95						990				
10.5 oz meal	298	28.0	23.0								540				
beef, chopped, Banquet	434		30.0				9	.20			1199	69			
11 oz meal	312	19.0	23.0			6803	.12	4.1			658		3.00		
beef pepper steak, rice & green beans, Armour Classic Lites—*10 oz meal*	240		7.0		60						1020				
	284	16.0	28.0								380				
beef short ribs															
boneless w/ bbq sce, Armour Dinner Classics—*10.5 oz meal*	460		26.0		95						1180				
	298	25.0	33.0								870				
boneless w/ horseradish sce, Armour Dinner Classics—*10.5 oz meal*	370		21.0		100						1160				
	298	25.0	21.0								720				
beef stroganoff, Armour Dinner Classics	370		17.0		85						1330				
11.25 oz meal	319	30.0	25.0								590				
beef w/ gravy, Banquet	345		19.0				8	.22			1009	59			
10 oz meal	284	23.0	19.0			6531	.13	5.0			624		3.00		
burrito, grande w/ rice & corn, Van de Kamp's—*14.75 oz meal*	530		20.0								1210				
	418	20.0	70.0												
chicken a la king, rice & green beans almondine, Le Menu—*10.25 oz meal*	320		13.0								1170				
	291	22.0	29.0												
chicken & dumplings, Banquet	286		13.0				7	.12			944	55			
9 oz meal	255	13.0	28.0			3101	.11	4.2			396		1.10		

[a] Ground veal, spinach & rice w/ cream sce.
[b] Veal & beef patties w/ tomato sce & cheese.
[c] Veal, carrots, green peas & mushrooms.

	KCAL / WT (g)	H₂O (g) / PRO (g)	FAT (g) / CHO (g)	PUFA (g) / SFA (g)	CHOL (mg) / DFIB (g)	A (RE) / A (IU)	C (mg) / B-1 (mg)	B-2 (mg) / NIA (mg)	B-6 (mg) / B-12 (mcg)	FOL (mcg) / PANT (mg)	Na (mg) / K (mg)	Ca (mg) / P (mg)	Mg (mg) / Fe (mg)	Zn (mg) / Cu (mg)	Mn (mg)
chicken breast medallions marsala,	270		7.0		85						970				
Armour Classic Lites—*11 oz meal*	312	22.0	28.0								470				
chicken breast, sherried, Armour Dinner	300		12.0		90						830				
Classics—*10 oz meal*	284	29.0	19.0								390				
chicken burgundy, Armour Classic Lites	230		5.0		70						1220				
11.25 oz meal	319	21.0	26.0								530				
chicken fricassee, rice & broccoli, Armour	340		12.0		70						1210				
Dinner Lites—*11.75 oz meal*	333	23.0	35.0								440				
chicken fricassee, veg w/ sce, rice & green	349		9.5		58		12	.36			167	166			
beans, Nutra Choice—*12.5 oz meal*	354	27.3	36.8			2746	.19	6.5			580	279	2.26		
chicken															
fried, Banquet	359		11.0				11	.18			1831	89			
11 oz meal	312	18.0	46.0			7589	.16	5.5			674		2.00		
hawaiian, Armour Dinner Classics—	360		10.0		75						700				
11.5 oz meal	326	21.0	47.0								540				
marsala, potatoes, carrots & green	304		8.6		68		11	.31			193	133			
beans, Nutra Choice—*12.25 oz meal*	347	26.2	29.2			9497	.16	7.7			728	248	2.04		
milan, Armour Dinner Classics	350		12.0		75						1360				
11.5 oz meal	326	24.0	38.0								540				
oriental, Armour Classic Lites	250		6.0		65						880				
10 oz meal	284	24.0	26.0								330				
parmigiana, fettucini alfredo & green	380		18.0								890				
beans, Le Menu—*11.5 oz meal*	326	27.0	28.0												
roasted breast, Armour Classic Lites	270		9.0		85						1220				
11 oz meal	312	22.0	26.0								560				
supreme w/ sce, parsleyed ndls, mxd	377		11.4		75		23	.31			176	141			
veg, Nutra Choice—*12 oz meal*	340	34.5	32.2			6678	.15	12.8			652	353	2.17		
tetrazzini, Armour Dinner Classics	320		14.0		75						850				
11.5 oz meal	326	23.0	26.0								490				
chopped sirloin w/ mushroom gravy															
potatoes & green beans, Le Menu	410		23.0								1080				
12.25 oz meal	347	26.0	28.0												
chow mein, chicken, Armour Dinner	220		4.0		60						1180				
Classics—*10.5 oz meal*	298	20.0	25.0								240				
cod															
almondine, Armour Dinner Classics	360		15.0		95						1440				
12 oz meal	340	23.0	33.0								590				
w/ lemon sce, rice, green peas &	279		8.0		85						196				
carrots, Nutra Choice—*12 oz meal*	340	23.0	27.0								632				
enchilada															
beef & cheese w/ rice & beans, Van de	540		20.0								1380				
Kamp's—*14.75 oz meal*	418	25.0	60.0												
beef, Banquet	497		15.0				7	.19			1805	109			
12 oz meal	340	19.0	72.0			376	.44	3.0			419		4.00		
beef, El Charrito	620		31.0		45						2130				
13.75 oz meal	390	19.0	63.0								790				
beef grande, El Charrito	950		49.0		70						2730				
21 oz meal	595	29.0	93.0								1290				
beef, 4 grande, El Charrito	890		47.0		65						2250				
16.5 oz meal	468	28.0	83.0								1140				
beef, 2 grande, 2 cheese tacos & beans,	800		35.0		45						2460				
El Charrito—*16.5 oz meal*	468	23.0	92.0								1220				
beef, shredded w/ rice & corn, Van de	490		15.0								1170				
Kamp's—*14.75 oz meal*	418	25.0	60.0												
beef, Van de Kamp's	390		15.0								2175				
12 oz meal	340	20.0	45.0												
cheese, Banquet	543		19.0				7	.27			2166	281			
12 oz meal	340	22.0	71.0			711	.44	2.5			429		3.00		

	KCAL / WT (g)	H_2O (g) / PRO (g)	FAT (g) / CHO (g)	PUFA (g) / SFA (g)	CHOL (mg) / DFIB (g)	A (RE) / A (IU)	C (mg) / B-1 (mg)	B-2 (mg) / NIA (mg)	B-6 (mg) / B-12 (mcg)	FOL (mcg) / PANT (mg)	Na (mg) / K (mg)	Ca (mg) / P (mg)	Mg (mg) / Fe (mg)	Zn (mg) / Cu (mg)	Mn (mg)
cheese, El Charrito	570		24.0		30						2330				
13.75 oz meal	390	16.0	72.0								860				
cheese, Van de Kamp's	450		20.0								1665				
12 oz meal	340	20.0	45.0												
cheese w/ rice & beans, Van de Kamp's	620		30.0								1460				
14.75 oz meal	418	25.0	60.0												
chicken, El Charrito	510		17.0		50						2150				
13.75 oz meal	390	22.0	69.0								940				
chicken suiza w/ rice & beans, Van de Kamp's—*14.75 oz meal*	550		20.0								1210				
	418	25.0	65.0												
fillet of fish, Van de Kamp's	300		10.0								1820				
12 oz meal	340	25.0	25.0												
fish, Banquet	553		33.0				18	.26			927	62			
8.75 oz meal	248	18.0	45.0			5305	.34	3.7			601		6.00		
grande mexican style, El Charrito	850		47.0		65						3100				
20 oz meal	567	26.0	87.0								1200				
grand satillo, El Charrito	820		34.0		45						3160				
20.75 oz meal	588	22.0	103.0								1380				
ham, Banquet	532		22.0				10	.28			1148	80			
10 oz meal	284	21.0	61.0			7176	.51	4.1			530		4.00		
italian style, Banquet	597		26.0				23	.24			1783	79			
12 oz meal	340	21.0	71.0			1391	.34	4.0			527		4.00		
imperial chicken w/ mushrooms in sce & veg, Weight Watchers[a]	240		4.0								840				
9.25 oz meal	262	22.0	29.0												
lasagna, Armour Dinner Classics	380		16.0		70						1120				
10 oz meal	284	17.0	42.0								580				
macaroni & cheese, Banquet	334		16.0				12	.16			940	184			
9 oz meal	255	11.0	37.0			5109	.19	1.5			295		2.60		
meat loaf, Banquet	437		27.0				18	.21			1525	80			
11 oz meal	312	20.0	30.0			5044	.19	4.0			702		3.00		
mexican style															
Banquet	483		18.0				8	.16			1995	110			
12 oz meal	340	18.0	62.0			358	.39	2.8			439		3.00		
Banquet Extra Helping	777		27.0				13	.43			4778	411			
21.25 oz meal	602	31.0	105.0			1261	.65	3.8			712		5.00		
combination, Banquet	518		17.0				7	.23			1978	194			
12 oz meal	340	20.0	72.0			539	.44	2.7			423		3.00		
El Charrito	690		35.0		45						2430				
14.25 oz meal	404	30.0	70.0								870				
Van de Kamp's	420		20.0								1040				
11.5 oz meal	326	20.0	45.0												
noodles & chicken, Banquet	361		11.0				4	.16			964	62			
9.5 oz meal	269	13.0	50.0			4923	.16	3.1			295		2.60		
oriental beef, veg & rice, Lean Cuisine	270		8.0								1150				
8.6 oz meal	245	20.0	30.0												
pork sliced w/ orange sce, rice & veg medley, Nutra Choice	354		10.3		81		34	.40			99	46			
11 oz meal	312	26.4	38.4			812	.53	5.6			525	230	2.68		
queso, El Charrito	490		16.0		15						2540				
13.25 oz meal	376	14.0	74.0								970				
salisbury steak															
Banquet	395		26.0				10	.15			1333	62			
11 oz meal	312	17.0	24.0			4203	.14	3.1			514		2.00		
Banquet Extra Helping	1024		65.0				13	.35			2175	127			
19 oz meal	539	39.0	72.0			4662	.22	6.0			851		4.00		
italian style sce & veg, Lean Cuisine	270		13.0								820				
9.5 oz meal	269	25.0	14.0												

[a] Vegetables include rice & broccoli.

	KCAL	H₂O (g)	FAT (g)	PUFA (g)	CHOL (mg)		A (RE)	C (mg)	Vitamins B-2 (mg)	B-6 (mg)	FOL (mcg)		Na (mg)	Ca (mg)	Minerals Mg (mg)	Zn (mg)	Mn (mg)
	WT (g)	PRO (g)	CHO (g)	SFA (g)	DFIB (g)		A (IU)	B-1 (mg)	NIA (mg)	B-12 (mcg)	PANT (mg)		K (mg)	P (mg)	Fe (mg)	Cu (mg)	
roman w/ rotini noodles, Weight	300		12.0										990				
Watchers—*8.75 oz meal*	248	25.0	24.0														
whipped potatoes, corn & cherry apple	350		14.0										1025				
crmb cke, Swanson's—*10.75 oz meal*	305	18.0	38.0														
salisbury steak w/ mushroom gravy,																	
broccoli & potatoes, Armour Classic	290		13.0		75								870				
Lites—*10 oz meal*	284	20.0	25.0										780				
potatoes, peas & corn, Armour Dinner	480		27.0		110								1400				
Classics—*11 oz meal*	312	20.0	39.0										740				
salisbury steak w/ mushroom sce,																	
escalloped potatoes & peas w/ sce,	378		19.0		97			19	.40				244	135			
Nutra Choice—*12 oz meal*	340	26.2	35.7				891	.32	5.3				824	284	4.57		
satillo, El Charrito	570		24.0		30								2200				
13.5 oz meal	383	16.0	70.0										860				
seafood newburg, Armour Dinner Classics	280		10.0		70								1500				
10.5 oz meal	298	11.0	36.0										330				
shrimp americana, Armour Dinner	280		10.0		110								1210				
Classics—*11.5 oz meal*	326	18.0	31.0										490				
shrimp creole, peas w/ sce & rice w/ sce,	261		3.9		80			32	.17				142	83			
Nutra Choice—*12 oz meal*	340	18.3	38.2				1333	.27	4.7				558	214	3.41		
shrimp in sherried crm sce, Armour	280		8.0		110								1220				
Classic Lites—*10.5 oz meal*	298	17.0	34.0										490				
sirloin roast, Armour Dinner Classics	280		9.0		100								780				
11 oz meal	312	25.0	24.0										730				
sirloin tips w/ mushroom gravy,																	
potatoes, grn beans & carrots, Armour	370		16.0		100								1180				
Dinner Classics—*11 oz meal*	312	29.0	27.0										710				
potatoes & broccoli w/ cheese sce, Le	390		19.0										825				
Menu—*11.5 oz meal*	326	29.0	26.0														
sirloin tips w/ mushroom sce,																	
parsleyed noodles & mixed veg,	297		7.6		69			29	.28				177	52			
Nutra Choice—*11 oz meal*	312	27.9	27.2				5565	.12	5.7				529	265	4.01		
spaghetti & meatballs																	
Armour Dinner Classics	350		21.0		95								1130				
11 oz meal	312	14.0	27.0										760				
Banquet	418		14.0					14	.20				1317	73			
9.5 oz meal	269	15.0	57.0				4918	.22	2.8				473		3.60		
steak diane mignonettes, Armour Classic	290		9.0		90								770				
Lites—*10 oz meal*	284	29.0	23.0										760				
stuffed cabbage, Armour Classic Lites	290		8.0		40								600				
12 oz meal	340	13.0	43.0										780				
stuffed green peppers, Armour Dinner	390		18.0		70								1750				
Classics—*12 oz meal*	340	16.0	41.0										580				
swedish meatballs, Armour Dinner	470		28.0		105								1560				
Classics—*11.5 oz meal*	326	23.0	32.0										760				
sweet & sour chicken																	
Armour Classic Lites	250		3.0		70								640				
11 oz meal	312	23.0	33.0										400				
Armour Dinner Classics	450		20.0		65								1240				
11 oz meal	312	16.0	51.0										450				
szechuan beef, Armour Classic Lites	280		9.0		70								1010				
10 oz meal	284	23.0	26.0										520				
teriyaki chicken, Armour Dinner Classics	250		5.0		80								1120				
10.5 oz meal	298	22.0	28.0										350				
teriyaki steak, Armour Dinner Classics	360		16.0		95								1445				
10 oz meal	284	24.0	32.0										540				

	KCAL	H₂O (g)	FAT (g)	PUFA (g)	CHOL (mg)	A (RE)	C (mg)	B-2 (mg)	B-6 (mg)	FOL (mcg)	Na (mg)	Ca (mg)	Mg (mg)	Zn (mg)	Mn (mg)
	WT (g)	PRO (g)	CHO (g)	SFA (g)	DFIB (g)	A (IU)	B-1 (mg)	NIA (mg)	B-12 (mcg)	PANT (mg)	K (mg)	P (mg)	Fe (mg)	Cu (mg)	
turf & surf, Armour Classic Lites	250		7.0		80						890				
10 oz meal	284	31.0	14.0								520				
turkey															
Banquet	320		9.0				8	.25			1416	98			
11 oz meal	312	19.0	41.0			5656	.20	5.1			639		2.00		
Banquet Extra Helping	723		23.0				14	.44			2165	176			
19 oz meal	539	31.0	98.0			4940	.34	7.6			946		4.00		
turkey parmesan w/ veg, butter sce & spaghetti, Armour Classic Lites	240		8.0		70						480				
11 oz meal	312	19.0	25.0								770				
veal parmigiana															
Banquet	413		21.0				19	.18			1310	121			
11 oz meal	312	14.0	43.0			889	.26	3.5			468		3.00		
fettucine alfredo, green beans & apple crumb, Swanson's—*12.25 oz meal*	490		19.0								965				
	347	23.0	56.0												
pasta primavera, Nutra Choice	354		16.4		76		25	.30			149	106			
11 oz meal	312	24.7	26.9			6534	.15	5.3			571	289	4.04		
spaghetti & italian mixed veg, Armour Dinner Classics—*10.75 oz meal*	400		22.0		75						1430				
	305	17.0	34.0								520				
w/ veg medley, Weight Weighters	230		9.0								1040				
8 oz meal	229	24.0	9.0												
western, Banquet	513		29.0				7	.16			1548	93			
11 oz meal	312	22.0	43.0			57	.14	3.6			643		4.00		
yankee pot roast															
w/ gravy & parsleyed potatoes, Armour Dinner Classics—*11 oz meal*	380		15.0		90						820				
	312	30.0	32.0								770				
w/ mushroom grvy, potatoes, carrots & pearl onions, Le Menu—*11 oz meal*	360		15.0								810				
	312	28.0	28.0												
w/ sce, parsleyed potatoes & veg, Nutra Choice—*12 oz meal*	281		5.5		67		24	.28			131	49			
	340	29.4	27.1			9562	.28	6.9			733	284	4.48		

10.5. HOMEMADE ENTREES

	KCAL	H₂O (g)	FAT (g)	PUFA (g)	CHOL (mg)	A (RE)	C (mg)	B-2 (mg)	B-6 (mg)	FOL (mcg)	Na (mg)	Ca (mg)	Mg (mg)	Zn (mg)	Mn (mg)
	WT (g)	PRO (g)	CHO (g)	SFA (g)	DFIB (g)	A (IU)	B-1 (mg)	NIA (mg)	B-12 (mcg)	PANT (mg)	K (mg)	P (mg)	Fe (mg)	Cu (mg)	
beef & veg stew	218	201.9	10.5				17	.17			91[a]	29			
1 cup	245	15.7	15.2	4.9		2400	.15	4.7			613	184	2.90		
beef, dried, chipped, ckd, creamed	377	176.4	25.2				1	.47			1754	257			
1 cup	245	20.1	17.4	13.7		880	.15	1.5			375	343	2.00		
chicken a la king	468	167.1	34.3				12	.42			760	127			
1 cup	245	27.4	12.3	12.7		1130	.10	5.4			404	358	2.50		
chicken & noodles	367	170.6	18.5				0	.17			600	26			
1 cup	240	22.3	25.7	5.9		430	.05	4.3			149	247	2.20		
chicken fricassee	386	171.1	22.3					.17			370	14			
1 cup	240	36.7	7.4	7.2		170	.05	5.8			336	271	2.20		
chop suey	300	188.5	17.0				33	.38			1053	60			
1 cup	250	26.0	12.8	8.5		600	.28	5.0			425	248	4.80		
chow mein	255	195.0	10.0				10	.23			718	58			
1 cup	250	31.0	10.0	2.4		280	.08	4.3			473	293	2.50		
clam fritter	124	16.1	6.0					.05				30			
1 fritter	40	4.6	12.4				.01	0.4			59	78	1.40		
corn fritter	132	10.2	7.5				1	.07			167	22			
1 fritter	35	2.7	13.9	2.0		140	.06	0.6			47	54	.60		
crab, deviled[b]	451	151.9	22.6				14	.26			2081	113			
1 cup	240	27.4	31.9				.19	3.5			398	329	2.90		
crab, imperial[c]	323	158.2	16.7				11	.26			1602	132			
1 cup	220	32.1	8.6				.13	2.4			288	365	2.00		

[a] W/o added salt.

[b] Prepared w/ bread crumbs, butter or marg, parsley, eggs, lemon jce & catsup.

[c] Prepared w/ butter or margarine, flour, milk, onion, green pepper, eggs & lemon jce.

	KCAL	H₂O (g)	FAT (g)	PUFA (g)	CHOL (mg)		A (RE)	C (mg)	Vitamins B-2 (mg)	B-6 (mg)	FOL (mcg)		Na (mg)	Ca (mg)	Minerals Mg (mg)	Zn (mg)	Mn (mg)
	WT (g)	PRO (g)	CHO (g)	SFA (g)	DFIB (g)		A (IU)	B-1 (mg)	NIA (mg)	B-12 (mcg)	PANT (mg)		K (mg)	P (mg)	Fe (mg)	Cu (mg)	
fish cake, fried[a]	103	39.6	4.8														
1 piece	60	8.8	5.6														
fish loaf, ckd[b]	186	108.3	5.6														
1 slice	150	21.2	11.0														
green pepper stuffed w/ beef & crumbs	315	116.7	10.2					74	.81				581	78			
1 stuffed pepper	185	24.1	31.1	4.8			520	.17	4.6				477	224	3.90		
ham croquette	163	35.1	9.8					0	.14				222	45			
1 croquette	65	10.6	7.6	3.9			170	.18	1.6				54	104	1.40		
lobster thermidor	405		26.6					0	.51					290			
5.5 oz	157	28.5	14.8		0.5		984	.15	4.8					451	1.90		
lobster newburg[c]	485	160.0	26.5						.28				573	218			
1 cup	250	46.3	12.8					.18					428	480	2.30		
macaroni & cheese	430	116.4	22.2					0	.40				1086	362			
1 cup	200	16.8	40.2	11.9			860	.20	1.8				240	322	1.80		
oyster stew[d]	233	196.8	15.4						.43				812	274			
1 cup	240	12.5	10.8				820	.14	2.2				319	266	4.60		
pizza																	
cheese	153	31.4	5.4					5	.13				456	144			
1 piece	65	7.8	18.4	2.1			410	.04	0.7				85	127	.70		
sausage	157	33.9	6.2					6	.08				488	11			
1 piece	67	5.2	19.8	1.8			380	.06	1.0				113	62	.80		
pot pie																	
beef	517	115.7	30.5					6	.25				596	29			
1 piece	210	21.2	39.5	8.4			1720	.23	4.2				334	149	3.80		
chicken	545	131.3	31.3					5	.26				594	70			
1 piece	232	23.4	42.5	10.9			3090	.26	4.2				343	232	3.00		
turkey	550	130.4	31.3					5	.30				633	63			
1 piece	232	24.1	42.9	10.5			3090	.26	5.8				459	234	3.20		
salmon patty	239		12.4		64			4	.22	.07	13		96	78	34	.84	
3.5 oz	100	15.8	16.1		0.8		66	.12	4.0	3.00	.66		89	104	1.24		
salmon rice loaf	212	129.5	7.8														
1 slice	174	20.9	12.7														
spanish rice	213	192.3	4.2					37	.07				774	34			
1 cup	245	4.4	40.7				1620	.10	1.7				566	96	1.50		
spaghetti in tomato sce	260	192.5	8.8					13	.18				955	80			
1 cup	250	8.8	37.0	2.0			1080	.25	2.3				408	135	2.30		
spaghetti w/ meatballs & tomato sce	332	173.8	11.7					22	.30				1009	124			
1 cup	248	18.6	38.7	3.3			1590	.25	4.0				665	236	3.70		
tuna patty	209		10.6	0.7	47			1	.16				154	56	25		
3.5 oz	100	19.8	7.4				78	.08	7.4				64	173	1.70		
welsh rarebit	415	162.9	31.6					0	.53				770	582			
1 cup	232	18.8	14.6	17.3			1230	.09	0.2				320	432	.70		

11. FAST FOODS
11.1. GENERIC FAST FOODS[e]

	KCAL	H₂O (g)	FAT (g)	PUFA (g)	CHOL (mg)		A (RE)	C (mg)	B-2 (mg)	B-6 (mg)	FOL (mcg)		Na (mg)	Ca (mg)	Mg (mg)	Zn (mg)	Mn (mg)
bacon cheeseburger	464	65.4	27.3		68			2	.27	.24	26		660	116	35	5.25	
1 sandwich	150	25.1	29.1				368	.15	4.9	1.80	.27		339	302	3.74	.120	
burrito	392	89.3	13.6		56			2	.30	.17	29		1030	191	70	2.37	
1 burrito	174	15.5	51.9				209	.37	5.4	1.04	.31		501	235	2.70	.313	
cheeseburger	299	51.3	15.1		45			1	.24	.11	18		672	136	22	2.24	
1 reg sandwich	112	14.8	28.3				340	.26	3.7	.90	.27		220	174	2.33	.112	
cheeseburger w/ 4 oz meat	524	88.3	31.4		105			2	.49	.23	23		1224	237	43	5.23	
1 sandwich	194	30.1	39.6				671	.33	7.4	2.33	.64		407	320	4.46	.155	
chicken, fried																	
dark meat	234	43.4	14.6		51			0	.29				414	13	21	1.86	
3 oz	85	19.0	6.5				119	.07	4.5				230	138	1.02	.068	
light & dark meat	235	41.6	13.6		72			0	.15	.26	11		394	12	21	1.11	
3 oz	85	14.3	13.2				97	.08	4.8	.17			223	146	.78	.060	

[a] Prepared w/ cnd flaked fish, potato & egg.
[b] Prepared w/ cnd flaked fish, bread cubes, egg, tomatoes, onions & butter or margarine.
[c] Prepared w/ butter or margarine, egg yolks, sherry & cream.
[d] Approximately 6 med oysters/cup.
[e] Averages of data covering a number of chains sampled in different locations.

	KCAL / WT (g)	H₂O (g) / PRO (g)	FAT (g) / CHO (g)	PUFA (g) / SFA (g)	CHOL (mg) / DFIB (g)	A (RE) / A (IU)	C (mg) / B-1 (mg)	B-2 (mg) / NIA (mg)	B-6 (mg) / B-12 (mcg)	FOL (mcg) / PANT (mg)	Na (mg) / K (mg)	Ca (mg) / P (mg)	Mg (mg) / Fe (mg)	Zn (mg) / Cu (mg)	Mn (mg)
light meat	224	42.8	12.1		81		0	.09	.26	11	366	10	21	.61	
3 oz	85	12.6	16.1			66	.07	5.7	.17		218	149	.52	.051	
chili	268	198.6	9.1		26		3	1.17	.34	31	1061	70	52	3.90	
1 cup	260	23.9	23.9			1708	.16	2.5	1.30		731	187	4.16	.416	
coffee	4	179.3	0.0		0		0	.02		0	2	4	13	.05	
6 fl oz	180	0.2	0.0			0	.00	0.4	.00		124	2	.02	.018	
cola	159	328.0	0.0		0		0	.00	.00		18	11	4	.15	
12 fl oz	369	0.0	40.6			0	.00	0.0			7	63	.18	.111	
coleslaw	122	72.8	11.5				1	.02	.11	39	232	44	14	.21	
1/2 cup	100	1.7	11.7				.03	0.0	.20	.15	196		.54	.030	
cookies															
choc chip	307	3.3	13.2		18		1	.39	.03	6	309	29	29	.49	
13 cookies	68	3.9	46.4			75	.10	2.3	.14	.15	168	106	1.54	.170	
plain	313	2.8	10.7		12		1	.24	.02	6	359	12	11	.31	
21 cookies	68	4.8	48.8			27	.25	2.5	.07	.22	57	65	1.50	.054	
enchilada	396	157.1	20.9					.25	.25		1332	97	76	1.29	
1 enchilada	230	17.3	34.7				.18		2.07	.60	653	198	3.29	.299	
english muffin w/ egg, cheese & bacon	359	67.5	18.1		214		1	.50	.15	41	832	197	28	1.86	
1 sandwich	138	17.9	31.1			654	.47	3.7	.83	.69	201	290	3.11	.124	
fish, breaded & fried	196	38.0	11.4					.04	.10	9	334	9	16	.32	
1 portion	73	10.5	11.8			64	.05	1.1	.73	.14	230	124	.22		
fish sandwich	469	81.8	26.7		90		0	.24	.12	43	621	61	34	.88	
1 large sandwich	170	18.0	41.1			109	.34	3.5	1.53		374	247	2.23	.136	
fish sandwich w/ cheese	421	60.2	22.7		196		3	.27	.10	24	668	132	29	.95	
1 reg sandwich	140	15.8	38.5			160	.32	3.3	.84	.48	274	223	1.85	.098	
french fries	274	32.8	14.2		14		5	.03	.22	23	30[a]	14	29	.33	
3 oz	85	3.4	33.1			25	.11	1.8	.09	.37	598	100	.71	.085	
ham & cheese sandwich	372	78.6	16.0		70		2	.64			1118	217			
1 sandwich	155	24.2	32.4			186	.19	2.5			332		3.98		
hamburger	245	44.8	11.1		32		1	.25	.10	17	463	56	20	1.86	
1 reg sandwich	98	12.1	28.4			77	.24	3.8	.78	.30	202	107	2.16	.098	
w/ 4 oz meat	444	87.5	21.1		71		2	.38	.28	24	762	75	40		
1 sandwich	174	24.9	37.6			157	.38	7.8	2.26	.52	404	226	4.84		
w/ 4 oz meat on large roll	668	139.7	37.2		114		3	.37	.40	55	1019	106	48	4.62	
1 sandwich	264	26.1	56.8			507	.48	6.8	2.11		560	243	4.73	.264	
w/ double meat	370	67.8	20.8		81			.27			459	59	29	3.70	
1 sandwich	137	21.2	24.9			459	.22	4.3			336	181	3.53	.123	
w/ double meat & 3 piece roll	524	89.5	29.6		78		2	.36	.21	19	893	154	34	3.99	
1 sandwich	190	24.9	42.8			296	.95	7.0	1.71	.25	323	281	3.52	.152	
hash brown potatoes	282	52.9									73[b]	9	19	.26	
3 oz	85										282	70	.54	.093	
hot chocolate	47	177.8	0.4		0		0	.08	2.06	2	68	43	24	.39	
6 fl oz	206	1.6	11.1			2	.00	0.0	.41	.25	107	43	.14	.103	
hot dog	214	43.7	13.7		37		0	.23			636	27	13	2.05	
1 hot dog	82	9.1	13.6				.20	3.1			140	84	1.53	.107	
hush puppies	153	17.9	5.5				0	.01	.06	12					
3 pieces	45	3.2	17.5				.00	1.2	.09	.13					
lemon-lime soda	153	332.6	0.0		0		0	.00			33	7	4	.15	
12 fl oz	372	0.0	39.1			0	.00	0.0	.00		4	0	.45	.037	
omelet w/ ham & cheese	255	74.3	17.7		446		6	.55	.19	41	399	116	19	2.17	
2 eggs	118	17.1	6.7			743	.14	0.7	1.18	1.01	189	287	2.94	.094	
onion rings	285	30.4	15.6		16		1	.10	.07	11	485	23	16	.37	
3 oz	85	3.8	33.5			9	.09	0.9	.26	.20	141	81	.78	.077	
orange jce	56	110.1	0.1		0		49	.03	.05	55	1	11	13	.05	
1/2 cup	125	0.9	13.5			98	.10	0.3	.00	.20	238	20	.13	.063	
orange soda	179	325.9	0.0		0		0	.00	.00		52	15	4	.26	
12 fl oz	372	0.0	45.8			0	.00	0.0			37	11	.26	.074	
pancakes w/ butter & syrup	468	97.8	10.4		46		4	.38	.08	8	1020	122	28	.62	
3 pancakes	200	7.0	82.2			244	.28	3.0	.20	.22	216	430	2.20	.100	

[a] 152 mg if salted.   [b] 453 mg if salted.

	KCAL / WT (g)	H₂O (g) / PRO (g)	FAT (g) / CHO (g)	PUFA (g) / SFA (g)	CHOL (mg) / DFIB (g)	A (RE) / A (IU)	C (mg) / B-1 (mg)	B-2 (mg) / NIA (mg)	B-6 (mg) / B-12 (mcg)	FOL (mcg) / PANT (mg)	Na (mg) / K (mg)	Ca (mg) / P (mg)	Mg (mg) / Fe (mg)	Zn (mg) / Cu (mg)	Mn (mg)
pizza															
cheese	290	54.6	8.6		56		2	.29	.12	55	698	220	31	1.67	
1 slice	120	14.6	39.1			750	.34	4.2	.48	.42	230	216	1.61	.144	
pepperoni	306	56.2	11.5				2	.29	.10	78	817	196			
1 slice	120	13.0	36.7			532	.32	5.1	.36	.42	216		2.52		
roast beef sandwich	347	78.0	13.4		56		3	.33	.29	42	756	60	38	3.66	
1 sandwich	150	22.4	33.8			240	.39	6.0	1.35	.75	338	222	4.04	.195	
salad bar items															
alfalfa sprouts, raw	2	5.5	0.0		0		0	.01	.00	2	0	2	2	.06	
2 T	6	0.2	0.2			9	.00	0.0	.00	.03	5	4	.06	.010	
bacon bits	27	0.2	1.6		0		0	.02			165	8			
1 T	6	1.9	1.7			0	.02	0.1				18	.30		
beets, pickled, cnd	18	22.9	0.0				1	.01			74	3	4	.07	
1/8 cup	28	0.2	4.6			0	.01	0.1	.00		41	5	.11	.034	
broccoli, raw	3	10.0	0.0		0		10	.01	.02	8	3	5	3	.04	
1/8 cup	11	0.3	0.6			170	.01	0.1	.00	.06	36	7	.10	.004	
carrots, raw	6	12.3	0.0		0		1	.01	.02	2	5	4	2	.03	
1/8 cup	14	0.1	1.4			394	.01	0.1	.00	.03	45	6	.07	.007	
cheese, cheddar, shredded	56	5.2	4.6		15		0	.05	.01	3	87	101	4	.44	
2 T	14	3.5	0.2			148	.00	0.0	.11	.06	14	72	.10		
cheese, cottage	116	89.3	5.1		17		0	.18	.08	14	458	68	6	.42	
1/2 cup	113	14.1	3.1			184	.02	0.1	.68	.24	95	149	.16		
cheese, parmesan	46	1.8	3.0		1		0	.04	.01	1	186	138	5	.32	
2 T	10	4.2	0.4			70	.00	0.0		.05	11	81	.10		
chickpeas (garbanzo beans)	21	19.6	0.4		0		0	.02			75	10			
2 T	25	1.2	3.5			16	.02	0.1	.00		43	23	.30		
croutons	60		3.0								155				
1/2 oz	14	2.0	8.0								30				
cucumber, raw	2	12.5	0.0		0		0	.00	.01	2	0	2	1	.03	
2 T	13	0.1	0.4			6	.00	0.0	.00	.03	19	2	.04	.005	
eggs, ckd, chopped	27	12.7	1.9		93		0	.05	.02	8	23	10	2	.24	
2 T	17	2.1	0.2			88	.01	0.0	.22	.29	22	31	.36		
lettuce	2	13.4	0.0		0		1	.00	.01	8	1	3	1		
1/4 cup	14	0.1	0.3			46	.01	0.0	.00	.01	22	3	.07		
mushrooms, raw	2	8.3	0.0		0		0	.04	.01	2	0	0	1	.07	
2 T	9	0.2	0.4			0	.01	0.4	.00	.20	33	9	.11	.044	
onion, raw	7	18.2	0.1		0		2	.00	.03	4	0	5	2	.04	
2 T	20	0.2	1.5				.01	0.0	.00	.03	31	6	.07	.008	
pepper, green, raw	3	11.1	0.0		0		23	.01	.02	2	0	1	2	.02	
2 T	12	0.1	0.6			684	.01	0.1	.00	.00	23	3	.15	.012	
potato salad	179	95.0	10.3		85		13	.08	.18	9	661	24	19	.39	
1/2 cup	125	3.4	14.0			261	.10	1.1	.00	.66	318	65	.81	.150	
salad dressing															
blue cheese	76	4.8	7.8				0	.02				12			
1 T	15	0.7	1.1			32	.00	0.0				11	.03		
french	79	5.2	8.7		0		0	.00	.00	0	176	2	0	.01	
1 T	15	0.0	0.6			1	.00	0.0	.00	.00	2	1	.03	.002	
italian	79	5.2	8.6		0		0	.00			162	1			
1 T	15	0.1	0.9			29	.00	0.0			5	1	.03		
italian, low calorie	7	12.9	0.0		0		0	.00			136	1			
1 T	15	0.0	1.7			0	.00	0.0			4	1	.02		
oil & vinegar	72	7.6	8.0								0				
1 T	16	0.0	0.4								1				
thousand island	57	6.9	5.4								105	2		.02	
1 T	15	0.1	2.3			48					17	3	.09		
tomato, raw	2	11.3	0.0		0		2	.01	.00	1	1	1	1	.01	
2 slices	12	0.1	0.5			136	.01	0.1	.00	.03	25	3	.06	.010	
sausage	198	17.1	17.2		39		0	.11	.13	1	418	8	9	1.30	
1 patty	46	8.4	1.9			31	.20	2.3	.51	.30	136	74	.81	.046	

	KCAL / WT (g)	H₂O (g) / PRO (g)	FAT (g) / CHO (g)	PUFA (g) / SFA (g)	CHOL (mg) / DFIB (g)	A (RE) / A (IU)	C (mg) / B-1 (mg)	B-2 (mg) / NIA (mg)	B-6 (mg) / B-12 (mcg)	FOL (mcg) / PANT (mg)	Na (mg) / K (mg)	Ca (mg) / P (mg)	Mg (mg) / Fe (mg)	Zn (mg) / Cu (mg)	Mn (mg)
scrambled eggs	175	63.5	14.7		346		1	.60	.19	63	226	60	13	1.64	
2 eggs	95	13.2	2.5			632	.08	0.3	.95	.72	154	230	2.59	.067	
shake															
choc	360	202.3	10.5	0.4	37	64	1	.69	.14	10	273	319	47	1.15	.110
10 fl oz	283	9.6	57.9	6.5		263	.16	0.5	.97	1.10	567	288	.88	.184	
strawberry	319	209.7	8.0		31	83	2	.55	.13	9	234	320	36	1.00	.042
10 fl oz	283	9.5	53.4			340	.13	0.5	.88	1.39	516	283	.30	.062	
van	314	211.5	8.4	0.3	32	90	2	.52	.15	9	232	344	35	1.01	.040
10 fl oz	283	9.8	50.8	5.3		368	.13	0.5	1.01	1.19	492	289	.26	.144	
sundae															
caramel	328	93.2	9.9		26		3	.31	.05	13	196	201	30	.87	
1 sundae	165	7.3	53.0			281	.07	1.0	.66	.40	338	231	.23	.083	
hot fudge	312	98.5	10.9		18		3	.31	.13	10	176	85	35	.99	
1 sundae	165	7.3	46.5			231	.07	1.1	.66	.35	413	238	.24	.132	
strawberry	285	101.6	8.9		20		3	.30	.05	20	97	175	28	.31	
1 sundae	165	6.9	46.0			231	.07	1.0	.66	.41	292	182	.38	.099	
taco	187	45.5	10.4		21		1	.06	.12	11	456	109	36	1.56	
1 taco	81	10.6	12.7			420	.09	1.4	.41	.11	263	134	1.15	.105	
tea	2	239.0	0.0		0		0	.00			2	0	2	.05	
8 fl oz	240	0.0	0.0				.02	0.0	.00		35	2	.07	.000	
tomato jce	21	114.6	0.1		0		22	.04	.13	24	440	11	13	.17	
1/2 cup	122	1.0	5.1			678	.06	0.8	.00	.31	268	23	.71	.122	
tuna submarine sandwich	566	142.3	27.8								1288	74	79	1.86	
1 sandwich	255	28.8	52.0								334	219	2.63	.434	
turnover															
apple	255	36.5	14.1		4ᵃ		1	.06	.03	3	326	12	8	.16	
3 oz	85	2.2	31.3			34	.09	1.0	.09	.13	42	34	.94	.043	
cherry	251	35.4	14.2		13		1	.06	.02	3	371	11	8	.17	
3 oz	85	2.0	32.2			186	.06	0.6	.00	.26	61	41	.70	.051	
lemon	279	32.0	15.0		20		0	.09	.02	3	286	11	9	.24	
3 oz	85	2.7	34.3				.18	1.7	.17	.10	43	53	1.17	.043	

11.2. BURGER KING

	KCAL / WT (g)	H₂O (g) / PRO (g)	FAT (g) / CHO (g)	PUFA (g) / SFA (g)	CHOL (mg) / DFIB (g)	A (RE) / A (IU)	C (mg) / B-1 (mg)	B-2 (mg) / NIA (mg)	B-6 (mg) / B-12 (mcg)	FOL (mcg) / PANT (mg)	Na (mg) / K (mg)	Ca (mg) / P (mg)	Mg (mg) / Fe (mg)	Zn (mg) / Cu (mg)	Mn (mg)
cheeseburger	317		15.0		48		3	.29			651	102	26	2.63	
1 sandwich	120	17.3	30.0	7.0		341	.23	3.9			247	186	2.74	.060	
cheeseburger, double w/ bacon	510		31.0		104			.42			728	168	37	5.06	
1 sandwich	159	31.9	27.0	15.0		384	.31	6.3			314	328	3.80	.090	
chicken sandwich	688		40.0		82			.31			1423	79	54	1.15	
1 sandwich	230	26.0	56.0	8.0		126	.45	9.6			375	274	3.30	.160	
chicken tenders	204		10.0		47			.08			636	18	24	.56	
6 pieces	95	20.2	10.0	2.0		95	.08	7.3			200	236	.67	.070	
croissandwich															
w/ bacon, egg & cheese	355		24.0					.30			762	136	20	1.51	
1 croissandwich	119	14.6	20.0	8.0		426	.32	2.1			182	249	2.01		
w/ ham, egg & cheese	335		20.0					.32			987	136	24	1.87	
1 croissandwich	145	17.5	20.0	6.0		426	.49	3.2			256	317	2.16		
w/ sausage, egg & cheese	538		41.0					.32			1042	145	19	2.43	
1 croissandwich	163	19.4	20.0	14.0		426	.36	4.2			285	292	2.89		
french fries	227		13.0		14			.30			160		21		
1 reg serving	74	2.6	24.0	7.0			.10	1.5			360	114	.52	.080	
french toast platter															
w/ bacon	469		30.0		73			.24			448	59	19	1.30	
1 platter	117	11.0	41.0	7.0			.24	3.3			151	118	2.72	.010	
w/ sausage	635		46.0		115			.27			686	70	18	2.32	
1 platter	158	16.1	41.0	13.0			.29	5.5			242	164	3.71		
ham & cheese sandwich	471		24.0		70		7	.42			1534	195	42	2.36	
1 sandwich	230	23.7	44.0	9.0		725	.87	6.1			419	384	3.19	.120	

ᵃ If fried, cholesterol is 14 mg.

	KCAL / WT (g)	H₂O (g) / PRO (g)	FAT (g) / CHO (g)	PUFA (g) / SFA (g)	CHOL (mg) / DFIB (g)	A (RE) / A (IU)	C (mg) / B-1 (mg)	B-2 (mg) / NIA (mg)	B-6 (mg) / B-12 (mcg)	FOL (mcg) / PANT (mg)	Na (mg) / K (mg)	Ca (mg) / P (mg)	Mg (mg) / Fe (mg)	Zn (mg) / Cu (mg)	Mn (mg)
hamburger	275		12.0		37		3	.25			509	37	23	2.36	
1 sandwich	109	15.0	29.0	5.0		150	.23	3.9			235	124	2.73	.060	
hot chocolate	131		4.0		1						172	27	17		
1 serving	244	1.3	22.0								140	59	.71	.070	
onion rings	274		16.0		0						665	124	18	.37	
1 reg serving	79	3.6	28.0	3.0							173	195	.80	.090	
pie															
apple	305		12.0		4		5	.16			412				
1 snack pie	125	3.1	44.0	4.0			.27	0.6			122	31	1.18		
cherry	357		13.0		6		8	.16			204		12		
1 snack pie	128	3.6	55.0	4.0		370	.24	0.5			166	37	1.12		
pecan	459		20.0		4			.18			374		24	16	
1 snack pie	113	5.0	64.0	16.0			.28				204	84	1.13	.140	
salad w/ dressing[a]	160		13.2		10		42	.13			322	45	28	.49	
1 salad	176	2.6	8.5	2.0		1630	.06	1.3			396	68	1.34	.120	
scrambled egg platter	468		30.0		370		3	.35			808	101	32	1.51	
1 platter	195	14.2	33.0			375	.31	2.9			487	271	2.69	.060	
w/ bacon	536		36.0		378		3	.38			975	103	35	1.85	
1 platter	206	18.0	33.0			375	.39	3.6			532	299	2.82	.060	
w/ sausage	702		52.0		420		3	.40			1213	112	33	2.69	
1 platter	247	21.9	33.0			375	.42	5.5			623	335	3.66	.060	
shake															
choc	320		12.0					.55			202	260	46	.95	
1 med shake	273	8.0	46.0				.13				567	262	1.61	.090	
choc w/ added syrup	374		11.0					.51			225	248	56	1.05	
1 med shake	284	8.3	60.0				.12				590	264	1.55	.160	
van	321		10.0					.57			205	295	32	.98	
1 med shake	273	8.7	49.0				.11				508	284			
whaler sandwich	488		27.0		84			.21			592	46	40	.82	
1 sandwich	189	18.6	45.0	6.0		36	.28	4.0			369	249	2.22	.110	
w/ cheese	530		30.0		95			.24			734	112	43	1.10	
1 sandwich	201	20.9	46.0	8.0		227	.27	4.0			381	311	2.24	.110	
whopper jr sandwich	322		17.0		41		6	.25			486	40	24	2.34	
1 sandwich	136	15.2	30.0	6.0		296	.23	4.0			278	127	2.81	.070	
w/ cheese	364		20.0		52		6	.29			628	105	27	2.62	
1 sandwich	147	17.4	31.0	8.0		488	.23	4.0			290	189	2.82	.070	
whopper sandwich	640		41.0		94		14	.41			842	80	43	4.54	
1 sandwich	265	26.9	42.0	14.0		618	.33	6.8			547	237	4.88	.140	
double beef	850		52.0				14	.56				91	60	8.54	
1 sandwich	351	46.4	52.0			617	.34	10.4				387	7.31	.180	
double beef w/ cheese	950		60.0				14	.63				222	65	9.07	
1 sandwich	374	50.8	54.0			1001	.35	10.4				510	7.34	.180	
w/ cheese	723		48.0		117		14	.48			1126	210	48	5.09	
1 sandwich	289	31.4	43.0	19.0		1001	.34	6.8			570	361	4.91	.140	

11.3. JACK-IN-THE-BOX

	KCAL / WT (g)	H₂O (g) / PRO (g)	FAT (g) / CHO (g)	PUFA (g) / SFA (g)	CHOL (mg) / DFIB (g)	A (RE) / A (IU)	C (mg) / B-1 (mg)	B-2 (mg) / NIA (mg)	B-6 (mg) / B-12 (mcg)	FOL (mcg) / PANT (mg)	Na (mg) / K (mg)	Ca (mg) / P (mg)	Mg (mg) / Fe (mg)	Zn (mg) / Cu (mg)	Mn (mg)
bacon cheeseburger	667		39.0	7.0	65						1127				
1 sandwich	230	11.0	67.0	15.0											
breakfast jack	307		13.0	2.5	203						871				
1 sandwich	126	18.0	30.0	5.2											
cheeseburger	325		17.0	2.7	41						746				
1 sandwich	113	15.0	28.0	7.5											

[a] Avg values for 6 types of dressing. If reduced calorie dressing is used, SFA is 0.0 g, fat is 1.0 g, chol is 0 mg & Kcal are 42.

	KCAL	H₂O (g)	FAT (g)	PUFA (g)	CHOL (mg)	A (RE)	C (mg)	B-2 (mg)	B-6 (mg)	FOL (mcg)	Na (mg)	Ca (mg)	Mg (mg)	Zn (mg)	Mn (mg)
	WT (g)	PRO (g)	CHO (g)	SFA (g)	DFIB (g)	A (IU)	B-1 (mg)	NIA (mg)	B-12 (mcg)	PANT (mg)	K (mg)	P (mg)	Fe (mg)	Cu (mg)	
cheesecake	309		17.5	1.0	63						208				
1 serving	99	8.0	29.0	9.0											
chef salad	295		18.0	1.0	107						812				
1 salad	369	32.0	3.0	9.4											
chicken breast tenderloins, wedge fries, garlic roll & dressing—1 dinner	689		30.0	7.3	100						1213				
	321	40.0	65.0	11.5											
chicken supreme	575		36.0	7.6	62						1525				
1 sandwich	231	27.0	34.0	14.3											
club pita	277		8.0	1.0	43						931				
1 sandwich	179	23.0	28.0	3.6											
crescent, canadian	452		31.0	7.0	226						851				
1 sandwich	134	19.0	25.0	9.7											
crescent, sausage	584		43.0	5.7	187						1012				
1 sandwich	156	22.0	28.0	15.5											
crescent, supreme	547		40.0	7.8	178						1053				
1 sandwich	146	20.0	27.0	13.2											
dressing															
blue cheese	131		11.0	6.3	9						459				
1 pkt	35	1.0	7.0	2.0											
buttermilk house	181		18.0	10.8	10						347				
1 pkt	35	1.0	4.0	2.9											
french, low calorie	80		4.0	2.4	0						300				
1 pkt	35	1.0	13.0	0.6											
thousand island	156		15.0	8.8	11						350				
1 pkt	35	1.0	6.0	2.5											
french fries	221		12.0	1.7	8						164				
reg serving	68	2.0	27.0	5.0											
french fries	353		19.0	2.7	13						262				
large serving	109	3.0	43.0	7.9											
grape jelly	38		0.0	0.0	0						3				
1 pkt	14	0.0	9.0	0.0											
ham & swiss burger	754		49.0	7.0	106						1217				
1 sandwich	259	13.0	65.0	21.0											
hamburger	288		13.0	2.4	26						556				
1 sandwich	103	13.0	29.0	5.1											
hot chocolate	133		4.0								180				
8 fl oz		1.3	24.0												
jumbo jack	573		34.0	8.0	73						733				
1 sandwich	222	16.0	50.0	11.0											
jumbo jack w/ cheese	665		40.0	9.0	102						1090				
1 sandwich	242	19.0	58.0	14.0											
moby jack	444		25.0	8.7	47						820				
1 sandwich	137	16.0	39.0	8.4											
monterey burger	865		57.0	4.0	152						1124				
1 sandwich	281	17.0	72.0	28.0											
mushroom burger	470		24.0	2.0	64						910				
1 sandwich	182	10.0	54.0	10.0											
nachos, cheese	571		35.0	8.0	37						1154				
1 serving	170	15.0	49.0	12.7											
nachos, supreme	639		36.0	6.5	37						2187				
1 serving	284	21.0	59.0	13.6											
onion rings	382		23.0	1.3	27						407				
1 serving	108	5.0	39.0	11.1											
pancake platter	630		27.0	3.3	85						1670				
1 platter	232	16.0	79.0	12.6											
pancake syrup	121		0.0	0.0	0						6				
1 pkt	42	0.0	30.0	0.0											
pasta & seafood salad	394		22.0	11.9	48						1570				
1 salad	417	15.0	32.0	4.0											
pizza pocket sandwich	497		28.0	2.0	32						940				
1 sandwich	162	19.0	42.0	13.1											

	KCAL	H₂O (g)	FAT (g)	PUFA (g)	CHOL (mg)	A (RE)	C (mg)	B-2 (mg)	B-6 (mg)	FOL (mcg)	Na (mg)	Ca (mg)	Mg (mg)	Zn (mg)	Mn (mg)
	WT (g)	PRO (g)	CHO (g)	SFA (g)	DFIB (g)	A (IU)	B-1 (mg)	NIA (mg)	B-12 (mcg)	PANT (mg)	K (mg)	P (mg)	Fe (mg)	Cu (mg)	
sauce															
A-1 steak sauce	35		0.0	0.0	0						809				
1 pkt	50	0.0	9.0	0.0											
barbeque	78		0.0	0.0	0						535				
1 pkt	50	0.7	19.0	0.0											
mayo-mustard	124		13.0		10						247				
1 pkt	21	0.5	2.0												
mayo-onion	143		15.0		20						140				
1 pkt	21	0.3	1.0												
seafood cocktail	57		0.0	0.0	0						367				
1 pkt	50	1.5	12.2	0.0											
scrambled egg platter	720		44.0	1.6	260						1110				
1 platter	267	26.0	55.0	21.1											
shake															
choc	330		7.0	1.0	25						270				
1 shake	322	11.0	55.0	4.3											
strawberry	320		7.0	1.0	25						240				
1 shake	328	10.0	55.0	4.3											
van	320		6.0	1.0	25						230				
1 shake	317	10.0	57.0	3.6											
side salad	51		3.0	0.0	1						84				
1 salad	111	7.0	1.0	2.0											
sirloin steak, wedge fries, garlic roll & sce	699		27.0	6.1	75						969				
1 dinner	334	38.0	75.0	10.0											
shrimp, breaded, wedge fries, garlic roll &	731		37.0	4.8	157						1510				
seafood cocktail sce—*1 dinner*	301	22.0	77.0	16.6											
swiss & bacon burger	681		47.0	7.0	92						1458				
1 sandwich	187	5.0	61.0	20.0											
taco	191		11.0	1.0	21						406				
1 taco	81	8.0	16.0	5.2											
taco salad	377		24.0	2.5	102						1436				
1 salad	358	31.0	10.0	10.4											
taco, super	288		17.0	1.2	37						765				
1 taco	135	12.0	21.0	8.0											
turnover, hot apple	410		24.0	2.0	15						350				
1 turnover	119	4.0	45.0	10.8											

11.4. KENTUCKY FRIED CHICKEN

	KCAL	H₂O (g)	FAT (g)	PUFA (g)	CHOL (mg)	A (RE)	C (mg)	B-2 (mg)	B-6 (mg)	FOL (mcg)	Na (mg)	Ca (mg)	Mg (mg)	Zn (mg)	Mn (mg)
beans, baked	105		1.2	0.3	0		2	.04			387	54			
1 serving	89	5.1	18.4	0.4			.06	0.5					1.43		
biscuit, buttermilk	269		13.6	1.1	0		0	.13			521	77			
1 biscuit	75	5.1	31.6	3.5			.28	1.8					1.22		
coleslaw	103		5.7	2.9	4		19	.03			171	29			
1 serving	79	1.3	11.5	0.8		269	.03	0.2					.19		
corn-on-the-cob	176		3.1	1.5	0		2	.11				7			
1 ear	143	5.1	31.9	0.5		272	.14	1.8					.79		
chicken, fried, extra crispy															
breast, center	353		20.9	2.0	93			.16			842	35			
1 center breast	120	26.9	14.4	5.3			.10	10.0					.86		
breast, side	354		23.7	2.3	66			.13			797	32			
1 side breast	98	17.7	17.3	6.0			.08	6.5					.86		
drumstick	173		10.9	1.3	65			.14			346	15			
1 drumstick	60	12.7	5.9	2.8			.05	2.8					.61		
thigh	371		26.3	3.4	121			.27			766	46			
1 thigh	112	19.6	13.8	6.9			.09	5.2					1.21		
wing	218		15.6	1.8	63			.07			437	21			
1 wing	57	11.5	7.8	4.0			.03	2.8					.52		

	KCAL / WT (g)	H₂O (g) / PRO (g)	FAT (g) / CHO (g)	PUFA (g) / SFA (g)	CHOL (mg) / DFIB (g)	A (RE) / A (IU)	C (mg) / B-1 (mg)	B-2 (mg) / NIA (mg)	B-6 (mg) / B-12 (mcg)	FOL (mcg) / PANT (mg)	Na (mg) / K (mg)	Ca (mg) / P (mg)	Mg (mg) / Fe (mg)	Zn (mg) / Cu (mg)	Mn (mg)
chicken, fried, original recipe															
breast, center	257		13.7	1.6	93			.14			532	39			
1 center breast	107	25.5	8.0	3.6			.09	10.0					.63		
breast, side	276		17.3	2.3	96			.18			654	48			
1 side breast	95	20.0	10.1	4.6			.07	6.8					.79		
drumstick	147		8.8	1.1	81			.13			269	13			
1 drumstick	58	13.6	3.4	2.3			.06	2.9					.60		
thigh	278		19.2	2.6	122			.28			517	28			
1 thigh	96	18.0	8.4	5.3			.08	4.6					1.05		
wing	181		12.3	1.7	67			.06			387	38			
1 wing	56	11.8	5.8	3.3		56	.03	3.2					.45		
chicken nuggets	46		2.9	0.3	12			.03			140	2			
1 nugget	16	2.8	2.2	0.7			.02	1.0					.13		
chicken nugget sce															
barbeque	35		0.6	0.3	0		0	.01			450	6			
1 oz		0.3	7.1	0.1		370	.01	0.2					.24		
honey	49		0.0	0.0	0			.00				1			
1/2 oz		0.0	12.1	0.0			.00	0.0					.11		
mustard	36		0.9		0		0	.01			346	10			
1 oz		0.9	6.0	0.1			.02	0.2					.26		
sweet & sour	58		0.6	0.3	0		0				148	5			
1 oz		0.1	13.0	0.1		60	.01	0.0					.16		
fries	268		12.8	0.7	2		3	.06			81	24			
1 serving	119	4.8	33.3	3.1			.17	2.7					.94		
gravy, chicken	59		3.7	0.3	2			.03			398	9			
1 serving	78	2.0	4.4	1.0			.00	0.5					.48		
potatoes, mashed	59		0.6	0.0	0		0	.04				21			
1 serving	80	1.9	11.6	0.2			.00	1.0					.28		
potatoes, mashed w/ gravy	62		1.4	0.1	0		0	.04			297	19			
1 serving	86	2.1	10.3	0.4			.00	1.0					.35		
potato salad	141		9.3	4.9	11		3	.02			396	10			
1 serving	90	1.8	12.6	1.4		90	.07	0.6					.32		

11.5. MCDONALDS

	KCAL / WT (g)	H₂O (g) / PRO (g)	FAT (g) / CHO (g)	PUFA (g) / SFA (g)	CHOL (mg) / DFIB (g)	A (RE) / A (IU)	C (mg) / B-1 (mg)	B-2 (mg) / NIA (mg)	B-6 (mg) / B-12 (mcg)	FOL (mcg) / PANT (mg)	Na (mg) / K (mg)	Ca (mg) / P (mg)	Mg (mg) / Fe (mg)	Zn (mg) / Cu (mg)	Mn (mg)
big mac	570		35.0		83		3	.38			979	203			
1 sandwich	200	24.6	39.2			380	.48	7.2					4.90		
biscuit															
plain	330		18.2		9			.15			786	74			
1 biscuit	85	4.9	36.6			179	.21	1.7					1.30		
w/ bacon, egg & cheese	483		31.6		263		2	.43			1269	2			
1 biscuit	145	16.5	33.2			653	.30	2.3					2.57		
w/ sausage	467		30.9		48			.22			1147	82			
1 biscuit	121	12.1	35.3			61	.56	3.4					2.05		
w/ sausage & egg	585		39.9		285			.49			1301	119			
1 biscuit	175	19.8	36.4			420	.53	3.9					3.43		
cheeseburger	318		16.0		41		2	.24			743	169			
1 sandwich	114	15.0	28.5			353	.30	4.3					2.84		
chicken mcnuggets	323		21.3		73		2	.14			512	11			
1 serving	109	19.1	13.7				.16	7.5					1.25		
chicken mcnugget sce															
barbeque	60		0.4		0		0	.01			309	4			
1 container	32	0.4	13.7			45	.01	0.1					.12		
honey	50		0.0		0		0	.00			2	1			
1 container	14	0.0	12.4				.00	0.0					.02		
hot mustard	63		2.1		3		0	.00			259	8			
1 container	30	0.6	10.5			9	.01	0.1					.17		
sweet & sour	64		0.3		0		0	.00			186	2			
1 container	32	0.2	15.0			200	.01	0.1					.08		

	KCAL / WT (g)	H₂O (g) / PRO (g)	FAT (g) / CHO (g)	PUFA (g) / SFA (g)	CHOL (mg) / DFIB (g)	A (RE) / A (IU)	C (mg) / B-1 (mg)	B-2 (mg) / NIA (mg)	B-6 (mg) / B-12 (mcg)	FOL (mcg) / PANT (mg)	Na (mg) / K (mg)	Ca (mg) / P (mg)	Mg (mg) / Fe (mg)	Zn (mg) / Cu (mg)	Mn (mg)
cookies															
chocolaty chip	342		16.3		18		1	.21			313	29			
1 box	69	4.2	44.8			76	.12	1.7					1.56		
mcdonaldland	308		10.8		10			.23			358	12			
1 box	67	4.2	48.7				.23	2.9					1.47		
egg mcmuffin	340		15.8		259			.44			885	226			
1 mcmuffin	138	18.5	31.0			591	.47	3.8					2.93		
english muffin w/ butter	186		5.3		15		1	.49			310	117			
1 muffin	63	5.0	29.5			164	.28	2.6					1.51		
filet-o-fish sandwich	435		25.7		45			.23			799	133			
1 sandwich	143	14.7	35.9			186	.36	3.0					2.47		
french fries	220		11.5		9		13	.02			109	9			
1 reg serving	68	3.0	26.1				.12	2.3					.61		
hamburger	263		11.3		29		2	.22			506	84			
1 sandwich	100	12.4	28.3			100	.31	4.1					2.85		
hash brown potatoes	125		7.0		7		4	.00			325	5			
1 serving	55	1.5	14.0				.06	0.8					.40		
hotcake w/ butter & syrup	500		10.3		47		5	.36			1070	103			
1 serving	214	7.9	93.9			257	.26	2.3					2.23		
ice cream in cone	185		5.2		24			.36			109	183			
1 serving	115	4.3	30.2			218	.06	0.4					.12		
Mc D.L.T.	680		44.0		101		8	.46			1030	230			
1 sandwich	254	30.0	40.0			508	.56	8.0					6.60		
pie															
apple	253		14.3		12			.02			398	14			
1 snack pie	85	1.9	29.3				.02	0.2					.62		
cherry	260		13.6		13			.02			427	12			
1 snack pie	88	2.0	32.1			114	.03	0.3					.59		
quarter pounder	427		23.5		81		3	.32			718	98			
1 sandwich	160	24.6	29.3			128	.35	7.2					4.30		
w/ cheese	525		31.6		107		3	.41			1220	255			
1 sandwich	186	29.6	30.5			614	.37	7.1					4.84		
sausage	210		18.6		39			.11			423	16			
1 serving	53	9.8	0.6				.27	2.1					.82		
sausage mcmuffin	427		26.3		59		1	.25			942	168			
1 mcmuffin	115	17.6	30.0			380	.70	4.1					2.25		
sausage mcmuffin w/ egg	517		32.9		287			.50			1044	196			
1 mcmuffin	165	22.9	32.2			660	.84	4.5					3.47		
scrambled eggs	180		13.0		514		1	.47			205	61			
1 serving	98	13.2	2.5			652	.08	0.2					2.53		
shake															
choc	383		9.0		30			.44			300	320			
1 shake	291	9.9	65.5			349	.12	0.5					.84		
strawberry	362		8.7		32		4	.44			207	322			
1 shake	290	9.0	62.1			377	.12	0.4					.17		
van	352		8.4		31		3	.70			201	329			
1 shake	291	9.3	59.6			349	.12	0.4					.18		
sundae															
caramel	361		10.0		31		4	.31			145	200			
1 sundae	165	7.2	60.8			279	.07	1.0					.23		
hot fudge	357		10.8		27		2	.31			170	215			
1 sundae	164	7.0	58.0			230	.07	1.1					.61		
strawberry	320		8.7		25		3	.30			90	174			
1 sundae	164	6.0	54.0			230	.07	1.0					.38		

	KCAL	H₂O (g)	FAT (g)	PUFA (g)	CHOL (mg)	A (RE)	C (mg)	Vitamins B-2 (mg)	B-6 (mg)	FOL (mcg)	Na (mg)	Ca (mg)	Minerals Mg (mg)	Zn (mg)	Mn (mg)
	WT (g)	PRO (g)	CHO (g)	SFA (g)	DFIB (g)	A (IU)	B-1 (mg)	NIA (mg)	B-12 (mcg)	PANT (mg)	K (mg)	P (mg)	Fe (mg)	Cu (mg)	

11.6. ROY ROGERS

bacon cheeseburger	581		39.2		103		1536
1 sandwich	180	32.3	25.0				
bar burger	611		39.4		115		1826
1 sandwich	208	36.1	28.0				
biscuit	231		12.1				575
1 biscuit	63	4.4	26.2				
breakfast crescent sandwich	401		27.3		148		867
1 sandwich	127	13.1	25.3				
w/ bacon	431		29.7		156		1035
1 sandwich	133	15.4	25.5				
w/ ham	557		41.7		189		1192
1 sandwich	165	19.8	25.3				
w/ sausage	449		29.4		168		1289
1 sandwich	162	19.9	25.9				
brownie	264		11.4		10		150
1 brownie	64	3.3	37.3				
cheeseburger	563		37.3		95		1404
1 sandwich	173	29.5	27.4				
chicken, fried							
breast	412		23.7		118		609
1 breast	144	33.0	16.9				
leg	140		8.0		40		190
1 leg	53	11.5	5.5				
thigh	296		19.5		85		406
1 thigh	98	18.4	11.7				
wing	192		12.8		47		285
1 wing	52	10.5	8.5				
coleslaw	110		6.9				261
1 serving	99	1.0	11.0				
crescent roll	287		17.7				547
1 roll	70	4.7	27.2				
danish							
apple	249		11.6		15		255
1 danish	71	4.5	31.6				
cheese	254		12.2		11		260
1 danish	71	4.9	31.4				
cherry	271		14.4		11		242
1 danish	71	4.4	31.7				
egg & biscuit platter	394		26.5		284		734
1 platter	165	16.9	21.9				
w/ bacon	435		29.6		294		957
1 platter	173	19.7	22.1				
w/ ham	442		28.6		304		1156
1 platter	200	23.5	22.5				
w/ sausage	550		40.9		325		1059
1 platter	203	23.4	21.9				
french fries	268		13.5		42		165
1 reg serving	85	3.9	32.0				
french fries	357		18.4		56		221
1 large serving	113	5.3	42.7				
hamburger	456		28.3		73		495
1 sandwich	143	23.8	26.6				

	KCAL	H₂O (g)	FAT (g)	PUFA (g)	CHOL (mg)	Vitamins					Minerals				
						A (RE)	C (mg)	B-2 (mg)	B-6 (mg)	FOL (mcg)	Na (mg)	Ca (mg)	Mg (mg)	Zn (mg)	Mn (mg)
	WT (g)	PRO (g)	CHO (g)	SFA (g)	DFIB (g)	A (IU)	B-1 (mg)	NIA (mg)	B-12 (mcg)	PANT (mg)	K (mg)	P (mg)	Fe (mg)	Cu (mg)	
hot chocolate	123		2.0		35						125				
6 fl oz		3.0	22.0												
macaroni salad	186		10.7								603				
1 serving	100	3.1	19.4												
pancake platter w/ butter & syrup	452		15.2		53						842				
1 platter	165	7.7	71.8												
w/ bacon	493		18.3		63						1065				
1 platter	173	10.4	72.0												
w/ ham	506		17.3		73						1264				
1 platter	200	14.3	72.4												
w/ sausage	608		29.6		94						1167				
1 platter	203	14.2	71.8												
potato, baked															
w/ bacon & cheese	397		21.7		34						778				
1 potato	248	17.1	33.3												
w/ broccoli & cheese	376		18.1								523				
1 potato	312	13.7	39.6												
w/ sour cream & chives	408		20.9		31						138				
1 potato	297	7.3	47.6												
w/ taco beef & cheese	463		21.8		37						726				
1 potato	359	21.8	45.0												
potato salad	107		6.1								696				
1 serving	100	2.0	10.9												
roast beef sandwich	317		10.1		55						785				
1 sandwich	154	27.2	29.1												
	360		11.9		73						1044				
1 large sandwich	182	33.9	29.6												
w/ cheese	424		19.2		77						1694				
1 sandwich	182	32.9	29.9												
w/ cheese	467		20.9		95						1953				
1 large sandwich	211	39.6	30.3												
shake															
choc	358		10.2		37						290				
1 shake	319	7.9	61.3												
strawberry	315		10.2		37						261				
1 shake	312	7.6	49.4												
van	306		10.7		40						282				
1 shake	306	8.0	45.0												
strawberry shortcake	447		19.2		28						674				
1 serving	205	10.1	59.3												
sundae															
caramel	293		8.5		23						193				
1 sundae	145	7.0	51.5												
hot fudge	337		12.5		23						186				
1 sundae	151	6.5	53.3												
strawberry	216		7.1		23						99				
1 sundae	142	5.7	33.1												

11.7. WENDY'S

	KCAL	H₂O (g)	FAT (g)	PUFA (g)	CHOL (mg)	A (RE)	C (mg)	B-2 (mg)	B-6 (mg)	FOL (mcg)	Na (mg)	Ca (mg)	Mg (mg)	Zn (mg)	Mn (mg)
bacon cheeseburger	455		25.0		98						843				
1 sandwich	154	31.0	26.0												
big classic hamburger	470		25.0		80						900				
1 sandwich	241	26.0	36.0								470				
big classic double hamburger	680		39.0		155						1005				
1 sandwich	317	46.0	36.0								740				

	KCAL / WT (g)	H₂O (g) / PRO (g)	FAT (g) / CHO (g)	PUFA (g) / SFA (g)	CHOL (mg) / DFIB (g)	A (RE) / A (IU)	C (mg) / B-1 (mg)	B-2 (mg) / NIA (mg)	B-6 (mg) / B-12 (mcg)	FOL (mcg) / PANT (mg)	Na (mg) / K (mg)	Ca (mg) / P (mg)	Mg (mg) / Fe (mg)	Zn (mg) / Cu (mg)	Mn (mg)
breakfast potatoes	360		22.0		20						745				
1 serving	103	4.0	37.0								615				
breakfast sandwich (white bread, egg & cheese)—*1 sandwich*	370		19.0		200						770				
	129	17.0	33.0								155				
buttermilk biscuit	320		17.0		0						860				
1 biscuit	94	5.0	37.0								125				
chicken club sandwich	479		25.0		78						813				
1 sandwich	187	29.0	30.0								365				
chicken fried steak	580		41.0		95						1040				
1 serving	176	28.0	25.0								390				
chicken nuggets	290		21.0		55						615				
6 pieces	95	16.0	11.0								165				
chicken nugget sce															
barbeque	50		0.0		0						100				
1 pkt	28	0.0	11.0								95				
honey	45		0.0		0						0				
1 pkt	14	0.0	12.0								0				
sweet & sour	45		0.0		0						55				
1 pkt	28	0.0	11.0								40				
sweet mustard	50		1.0		0						140				
1 pkt	28	0.0	9.0								30				
chili	240		8.0		25						990				
1 reg serving	236	19.0	24.0								540				
chili	360		12.0		38						1485				
1 large serving	354	29.0	36.0								810				
coleslaw	80		5.0		40						165				
1/4 cup	57	0.0	9.0								85				
cookie, choc chip	320		17.0		5						235				
1 cookie	64	3.0	40.0								100				
danish, apple/cheese/cinn raisin	400		17.7								437				
1 danish	95	7.0	53.3												
double cheeseburger	620		36.0		165						760				
1 sandwich	221	48.0	26.0												
double hamburger	560		30.0		150						465				
1 sandwich	203	44.0	26.0								590				
egg, fried	90		6.0		230						95				
1 egg	44	6.0	1.0								65				
eggs, scrambled	190		12.0		450						160				
2 eggs	91	14.0	7.0								130				
fish fillet	210		11.0		45						475				
1 serving	92	14.0	13.0								255				
french fries	310		15.0		15						105				
1 reg serving	106	4.0	38.0								675				
french fries	403		19.5		20						137				
1 large serving	138	5.2	49.4								878				
french toast	400		19.0		115						850				
2 slices	135	11.0	45.0								175				
frosty dairy dessert	400		14.0		50						220				
1 small	243	8.0	59.0								585				
	520		18.2		65						286				
1 med	316	10.4	76.7								761				
	680		23.8		85						374				
1 large	413	13.6	100.3								995				
kid's hamburger	200		9.0		35						225				
1 sandwich	72	13.0	17.0								170				
omelet															
w/ ham & cheese	290		21.0		355						570				
1 serving	118	18.0	7.0								190				

	KCAL	H₂O (g)	FAT (g)	PUFA (g)	CHOL (mg)	A (RE)	C (mg)	B-2 (mg)	B-6 (mg)	FOL (mcg)	Na (mg)	Ca (mg)	Mg (mg)	Zn (mg)	Mn (mg)
	WT (g)	PRO (g)	CHO (g)	SFA (g)	DFIB (g)	A (IU)	B-1 (mg)	NIA (mg)	B-12 (mcg)	PANT (mg)	K (mg)	P (mg)	Fe (mg)	Cu (mg)	
w/ ham, cheese & mushrooms	250		17.0		450						405				
1 serving	114	18.0	6.0								180				
w/ ham, cheese, onion & green pepper	280		19.0		525						485				
1 serving	128	19.0	7.0								200				
w/ mushrooms, green pepper & onion	210		15.0		460						200				
1 serving	114	14.0	7.0								190				
pasta salad	130		6.0		5						190				
1/4 cup	57	3.0	18.0								70				
potato, baked	250		2.0		0						60				
1 potato	250	6.0	52.0								1360				
w/ bacon & cheese	570		30.0		22						1180				
1 potato	350	19.0	57.0								1380				
w/ broccoli & cheese	500		25.0		22						430				
1 potato	365	13.0	54.0								1550				
w/ cheese	590		34.0		22						450				
1 potato	350	17.0	55.0								1380				
w/ chili & cheese	510		20.0		22						610				
1 potato	400	22.0	63.0								1590				
w/ sour cream & chives	460		24.0		15						230				
1 potato	310	6.0	53.0								1420				
sausage	200		18.0		45						405				
1 patty	45	8.0	0.0								115				
sausage gravy	440		36.0		85						1300				
6 oz	214	17.0	13.0								340				
single cheeseburger	410		22.0		90						655				
1 sandwich	145	28.0	26.0												
single hamburger	350		16.0		75						360				
1 sandwich	127	24.0	26.0								320				
syrup	140		0.0		0						5				
1 pkt	43	0.0	37.0								0				
taco salad	430		19.0		45						1260				
1 serving	398	22.0	43.0								760				
tartar sce	80		9.0								75				
1 T	14	0.0	0.0												

12. FATS, OILS & SHORTENINGS
12.1. ANIMAL FATS

	KCAL	H₂O (g)	FAT (g)	PUFA (g)	CHOL (mg)	A (RE)	C (mg)	B-2 (mg)	B-6 (mg)	FOL (mcg)	Na (mg)	Ca (mg)	Mg (mg)	Zn (mg)	Mn (mg)
	WT (g)	PRO (g)	CHO (g)	SFA (g)	DFIB (g)	A (IU)	B-1 (mg)	NIA (mg)	B-12 (mcg)	PANT (mg)	K (mg)	P (mg)	Fe (mg)	Cu (mg)	
beef suet, raw	242	1.1	26.7	0.9	19										
1 oz	28	0.4	0.0	14.8							5				
beef tallow, raw	116	0.0	12.8	0.5	14						0				
1 T	13	0.0	0.0	6.4	0.0						0				
chicken fat, raw	115	0.0	12.8	2.7	11									.002	
1 T	13	0.0	0.0	3.8											
duck fat, raw	115	0.0	12.8	1.7	13										
1 T	13	0.0	0.0	4.3	0.0										
goose fat, raw	115	0.0	12.8	1.4	13										
1 T	13	0.0	0.0	3.5	0.0										
mutton tallow, raw	116	0.0	12.8	1.0	13										
1 T	13	0.0	0.0	6.1	0.0										
pork fat (lard), raw	115	0.0	12.8	1.4	12						0	0	0	.01	
1 T	13	0.0	0.0	5.0	0.0						0				
pork separable fat	200	5.1	21.3	2.3	26	1	0	.03	.17	0	9	1	2	.17	.001
1 oz	28	1.9	0.0	7.8	0.0	4	.09	0.7	.17	.04	23	17	.09	.015	
salt pork, raw	212	3.1	22.8	2.7	25	0		.02	.02	0	404	2	2	.26	.001
1 oz	28	1.4	0.0	8.3	0.0	0	.06	0.5	.08	.06	19	15	.12	.014	
turkey fat, raw	115	0.0	12.8	3.0	13										
1 T	13	0.0	0.0	3.8	0.0										

	KCAL / WT (g)	H₂O (g) / PRO (g)	FAT (g) / CHO (g)	PUFA (g) / SFA (g)	CHOL (mg) / DFIB (g)	A (RE) / A (IU)	C (mg) / B-1 (mg)	B-2 (mg) / NIA (mg)	B-6 (mg) / B-12 (mcg)	FOL (mcg) / PANT (mg)	Na (mg) / K (mg)	Ca (mg) / P (mg)	Mg (mg) / Fe (mg)	Zn (mg) / Cu (mg)	Mn (mg)

12.2. VEGETABLE OILS

	KCAL/WT	H₂O/PRO	FAT/CHO	PUFA/SFA	CHOL/DFIB	A RE/A IU	C/B-1	B-2/NIA	B-6/B-12	FOL/PANT	Na/K	Ca/P	Mg/Fe	Zn/Cu	Mn
almond oil	120	0.0	13.6	2.4											
1 T	14	0.0	0.0	1.1	0.0										
coconut oil	120	0.0	13.6	0.2											
1 T	14	0.0	0.0	11.8	0.0										
corn oil	120	0.0	13.6	8.0											
1 T	14	0.0	0.0	1.7	0.0										
corn oil, Fleischmann's	120		14.0	8.0	0						0				
1 T		0.0	0.0	2.0							0				
corn oil, Mazola	125	0.0	14.0	8.3	0						0				
1 T	14	0.0	0.0	1.8											
cottonseed oil	120	0.0	13.6	7.1											
1 T	14	0.0	0.0	3.5	0.0										
Crisco oil	124	0.0	14.0	7.3	0										
1 T	14	0.0	0.0	1.8	0.0										
olive oil	119	0.0	13.5	1.1							0	0	0	.01	
1 T	14	0.0	0.0	1.8	0.0						0	.05			
palm oil	120	0.0	13.6	1.3											
1 T	14	0.0	0.0	6.7	0.0						0	.00			
palm kernel oil	120	0.0	13.6	0.2											
1 T	14	0.0	0.0	11.1	0.0										
peanut oil	119	0.0	13.5	4.3							0	0	0	.00	
1 T	14	0.0	0.0	2.3	0.0						0	0	.00		
peanut oil, Planters	130		14.0	4.0	0						0				
1 T		0.0	0.0	2.0							0				
popcorn oil, Planters	130		14.0	4.0	0						0				
1 T		0.0	0.0	0.2							0				
Puritan oil	124	0.0	14.0	9.5	0										
1 T	14	0.0	0.0	1.7	0.0										
safflower oil	120	0.0	13.6	10.1											
1 T	14	0.0	0.0	1.2	0.0										
sesame oil	120	0.0	13.6	5.7											
1 T	14	0.0	0.0	1.9	0.0										
soybean oil	120	0.0	13.6	7.9							0	0	0	.00	
1 T	14	0.0	0.0	2.0	0.0						0	.00			
soybean oil, hydrogenated	120	0.0	13.6	5.1											
1 T	14	0.0	0.0	2.0	0.0										
soybean (hydrogenated) & cottonseed oil	120	0.0	13.6	6.5											
1 T	14	0.0	0.0	2.4	0.0										
soybean lecithin[a]	120	0.0	13.6	6.1											
1 T	14	0.0	0.0	2.1	0.0										
sunflower oil	120	0.0	13.6	8.9											
1 T	14	0.0	0.0	1.4	0.0										
veg oil spray, Mazola No Stick	6	0.0	0.7	0.4	0						0				
2.5 sec spray	0.7	0.0	0.0	0.1											
wheat germ oil	120	0.0	13.6	8.5										.50	
1 T	14	0.0	0.0	2.6	0.0										

12.3. SHORTENINGS

	KCAL/WT	H₂O/PRO	FAT/CHO	PUFA/SFA	CHOL/DFIB	A RE/A IU	C/B-1	B-2/NIA	B-6/B-12	FOL/PANT	Na/K	Ca/P	Mg/Fe	Zn/Cu	Mn
Crisco, reg/butter flavor	106	0.0	12.0	3.6	0										
1 T	12	0.0	0.0	3.1	0.0	0[b]									
lard & veg oil	115	0.0	12.8	1.4											
1 T	13	0.0	0.0	5.2	0.0										
soybean & cottonseed	113	0.0	12.8	3.3											
1 T	13	0.0	0.0	3.2	0.0										
soybean & palm	113	0.0	12.8	1.8											
1 T	13	0.0	0.0	3.9	0.0										

[a] 70% soybean phosphatide in 30% soybean oil.

[b] Butter flavor contains 190 IU Vit A.

	KCAL	H₂O (g)	FAT (g)	PUFA (g)	CHOL (mg)		A (RE)	C (mg)	**Vitamins** B-2 (mg)	B-6 (mg)	FOL (mcg)		Na (mg)	Ca (mg)	**Minerals** Mg (mg)	Zn (mg)	Mn (mg)
	WT (g)	PRO (g)	CHO (g)	SFA (g)	DFIB (g)		A (IU)	B-1 (mg)	NIA (mg)	B-12 (mcg)	PANT (mg)		K (mg)	P (mg)	Fe (mg)	Cu (mg)	

13. FISH, SHELLFISH & CRUSTACEA

	KCAL	H₂O	FAT	PUFA	CHOL		A(RE)	C	B-2	B-6	FOL		Na	Ca	Mg	Zn	Mn
abalone																	
raw	89	63.4	0.6	0.1	72						4		255	27	41	.69	.034
3 oz	85	14.5	5.1	0.1							2.55				2.71	.167	
fried[a]	161	51.1	5.8	1.4	80						5		502	32	47	.80	
3 oz	85	16.7	9.4	1.4											3.23	.194	
alewife																	
raw	127	79.4	4.9														
3.5 oz	100	19.4	0.0		0.0									218			
cnd	141	73.0	8.0														
3.5 oz	100	16.2	0.0		0.0												
anchovy																	
raw	111	62.4	4.1	1.4					.22	.12			88	125	35	1.46	
3 oz	85	17.3	0.0	1.1				.05	11.9	.53			325	148	2.76	.179	
cnd in olive oil	42	10.1	1.9	0.5					.07	.04			734	46	14	.49	
5 anchovies	20	5.8	0.0	0.4				.02	4.0	.18			109	50	.93	.068	
paste	14		0.8	0.5													
1 t	7	1.4	0.3		0.0												
pickled	49	16.4	2.9											47			
1 oz	28	5.4	0.1		0.0									59			
barracuda, pacific, raw	113	75.4	2.6														
3.5 oz	100	21.0	0.0		0.0												
bass, black																	
raw	93	79.3	1.2										68				
3.5 oz	100	19.2	0.0		0.0								256				
baked	287		19.4					0	.16				68	96			
4 oz	113	23.6	3.0		0.0		97	.07	3.5				256	269	1.20		
stuffed, baked[b]	259	52.9	15.8														
3.5 oz	100	16.2	11.4		0.0												
bass, freshwater, raw	97	64.3	3.1	0.9	58								59	68	26	.56	.756
3 oz	85	16.0	0.0	0.7									303	170	1.27	.079	
bass, striped, raw	82	67.3	2.0	0.7	68								59			.34	.013
3 oz	85	15.1	0.0	0.4					3.25						.71	.026	
bluefish, raw	105	60.2	3.6	0.9	50		338		.07	.34	1		51	6	28	.69	.018
3 oz	85	17.0	0.0	0.8			101	.05	5.1	4.58	.70		316	193	.41	.045	
bonito, cnd	257		19.1						.09				514	8	28		
3.5 oz	100	19.8	0.0		0.0			.01	9.8				302	193	1.00		
bullhead, black, raw	84	81.3	1.6														
3.5 oz	100	16.3	0.0		0.0												
burbot, raw	76	67.4	0.7	0.3	51								82	43	27	.64	.595
3 oz	85	16.4	0.0	0.1									343	170	.77	.170	
butterfish, raw	124	63.0	6.8		55								75			.65	.013
3 oz	85	14.7	0.0										318		.42	.046	
carp																	
raw	108	64.9	4.8	1.2	56		25	1		.16			42	35	25	1.26	
3 oz	85	15.2	0.0	0.9			7		1.30				283	352	1.05	.048	
ckd by dry heat	138	59.2	6.1	1.6	72		27	1		.19	1		54	44	32	1.62	
3 oz	85	19.4	0.0	1.2			8						363	452	1.35	.062	
catfish, channel																	
raw	99	64.9	3.6	0.9	49				.09				54	34	21	.61	.013
3 oz	85	15.5	0.0	0.8				.04	1.8				296	181	.83	.080	
breaded & fried[c]	194	50.0	11.3	2.8	69		24	0	.11				238	37	23	.73	
3 oz	85	15.4	6.8	2.8			7	.06	1.9				289	183	1.22	.086	

[a] Dipped in flour & salt before frying.
[b] Stuffed w/ bacon, butter, celery, onion & bread cubes.
[c] Breading consists of cornmeal, egg, milk & salt.

	KCAL / WT (g)	H₂O (g) / PRO (g)	FAT (g) / CHO (g)	PUFA (g) / SFA (g)	CHOL (mg) / DFIB (g)	A (RE) / A (IU)	C (mg) / B-1 (mg)	B-2 (mg) / NIA (mg)	B-6 (mg) / B-12 (mcg)	FOL (mcg) / PANT (mg)	Na (mg) / K (mg)	Ca (mg) / P (mg)	Mg (mg) / Fe (mg)	Zn (mg) / Cu (mg)	Mn (mg)
caviar, black & red, granular	40	7.6	2.9		94						240				
1 T	16	3.9	0.6												
chub, raw	145	74.9	8.8		50										
3.5 oz	100	15.3	0.0		0.0										
cisco															
raw	84	67.1	1.6	0.5							47				
3 oz	85	16.1	0.0	0.4							301		.31		
smoked	151	59.3	10.1	1.9	27	802		.14	.23	2	409	22	14	.25	.018
3 oz	85	13.9	0.0	1.5		240	.04	2.0	3.62	.26	249	128	.41	.183	
clam liquid, cnd	6	234.5	0.1								516	31	26	.24	.178
1 cup	240	1.0	0.2											.934	
clams															
raw	63	69.6	0.8	0.2	29	255		.18			47	39	8	1.16	.425
3 oz (4 large or 9 small)	85	10.9	2.2	0.1		77		1.5	42.03	.31	267	144	11.88	.292	
breaded & fried[a]	171	52.3	9.5	2.4	52	257		.21			309	54	12	1.24	
3 oz (9 small)	85	12.1	8.8	2.3		77		1.8	34.23		277	160	11.83	.303	
cnd	126	54.1	1.7	0.5	57	484		.36			95	78	16	2.32	
3 oz	85	21.7	4.4	0.2		145		2.9	84.06		534	287	23.76	.585	
ckd by moist heat	126	54.1	1.7	0.5	57	484		.36			95	78	.16	2.32	
3 oz (9 small)	85	21.7	4.4	0.2		145		2.9	84.06		534	287	23.76	.585	
cod, atlantic															
raw	70	69.0	0.6	0.2	37	34	1	.06	.21		46	13	27	.38	.013
3 oz	85	15.1	0.0	0.1		10	.07	1.8	.77	.13	351	173	.32	.024	
cnd	89	64.3	0.7	0.2	47	39	1	.07	.24		185	18	35	.49	
3 oz	85	19.4	0.0	0.1		12	.07	2.1	.89		449	221	.41	.031	
ckd by dry heat	89	64.5	0.7	0.2	47	39	1	.07	.24		66	12	36	.49	
3 oz	85	19.4	0.0	0.1		12	.08	2.1	.89		208	117	.41	.031	
dried & salted	246	13.7	2.0	0.7	129	120	3	.20	.73		5973	136	113	1.35	
3 oz	85	53.4	0.0	0.4		36	.23	6.4	8.50	1.42	1239	808	2.13	.150	
cod, frzn															
almondine, Am Hosp Co	347		21.7		49		1	.18			312	91			
5.5 oz	156	26.4	11.4			307	.13	4.3			511		1.70		
breaded, Van de Kamp's	290		20.0								370				
5 oz piece	142	20.0	10.0												
w/ creole sce, Am Hosp Co	158		5.6		38		28	.14			498	42			
6.44 oz	183	20.5	6.4			875	.12	3.4			621		1.42		
w/ fish-based crm sce, Am Hosp Co	188		9.3		72		1	.05			497	81			
6 oz	170	15.8	10.2			343	.05	1.7			282		1.08		
w/ lemon sce, Am Hosp Co	237		11.4		63		4	.13			165	42			
6.5 oz	184	29.2	4.4			431	.09	3.0			436		1.01		
cod, pacific, raw	70	69.1	0.5	0.2	31	24		.04			60	6	21	.34	.010
3 oz	85	15.2	0.0	0.1		7	.02	1.7		.12	342	148	.22	.022	
crab, alaska king															
raw	71	67.6	0.5		35	20		.04			711	39		5.05	.030
3 oz	85	15.6	0.0			6	.04	0.9			173	186	.50	.784	
ckd by moist heat	82	65.9	1.3	0.5	45	25		.05			911	50		6.48	
3 oz	85	16.5	0.0	0.1		7	.05	1.1			222	238	.64	1.01	
imitation, made from surimi	87	62.3	1.1		17			.02			715	11			
3 oz	85	10.2	8.7				.03	0.2			77		.33		
crab, blue															
raw	74	67.2	0.9	0.3	66						249	76	29	3.00	1.28
3 oz	85	15.4	0.0	0.2							280	195	.63	.569	
cnd	84	64.7	1.0	0.4	76			.07			283	86	33	3.41	
3 oz	85	17.4	0.0	0.2				1.2	.39		318	221	.71	.646	
ckd by moist heat	87	65.8	1.5			0.6	85				237	88	28	3.58	
3 oz	85	17.2	0.0			0.2			6.21		275	175	.77	.548	

[a] Prepared w/ bread crumbs, egg, milk & salt.

	KCAL / WT (g)	H₂O (g) / PRO (g)	FAT (g) / CHO (g)	PUFA (g) / SFA (g)	CHOL (mg) / DFIB (g)	A (RE) / A (IU)	C (mg) / B-1 (mg)	B-2 (mg) / NIA (mg)	B-6 (mg) / B-12 (mcg)	FOL (mcg) / PANT (mg)	Na (mg) / K (mg)	Ca (mg) / P (mg)	Mg (mg) / Fe (mg)	Zn (mg) / Cu (mg)	Mn (mg)
crab cakes[a]	93	42.6	4.5	1.4	90						198	63	20	2.46	
1 cake	60	12.1	0.3	0.9				3.56			195	128	.65	.366	
crab, dungeness, raw	73	67.3	0.8	0.3	50			.14			251	39	39	3.63	.068
3 oz	85	14.8	0.6	0.1			.04	2.7			301	154	.31	.573	
crab, queen, raw	76	68.5	1.0	0.4	47						458	22			
3 oz	85	15.7	0.0	0.1							147	113			
crayfish															
raw	76	68.7	0.9	0.3	118		3	.05			45	20	21	1.11	
3 oz	85	15.9	0.0	0.2				2.0	2.30	.35	233	219	2.08	.371	
ckd by moist heat	97	64.1	1.2	0.3	151		3	.07			58	26	27	1.42	
3 oz	85	20.3	0.0	0.2				2.5	2.94		298	280	2.67	.475	
croaker, atlantic															
raw	89	66.3	2.7	0.4	52						47	13	34	.35	.021
3 oz	85	15.1	0.0	0.9							293	178	.31	.036	
breaded & fried[b]	188	50.8	10.8	2.5	71						296	27	35	.44	
3 oz	85	15.5	6.4	3.0							289	184	.73	.055	
cusk, raw	74	64.9	0.6		35			.11	.33		27	9	27	.32	.013
3 oz	85	16.1	0.0					2.3	.89	.24	333	173	.71	.015	
cuttlefish, raw	67	68.5	0.6	0.1	95		5	.77			316	77		1.47	
3 oz	85	13.8	0.7	0.1			.01	1.0	2.55		301	329	5.11	.499	
dogfish, spiny, raw	156	72.3	9.0												
3.5 oz	100	17.6	0.0		0.0		.05								
dolphinfish, raw	73	65.9	0.6	0.1	62						74			.39	.013
3 oz	85	15.7	0.0	0.2							354		.96	.035	
drum, freshwater, raw	101	65.7	4.2	1.0	54						64	51	26	.56	.595
3 oz	85	14.9	0.0	1.0							234	153	.77	.197	
fish fillets, frzn															
batter-dipped, Van de Kamp's	180		10.0								230				
3 oz piece	85	10.0	15.0												
country seasoned, Van de Kamp's	200		10.0								335				
2 oz piece	57	10.0	10.0												
light & crispy, Van de Kamp's	180		15.0								175				
2 oz piece	57	5.0	10.0												
fish kabobs, frzn, batter-dipped, Van de Kamp's—4 oz	240		15.0								430				
	113	10.0	15.0												
fish nuggets, frzn, light & crispy, Van de Kamp's—2 oz	130		10.0								160				
	57	10.0	10.0												
fish pieces, frzn, reheated[c]	155	26.4	7.0	1.8	64	60		.10	.03	10	332	11	14	.38	.135
1 piece (4"x2"x1/2")	57	8.9	13.5	1.8		18	.07	1.2	1.02		149	103	.42	.058	
fish sticks															
frzn[c]	76	13.0	3.4	0.9	31	30		.05	.02	5	163	6	7	.19	.066
1 stick (4"x2"x1/2")	28	4.4	6.7	0.9		9	.04	0.6	.50		73	51	.21	.028	
frzn, batter-dipped, Van de Kamp's	220		15.0								330				
4 pieces (4 oz)	113	10.0	15.0												
frzn, light & crispy, Van de Kamp's	270		20.0								300				
4 pieces (3.75 oz)	106	10.0	15.0												
flatfish															
raw	78	67.2	1.0	0.3	41	28		.07	.18		69	15	27	.39	.014
3 oz	85	16.0	0.0	0.2		8	.08	2.5	1.29	.43	307	156	.30	.027	
ckd by dry heat	99	62.2	1.3	0.4	58	32		.10	.20		89	16	50	.53	
3 oz	85	20.5	0.0	0.3		10	.07	1.9	2.13		292	246	.28	.022	
flounder/sole															
raw	68		0.5					.05			56	61	30		
3.5 oz	100	14.9	0.0		0.0		.06	1.7			366	195	.80		
baked	202	58.1	8.2				2	.08			237	23	30		
3.5 oz	100	30.0	0.0		0.0		.07	2.5			587	344	1.40		

[a] Prepared w/ crab meat, egg, onion & margarine.
[b] Breading consists of bread crumbs, egg, milk & salt.
[c] Prepared from walleye pollock, bread crumbs, egg, milk & salt.

	KCAL	H₂O (g)	FAT (g)	PUFA (g)	CHOL (mg)	A (RE)	C (mg)	Vitamins B-2 (mg)	B-6 (mg)	FOL (mcg)	Minerals Na (mg)	Ca (mg)	Mg (mg)	Zn (mg)	Mn (mg)
	WT (g)	PRO (g)	CHO (g)	SFA (g)	DFIB (g)	A (IU)	B-1 (mg)	NIA (mg)	B-12 (mcg)	PANT (mg)	K (mg)	P (mg)	Fe (mg)	Cu (mg)	
frzn, breaded, Van de Kamp's	300		15.0								410				
5 oz piece	142	15.0	15.0												
gefiltefish w/ broth, sweet	35	33.7	0.7	0.1	12	37		.03	.08	1	220	10	4	.35	.031
1 piece	42	3.8	3.1	0.2		11	.03		.35		38	31	1.04	.082	
grouper															
raw	78	67.3	0.9	0.3	31			.00			45	23	26	.41	.012
3 oz	85	16.5	0.0	0.2			.06	0.3	51		410	138	.75	.017	
ckd by dry heat	100	62.4	1.1	0.3	40			.01			45	18	32	.43	.010
3 oz	85	21.1	0.0	0.3			.07	0.3	.59		403	121	.96	.038	
haddock															
raw	74	67.9	0.6	0.2	49	47		.03	.26		58	28	33	.32	.021
3 oz	85	16.1	0.0	0.1		14	.03	3.2	1.02	.11	264	160	.89	.022	
ckd by dry heat	95	63.1	0.8	0.3	63	54		.04	.29		74	36	43	.41	
3 oz	85	20.6	0.0	0.1		16	.03	3.9	1.18		339	205	1.14	.028	
frzn, batter-dipped, Van de Kamp's	240		10.0								430				
2 pieces (4 oz)	113	10.0	20.0												
frzn, breaded, Van de Kamp's	300		20.0								310				
5 oz piece	142	15.0	15.0												
frzn fillet, light & crispy, Van de Kamp's—*2 oz piece*	180		15.0								160				
	57	5.0	10.0												
smoked	99	60.8	0.8	0.3	65	62		.04	.34		649	41	46	.42	
3 oz	85	21.4	0.0	0.1		19	.04	4.3	1.36		353	214	1.19	.036	
halibut, atlantic & pacific															
raw	93	66.2	2.0	0.7	27	132		.06	.29		46	40	71	.35	.013
3 oz	85	17.7	0.0	0.3		40	.05	5.0	1.01	.28	382	189	.71	.023	
ckd by dry heat	119	60.9	2.5	0.8	35	152		.08	.34		59	51	91	.45	
3 oz	85	22.7	0.0	0.4		46	.06	6.1	1.16		490	242	.91	.030	
halibut, frzn															
batter-dipped, Van de Kamp's	260		15.0								440				
3 pieces (4 oz)	113	10.0	15.0												
breaded, Van de Kamp's	220		10.0								520				
4 oz piece	113	15.0	15.0												
halibut, greenland, raw	158	59.7	11.8	1.2	39	47		.07		1	68	3			
3 oz	85	12.2	0.0	2.1		14	.05	1.3	1.00	.21	228	139	.56		
herring, atlantic															
raw	134	61.2	7.7	1.8	51	80	1	.20	.26		76	49	27	.84	.030
3 oz	85	15.3	0.0	1.7		24	.08	2.7	11.62	.55	278	201	.94	.078	
ckd by dry heat	172	54.9	9.9	2.3	65	87	1	.25	.30		98	63	35	1.08	
3 oz	85	19.6	0.0	2.2		26	.10	3.5	11.17		356	258	1.20	.100	
kippered	87	23.9	5.0	1.2	33	51	0	.13	.17		367	33	18	.54	
1 piece (4 3/8"x1 3/4"x 1/4")	40	9.8	0.0	1.1		15	.05	1.8	7.48		179	130	.60	.054	
pickled	39	8.3	2.7	0.3	2	129		.02		0	131	12	1	.08	.006
1 piece (1 3/4"x7/8"x1/2")	15	2.1	1.5	0.4		39	.01		.64	.01	10	13	.18	.016	
herring, pacific, raw	166	60.8	11.8	2.1	65	90					63			.45	.038
3 oz	85	13.9	0.0	2.8		27					359		.95	.066	
inconnu, raw	146	72.0	6.8												
3.5 oz	100	19.9	0.0		0.0										
kingfish															
raw	105	77.3	3.0								83				
3.5 oz	100	18.3	0.0		0.0						250				
ckd	255		13.4				0	.13			101	80	56		
3.5 oz	100	22.3	11.7		0.0	93	.11	2.9			293	287	1.90		
ling, raw	74	67.7	0.5			85		.16	.26		115	29	53		
3 oz	85	16.1	0.0			26	.09	2.0	.48	.27	322	169	.55		
lingcod, raw	72	68.9	0.9	0.3	44			.10			50	12	22	.39	.017
3 oz	85	15.0	0.0	0.2			.03	1.6			371	171	.27	.023	

	KCAL / WT (g)	H₂O (g) / PRO (g)	FAT (g) / CHO (g)	PUFA (g) / SFA (g)	CHOL (mg) / DFIB (g)	A (RE) / A (IU)	C (mg) / B-1 (mg)	B-2 (mg) / NIA (mg)	B-6 (mg) / B-12 (mcg)	FOL (mcg) / PANT (mg)	Na (mg) / K (mg)	Ca (mg) / P (mg)	Mg (mg) / Fe (mg)	Zn (mg) / Cu (mg)	Mn (mg)
lobster, northern															
raw	77	65.2	0.8		81			.04			1			2.57	.047
3 oz	85	16.0	0.4					1.2	.79	1.39				1.414	
ckd by moist heat	83	64.6	0.5	0.1	61	74		.06	.07	9	323	52	30	2.48	.052
3 oz	85	17.4	1.1	0.1		22	.01	0.9	2.64	.24	299	157	.33	1.649	
lobster paste	13	4.3	0.7					.02				5			
1 t	7	1.5	0.1				.01					13	.10		
lobster salad[a]	110	80.3	6.4	0.9			18	.08			124	36			
3.5 oz	100	10.1	2.3				.09				264	95	.90		
mackerel, atlantic															
raw	174	54.0	11.8	4.0	60	140	0	.27	.34		76	10	64	.53	.013
3 oz	85	15.8	0.0	2.8		42	.15	7.7	7.40	.73	267	184	1.38	.063	
ckd by dry heat	223	45.3	15.1	3.7	64	153	0	.35	.39		71	13	83	.80	
3 oz	85	20.3	0.0	3.6		46	.14	5.8	16.15		341	236	1.33	.080	
mackerel, jack, cnd	296	131.4	12.0	0.2	150	825	2	.40	.40	10	720	458	70	1.94	.076
1 cup	190	44.1	0.0	3.4		247	.08	11.7	13.19	.58	369	572	3.88	.279	
mackerel, king, raw	89	64.5	1.7	0.4	45	618		.41	.38	6	134	26	27	.48	.004
3 oz	85	17.2	0.0	0.3		185	.09	7.3	13.26	.71	370	211	1.51	.220	
mackerel, pacific & jack, raw	133	59.6	6.7	2.1	40	37		.36	.28	2	73	19	24	.57	.013
3 oz	85	17.1	0.0	1.9		11	.09	7.1	3.74	.27	345	106	.99	.079	
mackerel, spanish															
raw	118	60.9	5.4	1.5	65			.15			50	10	28	.42	.012
3 oz	85	16.4	0.0	1.6			.11	2.0	2.04		379	175	.37	.047	
ckd by dry heat	134	58.2	5.4	1.5	62			.18			56	11	32	.53	.010
3 oz	85	20.1	0.0	1.5			.11	4.3	5.95		471	231	.63	.055	
menhaden, atlantic, cnd	172	67.9	10.2												
3.5 oz	100	18.7	0.0		0.0										
milkfish, raw	126	60.2	5.7		44			.05	.36			43		.70	
3 oz	85	17.5	0.0				.01	5.5	2.89			138	.27	.029	
monkfish, raw	64	70.8	1.3		21			.05			16	7	18	.35	.020
3 oz	85	12.3	0.0				.02						.27	.024	
mullet, striped															
raw	99	65.5	3.2	0.6	42	104	1		.36	7	55	34	24	.44	.014
3 oz	85	16.5	0.0	0.9		31				.65	304	188	.87	.043	
ckd by dry heat	127	59.9	4.1	0.8	54	120	1		.42	8	61	26	28	.75	.019
3 oz	85	21.1	0.0	1.2		36					389	207	1.20	.120	
mussels, blue															
raw	73	68.5	1.9	0.5	24						243	22	29	1.36	
3 oz	85	10.1	3.1	0.4							272	168	3.36	.080	
ckd by moist heat	147	52.0	3.8	1.0	48						313	28	32	2.27	
3 oz	85	20.2	6.3	0.7							228	242	5.71	1.270	
ocean perch, atlantic															
raw	80	66.9	1.4	0.4	36	34		.09			64	91	26	.41	.013
3 oz	85	15.8	0.0	0.2		10		1.7	.85	.31	232	184	.78	.022	
ckd by dry heat	103	61.8	1.8	0.5	46	39		.11			82	117	33	.52	
3 oz	85	20.3	0.0	0.3		12		2.1	.98		298	235	1.00	.028	
oysters, eastern															
raw	58	71.5	2.1	0.6	46			.14	.04	8	94	38	46	76.4	.378
6 med	84	5.9	3.3	0.5				1.1	16.07	.16	192	117	5.63	3.747	
breaded & fried[b]	167	55.0	10.7	2.8	69			.17	.05	12	355	53	49	74.06	
3 oz (about 6 med)	85	7.5	9.9	2.7				1.4	13.29		208	135	5.91	3.650	
cnd	58	72.4	2.1	0.6	4.6			.14	.08	8	95	38	46	77.31	
3 oz	85	6.0	3.3	0.5				1.1	16.26		195	118	5.70	3.792	
ckd by moist heat	117	59.7	4.2	1.3	93			.28	.08	15	190	76	92	154.62	
3 oz (12 med)	85	12.0	6.7	1.1				2.1	32.53		389	236	11.39	7.584	

[a] Prepared w/ onion, sweet pickle, celery, egg, mayonnaise & tomato. [b] Prepared w/ bread crumbs, egg, milk & salt.

	KCAL	H₂O (g)	FAT (g)	PUFA (g)	CHOL (mg)	A (RE)	C (mg)	B-2 (mg)	B-6 (mg)	FOL (mcg)	Na (mg)	Ca (mg)	Mg (mg)	Zn (mg)	Mn (mg)
	WT (g)	PRO (g)	CHO (g)	SFA (g)	DFIB (g)	A (IU)	B-1 (mg)	NIA (mg)	B-12 (mcg)	PANT (mg)	K (mg)	P (mg)	Fe (mg)	Cu (mg)	
oysters, pacific, raw	69	69.8	2.0	0.4				.20			90	7	19	14.13	.547
3 oz	85	8.0	4.2	0.4			.06	1.7			143	138	4.34	1.340	
perch															
raw	77	67.3	0.8	0.3	76						52	68	26	.95	.595
3 oz	85	16.5	0.0	0.2							228	170	.77	.128	
ckd by dry heat	99	62.3	1.0	0.4	98						67	87	33	1.21	
3 oz	85	21.0	0.0	0.2							293	218	.98	.163	
frzn															
batter-dipped, Van de Kamp's	270		15.0								510				
2 pieces (4 oz)	113	10.0	20.0												
fillet, light & crispy, Van de Kamp's	170		10.0								115				
2 oz piece	57	5.0	10.0												
pike, northern															
raw	75	67.1	0.6	0.2	33	60	3	.05	.10		33	48		.57	
3 oz	85	16.4	0.0	0.1		18	.05				220	187	.47	.043	
ckd by dry heat	96	62.0	0.8	0.2	43	69	3	.07	.12		42	62		.73	
3 oz	85	21.0	0.0	0.1		21	.06				282	239	.60	.055	
pike, pickerel, raw	84	79.7	0.5												
3.5 oz	100	18.7	0.0		0.0								.70		
pike, walleye, raw	79	67.4	1.0	0.4	73						43	94	26	.53	.680
3 oz	85	16.3	0.0	0.2							331	179	1.11	.151	
pollock, atlantic, raw	78	66.5	0.8	0.4	60	30		.16	.24		73	51	57	.40	.013
3 oz	85	16.5	0.0	0.1		9	.04	2.8	2.71	.30	302	188	.39	.043	
pollock, ckd, creamed[a]	128	74.7	5.9				0	.13			111				
3.5 oz	100	13.9	4.0		0.0		.03	0.7			238				
pollock, walleye															
raw	68	69.3	0.7	0.4	61	56		.05	.05	3	84	4		.37	.013
3 oz	85	14.6	0.0	0.1		17	.06	1.1	2.64	.12	277		.20	.037	
ckd by dry heat	96	63.0	1.0	0.4	82	65		.07	.06	3	98	5		.51	
3 oz	85	20.0	0.0	0.2		19	.06	1.4	3.57		329		.24	.047	
pompano, florida															
raw	140	60.5	8.1	1.0	43						55	19	23	.61	.011
3 oz	85	15.7	0.0	3.0							324	166	.51	.032	
ckd by dry heat	79	53.5	10.3	1.2	54						65	36	27	.59	.021
3 oz	85	20.1	0.0	3.8							541	290	.57	.066	
porgy/scup															
raw	112	76.2	3.4								63	54			
3.5 oz	100	19.0	0.0		0.0						287	250			
fried	279		15.5				0	.08			58	24			
3.3 oz	93	22.7	10.8			75	.07	3.7			266	258	1.50		
pout, ocean, raw	67	69.2	0.8	0.0	44						52	8	11	.88	.013
3 oz	85	14.1	0.0	0.3									.24	.027	
rockfish, pacific															
raw	80	67.4	1.3	0.4	29	162		.06			51	8	22	.35	.014
3 oz	85	15.9	0.0	0.3		48	.03	2.7			344	151	.35	.025	
ckd by dry heat	103	62.4	1.7	0.5	38	186		.07			65	10	29	.45	
3 oz	85	20.4	0.0	0.4		56	.04	3.3			442	194	.45	.031	
roe, mixed species, raw	39	19.0	1.8	0.7	105										
1 oz	28	6.3	0.4	0.4											
roughy, orange, raw	107	64.5	6.0	0.1	17						54				
3 oz	85	12.5	0.0	0.1										.15	
sablefish															
raw	166	60.4	13.0	1.7	42						48			.28	.013
3 oz	85	11.4	0.0	2.7							304		1.09	.019	

[a] Prepared w/ flour, butter & milk.

	KCAL / WT (g)	H₂O (g) / PRO (g)	FAT (g) / CHO (g)	PUFA (g) / SFA (g)	CHOL (mg) / DFIB (g)	A (RE) / A (IU)	C (mg) / B-1 (mg)	B-2 (mg) / NIA (mg)	B-6 (mg) / B-12 (mcg)	FOL (mcg) / PANT (mg)	Na (mg) / K (mg)	Ca (mg) / P (mg)	Mg (mg) / Fe (mg)	Zn (mg) / Cu (mg)	Mn (mg)
smoked	218	51.1	17.1	2.3	55						626			.36	
3 oz	85	15.0	0.0	3.6							401			.031	
salmon, atlantic, raw	121	58.2	5.4	2.2	47	34		.32	.70		37	10			
3 oz	85	16.9	0.0	0.8		10	.19	6.7	2.70	1.41	417	170	.68		
salmon, chinook															
raw	153	62.2	8.9	1.8	56		3	.10			40	19		.38	.013
3 oz	85	17.1	0.0	2.1			.03	6.7			335		.60	.035	
smoked	99	61.2	3.7	0.8	20	75		.09	.24	2	666	9	15	.26	.014
3 oz	85	15.5	0.0	0.8		22	.02	4.0	2.77	.74	149	139	.72	.196	
salmon, chum, raw															
raw	102	64.1	3.2	0.8	63	84		.15			42	9		.40	.013
3 oz	85	17.1	0.0	0.7		25	.07				365	241	.47	.047	
cnd w/ bone	120	60.2	4.7	1.3	33	51					414ᵃ	212ᵇ			
3 oz	85	18.2	0.0	1.3		15						301			
salmon, coho															
raw	124	61.7	5.1	1.5	33		1				39			.35	.013
3 oz	85	18.4	0.0	0.9							359		.60	.043	
ckd by moist heat	157	55.6	6.4	1.9	42		1				50			.44	
3 oz	85	23.3	0.0	1.2							454		.76	.055	
salmon, pink															
raw	99	64.9	2.9	1.2	44	100					57			.46	.013
3 oz	85	17.0	0.0	0.5		30					274		.65	.065	
cnd w/ bone	118	58.5	5.1	1.7		47	0	.16		13	471ᵃ	181ᵇ	29	.78	
3 oz	85	16.8	0.0	1.3		14	.02	5.6			277	279	.72	.087	
salmon, sockeye															
raw	143	59.7	7.3	1.6	53	163		.13	.16		40	5	20	.45	.012
3 oz	85	18.1	0.0	1.3		49	.17	4.9	.24	.52	332	183	.40	.044	
cnd w/ bone	130	58.4	6.2	1.9	37	149	0	.16		8	458ᵃ	203ᵇ	25	.86	
3 oz	85	17.4	0.0	1.4		45	.01	4.7			321	277	.90	.071	
ckd by dry heat	83	52.6	9.3	2.0	74	178		.15	.19		56	6	26	.43	
3 oz	85	23.2	0.0	1.6		53	.18	5.7	4.93		319	234	.47	.057	
sardines															
atlantic, cnd in soybean oil	50	14.3	2.8	1.2	34	54		.05	.04	3	121	92ᵇ	9	.31	.026
2 sardines (3″x1″x1/2″)	24	5.9	0.0	0.4		16	.02	1.3	2.15	.15	95	118	.70	.045	
pacific, cnd in tomato sce	68	26.0	4.6	1.6	23	139	0	.07	.05	9	157	91ᵇ	13	.53	.078
1 sardine (4 3/4″x1 1/8″x5/8″)	38	6.2	0.0	1.2		26	.02	1.6	3.42	.28	130	139	.87	.103	
scallops															
raw	75	66.8	0.6	0.2	28			.06			137	21	48	.81	.077
3 oz (6 large or 14 small)	85	14.3	2.0	0.1			.01	1.0	1.30	.12	274	186	.25	.045	
breaded & friedᶜ	67	18.1	3.4	0.9	19			.03			144	13	18	.33	
2 large	31	5.6	3.1	0.8			.01	0.5	.41		103	73	.25	.024	
imitation, made from surimi	84	62.8	0.4		18			.01			676	7			
3 oz	85	10.9	9.0				.01	0.3			88		.26		
scup, raw	89	64.1	2.3								36	34	19	.41	.030
3 oz	85	16.0	0.0								244		.45	.043	
sea bass															
raw	82	66.5	1.7	0.6	35	157					58	9	35	.34	.013
3 oz	85	15.7	0.0	0.4		47					218	165	.25	.016	
ckd by dry heat	105	61.3	2.2	0.8	45	181					74	11	45	.44	
3 oz	85	20.1	0.0	0.6		54					279	211	.32	.020	
seafood (crab, langostinos & gulf shrimp)	132		3.5		65		1	.11			342	132			
w/ crm sce, frzn, Am Hosp Co—6 oz	170	16.2	10.8			654	.08	1.8			288		1.20		
sea nuggets/strips, frzn, Fishery	171		4.6		35						498				
Productsᵈ—4 oz	113	15.0	16.4												

ᵃ Value is for product w/ salt added. If no salt is added, sodium is 64 mg.

ᵇ If bones are discarded, calcium is greatly reduced.

ᶜ Prepared w/ bread crumbs, egg, milk & salt.

ᵈ Values reflect averages for natural fillet, natural shaped & formed sea nuggets and prime sea strips.

	KCAL	H_2O (g)	FAT (g)	PUFA (g)	CHOL (mg)	A (RE)	C (mg)	B-2 (mg)	B-6 (mg)	FOL (mcg)	Na (mg)	Ca (mg)	Mg (mg)	Zn (mg)	Mn (mg)
	WT (g)	PRO (g)	CHO (g)	SFA (g)	DFIB (g)	A (IU)	B-1 (mg)	NIA (mg)	B-12 (mcg)	PANT (mg)	K (mg)	P (mg)	Fe (mg)	Cu (mg)	
seatrout, raw	88	66.4	3.1	0.6	71						49	15	27	.38	.013
3 oz	85	14.2	0.0	0.9							290	212	.23	.026	
shad, american, raw	167	58.0	11.7								44	40	26	.31	.036
3 oz	85	14.4	0.0								326	231	.82	.054	
sheepshead															
raw	92	66.3	2.1	0.4							61	18	27	.33	.011
3 oz	85	17.2	0.0	0.5							344	266	.39	.026	
ckd by dry heat	107	58.7	1.4	0.3							62	32	30	.54	.018
3 oz	85	22.1	0.0	0.3							435	297	.57	.104	
shrimp															
raw	90	64.5	1.5	0.6	130			.03	.09	3	126	44	31	.94	.043
3 oz (12 large)	85	17.3	0.8	0.3			.02	2.2	.99	.24	157	175	2.05	.224	
breaded & fried[a]	206	44.9	10.4	4.3	150			.12	.08	7	292	57	34	1.17	
3 oz (11 large)	85	18.2	9.8	1.8			.11	2.6	1.59		191	185	1.07	.233	
cnd	102	61.7	1.7	0.6	147			.03	.09	2	143	50	35	1.07	
3 oz	85	19.6	0.9	0.3			.02	2.3	.95		179	198	2.32	.255	
ckd by moist heat	84	65.7	0.9	0.4	166			.03	.11	3	190	33	29	1.33	.029
3 oz (15 1/2 large)	85	17.8	0.0	0.2			.03	2.2	1.27		154	116	2.62	.164	
french fried[b]	225	56.9	10.8		120			.08			186	72	51		
3.5 oz	100	20.3	10.0				.04	2.7			229	191	2.00		
frzn, breaded, raw	139	65.0	0.7					.03				38			
3.5 oz	100	12.3	19.9	0.1			.03	2.0				111	1.00		
imitation, made from surimi	86	63.7	1.3		31			.03			599	16			
3 oz	85	10.5	7.8				.02	0.1			76		.51		
shrimp paste	180	61.3	9.4												
3.5 oz	100	20.8	1.5											.26	
skate (rajah fish), raw	98	77.8	0.7												
3.5 oz	100	21.5	0.0		0.0		.02								
smelt, atlantic															
raw	98	79.0	0.0					.12							
4-5 med	100	18.6	2.1		0.0		.01	1.4				272	.40		
cnd	200	62.7	13.5									358			
4-5 med	100	18.4	0.0									370	1.70		
smelt, rainbow															
raw	83	67.0	2.1	0.8	60			.10			51	51	26	1.40	.595
3 oz	85	15.0	0.0	0.4				1.2	2.92	.54	247	196	.77	.118	
ckd by dry heat	106	61.9	2.6	1.0	76			.12			65	65	33	1.80	
3 oz	85	19.2	0.0	0.5				1.5	3.37		316	251	.98	.151	
snapper															
raw	85	65.3	1.1	0.4	31			.00			54	27	27	.30	.011
3 oz	85	17.4	0.0	0.2			.04	0.2			355	169	.15	.024	
ckd by dry heat	109	59.8	1.5	0.5	40			.00			48	34	31	.37	.014
3 oz	85	22.4	0.0	0.3			.05	0.3			444	171	.20	.039	
sole															
raw	68		0.5					.05			56	61	30		
3.5 oz	100	14.9	0.0		0.0		.06	1.7			366	195	.80		
fillet	80		0.8								162				
1 serving	113	17.7	0.6								281				
frzn fillet															
Am Hosp Co	82		0.8		80			.06			88	70			
4 oz	113	18.0	0.6				.07	2.0			282		.88		
batter-dipped, Van de Kamp's	250		15.0								575				
2 pieces (4 oz)	113	15.0	25.0												
breaded, Van de Kamp's	300		15.0								410				
5 oz piece	142	15.0	15.0												

[a] Prepared w/ bread crumbs, egg, milk & salt.

[b] Dipped in egg & flour/bread crumbs or in batter.

	KCAL / WT (g)	H₂O (g) / PRO (g)	FAT (g) / CHO (g)	PUFA (g) / SFA (g)	CHOL (mg) / DFIB (g)	A (RE) / A (IU)	C (mg) / B-1 (mg)	B-2 (mg) / NIA (mg)	B-6 (mg) / B-12 (mcg)	FOL (mcg) / PANT (mg)	Na (mg) / K (mg)	Ca (mg) / P (mg)	Mg (mg) / Fe (mg)	Zn (mg) / Cu (mg)	Mn (mg)
w/ crabmeat stuffing & herb sce, Am Hosp Co—4.44 oz	100		6.7		48		0	.04			194	18			
	126	7.6	2.4			278	.03	0.7			155		.38		
w/ herb sce, Am Hosp Co	110		6.7		57		0	.05			150	19			
3.5 oz	100	10.3	2.3			245	.03	1.0			214		.50		
spiny lobster, raw	95	63.0	1.3	0.5	60			.04			150	41		4.82	.013
3 oz	85	17.5	2.1	0.2			.01	3.6				202	1.04	.324	
spot, raw	105	64.6	4.2	0.9							24	12	36	.43	.030
3 oz	85	15.7	0.0	1.2							422	158	.27	.039	
sturgeon															
raw	90	65.1	3.4	0.6		595								.36	.021
3 oz	85	13.7	0.0	0.8		179					241			.035	
ckd by dry heat	115	59.5	4.4	0.8		687								.46	
3 oz	85	17.6	0.0	1.0		206					309			.045	
smoked	147	53.1	3.7	0.4											
3 oz	85	26.5	0.0	0.9											
sucker, white, raw	79	67.8	2.0	0.7	35						34	60	26	.64	.500
3 oz	85	14.2	0.0	0.4							323	179	1.11	.166	
sunfish, pumpkinseed, raw	76	67.6	0.6	0.2	57						68	68	26	1.32	.595
3 oz	85	16.5	0.0	0.1							298	153	1.02	.255	
surimi[a]	84	64.9	0.8		25			.02			122	7			
3 oz	85	12.9	5.8				.02	0.2			95		.22		
swordfish															
raw	103	64.3	3.4	0.8	33	101	1	.08	.28		76	4	23	.97	.016
3 oz	85	16.8	0.0	0.9		30	.03	8.2	1.49	.35	245	224	.69	.107	
ckd by dry heat	132	58.4	4.4	1.0	43	117	1	.10	.32		98	5	29	1.25	
3 oz	85	21.6	0.0	1.2		35	.04	10.0	1.72		314	287	.88	.138	
tautog (blackfish), raw	89	79.3	1.1												
3.5 oz	100	18.6	0.0		0.0							227			
tilefish															
raw	81	67.1	2.0	0.5							45	22	24	.31	.009
3 oz	85	14.9	0.0	0.4							368	159	.21	.035	
ckd by dry heat	125	59.7	4.0	1.1							50	22	28	.45	.013
3 oz	85	20.8	0.0	0.7							435	201	.26	.044	
tomcod, atlantic, raw	77	81.5	0.4					.17							
3.5 oz	100	17.2	0.0		0.0										
trout, dolly varden	144	73.1	6.5					.06							
3.5 oz	100	19.9	0.0		0.0		.06								
trout, mixed species, raw	126	60.7	5.6	2.0	49	49	0	.28		11	44	36	19	.56	.723
3 oz	85	17.7	0.0	1.0		15	.30		6.62	1.65	307	208	1.27	.160	
trout, rainbow															
raw	100	60.8	2.9	1.0	48	55	3	.16			23	57	26	.92	.595
3 oz	85	17.5	0.0	0.6		17	.06				421	213	1.62	.094	
ckd by dry heat	129	53.9	3.7	1.3	62	63	3	.19			29	73	33	1.18	
3 oz	85	22.4	0.0	0.7		19	.07				539	272	2.07	.120	
tuna															
cnd in oil															
light	169	50.9	7.0	2.5	15	66		.09		5	301[b]	11	26	.77	.013
3 oz	85	24.8	0.0	1.3		20	.03				176	265	1.18	.060	
light, chunk, Star-Kist	170		13.0	7.0	20						310		54		
2 oz	58	13.0	0.0	1.0	0.0						79	83			
light, chunk, 60% less salt, Star-Kist	170		13.0	7.0	20						135		54		
2 oz	58	14.0	0.0	1.0	0.0						79	83			
light, solid, Star-Kist	150		13.0	7.0	20						310		54		
2 oz	58	13.0	0.0	1.0	0.0						79	83			
white	158	54.4	6.9		26			.07		4	336[b]	4	29	.40	
3 oz	85	22.6	0.0				.01	9.9			283	227	.56	.111	

[a] Prepared from walleye pollock. [b] Value reflects added salt. If no salt is added, sodium is 43 mg.

	KCAL	H₂O (g)	FAT (g)	PUFA (g)	CHOL (mg)	A (RE)	C (mg)	B-2 (mg)	B-6 (mg)	FOL (mcg)	Na (mg)	Ca (mg)	Mg (mg)	Zn (mg)	Mn (mg)
	WT (g)	PRO (g)	CHO (g)	SFA (g)	DFIB (g)	A (IU)	B-1 (mg)	NIA (mg)	B-12 (mcg)	PANT (mg)	K (mg)	P (mg)	Fe (mg)	Cu (mg)	
white, chunk, Star-Kist	140		10.0	7.0	20						310		54		
2 oz	58	14.0	0.0	1.0	0.0						79	83			
white, solid, Star-Kist	140		10.0	7.0	20						310		54		
2 oz	58	14.0	0.0	1.0	0.0						79	83			
cnd in spring water															
light	111	60.0	0.4	0.1					.32	4	303ᵃ	10	25	.37	
3 oz	85	25.1	0.0	0.1							267	158	2.72	.009	
light, chunk, Star-Kist	60		1.0	0.8	20						310		54		
2 oz	58	13.0	0.0	0.2	0.0						85	83			
light, chunk, 60% less salt, Star-Kist	65		1.0	0.8	20						135		54		
2 oz	58	14.0	0.0	0.2	0.0						85	83			
light, solid, Star-Kist	60		1.0	0.8	20						310		54		
2 oz	58	14.0	0.0	0.2	0.0						85	83			
white	116	59.1	2.1	0.8	35			.04		4	333ᵃ				
3 oz	85	22.7	0.0	0.6			.00	4.9			241		.51	.163	
white, chunk, dietetic, Star-Kist	70		1.0	0.8	20						30		54		
2 oz	58	15.0	0.0	0.2	0.0						85	83			
white, chunk, 60% less salt, Star-Kist	70		1.0	0.8	20						120		54		
2 oz	58	15.0	0.0	0.2	0.0						85	83			
white, solid, import albacore, Star-Kist—2 oz	70		1.0	0.8	20						310		54		
2 oz	58	15.0	0.0	0.2	0.0						85	83			
white, solid, local albacore, Star-Kist	100		5.0		20						310		54		
2 oz	58	14.0	0.0		0.0						85	83			
saladᵇ	383	129.5	19.0	8.5	27	199	5		.17	15	824	35	40	1.15	
1/2 cup	205	32.9	19.3	3.2		55	.06				365	365	2.04	.297	
tuna, bluefin															
raw	122	57.9	4.2	1.4	32	1856		.21	.39		33			.51	.013
3 oz	85	19.8	0.0	1.1		557	.21	7.4	8.01	.90	214		.87	.073	
ckd in dry heat	157	50.2	5.3	1.6	42	2142		.26	.45		43			.65	
3 oz	85	25.4	0.0	1.4		642	.24	9.0	9.25		275		1.11	.094	
tuna, skipjack, raw	88	60.0	0.9	0.3	40	44			.72		31	24	29	.70	.013
3 oz	85	18.7	0.0	0.3		13	.03				346	188	1.06	.073	
tuna, yellowfin, raw	92	60.3	0.8	0.2	38	30		.04			31	14		.45	.013
3 oz	85	19.9	0.0	0.2		15	.37	8.3				163	.62	.054	
turbot, european, raw	81	65.4	2.5			30		.07			127	15	44	.19	
3 oz	85	13.6	0.0			9	.06	1.9	1.87	.49	202	1.9		.031	
weakfish (sea trout)															
raw	121	76.5	5.6					.06			75				
3.5 oz	100	16.5	0.0		0.0		.09	2.7			317				
broiled	208	61.4	11.4					.08			560				
3.5 oz	100	24.6	0.0		0.0		.10	3.5			465				
whelk															
raw	117	56.1	0.3	0.0	55	72		.09	.29	5	175	48	73	1.39	.380
3 oz	85	20.3	6.6	0.0		22	.02	0.9	7.71	.18	295	120	4.28	.876	
ckd by moist heat	233	27.2	0.7	0.0	110	137		.18	.55	10	350	96	147	2.77	
3 oz	85	40.5	13.2	0.1		41	.04	1.7	15.42		590	240	8.55	1.751	
whitefish															
raw	114	61.9	5.0	1.8	51						43		28	.84	
3 oz	85	16.2	0.0	0.8							269		.31	.061	
baked, stuffedᶜ	215	63.2	14.0				0	.11			195				
3.5 oz	100	15.2	5.8				.11	2.3			291	246	.50		
smoked	92	60.2	0.8	0.2	28	162		.09	.33	6	866	15	19	.41	.029
3 oz	85	19.9	0.0	0.2		48	.03	2.0	2.77	.09	360	112	.43	.268	
white perch															
raw	118	75.7	4.0												
3.5 oz	100	19.3	0.0		0.0							192			

ᵃ Value reflects added salt. If no salt is added, sodium is 43 mg.
ᵇ Prepared w/ light tuna cnd in oil, pickle relish, salad dressing, onion & celery.
ᶜ Stuffed w/ bacon, onion, celery & bread crumbs.

	KCAL	H₂O (g)	FAT (g)	PUFA (g)	CHOL (mg)	Vitamins A (RE)	C (mg)	B-2 (mg)	B-6 (mg)	FOL (mcg)	Minerals Na (mg)	Ca (mg)	Mg (mg)	Zn (mg)	Mn (mg)
	WT (g)	PRO (g)	CHO (g)	SFA (g)	DFIB (g)	A (IU)	B-1 (mg)	NIA (mg)	B-12 (mcg)	PANT (mg)	K (mg)	P (mg)	Fe (mg)	Cu (mg)	
fried fillet	108		5.3				0	.05				9			
2.3 oz	65	12.5	0.0		0.0	0	.04	2.7				113	.70		
whiting															
raw	77	68.2	1.1	0.4	57	84		.04	.13	11	61	41	18	.75	.088
3 oz	85	15.6	0.0	0.2		25	.05	1.1	1.96	.18	212	189	.29	.026	
ckd by dry heat	98	63.5	1.4	0.5	71	97		.05	.15	13	113	53	23	.45	
3 oz	85	20.0	0.0	0.3		29	.06	1.4	2.21		369	242	.36	.034	
wolffish, atlantic, raw	82	67.9	2.0	0.7	39	319		.07			72			.66	.013
3 oz	85	14.9	0.0	0.3		96	.15	1.8	1.73	.49			.08	.025	
yellowtail, raw	124	63.3	4.5			81	2	.03	.14	3	33			.44	.013
3 oz	85	19.7	0.0			24	.12	5.8	1.11	.50		133	.41	.038	

14. FRUIT & VEGETABLE JUICES

	KCAL	H₂O (g)	FAT (g)	PUFA (g)	CHOL (mg)	A (RE)	C (mg)	B-2 (mg)	B-6 (mg)	FOL (mcg)	Na (mg)	Ca (mg)	Mg (mg)	Zn (mg)	Mn (mg)
acerola jce, fresh	51	228.2	0.7			123	3872	.15	.01		7	24	29		
8 fl oz	242	1.0	11.6			1232	.05	1.0	.00	.50	235	22	1.21		
apple jce, cnd/bottled	116	218.1	0.3	0.1	0	0	2	.04	.07	0	7	16	8	.07	.280
8 fl oz	248	0.2	29.0	0.0		2	.05	0.2	.00		296	18	.92	.055	
apple jce, from frzn conc	111	210.1	0.3	0.1	0		1	.04	.08	1	17	14	12	.09	.151
8 fl oz	239	0.3	27.6	0.0			.01	0.1	.00	.15	301	16	.61	.033	
apricot nectar, cnd	141	213.0	0.2	0.0	0	330	1	.04			9	17	13	.23	
8 fl oz	251	0.9	36.1	0.0	0.8	3304	.02	0.7	.00	3	286	23	.96	.183	
beef broth & tomato jce	61	151.0	0.2	0.0		21	2	.05			220	19			
5.5 fl oz can	168	1.0	14.3	0.1		215	.00	0.3			162		.98		
carrot jce, cnd	73	163.5	0.3	0.1	0	4738	16	.10	.40	7	54	44	26	.33	.239
6 fl oz	184	1.7	17.1	0.1		47381	.17	0.7	.00	.42	538	77	.85	.085	
clam & tomato jce, cnd	77	145.3	0.1	0.0		36	7	.05			664	21			
5.5 fl oz can	166	1.1	18.1	0.0		357	.07	0.3			149		1.00		
cranberry jce cocktail, bottled	147	215.1	0.1		0		108	.04		1	10	8	8	.05	.397
8 fl oz	253	0.1	37.7				.01	0.1	.00	.17	61	3	.40	.033	
grape jce, cnd/bottled	155	212.8	0.2	0.1	0	2	0	.09	.16	7	7	22	24	.13	.911
8 fl oz	253	1.4	37.9	0.1		20	.07	0.7	.00	.10	334	27	.60	.071	
grape jce, from frzn conc, sweetened	128	217.3	0.2	0.1	0	2	60	.07	.11	3	5	9	11	.10	.443
8 fl oz	250	0.5	31.9	0.1		19	.04	0.3	.00	.06	53	11	.26	.033	
grapefruit jce, fresh	96	222.3	0.3	0.1	0		94	.05			2	22	30	.13	.049
8 fl oz	247	1.2	22.7	0.0		ª	.10	0.5	.00		400	37	.49	.082	
grapefruit jce, cnd	93	222.6	0.2	0.1	0	2	72	.05	.05	26	3	18	24	.21	.049
8 fl oz	247	1.3	22.1	0.0		18	.10	0.6	.00	.32	378	27	.50	.094	
grapefruit jce, cnd, sweetened	116	218.4	0.2	0.1	0	0	67	.06	.05	26	4	20	24	.15	.050
8 fl oz	250	1.5	27.8	0.0		0	.10	0.8	.00	.33	405	27	.89	.120	
grapefruit jce, from frzn conc	102	220.6	0.3	0.1	0	2	83	.05	.11	9	2	19	26	.13	.049
8 fl oz	247	1.4	24.0	0.0		22	.10	0.5	.00	.47	337	34	.34	.082	
lemon jce, fresh	4	13.8	0.0		0	0	7	.00	.01	2	0	1	1	.01	.001
1 T	15	0.1	1.3			3	.01	0.0	.00	.02	19	1	.00	.004	
lemon jce, fresh	60	221.4	0.0		0	5	112	.02	.12	32	2	18	16	.12	.020
8 fl oz	244	0.9	21.1			49	.07	0.2	.00	.25	303	14	.08	.071	
lemon jce, cnd/bottled	3	14.1	0.0	0.0	0	0	4	.00	.01	2	3	2	1	.01	.003
1 T	15	0.1	1.0	0.0		2	.01	0.0	.00	.01	15	1	.02	.006	
lemon jce, frzn, single-strength	3	14.0	0.1	0.0	0	0	5	.00	.01	1	0	1	1	.01	.005
1 T	15	0.1	1.0	0.0		2	.01	0.0	.00	.02	14	1	.02	.005	
lime jce, fresh	4	13.9	0.0	0.0	0	0	5	.00	.01		0	1	1	.01	.001
1 T	15	0.1	1.4	0.0		2	.00	0.0	.00	.02	17	1	.00	.005	
lime jce, fresh	66	221.9	0.3	0.1	0	2	72	.03	.11		2	22	14	.15	.020
8 fl oz	246	1.1	22.2	0.0		25	.05	0.2	.00	.34	268	18	.08	.074	
lime jce, cnd/bottled	3	14.3	0.0	0.0	0	0	1	.00	.00	1	2	2	1	.01	.001
1 T	15	0.0	1.0	.0		3	.01	0.0	.00	.01	12	1	.03	.005	
new breakfast jce, cnd, Dole[b]	90		0.0				60				8				
6 fl oz		1.0	22.1								59				
orange jce, fresh	111	219.0	0.5	0.1	0	50	124	.07	.01		2	27	27	.13	.035
8 fl oz	248	1.7	25.8	0.1		496	.22	0.1	.00	.47	496	42	.50	.109	

ª 25 IU for white grapefruit jce; 1087 IU for pink or red grapefruit jce.　　　　ᵇ Avg of 3 flavors - pineapple, pineapple-orange & pineapple-grapefruit.

	KCAL / WT (g)	H₂O (g) / PRO (g)	FAT (g) / CHO (g)	PUFA (g) / SFA (g)	CHOL (mg) / DFIB (g)	A (RE) / A (IU)	C (mg) / B-1 (mg)	B-2 (mg) / NIA (mg)	B-6 (mg) / B-12 (mcg)	FOL (mcg) / PANT (mg)	Na (mg) / K (mg)	Ca (mg) / P (mg)	Mg (mg) / Fe (mg)	Zn (mg) / Cu (mg)	Mn (mg)
orange jce, cnd	104	221.6	0.4	0.1	0	44	86	.07	.22		6	21	27	.17	.035
8 fl oz	249	1.5	24.5	0.0		437	.15	0.8	.00	.37	436	36	1.10	.142	
orange jce, from frzn conc	112	219.4	0.1	0.0	0	19	97	.05	.11	109	2	22	24	.13	.035
8 fl oz	249	1.7	26.8	0.0		194	.20	0.5	.00	.39	474	40	.24	.110	
orange jce, unsweetened, Kraft	83	158.8	0.2	0.0	0		69	.04	.05	5	2	12	18		
6 fl oz	180	1.2	19.1	0.0		47	.13	0.4	.00	.28	335	30	.20		
orange grapefruit jce, cnd	107	218.9	0.2	0.0	0	29	72	.07	.06		8	21	24	.18	.042
8 fl oz	247	1.5	25.4	0.0		293	.14	0.8	.00	.35	390	34	1.15	.188	
papaya nectar, cnd	142	212.5	0.4	0.1	0	28	8	.01	.02	5	14	24	8	.38	.033
8 fl oz	250	0.4	36.3	0.1		277	.02	0.4	.00	.14	78	1	.86	.033	
passion fruit jce, purple, fresh	126	211.5	0.1		0	177	74	.32				9			
8 fl oz	247	1.0	33.6			1771		3.6	.00			31	.59		
passion fruit jce, yellow, fresh	149	208.0	0.4		0	595	45	.25			15	9	41		
8 fl oz	247	1.7	35.7			5953		5.5	.00		687	61	.89		
peach nectar, cnd	134	213.2	0.1	0.0	0	64	13	.04			17	13	11	.20	.047
8 fl oz	249	0.7	34.7	0.0	0.4	643	.01	0.7	.00		101	16	.47	.172	
pear nectar, cnd	149	210.0	0.0	0.0	0	0	3	.03			9	11	6	.16	.075
8 fl oz	250	0.3	39.4	0.0	1.6	1	.01	0.3	.00		33	7	.65	.168	
pineapple jce, cnd	139	213.8	0.2	0.1	0	1	27	.06	.24	58	2	42	34	.29	2.475
8 fl oz	250	0.8	34.4	0.0		12	.14	0.6	.00	.25	334	20	.65	.225	
pineapple jce, cnd, Dole	103		0.2					.04			2	28			
6 fl oz	188	0.8	25.4			90	.09	0.4			280	17	.60		
pineapple jce, from frzn conc	129	216.3	0.1	0.0	0	3	30	.05	.19		3	28	23	.29	2.475
8 fl oz	250	1.0	31.9	0.0		25	.18	0.5	.00	.31	340	20	.75	.225	
prune jce, cnd	181	208.0	0.1	0.0	0	1	11	.18		1	11	30	36	.52	.387
8 fl oz	256	1.6	44.7	0.0		9	.04	2.0	.00		706	64	3.03	.174	
tangerine jce, fresh	106	219.6	0.5	0.1	0	104	77	.05			2	44	20	.06	.091
8 fl oz	247	1.2	25.0	0.0		1037	.15	0.2	.00		440	35	.49	.062	
tangerine jce, cnd, sweetened	125	216.6	0.5	0.1	0	105	55	.05	.80		2	45	20	.06	.092
8 fl oz	249	1.3	29.9	0.0		1046	.15	0.2	.00		443	35	.50	.062	
tangerine jce, from frzn conc, sweetened	110	212.3	0.3	0.0	0	138	58	.05	.10	11	2	18	19	.06	.089
8 fl oz	241	1.0	27.7	0.0		1382	.13	0.2	.00	.30	273	20	.23	.060	
tomato jce	32	170.9	0.1	0.0	0	101	33	.06	.20	36	658[a]	16	20	.26	.140
6 fl oz	182	1.4	7.7	0.0		1012	.09	1.2	.00	.46	400	34	1.06	.184	
tomato jce cocktail, cnd	51	226.0	0.2				39	.05			486	24			
8 fl oz	243	1.7	12.2			1940	.12	1.5			537	44	2.20		
tomato jce, from concentrate, Campbell's	39		0.0				25	.05			589	20			
6 fl oz	182	1.6	8.0			1148	.06	1.1			376		2.20		
veg jce cocktail, cnd	34	170.2	0.2	0.1	0	213	50	.05	.26		664	20	20	.36	.182
6 fl oz	182	1.1	8.3	0.0		2130	.08	1.3	.00		351	31	.77	.364	
V-8 vegetable cocktail, Campbell's	37		0.1				37	.04			593	23			
6 fl oz	182	1.3	7.8			2522	.04	1.3			383		1.00		
no salt added	40		0.0				39	.05			41	31			
6 fl oz	182	1.6	8.2			3204	.04	1.2			439		1.20		
spicy hot	37		0.0				38	.04			634	24			
6 fl oz	182	1.5	7.8			2680	.03	1.2			408		1.20		

15. FRUITS

	KCAL / WT (g)	H₂O (g) / PRO (g)	FAT (g) / CHO (g)	PUFA (g) / SFA (g)	CHOL (mg) / DFIB (g)	A (RE) / A (IU)	C (mg) / B-1 (mg)	B-2 (mg) / NIA (mg)	B-6 (mg) / B-12 (mcg)	FOL (mcg) / PANT (mg)	Na (mg) / K (mg)	Ca (mg) / P (mg)	Mg (mg) / Fe (mg)	Zn (mg) / Cu (mg)	Mn (mg)
acerola, raw	31	89.6	0.3		0	75	1644	.06	.01		7	12	18		
1 cup	98	0.4	7.5			751	.02	0.4	.00	.30	143	11	.20		
apple															
raw, w/ skin	81	115.8	0.5	0.1	0	7	8	.02	.07	4	1	10	6	.05	.062
1 med	138	0.3	21.1	0.1	2.8	74	.02	0.1	.00	.08	159	10	.25	.057	
raw, w/o skin	72	108.1	0.4	0.1	0	6	5	.01	.06	1	0	5	4	.05	.029
1 med	128	0.2	19.0	0.1	2.9	56	.02	0.1	.00	.07	144	9	.09	.040	
boiled, w/o skin	91	146.2	0.6	0.2	0	7	0	.02	.08	1	1	8	5	.07	.202
1 cup	171	0.5	23.3	0.1	3.3	75	.03	0.2	.00	.08	150	13	.32	.060	

[a] If no salt added, Na is 18 mg.

	KCAL / WT (g)	H₂O (g) / PRO (g)	FAT (g) / CHO (g)	PUFA (g) / SFA (g)	CHOL (mg) / DFIB (g)	A (RE) / A (IU)	C (mg) / B-1 (mg)	B-2 (mg) / NIA (mg)	B-6 (mg) / B-12 (mcg)	FOL (mcg) / PANT (mg)	Na (mg) / K (mg)	Ca (mg) / P (mg)	Mg (mg) / Fe (mg)	Zn (mg) / Cu (mg)	Mn (mg)
cnd, sliced, sweetened	68	84.0	0.5	0.1	0	5	0	.01	.05	0	3	4	2	.03	.156
1/2 cup	102	0.2	17.0	0.1	1.9	52	.01	0.1	.00	.03	69	6	.23	.054	
dried, sulfured	155	20.3	0.2	0.1	0	0	3	.10	.08		56	9	10	.13	.058
10 rings	64	0.6	42.4	0.0		0	.00	0.6	.00		288	25	.90	.122	
micro ckd w/o skin	96	143.9	0.7	0.2	0	7	1	.02	.08	1	1	8	6	.06	.241
1 cup	170	0.5	24.5	0.1	4.0	68	.03	0.1	.00	.08	159	14	.28	.078	
applesce, cnd, sweetened	97	101.9	0.2	0.1	0	1	2	.04	.03	1	4	5	4	.05	.096
1/2 cup	128	0.2	25.5	0.0	1.4	14	.02	0.2	.00	.07	78	9	.45	.055	
applesce, cnd, unsweetened	53	107.8	0.1				2	.02			2	4	4	.03	.092
1/2 cup	122	0.2	13.8		1.4	29	.01	0.0			91	9	.15	.032	
apricots															
raw	51	91.5	0.4	0.1	0	277	11	.04	.06	9	1	15	8	.28	.084
3 med	106	1.5	11.8	0.0	1.4	2769	.03	0.6	.00	.25	313	21	.58	.094	
cnd, heavy syrup	75	69.9	0.1	0.0	0	112	3	.02	.05	2	9	8	7	.09	.046
4 halves	90	0.5	19.3	0.0		1116	.02	0.4	.00	.08	120	12	.38	.059	
cnd, jce pack	40	72.8	0.0	0.0	0	142	4	.02			3	10	8	.09	.044
3 halves	84	0.5	10.4	0.0	0.4	1421	.02	0.3	.00		139	17	.25	.045	
cnd, light syrup	54	70.2	0.0	0.0	0	112	2	.02	.05	1	3	10	7	.09	.044
3 halves	85	0.5	14.0	0.0		1124	.01	0.3	.00	.08	117	11	.33	.067	
cnd, water pack	20	84.1	0.0	0.0	0	163	2	.02	.05	2	10	8	8	.10	.048
4 halves	90	0.6	4.9	0.0		1629	.02	0.4	.00	.08	139	15	.49	.061	
dried, sulfured	83	10.9	0.2	0.0	0	253	1	.05	.06	4	3	16	16	.26	.096
10 halves	35	1.3	21.6	0.0		2534	.00	1.0	.00	.26	482	41	1.65	.150	
frzn, sweetened	119	88.7	0.1	0.0	0	203	11	.05	.07		5	12	11	.12	.061
1/2 cup	121	0.9	30.4	0.0		2033	.02	1.0	.00	.24	277	23	1.09	.077	
banana, raw	105	84.7	0.6	0.1	0	9	10	.11	.66	22	1	7	33	.19	.173
1 med	114	1.2	26.7	0.2	1.6	92	.05	0.6	.00	.30	451	22	.35	.119	
blackberries															
raw	37	61.7	0.3		0	12	15	.03	.04		0	23	14	.20	.930
1/2 cup	72	0.5	9.2		3.3	119	.02	0.3	.00	.17	141	15	.41	.101	
cnd, heavy syrup	118	96.1	0.2		0	28	4	.05	.05	34	3	27	22	.23	.892
1/2 cup	128	1.7	29.6			280	.04	0.4	0	.19	127	18	.83	.170	
frzn, unsweetened	97	124.1	0.7		0	17	5	.07	.09	51	2	44	33	.37	1.847
1 cup	151	1.8	23.7			172	.04	1.8	.00	.23	211	46	1.21	.181	
blueberries															
raw	82	122.7	0.6		0	15	19	.07	.05	9	9	9	7	.16	.409
1 cup	145	1.0	20.5		4.4	145	.07	0.5	.00	.14	129	15	.24	.088	
cnd, heavy syrup	112	98.3	0.4		0	8	1	.07	.05	2	4	7	4	.09	.260
1/2 cup	128	0.8	28.2			82	.04	0.1	0	.11	51	13	.42	.068	
frzn, sweetened	187	178.0	.3		0	10	2	.12	.14	16	3	13	6	.14	.603
1 cup	230	0.9	50.5			102	.05	0.6	.00	.29	137	16	.90	.090	
boysenberries															
cnd, heavy syrup	113	97.6	0.2		0	5	8	.04	.05	44	4	23	14	.24	.320
1/2 cup	128	1.3	28.6			21	.03	0.3	.00	.17	115	13	.55	.090	
frzn, unsweetened	66	113.4	0.4		0	9	4	.05	.07	84	2	36	21	.29	.722
1 cup	132	1.5	16.1			89	.07	1.0	.00	.33	183	36	1.12	.106	
breadfruit, raw	99	67.8	0.2		0	4	28	.03			2	17	24	.11	.058
1/4 small	96	1.0	26.0			38	.11	0.9	.00	.44	470	29	.52	.081	
cantaloupe, raw	57	143.6	0.4		0	516	68	.03	.18	27	14	17	17	.25	.075
1 cup pieces	160	1.4	13.4		0.5	5158	.06	0.9	.00	.21	494	27	.34	.067	
carambola, raw	42	115.5	0.4		0	62	27	.03			2	6	12	.14	.104
1 med	127	0.7	9.9		1.5	626	.04	0.5	.00		207	20	.33	.152	
carissa, raw	12	16.8	0.3		0	1	8	.01			1	2	3		
1 med	20	0.1	2.7			8	.01	0.0	.00		52	1	.26	.042	
casaba melon, raw	45	156.4	0.2		0	5	27	.03			20	9	14		
1 cup pieces	170	1.5	10.5			51	.10	0.7	.00		357	12	.68		
cherimoya, raw	515	402.1	2.2		0	5	49	.60				126			
1 med	547	7.1	131.3			55	.55	7.1	.00			219	2.74		

	KCAL / WT (g)	H₂O (g) / PRO (g)	FAT (g) / CHO (g)	PUFA (g) / SFA (g)	CHOL (mg) / DFIB (g)	A (RE) / A (IU)	C (mg) / B-1 (mg)	B-2 (mg) / NIA (mg)	B-6 (mg) / B-12 (mcg)	FOL (mcg) / PANT (mg)	Na (mg) / K (mg)	Ca (mg) / P (mg)	Mg (mg) / Fe (mg)	Zn (mg) / Cu (mg)	Mn (mg)
cherries															
sour, cnd, heavy syrup	116	96.8	0.1	0.0	0	91	3	.05	.06	10	9	13	7	.08	.092
1/2 cup	128	0.9	29.8	0.0		914	.02	0.2	.00	.13	119	12	1.70	.084	
sour, cnd, water pack	43	109.7	0.1	0.0	0	92	3	.05	.05	10	9	13	7	.08	.093
1/2 cup	122	0.9	10.9	0.0		920	.02	0.2	.00	.13	120	12	1.67	.085	
sweet, raw	49	54.9	0.7	0.2	0	15	5	.04	.02	3	0	10	8	.04	.063
10 cherries	68	0.8	11.3	0.1	1.1	146	.03	0.3	.00	.09	152	13	.26	.065	
sweet, cnd, heavy syrup	107	100.1	0.2	0.1	0	20	5	.05	.04		3	12	11	.13	.076
1/2 cup	129	0.8	27.4	0.0		199	.03	0.5	.00		187	23	.46	.183	
sweet, cnd, jce pack	68	106.2	0.0	0.0	0	16	3	.03			3	17	16	.12	.076
1/2 cup	125	1.1	17.3	0.0	0.3	156	.02	0.5	.00		163	27	.73	.091	
sweet, cnd, water pack	57	107.9	0.2	0.0	0	20	3	.05	.04		2	13	11	.09	.077
1/2 cup	124	1.0	14.6	0.0		198	.03	0.5	.00		162	19	.45	.093	
sweet, frzn, sweetened	232	195.6	0.3	0.1	0	49	3	.12			3	31	26	.10	.282
1 cup	259	3.0	57.9	0.1		489	.07	0.5	.00		514	41	.90	.062	
crabapples, raw	83	86.8	0.3	0.1	0	4	9	.02			1	20	7		.127
1 cup slices	110	0.4	21.9	0.1		44	.03	0.1	.00		213	17	.39	.074	
cranberries, raw	46	82.2	0.2		0	4	13	.02	.06	2	1	7	5	.12	.149
1 cup whole	95	0.4	12.1			44	.03	0.1	.00	.21	67	8	.19	.055	
cranberry sce, jelled, cnd	209	83.7	0.2		0	3	3	.03	.02		40	5	4	.07	.083
1/2 cup	138	0.3	53.7			28	.02	0.1	.00		35	8	.30	.028	
cranberry orange relish, cnd	246	73.4	0.1			10	25	.03		0	44	15	6		
1/2 cup	138	0.4	63.8			97	.04	0.1	.00		53	11	.28	.055	
currants, european black, raw	36	45.9	0.2	0.1	0	13	101	.03	.04		1	31	14	.15	.143
1/2 cup	56	0.8	8.6	0.0	3.0	129	.03	0.2	.00	.22	180	33	.86	.048	
currants, red & white, raw	31	47.0	0.1	0.0	0	7	23	.03	.04		1	18	7	.13	.104
1/2 cup	56	0.8	7.7	0.0		67	.02	0.1	.00	.04	154	24	.56	.060	
currants, zante, dried[a]	204	13.8	0.2	0.1	0	5	3	.10	.21	7	6	62	30	.47	.338
1/2 cup	72	2.9	53.3	0.0		52	.12	1.2	.00	.03	642	90	2.34	.337	
custard apple, raw	101	71.5	0.6		0		19	.10	.22		4	30	18		
3.5 oz	100	1.7	25.2				.08	0.5	.00	.14	382	21	.71		
dates, dried	228	18.7	0.4		0	4	0	.08	.16	10	2	27	29	.24	.247
10 dates	83	1.6	61.0		4.2	42	.08	1.8	.00	.65	541	33	.96	.239	
dates, dried, chopped, Dromedary	130		0.0								0				
1/4 cup		1.0	31.0								190				
elderberries, raw	105	115.7	0.7		0	87	52	.09	.33			55			
1 cup	145	1.0	26.7			870	.10	0.7	.00	.20	406	57	2.32		
figs, raw	37	39.6	0.2	0.1	0	7	1	.03	.06		1	18	8	.07	.064
1 med	50	0.4	9.6	0.0		71	.03	0.2	.00	.15	116	7	.18	.035	
figs, cnd, heavy syrup	75	64.9	0.1	0.0	0	3	1	.03			1	23	8	.09	.071
3 figs	85	0.3	19.5	0.0		31	.02	0.4	.00	.06	85	9	.24	.090	
figs, dried	477	53.2	2.2	1.0	0	25	2	.17	.42	14	20	269	111	.94	.726
10 figs	187	5.7	122.2	0.4		248	.13	1.3	.00	.81	1332	128	4.18	.585	
fruit bars, General Mills[b]	90		2.0								10				
1 bar	23	1.0	17.2								37				
fruit cocktail[c]															
cnd, heavy syrup	93	102.9	0.1	0.0	0	26	2	.02	.06		7	8	7	.11	
1/2 cup	128	0.5	24.2	0.0		262	.02	0.5	.00	.08	112	14	.36	.088	
cnd, jce pack	56	108.4	0.0	0.0	0	38	3	.02			4	10	9	.11	
1/2 cup	124	0.6	14.7	0.0		376	.02	0.5	.00		118	17	.26	.077	
cnd, water pack	40	110.7	0.1	0.0	0	30	3	.01	.06		5	6	8	.11	
1/2 cup	122	0.5	10.4	0.0		305	.02	0.4	.00	.08	115	14	.31	.087	

[a] Dried black Corinth grapes: not related to European black, red or white currants.

[b] Values are averages for cherry, grape, orange-pineapple, strawberry & tropical flavors.

[c] Peaches, pears, grapes, pineapples & cherries.

	KCAL	H₂O (g)	FAT (g)	PUFA (g)	CHOL (mg)	A (RE)	C (mg)	B-2 (mg)	B-6 (mg)	FOL (mcg)	Na (mg)	Ca (mg)	Mg (mg)	Zn (mg)	Mn (mg)
	WT (g)	PRO (g)	CHO (g)	SFA (g)	DFIB (g)	A (IU)	B-1 (mg)	NIA (mg)	B-12 (mcg)	PANT (mg)	K (mg)	P (mg)	Fe (mg)	Cu (mg)	
fruit roll-ups, General Mills[a]	50		1.0								7				
1 roll	15	0.0	12.0								43				
fruit salad, cnd															
heavy syrup[b]	94	102.7	0.1	0.0	0	65	3	.03	.04		7	8	7	.09	
1/2 cup	128	0.4	24.5	0.0		646	.02	0.4	.00		103	12	.36	.082	
jce pack[b]	62	106.8	0.0	0.0	0	74	4	.02			7	14	10	.18	
1/2 cup	124	0.6	16.2	0.0		744	.01	0.4	.00		144	18	.31	.062	
tropical, heavy syrup[c]	110	98.3	0.1		0	16	22	.06			3	17	17	.14	
1/2 cup	128	0.5	28.6			162	.07	0.7	.00		168	10	.66	.102	
fruit salad, Fresh Chef															
fruit cooler	67		0.1				2	.04			121	22			
2/5 cup	99	0.8	15.7			78	.02	0.3			66		.40		
tropical delight	235		10.8				7	.12			253	65			
4/5 cup	198	2.4	32.1			803	.08	0.3			147		1.30		
fruit wrinkles, General Mills[d]	100		2.0								105				
1 pouch	26	1.0	21.0								38				
gooseberries, raw	67	131.8	0.9	0.0	0	44	42	.05	.12		1	38	15	.18	.216
1 cup	150	1.3	15.3	0.1		435	.06	0.5	.00	.43	297	40	.47	.105	
gooseberries, cnd, light syrup	93	100.9	0.3	0.1	0	17	13	.07	.02	4	3	20	8	.14	.223
1/2 cup	126	0.8	23.6			174	.03	0.2	.00	.17	97	9	.42	.273	
grapefruit															
raw, pink & red	37	112.4	0.1	0.0	0	32	47	.03	.05	15	0	13	10	.09	.012
1/2 med	123	0.7	9.5	0.0		318	.04	0.2	.00	.35	158	11	.15	.054	
raw, white	39	106.8	0.1	0.0	0	1	39	.02	.05	12	0	14	11	.08	.015
1/2 med	118	0.8	9.9	0.0		12	.04	0.3	.00	.33	175	9	.07	.059	
cnd, jce pack	46	111.2	0.1	0.0	0	0	42	.02			9	19	13	.09	.009
1/2 cup	124	0.9	11.4	0.0	0.3	0	.04	0.2	.00		209	15	.26	.046	
cnd, light syrup	76	106.2	0.1	0.0	0	0	27	.03	.03	11	2	18	13	.11	.009
1/2 cup	127	0.7	19.6	0.0		0	.05	0.3	.00	.15	164	13	.51	.084	
grapes															
american (slip skin), raw	58	74.8	0.3	0.1	0	9	4	.05	.10	4	2	13	5	.04	.661
1 cup	92	0.6	15.8	0.1		92	.09	0.3	.00	.02	176	9	.27	.037	
european (adherent skin), raw	114	128.9	0.9	0.3	0	12	17	.09	.18	6	3	17	10	.09	.093
1 cup	160	1.1	28.4	0.3	2.6	117	.15	0.5	.00	.04	296	21	.41	.144	
thompson seedless, cnd, heavy syrup	94	101.8	0.1	0.0	0	8	1	.03			7	13	8	.06	.049
1/2 cup	128	0.6	25.2	0.0		81	.04	0.2	.00		132	22	1.20	.069	
groundcherries, raw[e]	74	119.6	1.0			101	15	.06		0		13			
1 cup	140	2.7	15.7			1008	.15	3.9	.00			56	1.40		
guava, raw	45	77.5	0.5	0.2	0	71	165	.05	.13		2	18	9	.21	.130
1 med	90	0.7	10.7	0.2		713	.05	1.1	.00	.14	256	23	.28	.093	
guava, strawberry, raw	169	196.8	1.5	0.6	0	22	90	.07			89	52	41		
1 cup	244	1.4	42.4	0.4		220	.07	1.5	.00		713	67	.53		
honeydew melon, raw	33		0.3				23	.03			12	14			
1/4 cup	100	0.8	7.7			40	.04	0.6			251	16	.40		
jackfruit, raw	94	72.3	0.3			30	7		.11	0	3	34	37	.42	.197
3.5 oz	100	1.5	24.0			297	.03	0.4	.00		303	36	.60	.187	
java plum, raw	82	112.2	0.3			0	19	.02	.05		18	25	21		
1 cup	135	1.0	21.0			5	.01	0.4	.00		106	23	.25		
jujube															
raw	79	77.9	0.2			4	69	.04	.08	0	3	21	10	.05	.084
3.5 oz	100	1.2	20.2			40	.02	0.9	.00		250	23	.48	.073	
dried	287	19.7	1.1				13	.36		0	9	79	37	.19	.305
3.5 oz	100	3.7	73.6				.21	0.5	.00		531	100	1.80	.265	

[a] Values are averages for apple, apricot, cherry, grape, orange, raspberry & strawberry flavors.
[b] Peaches, pears, apricots, pineapples & cherries.
[c] Pineapples, papayas, pineapple jce, bananas, guava puree, cherries & passion fruit jce.
[d] Values are averages for cherry, lemon, orange & strawberry flavors.
[e] Roundish yellow berries 3/4 inch across, sweet & slightly acid: enclosed within the laternlike, light brown calyx or husk; native to eastern & central North America.

	KCAL / WT (g)	H₂O (g) / PRO (g)	FAT (g) / CHO (g)	PUFA (g) / SFA (g)	CHOL (mg) / DFIB (g)	A (RE) / A (IU)	C (mg) / B-1 (mg)	B-2 (mg) / NIA (mg)	B-6 (mg) / B-12 (mcg)	FOL (mcg) / PANT (mg)	Na (mg) / K (mg)	Ca (mg) / P (mg)	Mg (mg) / Fe (mg)	Zn (mg) / Cu (mg)	Mn (mg)
kiwifruit, raw	46	63.1	0.3			13	75	.04		0	4	20	23		
1 med	76	0.8	11.3			133	.02	0.4	.00		252	31	.31		
kumquats, raw	12	15.5	0.0		0	6	7	.02			1	8	2	.02	.016
1 med	19	0.2	3.1			57	.02		.00		37	4	.07	.020	
lemon peel	a	4.9	0.0	0.0	0	0	8	.01	.01		0	8	1		
1 T	6	0.1	1.0	0.0		3	.00	.00	.00	.02	10	1	.05		
lemon, raw	17	51.6	0.2	0.0	0	2	31	.01	.05	6	1	15		.04	
1 med	58	0.6	5.4	0.0		17	.02	.06	.00	.11	80	9	.35	.021	
lime, raw	20	59.1	0.1	0.0	0	1	20	.01		6	1	22		.07	
1 med	67	0.5	7.1	0.0		7	.02	0.1	.00	.15	68	12	.40	.044	
loganberries, frzn	80	124.4	0.5		0	5	23	.05	.10	38	1	38	32	.50	1.833
1 cup	147	2.2	19.1			52	.07	1.2	.00	.36	213	38	.94	.172	
longans															
raw	60	82.8	0.1		0		84	.14			0	1	10	.05	.052
31 fruits	100	1.3	15.1				.03	0.3	.00		266	21	.13	.169	
dried	286	17.6	0.4		0	0	28				48	45	46	.22	.248
3.5 oz	100	4.9	74.0			0	.04		.00		658	196	5.40	.807	
loquats, raw	47	86.7	0.2	0.1	0	153	1	.02			1	16	13	.05	.148
10 med	100	0.4	12.1	0.0		1528	.02	0.2	.00		266	27	.28	.040	
lychees															
raw	66	81.8	0.4		0	0	72	.07			1	5	10	.07	.055
10 med	100	0.8	16.5			0	.01	0.6	.00		171	31	.31	.148	
dried	277	22.3	1.2		0	0	183	.57			3	33	42	.28	.234
3.5 oz	100	3.8	70.7			0	.01	3.1	.00		1110	181	1.70	.631	
mammy apple, raw	51	86.2	0.5		0	23	14	.04			15	11			
1/8 med	100	0.5	12.5			230	.02	0.4	.00	.10	47	11	.70		
mandarin oranges, cnd															
jce pack	46	111.0	0.0	0.0	0	106	43	.04			7	14	14	.63	
1/2 cup	124	0.8	11.9	0.0		1056	.10		.00		165	13	.33	.041	
light syrup	76	104.7	0.1	0.0	0	106	25	.06			8	9	10	.30	
1/2 cup	126	0.6	20.4	0.0		1058	.07	0.6	.00		99	12	.46	.055	
mango, raw	135	169.1	0.6	0.1	0	806	57	.12	.28		4	21	18	.07	.056
1 med	207	1.1	35.2	0.1	2.2	8060	.12	1.2	.00	.33	322	22	.26	.228	
melon balls (cantaloupe & honeydew),	55	156.1	0.4		0	307	11	.04	.18	45	53	17	24	.29	.069
frzn—1 cup	173	1.5	13.7			3069	.29	1.1	.00	.28	484	22	.50	.104	
mixed fruit															
cnd, heavy syrup[b]	92	103.1	0.1	0.0	0	25	88[c]	.05			5	1	6	.09	
1/2 cup	128	0.5	24.0	0.0		248	.02	0.8	.00		108	13	.46	.074	
dried	243	31.2	0.5	0.1	0	244	4	.16	.16		18	38	39	.50	.227
3.5 oz	100	2.5	64.1	0.0		2442	.04	1.9	.00		796	77	2.71	.385	
dried, diced, Del Monte	74	7.0	0.1				3		.10		26	15	11	.11	
1 oz	28		19.5			743	.01	0.4			292	5	1.00	.080	
frzn, in syrup, Birds Eye[d]	123	109.2	0.4	0.1	0		27	.05	.03	8	4	9	8	.09	
1/2 cup	142	0.8	31.0	0.1		348	.03	0.6	.00	.14	161	14	.42	.058	
frzn, sweetened[d]	245	184.3	0.5	0.2	0	81	188[c]	.09	.25	9	8	18	14	.12	.160
1 cup	250	3.5	60.6	0.1		806	.04	1.0	.00	.41	327	30	.70	.085	
mulberries, raw	61	122.8	0.6		0	4	51	.14			14	55	25		
1 cup	140	2.0	13.7			35	.04	0.9	.00		271	53	2.59		
nectarine, raw	67	117.3	0.6		0	100	7	.06	.03	5	0	6	11	.12	.060
1 med	136	1.3	16.0			1001	.02	1.3	.00	.22	288	22	.21	.099	
oheloberries, raw	39	129.2	0.3		0	116	8	.05			2	10	9		
1 cup	140	0.5	9.6			1162	.02	0.4	.00		54	14	.13		
orange															
navel, raw	65	121.5	0.1	0.0	0	26	80	.06	.10	47	1	56	15	.08	.038
1 med	140	1.4	16.3	0.0		256	.12	0.4	.00	.35	250	27	.17	.078	
valencia, raw	59	104.5	0.4	0.1	0	28	59	.05	.08	47	0	48	12	.07	.028
1 med	121	1.3	14.4	0.0		278	.11	0.3	.00	.30	217	21	.11	.045	

a Cannot be calculated; no digestibility value for peel.
b Peaches, pears & pineapple.
c Added vitamin C.
d Peaches, sweet cherries, red sour cherries, red raspberries, boysenberries & grapes.

	KCAL / WT (g)	H₂O (g) / PRO (g)	FAT (g) / CHO (g)	PUFA (g) / SFA (g)	CHOL (mg) / DFIB (g)	A (RE) / A (IU)	C (mg) / B-1 (mg)	B-2 (mg) / NIA (mg)	B-6 (mg) / B-12 (mcg)	FOL (mcg) / PANT (mg)	Na (mg) / K (mg)	Ca (mg) / P (mg)	Mg (mg) / Fe (mg)	Zn (mg) / Cu (mg)	Mn (mg)
orange peel	a	4.4	0.0	0.0	0	3	8	.01	.01		0	10	1		
1 T	6	0.1	1.5	0.0		25	.01	0.1	.00	.03	13	1	.05		
papaya, raw	117	270.0	0.4	0.1	0	612	188	.10	.06		8	72	31	.22	.033
1 med	304	1.9	29.8	0.1	2.8	6122	.08	1.0	.00	.66	780	16	.30	.049	
passion fruit (purple grandilla), purple,	18	13.1	0.1		0	13	5	.02			5	2	5		
raw—*1 med*	18	0.4	4.2			126		0.3	.00		63	12	.29		
peach															
raw	37	76.3	0.1	0.0	0	47	6	.04	.02	3	0	5	6	.12	.041
1 med	87	0.6	9.7	0.0	0.5	465	.02	0.9	.00	.15	171	11	.10	.059	
cnd, heavy syrup	190	203.0	0.3	0.1	0	85	7	.06	.05	8	16	8	13	.22	.115
1 cup	256	1.2	51.0	0.0		849	.03	1.6	.00	.13	235	29	.69	.131	
cnd, heavy syrup, spiced	66	69.7	0.1	0.0	0	28	5	.03			3	5	6	.07	
1 med	88	0.4	17.7	0.0		279	.01	0.5	.00		75	8	.25	.086	
cnd, jce pack	109	217.0	0.1	0.0	0	95	9	.04			11	15	18	.26	
1 cup	248	1.6	28.7	0.0	1.1	945	.02	1.4	.00		317	43	.66	.124	
cnd, light syrup	136	212.6	0.1	0.0	0	89	6	.06	.05	8	13	9	12	.22	.115
1 cup	251	1.1	36.5	0.0		888	.02	1.5	.00	.13	244	27	.90	.131	
cnd, water pack	58	227.2	0.1	0.0	0	130	7	.05	.05	8	8	6	12	.22	.117
1 cup	244	1.1	14.9	0.0		1298	.02	1.3	.00	.12	241	25	.77	.132	
dried, sulfured	311	41.3	1.0	0.5	0	281	6	.28	.09		9	37	54	.75	.397
10 halves	130	4.7	79.7	0.1		2812	.00	5.7	.00		1295	155	5.28	.473	
frzn, sweetened	235	186.8	0.3	0.2	0	71	235[b]	.09	.05	24	16	6	12	.13	.073
1 cup	250	1.6	59.9	0.0		709	.03	1.6	.00	.33	325	28	.93	.060	
pear															
raw	98	139.1	0.7	0.2	0	3	7	.07	.03	12	1	19	9	.20	.126
1 med	166	0.7	25.1	0.0	4.1	33	.03	0.2	.00	.12	208	18	.41	.188	
cnd, heavy pack	188	204.9	0.3	0.1	0	0	3	.06	.04	3	13	12	11	.21	.082
1 cup	255	0.5	48.9	0.0		0	.03	0.6	.00	.06	165	17	.56	.125	
cnd, jce pack	123	214.5	0.2	0.0	0	1	4	.03			10	21	17	.22	
1 cup	248	0.9	32.1	0.0	2.3	14	.03	0.5	.00		238	29	.71	.131	
cnd, light syrup	144	212.0	0.1	0.0	0	0	2	.04	.04	3	13	13	11	.21	.083
1 cup	251	0.5	38.1	0.0		0	.03	0.4	.00	.06	165	17	.70	.123	
cnd, water pack	71	224.0	0.1	0.0	0	0	3	.02	.03	3	5	9	9	.21	.083
1 cup	244	0.5	19.1	0.0		0	.02	0.1	.00	.05	130	17	.52	.124	
dried, sulfured	459	46.7	1.1	0.3	0	1	12	.25			10	59	58	.68	.572
10 halves	175	3.3	122.0	0.1		6	.01	2.4	.00		932	103	3.68	.649	
persimmon, raw	32	16.1	0.1		0		17				0	7			
1 med	25	0.2	8.4						.00		78	7	.63		
persimmon, japanese															
raw	118	134.9	0.3		0	364	13	.03		13	3	13	15	.18	.596
1 med	168	1.0	31.2			3640	.05	0.2	.00		270	28	.26	.190	
dried	93	7.8	0.2		0	19	0	.01			1	8	11	.14	.473
1 med	34	0.5	25.0			190		0.1	.00		273	27	.25	.150	
pineapple															
raw	77	135.1	0.7	0.2	0	4	24	.06	.14	16	1	11	21	.12	2.556
1 cup pieces	155	0.6	19.2	0.0	2.4	35	.14	0.7	.00	.25	175	11	.57	.171	
cnd, heavy syrup	199	201.4	0.3	0.1	0	4	19	.06	.19	12	3	35	40	.29	2.754
1 cup pieces	255	0.9	51.5	0.0		37	.23	0.7	.00	.26	264	17	.98	.258	
cnd, jce pack	150	208.8	0.2	0.1	0	9	24	.05			4	34	35	.24	
1 cup pieces	250	1.0	39.2	0.0	1.9	95	.24	0.7	.00		304	16	.70	.215	
pitanga, raw	57	157.1	0.7		0	260	46	.07			5	16	21		
1 cup	173	1.4	13.0			2595	.05	0.5	.00		178	19	.35		
plantain, ckd	179	103.6	0.3		0	140	17	.08	.37	40	8	3	49	.20	
1 cup slices	154	1.2	48.0			1400	.07	1.2	.00	.36	716	43	.89	.102	
plum															
raw	36	56.2	0.4	0.1	0	21	6	.06	.05	1	0	2	4	.06	.032
1 med	66	0.5	8.6	0.0		213	.03	0.3	.00	.12	113	7	.07	.028	

a Cannot be calculated; no digestibility value for peel.   b Contains added ascorbic acid.

	KCAL	H₂O (g)	FAT (g)	PUFA (g)	CHOL (mg)	A (RE)	C (mg)	B-2 (mg)	B-6 (mg)	FOL (mcg)	Na (mg)	Ca (mg)	Mg (mg)	Zn (mg)	Mn (mg)
	WT (g)	PRO (g)	CHO (g)	SFA (g)	DFIB (g)	A (IU)	B-1 (mg)	NIA (mg)	B-12 (mcg)	PANT (mg)	K (mg)	P (mg)	Fe (mg)	Cu (mg)	
heavy syrup	119	101.2	0.1	0.0	0	34	1	.05	.04	3	26	12	7	.10	.041
3 plums	133	0.5	30.9	0.0		344	.02	0.4	.00	.10	121	17	1.12	.049	
cnd, jce pack	55	79.8	0.0	0.0	0	96	3	.06			1	9	8	.10	
3 plums	95	0.5	14.4	0.0	0.4	958	.02	0.5	.00		147	15	.32	.051	
pomegranate, raw	104	124.7	0.5		0		9	.05	.16		5	5			
1 med	154	1.5	26.4				.05	0.5	.00	.92	399	12	.46		
pricklypear, raw	42	90.2	0.5		0	5	14	.06			6	58	88		
1 med	103	0.8	9.9			53	.01	0.5	.00		226	25	.31		
prunes															
cnd, heavy syrup	90	60.8	0.2	0.0	0	69	2	.11			2	15	13	.16	.084
5 prunes	86	0.8	23.9	0.0		686	.03	0.7	.00		194	22	.35	.101	
dried	201	27.2	0.4	0.1	0	167	3	.14	.22	3	3	43	38	.45	.185
10 prunes	84	2.2	52.7	0.0		1669	.07	1.6	.00	.39	626	66	2.08	.361	
dried, ckd	113	73.9	0.2	0.0	0	32	3	.11	.23	0	2	24	21	.25	.104
1/2 cup	106	1.2	29.8	0.0		324	.03	0.8	.00	.11	354	37	1.18	.205	
pummelo, raw	71	169.3	0.1		0	0	116	.05	.07		2	7	12	.15	.032
1 cup pieces	190	1.4	18.3			0	.07	0.4	.00		411	32	.22	.091	
quince, raw	53	77.1	0.1	0.0	0	4	14	.03	.04		4	10	7		
1 med	92	0.4	14.1	0.0		38	.02	0.2	.00	.08	181	16	.64	.120	
raisins															
golden seedless	302	14.9	0.5	0.1	0	4	3	.19	.32	3	12	53	35	.32	.308
2/3 cup	100	3.4	79.5	0.2		44	.01	1.1	.00	.14	746	115	1.79	.363	
seeded	296	16.6	0.5	0.2	0	0	5	.18	.19	3	28	28	30	.18	.267
2/3 cup	100	2.5	78.5	0.2		0	.11	1.1	.00		825	75	2.59	.302	
seedless	300	15.4	0.5	0.1	0	1	3	.09	.25	3	12	49	33	.27	.308
2/3 cup	100	3.2	79.1	0.2		8	.16	0.8	.00	.05	751	97	2.08	.309	
w/ van yogurt coating, Del Monte	134	0.8	5.5				1	.02			27	51	10	.15	
1 oz	28	1.6	19.5			3	.03	0.0			239	38	.15	.032	
raspberries															
raw	61	106.5	0.7	0.4	0	16	31	.11	.07		0	27	22	.57	1.246
1 cup	123	1.1	14.2	0.0	5.8	160	.04	1.1	.00	.30	187	15	.70	.091	
cnd, heavy syrup	117	96.4	0.2	0.1	0	4	11	.04	.05	13	4	14	16	.20	.298
1/2 cup	128	1.1	29.9	0.0		43	.03	0.6	.00	.31	120	12	.54	.073	
frzn, in lite syrup, Birds Eye	99	115.7	0.5	0.3	0		23	.08	.05	24	0	20	17	.43	
1/2 cup	142	0.8	24.6	0.0	4.3	120	.03	0.8	.00	.22	141	11	.54	.071	
frzn, sweetened	103	72.8	0.2	0.1	0	6	17	.05	.03	26	1	15	13	.18	.650
2/5 cup	100	0.7	26.2	0.0		60	.02	0.2	.00	.15	114	17	.65	.110	
rose apple, raw	25	93.0	0.3			34	22	.03			0	29	5	.06	.029
3.5 oz	100	0.6	5.7			339	.02	0.8	.00		123	8	.07	.016	
roselle, raw	28	49.4	0.4		0	16	7	.02			3	123	29		
1 cup	57	0.6	6.5			163	.01	0.2	.00		118	21	.84		
sapodilla, raw	140	132.6	1.9		0	10	25	.03	.06		20	36			
1 med	170	0.7	33.9		9.0	102		0.3	.00	.43	328	20	1.36		
sapote, raw	301	140.5	1.4		0	92	45	.05			21	88	68		
1 med	225	4.8	76.0			923	.02	4.1	.00		773	63	2.25		
soursop, raw	150	182.6	0.7		0	1	46	.11	.13		31	32	46		
1 cup	225	2.3	37.9			5	.16	2.0	.00	.57	626	61	1.35		
strawberries															
raw	45	136.4	0.6	0.3	0	4	85	.10	.09	26	2	21	16	.19	.432
1 cup	149	0.9	10.5	0.0	2.8	41	.03	0.3	.00	.51	247	28	.57	.073	
frzn, in lite syrup, Birds Eye	87	118.8	0.4	0.2	0		64	.07	.07	20	1	16	11	.15	
1/2 cup	142	0.7	21.6	0.2	2.2	31	.02	0.3	.00	.39	189	22	.44	.058	
frzn, sweetened	245	186.6	0.3	0.2	0	6	106	.13	.08	38	8	28	18	.14	.638
1 cup	255	1.4	66.1	0.0		61	.04	1.02	.00	.28	249	32	1.49	.051	
frzn, unsweetened	52	134.1	0.2	0.1	0	7	61	.06	.04	25	3	23	16	.19	.432
1 cup	149	0.6	13.6	0.0		66	.03	0.7	.00	.16	220	20	1.12	.073	
sugar apple, raw	146	113.5	0.5		0	1	56	.18	.31		15	37	33		
1 med	155	3.2	36.6			9	.17	1.4	.00	.35	384	50	.93		

	KCAL	H₂O (g)	FAT (g)	PUFA (g)	CHOL (mg)		**Vitamins**					**Minerals**			
						A (RE)	C (mg)	B-2 (mg)	B-6 (mg)	FOL (mcg)	Na (mg)	Ca (mg)	Mg (mg)	Zn (mg)	Mn (mg)
	WT (g)	PRO (g)	CHO (g)	SFA (g)	DFIB (g)	A (IU)	B-1 (mg)	NIA (mg)	B-12 (mcg)	PANT (mg)	K (mg)	P (mg)	Fe (mg)	Cu (mg)	
tamarind, raw	287	37.7	0.7	0.1	0	4	4	.18	.08		33	89	110		
1 cup	120	3.4	75.0	0.3		36	.51	2.3	.00	.17	753	136	3.36		
tangerine, raw	37	73.6	0.2	0.0	0	77	26	.02	.06	17	1	12	10		.027
1 med	84	0.5	9.4	0.0		773	.09	0.1	.00	.17	132	8	.09	.024	
watermelon, raw	50	146.4	0.7		0	58	15	.03	.23	3	3	13	17	.11	.059
1 cup	160	1.0	11.5		0.3	585	.13	0.3	.00	.34	186	14	.28	.051	

16. GRAIN FRACTIONS

	KCAL	H₂O (g)	FAT (g)	PUFA (g)	CHOL (mg)	A (RE)	C (mg)	B-2 (mg)	B-6 (mg)	FOL (mcg)	Na (mg)	Ca (mg)	Mg (mg)	Zn (mg)	Mn (mg)
barley, pearled															
light	698	22.2	2.0				0	.10			6	32			
1 cup	200	16.4	157.6			0	.24	6.2			320	378	4.00		
med, Scotch Brand	172	5.0	0.5					.06	.13	32	5	11	34	.00	.000
1/4 cup	48	5.5	36.3			11	.11	2.2	.00	.00	130	113	1.24	.210	
pot/scotch	696	21.6	2.2				0	.14				68			
1 cup	200	19.2	154.4			0	.42	7.4			592	580	5.40		
quick, Scotch Brand	172	5.0	0.5					.06	.13	32	5	11	34	.00	.000
1/4 cup	48	5.5	36.3			11	.11	2.2	.00	.00	130	113	1.24	.210	
bisquick, General Mills	240		8.0								700				
2 oz	57	4.0	37.0								80				
buckwheat flour															
dark	326	11.8	2.5				0	.15				32			
1 cup	98	11.5	70.6			0	.57	2.8				340	2.70		
light	340	11.8	1.2				0	.04				11			
1 cup	98	6.3	77.9			0	.08	0.4			314	86	1.00		
carob (St. Johnsbread) flour	185	3.7	0.7	0.2	0	1	0	.48	.38	30	36	359	56	.94	.523
1 cup	103	4.8	91.6	0.1	10.9	15	.06	2.0	.00	.05	852	81	3.03	.588	
corn flour	431	14.0	3.0				0	.07			1	7			
1 cup	117	9.1	89.9	0.3		400ᵃ	.23	1.6				92	2.10		
corn germ, Ener-G Foods	490		25.0				4	.75	1.41	90	31	0	672	10.60	
1 cup	100	17.0	42.0		20.8	60	1.70	2.2			1420	1587	7.80	.500	
corn meal mix															
white, bolted, Aunt Jemima	99	2.9	0.7					.14	.14	19	337	60	21	.00	.000
1/6 cup (1 oz)	28	2.4	20.8			0	.22	1.6	.00	.00	68	181	1.08	.040	
white, buttermilk, self-rising, Aunt	101	2.6	1.1					.14	.12	10	439	60	25	1.00	.000
Jemima—*3 T (1 oz)*	28	2.5	20.2			0	.22	1.6	.00	.00	87	207	1.08	.080	
yellow, bolted, Aunt Jemima	97	2.9	0.4					.14	.07	14	369	60	10	.00	.000
1/6 cup (1 oz)	28	2.4	20.9			81	.22	1.6	.00	.00	40	156	1.08	.020	
corn meal, white, bolted	442	14.6	4.1				0	.10			1	21			
*1 cup*ᵇ	122	11.0	90.9	0.5		0	.37	2.3			303	272	2.20		
enr, self-rising, Aunt Jemima	99	3.0	0.9					.07	.15	16	382	109	24	1.00	.000
1/6 cup (1 oz)	28	2.3	20.4			0	.12	1.0	.00	.00	78	207	.81	.040	
corn meal, white, degermed															
enr	502	16.6	1.7				0	.36			1	8			
1 cup	138	10.9	108.2			0	.61	4.8			166	137	4.00		
enr, ckd	120	210.5	0.5				0	.10			264	2			
1 cup	240	2.6	25.7			0	.14	1.2			38	34	1.00		
enr, Quaker/Aunt Jemima	102	3.0	0.5					.07	.10	15	1	1	12	.00	.000
3 T (1 oz)	28	2.4	22.2			0	.12	1.0	.00	.00	51	37	.81	.020	
enr, self-rising, Aunt Jemima	98	3.0	0.5					.07	.11	9	381	109	14	.00	.000
1/6 cup (1 oz)	28	2.3	21.1			0	.12	1.0	.00	.00	45	179	.81	.040	
unenr	502	16.6	1.7				0	.07			1	8			
1 cup	138	10.9	108.2			0	.19	1.4			166	137	1.50		

ᵃ For yellow varieties; white varieties have only a trace of Vit A activity.   ᵇ Nearly whole grain.

	KCAL	H₂O (g)	FAT (g)	PUFA (g)	CHOL (mg)	A (RE)	C (mg)	B-2 (mg)	B-6 (mg)	FOL (mcg)	Na (mg)	Ca (mg)	Mg (mg)	Zn (mg)	Mn (mg)
	WT (g)	PRO (g)	CHO (g)	SFA (g)	DFIB (g)	A (IU)	B-1 (mg)	NIA (mg)	B-12 (mcg)	PANT (mg)	K (mg)	P (mg)	Fe (mg)	Cu (mg)	
unenr, ckd	120	210.5	0.5				0	.02			264	2			
1 cup	240	2.6	25.7			0	.05	0.2			38	34	.50		
unenr, self-rising	491	15.9	1.6				0	.07			1946	409			
1 cup	141	10.6	106.2			0	.18	1.3			159	739	1.40		
corn meal, white, whole ground	433	14.6	4.8				0	.13			1	24			
1 cup	122	11.2	89.9	0.5		0	.46	2.4			346	312	2.90		
self-rising	465	15.1	4.3				0	.11			1849	402			
1 cup	134	11.4	95.9	0.5		0	.38	2.4			314	859	2.30		
corn meal, yellow, bolted[a]	442	14.6	4.1				0	.10			1	21			
1 cup	122	11.0	90.0	0.5		590	.37	2.3			303	272	2.20		
corn meal, yellow, degermed															
enr	502	16.6	1.7				0	.36			1	8			
1 cup	138	10.9	108.2			610	.61	4.8			166	137	4.00		
enr, ckd	120	210.5	0.5				0	.10			264	2			
1 cup	240	2.6	25.7			140	.14	1.2			38	34	1.00		
enr, Quaker/Aunt Jemima	102	3.0	0.5					.07	.10	15	1	1	12	.00	.000
3 T (1 oz)	28	2.4	22.2			115	.12	1.0	.00	.00	51	37	.81	.020	
unenr	502	16.6	1.7				0	.07			1	8			
1 cup	138	10.9	108.2			610	.19	1.4			166	137	1.50		
unenr, ckd	120	210.5	0.5				0	.02			264	2			
1 cup	240	2.6	25.7			140	.05	0.2			38	34	.50		
unenr, self-rising	491	15.9	1.6				0	.07			1946	409			
1 cup	141	10.6	106.2			590	.18	1.3			159	739	1.40		
corn meal, yellow, whole ground	433	14.6	4.8				0	.13			1	24			
1 cup	122	11.2	89.9	0.5		620	.46	2.4			346	312	2.90		
self-rising	465	15.1	4.3				0	.11			1849	402			
1 cup	134	11.4	95.9	0.5		600	.38	2.4			314	859	2.30		
masa harina de maiz, Quaker	137	3.2	1.5					.29	.19	17	2	77	41	1.00	.000
1/3 cup	37	3.5	27.4			0	.52	3.5	.00	.00	115	86	2.93	.080	
masa trigo, Quaker	149	3.7	4.0					.24	.01	29	794	66	8	.00	.000
1/3 cup	37	3.5	24.7			0	.39	3.1	.00	.00	36	77	2.60	.040	
millet, proso (broomcorn/hogmillet),	327	11.8	2.9				0	.38				20	162		
whole grain—*3.5 oz*	100	9.9	72.9			0	.73	2.3			430	311	6.80		
oat bran, Quaker	110	2.6	2.5					.12	.04	20	1	20	67	1.00	2.000
1/3 cup (1 oz)	28	5.7	16.2			0	.33	.23	.00	.00	170	200	1.63	.080	
potato flour	316	6.8	0.7	0.3	0		0	17	.13			31	30			
1/2 cup	90	7.2	71.9	0.2		0	.38	3.1	.00		1429	160	15.48		
rice bran, Ener-G Foods	438	5.2	21.7				0	.26			6	355		5.98	
3.5 oz	100	11.9	45.6			0	2.21	31.2			1370		7.88	.750	
rice flour, Ener-G Foods	398	13.6	0.9				0	.59			45	13	60		
1 cup	113	7.1	91.0			0	.52	2.9			168	249	3.28		
rice polish, Ener-G Foods	220	5.5	7.0				0	.22			21	34			
1/2 cup	56	7.0	33.0			0	1.79	0.2			829			.08	
rye & wheat flour, bohemian style,	400	12.9	1.5				0	.29			2	24			
Pillsbury—*1 cup*	112	12.3	84.4			0	.56	4.4			167	196	4.03		
rye flour															
dark	419	14.1	3.3				0	.28			1	69			
1 cup	128	20.9	87.2			0	.78	3.5			1101	686	5.80		
light	364	11.2	1.0				0	.07			1	22			
1 cup	102	9.6	79.5			0	.15	0.6			159	189	1.10		
med, Pillsbury's Best	400	11.1	2.2				0	.11			5	22			
1 cup	111	12.2	81.0			0	.33	1.3			322	152	2.78		
sorghum grain	332	11.0	3.3				0	.15				28			
3.5 oz	100	11.0	73.0	0.0		0	.38	3.9			350	287	4.40		

[a] Nearly whole grain.

	KCAL	H₂O (g)	FAT (g)	PUFA (g)	CHOL (mg)	A (RE)	C (mg)	B-2 (mg)	B-6 (mg)	FOL (mcg)	Na (mg)	Ca (mg)	Mg (mg)	Zn (mg)	Mn (mg)
	WT (g)	PRO (g)	CHO (g)	SFA (g)	DFIB (g)	A (IU)	B-1 (mg)	NIA (mg)	B-12 (mcg)	PANT (mg)	K (mg)	P (mg)	Fe (mg)	Cu (mg)	
soybean flour															
defatted	327	7.3	1.2	0.5	0	4	0	.25	.57	305	20	241	290	2.46	3.018
1 cup	100	51.5	33.9	0.1		40	.70	2.6	.00	2.00	2384	674	9.24	4.065	
full fat	368	4.4	17.6	9.9	0	10	0	.99	.39	293	11	175	364	3.33	1.934
1 cup	85	32.1	27.1	2.5		102	.49	3.7	.00	1.35	2138	420	5.42	2.482	
full fat, roasted	373	3.2	18.6	10.5	0	9	0	.80	.30	193	11	160	314	3.04	1.765
1 cup	85	32.4	25.8	2.7		93	.35	2.8	.00	1.03	1734	405	4.94	1.888	
low fat	326	2.4	5.9	3.3	0	3	0	.25	.46	361	16	165	202	1.04	2.710
1 cup	88	44.8	29.6	0.9		35	.33	1.9	.00	1.60	2262	522	5.27	4.470	
soy meal, defatted	411	8.5	2.9	1.3	0	5	0	.31	.69	369	3	297	373	6.17	4.636
1 cup	122	60.0	43.8	0.3	14.0	48	.84	3.2	.00	2.41	3038	855	16.71	2.440	
wheat bran															
toasted, Kretschmer	60		1.0		0						0				
1 oz	28	5.0	18.0		12.0						410				
unprocessed, Quaker	21	0.8	0.2					.03	.04	11	0	6	46	1.00	1.000
2 T	7	1.1	3.6			0	.04	1.5	.00	.00	106	115	1.02	.080	
wheat flour															
all purpose, enr	499	16.4	1.4				0	.36			3	22			
1 cup	137	14.4	104.3			0	.60	4.8			130	119	4.00		
all purpose, enr, Ballard	399	10.1	1.0				0	.44			7	230			
1 cup	109	10.7	86.1			0	.76	4.8			132	110	4.91		
all purpose, enr, Pillsbury's Best	401	13.3	0.9				0	.34			2	238			
1 cup	113	10.8	87.0			0	.79	6.2			148	113	5.09		
all purpose, enr, unbleached, Pillsbury's Best—*1 cup*	401	10.9	2.2				0	.37			3	230			
	109	12.3	82.8			0	.76	5.5			136	518	4.91		
bread, enr	500	16.4	1.5				0	.36			3	22			
1 cup	137	16.2	102.3			0	.60	4.8			130	130	4.00		
bread, Pillsbury's Best	401	14.6	3.0				0	.56			9	236			
1 cup	112	13.4	79.5			0	1.01	2.7			91	115	5.04		
cake/pastry	430	14.2	0.9				0	.04			2	20			
1 cup	118	8.9	93.7			0	.04	0.8			112	86	.60		
gluten (45% gluten & 55% all-purpose)	529	11.9	2.7				0				3	56			
1 cup	140	58.0	66.1			0					84	196			
sauce 'n gravy, Pillsbury's Best	50	1.5	0.1				0	.05			1	30			
2 T	14	1.5	10.8			0	.08	0.7			12	14	.41		
self-rising, enr	440	14.4	1.3				0	.33			1349	331			
1 cup	125	11.6	92.8			0	.55	4.4				583	3.60		
self-rising, enr, Aunt Jemima	109	3.1	0.3					.11	.01	12	368	60	6	.00	.000
1/4 cup (1 oz)	28	3.0	23.6			0	.18	1.5	.00	.00	32	153	1.25	.020	
self-rising, enr, bleached/unbleached, Pillsbury's Best/Ballard—*1 cup*	380	15.0	1.2				0	.39			1309	243			
	115	8.7	84.0			0	.74	5.8			104	546	5.18		
whole wheat	400	14.4	2.4				0	.14			4	49			
1 cup	120	16.0	85.2			0	.66	5.2			444	446	4.00		
whole wheat, Pillsbury's Best	398	12.2	2.1				0	.20			9	30			
1 cup	111	16.7	77.7			0	.43	7.3			487	382	4.33		
wheat germ															
toasted	108	1.6	3.0	1.8			2	.23	.28	100	1	13	91	4.73	5.658
1/4 cup (1 oz)	28	8.3	14.1	0.5	3.0		.47	1.6		.39	268	325	2.58	.176	
toasted w/ brown sugar & honey	107	1.6	2.3	1.4				.18	.21	75	1	9	68	3.54	4.243
1/4 cup (1 oz)	28	6.2	17.2	0.4	2.0		.35	1.2		.30	201	244	1.93	.132	

	KCAL / WT (g)	H2O (g) / PRO (g)	FAT (g) / CHO (g)	PUFA (g) / SFA (g)	CHOL (mg) / DFIB (g)	A (RE) / A (IU)	C (mg) / B-1 (mg)	B-2 (mg) / NIA (mg)	B-6 (mg) / B-12 (mcg)	FOL (mcg) / PANT (mg)	Na (mg) / K (mg)	Ca (mg) / P (mg)	Mg (mg) / Fe (mg)	Zn (mg) / Cu (mg)	Mn (mg)

17. GRAIN PRODUCTS
17.1. BISCUITS

	KCAL	H2O	FAT	PUFA	CHOL	A(RE)/A(IU)	C/B-1	B-2/NIA	B-6/B-12	FOL/PANT	Na/K	Ca/P	Mg/Fe	Zn/Cu	Mn
from mix	93	8.0	3.3				0	.11	.01	2	262	58	7	.18	
1 biscuit	28	2.1	13.6			16	.12	0.8	.05	.10	56	128	.57	.031	
from refrig dough	91	8.4	3.1				0	.07	.01		349	6	4	.12	
1 biscuit	28	2.0	13.6			0	.11	0.9	.00	.07	25	110	.66	.020	
homemade, enr	103	7.7	4.8				0	.06			175	34			
1 biscuit	28	2.1	12.8	1.2[a]		0	.06	0.5			23	49	.40		
homemade, unenr	102	7.7	4.8				0	.03			175	34			
1 biscuit	28	2.1	12.8	1.2[a]		0	.01	0.1			33	49	.10		
refrig dough															
baking powder, 1869 Brand	207	21.4	10.1				0	.20			616	24			
2 biscuits	63	3.8	25.2			2	.20	1.9			60	265	1.51		
baking powder/buttermilk, tenderflake, Pillsbury—2 biscuits	110	12.6	5.3				0	.07			333	5			
	35	1.9	14.0			0	.10	0.9			19	139	.77		
butter/buttermilk/country, Pillsbury	101	17.2	1.5				0	.11			361	10			
2 biscuits	43	2.7	19.8			0	.15	1.4			209	196	1.18		
buttermilk/buttertastin', 1869 Brand	209	21.4	10.1				0	.17			278	23			
2 biscuits	63	3.8	25.8			3	.18	1.7			57	260	1.39		
butter flavored, Continental	190		8.0								520				
2 biscuits	57	3.0	25.0								45				
butter-me-not, Continental	190		9.0								530				
2 biscuits	57	3.0	22.0								40				
buttermilk, big country, Pillsbury	201	24.1	7.6				0	.11			633	15			
2 biscuits	67	4.3	28.8			0	.32	1.2			176	304	2.01		
buttermilk, deluxe heat 'n eat, Pillsbury	281	17.3	14.8				0	.32			614	37			
2 biscuits	72	4.7	32.4			5	.19	1.8			78	228	1.44		
buttermilk, extra rich, Hungry Jack	110	17.6	2.6				0	.11			360	11			
2 biscuits	44	2.6	19.4			0	.15	1.4			209	194	1.14		
buttermilk, flaky, extra lights, Pillsbury	109	16.0	3.4				0	.09			323	9			
2 biscuits	41	2.4	17.2			0	.15	1.2			186	175	.78		
buttermilk, flaky, Hungry Jack	171	19.6	6.5				0	.12			582	8			
2 biscuits	56	3.4	24.6			0	.17	1.8			41	230	1.06		
buttermilk, fluffy, Hungry Jack	179	20.3	8.1				0	.13			571	9			
2 biscuits	58	3.5	24.1			0	.19	1.7			34	238	1.39		
buttermilk, heat 'n eat, Pillsbury	170	14.6	5.2				0	.14			532	24			
2 biscuits	52	3.9	26.5			3	.19	1.6			55	237	1.04		
buttermilk/plain, Ballard Ovenready	99	17.2	1.3				0	.11			361	10			
2 biscuits	43	2.7	19.8			0	.15	1.4			209	196	1.12		
buttermilk, Weight Watchers	90		1.0								420				
2 biscuits	40	2.0	18.0								30				
buttertastin', big country, Pillsbury	191	26.9	7.9				0	.16			658	13			
2 biscuits	69	3.7	27.6			0	.26	2.0			184	295	1.24		
buttertastin', flaky, Hungry Jack	180	20.0	7.4				0	.32			307	11			
2 biscuits	57	3.4	24.5			0	.23	1.5			25	250	1.60		
flaky, Hungry Jack	169	20.2	7.3				0	.13			580	9			
2 biscuits	56	3.1	23.5			0	.18	1.6			32	240	1.34		
good 'n buttery, fluffy, Pillsbury	179	19.8	9.5				0	.13			727	12			
2 biscuits	56	2.7	21.0			0	.18	1.6			143	238	1.34		
homestyle, Continental	100		2.0								340				
2 biscuits	43	3.0	18.0								115				
mountain man, Continental	190		8.0								630				
2 biscuits	57	3.0	25.0								75				
southern style, Big Country	201	24.1	8.0				0	.16			624	16			
2 biscuits	67	3.7	28.8			0	.23	2.0			40	291	1.74		
texas style, butter flavored, Continental	210		9.0								660				
2 biscuits	68	4.0	28.0								70				
texas style, Continental	190		5.0								760				
2 biscuits	68	4.0	32.0								65				

[a] Made w/ veg shortening; 1.8 g if made w/ lard.

	KCAL	H₂O (g)	FAT (g)	PUFA (g)	CHOL (mg)	A (RE)	C (mg)	B-2 (mg)	B-6 (mg)	FOL (mcg)	Na (mg)	Ca (mg)	Mg (mg)	Zn (mg)	Mn (mg)
	WT (g)	PRO (g)	CHO (g)	SFA (g)	DFIB (g)	A (IU)	B-1 (mg)	NIA (mg)	B-12 (mcg)	PANT (mg)	K (mg)	P (mg)	Fe (mg)	Cu (mg)	
texas style, flaky, Continental	220		9.0								640				
2 biscuits	68	4.0	30.0								50				
wheat, Weight Watchers	90		1.0								440				
2 biscuits	40	2.0	18.0								50				

17.2. BREADS & BREAD PRODUCTS

	KCAL	H₂O (g)	FAT (g)	PUFA (g)	CHOL (mg)	A (RE)	C (mg)	B-2 (mg)	B-6 (mg)	FOL (mcg)	Na (mg)	Ca (mg)	Mg (mg)	Zn (mg)	Mn (mg)
	WT (g)	PRO (g)	CHO (g)	SFA (g)	DFIB (g)	A (IU)	B-1 (mg)	NIA (mg)	B-12 (mcg)	PANT (mg)	K (mg)	P (mg)	Fe (mg)	Cu (mg)	
bagel	163	16.0	1.4				0	.16	.02	13	198	23	11	.29	
1 bagel	55	6.0	30.9		0.6	0	.21	1.9	.00	.20	41	37	1.46	.046	
boston brown															
cnd	95	20.3	0.6				0	.03			113	41			
1 piece	45	2.5	20.5			0	.05	0.5			131	72	.90		
plain/raisin, cnd, B&M/Friends	80		0.0								220				
1/2" slice	45	2.0	18.0								45				
bran 'n honey, Country Hearth	70		1.0								160				
1 slice	28	3.0	13.0								50				
bread crumbs															
dry, grated, enr	392	6.5	4.6				0	.30			736	122			
1 cup	100	12.6	73.4	1.1		0	.22	3.5			152	141	3.60		
seasoned, Contadina	35	0.6	0.3	0.0	1		0	.04			250	9	3	.14	
1 rd T	10	1.4	6.8	0.1		9	.04	0.5			14	14	.42		
bread sticks	77	1.0	0.6				0	.01			140	6			
2 sticks	20	2.4	15.1	0.1		0	.01	0.2			18	20	.18		
bread sticks, soft, refrig dough, Pillsbury	100	14.4	2.5				1	.73			238	4			
1 stick	37	3.1	16.2			0		1.2			297	35	1.04		
buttermilk, Grant's Farm	80		1.0		0						190				
1 slice	28	3.0	14.0								50				
buttermilk, old fashioned, Country Hearth	70		1.0								190				
1 slice	28	3.0	13.0								35				
cinnamon raisin	80		1.0		5						140				
1 slice	28	2.0	15.0												
corn bread															
from mix	178	16.7	5.8				0	.10			263	133			
1 piece	55	3.8	27.5	1.7		130	.10	0.8			61	209	.80		
from mix, Dromedary	130		3.0								480				
2" square		3.0	20.0								65				
mix, Aunt Jemima	205	2.5	6.1					.20	.08	46	519	15	13	.00	.000
1.7 oz dry mix	48	3.6	33.9			39	.26	2.1	.00	.00	55	270	1.48	.040	
w/ enr corn meal, homemade	198	41.7	7.3				0	.15	.06	8	232	90	15	.39	
1 piece	83	4.1	28.7			114	.15	1.2	.14	.32	78	81	1.24	.033	
w/ whole ground corn meal,	172	42.0	6.9				0	.11	.08	6	209	83	20	.46	
homemade—*1 piece*	78	3.7	24.1			104	.11	0.8	.13	.28	88	90	.83	.047	
corn pone w/ white whole ground corn	122	31.1	3.2				0	.03			238	37			
meal[a]—*1 piece*	60	2.7	21.7	0.6		0	.09	0.5			37	98	.70		
country meal, Continental	80		1.0		0						135				
1 slice	28	3.0	14.0		1.3						60				
cracked wheat	66	8.8	0.9				0	.10	.02		108	16	9		
1 slice	25	2.3	12.5		1.1	0	.10	0.8	.00	.15	33	32	.67		
mini loaf, Earth Grains	70		1.0								170				
1 oz	28	2.0	14.0								55				
toasted	63	5.5	0.8				0	.09	.02		104	16	8	.39	.403
1 slice	21	2.2	12.0		1.1	0	.07	0.8	.00	.14	32	30	.64	.067	
Wonder	70		1.0		5						180				
1 slice	28	3.0	13.0												
d'italia, Country Hearth	70		1.0								200				
1 slice	28	3.0	13.0								25				

[a] Cornbread often made w/o milk or eggs, shaped in irregular ovals by the palm of the hand & baked or fried on a griddle.

	KCAL / WT (g)	H₂O (g) / PRO (g)	FAT (g) / CHO (g)	PUFA (g) / SFA (g)	CHOL (mg) / DFIB (g)	A (RE) / A (IU)	C (mg) / B-1 (mg)	B-2 (mg) / NIA (mg)	B-6 (mg) / B-12 (mcg)	FOL (mcg) / PANT (mg)	Na (mg) / K (mg)	Ca (mg) / P (mg)	Mg (mg) / Fe (mg)	Zn (mg) / Cu (mg)	Mn (mg)
european, butter sesame, Country Hearth	70		1.0								190				
1 slice	28	3.0	12.0								25				
french	81	8.3	1.1					.07	.01	9	163	22	6	.24	.154
1 slice	28	2.7	14.8		0.6		.11	1.2		.12	32	30	.92	.056	
DiCarlo Parisian	70		1.0		5						180				
1 slice	28	3.0	13.0												
from mix, Home Hearth	170		3.0								320				
two 3/8" slices		6.0	29.0								70				
refrig dough, Pillsbury	61	10.7	0.3				0	.07			129	5			
1" slice	26	2.6	11.4			0	.12	1.0			16	15	.86		
vienna	70	8.5	1.0		0		0	.09	.01	9	138	28	5	.16	
1 slice	25	2.4	12.6			0	.12	1.0	.00	.09	22	20	.77	.036	
Wonder	70		1.0		0						180				
1 slice	28	3.0	13.0												
fruit quick bread, from mix	118	13.0	2.4					.09			126	12	6	.16	
1 slice	40	1.8	22.3				.10	1.1			40	48	.69	.041	
fruit/nut quick bread, from mix, Pillsbury[a]—1/12 loaf	161		4.4								162				
		2.4	28.0								51				
grainola, Country Hearth	70		1.0								240				
1 slice	28	3.0	12.0								60				
hillbilly, Continental	70		1.0		5						140				
1 slice	28	3.0	14.0												
hollywood, Continental															
dark	70		1.0		5						160				
1 slice	28	3.0	13.0												
light	70		1.0		5						150				
1 slice	28	3.0	13.0												
honey & oat, Country Hearth	70		1.0								180				
1 slice	28	3.0	13.0								40				
honey grain, Continental	70		1.0		0						170				
1 slice	28	3.0	13.0	1.0							45				
italian	78	8.9	0.6					.06	.01	9	151	14	7	.24	.134
1 slice	28	2.8	14.9		0.3		.10	1.2		.11	31	26	.74	.056	
Wonder Family	70		1.0		5						160				
1 slice	28	2.0	13.0												
mixed grain	64	9.4	0.9				0	.10	.03	16	103	26	12	.30	.350
1 slice	25	2.5	11.7		1.0	0	.10	1.0	.00	.16	55	53	.82	.071	
toasted	63	5.8	0.9				0	.09	.03	16	100	25	12	.29	
1 slice	21	2.4	11.4			0	.08	1.0	.00	.15	53	52	.79	.069	
nature's wheat, Country Hearth	70		1.0								150				
1 slice	28	3.0	13.0								50				
oatmeal	71	10.3	1.2					.06	.01	9	138	17	10	.27	.241
1 slice	28	2.4	13.0		0.6		.10	0.9		.10	44	35	.74	.062	
pita pocket	106	11.9	0.6				0	.10	.01	8	215	31	10	.30	.186
1 pocket	38	4.0	20.6		0.3	0	.18	1.6		.16	45	38	.92	.068	
raisin	70	8.3	1.0				0	.16	.01	9	94	26	4	.16	.126
1 slice	25	2.1	13.2		0.6	0	.08	1.0	.00	.11	60	22	.78	.035	
toasted	66	5.1	0.9				0	.15	.01	8	90	24	6	.15	
1 slice	21	1.9	12.6			0	.06	1.0	.00	.10	57	21	.74	.033	
roman meal	70		1.0		5						140				
1 slice	28	3.0	13.0												

[a] Values are averages for applesce spice, apricot nut, banana, blueberry nut, carrot nut, cherry nut, cranberry, date, honey granola & nut.

	KCAL	H_2O (g)	FAT (g)	PUFA (g)	CHOL (mg)	Vitamins A (RE)	C (mg)	B-2 (mg)	B-6 (mg)	FOL (mcg)	Minerals Na (mg)	Ca (mg)	Mg (mg)	Zn (mg)	Mn (mg)
	WT (g)	PRO (g)	CHO (g)	SFA (g)	DFIB (g)	A (IU)	B-1 (mg)	NIA (mg)	B-12 (mcg)	PANT (mg)	K (mg)	P (mg)	Fe (mg)	Cu (mg)	
rye															
am	66	9.4	0.9				0	.08	.02	10	174	20	6	.32	
1 slice	25	2.1	12.0			0	.10	0.8	.00	.11	51	36	.68	.026	
am, toasted	66	6.3	0.9				0	.08	.02	10	175	20	6	.32	
1 slice	22	2.1	12.0			0	.08	0.8	.00	.11	51	37	.68	.026	
deli, Country Hearth	70		1.0								160				
1 slice	28	3.0	14.0								40				
from mix, Home Hearth	150		1.0								360				
two 3/8″ slices		6.0	28.0								90				
hearty/mild	70		1.0		5						180				
1 slice	28	3.0	13.0												
honey cracked, Grant's Farm	70		1.0		0						190				
1 slice	28	3.0	13.0								55				
pumpernickel	82	11.9	0.8				0	.17	.05		173	23	22	.36	
1 slice	32	2.9	15.4			0	.11	1.1	.00	.15	139	70	.88		
pumpernickel, toasted	82	7.9	1.1				0	.17	.05		174	23	22	.36	
1 slice	28	2.9	15.4			0	.09	1.1	.00	.15	139	70	.88		
seven grain, Grant's Farm	70		1.0		0						140				
1 slice	28	3.0	13.0								60				
seven grain, Home Pride	70		1.0		5						140				
1 slice	28	3.0	13.0												
seven whole grain, Country Hearth	80		1.0								190				
1 slice	28	3.0	14.0								45				
sourdough															
DiCarlo	70		1.0		0						140				
1 slice	28	3.0	12.0												
mini loaf, Earth Grains	80		1.0								210				
1 oz	28	2.0	14.0								40				
spoon bread w/ white whole ground	468	151.2	27.4				1	.43			1157	230			
cornmeal[a]—*1 cup*	240	16.1	40.6	8.7		700	.22	1.0			317	394	2.40		
wheat	61	8.9	1.0		0		0	.08	.03	11	129	30	11	.25	
1 slice[b]	24	2.3	11.3			0	.11	1.1	1.1	.10	33	44	.84	.058	
toasted[b]	61	5.9	1.0				0	.08	.03	11	129	30	11	.25	
1 slice	21	2.3	11.3			0	.09	1.1	.00	.10	33	44	.84	.058	
butter split top, Country Hearth	70		1.0								160				
1 slice	28	3.0	13.0								45				
butter top, Home Pride	70		1.0		5						140				
1 slice	28	3.0	13.0												
Fresh & Natural	70		1.0		0						140				
1 slice	28	3.0	13.0												
Fresh Horizons	50		1.0		0						140				
1 slice	28	3.0	10.0												
honey buttered, split top, Family	70		1.0		0						150				
Recipe—*1 slice*	28	3.0	14.0		1.2						55				
refrig dough, pipin' hot, Pillsbury	81	12.1	1.9				0	.07			185	5			
1″ slice	31	2.9	13.0			0	.65	1.0			21	26	.90		
sandwich, Country Hearth	70		1.0								150				
1 slice	28	3.0	13.0								50				
sandwich, Family Recipe	70		1.0		0						150				
1 slice	28	3.0	14.0		1.0						45				
stone ground, Grant's Farm	70		1.0		0						150				
1 slice	28	3.0	12.0								60				
wheat berry, Country Hearth	70		1.0								160				
1 slice	28	3.0	13.0								50				
wheat berry, Grant's Farm	70		1.0		0						170				
1 slice	28	3.0	13.0								45				
wheat berry, Home Pride	70		1.0		5						160				
1 slice	28	3.0	12.0												

[a] Bread made of cornmeal w/ or w/o added rice & hominy & mixed w/ milk, eggs, shortening & leavening to a consistency that is served w/ a spoon.

[b] Made w/ white enr flour & colored brown; not the same as whole wheat bread.

	KCAL	H₂O (g)	FAT (g)	PUFA (g)	CHOL (mg)	A (RE)	C (mg)	B-2 (mg)	B-6 (mg)	FOL (mcg)	Na (mg)	Ca (mg)	Mg (mg)	Zn (mg)	Mn (mg)
	WT (g)	PRO (g)	CHO (g)	SFA (g)	DFIB (g)	A (IU)	B-1 (mg)	NIA (mg)	B-12 (mcg)	PANT (mg)	K (mg)	P (mg)	Fe (mg)	Cu (mg)	
Wonder Family	70		1.0		5						140				
1 slice	28	3.0	13.0												
white	64	8.9	0.9				0	.07	.01	8	123	30	5	.15	
1 slice	24	2.0	11.7			0	.11	0.9	.00	.10	27	26	.68	.033	
toasted	64	5.9	0.9				0	.08	.01	8	123	30	5	.15	
1 slice	21	2.0	11.7			0	.03	0.9	.00	.10	27	26	.68	.033	
homemade	72	9.0	1.7				0	.08	.01	9	102	16	6	.16	
1 slice	25	1.9	12.0			13	.08	0.7	.03	.12	33	25	.54	.019	
homemade, toasted	70	5.5	1.6				0	.08	.01	9	98	16	5	.16	
1 slice	21	1.8	11.7			12	.06	0.7	.03	.11	32	24	.52	.018	
butter split, Country Hearth	70		1.0								160				
1 slice	28	3.0	13.0								30				
butter top, Home Pride	70		1.0		5						160				
1 slice	28	3.0	13.0								30				
country mini loaf, Earth Grains	70		1.0								150				
1 oz	28	2.0	14.0								40				
Fresh Horizons	50		1.0		0						140				
1 slice	28	3.0	10.0												
from mix, Home Hearth	150		1.0								260				
two 3/8″ slices		6.0	29.0								70				
honey-buttered split top, Family Recipe	80		1.0		0						150				
1 slice	28	2.0	14.0								40				
old fashioned, Country Hearth	70		1.0								160				
1 slice	28	3.0	13.0								30				
refrig dough, pipin' hot, Pillsbury	81	12.0	1.7				0	.08			179	5			
1″ slice	30	2.8	12.9			0	.62	1.0			18	23	.84		
Rich's dough, baked	116	16.5	1.1		0			.14			300	110	11	.30	
2 slices	46	3.9	22.7				.18	1.7			51	55	2.10	.070	
w/ buttermilk, Wonder	70		1.0		5						160				
1 slice	28	2.0	13.0												
Wonder	70		1.0		5						140				
1 slice	28	3.0	13.0												
whole wheat	61	9.6	1.1				0	.05	.05	14	159	18	23	.42	.594
1 slice	25	2.4	11.4		1.6	0	.09	1.0	.00	.18	44	65	.86	.086	
toasted	59	6.1	1.1				0	.05	.04	13	153	17	23	.40	
1 slice	21	2.3	10.9			0	.07	0.9	.00	.18	42	63	.82	.082	
homemade	67	9.0	1.6				0	.04	.05	12	89	20	23	.56	
1 slice	25	2.2	11.6			11	.07	0.8	.03	.20	85	63	.67	.064	
homemade, toasted	65	5.5	1.6				0	.04	.05	12	87	19	23	.55	
1 slice	21	2.2	11.3			11	.05	0.8	.03	.20	83	61	.65	.062	
stone ground, 100%, Country Hearth	70		1.0								160				
1 slice	28	3.0	12.0								75				
Wonder	70		1.0		5						160				
1 slice	28	3.0	12.0												
Wonder, soft	70		1.0		0						140				
1 slice	28	4.0	10.0												

17.3. CRACKERS

	KCAL	H₂O (g)	FAT (g)	PUFA (g)	CHOL (mg)	A (RE)	C (mg)	B-2 (mg)	B-6 (mg)	FOL (mcg)	Na (mg)	Ca (mg)	Mg (mg)	Zn (mg)	Mn (mg)
	WT (g)	PRO (g)	CHO (g)	SFA (g)	DFIB (g)	A (IU)	B-1 (mg)	NIA (mg)	B-12 (mcg)	PANT (mg)	K (mg)	P (mg)	Fe (mg)	Cu (mg)	
bacon flavored thins, Nabisco	70		4.0								210				
7 crackers	14	1.0	8.0								20				
bacon, Great Crisps	70		4.0								230				
9 crackers	14	2.0	8.0								20				
blue cheese/cheddar/nacho/swiss thins,	70		4.0								233				
Nabisco—*10 crackers*	14	1.3	8.0								15				
cheddar cheese, Safari	130		10.0								450				
1 oz	28	3.0	14.0												
cheese	81	0.6	4.9				0	.06			180	16	3	.15	
5 pieces	15	1.4	7.8				.06	1.2			28	32	.53	.070	

						Vitamins					Minerals				
	KCAL	H₂O (g)	FAT (g)	PUFA (g)	CHOL (mg)	A (RE)	C (mg)	B-2 (mg)	B-6 (mg)	FOL (mcg)	Na (mg)	Ca (mg)	Mg (mg)	Zn (mg)	Mn (mg)
	WT (g)	PRO (g)	CHO (g)	SFA (g)	DFIB (g)	A (IU)	B-1 (mg)	NIA (mg)	B-12 (mcg)	PANT (mg)	K (mg)	P (mg)	Fe (mg)	Cu (mg)	
cheese 'n chives, Dip in a Chip	70		4.0								130				
8 crackers	14	1.0	8.0								40				
cheese 'n chives, Great Crisps	70		4.0								170				
9 crackers	14	2.0	8.0								20				
cheese nips, Nabisco	70		3.0								130				
13 crackers	14	1.0	9.0								25				
cheese, Ritz	70		3.0								120				
5 crackers	14	1.0	8.0								15				
cheese tidbits	70		4.0								200				
16 crackers	14	1.0	8.0								20				
cheese wheat thins	70		3.0								220				
9 crackers	14	2.0	9.0								35				
cheese w/ peanut butter filling	209	1.0	10.2				0	.03			422	24			
6 sandwiches	42	6.5	23.8	2.7		20	.01	1.5			96	76	.30		
cheese w/ peanut butter filling, Nabisco	70		3.0								150				
2 sandwiches	14	2.0	8.0								25				
chicken in a biskit	70		4.0								115				
7 crackers	14	1.0	8.0								20				
cinnamon treats, Nabisco	60		1.0								80				
2 crackers	14	1.0	11.0								20				
cracker meal, Nabisco	50		0.0								0				
2 T	14	1.0	12.0								20				
escort, Nabisco	80		4.0								110				
3 crackers	14	1.0	9.0								20				
french onion, Great Crisps	70		4.0								90				
7 crackers	14	1.0	8.0								20				
garlic butter n' cheese, Safari	150		9.0								690				
1 oz	28	2.0	15.0												
garlic, Great Crisps	70		3.0								190				
8 crackers	14	1.0	9.0								20				
graham	60	0.5	1.5				0	.04	.01	2	66	5	5	.11	
2 squares	14	1.0	10.8			0	.05	0.4	.00	.07	23	17	.37	.017	
graham crumbs, Nabisco	60		1.0								90				
2 T	14	1.0	11.0								15				
graham, Nabisco	60		1.0								115				
2 crackers	14	1.0	11.0								20				
holland rusk	60		1.0								35				
1 cracker	14	2.0	10.0								35				
meal mates sesame wafers	70		3.0								140				
3 crackers	14	1.0	9.0								30				
nacho, Great Crisps	70		4.0								250				
8 crackers	14	1.0	8.0								30				
nutty wheat thins, Nabisco	80		5.0								250				
7 crackers	14	1.0	8.0								35				
onion 'n cheese, Safari	150		9.0								750				
1 oz	28	2.0	15.0												
oyster	33	0.3	1.0				0	.00			83	2			
10 crackers	8	0.7	5.3	0.2		0	.00	0.1			9	7	.10		
oyster, Dandy	60		1.0								220				
20 crackers	14	1.0	10.0								20				
pizza, Safari	150		9.0								480				
1 oz	28	3.0	15.0												
premium unsalted tops	60		2.0								115				
5 crackers	14	1.0	10.0								20				
rice cakes, Chico San	35		0.2				0	.01			ᵃ	0			
1 cake		0.4	8.0	0.4		14	.00	0.6			25		.10		
ritz	70		4.0								120ᵇ				
4 crackers	14	1.0	9.0								15				
round toast, Planters	140		7.0								270				
1 oz	28	4.0	15.0								85				
royal lunch, Nabisco	60		2.0								80				
1 cracker	14	1.0	10.0								15				

ᵃ 0 mg in sodium free varieties; 10 mg in very low sodium varieties.　　ᵇ Low sodium Ritz crackers contain 60 mg sodium.

	KCAL / WT (g)	H₂O (g) / PRO (g)	FAT (g) / CHO (g)	PUFA (g) / SFA (g)	CHOL (mg) / DFIB (g)	A (RE) / A (IU)	C (mg) / B-1 (mg)	B-2 (mg) / NIA (mg)	B-6 (mg) / B-12 (mcg)	FOL (mcg) / PANT (mg)	Na (mg) / K (mg)	Ca (mg) / P (mg)	Mg (mg) / Fe (mg)	Zn (mg) / Cu (mg)	Mn (mg)
rusk	38	0.4	0.8				0	.02			22	2			
1 rusk	9	1.2	6.4	0.2		20	.01	0.1			14	11	.10		
rye, whole grain	45	0.8	0.2				0	.03			115	7			
2 wafers	13	1.7	9.9			0	.04	0.2			78	50	.50		
ry-krisp	40		0.2					.07	.06	7	112	12	34	.79	
1/4 large square	14	1.5	13.0		2.5		.04	0.3		.16	63	93	.44	.135	
seasoned	50		1.1					.11	.05	27	147	11	32	.75	
1/4 large square	14	1.5	10.0		2.4		.03	0.6		.12	61	47	.45	.120	
sesame	60		1.4					.06	.05	9	148	5	36	.80	
2 triple crackers	14	1.6	12.0		2.1		.04	0.4		.12	61	88	.45	.140	
saltines	26	0.2	0.6				0	.03	.00	1	80	4	2	.04	
2 crackers	6	0.6	4.4			0	.02	0.4	.00	.02	8	6	.16	.010	
saltines, Premium	60		2.0								180				
5 crackers	14	1.0	10.0								20				
sea round, Nabisco	60		2.0								140				
1 cracker	14	1.0	10.0								20				
sesame, Great Crisps	70		4.0								190				
9 crackers	14	2.0	8.0								25				
sociables, Nabisco	70		3.0								130				
6 crackers	14	1.0	10.0								30				
soda	125	1.1	3.7				0	.01			312	6			
10 crackers	28	2.6	20.1	0.9		0	.00	0.3			34	25	.40		
soda, Sultana	60		1.0								115				
4 crackers	14	1.0	11.0								20				
square cheese, Planters	140		7.0								270				
1 oz	28	4.0	15.0								85				
tomato & celery, Great Crisps	70		4.0								160				
9 crackers	14	1.0	8.0								35				
triscuit	60		2.0								90[a]				
3 crackers	14	1.0	10.0								40				
twigs, sesame & cheese, Nabisco	70		4.0								200				
7 crackers	14	1.0	8.0								25				
uneeda unsalted tops, Nabisco	60		2.0								100				
3 crackers	14	1.0	10.0								20				
vegetable thins, Nabisco	70		4.0								100				
7 crackers	14	1.0	8.0								30				
waverly, Nabisco	70		3.0								160				
4 crackers	14	1.0	10.0								15				
wheatsworth stone ground wheat	70		3.0								135				
5 crackers	14	1.0	9.0								35				
wheat thins, Nabisco	70		3.0								120[b]				
8 crackers	14	1.0	9.0								25				
zwieback	30	0.4	0.6				0	.00			18	1			
1 piece	7	0.7	5.2	0.2		0	.00	0.1			11	5	.00		
zwieback, Nabisco	60		1.0								20				
2 pieces	14	2.0	10.0								20				

17.4. FRENCH TOAST

	KCAL / WT (g)	H₂O (g) / PRO (g)	FAT (g) / CHO (g)	PUFA (g) / SFA (g)	CHOL (mg) / DFIB (g)	A (RE) / A (IU)	C (mg) / B-1 (mg)	B-2 (mg) / NIA (mg)	B-6 (mg) / B-12 (mcg)	FOL (mcg) / PANT (mg)	Na (mg) / K (mg)	Ca (mg) / P (mg)	Mg (mg) / Fe (mg)	Zn (mg) / Cu (mg)	Mn (mg)
frzn															
Am Hosp Co	71		0.8		7		0	.02			33	24			
1.3 oz	37	4.3	11.6			436	.03				48		.55		
Aunt Jemima	168	46.2	3.9					.17	.30	37	430	94	12	1.00	.000
2 slices	85	6.4	26.8			173	.15	1.6	.90	1.00	97	120	1.44	.080	
cinn swirl, Aunt Jemima	190	43.4	6.0					.17	.30	48	359	94	16	1.00	.000
2 slices	85	6.4	27.6			183	.15	1.6	.90	1.00	126	126	1.44	.060	
raisin, Aunt Jemima	185	42.3	4.2					.17	.20	18	423	96	17	1.00	.000
2 slices	85	7.3	29.4			120	.15	1.2	.60	1.00	160	99	1.44	.100	

[a] Low salt triscuits contain 35 mg sodium. [b] Low salt wheat thins contain 35 mg sodium.

	KCAL	H₂O (g)	FAT (g)	PUFA (g)	CHOL (mg)	A (RE)	C (mg)	B-2 (mg)	B-6 (mg)	FOL (mcg)	Na (mg)	Ca (mg)	Mg (mg)	Zn (mg)	Mn (mg)
	WT (g)	PRO (g)	CHO (g)	SFA (g)	DFIB (g)	A (IU)	B-1 (mg)	NIA (mg)	B-12 (mcg)	PANT (mg)	K (mg)	P (mg)	Fe (mg)	Cu (mg)	
homemade	153	34.3	6.7				0	.16	.04	18	257	72	12	.55	
1 slice	65	5.7	17.2			111	.12	1.0	.29	.53	86	87	1.34	.059	

17.5. MUFFINS

	KCAL	H₂O (g)	FAT (g)	PUFA (g)	CHOL (mg)	A (RE)	C (mg)	B-2 (mg)	B-6 (mg)	FOL (mcg)	Na (mg)	Ca (mg)	Mg (mg)	Zn (mg)	Mn (mg)
	WT (g)	PRO (g)	CHO (g)	SFA (g)	DFIB (g)	A (IU)	B-1 (mg)	NIA (mg)	B-12 (mcg)	PANT (mg)	K (mg)	P (mg)	Fe (mg)	Cu (mg)	
apple steusel, frzn, Am Hosp Co	225		10.6		37		0	.03			517	20			
3 oz	85	3.0	29.4			269	.03	0.1			243		2.00		
banana, frzn, Am Hosp Co	242		10.3		51			.12			251	65			
3 oz	85	5.8	27.2			178	.01	1.2			90		.60		
banana nut, frzn, Am Hosp Co	371		15.8		64		0	.13			676	21			
4 oz	113	6.4	52.2			90	.14	0.0			219		1.60		
blueberry															
from mix	126	13.0	4.3				0	.07			200	14	4	.20	
1 muffin	40	2.4	19.5			40	.10	1.0			48	80	.48	.032	
homemade, enr	112	15.6	3.7				0	.08			253	34			
1 muffin	40	2.9	16.8	1.1		90	.06	0.5			46	53	.60		
muffin mix, Duncan Hines	99	5.1	2.7		0			.06			147	9	3	.11	
1/12 pkg (1 muffin)	27	1.5	17.0				.07	1.0			30	57	.35	.022	
bran & honey muffin mix, Duncan Hines	98	1.1	2.8		0			.06			161	8	20	.35	
1/12 pkg (1 muffin)	23	1.7	16.3				.08	1.2			46	110	.83	.048	
bran, homemade	112	14.2	5.1				2	.11	.11	17	168	54	35	1.08	
1 muffin	40	3.0	16.7			206	.10	1.3	.09	.24	99	111	1.26	.085	
carrot, frzn, Am Hosp Co	260		13.7		53		0	.09			248	22			
3 oz	85	5.1	25.5			26	.04				109		.34		
corn															
from mix	130	12.2	4.2				0	.08			192	96			
1 muffin	40	2.8	20.0	1.2		100	.07	0.6			44	152	.60		
from mix, Dromedary	120		4.0								270				
1 muffin		3.0	20.0								50				
homemade w/ enr degermed corn meal	126	13.1	4.0				0	.09			192	42			
1 muffin	40	2.8	19.2	1.2		120	.08	0.6			54	68	.70		
homemade w/ whole ground corn meal	115	15.1	4.1				0	.07			198	45			
1 muffin	40	2.9	17.0	1.2		120	.07	0.4			53	86	.60		
muffin mix, Flako	116	2.2	2.2					.03	.02	3	317	18	5	.00	.000
1 oz dry mix	28	1.9	19.5			32	.05	0.6	.00	.00	26	25	.38	.020	
date nut, frzn, Am Hosp Co	427		16.7		73		9	.02			612	45			
4 oz	113	5.7	63.5			124	.01	1.1			256		2.72		
english muffin															
cinn raisin, refrig dough	140		1.0								250				
1 muffin	57	4.0	30.0								105				
multi grain, refrig dough	130		1.0								270				
1 muffin	57	4.0	26.0								80				
plain	135	23.9	1.1		0		0	.18	.02	18	364	92	11	.41	
1 muffin	57	4.5	26.2			0	.26	2.1	.00	.29	319	64	1.61	.177	
plain, Earth Grains	150		1.0								450				
1 muffin	65	5.0	30.0								70				
plain, refrig dough	120		1.0								280				
1 muffin	57	4.0	24.0								45				
plain, toasted	145	14.5	1.2		0		0	.19	.02	19	391	99	12	.44	
1 muffin	50	4.8	28.1			0	.22	2.3	.00	.31	343	68	1.73	.190	
plain, Wonder	130		1.0		0						280				
1 muffin	56	4.0	26.0												
sour dough, Wonder	130		1.0		0						250				
1 muffin	56	4.0	27.0												
wheatberry, Earth Grains	150		1.0								460				
1 muffin	65	6.0	28.0								110				
whole wheat, Earth Grains	170		2.0								540				
1 muffin	82	7.0	33.0								190				
w/ raisins, Earth Grains	150		1.0								390				
1 muffin	65	5.0	31.0								135				

	KCAL	H₂O (g)	FAT (g)	PUFA (g)	CHOL (mg)	A (RE)	C (mg)	B-2 (mg)	B-6 (mg)	FOL (mcg)	Na (mg)	Ca (mg)	Mg (mg)	Zn (mg)	Mn (mg)
	WT (g)	PRO (g)	CHO (g)	SFA (g)	DFIB (g)	A (IU)	B-1 (mg)	NIA (mg)	B-12 (mcg)	PANT (mg)	K (mg)	P (mg)	Fe (mg)	Cu (mg)	
w/ raisins, Wonder	140		2.0		0						280				
1 muffin	56	4.0	27.0												
plain, homemade, enr	118	15.2	4.0				0	.09			176	42			
1 muffin	40	3.1	16.9	1.0		40	.07	0.6			50	60	.60		
toaster muffin, Pillsbury															
apple spice	138		4.5				0	.05			123	16			
1 muffin	39	1.9	24.1			12	.07	0.6			50	75	.62		
banana nut	141		6.2				1	.06			83	10			
1 muffin	39	2.1	19.7			28	.08	0.7			69	35	.66		
blueberry	131		4.1				0	.05			122	9			
1 muffin	39	1.6	23.0			10	.07	0.7			26	55	.53		
corn, old fashioned	141		5.1				0	.08			253	20			
1 muffin	38	2.7	20.9			42	.09	0.7			52	130	.49		
raisin bran	121		4.1				0	.07			222	27			
1 muffin	39	2.4	19.9			15	.07	0.9			118	108	1.01		

17.6. PANCAKES

	KCAL	H₂O (g)	FAT (g)	PUFA (g)	CHOL (mg)	A (RE)	C (mg)	B-2 (mg)	B-6 (mg)	FOL (mcg)	Na (mg)	Ca (mg)	Mg (mg)	Zn (mg)	Mn (mg)
	WT (g)	PRO (g)	CHO (g)	SFA (g)	DFIB (g)	A (IU)	B-1 (mg)	NIA (mg)	B-12 (mcg)	PANT (mg)	K (mg)	P (mg)	Fe (mg)	Cu (mg)	
from frzn batter															
blueberry, Aunt Jemima	205	61.2	1.6					.23	.11	9	698	68	17	.00	.000
three 4" pancakes	113	6.2	41.5			54	.32	2.0	.89	.00	93	329	1.44	.090	
buttermilk, Aunt Jemima	212	59.2	1.5					.28	.06	14	733	91	18	.00	.000
three 4" pancakes	113	7.1	42.6			56	.34	2.2	.28	.00	129	340	1.70	.090	
plain, Aunt Jemima	210	60.2	1.6					.24	.05	19	857	68	16	.00	.000
three 4" pancakes	113	6.6	42.2			53	.35	2.1	.37	.00	92	329	1.51	.090	
from mix															
buttermilk complete, Hungry Jack	180		1.0				0	.20			710	154			
three 4" pancakes		4.2	38.2			51	.24	2.1			80	368	2.65		
extra lights complete, Hungry Jack	190		2.5				0	.18			698	133			
three 4" pancakes		4.2	37.1			42	.28	2.1			32	357	2.86		
plain	159	39.1	5.9				0	.16	.15	8	431	96	14	.52	
1 large (7 T batter)	73	5.0	21.3			104	.10	0.7	.96	.26	117	191	.72	.044	
frzn															
apple, Am Hosp Co	90		0.4		0		2	.07			262	11			
2 oz	57	1.6	20.3			6	.08	0.6			48				
blueberry, Am Hosp Co	93		0.4		0		2	.06			256	11			
2 oz	57	1.6	20.8			11	.08	0.5			44		.70		
blueberry, Aunt Jemima	249	45.8	4.0					.34	.05	55	789	67	20	1.00	.000
three 4" pancakes	106	6.8	46.4			131	.24	2.1	.64	1.00	136	380	1.95	.100	
buttermilk, Aunt Jemima	240	47.6	3.7					.33	.04	29	778	76	20	1.00	.000
three 4" pancakes	106	7.0	44.7			0	.26	2.1	.59	1.00	103	309	2.06	.080	
buttermilk, Pillsbury Microwave	260	44.3	3.6				0	.22			592	90			
3 pancakes	108	5.6	50.8			26	.32	2.4			85	162	1.30		
plain, Aunt Jemima	246	46.0	3.7					.41	.05	26	777	66	19	1.00	.000
three 4" pancakes	106	6.7	46.6			83	.25	2.2	.32	1.00	133	367	1.17	.070	
plain, Pillsbury Microwave	241	47.7	3.5				0	.21			542	85			
3 pancakes	106	5.3	46.6			3	.21	2.1			85	153	1.17		
homemade, enr, plain	62	13.5	1.9				0	.06			115	27			
one 4" pancake	27	1.9	9.2	0.5		30	.05	0.4			33	38	.40		
pancake batter, panshakes, Hungry Jack	197	5.0	3.2				0	.27			834	45			
amt for three 4" pancakes	54	4.2	38.8			0	.41	3.4			38	400	2.16		
pancake/waffle mix															
blueberry, Hungry Jack	171	22.4	1.0				1	.14			772	52			
amt for three 4" pancakes	68	3.2	37.4			33	.20	1.8			51	116	1.63		
buckwheat, Aunt Jemima	107	2.9	0.8					.12	.05	11	432	150	42	.00	.000
1/4 cup dry mix	31	3.4	21.3			0	.17	1.4	.00	.00	99	273	1.66	.110	

	KCAL	H₂O (g)	FAT (g)	PUFA (g)	CHOL (mg)	A (RE)	C (mg)	B-2 (mg)	B-6 (mg)	FOL (mcg)	Na (mg)	Ca (mg)	Mg (mg)	Zn (mg)	Mn (mg)
WT (g)	PRO (g)	CHO (g)	SFA (g)	DFIB (g)	A (IU)	B-1 (mg)	NIA (mg)	B-12 (mcg)	PANT (mg)	K (mg)	P (mg)	Fe (mg)	Cu (mg)		
buttermilk, Aunt Jemima	175	5.1	0.7					.28	.05	12	832	230	28	.00	.000
1/3 cup dry mix	51	5.7	36.5			0	.34	2.4	.00		114	342	1.85	.100	
buttermilk, complete, Aunt Jemima	264	5.8	3.0					.42	.09	19	858	403	37	1.00	.000
1/2 cup dry mix	71	8.1	51.2			0	.39	2.8	.00	1.00	236	514	2.40	.140	
buttermilk, Hungry Jack	131	3.0	0.4				0	.15			514	26			
amt for three 4″ pancakes	37	3.0	28.5			0	.19	1.1			61	226	1.52		
extra lights, Hungry Jack	119	3.1	0.3				0	.09			429	99			
amt for three 4″ pancakes	34	2.6	26.2			0	.26	1.1			102	228	1.22		
plain, Aunt Jemima	108	2.9	0.6					.06	.06	4	450	140	19	.00	.000
1/4 cup dry mix	31	3.0	22.5			0	.15	0.9	.00	.00	61	242	1.17	.060	
plain, complete, Aunt Jemima	272	7.0	3.6					.36	.16	86	881	155	33	1.00	.000
1/2 cup dry mix	74	7.5	52.3			50	.39	2.5	.00	1.00	182	503	2.43	.100	
whole wheat, Aunt Jemima	142	4.0	0.5					.17	.06	21	587	181	51	1.00	1.000
1/3 cup dry mix	43	5.9	28.5			0	.15	2.0	.00	.00	184	311	1.78	.210	

17.7. PASTA

	KCAL	H₂O (g)	FAT (g)	PUFA (g)	CHOL (mg)	A (RE)	C (mg)	B-2 (mg)	B-6 (mg)	FOL (mcg)	Na (mg)	Ca (mg)	Mg (mg)	Zn (mg)	Mn (mg)
macaroni, enr, ckd	159	100.1	0.7		0		0	.11	.01	3	1	11	25	.70	.280
1 cup	140	5.2	33.7	0.6		0	.20	1.5	.00	.15	85	70	2.25	.028	
macaroni salad, cnd, Joan of Arc	200	75.0	12.8					.09			830	13			
1/2 cup	112	2.9	18.5			22	.09	0.9				47	.90		
noodles, chow mein, cnd	220	0.5	10.6												
1 cup	45	5.9	26.1												
noodles															
enr, ckd	200	112.6	2.4		50		0	.19	.01	7	3	16	28	.20	.320
1 cup	160	6.6	37.3	0.2		110	.26	1.9	.50	.25	70	94	2.40	.136	
Mueller's	220	5.5	2.9		55			.30			9				
2 oz dry	57	8.0	39.8				.60	4.2					2.00		
Mueller's golden rich	220	6.2	3.2		70			.30			5				
2 oz dry	57	8.0	38.8				.60	4.2					2.00		
lasagne, Mueller's	210	5.5	1.4		0			.30			3				
2 oz dry	57	7.8	41.6				.60	4.2					2.00		
parsleyed, frzn, Banquet	186		10.0		35		0	.07			568	40			
4 oz	113	5.0	19.0			270	.07	0.4			21		.60		
romanoff w/ cheese sce, frzn, Am Hosp	198		5.6		9		0	.15			124	199			
Co—4 oz	113	16.8	20.4			264	.09				256		.60		
pasta salad, italian, Fresh Chef	111		5.3				1	.06			334	25			
3.2 oz	92	2.4	13.4			234	.06	0.7			47		1.60		
pasta salad, seafood, Fresh Chef	229		16.2				1	.12			426	41			
4.3 oz	123	6.0	14.7			176	.09	1.1			117		1.80		
spaghetti, enr, ckd	159	100.1	0.7		0		0	.11	.01	3	1	11	25	.70	.322
1 cup	140	5.2	33.7	0.6		0	.20	1.5	.00	.15	85	70	2.25	.028	
spaghetti/macaroni, Mueller's	210	5.8	1.1		0			.30			2	8			
2 oz dry	57	7.3	42.1				.60	4.2					2.00		

17.8. ROLLS

	KCAL	H₂O (g)	FAT (g)	PUFA (g)	CHOL (mg)	A (RE)	C (mg)	B-2 (mg)	B-6 (mg)	FOL (mcg)	Na (mg)	Ca (mg)	Mg (mg)	Zn (mg)	Mn (mg)
brown 'n serve, Wonder															
gem style	80		2.0		5						140				
1 roll	28	2.0	13.0												
half & half	80		2.0		5						140				
1 roll	28	2.0	13.0												
home bake	80		2.0		5						130				
1 roll	28	2.0	13.0												
w/ buttermilk	80		2.0		5						140				
1 roll	28	2.0	13.0												
dinner, Home Pride	80		2.0		5						170				
1 roll	28	2.0	14.0												
dinner/pan	85	9.0	2.1				0	.09	.02	11	155	33	6	.20	
1 roll	28	2.4	14.0			0	.14	1.1		.15	36	44	.83	.023	

	KCAL	H_2O (g)	FAT (g)	PUFA (g)	CHOL (mg)	A (RE)	C (mg)	B-2 (mg)	B-6 (mg)	FOL (mcg)	Na (mg)	Ca (mg)	Mg (mg)	Zn (mg)	Mn (mg)
	WT (g)	PRO (g)	CHO (g)	SFA (g)	DFIB (g)	A (IU)	B-1 (mg)	NIA (mg)	B-12 (mcg)	PANT (mg)	K (mg)	P (mg)	Fe (mg)	Cu (mg)	
french															
Earth Grains	100		1.0								270				
1 roll	34	4.0	19.0								45				
enr	137	16.0	0.4				0	.18	.03	29	287	8	12	.30	
1 roll	50	4.3	28.3			0	.20	2.0	.00	.24	50	41	1.48	.050	
sourdough, Earth Grains	100		1.0								160				
1 roll	34	3.0	19.0								40				
unenr	137	16.0	0.4				0	.06	.03	29	287	8	12	.30	
1 roll	50	4.3	28.3			0	.05	0.6	.00	.25	50	41	.63	.050	
from mix, Pillsbury	240		4.0								430				
2 rolls		7.0	42.0								90				
hamburger	114	13.6	2.1				0	.13	.01	15	241	54	8	.25	
1 roll	40	3.4	20.1			0	.20	1.6	.00	.21	37	33	1.19	.066	
hamburger, Wonder	80		1.0		5						150				
1 roll	28	2.0	14.0												
hoagie	400		7.0		5						840				
1 roll	142	13.0	73.0												
home style, Rich's dough, baked	152	20.8	2.4					.20			335	27	12	.34	
2 rolls	50	4.5	28.1				.32	2.1			68	76	2.04	.180	
hot dog	114	13.6	2.1				0	.13	.01	15	241	54	8	.25	
1 roll	40	3.4	20.1			0	.20	1.6	.00	.21	37	33	1.19	.070	
hot dog, Wonder	80		1.0		5						150				
1 roll	28	2.0	14.0												
kaiser, Earth Grains	190		2.0								520				
1 roll	65	7.0	37.0								85				
onion, Earth Grains	190		2.0								530				
1 roll	65	7.0	35.0								100				
pan, Wonder	80		1.0		5						140				
1 roll	28	2.0	14.0												
popover, enr, homemade	90	22.0	3.7				0	.10			88	38			
1 popover	40	3.5	10.3	1.3		130	.06	0.4			60	56	.60		
popover mix, Flako	102	3.3	1.2					.02	.02	6	273	9	8	.00	.000
1 oz dry mix	28	3.2	19.6			0	.07	0.5	.00	.00	31	28	.45	.040	
refrig dough															
butterflake, Pillsbury	110	14.1	3.8				0	.08			415	5			
1 roll	38	2.5	16.3			0	.13	1.0			20	165	.66		
crescent, Continental	190		8.0								600				
2 rolls	57	3.0	25.0								40				
crescent, Pillsbury	200	17.7	11.2				0	.12			463	11			
2 rolls	57	3.4	22.2			0	.17	1.5			127	215	1.31		
submarine, Earth Grains	180		1.0								500				
1/2 roll	64	7.0	36.0								85				
wheat, Earth Grains	110		1.0								260				
1 roll	40	4.0	22.0								65				
whole wheat	72	9.0	0.8				0	.04			158	30			
1 roll	28	2.8	14.6			0	.10	0.8			82	79	.67		
weiner wrap, Pillsbury	60	7.9	1.4				0	.05			373	4			
1 wrap	22	1.5	10.5			0	.18	0.8			23	112	.75		

17.9. STUFFING

	KCAL	H_2O (g)	FAT (g)	PUFA (g)	CHOL (mg)	A (RE)	C (mg)	B-2 (mg)	B-6 (mg)	FOL (mcg)	Na (mg)	Ca (mg)	Mg (mg)	Zn (mg)	Mn (mg)
	WT (g)	PRO (g)	CHO (g)	SFA (g)	DFIB (g)	A (IU)	B-1 (mg)	NIA (mg)	B-12 (mcg)	PANT (mg)	K (mg)	P (mg)	Fe (mg)	Cu (mg)	
from mix															
americana san francisco style, Stove	173	72.0	8.7	0.6	21		0	.11	.04	29	638	34	15	.27	
Top—*1/2 cup*	107	4.1	19.9	5.0		293	.15	1.4	.02	.22	77	47	1.17	.079	
bread	416	122.8	25.6				0	.18			1008	80			
1 cup	200	8.8	39.4	13.1		840	.10	1.6			116	132	2.00		
chicken flavored, Stove Top	176	72.0	8.8	0.6	21		1	.10	.04	24	560	33	12	.25	
1/2 cup	107	3.9	20.5	5.0		319	.15	1.3	.02	.20	86	41	1.23	.081	
chicken flavored w/ rice, Stove Top	182	72.3	8.8	0.6	22		1	.09	.04	23	557	32	12	.28	
1/2 cup	109	4.0	22.0	5.0		324	.16	1.3	.02	.17	84	42	1.27	.082	

	KCAL	H₂O (g)	FAT (g)	PUFA (g)	CHOL (mg)	A (RE)	C (mg)	Vitamins B-2 (mg)	B-6 (mg)	FOL (mcg)	Na (mg)	Ca (mg)	Minerals Mg (mg)	Zn (mg)	Mn (mg)
	WT (g)	PRO (g)	CHO (g)	SFA (g)	DFIB (g)	A (IU)	B-1 (mg)	NIA (mg)	B-12 (mcg)	PANT (mg)	K (mg)	P (mg)	Fe (mg)	Cu (mg)	
chicken florentine, Stove Top	207	62.9	12.7	0.7	32		4	.10	.06	37	670	56	26	.32	
1/2 cup	102	4.2	19.7	7.4		1119	.11	1.1	.03	.12	178	66	1.33	.097	
cornbread, Stove Top	175	72.0	8.5	0.5	21		1	.07	.04	12	568	24	13	.21	
1/2 cup	107	3.2	21.5	5.0		334	.10	0.8	.01	.09	84	34	.94	.066	
for beef, Stove Top	178	71.9	8.6	0.6	21		1	.10	.05	25	581	33	11	.26	
1/2 cup	108	4.1	21.4	5.0		815	.16	1.3	.02	.18	82	42	1.30	.085	
for pork, Stove Top	174	72.0	8.7	0.6	21		1	.11	.05	27	620	36	13	.24	
1/2 cup	107	3.8	20.4	5.0		353	.23	1.4	.02	.19	91	45	1.40	.100	
garden herb, Stove Top	208	56.3	12.6	0.7	32		2	.09	.05	19	416	41	16	.30	
1/2 cup	95	3.7	20.8	7.4		576	.13	1.2	.03	.13	151	48	1.28	.099	
homestyle herb, Stove Top	173	61.7	8.6	1.2	16		1	.09	.03	19	530	37	11	.25	
1/2 cup	96	3.8	20.1	4.1		314	.11	1.2	.02	.11	74	39	1.28	.081	
long grain & wild rice flavor, Stove Top	183	72.3	8.7	0.6	21		0	.10	.03	22	552	32	12	.28	
1/2 cup	109	3.9	22.2	5.0		325	.16	1.3	.02	.15	70	41	1.33	.079	
savory herbs flavor, Stove Top	175	72.0	8.8	0.6	22		1	.10	.04	24	579	36	13	.26	
1/2 cup	107	4.0	20.3	5.0		345	.15	1.3	.02	.16	89	41	1.45	.082	
turkey flavor, Stove Top	174	72.0	8.7	0.6	21		1	.09	.04	24	628	33	13	.24	
1/2 cup	107	3.7	20.6	5.0		318	.15	1.2	.02	.16	83	41	1.21	.080	
veg & almond, Stove Top	215	56.2	14.2	1.2	32		9	.12	.06	22	439	47	23	.40	
1/2 cup	95	4.0	18.8	7.6		1376	.11	1.2	.03	.15	142	59	1.44	.122	
wild rice & mushroom, Stove Top	205	56.2	12.4	0.7	31		17	.12	.05	17	582	31	15	.33	
1/2 cup	95	3.7	20.6	7.3		857	.12	1.5	.03	.36	140	57	1.20	.135	
frzn															
apple, Am Hosp Co	84		2.1		9		1	.02			38	11			
1 oz	28	1.4	12.8			11	.06	0.2			20		.27		
chicken flavored, Banquet	281		8.0		15		0	.27			693	65			
4 oz	113	8.0	44.0			127	.55	3.3			99		3.50		
chicken, Stuffing Originals	170	61.1	7.5				0	.12			686	27			
1/2 cup	96	3.6	21.8			127	.32	1.7			93	42	1.44		
cornbread, Stuffing Originals	170	58.9	6.4				0	.10			662	20			
1/2 cup	95	3.3	24.7			125	.28	1.5			100	40	1.02		
mushroom, Stuffing Originals	150	60.7	6.4				0	.12			763	25			
1/2 cup	92	4.0	18.7			98	.18	1.7			100	46	1.29		

17.10. TORTILLAS

	KCAL	H₂O	FAT	PUFA	CHOL	A	C	B-2	B-6	FOL	Na	Ca	Mg	Zn	Mn
corn															
enr	67	13.6	1.1				0	.14	.09	6	53	42	20	.43	.123
1 tortilla	30	2.1	12.8		1.0		.20	1.5	.00	.06	52	55	1.42	.090	
taco/tostada shell	50	0.4	2.2					.02			72	16	11	.14	
1 shell	11	1.0	7.2				.03	0.2	.00		25	25	.29	.035	
unenr	67	13.6	1.1				0	.03	.09	6	53	42	20	.43	
1 tortilla	30	2.1	12.8				.05	0.4	.00	.06	52	55	.57	.090	
flour, El Charrito	170		4.0								310				
2 tortillas		5.0	30.0								60				

17.11. WAFFLES

	KCAL	H₂O	FAT	PUFA	CHOL	A	C	B-2	B-6	FOL	Na	Ca	Mg	Zn	Mn
homemade, plain	245	27.6	12.6				0	.24	.05	14	445	154	17	.65	
1 large waffle	75	6.9	25.7			140	.18	1.5	.36	.60	129	135	1.48	.050	
from mix, plain, enr	206	31.2	8.0				0	.17			515	179			
1 waffle (7″ diameter)	75	6.6	27.2	2.7		170	.11	0.7			146	257	1.00		
frzn															
apple & cinn, Aunt Jemima	173	31.0	4.3					.34	.40	30	503	100	14	1.00	.000
2 waffles	71	4.6	29.0			0	.30	3.0	1.50	.00	87	270	3.60	.050	
blueberry, Aunt Jemima	173	31.0	4.3					.34	.40	30	503	100	14	1.00	.000
2 waffles	71	4.6	29.0			0	.30	3.0	1.50	.00	87	270	3.60	.050	
blueberry/raisin & bran, nutri-grain,	130		5.0				0	.17			300	20			
Eggo—*1 rd waffle*	39	3.0	18.0			500	.15	2.0					1.80		
buttermilk, Aunt Jemima	175	31.0	4.3					.34	.40	30	550	100	14	1.00	.000
2 waffles	71	4.6	29.0			0	.30	3.0	1.50	.00	79	270	3.60	.050	

	KCAL	H₂O (g)	FAT (g)	PUFA (g)	CHOL (mg)	A (RE)	C (mg)	Vitamins B-2 (mg)	B-6 (mg)	FOL (mcg)	Na (mg)	Ca (mg)	Minerals Mg (mg)	Zn (mg)	Mn (mg)
	WT (g)	PRO (g)	CHO (g)	SFA (g)	DFIB (g)	A (IU)	B-1 (mg)	NIA (mg)	B-12 (mcg)	PANT (mg)	K (mg)	P (mg)	Fe (mg)	Cu (mg)	
buttermilk/homestyle, Eggo	120		5.0				0	.17			300	20			
1 rd waffle	39	3.0	16.0			500	.15	2.0					1.80		
plain	95	13.7	3.2				0	.18	.09	1	235	28	7	.28	
1 waffle	34	2.0	14.2			436	.15	1.8		.12	71	130	1.65	.021	
plain, Aunt Jemima	173	31.0	4.3					.34	.40	30	539	100	14	1.00	.000
2 waffles	71	4.6	29.0			0	.30	3.0	1.50	.00	86	270	3.60	.050	
raisin, Aunt Jemima	200	34.6	3.9					.34	.40	11	526	100	19	.00	.000
2 waffles	82	4.7	36.4			0	.30	3.0	1.50	.00	163	257	3.60	.110	

18. INFANT FORMULA

						Vitamins					Minerals				
	KCAL	H_2O (g)	FAT (g)	PUFA (g)	CHOL (mg)	A (RE)	C (mg)	B-2 (mg)	B-6 (mg)	FOL (mcg)	Na (mg)	Ca (mg)	Mg (mg)	Zn (mg)	Mn (mg)
	WT (g)	PRO (g)	CHO (g)	SFA (g)	DFIB (g)	A (IU)	B-1 (mg)	NIA (mg)	B-12 (mcg)	PANT (mg)	K (mg)	P (mg)	Fe (mg)	Cu (mg)	
					INOS (mg)	D (IU)	E (IU)	K (mcg)	BIO (mcg)	CHLN (mg)	Cl (mg)	I (mcg)	Mo (mg)		
	TRY (mg)	THR (mg)	ISO (mg)	LEU (mg)	LYS (mg)	MET (mg)	CYS (mg)	PHE (mg)	TYR (mg)	VAL (mg)	ARG (mg)	HIS (mg)			
enfamil, Mead Johnson	20	26.8	1.1				2	.03	.01	3	5	14	2	.16	.003
1 fl oz	30	0.4	2.1			62	.02	0.3	.05	.09	22	9	.03[a]	.019	
(whey, nonfat milk, coconut oil, corn					0.9	12	0.6	2	0.5	3.1	12	2			
oil & lactose)	7	23	27	46	31	9	5	17	20	27		9			
enfamil human milk fortifier, Mead	14	0.2	0.0								7	60	4	.80	
Johnson[b]—*4 packets*	3.8	0.7	2.7								16	33		.040	
(corn syrup solids, whey protein											18				
concentrate & casein)															
enfamil premature formula, 20 cal/fl oz,	20	26.6	1.0				7	.07	.05	7	8	23	1	.20	.003
Mead Johnson—*1 fl oz* (corn syrup	30	0.6	2.2			240	.05	0.8	.06	.24	22	12	.05	.032	
solids, whey protein concentrate, nonfat					0.9	66	0.9	3	0.4	1.5	17	2			
milk, soy oil, MCTs & coconut oil)	9	33	35	58	46	14	6	23	27	36		12			
enfamil premature formula, 24 cal/fl oz,	24	25.7	1.2				8	.08	.06	8	9	28	1	.24	.003
Mead Johnson—*1 fl oz* (corn syrup	30	0.7	2.6			286	.06	1.0	.07	.29	26	14	.06	.038	
solids, whey protein concentrate, nonfat					1.1	79	1.1	3	0.5	1.8	20	2			
milk, soy oil, MCTs & coconut oil)	11	40	42	70	55	16	8	28	32	44		12			
isomil, 20 cal/fl oz, Ross Labs	20	27.0	1.1	0.4	0	18	2	.02	.01	3	9	21	2	.15	.006
1 fl oz	30	0.5	2.0	0.5	0.0	60	.01	0.3	.09	.15	28	15	.36	.015	
(soy protein isolate, corn syrup,					1.0	12	0.6[c]	3.0	0.9	1.6	13	3			
sucrose, soy & coconut oils)	6	20	23	42	32	12	6	27	19	23	38	13			
isomil, SF, 20 cal/fl oz, Ross Labs	20	27.0	1.1	0.4	0	18	2	.02	.01	3	9	21	2	.15	.006
1 fl oz	30	0.6	2.0	0.5	0.0	60	.01	0.3	.09	.15	23	15	.36	.015	
(soy protein isolate, corn syrup solids,					1.0	12	0.6[c]	3.0	0.9	1.6	18	3			
soy & coconut oils)	6	23	26	47	36	13	7	30	21	25	43	14			
lofenalac, Mead Johnson[d]—*1 fl oz*	20		0.8				2	.02	.01	3	9	19	2	.12	.032
(corn syrup solids, enzymatic digest of	30	0.7	2.7			50	.02	0.3	.06	.09	20	14	.38		
casein processed to remove phe, corn					0.9	12	0.3	3	1.6	2.7	14	1	.018		
oil & modified tapioca starch)	9	34	38	74	72	24	3	3	35	60		21			
low met diet, Mead Johnson[e]	20	27.0	1.1				2	.02	.01	3	9	19	2	.16	.006
1 fl oz	30	0.6	2.0			50	.02	0.2	.06	.09	23	15	.38	.019	
(corn syrup solids, soy protein isolate,					0.9	12	0.3	3	1.6	1.6	16	2			
coconut oil & corn oil)	7	19	28	46	36	6	6	29	20	28		14			
low phe/try diet, Mead Johnson[f]—*5 fl oz*	100	134.0	3.9				8	.09	.07	16	47	94	11	.63	.152
(corn syrup solids, specially processed	150	3.3	13.0			250	.08	1.3	.30	.48	102	70	1.87	.087	
casein hydrolysate, corn oil, modified					4.8	63	1.5	16	7.8	13.3	70	7			
tapioca starch)	43	172	191	370	360	119	13	16	<8	300		99			
mono- and disaccharide-free diet, Mead	100	127.0	4.2				12	.09	.06	16	43	94	11	.62	.031
Johnson[g]—*5 fl oz* (modified tapioca	150	2.8	13.4			380	.08	1.3	.31	.47	109	62	1.88	.094	
starch, MCTs, casein enzymically hydro-					4.7	75	3.8	19	7.8	13.3	86	7			
lyzed & charcoal treated & corn oil)	70	220	260	450	380	136	70	210	101	330		132			
msud diet, Mead Johnson[h]—*5 fl oz*	100	134.0	3.9				8	.09	.06	16	47	94	11	.78	.031
(corn syrup solids, casein enzymically	150	3.3	13.0			310	.08	1.3	.31	.47	102	70	1.88	.094	
hydrolyzed, corn oil & modified					4.7	62	3.1	16	7.8	13.3	70	7			
tapioca starch)	41	114			105	51	51	114	134			51			
nursoy, 20 cal/fl oz, Wyeth—*1 fl oz*	20	26.4	1.1	0.1	0	18	2	.03	.01	1	6	18	2	.15	.006
	30	0.6	2.0	0.4	0.0	59	.02	0.1	.06	.09	21	12	.34	.014	
					0.8	12	0.2	3	0.4	3		2			
	7	21	34	52	39	16	6	36	26	34	49	17			

[a] Enfamil w/ iron contains .38 mg.
[b] Nutritional supplement to be added to mother's milk for premature infants.
[c] Alpha-tocopherol is 0.4 mg.
[d] Low phenylalanine formula for infants & children w/ phenylketonuria.
[e] Formula w/o added methionine for management of infants w/ homocystinuria.
[f] Formula low in phenylalanine & tyrosine for use in tyrosinemia.
[g] Protein hydrolysate formula base for use w/ added carbohydrate.
[h] Free of branched chain amino acids for use w/ maple syrup urine disease.

Item	KCAL WT (g) — TRY (mg)	H2O (g) PRO (g) — THR (mg)	FAT (g) CHO (g) — ISO (mg)	PUFA (g) SFA (g) — LEU (mg)	CHOL (mg) DFIB (g) INOS (mg) LYS (mg)	A (RE) A (IU) D (IU) MET (mg)	C (mg) B-1 (mg) E (IU) CYS (mg)	B-2 (mg) NIA (mg) K (mcg) PHE (mg)	B-6 (mg) B-12 (mcg) BIO (mcg) TYR (mg)	FOL (mcg) PANT (mg) CHLN (mg) VAL (mg)	Na (mg) K (mg) Cl (mg) ARG (mg)	Ca (mg) P (mg) I (mcg) HIS (mg)	Mg (mg) Fe (mg) Mo (mg)	Zn (mg) Cu (mg)	Mn (mg)
nutramigen, Mead Johnson[a]—*1 fl oz*	20	26.6	0.8				2	.02	.01	3	9	19	2	.16	.006
(corn syrup solids, corn oil, casein	30	0.6	2.7			62	.02	0.3	.06	.09	22	12	.38	.019	
enzymatically hydrolyzed & charcoal-					0.9	12	0.6	3	1.6	2.7	17	1			
treated & modified corn starch)	7	32	40	68	56	20	3	32	7	50		20			
portagen, 20 cal/fl oz, Mead Johnson[b]	20	26.8	1.0				2	.04	.04	3	10	19	4	.19	.025
1 fl oz	30	0.7	2.3			156	.03	0.4	.12	.21	25	14	.38	.031	
(corn syrup solids, MCTs, sodium						16	0.6	3	1.6	2.7	17	1			
caseinate, sucrose & corn oil)	9	29	40	70	60	20	2	38	40	50		22			
pregestimil, Mead Johnson[c]—*1 fl oz*	20	25.4	0.8				2	.02	.01	3	9	19	2	.12	.006
(corn syrup solids, casein enzymatically	30	0.6	2.7			62	.02	0.3	.06	.09	22	12	.38	.019	
hydrolyzed & charcoal-treated, corn					0.9	12	0.5	3	1.6	2.7	17	1			
oil, modified tapioca starch & MCTs)	9	27	34	58	48	17	9	27	13	42		17			
prosobee, Mead Johnson[d]	20	26.8	1.1				2	.02	.01	3	7	19	2	.16	.005
1 fl oz	30	0.6	2.0			62	.02	0.3	.06	.09	24	15	.38	.019	
(corn syrup solids, soy protein isolate,					0.9	12	0.6	3	1.6	1.6	17	2			
coconut oil & corn/soy oil)	7	19	28	46	36	11	5	29	20	27		14			
protein-free diet powder, Mead	490	3.0	22.5				45	.54	.36	90	540				
Johnson[e]—*100 g dry powder*	100	0.0	71.8			1440	.45	7.2	1.80	2.70					
(corn syrup solids, corn oil & modified					27.0	360	9.0	90	45.0	76					
tapioca starch)															
RCF (Ross carbohydrate free), Ross	12	26.0	1.1	0.4	0	18	2	.02	.01	3	9	21	2	.15	.006
Labs[f]—*1 fl oz*	30	0.6	0.0	0.4	0.0	60	.01	0.3	.09	.15	23	15	.05	.015	
(soy protein isolate, soy & coconut oils)					1.0	12	0.6[f]	3.0	0.9	1.6	13	3			
	6	23	26	47	36	13	7	30	21	25	43	14			
Similac, 13 cal/fl oz, Ross Labs	13	28.0	0.7	0.3	0	12	1	.02	.01	2	6	12	1	.10	.001
1 fl oz	30	0.4	1.4	0.3	0.0	39	.01	0.1	.03	.06	18	9	g	.012	
(cow milk, lactose, soy & coconut oils)					0.6	8	0.4[h]	1.0	0.6	2.1	12	2			
	5	15	17	34	26	10	3	17	15	20	11	8			
Similac, 20 cal/fl oz, Ross Labs	20	27.0	1.1	0.4	0	18	2	.03	.01	3	6	15	1	.15	.001
1 fl oz	30	0.4	2.1	0.5	0.0	60	.02	0.2	.05	.09	24	12	i	.018	
(cow milk, lactose, soy & coconut oils)					0.9	12	0.6[f]	1.6	0.9	3.2	15	3			
	6	19	22	42	32	13	4	22	20	24	15	9			
Similac, 24 cal/fl oz, Ross Labs	24	26.0	1.3	0.5	1	22	2	.04	.01	4	10	21	2	.18	.001
1 fl oz	30	0.7	2.5	0.5	0.0	72	.02	0.3	.06	.11	32	17	j	.021	
(cow milk, lactose, soy & coconut oils)					1.1	14	0.7[k]	1.9	1.0	3.8	21	4			
	8	28	32	62	47	18	6	32	29	36	22	14			
Similac, 27 cal/fl oz, Ross Labs	27	26.0	1.4	0.5	1	24	2	.04	.02	4	12	24	2	.20	.001
1 fl oz	30	0.7	2.8	0.6	0.0	81	.03	0.3	.07	.12	36	19	.06	.024	
(cow milk, lactose, soy & coconut oils)					1.3	16	0.8[k]	2.2	1.2	4.3	24	4			
	10	32	36	70	53	20	6	36	34	40	25	15			
Similac LBW (low birth weight), 24 cal/fl	24	26.6	1.3	0.2	0	22	3	.04	.01	4	10	21	2	.24	.001
oz, Ross Labs—*1 fl oz*	30	0.7	2.5	0.9	0.0	72	.03	0.3	.06	.11	36	17	.09	.024	
(cow milk, lactose, corn syrup solids,					1.1	14	0.7[k]	1.9	1.0	3.8	26	4			
MCT, soy & coconut oils)	8	28	32	62	47	18	6	32	29	36	22	14			
Similac Natural Care, 24 cal/fl oz, Ross	24	26.0	1.3	0.2	1	49	9	.15	.06	9	12	50	3	.36	.003
Labs—*1 fl oz*	30	0.7	2.5	0.8	0.0	162	.06	1.2	.13	.45	33	25	.09	.060	
(cow milk, whey, lactose, corn syrup					1.3	36	1.0[l]	2.9	8.8	2.4	21	5			
solids, MCT, soy & coconut oils)	8	38	35	65	52	16	11	22	22	36	18	13			
Similac PM 60/40, 20 cal/fl oz, Ross	20	27.0	1.1	0.4	1	18	2	.03	.01	3	5	11	1	.15	.001
Labs—*1 fl oz*	30	0.5	2.0	0.5	0.0	60	.02	0.2	.05	.09	17	6	.04	.018	
(whey, caseinate, lactose, soy & coconut					0.9	12	0.6[f]	1.6	1.8	3.2	12	1			
oils)	7	28	29	49	41	12	8	18	17	30	14	11			

[a] Hypoallergenic formula supplying protein nutrients in hydrolyzed form for infants & children sensitive to intact proteins of milk & other foods.

[b] Nutritionally complete formula for infants, children & adults who do not efficiently digest conventional food fat or absorb the resulting long chain fatty acids.

[c] Formula w/ readily digestible sources of protein, fat & carbohydrate in hypoallergenic, low osmolality form.

[d] Milk-free, lactose-free & sucrose-free formula w/ soy protein for infants w/ milk sensitivity.

[e] Formula base for use in making diets for infants requiring specific mixtures of amino acids.

[f] Alpha-tocopherol is 0.4 mg.

[g] .03 mg iron for Similac 13 and .23 mg iron for Similac 13 w/ iron.

[h] Alpha-tocopherol is 0.3 mg.

[i] .04 mg iron for Similac 20 and .36 mg iron for Similac 20 w/ iron.

[j] .05 mg iron for Similac 24 and .43 mg iron for Similac 24 w/ iron.

[k] Alpha-tocopherol is 0.5 mg.

[l] Alpha-tocopherol is 0.7 mg.

	KCAL	H₂O (g)	FAT (g)	PUFA (g)	CHOL (mg)	A (RE)	C (mg)	B-2 (mg)	B-6 (mg)	FOL (mcg)	Na (mg)	Ca (mg)	Mg (mg)	Zn (mg)	Mn (mg)
	WT (g)	PRO (g)	CHO (g)	SFA (g)	DFIB (g)	A (IU)	B-1 (mg)	NIA (mg)	B-12 (mcg)	PANT (mg)	K (mg)	P (mg)	Fe (mg)	Cu (mg)	
					INOS (mg)	D (IU)	E (IU)	K (mcg)	BIO (mcg)	CHLN (mg)	Cl (mg)	I (mcg)	Mo (mg)		
	TRY (mg)	THR (mg)	ISO (mg)	LEU (mg)	LYS (mg)	MET (mg)	CYS (mg)	PHE (mg)	TYR (mg)	VAL (mg)	ARG (mg)	HIS (mg)			
Similac Special Care, 20 cal/fl oz, Ross Labs—*1 fl oz* (cow milk, whey, lactose, corn syrup solids, MCT, soy & coconut oils)	20	27.0	1.1	0.2	1	41	7	.12	.05	7	10	36	2	.30	.002
	30	0.5	2.1	0.6	0.0	136	.05	1.0	.11	.38	28	18	.07	.050	
					0.9	8	0.8ᵃ	2.4	0.9	3.2	18	4			
	6	32	29	54	44	14	9	18	19	30	15	11			
Similac Special Care, 24 cal/fl oz, Ross Labs—*1 fl oz* (cow milk, whey, lactose, corn syrup solids, MCT, soy & coconut oils)	24	26.0	1.3	0.2		41	9	.15	.06	9	10	36	2	.30	.002
	30	0.7	2.1	0.8		136	.06	1.2	.13	.45	28	18	.07	.050	
					1.1	30	1.0ᵃ	2.9	1.0	3.8	21	4			
	8	38	35	65	52	16	11	22	22	36	18	13			
Similac w/ whey & iron, 20 cal/fl oz, Ross Labs—*1 fl oz* (cow milk, whey, lactose, soy & coconut oils)	20		1.1			18	2	.03	.01	3	7	12	2	.15	.001
	30	0.4	2.1			60	.02	0.2	.05	.09	22	9	.36	.018	
					0.9	12	0.6ᵇ	1.6	0.9	3.2	13	3			
	6	28	27	48	38	14	7	20	18	29	14	11			
SMA preemie, 24 cal/fl oz, Wyeth *1 fl oz*	24	25.9	1.3	0.2		21	2	.04	.01	3	9	22	2	.24	.006
	30	0.6	2.5	0.6		71	.02	0.2	.06	.11	22	12	.09	.021	
					0.9	14	0.3	2	1.0	4		2			
	10	41	40	73	61	14	11	32	32	43	21	15			
SMA w/ iron, 20 cal/fl oz, Wyeth *1 fl oz*	20	26.6	1.1	0.1	1	18	2	.03	.01	1	4	12	1	.15	.004
	30	0.4	2.1	0.4	0.0	59	.02	0.1	.04	.06	16	8	.35	.014	
					0.9	12	0.2	2	0.5	3		2			
	8	31	30	54	46	10	8	24	24	32	16	11			

ᵃ Alpha-tocopherol is 0.5 mg.

ᵇ Alpha-tocopherol is 0.4 mg.

	KCAL	H₂O (g)	FAT (g)	PUFA (g)	CHOL (mg)	A (RE)	C (mg)	B-2 (mg)	B-6 (mg)	FOL (mcg)	Na (mg)	Ca (mg)	Mg (mg)	Zn (mg)	Mn (mg)
	WT (g)	PRO (g)	CHO (g)	SFA (g)	DFIB (g)	A (IU)	B-1 (mg)	NIA (mg)	B-12 (mcg)	PANT (mg)	K (mg)	P (mg)	Fe (mg)	Cu (mg)	

19. INFANT, JUNIOR & TODDLER FOODS

19.1. BAKED PRODUCTS

	KCAL/WT	H₂O/PRO	FAT/CHO	PUFA/SFA	CHOL/DFIB	A(RE)/A(IU)	C/B-1	B-2/NIA	B-6/B-12	FOL/PANT	Na/K	Ca/P	Mg/Fe	Zn/Cu	Mn
arrowroot cookie	24	0.3	0.9	0.0	0.1		0	.03	.00		22	2	1	.03	
1 cookie	6	0.4	4.3	0.2	0.0		.03	0.3	.00		9	7	.18		
baby cookie	28	0.4	0.9	0.0		0	1	.21	.38		12	7	3	.07	
1 cookie	7	0.8	4.4	0.2	0.1	2	.10	1.0	.30		33	12	.27		
baby pretzel	24	0.2	0.1					.03	.00		16	1	2	.05	
1 pretzel	6	0.7	4.9		0.1		.04	0.4			8	7	.23		
teething biscuit	43	0.7	0.5			1	1	.06	.01		40	29	4	.10	
1 biscuit	11	1.2	8.4		0.1	13	.03	0.5	.01		35	18	.39		
zwieback	30	0.3	0.7	0.1	1	0	0	.02	.01		16	1	1	.04	
1 piece	7	0.7	5.2	0.3	0.0	4	.02	0.1			21	4	.04		

19.2. CEREALS

	KCAL/WT	H₂O/PRO	FAT/CHO	PUFA/SFA	CHOL/DFIB	A(RE)/A(IU)	C/B-1	B-2/NIA	B-6/B-12	FOL/PANT	Na/K	Ca/P	Mg/Fe	Zn/Cu	Mn
barley															
dry	9	0.2	0.1			0	0	.07	.01	1	1	19	3	.08	
1 T	2.4	0.3	1.8		0.1		.07	0.9			9	11	1.80	.011	
prep w/ whole milk	31	21.2	0.9				0	.16	.03	3	14	65	8	.24	
1 oz	28	1.3	4.6			30	.14	1.7			54	43	3.50		
cereal & egg yolks															
jr	110	189.0	3.8	0.3		88	2	.10	.04	7	70	51	6	.62	
1 jar	213	4.1	15.1	1.3		306	.02	0.1	.13		75	84	1.09	.047	
str	66	113.7	2.3	0.2	81	52	1	.06	.03	4	42	30	4	.37	
1 jar	128	2.5	9.0	0.8		181	.01	0.1	.09		50	51	.60	.028	
cereal, egg yolks & bacon															
jr	178	180.9	11.0				3	.11		9	97	54	10	.61	
1 jar	213	5.4	15.1			142	.04	0.3			79		1.06	.045	
str	101	109.9	6.4				1	.10		5	62	36	6	.35	
1 jar	128	3.2	8.0			121	.06	0.3			44		.60	.026	
grits & egg yolks, str	76	112.7	2.9			39	1	.09	.03	4	71	46	6	.64	
1 jar	128	2.3	10.1			156	.04	0.4	.05			36	.29		
high protein															
dry	9	0.1	0.1			0	0	.07	.01	5	1	17	5	.11	
1 T	2.4	0.9	1.1		0.1		.06	0.8			32	15	1.77	.031	
prep w/ whole milk	31	21.1	1.1			0	0	.16	.03	10	14	62	14	.30	
1 oz	28	2.5	3.3			30	.13	1.6			99	50	3.44		
w/ apple & orange, dry	9	0.1	0.2			0	0	.10	.01		2	18	4	.07	
1 T	2.4	0.6	1.4			1	.09	0.6	.01		32	13	2.10		
w/ apple & orange, prep w/ whole milk	32	21.1	1.1			8	0	.24	.03		16	63	10	.22	
1 oz	28	2.0	3.8			27	.19	1.1	.10		98	47	4.09		
mixed															
dry	9	0.2	0.1			0	0	.07	.01	1	1	18	2	.06	
1 T	2.4	0.3	1.8		0.1		.06	0.8			10	9	1.52	.013	
prep w/ whole milk	32	21.2	1.0			0	0	.17	.02	3	13	62	8	.20	
1 oz	28	1.3	4.5			30	.12	1.6			56	40	2.96		
w/ applesce & bananas, jr	183	175.1	0.9			4	20	.78	.31	8	78	9	16	.48	
1 jar	220	2.6	40.5		1.3	41	.63	8.9			70	63	12.33		
w/ applesce & bananas, str	111	108.0	0.7			2	35	.47	.19	5	3	9	11	.26	
1 jar	135	1.6	24.2		0.8	24	.38	5.4			55	32	8.94		
w/ bananas, dry	9	0.1	0.1			0	0	.09	.10		3	17	2	.03	
1 T	2.4	0.3	1.9		0.1	3	.09	0.5	.01		16	9	1.62		
w/ bananas, prep w/ whole milk	33	21.1	1.0			8	0	.20	.03		17	61	7	.16	
1 oz	28	1.3	4.7			31	.19	1.0	.10		67	39	3.16		
w/ bananas & apple jce, inst, Heinz	50		0.0				17	.28	.06		2	6		.18	
4 T dry	14	1.0	12.0			10	.24	3.6			80	26	7.00	.040	

	KCAL	H₂O (g)	FAT (g)	PUFA (g)	CHOL (mg)	A (RE)	C (mg)	B-2 (mg)	B-6 (mg)	FOL (mcg)	Na (mg)	Ca (mg)	Mg (mg)	Zn (mg)	Mn (mg)
	WT (g)	PRO (g)	CHO (g)	SFA (g)	DFIB (g)	A (IU)	B-1 (mg)	NIA (mg)	B-12 (mcg)	PANT (mg)	K (mg)	P (mg)	Fe (mg)	Cu (mg)	
w/ honey, dry	9	0.1	0.1			0	0	.07			1	28			
1 T	2.4	0.3	1.8			1	.06	0.9			6	16	1.64		
w/ honey, prep w/ whole milk	33	21.0	1.0			7	0	.17			14	83			
1 oz	28	1.4	4.5			26	.13	1.8			49	52	3.20		
oatmeal															
dry	10	0.1	0.2				0	.06	.00	1	1	18	3	.09	
1 T	2.4	0.3	1.7	0.1			.07	0.9			11	12	1.77	.013	
prep w/ whole milk	33	21.1	1.2				0	.16	.02	3	13	62	10	.26	
1 oz	28	1.4	4.3			30	.14	1.7			58	45	3.44		
w/ applesce & bananas, jr	165	179.9	1.6				42	.11	.53	8	69	12	24	.73	
1 jar	220	2.9	34.6	1.5			.53	7.4			106	89	12.13	.165	
w/ applesce & bananas, str	99	110.9	0.9			4	29	.49	.27	5	2	11	15	.47	
1 jar	135	1.8	20.8	1.1		40	.57	6.8			63	56	7.63	.099	
w/ bananas, dry	9	0.1	0.1			0	0	.09	.01		3	16	3	.05	
1 T	2.4	0.3	1.8	0.1		2	.09	0.5	.01		18	11	1.63		
w/ bananas, prep w/ whole milk	33	21.1	1.1			8	0	.21	.03		17	59	9	.18	
1 oz	28	1.3	4.5			28	.18	1.0	.09		70	43	3.18		
w/ bananas & apple jce, inst, Heinz	60		0.0				17	.28	.07		1	12		.25	
4 T dry	14	1.0	12.0			4	.24	3.6			112	38	7.00	.050	
w/ honey, dry	9	0.1	0.2			0	0	.07			1	28			
1 T	2.4	0.3	1.7			1	.07	0.9			6	18	1.61		
w/ honey, prep w/ whole milk	33	21.1	1.1			7	0	.17			14	82			
1 oz	28	1.4	4.3			26	.14	1.7			48	56	3.14		
rice															
dry	9	0.2	0.1				0	.05	.01	1	1	20	5	.05	
1 T	2.4	0.2	1.9	0.0			.06	0.8			9	14	1.77	.008	
prep w/ whole milk	33	21.2	1.0				0	.14	.03	2	13	68	13	.18	
1 oz	28	1.1	4.7			30	.13	1.5			54	50	3.46		
w/ applesce & bananas, str	107	109.4	0.5			3	43	.57	.32	3	38	23	4	.11	
1 jar	135	1.6	23.1	0.4		28	.35	5.4			38	16	9.09		
w/ bananas, dry	10	0.1	0.1			0	0	.09	.02		2	17	3	.04	
1 T	2.4	0.2	1.9	0.0		1	.10	0.6	.01		18	10	1.61		
w/ bananas, prep w/ whole milk	33	21.1	1.0			7	0	.22	.04		16	60	10	.16	
1 oz	28	1.2	4.8			26	.19	1.1	.09		72	41	3.13		
w/ bananas & apple jce, inst, Heinz	50		0.0				17	.28	.07		1	5		.16	
4 T dry	14	1.0	12.0			10	.24	3.6			70	20	7.00	.040	
w/ honey, dry	9	0.1	0.1			0	0	.07			1	28			
1 T	2.4	0.2	1.9			1	.07	0.9			2	15	1.57		
w/ honey, prep w/ whole milk	33	21.1	0.9			7	0	.17			14	83			
1 oz	28	1.1	4.9			26	.14	1.7			40	51	3.07		
w/ mixed fruit, jr	186	175.2	0.5			3	45	1.31	.54		24	43	10	.40	
1 jar	220	2.3	41.0	0.7		33	.55	6.0			72	51	10.36		
w/ pears & apple jce, inst, Heinz	50		0.0				17	.28	.02		8	8		.17	
4 T dry	14	1.0	13.0			12	.24	3.6			43	22	7.00	.060	

19.3. DESSERTS

	KCAL	H₂O (g)	FAT (g)	PUFA (g)	CHOL (mg)	A (RE)	C (mg)	B-2 (mg)	B-6 (mg)	FOL (mcg)	Na (mg)	Ca (mg)	Mg (mg)	Zn (mg)	Mn (mg)
	WT (g)	PRO (g)	CHO (g)	SFA (g)	DFIB (g)	A (IU)	B-1 (mg)	NIA (mg)	B-12 (mcg)	PANT (mg)	K (mg)	P (mg)	Fe (mg)	Cu (mg)	
apple betty															
jr	153	177.1	0.0			3	60	.11		1	19	36			
1 jar	220	0.8	41.7			35	.02	0.1			117		.44		
str	97	107.9	0.0			2	47	.05		1	14	25			
1 jar	135	0.5	26.5			23	.02	0.1			68		.24		
banana apple															
jr, Gerber	143		0.4			6	16	.04	.15		11	6			
1 jar	213	0.6	33.9	0.6			.02	0.3			153	13	.21		
str, Gerber	86		0.3			4	16	.03	.09		9	4			
1 jar	128	0.4	21.0	0.4			.01	0.2			91	12	.13		
banana pudding, str, Heinz	96		0.6				15	.04	.14		15	13		.15	
1 jar	128	1.0	21.6			59	.01	0.2			99	40	.51	.051	

	KCAL	H₂O (g)	FAT (g)	PUFA (g)	CHOL (mg)	A (RE)	C (mg)	B-2 (mg)	B-6 (mg)	FOL (mcg)	Na (mg)	Ca (mg)	Mg (mg)	Zn (mg)	Mn (mg)
	WT (g)	PRO (g)	CHO (g)	SFA (g)	DFIB (g)	A (IU)	B-1 (mg)	NIA (mg)	B-12 (mcg)	PANT (mg)	K (mg)	P (mg)	Fe (mg)	Cu (mg)	
caramel pudding															
jr	167	171.2	1.9			7	5	.15		2	60	116			
1 jar	213	2.9	36.2			69	.02	0.1			124		.34		
str	104	108.5	0.9			5	3	.11		1	37	60			
1 jar	135	1.8	23.2			49	.01	.05			70		.23		
cherry van pudding															
jr	152	178.2	0.4			44	2	.02	.03	1	32	11	5	.07	
1 jar	220	0.4	40.5		0.2	440	.02	0.1			73	15	.38		
str	91	110.0	0.4			27	2	.02	.02	0	22	7	3	.05	
1 jar	135	0.3	24.1		0.1	270	.01	0.1			46	10	.26		
choc custard pudding															
jr	195	172.7	3.5			10	2	.24	.03	11	55	134	23	.73	
1 jar	220	4.2	38.3			101	.03	0.2			196	112	.88		
str	107	102.2	2.1			6	2	.13	.02	6	30	78	13	.41	
1 jar	135	2.4	20.6		0.1	58	.02	0.1			110	63	.48		
cottage cheese w/ pineapple															
jr	172	176.5	1.5			3	52	.11	.02	11	113	68	8	.39	
1 jar	220	6.5	35.1			35	.03	0.1	.16		92	86	.28		
str	94	111.5	1.1			4	31	.07	.02	6	70	35	5	.21	
1 jar	135	4.0	17.8			39	.02	0.1	.10		58	51	.14		
dutch apple															
jr	151	180.6	2.1	0.0		11	47	.03	.03	2	36	10	4	.06	
1 jar	220	0.0	37.0	1.4	0.9	110	.03	0.1			82	8	.44		
str	92	111.0	1.2	0.0		7	29	.02	.02	1	21	6	2	.01	
1 jar	135	0.0	22.6	0.8	0.5	66	.02	0.1			44	5	.27		
fruit dessert															
jr	138	180.9	0.0			53	66	.03	.07	8	29	19	11	.11	
1 jar	220	0.6	37.9			525	.05	0.3			208	17	.46	.066	
str	79	112.6	0.0			34	3	.01	.05	4	18	11	7	.05	
1 jar	135	0.4	21.6		0.3	338	.02	0.2			127	9	.29	.041	
hawaiian delight															
jr, Gerber	190		0.6			4	16	.11	.09		40	81			
1 jar	213	3.0	42.2		0.2		.09	0.3			181	72	.21		
str, Gerber	113		0.5			1	16	.06	.05		23	49			
1 jar	128	1.9	24.8				.04	0.2			104	44	.13		
orange pudding, str	108	107.7	1.2			16	12	.08	.04	11		43	7	.23	
1 jar	135	1.5	23.8		0.1	155	.05	.02			117	38	.14		
peach cobbler															
jr	147	178.6	0.0			31	45	.03	.02	3	20	9			.07
1 jar	220	0.7	40.3		0.7	312	.02	0.6			123	13	.22		
str	88	110.4	0.0			19	28	.02	.01	2	10	6		.41	
1 jar	135	0.4	24.0		0.4	192	.01	0.4			73	7		.256	
peach melba															
jr	132	182.5	0.0			43	57	.07		4	19	23			
1 jar	220	0.6	36.1			430	.02	0.6			205		.66		
str	81	111.9	0.0			25	42	.05		3	12	13			
1 jar	135	0.3	22.3			249	.01	0.5			112		.45		
pineapple orange dessert, str	89	103.0	0.0			7	18	.03		3	13	14	5		
1 jar	128	0.3	24.4			73	.02	0.1			61	5	.23	.026	
pineapple pudding															
jr	192	167.4	0.9			8	59	.11	.09	12	48	75	20	.42	
1 jar	220	3.1	47.4			80	.09	0.3	.14		198	67	.41		

	KCAL / WT (g)	H₂O (g) / PRO (g)	FAT (g) / CHO (g)	PUFA (g) / SFA (g)	CHOL (mg) / DFIB (g)	A (RE) / A (IU)	C (mg) / B-1 (mg)	B-2 (mg) / NIA (mg)	B-6 (mg) / B-12 (mcg)	FOL (mcg) / PANT (mg)	Na (mg) / K (mg)	Ca (mg) / P (mg)	Mg (mg) / Fe (mg)	Zn (mg) / Cu (mg)	Mn (mg)
str	104	99.3	0.4			5	35	.06	.05	7	24	40	11	.26	
1 jar	128	1.6	26.0			52	.05	0.1	.08		104	39	.23		
tropical fruit dessert, jr	131	183.1	0.0			4	41	.07			16	22			
1 jar	220	0.4	36.1			43	.02	0.2			128		.57		
tutti frutti															
jr, Heinz	149		0.9				15	.04	.04		36	32		.40	
1 jar	213	0.9	34.3			430	.06	0.2			92	34	.64	.426	
str, Heinz	88		0.4				16	.03	.03		19	15		.10	
1 jar	128	0.5	20.6			166	.03	0.1			51	15	.26	.026	
van custard pudding															
jr	196	174.7	5.0	0.2		8	2	.17	.04	14	64	123	11	.61	
1 jar	220	3.5	35.7	2.6		79	.03	0.1		.56	136	99	.57	.112	
str	109	102.2	2.5	0.1		8	1	.10	.03	8	36	71	7	.36	
1 jar	128	2.0	20.6	1.3		82	.02	0.1		.32	85	57	.31	.064	

19.4. DINNERS

	KCAL / WT (g)	H₂O (g) / PRO (g)	FAT (g) / CHO (g)	PUFA (g) / SFA (g)	CHOL (mg) / DFIB (g)	A (RE) / A (IU)	C (mg) / B-1 (mg)	B-2 (mg) / NIA (mg)	B-6 (mg) / B-12 (mcg)	FOL (mcg) / PANT (mg)	Na (mg) / K (mg)	Ca (mg) / P (mg)	Mg (mg) / Fe (mg)	Zn (mg) / Cu (mg)	Mn (mg)
beef & egg noodles															
jr	122	186.9	4.0			187	3	.08	.07	12	37	18	16	.85	
1 jar	213	5.4	15.7		0.9	1397	.06	1.2	.20	.49	99	64	.92	.068	
str	68	113.4	2.2			141	2	.05	.06	7	37	12	9	.48	
1 jar	128	2.9	9.0			1053	.05	0.9	.12	.27	61	37	.53	.038	
beef & egg noodles w/ veg, toddler, Gerber—*1 jar*	119 / 170	/ 6.8	3.9 / 15.8		/ 1.0	303	3 / .09	.09 / 1.5	.09		374 / 236	20 / 75	/ 1.02		
beef & egg rice, toddler	146	144.9	5.1				7	.12	.25		632	20	15	1.62	
1 jar	177	8.9	15.5			889	.03	2.4			212	62	1.23	.096	
beef lasagna, toddler	137	145.7	3.8				3	.16	.13		804	32	20	1.24	
1 jar	177	7.4	17.7			2058	.13	2.4			216	70	1.55		
beef stew, toddler	90	153.8	2.1	0.0	22		5	.12	.13		611	16	20	1.54	
1 jar	177	9.1	9.6	1.0		2918	.03	2.3			251	78	1.27		
chicken & noodles															
inst, Heinz	60		2.0				0	.06	.07		35	24		.40	
6 T dry	14	4.0	7.0			150	.08	1.1			161	70	.90	.080	
jr	109	189.0	3.0			229	3	.07	.06	11	36	36	19	.63	
1 jar	213	4.1	16.1		0.4	1907	.06	1.1			75	51	.84	.083	
str	67	113.3	1.9			144	1	.07	.04	7	20	29	11	.38	
1 jar	128	2.7	9.6			1158	.04	0.6			50	31	.58	.051	
chicken soup, crm of, str	74	111.5	2.0			149	2	.06	.06		24	44	7	.33	
1 jar	128	3.2	10.8			929	.01	0.5	.07		100	38	.38		
chicken soup, str	64	114.1	2.2				1	.04		7	20	47			
1 jar	128	2.0	9.2			1772	.02	0.4					.35		
chicken stew, toddler	132	141.5	6.4	0.1	49		3	.13	.08		683	60	18	.70	
1 jar	170	8.9	10.9	1.9		1718	.05	2.0			156	87	1.13	.034	
lamb & noodles, jr	138	184.3	4.7				4	.14			39	39			
1 jar	213	4.8	18.6			1667	.08	1.4			165		.78		
macaroni alphabets															
w/ beef & tomato sce, toddler, Gerber	140		3.4			11	3	.11	.16		356	21			
1 jar	177	7.1	20.4		0.9		.11	2.4			262	87	1.42		
w/ tomato sce & cheese, toddler, Gerber—*1 jar*	120 / 177	/ 3.7	1.8 / 22.5		/ 0.7	175	5 / .09	.12 / 1.5	.11		550 / 250	51 / 81	/ .71		
macaroni & bacon, jr	160	181.2	7.1				5	.18			166	152			
1 jar	213	5.4	18.2			2471	.10	1.3			180		.80		
macaroni & cheese															
jr	130	184.2	4.3			6	3	.14	.03	3	163	108	14	.68	
1 jar	213	5.5	17.5		0.4	28	.12	1.2	.06		94	125	.64	.045	

	KCAL / WT (g)	H₂O (g) / PRO (g)	FAT (g) / CHO (g)	PUFA (g) / SFA (g)	CHOL (mg) / DFIB (g)	A (RE) / A (IU)	C (mg) / B-1 (mg)	B-2 (mg) / NIA (mg)	B-6 (mg) / B-12 (mcg)	FOL (mcg) / PANT (mg)	Na (mg) / K (mg)	Ca (mg) / P (mg)	Mg (mg) / Fe (mg)	Zn (mg) / Cu (mg)	Mn (mg)
str	76	111.5	2.7			7	2	.09	.02	2	93	69	11	.44	
1 jar	128	3.3	9.6			35	.07	0.6	.04		59	73	.40	.026	
macaroni & ham, jr	127	184.2	2.9				5	.21			101	159			
1 jar	213	6.8	18.0			1120	.12	1.7			225		.81		
macaroni, tomato & beef															
jr	125	184.6	2.4			233	3	.12	.10		35	30	15	.77	
1 jar	213	5.3	20.1		1.1	1471	.10	1.6	.51		153	94	.76	.083	
str	71	111.7	1.4			116	2	.08	.06	26	21	21	12	.40	
1 jar	128	2.9	11.3			679	.09	1.1	.29		123	54	.63	.051	
mixed veg dinner															
jr	71	193.0	0.1			520	7	.05		14	19	37			
1 jar	213	2.1	16.8			5205	.03	0.9			240		.65		
str	52	113.6	0.1			349	4	.04		10	10	29			
1 jar	128	1.5	12.2			3492	.02	0.6			155		.42		
noodles & beef, homestyle, toddler, Gerber—*1 jar*	143		5.1			119	2	.12	.09		369	37			
1 jar	170	6.5	17.7		0.7		.09	1.8			167	97	1.19		
noodles & chicken w/ carrots & peas, toddler, Gerber—*1 jar*	109		2.9			332	3	.09	.10		316	19			
1 jar	170	6.5	14.1		1.0	.	.07	2.1			162	71	.68		
potatoes & ham, toddler, Gerber	111		4.3			10	4	.05	.12		272	27			
1 jar	170	3.9	14.6		0.5		.09	1.1			216	60	.34		
rice w/ beef & tomato sce, toddler, Gerber	135		3.7			117	3	.11	.16		372	23			
1 jar	177	5.7	19.5		0.9		.04	1.5			285	67	1.42		
saucy rice w/ chicken, toddler, Gerber	109		2.2			320	3	.07	.12		391	29			
1 jar	170	5.4	16.8		0.7		.03	1.6			170	85	.68		
spaghetti, tomato & meat															
jr	135	182.1	2.7			284	5	.15	.13		42	39		.90	
1 jar	213	5.4	21.6		1.1	1477	.14	2.3		.38	230	79	1.16		
toddler	133	144.4	1.8				7	.18	.15		634	39	26	.85	
1 jar	177	9.4	19.1			784	.11	2.8			289	79	1.59	.035	
spaghetti, tomato sce & beef, toddler, Gerber—*1 jar*	129		2.8			195	4	.12	.12		358	37			
1 jar	177	6.4	19.6		1.1		.12	2.0			281	74	1.06		
split peas & ham, jr	152	177.8	2.8			170	4	.10	.09		30	49			
1 jar	213	7.0	24.1		1.7	1284	.10	1.0			291	104	1.06		
turkey & rice															
jr	104	190.2	2.9	0.4		316	3	.06	.06	7	33	50			
1 jar	213	3.8	15.3	0.9	0.9	2256	.02	0.6			71	36	.62		
str	63	114.0	1.7	0.2	13	122	2	.03	.04	4	21	27			
1 jar	128	2.4	9.4	0.5		783	.01	0.4			53	26	.33		
veg & bacon															
jr	150	183.5	8.2	0.1		463	2	.06	.14	19	96	23			
1 jar	213	3.9	16.1	3.0	1.5	3356	.11	1.2			184	81	.88		
str	88	110.0	4.2	0.1	4	382	2	.04	.10	12	55	17			
1 jar	128	2.0	11.0	1.5		3409	.04	0.7	.13		115	40	.45		
veg & beef															
inst, Heinz	60		2.0				0	.04	.05		25	12		.51	
5 T dry	14	3.0	7.0			354	.04	0.8			142	47	.90	.070	
jr	113	187.3	3.6			410	3	.07	.14	10	52	22	13	.88	
1 jar	213	5.0	15.8		1.1	3011	.07	1.4	.56	.27	224	91	1.01	.083	
str	67	113.3	2.6			219	2	.04	.07	6	27	16	7	.42	
1 jar	128	2.5	9.0			1520	.03	0.7	.32	.15	129	52	.49	.013	
toddler, Gerber	120		3.7			250	4	.07	.12		317	19			
1 jar	177	6.2	15.2		0.9		.04	1.3			297	65	.89		
veg & chicken															
inst, Heinz	60		2.0				0	.02	.06		30	24		.36	
6 T dry	14	3.0	8.0			424	.05	1.1			130	58	.90	.060	

	KCAL / WT (g)	H₂O (g) / PRO (g)	FAT (g) / CHO (g)	PUFA (g) / SFA (g)	CHOL (mg) / DFIB (g)	A (RE) / A (IU)	C (mg) / B-1 (mg)	B-2 (mg) / NIA (mg)	B-6 (mg) / B-12 (mcg)	FOL (mcg) / PANT (mg)	Na (mg) / K (mg)	Ca (mg) / P (mg)	Mg (mg) / Fe (mg)	Zn (mg) / Cu (mg)	Mn (mg)
jr	106	187.9	2.3			316	3	.04	.08	8	18	30			
1 jar	213	4.0	18.0		0.6	2546	.03	0.7		.50	54	55	.64		
str	55	115.2	1.4			176	1	.02	.03	4	14	18			
1 jar	128	2.5	8.5			1416	.02	0.2		.26	38	32	.35		
toddler, Gerber	119		2.8			276	3	.05	.09		324	19			
1 jar	177	6.5	16.6		1.1		.05	1.5			211	78	.71		
veg & ham															
inst, Heinz	60		2.0				1	.04	.05		30	12		.47	
6 T dry	14	4.0	7.0			300	.08	0.6			86	47	.60	.060	
jr	110	188.2	3.6			173	3	.05	.07	11	37	16	15	.48	
1 jar	213	5.2	14.9		0.9	1310	.08	0.7			197	55	.47	.066	
str	62	114.2	2.2			122	2	.04	.04	6	15	11	6	.25	
1 jar	128	2.2	8.8			872	.04	0.5			109	29	.38	.037	
toddler	128	148.0	5.2	0.3	14		7	.10	.15		531	41	30	.83	
1 jar	177	7.4	13.9	1.9	0.9	629	.08	1.2			271	71	1.23	.071	
veg & lamb															
jr	108	188.8	3.7			423	4	.07	.09	8	28	27	16	.47	
1 jar	213	4.4	15.1		0.9	3158	.05	1.2	.34	.34	202	104	.72	.062	
str	67	113.4	2.6			363	2	.04	.06	5	26	15	9	.28	
1 jar	128	2.5	8.9			2554	.02	0.7	.20	.21	120	63	.44	.037	
veg & liver															
jr	93	189.4	1.2			1524	4	.48	.21	68	27	20			
1 jar	213	3.9	17.5			8368	.05	2.5			189	81	3.86		
str	50	115.2	0.6			940	2	.34	.11	37	24	9			
1 jar	128	2.8	8.8			3981	.02	1.5			120	51	2.83		
veg & turkey															
inst, Heinz	60		2.0				0	.02	.03		25	12		.42	
6 T dry	14	4.0	8.0			450	.05	1.1			140	57	.90	.070	
jr	101	189.5	2.7			244	2	.04	.05	6	36	27		.54	
1 jar	213	3.6	16.4		0.9	1909	.03	0.5		.45	53	40	.68	.055	
str	54	115.4	1.5			143	1	.02	.04	3	17	21		.29	
1 jar	128	2.1	8.4			1068	.01	0.4		.24	56	24	.34	.029	
toddler	141	145.7	6.1				6	.16	.11		591	82	29	.55	
1 jar	177	8.5	14.2		0.9	3715	.04	1.0			294	102	1.07		
veg, dumplings & beef															
jr	103	188.7	1.7				2	.08		16	110	30	14	.70	
1 jar	213	4.4	17.0			1407	.08	1.0					1.00		
str	61	113.8	1.2				1	.05		9	62	18	8	.51	
1 jar	128	2.6	9.8			533	.06	0.7					.49		
veg, noodles & chicken															
jr	137	183.7	4.8				2	.08	.05	7	54	54		.68	
1 jar	213	3.7	19.4			2238	.09	1.4	.19	.44	126	70	1.04	.124	
str	81	111.6	3.3				1	.06	.03	4	26	35		.32	
1 jar	128	2.6	10.1			1813	.04	0.5	.10	.24	70	39	.45	.069	
veg, noodles & turkey															
jr	110	189.0	3.2				2	.09	.04	6	37	67	20	.64	
1 jar	213	3.9	16.1			2118	.05	0.6	.25		156	62	.55		
str	56	115.6	1.6				1	.06	.02	3	27	41	10	.35	
1 jar	128	1.5	8.7			1268	.02	0.3	.13		81	32	.25		

	KCAL	H₂O (g)	FAT (g)	PUFA (g)	CHOL (mg)	A (RE)	C (mg)	B-2 (mg)	B-6 (mg)	FOL (mcg)	Na (mg)	Ca (mg)	Mg (mg)	Zn (mg)	Mn (mg)
	WT (g)	PRO (g)	CHO (g)	SFA (g)	DFIB (g)	A (IU)	B-1 (mg)	NIA (mg)	B-12 (mcg)	PANT (mg)	K (mg)	P (mg)	Fe (mg)	Cu (mg)	

19.5. DINNERS, HIGH MEAT/CHEESE

beef w/ veg

	KCAL	H₂O	FAT	PUFA	CHOL	A	C	B-2	B-6	FOL	Na	Ca	Mg	Zn	Mn
jr	108	106.5	5.9			126	2	.10	.11	8	42	15	10	1.79	
1 jar	128	8.1	6.7		0.5	1014	.05	1.8	.75	.35	191	66	1.01	.118	
str	96	109.3	5.3			141	3	.08	.11	7	46	15	10	1.66	
1 jar	128	7.3	5.3		0.4	1003	.04	1.7	.65	.31	179	62	.93	.102	

chicken w/ veg

jr	117	105.8	7.0			167	1	.09	.05	2	33	56	9	1.28	
1 jar	128	9.0	5.4		0.5	1071	.04	1.3	.20	.44	79	68	.95		
str	100	107.2	4.6			106	1	.09	.08	1	35	66	9	1.21	
1 jar	128	8.0	7.6		0.5	737	.04	1.3	.18	.41	76	69	1.27		
cottage cheese w/ pineapple, str	157	97.2	3.0			10	2	.19	.06		201	88	9	.39	
1 jar	135	8.5	25.4			104	.05	0.1	.30		128	98	.14		

ham w/ veg

jr	98	107.0	4.2	0.3	23	85	2	.11	.12	8	28	12	12	1.39	
1 jar	128	8.2	7.9	1.4	0.5	332	.14	1.5	.36	.52	208	71	.76		
str	97	107.6	4.4	0.3		52	2	.11	.14	8	29	14	11	1.28	
1 jar	128	8.0	7.1	1.5	0.4	214	.13	1.8	.34	.50	198	72	.75		

turkey w/ veg

jr	115	105.6	6.4			143	2	.09	.05	13	56	91	10	1.16	
1 jar	128	7.6	7.5		0.4	810	.01	1.0	.58		137	81	1.00	.054	
str	111	106.3	6.1			119	2	.09	.05	12	38	80	10	1.28	
1 jar	128	7.2	7.7		0.3	419	.01	1.3	.56		153	90	.88	.052	

veal w/ veg

jr	93	107.9	4.0			101	2	.11	.09		32	14	12	1.41	
1 jar	128	7.8	7.4		0.5	545	.03	2.0	.59	.32	201	69	1.13	.131	
str	89	108.2	3.4			96	2	.10	.11		30	12	10	1.28	
1 jar	128	7.6	7.8		0.3	349	.03	2.1	.58	.32	196	69	.76	.128	

19.6. FRUIT JUICES

apple

str	61	114.3	0.1			2	75	.02	.04	0	4	6	4	.04	
1 jar	130	0.0	15.2		0.1	24	.01	0.1			118	7	.73		
toddler, Gerber	116		0.2			2	76	.02	.07		5	7			
1 jar	236	0.5	27.6		0.2		.02	0.2			212	17	.47		

apple-apricot

str, Gerber	61		0.1			56	40	.02	.02		4	7			
1 jar	124	0.2	14.5		0.2		.01	0.2			140	10	.25		
str, Heinz	61		0.3				42	.01	.04		8	14		.09	
1 jar	130	0.1	14.7			553	.03	0.2			148	14	.65	.039	
apple-banana, str, Gerber	62		0.2			2	40	.02	.07		4	5			
1 jar	124	0.2	14.8		0.2		.01	0.2			143	9	.12		
apple-berry, toddler, Gerber	109		0.2				76	.05	.07		14	24			
1 jar	236	0.2	26.4		0.2		.02	0.3			264	19	.71		

apple-cherry

str	53	116.3	0.3			1	76	.02	.04	0	4	7	4	.04	
1 jar	130	0.2	12.9		0.1	7	.01	0.1			127	7	.86		
toddler, Gerber	113		0.2			2	76	.05	.05		12	14			
1 jar	236	0.5	27.1		0.2		.02	0.2			241	19	.71		

	KCAL	H₂O (g)	FAT (g)	PUFA (g)	CHOL (mg)	A (RE)	C (mg)	B-2 (mg)	B-6 (mg)	FOL (mcg)	Na (mg)	Ca (mg)	Mg (mg)	Zn (mg)	Mn (mg)
	WT (g)	PRO (g)	CHO (g)	SFA (g)	DFIB (g)	A (IU)	B-1 (mg)	NIA (mg)	B-12 (mcg)	PANT (mg)	K (mg)	P (mg)	Fe (mg)	Cu (mg)	
apple-grape															
str	60	114.5	0.2			1	70	.03	.04	0	4	7	4	.05	
1 jar	130	0.1	14.8		0.1	7	.01	0.1			118	7	.51		
toddler, Gerber	116		0.2			2	76	.05	.07		9	14			
1 jar	236	0.5	27.6		0.2		.02	0.2			212	17	.24		
apple-peach, str	55	115.7	0.1			8	76	.01	.03	2		4	4	.03	
1 jar	130	0.2	13.6		0.1	82	.01	0.3			126	6	.73		
apple-pear, toddler, Gerber	111		0.2			2	76	.02	.05		7	14			
1 jar	236	0.5	26.7		0.7		.02	0.2			222	14	.47		
apple-pineapple															
str, Gerber	60		0.2			1	40	.01	.04		2	10			
1 jar	124	0.2	14.5		0.1		.04	0.1			130	7	.12		
str, Heinz	61		0.3				42	.01	.06		7	16		.07	
1 jar	130	0.3	14.6			21	.03	0.1			108	10	.39	.026	
apple-plum	63	113.5	0.0			6	76	.02	.04	0		6	4	.04	
1 jar	130	0.1	16.0			55	.03	0.3			131	4	.81		
apple-prune	94	105.7	0.2				88	.00	.05	0	7	12			
1 jar	130	0.3	23.4		0.1		.01	0.4			192	20	1.24		
fruit-a-plenty, toddler, Gerber	113		0.2				76	.02	.07		31	21			
1 jar	236	0.2	27.6		0.2		.02	0.3			243	17	.71		
fruits of the sun, toddler, Gerber	113		0.2				76	.05	.09		26	26			
1 jar	236	0.5	26.9		0.2		.05	0.3			413	24	.24		
mixed fruit															
str	61	114.3	0.1			5	83	.02	.06	9	5	10	7	.04	
1 jar	130	0.2	15.1		0.1	54	.03	0.2			131	7	.44		
toddler, Gerber	113		0.2			7	76	.02	.07		9	21			
1 jar	236	0.7	26.4		0.2		.07	0.3			262	19	.47		
orange	58	115.0	0.3			7	81	.04	.07	34	2	16	11	.07	
1 jar	130	0.8	13.3		0.1	72	.06	0.3			240	14	.22		
orange-apple	56	115.5	0.3			9	100	.04	.05	16	4	13	6	.03	
1 jar	130	0.5	13.2		0.1	95	.05	0.2			179	9	.26		
orange-apple-banana	61	113.9	0.1			4	42	.04	.08	13	6	6	7	.03	
1 jar	130	0.5	15.0			35	.06	0.3			174	10	.45		
orange-apricot	60	114.1	0.1			28	112	.04	.07	26	7	8	9	.05	
1 jar	130	1.0	14.2			281	.08	0.3			259	15	.49		
orange-banana	65	113.0	0.1			6	44	.06		32	4	22			
1 jar	130	0.9	15.4			60	.07	0.2			261		.14		
orange-pineapple	63	113.5	0.1			4	69	.03	.08	24	2	10	12	.05	
1 jar	130	0.7	15.2			40	.07	0.3			183	11	.55		
prune-orange	91	106.5	0.4			17	83	.16	.08	17	2	16	10	.05	
1 jar	130	0.8	21.8			170	.06	0.5			236	14	1.13		

19.7. FRUITS

	KCAL	H₂O (g)	FAT (g)	PUFA (g)	CHOL (mg)	A (RE)	C (mg)	B-2 (mg)	B-6 (mg)	FOL (mcg)	Na (mg)	Ca (mg)	Mg (mg)	Zn (mg)	Mn (mg)
	WT (g)	PRO (g)	CHO (g)	SFA (g)	DFIB (g)	A (IU)	B-1 (mg)	NIA (mg)	B-12 (mcg)	PANT (mg)	K (mg)	P (mg)	Fe (mg)	Cu (mg)	
apple-blueberry															
jr	137	182.2	0.4			9	31	.10	.09	8	28	10			
1 jar	220	0.4	36.5		2.2	92	.04	0.2			143	15	.88		
str	82	112.2	0.3			3	38	.05	.05	5	2	5			
1 jar	135	0.3	22.0		1.4	27	.02	0.2			93	11	.27		
apple-raspberry															
jr	127	184.7	0.4			7	64	.06	.08	7	4	11			
1 jar	220	0.4	34.1			66	.03	0.2		.19	158	17	.48		
str	79	113.1	0.2			3	36	.04	.05	5	3	7			
1 jar	135	0.3	21.2			29	.02	0.1		.12	108	11	.30		
apples, inst, Heinz	50		0.0				16	.01	.03		4	8		.04	
4 T dry	14	0.0	13.0			28	.02	0.1			104	9	.20	.040	

	KCAL / WT (g)	H₂O / PRO (g)	FAT / CHO (g)	PUFA / SFA (g)	CHOL (mg) / DFIB (g)	A (RE) / A (IU)	C (mg) / B-1 (mg)	B-2 (mg) / NIA (mg)	B-6 (mg) / B-12 (mcg)	FOL (mcg) / PANT (mg)	Na (mg) / K (mg)	Ca (mg) / P (mg)	Mg (mg) / Fe (mg)	Zn (mg) / Cu (mg)	Mn (mg)
apples & apricots, inst, Heinz	50		1.0				17	.02	.04		2	11		.06	
3 T dry	14	0.0	13.0			566	.02	0.2			160	10		.040	
apples & apricots															
jr, Heinz	115		0.6				15	.04	.11		4	17		.15	
1 jar	213	0.9	26.6			694	.09	0.3			217	34	.43	.192	
str, Heinz	69		0.4				16	.03	.06		3	19		.09	
1 jar	128	0.5	16.0			417	.05	0.2			131	20	.26	.115	
apples & bananas, inst, Heinz	50		0.0				17	.01	.04		3	0		.04	
4 T dry	14	0.0	13.0			28	.02	0.1			143	7	.20	.010	
apples & cranberries w/ tapioca															
jr, Heinz	179		0.6				6	.02	.04		19	13		.06	
1 jar	213	0.2	43.0			60	.02	0.2			38	13	.43	.085	
str, Heinz	108		0.4				4	.01	.03		7	8		.04	
1 jar	128	0.1	25.9			36	.01	0.1			23	8	.26	.051	
apples & peaches, inst, Heinz	50		0.0				17	.02	.04		7	8		.06	
3 T dry	14	0.0	13.0			212	.02	0.2			128	10	.30	.030	
apples & pears															
inst, Heinz	50		1.0				17	.01	.02		2	9		.05	
3 T dry	14	0.0	13.0			28	.01	0.1			107	8	.20	.050	
jr, Heinz	121		0.4				15	.04	.06		4	17		.13	
1 jar	213	0.4	28.8			87	.04	0.3			222	19	.43	.128	
str, Heinz	73		0.3				16	.03	.04		3	10		.08	
1 jar	128	0.3	17.3			52	.03	0.2			133	12	.26	.077	
applesce															
jr	79	190.7	0.0			2	81	.06	.06	4	5	10	6	.09	
1 jar	213	0.1	21.9	1.7		20	.03	0.1		.21	164	13	.46	.075	
str	53	113.4	0.2			2	49	.04	.04	2	3	5	4	.03	
1 jar	128	0.2	14.0	1.0		22	.02	0.1		.14	91	9	.28	.049	
applesce & apricots															
jr	104	191.1	0.5			74	39	.07	.07	3	6	13	8	.07	
1 jar	220	0.5	27.3	2.0		745	.03	0.3			240	23	.57		
str	60	118.3	0.3			52	26	.04	.04	2	4	8	5	.06	
1 jar	135	0.3	15.7	1.4		523	.02	0.2			162	13	.34		
applesce & cherries															
jr	106	189.9	0.0			9	51	.10		1	6	20			
1 jar	220	0.6	28.9			91	.03	0.2			214		.89		
str	65	116.6	0.0			5	45	.06		1	3	14			
1 jar	135	0.4	17.7			53	.02	0.1			130		.52		
applesce & pineapple															
jr	83	189.8	0.2			4	57	.06	.08	4	4	8	7		.64
1 jar	213	0.2	22.3			45	.05	0.2			162	13	.21		
str	48	114.6	0.1			3	36	.03	.05	3	3	5	4	.02	
1 jar	128	0.1	12.9	1.2		26	.03	0.1			100	8	.13		
apricots w/ tapioca															
jr	139	180.5	0.0			159	39	.03	.06	4	14	18	9	.09	
1 jar	220	0.6	38.0	0.7		1590	.02	0.4		.30	274	22	.58		
str	80	112.2	0.0			98	29	.02	.04	2	11	12	5	.06	
1 jar	135	0.4	22.0	0.4		979	.01	0.3		.18	163	14	.40	.046	
bananas, inst, Heinz	50		0.0				16	.04	.24		4	4		.10	
3 T dry	14	1.0	13.0			60	.01	0.2			236	16	.20	.050	
bananas & pineapple w/ tapioca															
jr	143	179.8	0.0			9	42	.04	.18	12	13	15	14	.09	
1 jar	220	0.5	39.3	0.9		90	.04	0.4			149	11	.51	.086	

	KCAL	H₂O (g)	FAT (g)	PUFA (g)	CHOL (mg)	A (RE)	C (mg)	B-2 (mg)	B-6 (mg)	FOL (mcg)	Na (mg)	Ca (mg)	Mg (mg)	Zn (mg)	Mn (mg)
	WT (g)	PRO (g)	CHO (g)	SFA (g)	DFIB (g)	A (IU)	B-1 (mg)	NIA (mg)	B-12 (mcg)	PANT (mg)	K (mg)	P (mg)	Fe (mg)	Cu (mg)	
str	91	109.5	0.1			5	29	.03	.12	7	10	9	8	.04	
1 jar	135	0.3	24.8		0.4	55	.02	0.2			105	7	.18	.054	
bananas w/ tapioca															
jr	147	179.3	0.4			10	57	.04	.31	14	21	17	25	.15	
1 jar	220	0.8	39.1		0.7	97	.03	0.5		.38	237	20	.66	.101	
str	77	113.4	0.1			6	23	.04	.16	7	12	6	14	.08	
1 jar	135	0.5	20.6		0.5	58	.02	0.2		.20	118	9	.26	.054	
guava, str, Gerber	79		0.1			32	16	.03	.05		3	8			
1 jar	128	0.3	19.2		0.9		.01	0.5			114	6	.26		
guava & papaya w/ tapioca, str	80	105.6	0.1			24	104	.03	.02		5	9	6	.08	
1 jar	128	0.3	21.8			236	.01	0.3			95	7	.26		
guava w/ tapioca, str	86	103.9	0.0			38	97	.09	.05		2	9	3	.10	
1 jar	128	0.4	23.4			383	.02	0.5			94	6	.26		
mango, str, Gerber	86		0.3			110	16	.01			4	8			
1 jar	128	0.3	20.6		0.4		.03	0.3			81	6	.13		
mango w/ tapioca, str	109	104.9	0.3			90	168	.04	.16		6	5	5	.08	
1 jar	135	0.4	29.2			898	.03	0.3			80	8	.14		
mixed fruit, inst, Heinz	50		0.0				17	.02	.04		2	9		.08	
4 T dry	14	1.0	13.0			212	.01	0.5			167	9	.20	.060	
papaya, str, Gerber	74		0.5			12	16	.01	.01		9	8			
1 jar	128	0.3	17.0		0.4		.01	0.1			81	5	.13		
papaya & applesce w/ tapioca, str	89	103.2	0.1			10	145	.04	.03		6	9	7	.04	
1 jar	128	0.3	24.2			97	.01	0.1			101	7			
peaches															
inst, Heinz	50		1.0				16	.01	.04		2	12		.14	
3 T dry	14	1.0	13.0			224	.01	0.5			202	20	.40	.070	
jr	157	176.1	0.4			39	42	.07	.04	9	10	11	12	.13	
1 jar	220	1.3	41.6		1.3	392	.03	1.4		.29	342	24	.59	.110	
str	96	108.1	0.2			22	42	.05	.02	5	8	8	8	.12	
1 jar	135	0.7	25.5		0.8	217	.02	0.8		.18	219	16	.32	.072	
pears															
jr	93	187.0	0.2			7	47	.06	.02	8	4	18	19	.17	
1 jar	213	0.6	24.7		4.0	72	.03	0.4		.20	246	26	.54	.170	
str	53	113.2	0.2			4	31	.04	.01	5	3	11	10	.10	
1 jar	128	0.4	13.9		2.3	43	.02	0.2		.12	166	15	.30	.083	
pears & pineapple															
jr	93	187.0	0.4			7	36	.05	.03	6	2	21	16	.27	
1 jar	213	0.6	24.4		3.8	68	.05	0.4			251	21	.44	.224	
str	52	113.3	0.1			4	35	.04	.02	4	5	13	9	.08	
1 jar	128	0.4	13.9		1.5	37	.03	0.3			148	11	.32	.179	
plums w/ tapioca															
jr	163	174.3	0.0			21	2	.07	.06	2	18	12	9	.18	
1 jar	220	0.3	45.0		0.9	207	.01	0.5		.25	182	14	.48	.084	
str	96	108.0	0.0			13	1	.04	.03	1	8	8	5	.11	
1 jar	135	0.2	26.6		0.5	128	.01	0.3		.15	115	8	.27	.050	
prunes w/ tapioca															
jr	155	176.1	0.2			90	2	.18	.19	1	5	33	21	.22	
1 jar	220	1.3	41.2			895	.05	1.2		.31	357	32	.72		
str	94	108.4	0.1			61	1	.10	.11	0	6	20	14	.12	
1 jar	135	0.8	25.0		0.9	612	.03	0.7		.19	238	21	.47	.082	
tropical fruit medley, str, Gerber	86		0.9			37	16	.01	.03		6	6			
1 jar	128	0.3	19.1		0.3		.03	0.1			50	5	.13		

	KCAL	H₂O (g)	FAT (g)	PUFA (g)	CHOL (mg)	A (RE)	C (mg)	B-2 (mg)	B-6 (mg)	FOL (mcg)	Na (mg)	Ca (mg)	Mg (mg)	Zn (mg)	Mn (mg)
	WT (g)	PRO (g)	CHO (g)	SFA (g)	DFIB (g)	A (IU)	B-1 (mg)	NIA (mg)	B-12 (mcg)	PANT (mg)	K (mg)	P (mg)	Fe (mg)	Cu (mg)	

19.8. MEAT/EGG YOLKS

beef

jr	105	79.1	4.9	0.2		31	2	.16	.12	6	65	8	9	1.98	
1 jar	99	14.3	0.0	2.6	0.1	102	.01	3.3	.146	.35	189	71	1.63	.091	
str	106	79.8	5.3	0.2		55	2	.14	.14	6	80	7	17	2.43	
1 jar	99	13.5	0.0	2.6		183	.01	2.8	1.41	.34	218	83	1.46	.043	
beef w/ beef heart, str	93	81.7	4.4	0.2		37	2	.35	.12	5	62	4	12	1.82	
1 jar	99	12.6	0.0	2.1		124	.02	3.8			198	93	1.97	.128	

chicken

jr	148	75.2	9.5	2.3		55	2	.16	.19	11	50	54	11	1.00	
1 jar	99	14.6	0.0	2.4		184	.01	3.4		.72	121	89	.98	.045	
str	128	76.7	7.8	1.9		40	2	.15	.20	10	47	63	13	1.20	
1 jar	99	13.6	0.1	2.0		133	.01	3.2		.67	139	96	1.39	.045	
chicken sticks, jr	134	48.5	10.2			678	1	.14	.07		340	52			
1 jar	71	10.4	1.0		0.1	2257	.01	1.4			75	86	1.11		
egg yolks, str	191	66.3	16.3	1.8	739	353	1	.25	.15	.87	37	72	6	1.80	
1 jar	94	9.4	0.9	4.9		1176	.07	0	1.45	2.01	73	270	2.60	.066	

ham

jr	123	77.8	6.6	0.9		9	2	.19	.20	2	66	5	11	1.68	
1 jar	99	14.9	0.0	2.2	0.1	31	.14	2.8		.53	208	88	1.00		
str	110	78.6	5.7	0.8		11	2	.15	.25	2	40	6	13	2.22	
1 jar	99	13.7	0.0	1.9		38	.14	2.6		.51	202	80	1.02	.064	

lamb

jr	111	78.8	5.2	0.2		8	2	.20	.18	2	73	7	10	2.57	
1 jar	99	15.0	0.0	2.5		27	.02	3.2	2.25	.42	209	90	1.65	.056	
str	102	79.5	4.7	0.2		25	1	.20	.15	2	62	7	13	2.73	
1 jar	99	13.9	0.1	2.3		85	.02	2.9	2.17	.41	203	96	1.48	.054	
liver, beef, str	100	78.5	3.7	0.1	182	11337	19	1.8	.34	334	73	3	13	2.95	
1 jar	99	14.2	1.4	1.4		37754	.05	8.2	2.14		224	201	5.23	1.965	
meat sticks, jr	130	49.3	10.4	1.1		15	2	.12	.06		388	24			
1 jar	71	9.5	0.8	4.1	0.1	49	.04	1.1			81	73	.98		
pork, str	123	77.6	7.1	0.8		11	2	.20	.20	2	42	5	10	2.25	
1 jar	99	13.8	0.0	2.4		38	.15	2.2	98		221	93	.99	.071	

turkey

jr	128	76.7	7.0	1.7		169	2	.25	.16	12	72	28	11	1.78	
1 jar	99	15.2	0.0	2.3		563	.02	3.4	1.06	.60	178	94	1.34		
str	113	78.1	5.8	1.4		166	2	.21	.18	11	54	23	14	1.81	
1 jar	99	14.1	0.1	1.9		554	.02	3.6	.99	.56	229	125	1.19	.040	
turkey sticks, jr	129	49.6	10.1			49	1	.11	.05		343	51			
1 jar	71	9.7	1.0		0.2	162	.01	1.2			65	73	.88		

veal

jr	109	79.0	4.9	0.2		15	2	.18	.12	7	68	6	11	2.49	
1 jar	99	15.1	0.0	2.4	0.1	50	.02	3.8		.45	234	97	1.24	.074	
str	100	80.0	4.7	0.2		14	2	.16	.15	6	64	7	12	1.98	
1 jar	99	13.4	0.0	2.3		45	.02	3.5		.43	214	97	1.26	.040	

19.9. VEGETABLES

beans, green

jr	51	190.5	0.3			89	17	.21	.07	67	3	133	46	.40	
1 jar	206	2.5	11.8			892	.04	0.7		.31	263	39	2.23	.101	
str	32	117.8	0.1			57	7	.11	.05	44	2	49	30	.26	
1 jar	128	1.7	7.6		1.4	574	.03	0.4		.21	202	26	.96	.065	

	KCAL / WT (g)	H₂O (g) / PRO (g)	FAT (g) / CHO (g)	PUFA (g) / SFA (g)	CHOL (mg) / DFIB (g)	A (RE) / A (IU)	C (mg) / B-1 (mg)	B-2 (mg) / NIA (mg)	B-6 (mg) / B-12 (mcg)	FOL (mcg) / PANT (mg)	Na (mg) / K (mg)	Ca (mg) / P (mg)	Mg (mg) / Fe (mg)	Zn (mg) / Cu (mg)	Mn (mg)
beans, green, buttered															
jr	67	187.9	1.8			79	18	.23		56	4	143			
1 jar	206	2.7	12.5			787	.03	0.7			351		2.38		
str	42	116.3	1.0			58	11	.14		36	4	82			
1 jar	128	1.6	8.5			584	.02	0.4			205		1.64		
beans, green, creamed	68	193.6	0.9			32	6	.12	.03		26	68	16	.34	
1 jar	213	2.1	15.3		1.5	320	.05	0.5			139	39	.56		
beets, str	43	115.3	0.1			4	3	.06	.03	39	106	18	18	.15	
1 jar	128	1.7	9.8		0.8	43	.01	0.2			233	17	.41	.090	
carrots															
inst, Heinz	60		2.0				1	.05	.03		30	45		.22	
4 T dry	14	1.0	11.0			9056	.04	0.3			225	38	.60	.080	
jr	67	193.9	0.4			2516	12	.09	.17	37	104	49	23	.38	
1 jar	213	1.7	15.4		1.7	25156	.05	1.1		.59	429	43	.82	.100	
str	34	118.1	0.2			1467	7	.05	.09	19	48	29	12	.20	
1 jar	128	1.0	7.7		0.9	14670	.03	0.6		.31	251	25	.47	.052	
carrots, buttered															
jr	70	194.6	1.2			2096	16	.12		18	34	76			
1 jar	213	1.7	14.3			20960	.04	1.1			310		.66		
str	46	116.0	0.8			1386	12	.07		12	24	45			
1 jar	128	1.1	9.4			13860	.02	0.8			292		.36		
corn, creamed															
inst, Heinz	60		1.0				0	.01	.04		1	4		.24	
6 T dry	14	2.0	10.0			23	.02	0.3			91	30	.30	.020	
jr	138	173.5	0.8			16	5	.10	.09	27	111	39	17	.49	
1 jar	213	3.1	34.7		0.9	165	.03	1.1	.05	.70	172	71	.58	.081	
str	73	107.0	0.5			10	3	.06	.05	14	55	25	11	.24	
1 jar	128	1.8	18.1		0.5	96	.02	0.7	.02	.37	115	43	.36	.044	
garden veg, str	48	115.2	0.3			777	7	.09	.13	52	45	36	27	.33	
1 jar	128	2.9	8.7		1.7	7766	.08	1.0			215	35	1.06	.090	
mixed veg															
inst, Heinz	60		2.0				0	.04	.05		3	12		.25	
5 T dry	14	2.0	9.0			1556	.05	0.6			116	31	.30	.060	
jr	88	190.4	0.8			894	5	.07	.17	9	77	24	23	.58	
1 jar	213	3.1	17.4		1.7	8935	.06	1.4		.55	362	53	.87	.087	
str	52	114.9	0.6			511	2	.03	.07	5	6	17	13	.19	
1 jar	128	1.6	10.2		0.8	5109	.03	0.4		.32	163	29	.41	.051	
peas															
jr, Gerber	113		1.1			75	9	.13	.09		15	34			
1 jar	213	7.0	19.0		4.0		.15	2.0			198	104	1.91		
str	52	112.0	0.4			72	9	.08	.09	33	5	26	19	.45	
1 jar	128	4.5	10.4		2.3	723	.10	1.3		.36	144	55	1.23	.082	
peas, buttered															
jr	123	172.0	2.6			84	26	.16		75	11	93			
1 jar	206	7.3	23.3			845	.15	2.8			240		2.14		
str	72	107.8	1.4			42	15	.09		44	10	49			
1 jar	128	4.7	13.6			422	.10	1.8			124		1.41		
peas, creamed															
inst, Heinz	60		2.0				0	.05	.06		2	12		.46	
5 T dry	14	3.0	8.0			142	.08	0.5			91	49	.60	.070	
str	68	110.7	2.4			11	2	.07	.06	29	18	16		.50	
1 jar	128	2.8	11.4			110	.11	1.0	.10		113	40	.72	.065	

	KCAL	H₂O (g)	FAT (g)	PUFA (g)	CHOL (mg)	A (RE)	C (mg)	B-2 (mg)	B-6 (mg)	FOL (mcg)	Na (mg)	Ca (mg)	Mg (mg)	Zn (mg)	Mn (mg)
	WT (g)	PRO (g)	CHO (g)	SFA (g)	DFIB (g)	A (IU)	B-1 (mg)	NIA (mg)	B-12 (mcg)	PANT (mg)	K (mg)	P (mg)	Fe (mg)	Cu (mg)	
spinach, creamed															
jr	90	187.9	3.0			783	8	.18	.12	147	117	240	134	.85	
1 jar	213	6.4	13.7			7830	.05	0.6			471	104	2.98	.145	
str	48	114.7	1.7			534	11	.13	.10	78	62	113	71	.40	
1 jar	128	3.2	7.3		1.0	5338	.02	0.3			244	70	.80	.077	
squash															
inst, Heinz	60		0.0				1	.05	.05		1	12		.25	
5 T dry	14	1.0	10.0			2830	.04	0.6			330	32	.30	.050	
jr	51	197.6	0.4			429	17	.14	.15	33	3	50	25	.17	
1 jar	213	1.8	12.0		1.7	4291	.02	0.8		.47	393	34	.74	.115	
str	30	118.7	0.2			259	10	.07	.08	20	3	30	16	.18	
1 jar	128	1.1	7.2		1.3	2590	.01	0.5		.28	229	20	.38	.069	
squash, buttered															
jr	63	195.5	1.3			326	16	.14		25	3	65			
1 jar	213	1.5	13.7			3259	.02	0.7			288		.89		
str	37	117.4	0.4			212	10	.09		15	2	42			
1 jar	128	0.8	8.8			2123	.01	0.5			163		.53		
sweet potatoes															
inst, Heinz	50		0.0				1	.01	.05		15	21		.16	
4 T dry	14	1.0	11.0			6940	.03	0.2			246	24	.30	.100	
jr	133	185.0	0.3			1460	21	.08	.25	23	49	35	26	.24	
1 jar	220	2.4	30.7		1.8	14600	.06	0.8		.90	535	52	.85	.220	
str	77	114.5	0.2			869	13	.05	.13	13	27	21	18	.28	
1 jar	135	1.5	17.8		0.9	8691	.04	0.5		.53	355	32	.50	.109	
sweet potatoes, buttered															
jr	126	188.3	1.6			1330	21	.10		29	17	61			
1 jar	220	1.8	26.8			13295	.04	0.7			474		.87		
str	76	115.9	1.0			920	12	.05		18	11	28			
1 jar	135	1.3	15.9			9199	.02	0.4			281		.60		

20. MEAT ANALOGUES & MEAT ANALOGUE ENTREES

	KCAL	H₂O (g)	FAT (g)	PUFA (g)	CHOL (mg)	A (RE)	C (mg)	B-2 (mg)	B-6 (mg)	FOL (mcg)	Na (mg)	Ca (mg)	Mg (mg)	Zn (mg)	Mn (mg)
	WT (g)	PRO (g)	CHO (g)	SFA (g)	DFIB (g)	A (IU)	B-1 (mg)	NIA (mg)	B-12 (mcg)	PANT (mg)	K (mg)	P (mg)	Fe (mg)	Cu (mg)	
bacon, simulated meat product	25	3.9	2.4	1.2	0	1	0	.04	.04	3	117	2	2	.03	.016
1 strip	8	0.9	0.5	0.4	0.1	7	.35	0.6	.00	.01	14	6	.19	.008	
big franks, cnd, Loma Linda	100	30.0	5.0		0			.61	.50		220	17		.61	
1 frank	51	10.0	4.0				.56	8.2	1.20	.82	75		.92		
bologna, frzn, Loma Linda	150	26.0	9.0		0			.31	.64		490	23		.91	
2 slices	57	14.0	5.0				.54	6.4	2.00	2.00	160		1.30		
burger-like, Millstone	90	89.0	0.2		0			.03			440	38		1.10	
4 oz	113	13.0	8.0				.03	1.8			430		.57		
burger mix, Millstone	90	2.9	0.7		0			.17			368	66		.55	
1 oz	28	10.0	12.0				.23	0.3		1.70	34		2.30		
chicken, fried, Loma Linda	180	26.0	14.0		0			.51	.63		510	26		.30	
1 piece	57	13.0	2.0				.63	9.1	2.00	2.50	80		1.10		
chicken, fried w/ gravy, cnd, Loma Linda	140	60.0	10.0		0			.41	.43		340	20		.55	
2 pieces	85	9.0	4.0				.60	4.0	2.00	1.50	250		1.40		
chicken, frzn, Loma Linda	160	31.0	13.0		0			.41	.51		330	25		.63	
2 slices	57	10.0	1.0				.46	5.1	2.00	1.40	120		1.10		
chicken supreme dry mix, Loma Linda	50	1.0	0.0		0			.26	.40		450	20		.40	
1/4 cup	16	9.0	4.0				.61	3.0	1.00	1.50	190		.72		
chicken nuggets, frzn, Loma Linda	230	41.0	13.0		0			.40	.51		640	41		.72	
5 pieces	85	14.0	15.0				.49	6.6	3.00	1.90	180		1.90		
chik-patties, frzn, Loma Linda	230	41.0	13.0		0			.40	.51		640	41		.72	
1 patty	85	14.0	15.0				.49	6.6	3.00	1.90	180		1.90		

	KCAL / WT (g)	H₂O (g) / PRO (g)	FAT (g) / CHO (g)	PUFA (g) / SFA (g)	CHOL (mg) / DFIB (g)	A (RE) / A (IU)	C (mg) / B-1 (mg)	B-2 (mg) / NIA (mg)	B-6 (mg) / B-12 (mcg)	FOL (mcg) / PANT (mg)	Na (mg) / K (mg)	Ca (mg) / P (mg)	Mg (mg) / Fe (mg)	Zn (mg) / Cu (mg)	Mn (mg)
chik stiks, Worthington	112	25.5	6.3	3.4	0			.22	.29		325	22			
1 piece	47	9.8	3.9	1.0			.60	5.4	4.30		90		2.24		
choplets, Worthington	105	66.3	1.4								317	11			
2 slices	92	19.1	4.0								33		1.86		
corn dogs, frzn, Loma Linda	200	32.0	10.0		0			.62	.80		620	20		.44	
1 corn dog	71	7.0	21.0				.58	6.2	3.00	2.60	150		.73		
dinner cuts, cnd, Loma Linda	110	71.0	1.0		0			.27	.46		550ᵃ	20		1.30	
2 cuts	100	21.0	4.0				.20	1.0	1.00	.21	180ᵃ		2.00		
dinner loaf, dry pack, Loma Linda	50	1.0	1.0		0			.24	.58		380	26		.30	
1/4 cup	16	9.0	4.0				.61	4.0	1.00	1.90	140		.64		
frichik, Worthington	155	61.8	9.8	7.7				.21	.18		487	18			
2 pieces	90	12.1	4.5	1.5			.21	3.1	3.92		41		1.72		
fripats, Worthington	193	29.5	12.3	6.2				.32	.44		304	77			
1 piece	64	15.5	5.0	2.1				8.8	7.04		135		3.72		
griddle steaks, frzn, Loma Linda	160	19.0	11.0		0			.33	.56		385	33		.52	
1 steak	47	11.0	4.0				.52	4.1	2.00	1.70	250		.61		
grillers, Morningstar Farms	190	29.9	11.9	6.2	1			.38	.44		285	66			
1 patty	64	15.6	5.1	2.1				7.5	6.70		140		3.48		
leanies, Worthington	93	21.5	5.0	3.8				.18	.19		385	41			
1 link	40	8.0	4.0	0.7			.31	3.3	3.10		39		1.89		
linketts, cnd, Loma Linda	150	44.0	8.0		0			.44	.37		340	20		.89	
2 links	74	15.0	5.0				.37	6.0	2.00	.81	75		1.80		
link, Morningstar Farms	178	35.5	11.0	7.4	1			.34	.39		406	22			
3 links	68	14.5	5.2	1.9				7.5	6.75		85		3.90		
little links, cnd, Loma Linda	80	29.0	5.0		0			.41	.16		210	12		.60	
2 links	46	8.0	2.0				.55	6.0	2.00	1.50	40		1.10		
meat extender, simulated meat product	88	2.1	0.8	0.5	0	1		.25	.37		3	57	61		
1 oz	28	10.7	10.7	0.1		9	.20	6.2	1.76				3.36		
meatless loaf, Millstone	90	2.9	0.8		0			.16			380	67		.46	
1 oz	28	9.9	12.0				.29	0.6		1.20	290		2.20		
nuteena, cnd, Loma Linda	160	39.0	12.0		0			.58	.45		120	21		.87	
1/2" slice	67	8.0	5.0				.47	14.0	1.00	2.80	200		1.20		
nut meat, Millstone	150	44.0	10.0		0			.03			280	21		.61	
2.5 oz	71	7.2	6.6				.01	2.0		.92	260		.36		
ocean fillet, frzn, Loma Linda	130	21.0	8.0		0			.42	.71		230	45		.66	
1 fillet	47	11.0	4.0				.52	5.6	2.00	2.10	280		.94		
ocean platter, dry pack, Loma Linda	50	1.0	1.0		0			.26	.40		260	26		.35	
1/4 cup	16	8.0	5.0				.61	3.0	2.00	1.50	170		.72		
patties, Morningstar Farms	203	37.1	11.4	6.0	1			.38	.50		640	38			
2 patties	76	18.2	6.8	2.3				9.2	7.14		149		4.27		
patty mix, dry pack, Loma Linda	50	1.0	1.0		0			.34	.58		320	30		.29	
1/4 cup	16	9.0	4.0				.61	3.7	2.00	1.50	150		.72		
proteena, cnd, Loma Linda	140	40.0	6.0		0			.50	.50		460	22		1.20	
1/2" slice	71	17.0	5.0				.64	7.8	2.00	2.50	280		1.60		
redi-burger, cnd, Loma Linda	130	39.0	6.0		0			.40	.80		370	19		1.20	
1/2" slice	68	14.0	5.0				.60	6.7	2.00	1.50	120		1.40		
roast beef, frzn, Loma Linda	150	31.0	11.0		0			.29	.32		670	24		.80	
2 slices	57	11.0	2.0				.17	1.4	1.00	1.80	120		1.50		
salami, frzn, Loma Linda	130	33.0	7.0		0			.34	.63		640	13		.91	
2 slices	57	13.0	2.0				.38	3.0	2.00	2.10	70		1.10		
sandwich spread, cnd, Loma Linda	70	34.0	4.0		0			.34	.53		300	21		.62	
3 T	48	4.0	4.0				.29	6.2	2.00	1.20	160		.72		
sausage, simulated meat product	64	12.6	4.5	2.3	0	16	0	.10	.21	7	222	16	9	.37	.181
1 link	25	4.6	2.5	0.7	0.1	160	.59	2.8	.00	.08	58	56	.93	.063	
sausage, simulated meat product	97	19.2	6.9	3.5	0	24	0	.15	.32	10	337	24	14	.55	.276
1 patty	38	7.0	3.7	1.1		243	.89	4.3	.00	.12	88	85	1.41	.095	
sizzle burger, frzn, Loma Linda	210	28.0	11.0		0			.51	.78		320	49		.78	
1 burger	71	15.0	13.0				.64	8.5	2.00	1.50	180		1.60		
sizzle franks, cnd, Loma Linda	170	43.0	13.0		0			.35	.35		340	20		.63	
2 franks	70	10.0	3.0				.35	4.1	2.00	2.30	100		1.40		
stakelets, Worthington	143	41.4	6.5	3.4				.29	.32		431	63			
1 piece	71	14.4	6.5	1.2				5.8	5.16		143		3.28		

ᵃ Product w/ no added salt contains 30 mg Na and 75 mg K.

	KCAL	H₂O (g)	FAT (g)	PUFA (g)	CHOL (mg)	A (RE)	C (mg)	B-2 (mg)	B-6 (mg)	FOL (mcg)	Na (mg)	Ca (mg)	Mg (mg)	Zn (mg)	Mn (mg)
	WT (g)	PRO (g)	CHO (g)	SFA (g)	DFIB (g)	A (IU)	B-1 (mg)	NIA (mg)	B-12 (mcg)	PANT (mg)	K (mg)	P (mg)	Fe (mg)	Cu (mg)	
stew pac, cnd, Loma Linda	70	39.0	2.0		0			.34	.62		220	17		.62	
2 oz	57	10.0	4.0				.39	6.7	1.00	1.60	60		.73		
stripples, Worthington	110	14.5	8.6	5.8				.15	.16		362	17			
4 strips	33	4.2	3.9	1.1				2.7	1.78		35		.95		
strips, Morningstar Farms	83	11.0	6.5	4.1	0			.11	.12		274	13			
3 strips	25	3.2	2.9	0.8				2.0	1.35		26		.72		
swedish meatballs, frzn, Loma Linda	190	30.0	8.0		0			.42	.70		420	32		.65	
8 meatballs	70	22.0	7.0				.49	4.9	2.00	1.50	170		1.80		
swiss steak w/ gravy, cnd, Loma Linda	140	45.0	8.0		0			.44	.37		350	34		.59	
1 steak	74	9.0	8.0				.52	4.0	1.00	1.50	180		1.50		
tastee cuts, cnd, Loma Linda	70	54.0	1.0		0			.18	.04		230	13		1.40	
2 cuts	71	12.0	2.0				.37	1.3	.20	.16	40		1.10		
tender bits, cnd, Loma Linda	80	40.0	3.0		0			.29	.63		340	17		.63	
4 pieces	57	8.0	4.0				.51	5.6	1.00	2.00	45		.91		
tender cuts, Millstone	80	82.0	1.2		0						410	13		3.00	
3.6 oz	102	14.0	3.3				.10	4.4		2.30	50		1.30		
tender rounds w/ gravy, cnd, Loma Linda	120	46.0	4.0		0			.44	.88		310	29		.66	
6 rounds	75	15.0	7.0				.66	5.0	2.00	.80	115		1.80		
turkey, frzn, Loma Linda	160	31.0	12.0		0			.23	.59		1350	19		.86	
2 slices	57	10.0	3.0				.46	1.9	2.00	2.50	110		1.10		
vege-burger, cnd, Loma Linda	110	79.0	1.0		0			.68	.56		190	32		1.10	
1/2 cup	108	22.0	4.0				.53	5.0	2.00	2.90	110		2.70		
vege-burger, Millstone	110	76.0	1.7		0			.14			450	12		2.70	
1/2 cup	108	20.0	3.0				.03			1.80	75		1.10		
vege-burger, no salt added, cnd, Loma Linda—1/2 cup	140	72.0	2.0		0			.51	.93		55	43		1.70	
	108	27.0	4.0				.53	6.5	2.00	1.30	200		1.80		
vegelona, cnd, Loma Linda	100	41.0	1.0		0			.50	.50		210	15		1.20	
1/2" slice	67	18.0	6.0				.54	8.7	.90	1.50	100		1.50		
vege-scallops, cnd, Loma Linda	70	60.0	1.0		0			.08	.02		180	17		1.10	
6 pieces	78	14.0	2.0				.34	0.7	2.00	.18	30		1.10		
vegetarian burger, Worthington	140	79.8	3.0					.25	.44		520	8			
1/2 cup	113	21.8	6.5				.63	12.1	8.45		57		4.48		
veg skallops, Worthington	92	62.2	1.1								347	9			
1/2 cup	85	16.9	3.4								12		1.51		
veja-links, Worthington	114	40.6	6.6	2.8				.22	.16		297	7			
2 links	62	8.2	5.4	1.1			.29	3.1	4.04		35		1.46		
vita-burger granules, dry pack, Loma Linda[a]—3 T	70	2.0	0.0		0			.13	.29		150	42		1.20	
	21	10.0	7.0				.13	3.4	1.00	.42	460		1.80		
wheat fries, Millstone	60	56.0	0.9		0			.12			180	5		.90	
2.5 oz	71	12.0	2.4					.80		1.20	20		1.40		

21. MEATS
21.1. BEEF

	KCAL	H₂O (g)	FAT (g)	PUFA (g)	CHOL (mg)	A (RE)	C (mg)	B-2 (mg)	B-6 (mg)	FOL (mcg)	Na (mg)	Ca (mg)	Mg (mg)	Zn (mg)	Mn (mg)
	WT (g)	PRO (g)	CHO (g)	SFA (g)	DFIB (g)	A (IU)	B-1 (mg)	NIA (mg)	B-12 (mcg)	PANT (mg)	K (mg)	P (mg)	Fe (mg)	Cu (mg)	
breakfast strips															
ckd	153	8.9	11.7	0.5	40		12[b]	.09	.11		766		9	2.17	
3 slices	34	10.6	0.5	4.9			.03	2.2	1.17		140	80	1.07		
ckd, Lean 'n Tasty	46	4.6	3.8		13						190	1	2	.48	
1 slice	12	2.9	0.2								36	20	.23	.040	
brisket															
sep lean & fat, braised	391	43.0	32.4	1.2	93	0		.17	.25	6	61	9	18	5.01	.014
3.5 oz	100	23.0	0.0	13.2	0.0		.06	3.0	2.23	.29	229	184	2.19	.094	
sep lean, braised	241	55.7	12.8	0.4	93	0		.22	.30	8	72	6	23	6.88	.017
3.5 oz	100	29.4	0.0	4.6	0.0		.07	3.8	2.55	.36	287	239	2.77	.120	
brisket, flat half															
sep lean & fat, braised	408	41.5	34.9	1.2	92	0		.18	.23	6	65	9	16	5.30	.013
3.5 oz	100	22.0	0.0	14.5	0.0		.06	2.9	2.24	.27	219	173	2.18	.090	
sep lean, braised	263	53.8	15.9	0.4	91	0		.23	.28	8	77	6	22	7.39	.017
3.5 oz	100	28.1	0.0	6.3	0.0		.07	3.6	2.58	.34	273	226	2.79	.115	

[a] Values correspond to 1/4 cup of vita-burger chunks.   [b] Contains added sodium ascorbate.

	KCAL / WT (g)	H₂O (g) / PRO (g)	FAT (g) / CHO (g)	PUFA (g) / SFA (g)	CHOL (mg) / DFIB (g)	A (RE) / A (IU)	C (mg) / B-1 (mg)	B-2 (mg) / NIA (mg)	B-6 (mg) / B-12 (mcg)	FOL (mcg) / PANT (mg)	Na (mg) / K (mg)	Ca (mg) / P (mg)	Mg (mg) / Fe (mg)	Zn (mg) / Cu (mg)	Mn (mg)
brisket, point half															
sep lean & fat, braised	366	44.9	28.9	1.1	95		0	.17	.25	6	56	8	19	4.72	.014
3.5 oz	100	24.7	0.0	11.8	0.0		.06	3.1	2.29	.30	239	197	2.25	.098	
sep lean, braised	212	57.8	8.6	0.3	95		0	.22	.31	8	63	5	25	6.36	.017
3.5 oz	100	31.5	0.0	3.1	0.0		.07	3.9	2.63	.37	298	257	2.84	.124	
chuck arm pot roast															
sep lean & fat, braised	350	45.5	26.0	1.0	99		0	.24	.28	9	60	10	19	6.74	.016
3.5 oz	100	27.1	0.0	10.7	0.0		.07	3.1	2.93	.32	244	217	3.07	.132	
sep lean, braised	231	55.2	10.0	0.4	101		0	.29	.33	11	66	9	24	8.66	.019
3.5 oz	100	33.0	0.0	3.8	0.0		.08	3.7	3.40	.38	289	268	3.79	.164	
chuck blade roast															
sep lean & fat, braised	383	43.3	30.4	1.1	103		0	.23	.25	5	63	13	18	7.83	.015
3.5 oz	100	25.4	0.0	12.7	0.0		.07	2.4	2.23	.30	223	191	2.96	.119	
sep lean, braised	270	52.7	15.3	0.5	106		0	.28	.29	6	71	13	23	10.27	.018
3.5 oz	100	31.1	0.0	6.2	0.0		.08	2.7	2.47	.35	263	235	3.68	.148	
corned beef															
cured brisket, ckd	251	59.8	19.0	0.7	98		16[a]	.17	.23		1134	8	12	4.58	.022
3.5 oz	100	18.2	0.5	6.3			.03	3.0	1.63	.42	145	125	1.86	.154	
cured, cnd	250	57.7	14.9	0.6	86	0	2	.15	.13		1006		14	3.57	.014
3.5 oz	100	27.1	0.0	6.2	0.0	0	.02	2.4	1.62		136	111	2.08	.064	
dried (chipped)	47	16.0	1.1	0.1							984	2	9	1.49	
1 oz	28	8.3	0.4	0.5							126	49	1.28	.045	
flank															
sep lean & fat, braised	257	56.2	15.5	0.5	72		0	.18	.35	9	71	7	23	5.92	.019
3.5 oz	100	27.5	0.0	6.6	0.0		.14	4.5	3.36	.38	344	261	3.40	.123	
sep lean & fat, broiled	254	56.9	16.3	0.5	71		0	.19	.42	8	82	6	24	4.71	.020
3.5 oz	100	25.1	0.0	7.0	0.0		.11	4.8	3.02	.44	398	215	2.50	.098	
sep lean, braised	244	57.3	13.8	0.4	71		0	.19	.36	9	72	6	24	6.05	.019
3.5 oz	100	28.0	0.0	5.9	0.0		.14	4.6	3.41	.38	351	267	3.47	.124	
sep lean, broiled	243	57.8	15.0	0.5	70		0	.19	.42	8	83	6	25	4.79	.020
3.5 oz	100	25.4	0.0	6.4	0.0		.11	4.9	3.05	.44	405	218	2.53	.098	
ground, extra lean															
baked, med	250	58.6	16.1	0.6	82		0	.24	.22	9	49	7	17	5.34	.015
3.5 oz	100	24.5	0.0	6.3	0.0		.04	4.2	1.73	.27	224	124	2.28	.075	
baked, well done	274	52.7	16.0	0.6	107		0	.31	.29	11	64	9	22	6.94	.020
3.5 oz	100	30.3	0.0	6.3	0.0		.05	5.4	1.86	.35	291	162	2.96	.097	
broiled, med	256	57.3	16.3	0.6	84		0	.27	.27	9	70	7	21	5.45	.016
3.5 oz	100	25.4	0.0	6.4	0.0		.06	5.0	2.17	.35	313	161	2.35	.070	
broiled, well done	265	53.9	15.8	0.6	99		0	.32	.32	11	82	9	25	6.43	.019
3.5 oz	100	28.6	0.0	6.2	0.0		.07	5.9	2.56	.42	369	190	2.77	.083	
pan fried, med	255	57.6	16.4	0.6	81		0	.26	.27	9	70	7	21	5.42	.016
3.5 oz	100	25.0	0.0	6.5	0.0		.06	4.7	2.00	.25	312	160	2.36	.087	
pan fried, well done	263	53.9	16.0	0.6	93		0	.30	.31	10	81	8	24	6.27	.018
3.5 oz	100	28.0	0.0	6.3	0.0		.07	5.4	2.32	.29	360	185	2.73	.101	
ground, lean															
baked, med	268	56.9	18.3	0.7	78		0	.19	.20	9	56	9	17	5.10	.014
3.5 oz	100	23.9	0.0	7.2	0.0		.05	4.3	1.77	.27	224	128	2.09	.072	
baked, well done	292	51.3	18.4	0.7	99		0	.24	.26	12	71	12	21	6.51	.018
3.5 oz	100	29.6	0.0	7.2	0.0		.07	5.5	2.26	.34	286	164	2.66	.092	
broiled, med	272	55.7	18.5	0.7	87		0	.21	.26	9	77	11	21	5.36	.014
3.5 oz	100	24.7	0.0	7.3	0.0		.05	5.2	2.35	.38	301	158	2.11	.066	
broiled, well done	280	52.9	17.6	0.7	101		0	.24	.30	11	89	12	24	6.20	.017
3.5 oz	100	28.2	0.0	6.9	0.0		.06	6.0	2.72	.44	349	182	2.45	.077	
pan fried, med	275	55.6	19.1	0.7	84		0	.22	.28	9	77	10	20	5.20	.014
3.5 oz	100	24.2	0.0	7.5	0.0		.05	4.8	2.27	.32	299	159	2.18	.077	

[a] Contains added sodium ascorbate.

	KCAL / WT (g)	H₂O (g) / PRO (g)	FAT (g) / CHO (g)	PUFA (g) / SFA (g)	CHOL (mg) / DFIB (g)	A (RE) / A (IU)	C (mg) / B-1 (mg)	B-2 (mg) / NIA (mg)	B-6 (mg) / B-12 (mcg)	FOL (mcg) / PANT (mg)	Na (mg) / K (mg)	Ca (mg) / P (mg)	Mg (mg) / Fe (mg)	Zn (mg) / Cu (mg)	Mn (mg)
pan fried, well done	277	53.7	17.7	0.7	95		0	.24	.32	10	87	11	23	5.91	.016
3.5 oz	100	27.6	0.0	6.9	0.0		.06	5.5	2.58	.36	340	181	2.48	.088	
ground, regular															
baked, med	287	55.1	20.9	0.8	87		0	.16	.23	9	60	10	15	4.89	.017
3.5 oz	100	23.0	0.0	8.2	0.0		.03	4.8	2.34	.22	221	137	2.41	.070	
baked, well done	317	48.9	21.5	0.8	108		0	.20	.29	11	75	12	19	6.07	.021
3.5 oz	100	28.8	0.0	8.4	0.0		.04	5.9	2.90	.27	274	170	2.99	.086	
broiled, med	289	54.2	20.7	0.8	90		0	.19	.27	9	83	11	20	5.18	.017
3.5 oz	100	24.1	0.0	8.1	0.0		.03	5.8	2.93	.33	292	170	2.44	.082	
broiled, well done	292	52.0	19.5	0.7	101		0	.21	.30	10	93	12	22	5.81	.020
3.5 oz	100	27.2	0.0	7.7	0.0		.04	6.5	3.28	.37	327	191	2.74	.092	
pan fried, med	306	52.3	22.6	0.8	89		0	.20	.24	9	84	11	20	5.07	.017
3.5 oz	100	23.9	0.0	8.9	0.0		.03	5.8	2.71	.34	300	171	2.45	.081	
pan fried, well done	286	52.7	18.9	0.7	98		0	.21	.27	10	93	13	22	5.62	.019
3.5 oz	100	27.0	0.0	7.4	0.0		.04	6.5	3.00	.38	332	189	2.71	.090	
rib eye, small end (rib 10-12)															
sep lean & fat, broiled	295	52.3	20.6	0.7	83		0	.20	.36	7	64	13	24	6.13	.014
3.5 oz	100	25.4	0.0	8.7	0.0		.09	4.3	3.06	.31	352	188	2.34	.098	
sep lean, broiled	225	58.4	11.6	0.4	80		0	.22	.40	8	69	13	27	6.99	.016
3.5 oz	100	28.0	0.0	4.9	0.0		.11	4.8	3.32	.34	394	208	2.57	.100	
rib, large end (rib 6-9)															
sep lean & fat, broiled	378	46.7	32.3	1.2	87		0	.16	.26	6	60	10	18	4.68	.013
3.5 oz	100	20.1	0.0	13.7	0.0		.07	2.6	2.73	.32	288	159	2.02	.089	
sep lean & fat, roasted	368	46.1	30.0	1.1	85		0	.18	.22	7	64	10	19	5.73	.013
3.5 oz	100	22.7	0.0	12.7	0.0		.07	3.6	2.33	.36	289	171	2.32	.100	
sep lean, broiled	233	60.1	14.2	0.4	82		0	.20	.32	7	70	8	24	6.25	.016
3.5 oz	100	24.6	0.0	6.1			.09	3.1	3.26	.41	369	199	2.48	.089	
sep lean, roasted	244	57.0	14.0	0.4	81		0	.22	.26	9	73	8	25	7.46	.016
3.5 oz	100	27.5	0.0	5.9	0.0		.09	4.4	2.61	.45	357	209	2.82	.105	
rib, shortribs															
sep lean & fat, braised	471	35.7	42.0	1.5	94		0	.15	.22	5	50	12	15	4.88	.013
3.5 oz	100	21.6	0.0	17.8	0.0		.05	2.5	2.62	.25	224	162	2.31	.099	
sep lean, braised	295	50.2	18.1	0.6	93		0	.20	.28	7	58	11	22	7.80	.018
3.5 oz	100	30.8	0.0	7.7	0.0		.07	3.2	3.46	.34	313	235	3.36	.107	
rib, small end (ribs 10-12)															
sep lean & fat, broiled	326	49.5	24.8	0.9	84		0	.19	.34	7	62	13	22	5.70	.013
3.5 oz	100	24.0	0.0	10.5	0.0		.09	4.1	2.93	.30	331	177	2.22	.097	
sep lean & fat, roasted	359	47.5	29.3	1.0	85		0	.17	.31	7	65	13	20	4.86	.013
3.5 oz	100	22.3	0.0	12.4	0.0		.07	3.1	2.88	.36	325	179	1.95	.089	
sep lean, broiled	221	58.7	11.2	0.3	80		0	.22	.40	8	69	13	27	6.99	.016
3.5 oz	100	28.0	0.0	4.8	0.0		.11	4.8	3.32	.34	394	208	2.57	.100	
sep lean, roasted	236	58.7	13.5	0.4	80		0	.20	.37	8	75	13	25	6.19	.016
3.5 oz	100	26.7	0.0	5.7	0.0		.08	3.6	3.37	.44	404	219	2.30	.089	
rib, whole (ribs 6-12)															
sep lean & fat, broiled	362	47.4	30.0	1.0	86		0	.17	.29	6	61	11	20	5.03	.013
3.5 oz	100	21.5	0.0	12.7	0.0		.08	3.1	2.79	.31	302	165	2.08	.092	
sep lean & fat, roasted	381	45.2	31.8	1.1	85		0	.17	.25	7	63	11	19	5.17	.013
3.5 oz	100	21.9	0.0	13.4	0.0		.07	3.3	2.51	.35	294	169	2.11	.095	
sep lean, broiled	228	59.5	13.0	0.4	82		0	.21	.35	8	69	10	25	6.55	.016
3.5 oz	100	26.0	0.0	5.5	0.0		.09	3.8	3.28	.38	379	202	2.52	.094	
sep lean, roasted	240	57.7	13.8	0.4	81		0	.21	.30	8	74	10	25	6.94	.016
3.5 oz	100	27.2	0.0	5.8	0.0		.08	4.1	2.92	.44	376	213	2.61	.098	
round, bottom															
sep lean & fat, braised	261	53.9	14.8	0.6	96		0	.25	.34	11	51	6	23	5.13	.017
3.5 oz	100	29.8	0.0	5.7	0.0		.07	3.9	2.40	.40	292	255	3.25	.130	

	KCAL / WT (g)	H₂O (g) / PRO (g)	FAT (g) / CHO (g)	PUFA (g) / SFA (g)	CHOL (mg) / DFIB (g)	A (RE) / A (IU)	C (mg) / B-1 (mg)	B-2 (mg) / NIA (mg)	B-6 (mg) / B-12 (mcg)	FOL (mcg) / PANT (mg)	Na (mg) / K (mg)	Ca (mg) / P (mg)	Mg (mg) / Fe (mg)	Zn (mg) / Cu (mg)	Mn (mg)
sep lean, braised	222	57.2	9.7	0.4	96	0	.26	.36		11	51	5	25	5.48	.018
3.5 oz	100	31.6	0.0	3.4	0.0		.08	4.1	2.47	.42	308	272	3.46	.134	
round, eye of															
sep lean & fat, roasted	243	57.5	14.2	0.5	73	0	.16	.35		7	59	6	25	4.33	.015
3.5 oz	100	26.8	0.0	5.8	0.0		.08	3.5	2.10	.42	362	208	1.84	.099	
sep lean, roasted	183	62.8	6.5	0.2	69	0	.17	.38		7	62	5	27	4.74	.016
3.5 oz	100	29.0	0.0	2.5	0.0		.09	3.8	2.17	.46	395	226	1.95	.100	
round, full cut															
sep lean & fat, broiled	274	55.1	18.2	0.8	84	0	.21	.44		9	60	7	24	4.13	.015
3.5 oz	100	25.5	0.0	7.3	0.0		.09	3.7	2.75	.36	366	211	2.42	.097	
sep lean, broiled	184	63.4	7.0	0.3	82	0	.23	.50		10	64	5	28	4.68	.016
3.5 oz	100	28.5	0.0	2.5	0.0		.10	4.2	2.98	.41	415	237	2.69	.107	
round, tip															
sep lean & fat, roasted	251	57.3	15.3	0.6	83	0	.25	.37		7	62	6	25	6.37	.015
3.5 oz	100	26.5	0.0	6.1	0.0		.09	3.5	2.74	.43	353	221	2.71	.121	
sep lean, roasted	190	62.8	7.5	0.2	81	0	.27	.40		8	65	5	27	7.07	.017
3.5 oz	100	28.7	0.0	2.8	0.0		.10	3.7	2.89	.47	386	242	2.94	.125	
round, top															
sep lean & fat, broiled	211	59.2	8.8	0.4	85	0	.26	.54		12	60	6	30	5.40	.016
3.5 oz	100	30.8	0.0	3.3	0.0		.11	5.9	2.44	.48	429	239	2.81	.121	
sep lean & fat, pan fried	290	50.4	17.1	1.1	97	0	.25	.55		11	67	6	31	4.18	.018
3.5 oz	100	31.7	0.0	6.5	0.0		.10	4.9	3.18	.42	459	262	2.86	.125	
sep lean, broiled	191	60.9	6.2	0.2	84	0	.27	.56		12	61	6	31	5.57	.017
3.5 oz	100	31.7	0.0	2.2	0.0		.12	6.0	2.48	.49	442	246	2.88	.123	
sep lean, pan fried	227	55.5	8.6	0.9	97	0	.28	.61		13	71	5	35	4.62	.020
3.5 oz	100	35.1	0.0	2.8	0.0		.11	5.5	3.43	.46	513	392	3.15	.131	
sausage, smoked, cnd	89	15.2	7.6	0.3	19	3ᵃ	.04	.03			321	2	4	.79	
1 oz	28	4.0	0.7	3.2			.01	0.9	.53		50	30	.50		
	134	23.0	11.6	0.5	29	5ᵃ	.06	.05			486	4	6	1.20	
1 sausage (8 per 12 oz pkg)	43	6.1	1.0	4.9			.02	1.4	.80		76	45	.76		
shank, crosscuts															
sep lean & fat, simmered	244	54.6	12.1	0.5	79	0	.20	.34		9	62	30	28	9.67	.018
3.5 oz	100	31.6	0.0	4.8	0.0		.13	5.5	3.59	.38	417	246	3.61	.161	
sep lean, simmered	201	58.2	6.4	0.2	78	0	.21	.37		10	64	32	30	10.49	.020
3.5 oz	100	33.7	0.0	2.3	0.0		.14	5.9	3.79	.41	447	263	3.86	.172	
short loin porterhouse steak															
sep lean & fat, broiled	299	52.4	21.2	0.8	83	0	.22	.35		7	61	8	25	4.70	.014
3.5 oz	100	25.1	0.0	8.8	0.0		.10	4.1	2.16	.31	356	189	2.66	.133	
sep lean, broiled	218	59.5	10.8	0.4	80	0	.25	.40		8	66	7	29	5.40	.015
3.5 oz	100	28.2	0.0	4.3	0.0		.11	4.6	2.27	.34	407	213	3.08	.143	
short loin T-bone steak															
sep lean & fat, broiled	324	50.4	24.6	0.9	84	0	.21	.33		7	60	9	24	4.46	.013
3.5 oz	100	24.0	0.0	10.2	0.0		.09	3.9	2.12	.29	339	177	2.54	.130	
sep lean, broiled	214	60.3	10.4	0.4	80	0	.25	.39		8	66	7	29	5.40	.015
3.5 oz	100	28.1	0.0	4.2	0.0		.11	4.6	2.27	.33	407	208	3.00	.143	
short loin tenderloin															
sep lean & fat, broiled	266	55.0	17.2	0.7	86	0	.27	.40		7	61	8	27	5.04	.015
3.5 oz	100	26.0	0.0	7.0	0.0		.12	3.6	2.45	.35	380	217	3.25	.167	
sep lean & fat, roasted	303	52.5	22.0	0.9	87	0	.28	.34		7	56	8	23	4.48	.014
3.5 oz	100	24.5	0.0	9.0	0.0		.09	3.1	2.56	.40	343	207	3.19	.152	
sep lean, broiled	204	60.4	9.3	0.4	84	0	.30	.44		7	63	7	30	5.59	.017
3.5 oz	100	28.3	0.0	3.6	0.0		.13	3.9	2.57	.38	419	238	3.58	.179	

ᵃ Contains added sodium ascorbate.

	KCAL	H₂O (g)	FAT (g)	PUFA (g)	CHOL (mg)	A (RE)	C (mg)	B-2 (mg)	B-6 (mg)	FOL (mcg)	Na (mg)	Ca (mg)	Mg (mg)	Zn (mg)	Mn (mg)
	WT (g)	PRO (g)	CHO (g)	SFA (g)	DFIB (g)	A (IU)	B-1 (mg)	NIA (mg)	B-12 (mcg)	PANT (mg)	K (mg)	P (mg)	Fe (mg)	Cu (mg)	
sep lean, roasted	219	60.0	11.3	0.4	86		0	.32	.38	8	59	7	27	5.17	.016
3.5 oz	100	27.5	0.0	4.4	0.0		.11	3.4	2.77	.45	393	236	3.66	.167	
short loin, top loin															
sep lean & fat, broiled	280	53.5	18.8	0.7	79		0	.18	.37	7	63	9	24	4.60	.014
3.5 oz	100	25.7	0.0	7.8	0.0		.08	4.7	1.94	.33	351	195	2.24	.104	
sep lean, broiled	203	60.2	8.9	0.3	76		0	.20	.42	8	68	8	27	5.22	.016
3.5 oz	100	28.6	0.0	3.6	0.0		.09	5.3	2.00	.37	396	218	2.47	.107	
wedge-bone sirloin															
sep lean & fat, broiled	280	52.6	18.0	0.7	90		0	.26	.40	9	63	11	28	5.75	.015
3.5 oz	100	27.4	0.0	7.5	0.0		.11	3.9	2.66	.35	360	218	3.01	.137	
sep lean & fat, pan fried	339	47.3	24.6	1.3	98		0	.27	.42	9	68	11	27	5.25	.016
3.5 oz	100	27.5	0.0	10.1	0.0		.12	3.7	3.21	.37	385	223	3.25	.135	
sep lean, broiled	208	58.6	8.7	0.4	89		0	.30	.45	10	66	11	32	6.52	.017
3.5 oz	100	30.4	0.0	3.6	0.0		.13	4.3	2.85	.39	403	244	3.36	.146	
sep lean, pan fried	238	55.9	11.0	0.9	99		0	.33	.50	10	77	11	33	6.40	.019
3.5 oz	100	32.5	0.0	4.2	0.0		.14	4.3	3.69	.43	465	267	3.90	.149	

21.2. LAMB

	KCAL	H₂O (g)	FAT (g)	PUFA (g)	CHOL (mg)	A (RE)	C (mg)	B-2 (mg)	B-6 (mg)	FOL (mcg)	Na (mg)	Ca (mg)	Mg (mg)	Zn (mg)	Mn (mg)
	WT (g)	PRO (g)	CHO (g)	SFA (g)	DFIB (g)	A (IU)	B-1 (mg)	NIA (mg)	B-12 (mcg)	PANT (mg)	K (mg)	P (mg)	Fe (mg)	Cu (mg)	
leg															
sep lean & fat, roasted	237	45.9	16.1					.23			52	9			
3 oz	85	21.5	0.0	9.0			.13	4.7			241	177	1.40		
sep lean, roasted	158	52.9	6.0					.26			60	11			
3 oz	85	24.4	0.0	3.4			.14	5.3			273	202	1.90		
loin chop															
sep lean & fat, broiled	255	33.4	20.9		58			.16			38	6			
1 chop	71	15.6	0.0	11.7			.09	3.6			175	122	.90		
sep lean, broiled	92	30.4	3.7		39			.14			34	6			
1 chop	49	13.8	0.0	2.1			.07	3.0			155	107	1.00		
rib chop															
sep lean & fat, broiled	273	28.7	23.9					.14			33	6			
1 chop	67	13.5	0.0	13.4			.08	3.1			151	105	.70		
sep lean, broiled	91	25.9	4.5					.12			29	5			
1 chop	43	11.7	0.0	2.5			.06	2.5			131	91	.80		
shoulder															
sep lean & fat, roasted	287	42.2	23.1					.20			45	9			
3 oz	85	18.4	0.0	12.9			.11	4.0			206	146	1.00		
sep lean, roasted	174	52.2	8.5					.24			56	10			
3 oz	85	22.8	0.0	4.8			.13	4.8			255	186	1.60		

21.3. PORK

	KCAL	H₂O (g)	FAT (g)	PUFA (g)	CHOL (mg)	A (RE)	C (mg)	B-2 (mg)	B-6 (mg)	FOL (mcg)	Na (mg)	Ca (mg)	Mg (mg)	Zn (mg)	Mn (mg)
	WT (g)	PRO (g)	CHO (g)	SFA (g)	DFIB (g)	A (IU)	B-1 (mg)	NIA (mg)	B-12 (mcg)	PANT (mg)	K (mg)	P (mg)	Fe (mg)	Cu (mg)	
arm picnic															
sep lean & fat, braised	345	46.6	25.5	2.9	109	3	0	.31	.27	4	88	7	18	4.04	.014
3.5 oz	100	26.8	0.0	9.3	0.0	9	.54	5.2	.69	.56	336	191	1.61	.138	
sep lean & fat, roasted	331	51.1	26.1	2.9	94	2	0	.30	.28	4	70	8	17	3.31	.033
3.5 oz	100	22.3	0.0	9.5	0.0	8	.52	3.9	.74	.49	293	206	1.19	.112	
sep lean, braised	248	54.3	12.2	1.5	114	2	0	.36	.41	5	102	8	22	4.97	.017
3.5 oz	100	32.3	0.0	4.2	0.0	8	.60	5.9	.71	.67	405	226	1.95	.161	
sep lean, roasted	228	60.3	12.6	1.5	95	2	0	.36	.41	5	80	9	20	4.07	.041
3.5 oz	100	26.7	0.0	4.4	0.0	7	.58	4.3	.78	.59	351	247	1.42	.128	
arm picnic, cured															
sep lean & fat, roasted	280	54.7	21.4	2.3	58	0		.19	.28	3	1072	10	14	2.51	.024
3.5 oz	100	20.4	0.0	7.7	0.0	0	.61	4.1	.93	.56	258	221	.95	.113	

	KCAL / WT (g)	H₂O (g) / PRO (g)	FAT (g) / CHO (g)	PUFA (g) / SFA (g)	CHOL (mg) / DFIB (g)	A (RE) / A (IU)	C (mg) / B-1 (mg)	B-2 (mg) / NIA (mg)	B-6 (mg) / B-12 (mcg)	FOL (mcg) / PANT (mg)	Na (mg) / K (mg)	Ca (mg) / P (mg)	Mg (mg) / Fe (mg)	Zn (mg) / Cu (mg)	Mn (mg)
sep lean, roasted	170	63.9	7.0	0.8	48	0		.23	.37	4	1231	11	16	2.94	.030
3.5 oz	100	24.9	0.0	2.4	0.0	0	.73	4.8	1.11	.65	292	243	1.00	.128	
bacon, canadian style															
unheated[a]	89	38.0	4.0	0.4	28	0	12[b]	.10	.22	2	799	5	10	.79	.013
2 slices	57	11.7	1.0	1.3	0.0	0	.43	3.5	.38	.30	195	138	.38	.026	
grilled[a]	86	28.7	3.9	0.4	27	0	10[b]	.09	.21	2	719	5	10	.79	.013
2 slices	47	11.3	0.6	1.3	0.0	0	.38	3.2	.36	.24	181	138	.38	.025	
bacon, cured															
broiled/pan fried	109	2.5	9.4	1.1	16	0	6[b]	.05	.05	1	303	2	5	.62	.008
3 med pieces	19	5.8	0.1	3.3	0.0	0	.13	1.4	.33	.20	92	64	.31	.032	
broiled/pan fried	732	16.4	62.5	7.4	107	0	43[b]	.36	.35	6	2026	15	30	4.14	.052
4.48 oz[c]	127	38.7	0.8	22.1	0.0	0	.88	9.3	2.23	1.34	617	426	2.05	.216	
raw	378	21.5	39.1	4.6	46	0	15[b]	.07	.10	1	466	5	6	.78	.005
3 med slices	68	5.9	0.1	14.5	0.0	0	.25	1.9	.63	.24	95	97	.41	.044	
bacon pieces															
bacon bits, Oscar Mayer	21	2.6	1.0		6		1	.03	.02		181	1	2	.33	
1/4 cup	7	2.6	0.2				.05	0.7	.15	.08	38	40	.14	.020	
bacOs, General Mills	30		1.0								130				
1 T		3.0	2.0								210				
blade roll, cured, sep lean & fat, roasted	287	56.2	23.5	2.5	67	0	3	.29	.21	3	973	7	13	2.45	.023
3.5 oz	100	17.3	0.4	8.4	0.0	0	.46	2.4	1.05	.77	194	156	.89	.076	
boston blade															
sep lean & fat, braised	371	43.3	28.7	3.3	111	3	0	.33	.23	1	67	7	19	4.61	.013
3.5 oz	100	26.4	0.0	10.3	0.0	10	.51	4.0	.84	.13	348	184	1.72	.143	
sep lean & fat, broiled	350	48.5	28.5	3.3	103	3	0	.38	.27	4	75	5	24	3.62	.008
3.5 oz	100	21.9	0.0	10.2	0.0	9	.67	4.0	1.04	.69	351	211	1.17	.117	
sep lean & fat, roasted	321	52.0	25.3	2.9	97	2	0	.33	.27	4	67	6	18	3.82	.012
3.5 oz	100	21.8	0.0	9.1	0.0	8	.55	4.0	.91	.51	313	195	1.42	.118	
sep lean, braised	294	49.4	17.6	2.1	116	3	0	.39	.27	5	75	8	22	5.57	.016
3.5 oz	100	31.2	0.0	6.1	0.0	9	.55	4.3	.90	.64	412	214	2.05	.165	
sep lean, broiled	274	55.1	18.4	2.2	105	2	0	.44	.31	5	84	6	28	4.27	.009
3.5 oz	100	25.2	0.0	6.4	0.0	8	.75	4.3	1.13	.82	409	244	1.35	.131	
sep lean, roasted	256	57.7	16.8	2.0	98	2	0	.37	.30	5	73	7	20	4.37	.014
3.5 oz	100	24.4	0.0	5.8	0.0	7	.59	4.3	.96	.57	353	219	1.60	.129	
breakfast strips															
ckd	156	9.2	12.5	1.9	36	0	15[b]	.13	.12	1	714	5	9	1.25	.015
3 slices	34	9.8	0.4	4.3	0.0	0	.25	2.6	.60	.31	158	90	.67	.052	
ckd, Lean 'n Tasty	50	4.9	4.3		13						191	1	2	.33	
1 slice	12	2.8	0.1								40	23	.14	.040	
center loin															
sep lean & fat, braised	354	44.1	25.4	2.9	107	3	0	.24	.38	4	51	6	19	2.47	.027
3.5 oz	100	29.4	0.0	9.2	0.0	9	.77	6.0	.61	.62	317	215	.83	.094	
sep lean & fat, broiled	316	49.8	22.1	2.5	97	3	0	.27	.40	5	70	4	25	1.93	.009
3.5 oz	100	27.4	0.0	8.0	0.0	9	1.00	5.0	.71	.59	359	211	.81	.076	
sep lean & fat, pan fried	375	45.4	30.5	3.5	103	3	0	.28	.39	5	72	5	26	1.96	.009
3.5 oz	100	23.3	0.0	11.0	0.0	9	1.02	5.2	.77	1.01	363	214	.84	.080	
sep lean & fat, roasted	305	51.8	21.8	2.5	91	2	0	.24	.40	1	64	5	19	2.04	.014
3.5 oz	100	25.4	0.0	7.9	0.0	8	.83	5.0	.60	.77	322	196	.99	.077	
sep lean, braised	272	50.2	13.7	1.7	111	3	0	.27	.45	5	55	6	22	2.91	.032
3.5 oz	100	34.8	0.0	4.7	0.0	9	.88	6.8	.61	.74	372	251	.95	.104	
sep lean, broiled	231	56.8	10.5	1.3	98	2	0	.31	.47	6	78	5	30	2.23	.011
3.5 oz	100	32.0	0.0	3.6	0.0	8	1.15	5.5	.74	.69	420	244	.92	.082	
sep lean, pan fried	266	54.4	15.9	2.0	107	2	0	.33	.50	6	85	5	32	2.41	.012
3.5 oz	100	28.8	0.0	5.5	0.0	8	1.25	6.0	.73	.79	455	265	1.00	.089	
sep lean, roasted	240	57.3	13.1	1.6	91	2	0	.26	.45	1	69	6	21	2.28	.016
3.5 oz	100	28.5	0.0	4.5	0.0	8	.91	5.5	.60	.65	362	219	1.09	.081	

[a] Bacon is fully cooked as purchased.
[b] Contains added ascorbic acid or sodium ascorbate.
[c] Yield from 1 lb raw bacon.

	KCAL	H₂O (g)	FAT (g)	PUFA (g)	CHOL (mg)	A (RE)	C (mg)	B-2 (mg)	B-6 (mg)	FOL (mcg)	Na (mg)	Ca (mg)	Mg (mg)	Zn (mg)	Mn (mg)
	WT (g)	PRO (g)	CHO (g)	SFA (g)	DFIB (g)	A (IU)	B-1 (mg)	NIA (mg)	B-12 (mcg)	PANT (mg)	K (mg)	P (mg)	Fe (mg)	Cu (mg)	

center rib

sep lean & fat, braised	367	43.2	27.2	3.1	95	3	0	.27	.32	6	48	10	20	2.35	.010
3.5 oz	100	28.6	0.0	9.8	0.0	10	.53	5.6	.54	.55	413	210	1.07	.090	
sep lean & fat, broiled	343	47.5	26.4	3.0	93	2	0	.28	.34	7	61	13	25	2.03	.017
3.5 oz	100	24.6	0.0	9.5	0.0	7	.79	4.7	.68	.63	371	227	.72	.075	
sep lean & fat, pan fried	390	44.1	33.0	3.7	84	3	0	.28	.33	6	45	7	21	1.62	.006
3.5 oz	100	21.6	0.0	11.9	0.0	9	.64	4.4	.62	.60	352	208	.66	.068	
sep lean & fat, roasted	318	50.7	23.6	2.7	81	3	0	.28	.35	8	44	10	19	1.96	.009
3.5 oz	100	24.7	0.0	8.5	0.0	8	.59	4.9	.56	.68	368	224	.89	.074	
sep lean, braised	277	49.9	14.4	1.8	97	3	0	.32	.39	8	52	12	23	2.82	.012
3.5 oz	100	34.4	0.0	5.0	0.0	9	.59	6.5	.52	.66	501	250	1.26	.100	
sep lean, broiled	258	54.5	14.9	1.8	94	2	0	.32	.40	9	67	15	30	2.38	.020
3.5 oz	100	28.8	0.0	5.2	0.0	6	.89	5.2	.70	.75	439	266	.81	.080	
sep lean, pan fried	257	55.1	15.3	1.9	81	2	0	.35	.44	8	50	9	28	2.06	.007
3.5 oz	100	28.0	0.0	5.3	0.0	8	.77	5.2	.62	.79	465	271	.80	.075	
sep lean, roasted	245	57.0	13.8	1.7	79	3	0	.31	.40	9	46	11	21	2.22	.011
3.5 oz	100	28.2	0.0	4.8	0.0	8	.64	5.4	.55	.58	423	256	1.00	.078	

ham, cured, lean

cnd	120	73.5	4.6	0.4	38	0	27[a]	.23	.45	6	1255	6	17	1.93	.025
3.5 oz	100	18.5	0.0	1.5	0.0	0	.84	5.3	.82	.49	364	224	.94	.084	
cnd, roasted	136	69.5	4.9	0.4	30	0	28[a]	.25	.45	5	1135	6	21	2.23	.024
3.5 oz	100	21.2	0.5	1.6	0.0	0	1.04	4.9	.71	.57	348	209	.92	.050	
roasted[b]	145	67.7	5.5	0.5	53	0	21[a]	.20	.40	3	1203	8	14	2.88	.054
3.5 oz	100	20.9	1.5	1.8	0.0	0	.75	4.0	.65	.40	287	196	1.48	.079	
unheated[b]	131	70.5	5.0	0.5	47	0	26[a]	.22	.46	4	1429	7	17	1.93	.033
3.5 oz	100	19.4	1.0	1.6	0.0	0	.93	4.8	.75	.47	350	218	.76	.074	

ham, cured, regular

center slice, sep lean & fat, unheated[b]	203	63.5	12.9	1.4	54	0		.21	.47	4	1386	7	16	1.88	.031
3.5 oz	100	20.2	0.1	4.6	0.0	0	.85	4.8	.80	.50	337	215	.75	.074	
cnd	190	66.5	13.0	1.5	39	0	22[a]	.23	.48	5	1240	6	14	1.66	.023
3.5 oz	100	17.0	0.0	4.3	0.0	0	.96	3.2	.78	.39	316	175	.83	.067	
cnd, roasted	226	60.9	15.2	1.8	62	0	14[a]	.26	.30	5	941	8	17	2.50	.029
3.5 oz	100	20.5	0.4	5.0	0.0	0	.82	5.3	1.06	.73	357	243	1.37	.130	
roasted[b]	178	64.5	9.0	1.4	59	0	23[a]	.33	.31		1500	8	22	2.47	.041
3.5 oz	100	22.6	0.0	3.1	0.0	0	.73	6.2	.70	.72	409	281	1.34	.145	
unheated[b]	182	64.6	10.6	1.2	57	0	28[a]	.25	.34	3	1317	7	19	2.14	.031
3.5 oz	100	17.6	3.1	3.4	0.0	0	.86	5.3	.83	.45	332	247	.99	.099	

ham patties

grilled[c]	203	30.6	18.4	2.0	43	0	0	.11	.10		632	5	6	1.13	
1 patty	60	7.9	1.0	6.6		0	.21	1.9	.42	.16	145	60	.96	.060	
unheated[c]	206	35.5	18.4	2.0	46	0	0	.10	.10		709	5	6	1.02	
1 patty	65	8.3	1.1	6.6		0	.30	2.0	.71	.20	156	97	.69	.046	

leg

sep lean & fat, roasted	294	52.4	20.7	2.3	93	2	0	.31	.39	10	59	6	22	2.86	.032
3.5 oz	100	25.0	0.0	7.5	0.0	8	.63	4.6	.70	.59	329	247	1.00	.100	
sep lean, roasted	220	59.7	11.0	1.3	94	2	0	.35	.45	12	64	7	25	3.26	.037
3.5 oz	100	28.3	0.0	3.8	0.0	7	.69	4.9	.72	.67	373	281	1.12	.108	

loin

sep lean & fat, braised	368	43.8	27.9	3.2	102	3	0	.30	.36	4	65	8	20	3.03	.015
3.5 oz	100	27.2	0.0	10.1	0.0	9	.61	6.0	.79	.57	345	199	1.16	.109	
sep lean & fat, broiled	346	48.0	27.2	3.1	94	3	0	.36	.38	5	66	7	25	2.45	.009
3.5 oz	100	23.6	0.0	9.8	0.0	9	.84	5.3	.98	.75	351	235	.81	.089	
sep lean & fat, roasted	319	51.3	24.3	2.8	90	2	0	.32	.38	5	63	8	19	2.62	.014
3.5 oz	100	23.4	0.0	8.8	0.0	8	.72	5.4	.87	.55	318	221	1.01	.093	
sep lean, braised	273	51.1	14.6	1.8	105	2	0	.36	.45	5	75	10	24	3.72	.018
3.5 oz	100	33.0	0.0	5.0	0.0	8	.69	6.9	.84	.70	419	239	1.40	.125	

[a] Contains added ascorbic acid or sodium ascorbate.
[b] Ham is fully cooked as purchased.
[c] Fully cooked as purchased.

	KCAL / WT (g)	H₂O (g) / PRO (g)	FAT (g) / CHO (g)	PUFA (g) / SFA (g)	CHOL (mg) / DFIB (g)	A (RE) / A (IU)	C (mg) / B-1 (mg)	B-2 (mg) / NIA (mg)	B-6 (mg) / B-12 (mcg)	FOL (mcg) / PANT (mg)	Na (mg) / K (mg)	Ca (mg) / P (mg)	Mg (mg) / Fe (mg)	Zn (mg) / Cu (mg)	Mn (mg)
sep lean, broiled	257	55.5	15.3	1.9	95	2	0	.42	.46	5	75	7	29	2.92	.010
3.5 oz	100	27.8	0.0	5.3	0.0	8	.97	6.0	1.08	.91	418	279	.93	.098	
sep lean, roasted	240	58.1	13.9	1.7	90	2	0	.36	.45	6	69	9	22	3.04	.016
3.5 oz	100	26.9	0.0	4.8	0.0	7	.80	6.0	.93	.64	367	254	1.15	.101	
loin blade															
sep lean & fat, braised	410	41.0	34.1	3.9	108	2	0	.28	.32	4	69	14	17	3.85	.013
3.5 oz	100	24.0	0.0	12.3	0.0	7	.49	4.8	.76	.52	334	176	1.30	.109	
sep lean & fat, broiled	393	44.4	33.8	3.9	98	3	0	.32	.34	4	67	11	21	3.05	.008
3.5 oz	100	20.7	0.0	12.2	0.0	9	.66	4.2	.94	.66	332	202	.91	.090	
sep lean & fat, pan fried	414	42.5	36.9	4.2	95	3	0	.29	.29	1	61	10	19	2.71	.007
3.5 oz	100	18.8	0.0	13.3	0.0	10	.61	3.9	.87	.59	297	182	.82	.083	
sep lean & fat, roasted	364	48.2	30.5	3.5	90	3	0	.30	.36	4	61	12	15	2.99	.011
3.5 oz	100	21.1	0.0	10.9	0.0	0	.52	4.2	.76	.60	293	173	1.07	.090	
sep lean, braised	313	48.6	20.6	2.5	113	2	0	.33	.41	5	81	17	20	4.93	.016
3.5 oz	100	29.7	0.0	7.1	0.0	5	.55	5.6	.82	.65	417	214	1.62	.128	
sep lean, broiled	300	52.4	21.5	2.6	100	3	0	.38	.43	5	77	13	25	3.79	.009
3.5 oz	100	24.9	0.0	7.4	0.0	8	.76	4.7	1.04	.82	408	245	1.08	.101	
sep lean, pan fried	283	53.5	19.8	2.5	97	2	0	.37	.40	0	74	13	24	3.67	.009
3.5 oz	100	24.3	0.0	6.8	0.0	8	.74	4.5	1.00	.80	394	237	1.04	.098	
sep lean, roasted	279	55.8	19.3	2.3	89	2	0	.34	.44	5	68	14	17	3.59	.013
3.5 oz	100	24.7	0.0	6.7	0.0	8	.57	4.7	.80	.53	346	201	1.25	.099	
rump															
sep lean & fat, roasted	274	54.7	17.8	2.0	95	3	0	.33	.27	6	61	7	26	2.74	.024
3.5 oz	100	26.6	0.0	6.5	0.0	10	.71	4.7	.71	.68	356	260	1.05	.103	
sep lean, roasted	221	59.3	10.7	1.3	96	3	0	.36	.30	6	65	7	29	3.01	.026
3.5 oz	100	29.1	0.0	3.7	0.0	9	.76	5.0	.73	.75	391	285	1.14	.109	
sausage, fresh															
ckd	48	5.8	4.1	0.5	11		0	.03	.04		168	4	2	.33	.009
1 link	13	2.6	0.1	1.4	0.0		.10	0.6	.22	.09	47	24	.16	.018	
ckd	100	12.0	8.4	1.0	22		1	.07	.09		349	9	5	.68	.019
1 patty	27	5.3	0.3	2.9	0.0		.20	1.2	.47	.20	97	50	.34	.038	
w/ beef, ckd	52	5.8	4.7	0.5				.02	.01		105		1	.24	
1 link	13	1.8	0.4	1.7			.05	0.4	.06	.06		14	.15	.000	
shank															
sep lean & fat, roasted	303	52.8	22.1	2.5	92	2	0	.30	.39	5	58	6	21	2.93	.028
3.5 oz	100	24.3	0.0	8.0	0.0	8	.58	4.5	.69	.59	310	239	.97	.098	
sep lean, roasted	215	60.4	10.5	1.3	92	2	0	.34	.46	6	64	7	25	3.45	.034
3.5 oz	100	28.2	0.0	3.6	0.0	7	.63	4.9	.71	.69	360	278	1.11	.108	
shoulder															
sep lean & fat, roasted	326	51.6	25.7	2.9	96	2	0	.32	.33	4	68	7	18	3.59	.022
3.5 oz	100	22.0	0.0	9.3	0.0	8	.54	4.0	.83	.50	304	200	1.31	.115	
sep lean, roasted	244	58.8	15.0	1.8	97	2	0	.36	.40	5	76	8	20	4.24	.026
3.5 oz	100	25.4	0.0	5.2	0.0	7	.58	4.3	.88	.58	352	231	1.52	.129	
sirloin															
sep lean & fat, braised	352	45.1	25.7	2.9	106	3	0	.30	.36	4	54	6	23	2.56	.014
3.5 oz	100	28.0	0.0	9.3	0.0	0	.66	5.0	.66	.59	369	196	1.13	.116	
sep lean & fat, broiled	331	49.3	25.3	2.9	97	3	0	.35	.44	5	55	5	29	2.03	.009
3.5 oz	100	24.2	0.0	9.1	0.0	9	.90	4.3	.80	.76	367	226	.78	.095	
sep lean & fat, roasted	291	54.0	20.4	2.3	91	2	0	.31	.38	5	59	9	22	2.27	.025
3.5 oz	100	25.0	0.0	7.4	0.0	8	.74	5.2	.76	.57	337	229	1.00	.099	
sep lean, braised	261	52.1	13.0	1.6	110	3	0	.35	.53	5	59	7	28	3.07	.017
3.5 oz	100	33.5	0.0	4.5	0.0	8	.74	5.7	.67	.71	443	232	1.33	.133	
sep lean, broiled	243	56.7	13.6	1.7	98	2	0	.40	.54	6	60	5	34	2.36	.010
3.5 oz	100	28.3	0.0	4.7	0.0	8	1.03	4.8	.84	.91	434	265	.89	.105	
sep lean, roasted	236	58.7	13.2	1.6	90	2	0	.33	.42	6	62	10	24	2.49	.028
3.5 oz	100	27.5	0.0	4.5	0.0	7	.80	5.6	.78	.63	370	252	1.09	.107	

	KCAL	H₂O (g)	FAT (g)	PUFA (g)	CHOL (mg)	A (RE)	C (mg)	B-2 (mg)	B-6 (mg)	FOL (mcg)	Na (mg)	Ca (mg)	Mg (mg)	Zn (mg)	Mn (mg)
	WT (g)	PRO (g)	CHO (g)	SFA (g)	DFIB (g)	A (IU)	B-1 (mg)	NIA (mg)	B-12 (mcg)	PANT (mg)	K (mg)	P (mg)	Fe (mg)	Cu (mg)	
spareribs, sep lean & fat, braised	397	40.4	30.3	3.5	121	3		.38	.35	4	93	47	24	4.60	.014
3.5 oz	100	29.1	0.0	11.8	0.0	10	.41	5.5	1.08	.75	320	261	1.85	.142	
tenderloin, sep lean, roasted	166	65.2	4.8	0.6	93	2	0	.39	.42	6	67	9	25	3.00	.039
3.5 oz	100	28.8	0.0	1.7	0.0	7	.94	4.7	.55	.69	538	288	1.54	.160	
top loin															
sep lean & fat, braised	381	42.1	29.2	3.3	95	3	0	.27	.31	6	47	10	19	2.28	.010
3.5 oz	100	27.7	0.0	10.6	0.0	10	.52	5.5	.54	.53	399	204	1.03	.088	
sep lean & fat, broiled	360	46.1	28.6	3.2	93	2	0	.27	.34	7	59	12	24	1.97	.016
3.5 oz	100	23.7	0.0	10.4	0.0	8	.76	4.6	.67	.61	357	219	.70	.074	
sep lean & fat, pan fried	392	44.0	33.2	3.7	84	3	0	.28	.33	6	45	7	21	1.62	.006
3.5 oz	100	21.5	0.0	12.0	0.0	9	.63	4.4	.61	.59	350	207	.66	.068	
sep lean & fat, roasted	330	49.7	25.1	2.8	82	3	0	.27	.34	7	44	10	18	1.92	.009
3.5 oz	100	24.2	0.0	9.1	0.0	8	.58	4.8	.56	.50	359	219	.88	.073	
sep lean, braised	277	49.9	14.4	1.8	97	3	0	.32	.39	8	52	12	23	2.82	.012
3.5 oz	100	34.4	0.0	5.0	0.0	9	.59	6.5	.52	.66	501	250	1.26	.100	
sep lean, broiled	258	54.5	14.9	1.8	94	2	0	.32	.40	9	67	15	30	2.38	.020
3.5 oz	100	28.8	0.0	5.2	0.0	6	.89	5.2	.70	.75	439	266	.81	.080	
sep lean, pan fried	257	55.1	15.3	1.9	81	2	0	.35	.44	8	50	9	28	2.06	.007
3.5 oz	100	28.0	0.0	5.3	0.0	8	.77	5.2	.62	.79	465	271	.80	.075	
sep lean, roasted	245	57.0	13.8	1.7	79	3	0	.31	.40	9	46	11	21	2.22	.011
3.5 oz	100	28.2	0.0	4.8	0.0	8	.64	5.4	.55	.58	423	256	1.00	.078	

21.4. VEAL

	KCAL	H₂O (g)	FAT (g)	PUFA (g)	CHOL (mg)	A (RE)	C (mg)	B-2 (mg)	B-6 (mg)	FOL (mcg)	Na (mg)	Ca (mg)	Mg (mg)	Zn (mg)	Mn (mg)
	WT (g)	PRO (g)	CHO (g)	SFA (g)	DFIB (g)	A (IU)	B-1 (mg)	NIA (mg)	B-12 (mcg)	PANT (mg)	K (mg)	P (mg)	Fe (mg)	Cu (mg)	
chuck, braised/pot roastd/stewed	200	49.7	10.9					.25			41	10			
3 oz	85	23.7	0.0	5.2			.08	5.4			190	128	3.00		
leg, sliced, frzn, Am Hosp Co	135		3.9		75			.20			42	6			
3 oz	85	24.9	0.1				.14	6.9			261		2.40		
loin, braised/broiled	199	50.1	11.4					.21			55	9			
3 oz	85	22.4	0.0	5.5			.06	4.6			251	191	2.70		
plate (breast), braised/stewed	258	44.3	18.0					.20			39	10			
3 oz	85	22.2	0.0	8.7			.04	3.9			178	117	2.81		
rib roast, roasted	229	46.4	14.4					.26			57	10			
3 oz	85	23.1	0.0	6.9			.11	6.6			259	211	2.90		
round w/ rump, braised/broiled	184	51.3	9.4					.21			56	9			
3 oz	85	23.0	0.0	4.5			.06	4.6			258	196	2.70		

21.5. VARIETY CUTS

	KCAL	H₂O (g)	FAT (g)	PUFA (g)	CHOL (mg)	A (RE)	C (mg)	B-2 (mg)	B-6 (mg)	FOL (mcg)	Na (mg)	Ca (mg)	Mg (mg)	Zn (mg)	Mn (mg)
	WT (g)	PRO (g)	CHO (g)	SFA (g)	DFIB (g)	A (IU)	B-1 (mg)	NIA (mg)	B-12 (mcg)	PANT (mg)	K (mg)	P (mg)	Fe (mg)	Cu (mg)	
brains															
beef, pan fried	196	70.8	15.8	2.3	1995	0	3	.26	.39	6	158	9	15	1.35	.032
3.5 oz	100	12.6	0.0	3.7	0.0	0	.13	3.8	15.2	.57	354	386	2.22	.220	
beef, simmered	160	73.3	12.5	1.4	2054	0	1	.17	.24	7	120	9	14	1.25	.035
3.5 oz	100	11.1	0.0	2.9	0.0	0	.08	2.2	8.60	.57	240	352	2.21	.240	
pork, braised	138	75.9	9.5	1.5	2552	0	14	.22	.14		91	9	12	1.48	.085
3.5 oz	100	12.1	0.0	2.2	0.0	0	.08	3.3	1.42	1.82	195	220	1.82	.263	
chitterlings, pork, simmered	303	62.4	28.8	7.2	143	0	0	.08			39	27	10	5.06	
3.5 oz	100	10.3	0.0	10.1	0.0	0	.00	0.1			8	47	3.70	.233	
ears, pork, simmered	183	79.9	11.9		99	0	0	.08			183	20	8		
1 ear	111	17.6	0.0		0.0	0	.02	0.6			44	27	1.65		
feet, pork															
cured, pickled	203	68.6	16.1	1.8	92	0	0	.04				32	4		
3.5 oz	100	13.5	0.0	5.6	0.0	0	.01	0.4				34			
simmered	194	66.0	12.4	1.4	100	0	0	.06				45	5		
3.5 oz	100	19.2	0.0	4.3	0.0	0	.01	0.5				48			
heart															
beef, simmered	175	64.1	5.6	1.4	193	0	2	1.54	.21	2	63	6	25	3.13	.059
3.5 oz	100	28.8	0.4	1.7	0.0	0	.14	4.1	14.30	.87	233	250	7.51	.740	

	KCAL / WT (g)	H₂O (g) / PRO (g)	FAT (g) / CHO (g)	PUFA (g) / SFA (g)	CHOL (mg) / DFIB (g)	A (RE) / A (IU)	C (mg) / B-1 (mg)	B-2 (mg) / NIA (mg)	B-6 (mg) / B-12 (mcg)	FOL (mcg) / PANT (mg)	Na (mg) / K (mg)	Ca (mg) / P (mg)	Mg (mg) / Fe (mg)	Zn (mg) / Cu (mg)	Mn (mg)
lamb, braised	377	78.4	20.9				0	1.49				20			
1 cup	145	42.8	1.5			150	.30	9.3				335			
pork, braised	191	87.8	6.5	1.7	285	9	3	2.20	.50	5	46	9	32	3.99	.094
1 heart	129	30.4	0.5	1.7	0.0	29	.72	7.8	4.89	3.19	266	230	7.52	.655	
jowl, pork, raw	655	22.2	69.6	8.1	90	3		.24	.09	1	25	4	3		.005
3.5 oz	100	6.4	0.0	25.3	0.0	9	.39	4.5	.82	.25	148	86	.42	.040	
kidneys															
beef, simmered	144	68.8	3.4	0.7	387	373	1	4.06	.52	98	134	17	18	4.22	.185
3.5 oz	100	25.5	1.0	1.1	0.0	1241	.19	6.0	51.30	1.69	179	306	7.31	.680	
pork, braised	151	68.7	4.7	0.4	480	78	11	1.59	.46	41	80	13	18	4.15	.149
3.5 oz	100	25.4	0.0	1.5	0.0	260	.40	5.8	7.79	2.87	143	240	5.29	.683	
liver															
beef, braised	161	65.9	4.9	1.1	389	10602	23	4.10	.91	217	70	7	20	6.07	.413
3.5 oz	100	24.4	3.4	1.9	0.0	35679	.20	10.7	71.00	4.57	235	404	6.77	2.789	
beef, pan fried	217	55.7	8.0	1.8	482	10728	23	4.14	1.43	220	106	11	23	5.45	.423
3.5 oz	100	26.7	7.9	2.8	0.0	36105	.21	14.4	111.80	5.92	364	461	6.28	2.822	
lamb, broiled	117	22.7	5.6				16	2.30			38	7			
1 slice	45	14.5	1.3			33530	.22	11.2			149	257	8.10		
pork, braised	165	64.3	4.4	1.1	355	5399	24	2.20	.57	163	49	10	14	6.72	.300
3.5 oz	100	26.0	3.8	1.4	0.0	17997	.26	8.4	18.67	4.77	150	241	17.92	.634	
lungs															
beef, braised	120	76.4	3.7	0.5	277	12	33	.14	.02	8	101	11	10	1.64	.015
3.5 oz	100	20.4	0.0	1.3	0.0	39	.04	2.5	2.59	.62	173	178	5.40	.221	
lamb, raw	103	76.7	2.3												
3.5 oz	100	19.3	0.0		0.0							180			
pork, braised	99	80.0	3.1	0.4	387	0	8	.32	.08		81	8	12	2.45	
3.5 oz	100	16.6	0.0	1.1	0.0	0	.08	1.9	2.03	.66	151	186	16.41		
pancreas															
beef, braised	271	55.6	17.2			0	20	.49	.18		60	16	21	4.60	.208
3.5 oz	100	27.1	0.0		0.0	0	.18	4.0	16.60	4.25	246	453	2.61	.089	
pork, braised	219	60.3	10.8		315	0	6	.66			42	16	23	4.29	.198
3.5 oz	100	28.5	0.0		0.0	0	.09	3.2	17.07	4.74	168	291	2.69	.110	
spleen															
beef, braised	145	70.0	4.2		347	0	50	.30	.04		57	12	19	2.79	.075
3.5 oz	100	25.1	0.0		0.0	0	.05	5.6	5.02		284	305	39.36	.924	
lamb, raw	115	74.4	3.9												
3.5 oz	100	18.8	0.0		0.0										
pork, braised	149	66.7	3.2	0.2	504	0	12	.26	.06			13		3.54	.045
3.5 oz	100	28.2	0.0	1.1	0.0	0	.14	5.9	2.76	.89	227	283	22.2	.133	
stomach, pork, raw	157	73.6	9.6		193			.12	.04		52	10		2.01	
3.5 oz	100	16.5	0.0		0.0		.09	4.5	.99		201	155	2.18	.365	
sweetbreads															
beef (yearlings), ckd	272	42.2	19.7								99				
3 oz	85	22.0	0.0								368	309			
calf, ckd	143	53.3	2.7					.14							
3 oz	85	27.7	0.0				.05	2.5							
lamb, ckd	149	54.9	5.2												
3 oz	85	23.9	0.0									173			
tail, pork, simmered	396	46.7	35.8	3.9	129	0	0	.07				14	7		
3.5 oz	100	17.0	0.0	12.5	0.0	0	.07	1.1				47			
thymus, beef, braised	319	52.8	25.0		294	0	30				116				
3.5 oz	100	21.9	0.0		0.0	0			1.51		423	364	1.49		
tongue															
beef, simmered	283	56.1	20.7	0.8	107		1	.35	.16	5	60	7	17	4.80	.026
3.5 oz	100	22.1	0.3	8.9	0.0		.03	2.2	5.90	.52	180	142	3.39	.220	

	KCAL / WT (g)	H₂O (g) / PRO (g)	FAT (g) / CHO (g)	PUFA (g) / SFA (g)	CHOL (mg) / DFIB (g)	A (RE) / A (IU)	C (mg) / B-1 (mg)	B-2 (mg) / NIA (mg)	B-6 (mg) / B-12 (mcg)	FOL (mcg) / PANT (mg)	Na (mg) / K (mg)	Ca (mg) / P (mg)	Mg (mg) / Fe (mg)	Zn (mg) / Cu (mg)	Mn (mg)
calf, braised	32	13.7	1.2												
1 slice	20	4.8	0.2												
lamb, braised	51	12.0	3.6												
1 slice	20	4.1	0.1									20			
pork, braised	271	56.9	18.6	1.9	146	0	2	.51	.23		109	19	20	4.53	
3.5 oz	100	24.1	0.0	6.5	0.0	0	.32	5.3	2.39		237	174	4.99		
sheep, braised	65	10.3	5.1												
1 slice	20	4.0	0.5										.70		
tripe, beef, raw	98	81.4	4.0	0.1	95	0	3	.17		2	46		8	2.47	
3.5 oz	100	14.6	0.0	2.0	0.0	0	.01	0.1	1.54		270	79	1.95	.090	

21.6. OTHER MEATS

	KCAL / WT (g)	H₂O (g) / PRO (g)	FAT (g) / CHO (g)	PUFA (g) / SFA (g)	CHOL (mg) / DFIB (g)	A (RE) / A (IU)	C (mg) / B-1 (mg)	B-2 (mg) / NIA (mg)	B-6 (mg) / B-12 (mcg)	FOL (mcg) / PANT (mg)	Na (mg) / K (mg)	Ca (mg) / P (mg)	Mg (mg) / Fe (mg)	Zn (mg) / Cu (mg)	Mn (mg)
beaver, roasted	211	47.8	11.6					.32							
3 oz	85	24.8	0.0				.07								
eel															
raw	156	58.0	9.9	0.8	107	2954		.03	.06		43	17		1.38	.030
3 oz	85	15.7	0.0	2.0		886	.13	3.0	2.55	.20	232	183	.43	.020	
ckd by dry heat	200	50.4	12.7	1.0	137	3219		.04	.07		55	22		1.76	
3 oz	85	20.1	0.0	2.6		966	.16	3.8	2.45		297	235	.54	.025	
smoked	330	50.2	27.8												
3.5 oz	100	18.6	0.0	6.0											
frog legs, raw	73	81.9	0.3					.25				18			
3.5 oz	100	16.4	0.0			0	.14	1.2				147	1.50		
muskrat, roasted	153	67.3	4.1					.21							
3.5 oz	100	27.2	0.0				.16								
octopus, raw	70	68.2	0.9	0.2	41			.03				45		1.43	.021
3 oz	85	12.7	1.9	0.2			.03	1.8				158	4.51	.370	
opossum, roasted	221	57.3	10.2					.38							
3.5 oz	100	30.2	0.0				.12								
rabbit, stewed	302	83.7	14.1					.10			57	29			
1 cup diced	140	41.0	0.0				.07	15.8			515	363	2.10		
racoon, roasted	255	54.8	14.5					.52							
3.5 oz	100	29.2	0.0				.59								
reindeer, lean, raw	127	73.3	3.8					.68							
3.5 oz	100	21.8	0.0				.33	5.5					5.30		
shark															
raw	111	62.5	3.8	1.3	43	198		.05			67	29	42	.36	.013
3 oz	85	17.8	0.0	0.8		60	.04	2.5	1.27	.59	136	179	.71	.028	
batter-dipped & fried[a]	194	51.1	11.8	3.1	50	153		.08			103	42	37	.41	
3 oz	85	15.8	5.4	2.7		46	.06	2.4	1.03		132	165	.94	.036	
snail, raw	90	79.2	1.4												
3.5 oz	100	16.1	2.0										3.50		
snail, giant african, raw	73	82.2	1.4												
3.5 oz	100	9.9	4.4												
squid															
raw	78	66.8	1.2	0.4	198		4	.35	.05		37	27	28	1.30	.030
3 oz	85	13.2	2.6	0.3			.02	1.8	1.10		209	188	.58	1.607	
fried[b]	149	54.9	6.4	1.8	221		4	.39	.05		260	33	33	1.48	
3 oz	85	15.3	6.6	1.6			.05	2.2	1.04		237	213	.86	1.797	
turtle, green															
cnd	106	75.0	0.7												
3.5 oz	100	23.4	0.0												
raw	89	78.5	0.5												
3.5 oz	100	19.8	0.0												
turtle, terrapin (diamond back), raw	111	77.0	3.5												
3.5 oz	100	18.6	0.0		0.0								3.20		
venison, lean, raw	107	62.9	3.4					.41				9			
3 oz	85	17.9	0.0	2.1			.20	5.4				212			

[a] Prepared w/ flour, oil, egg, milk, baking powder & salt. [b] Dipped in flour & salt before frying.

	KCAL / WT (g)	H₂O (g) / PRO (g)	FAT (g) / CHO (g)	PUFA (g) / SFA (g)	CHOL (mg) / DFIB (g)	A (RE) / A (IU)	C (mg) / B-1 (mg)	B-2 (mg) / NIA (mg)	B-6 (mg) / B-12 (mcg)	FOL (mcg) / PANT (mg)	Na (mg) / K (mg)	Ca (mg) / P (mg)	Mg (mg) / Fe (mg)	Zn (mg) / Cu (mg)	Mn (mg)
whale meat, raw	156	70.9	7.5				6	.08			78	12			
3.5 oz	100	20.6	0.0	1.0		1860	.09				22	144			

22. MEATS, LUNCHEON[a]

	KCAL / WT (g)	H₂O (g) / PRO (g)	FAT (g) / CHO (g)	PUFA (g) / SFA (g)	CHOL (mg) / DFIB (g)	A (RE) / A (IU)	C (mg) / B-1 (mg)	B-2 (mg) / NIA (mg)	B-6 (mg) / B-12 (mcg)	FOL (mcg) / PANT (mg)	Na (mg) / K (mg)	Ca (mg) / P (mg)	Mg (mg) / Fe (mg)	Zn (mg) / Cu (mg)	Mn (mg)
bbq loaf (pork & beef)	40	14.9	2.1	0.2	9		4	.06	.06		307	13	4	.60	.009
1 slice	23	3.6	1.5	0.7			.08	0.5	.39	.36	76	30	.27	.020	
bbq loaf (pork & beef), Oscar Mayer	48	18.0	2.5		13		5	.07	.07		346	16	5	.70	.012
1 oz slice	28	4.5	1.9			28	.11	0.6	.45	.44	94	42	.35	.020	
beef															
chopped, smoked	38	19.5	1.3	0.1	13		6	.05	.10		357		6	1.11	
1 oz	28	5.7	0.5	0.5			.02	1.3	.49	.17	107	51	.81		
italian style, Oscar Mayer	18	12.5	0.5		6						209	2	3	.52	
1 slice	17	3.2	0.2								54	35	.47	.050	
jellied lunch meat	31	21.2	0.9	0.1							375	3	5	1.01	
1 oz	28	5.4	0.0	0.4							114		.98	.034	
loaved lunch meat	87	14.9	7.4	0.3	18		4	.06	.05		377	3	4	.72	.013
1 slice	28	4.1	0.8	3.2	0.0		.03	1.0	1.10	.15	59	34	.66	.030	
summer sausage	77	11.7	6.8	0.3	17		5	.08	.06		286	3	3	.59	
1 slice	23	3.6	0.1	2.8			.04	1.0	1.27		62	26	.58	.035	
thin sliced lunch meat	37	12.2	0.8	0.0	9		3	.04	.07		302	2	4	.84	.008
5 slices	21	5.9	1.2	0.4			.02	1.1	.54	.12	90	35	.57	.007	
berliner (pork & beef)	53	14.0	4.0	0.4	11		2	.05	.05		298	3	3	.57	.009
1 slice	23	3.5	0.6	1.4	0.0		.09	0.7	.61		65	30	.27	.020	
blood sausage (blood pudding)[b]	95	11.8	8.6	0.9	30										
1 slice	25	3.7	0.3	3.3											
bockwurst (pork, veal, milk, etc.), raw	200	36.5	17.9	1.9											
1 link	65	8.7	0.3	6.6											
bologna															
beef	72	12.7	6.6	0.3	13		5	.03	.03	1	226	3	3	.50	.006
1 slice	23	2.8	0.2	2.8	0.0		.01	0.6	.33	.06	36	20	.38	.010	
beef & pork	73	12.5	6.5	0.6	13		5	.03	.04	1	234	3	3	.45	.01
1 slice	23	2.7	0.6	2.5	0.0		.04	0.6	.31	.06	41	21	.35	.020	
beef garlic, Oscar Mayer	73	12.5	6.8		13		4	.02	.03		241	3	3	.51	
1 slice	23	2.5	0.5				.01	0.5	.25		44	22	.31	.010	
pork	57	13.9	4.6	0.5	14		8	.04	.06	1	272	3	3	.47	.008
1 slice	23	3.5	0.2	1.6	0.0		.12	0.9	.21	.17	65	32	.18	.017	
turkey	60	18.5	4.5	1.1	20			.05	.05		222	26	4	.64	
1 slice	28	3.9	0.6	1.4			.01	1.1	.44		53	44	.36	.010	
turkey, Louis Rich	58	18.5	4.6	1.3	19						225	31	5	.59	
1 oz slice	28	3.6	0.5	1.5							48	45	.41		
turkey, Oscar Mayer	58	18.5	4.7		19		0	.05			225	31	5	.59	.008
1 oz slice	28	3.6	0.5				.01	1.1	.42		48	45	.41	.020	
w/ cheese, Oscar Mayer	74	12.2	6.7		14		4	.03	.04	1	242	13	3	.39	
1 slice	23	2.7	0.6				.05	0.6	.29	.08	40	41	.24		
bratwurst, pork, ckd	256	47.7	22.0	2.3	51		1	.16	.18		473	38	12	1.96	.039
1 link	85	12.0	1.8	7.9			.43	2.7	.81	.27	180	126	1.09	.079	
braunschweiger (pork liver sausage)	65	8.6	5.8	0.7	28	760	2	.28	.06		206	2	2	.51	.028
1 slice	18	2.4	0.6	2.0	0.0	2529	.05	1.5	3.62	.61	36	30	1.68	.043	
german brand, Oscar Mayer	94	14.2	8.4		46		10	.47	.09	26	341	3	4	.86	
1 oz	28	3.9	0.5			5113	.06	2.4	4.97	.83	60	57	2.79	.120	
in tube, Oscar Mayer	96	13.8	8.7		41		2	.43	.09	17	319	2	3	.83	
1 oz	28	3.8	0.8			4668	.06	2.3	5.19	.81	50	52	2.68	.080	
brotwurst (pork & beef w/ nfdm)	226	35.9	19.5	2.0	44		20	.16	.09		778	34	11	1.47	.027
1 link	70	10.0	2.1	7.0			.18	2.3	1.44	.04	197	94	.72	.050	

[a] Some luncheon meats contain added ascorbic acid or sodium ascorbate. [b] Sausage contains a large proportion of blood so that it is very dark in color.

	KCAL / WT (g)	H₂O (g) / PRO (g)	FAT (g) / CHO (g)	PUFA (g) / SFA (g)	CHOL (mg) / DFIB (g)	A (RE) / A (IU)	C (mg) / B-1 (mg)	B-2 (mg) / NIA (mg)	B-6 (mg) / B-12 (mcg)	FOL (mcg) / PANT (mg)	Na (mg) / K (mg)	Ca (mg) / P (mg)	Mg (mg) / Fe (mg)	Zn (mg) / Cu (mg)	Mn (mg)
chicken roll, light meat	90	38.9	4.2	0.9	28			.07			331	24	10	.41	
2 slices	57	11.1	1.4	1.2			.04	3.0			129	89	.55	.020	
chorizo (pork & beef)	265	19.1	23.0	2.1											
1 link	60	14.5	0.0	8.6											
corned beef loaf, jellied	43	19.6	1.7	0.1	13	2	.03	.03	1		270	3	3	1.16	.009
1 oz slice	28	6.5	0.0	0.7			.00	0.5	.36	.05	29	21	.58	.017	
corned beef, sliced, Oscar Mayer	16	12.7	0.3		5						201	1	3	.50	
1 thin slice	17	3.3	0.0								53	34	.48	.050	
frankfurter															
beef	180	31.2	16.3	0.8	35	14	.06	.07	2		585	11	2	1.24	.019
1 frank (8 per 1 lb pkg)	57	6.9	1.0	6.9			.03	1.4	.87	.17	94	50	.81	.030	
beef	142	24.6	12.8	0.6	27	11	.05	.05	2		462	9	2	.98	.015
1 frank (10 per 1 lb pkg)	45	5.4	0.8	5.4			.02	1.1	.69	.13	75	39	.64	.030	
beef & pork	144	24.2	13.1	1.2	22	12	.05	.06	2		504	5	5	.83	.01
1 frank	45	5.1	1.2	4.8	0.0		.09	1.2	.58	.16	75	38	.52	.040	
beef & pork, small, Oscar Mayer	28	4.9	2.6		5	2	.01	.01	0		92	1	1	.17	
1 frank	9	1.0	0.2				.02	0.2	.11	.04	15	8	.10	.010	
chicken	116	25.9	8.8	1.8	45			.05			617	43			
1 frank	45	5.8	3.1	2.5			.03	1.4					.90		
nacho style w/ cheese, Oscar Mayer	138	24.4	12.5		30						550	23	5	.66	
1 frank	45	5.4	1.1								75	77	.34	.070	
w/ bacon & cheddar cheese, Oscar Mayer—*1 frank*	143	23.5	12.7		30						509	17	5	.75	
	45	6.1	1.1								85	86	.42	.100	
turkey	100	28.8	8.1	2.1	39			.08	.09		472	58	8	1.00	
1 frank	45	5.8	0.6	2.7			.03	1.7	.58		72	83	.71	.020	
turkey, Louis Rich	103	28.3	8.5	2.2	39						482	61	7	.98	
1 frank	45	5.6	1.1	2.7							88	83	.75		
turkey, Oscar Mayer	103	28.3	8.5		40	0	.08	.09			484	61	7	.98	.017
1 frank	45	5.6	1.1				.03	1.7	.58		88	83	.75	.030	
turkey w/ cheese, Louis Rich	108	27.1	8.6	2.2	40						514	46	8	.83	
1 frank	45	6.0	1.6	3.1							73	95	.65		
turkey w/ cheese, Oscar Mayer	103	25.9	8.3		37						487	55	7	.85	
1 frank	43	5.8	1.5								76	92	.67	.030	
ham, cured															
chopped, cnd	50	12.8	4.0	0.4	10	0 / 0	0 / .11	.04 / 0.7	.07 / .15		287 / 60	1 / 29	3 / .20	.38 / .010	.005
1 slice	21	3.4	0.1	1.3	0.0										
chopped, packaged	48	13.4	3.6	0.4	11	0 / 0	4 / .13	.04 / 0.8	.07 / .19	0 / .06	288 / 67	1 / 33	3 / .17	.41 / .013	.009
1 slice	21	3.6	0.0	1.2	0.0										
minced	55	12.0	4.3	0.5	15	0 / 0	6 / .15	.04 / 0.9	.06 / .20	/ .04	261 / 65	2 / 33	3 / .16	.40 / .016	.007
1 slice	21	3.4	0.4	1.5	0.0										
sliced, lean (5% fat)	37	20.0	1.4	0.1	13	/	7 / .26	.06 / 1.4	.13 / .21	1 / .13	405 / 99	2 / 62	5 / .22	.55 / .020	.009
1 slice	28	5.5	0.3	0.5	0.0										
sliced, reg (11% fat)	52	18.3	3.0	0.3		/	8 / .24	.07 / 1.5	.10 / .24	1 / .13	373 / 94	2 / 70	5 / .28	.61 / .030	.009
1 slice	28	5.5	0.9	1.0	0.0										
w/ black cracked pepper, Oscar Mayer	24	15.4	0.9		1						269	1	4	.41	
1 slice	21	3.9	0.1								61	58	.23	.060	
ham & cheese loaf/roll	73	16.4	5.7	0.6	16	7 /	/ .17	.05 / 1.0	.07 / .23	/ .15	381 / 83	16 / 72	5 / .26	.57 / .021	.008
1 slice	28	4.7	0.4	2.1	0.0										
headcheese (pork)	60	18.4	4.5	0.5	23	6 /	/ .01	.05 / 0.3	.05 / .30	1 / .06	356 / 9	4 / 17	3 / .33	.37 / .035	.005
1 slice	28	4.5	0.1	1.4	0.0										
honey loaf (pork & beef)	36	20.0	1.3	0.1	10	6 /	/ .14	.07 / 0.9	.09 / .31	/ .19	374 / 97	5 / 41	5 / .38	.69 / .020	.009
1 slice	28	4.5	1.5	0.4											
honey roll sausage (beef)	42	14.9	2.4	0.1	12	4 /	/ .02	.04 / 1.0	.06 / .54	/ .11	304 / 67	2 / 31	4 / .51	.75 / .020	.008
1 slice	23	4.3	0.5	0.9	0.0										
italian sausage, pork, ckd	217	33.5	17.2	2.2	52	1 /	/ .42	.16 / 2.8	.22 / .87	/ .30	618 / 204	16 / 114	12 / 1.01	1.59 / .054	.055
1 link	67	13.4	1.0	6.1											
kielbasa/kolbassy (pork & beef w/ nfdm)	81	14.0	7.1	0.8	17	6 /	/ .06	.06 / 0.7	.05 / .42	/ .21	280 / 70	11 / 38	4 / .38	.53 / .030	.010
1 slice	26	3.5	0.6	2.6	0.0										
knackwurst/knockwurst (pork & beef)	209	37.7	18.9	2.0	39	18 /	/ .23	.10 / 1.9	.11 / .80	/ .22	687 / 136	7 / 67	8 / .62	1.13 / .040	
1 slice	68	8.1	1.2	6.9	0.0										

	KCAL / WT (g)	H₂O (g) / PRO (g)	FAT (g) / CHO (g)	PUFA (g) / SFA (g)	CHOL (mg) / DFIB (g)	A (RE) / A (IU)	C (mg) / B-1 (mg)	B-2 (mg) / NIA (mg)	B-6 (mg) / B-12 (mcg)	FOL (mcg) / PANT (mg)	Na (mg) / K (mg)	Ca (mg) / P (mg)	Mg (mg) / Fe (mg)	Zn (mg) / Cu (mg)	Mn (mg)
lebanon bologna (beef)	49	14.0	3.0	0.1	16	5	.04	.06		1	308	3	4	.92	.013
1 slice	23	4.5	0.6	1.3			.01	1.0	.59	.12	69	35	.57	.020	
livercheese, pork fat wrapped, Oscar Mayer—1 slice	116	20.2	10.0		76		1	.84	.17	43	433	3	5	1.43	
	38	5.8	0.7				.08	4.6	9.23	1.42	83	83	4.31	.160	
livercheese (pork liver)	115	20.4	9.7	1.3	66	1996	1	.85	.18		465	3	4	1.41	.076
1 slice	38	5.8	0.8	3.4		6646	.08	4.5	9.33	1.34	86	79	4.11	.146	
liver pate															
chicken, cnd	57		3.7				3	.40				3			
1 oz	28	3.8	1.9			205	.02	2.1					2.60		
goose (pate de fois gras), cnd	60	4.8	5.7				.04								
1 T	13	1.5	0.6				.01	0.3							
goose, smoked, cnd	131	10.5	12.4		43		.09								
1 oz	28	3.2	1.3		0.0		.03	0.7	2.66						
unspecified, cnd	90	15.3	7.9			0	.17	.02		17	198	20	4		.034
1 oz	28	4.0	0.4			936	.01	0.9	.91	.34	39	57	1.56	.110	
liver sausage/liverwurst (pork)	59	9.4	5.1	0.5	28			.19	.03	5		5			
1 slice	18	2.5	0.4	1.9			.05		2.42	.53		41	1.15		
luxury loaf (pork)	40	19.4	1.4	0.1	10	6	.08	.09		1	347	10	6	.86	.012
1 slice	28	5.2	1.4	0.5			.20	1.0	.39	.15	107	52	.30	.028	
meat spread															
chicken, Carnation	100	34.1	6.0	4.0	16	0	.03	.12		3	200	6	7	.28	
1.9 oz	53	6.0	2.0	1.2		9	.02	2.4	.07	.19	72	47	.30	.020	
chicken, cnd	55		3.3					.03				35			
1 oz	28	4.4	1.5				.00	0.8					.66		
chicken, cnd, Swanson	60		3.9			0	.03				136	68			
1.02 oz	29	4.4	1.8			0	.00	0.9			37		.60		
ham & cheese	69	16.8	5.3	0.4	17	2	.06	.04			339	62	5	.64	.010
1 oz	28	4.6	0.6	2.4			.09	0.6	.21	.17	46	140	.22	.026	
ham & cheese, Oscar Mayer	66	16.7	5.1		16	2	.07	.04			329	51	5	.60	
1 oz	28	4.5	0.6				.09	0.6	.19		47	134	.18	.030	
ham, Carnation	100	34.2	6.0	2.3	24	0	.05	.00		3	350	6	2	.05	
1.9 oz	53	5.0	4.0	1.8		4	.15	1.0	.00	.01	104	47	.82	.010	
ham salad	61	17.8	4.4	0.8	10	2	.03	.04			259	2	3	.31	
1 oz	28	2.5	3.0	1.4			.12	0.6	.22	.09	42	34	.17	.021	
ham salad, Oscar Mayer	62	17.1	4.2		11	2	.03	.04		0	265	2	3	.30	
1 oz	28	2.5	3.5				.13	0.7	.20	.09	39	34	.17	.020	
Oscar Mayer	67	16.7	4.9		10	0	.04	.03		1	261	3	3	.27	
1 oz	28	2.1	3.6				.05	0.5	.30	.12	29	17	.25	.030	
pork & beef	67	17.1	4.9	0.7	11	0	.04	.03			287	3	2	.29	.01
1 oz	28	2.2	3.4	1.7			.05	0.5	.32	.12	31	17	.22	.040	
poultry (chicken/turkey) salad	57		3.8	1.8	9	0	.02	.03		1	107	3			
1 oz	28	3.3	2.1	1.0		39	.01	0.5	.11	.08	52	9	.17		
tuna, Carnation	90	34.9	6.0	3.7	13	0	.02	.00		3	370	6	2	.05	
1.9 oz	53	6.0	2.0	0.9		22	.01	2.8	.00	.01	82	45	.43	.010	
turkey, Carnation	100	34.9	6.0	3.9	20	0	.04	.10		4	190	7	8	.67	
1.9 oz	53	6.0	2.0	1.1		3	.02	1.1	.08	.21	76	48	.50	.040	
mortadella (beef & pork)	47	7.9	3.8	0.5	8	4	.02	.02			187	3	2	.32	.004
1 slice	15	2.5	0.5	1.4			.02	0.4	.22		24	15	.21	.010	
mothers loaf (pork)	59	11.5	4.7	0.5	9	0	.04	.04			237	9	3	.30	.014
1 slice	21	2.5	1.6	1.7			.12	0.7	.22	.10	47	27	.28	.020	
new england brand sausage (pork & beef)	37	15.4	1.7	0.2	11	5	.06	.08		2	281	2	4	.62	.008
1 slice	23	4.0	1.1	1.6	0.0		.15	0.8	.31	.16	74	31	.22	.020	
old fashioned loaf (pork & beef)	68	16.8	5.1	0.5	13	5	.08	.06		1	354	24	6	.49	.009
1 slice	28	3.8	1.6	1.8			.09	0.7	.37	.17	107	46	.35	.020	
olive loaf (pork)	67	16.5	4.7	0.6	11	3	.07	.07		1	421	31	5	.39	.010
1 slice	28	3.4	2.6	1.7			.08	0.5	.36	.22	84	36	.15	.014	
pastrami															
beef	99	13.3	8.3	0.3	26		1	.05	.05		348	2	5	1.21	
1 oz	28	4.9	0.9	3.0			.03	1.4	.50		65	43	.54		

	KCAL	H₂O (g)	FAT (g)	PUFA (g)	CHOL (mg)	A (RE)	C (mg)	B-2 (mg)	B-6 (mg)	FOL (mcg)	Na (mg)	Ca (mg)	Mg (mg)	Zn (mg)	Mn (mg)
	WT (g)	PRO (g)	CHO (g)	SFA (g)	DFIB (g)	A (IU)	B-1 (mg)	NIA (mg)	B-12 (mcg)	PANT (mg)	K (mg)	P (mg)	Fe (mg)	Cu (mg)	
sliced, Oscar Mayer	17	12.6	0.4		5						216	1	3	.49	
1 slice	17	3.2	0.1								52	35	.43	.060	
peppered loaf (pork & beef)	42	19.1	1.8	0.1	13		7	.09	.08	1	432	15	6	.92	.018
1 slice	28	4.9	1.3	0.7			.11	0.9	.56	.15	112	48	.30	.030	
pepperoni (pork & beef)	27	1.5	2.4	0.2				.01	.01		112	1	1	.14	
1 slice	6	1.2	0.2	0.9	0.0		.02	0.3	.14	.10	19	7	.08	.000	
pickle & pimento loaf (pork)	74	16.2	6.0	0.7	10		4	.07	.05	1	394	27	5	.40	.008
1 slice	28	3.3	1.7	2.2		28	.08	0.6	.33	.23	96	40	.29	.035	
picnic loaf (pork & beef)	66	17.1	4.7	0.5	11		5	.07	.09	1	330	13	4	.62	.008
1 slice	28	4.2	1.4	1.7			.11	0.7	.42	.19	76	36	.29	.020	
polish sausage (pork)	92	15.1	8.1	0.9	20		0	.04	.05		248	3	4	.55	.014
1 oz	28	4.0	0.5	2.9	0.0		.14	1.0	.28	.13	67	39	.41	.026	
pork & beef lunch meat	100	14.0	9.1	1.1	15		4	.04	.06	2	367	3	4	.47	.008
1 slice	28	3.6	0.7	3.3	0.0		.09	0.8	.36	.18	57	24	.24	.010	
pork & beef luncheon sausage	60	13.5	4.8	0.5	15		4	.05	.05		272	3	3	.56	.01
1 slice	23	3.5	0.4	1.8	0.0		.05	0.8	.45	.09	56	28	.33	.020	
pork lunch meat, cnd	70	10.8	6.4	0.8	13		0	.04	.04	1	271	1	2	.31	.005
1 slice	21	2.6	0.4	2.3	0.0		.08	0.7	.19	.10	45	17	.15	.008	
salami															
beerwurst, beef	76	12.2	6.9	0.3	14		4	.03	.04	1	236	2	3	.56	
1 slice	23	2.9	0.4	3.0	0.0		.02	0.8	.45	.08	40	22	.35	.009	
beerwurst, pork	55	14.1	4.3	0.5	14		7	.04	.08	1	285	2	3	.40	.007
1 slice	23	3.3	0.5	1.4	0.0		.13	0.7	.20	.11	58	24	.17	.012	
ckd, beef	60	13.4	4.8	0.2	15		4	.04	.04	0	271	2	3	.50	.011
1 slice	23	3.5	0.7	2.1			.02	0.7	.70	.22	51	26	.50	.028	
ckd, beef & pork	57	13.9	4.6	0.5	15		3	.09	.05	0	245	3	3	.49	.013
1 slice	23	3.2	0.5	1.9	0.0		.06	0.8	0.8	.20	46	27	.61	.050	
dry/hard pork	41	3.6	3.4	0.4				.03	.06		226	1	2	.42	.007
1 slice	10	2.3	0.2	1.2	0.0		.09	0.6	.28			23	.13	.016	
dry/hard pork & beef	42	3.5	3.4	0.3	8		3	.03	.05		186	1	2	.32	.000
1 slice	10	2.3	0.3	1.2	0.0		.06	0.5	.19	.11	38	14	.15	.010	
smoked link sausage															
pork	265	26.7	21.6	2.6	46		1	.18	.24		1020	20	13	1.92	
1 link	68	15.1	1.4	7.7	0.0		.48	3.1	1.11	.53	228	110	.79	.049	
pork & beef	229	35.5	20.6	2.2	48		13	.12	.12		642	7	8	1.44	.03
1 link	68	9.1	1.0	9.2	0.0		.18	2.2	1.03	.30	129	73	.99	.040	
pork & beef, Oscar Mayer	28	4.8	2.5		5						92	1	2	.19	
1 small link	9	1.2	0.2								17	20	.11	.020	
pork & beef w/ american cheese	141	22.6	12.5	1.3	29		8	.07	.05		465	25	5	.97	.01
1 link	43	6.0	0.6	4.5			.11	1.2	.74	.33	89	76	.46	.030	
pork & beef w/ flour & nfdm	182	39.0	14.6	1.5	59		2	.12	.09		741	12	9	1.36	.035
1 link	68	9.5	2.7	5.3			.16	1.8	.89		105	74	1.05	.060	
pork & beef w/ nfdm	213	36.7	18.8	2.1	44		14	.15	.12		798	28	11	1.33	.026
1 link	68	9.0	1.3	6.6	0.0		.13	1.9	1.07	.41	194	92	1.00	.060	
thuringer (cervelat/summer sausage)															
beef	70	11.7	6.2	0.3	17		5	.07	.06	1	317	2	3	.48	
1 slice	23	3.4	0.7	2.8			.03	1.0	1.29		53	24	.52	.020	
beef & pork	80	11.0	6.9	0.4	16		5	.07	.07	1	334	2	3	.47	.007
1 slice	23	3.7	0.5	2.8	0.0		.04	0.9	1.06	.13	53	23	.47	.020	
turkey breast meat	23	15.1	0.3	0.1	9		0	.02	.07		301	1	4	.24	
1 slice	21	4.7	0.0	0.1			.01	1.7	.42	.12	58	48	.08	.010	
turkey cotto salami															
Louis Rich	52	18.8	3.7	0.9	22						257	5	5	.71	
1 oz slice	28	4.4	0.3	1.1							60	43	.49		
Oscar Mayer	52	18.8	3.7		22		0	.09	.07		256	5	5	.71	.014
1 oz slice	28	4.4	0.3				.03	1.2	.99		62	43	.49	.050	
turkey ham (cured thigh meat)	73	40.5	2.9	0.9	34			.14	.14		565	5	10	1.50	
2 slices	57	10.7	0.2	1.0			.03	2.0	1.02		184	108	1.57	.060	

	KCAL / WT (g)	H₂O (g) / PRO (g)	FAT (g) / CHO (g)	PUFA (g) / SFA (g)	CHOL (mg) / DFIB (g)	A (RE) / A (IU)	C (mg) / B-1 (mg)	B-2 (mg) / NIA (mg)	B-6 (mg) / B-12 (mcg)	FOL (mcg) / PANT (mg)	Na (mg) / K (mg)	Ca (mg) / P (mg)	Mg (mg) / Fe (mg)	Zn (mg) / Cu (mg)	Mn (mg)
turkey luncheon loaf, Louis Rich	43	19.5	2.6	0.7	14						278	2	5	.45	
1 oz slice	28	4.6	0.4	0.8							59	65	.15		
turkey luncheon loaf, Oscar Mayer	43	19.6	2.6		14	0		.03	.07		278	2	5	.45	.005
1 oz slice	28	4.6	0.4				.01	1.6	.19		60	65	.15		.010
turkey pastrami	80	40.1	3.5	0.9	30			.14	.16		593	5	8	1.22	
2 slices	57	10.4	0.9	1.0			.03	2.0	1.14		147	113	.94		.030
Louis Rich	33	20.3	1.2	0.4	17						276	2	6	.82	
1 oz slice	28	5.4	0.1	0.4							75	85	.40		
Oscar Mayer	33	20.3	1.2		17	0		.08	.08		275	2	6	.82	.017
1 oz slice	28	5.4	0.1				.03	1.5	.57		76	85	.40		.040
turkey roll															
light & dark meat	84	39.8	4.0	1.0	31			.16			332	18	10	1.13	
2 slices	57	10.3	1.2	1.2			.05	2.7			153	95	.76		.040
light meat	83	40.6	4.1	1.0	24			.13			277	23	9	.88	
2 slices	57	10.6	0.3	1.1			.05	4.0			142	104	.72		.020
turkey salami	111	37.3	7.8		46	0		.10			569	11	8	1.03	
2 slices	57	9.3	0.3				.04	2.0			138	60	.91		.031
ckd	50	19.0	3.4	1.1	20			.08	.07		251	7	5	.72	
1 slice	28	4.5	0.3	1.0			.02	1.2	.60		63	45	.38		.030
Louis Rich	52	18.9	3.7	1.0	19						245	6	5	.69	
1 oz slice	28	4.5	0.1	1.1							61	42	.35		
Oscar Mayer	52	18.9	3.7		19	0		.07	.07		244	6	5	.69	.007
1 oz slice	28	4.5	0.1				.02	1.3	.43		63	42	.35		.030
turkey summer sausage															
Louis Rich	52	18.3	3.5	0.9	22						311	4	5	.68	
1 oz slice	28	4.7	0.4	1.2							68	71	.56		
Oscar Mayer	52	18.3	3.5		23	0		.12	.07		312	4	5	.68	.035
1 oz slice	28	4.7	0.4				.03	1.4	1.15		69	71	.56		.040
vienna sausage, cnd (beef & pork)	45	9.6	4.0	0.3	8	0		.02	.02		152	2	1	.26	.005
1 sausage	16	1.7	0.3	1.5	0.0		.01	0.3	.16		16	8	.14		.000

23. MILK, YOGURT, MILK BEVERAGES & MILK BEVERAGE MIXES
23.1. COW MILK

	KCAL / WT (g)	H₂O (g) / PRO (g)	FAT (g) / CHO (g)	PUFA (g) / SFA (g)	CHOL (mg) / DFIB (g)	A (RE) / A (IU)	C (mg) / B-1 (mg)	B-2 (mg) / NIA (mg)	B-6 (mg) / B-12 (mcg)	FOL (mcg) / PANT (mg)	Na (mg) / K (mg)	Ca (mg) / P (mg)	Mg (mg) / Fe (mg)	Zn (mg) / Cu (mg)	Mn (mg)
buttermilk, cultured	99	220.8	2.2	0.0	9	20	2	.38	.08		257	285	27	1.03	
8 fl oz	245	8.1	11.7	1.3	0.0	81	.08	0.1	.54	.67	371	219	.12		
buttermilk, dry	25	0.2	0.4	0.0	5	4	0	.10	.02	3	34	77	7	.26	
1 T	7	2.2	3.2	0.2	0.0	14	.03	0.1	.25	.21	103	61	.02		
calcimilk (lactose reduced lowfat milk w/ added Ca)[a]—*8 fl oz*	102		3.0		10		2	.41			123	500			
		8.0	12.0			500	.10	0.2			381	340	.12		
condensed, sweetened, cnd	123	10.4	3.3	0.1	13	31	1	.16	.02	4	49	108	10	.49	
1 fl oz	38	3.0	20.8	2.1	0.0	125	.03	0.1	.17	.29	142	97	.07		
evaporated, lowfat, cnd, Carnation	110	97.3	3.0				1	.34	.06	10	138	318	32		
4 fl oz	127	9.0	12.0			630	.04	0.2	.27		416	267	.29		
evaporated, skim, cnd	25	25.3	0.1	0.0	1	37	0	.10	.02	3	37	92	9	.29	
1 fl oz	32	2.4	3.6	0.0	0.0	125	.01	0.1	.08	.24	106	62	.09		
evaporated, whole, cnd	42	23.3	2.4	0.1	9	17	1	.10	.02	2	33	82	8	.24	
1 fl oz	32	2.1	3.2	1.5	0.0	77	.02	0.1	.05	.20	95	64	.06		
evaporated, whole, cnd	169	93.3	9.5	0.3	37	68	2	.40	.06	10	133	329	30	.97	
4 fl oz	126	8.6	12.7	5.8	0.0	306	.06	0.2	.21	.80	382	255	.24		
lactaid (lactose reduced lowfat milk)[a]	102		3.0		10		2	.41			123	300			
8 fl oz		8.0	12.0			500	.10	0.2			381	235	.12		
lowfat 1% fat	102	219.8	2.6	0.1	10	145	2	.41	.11	12	123	300	34	.95	
8 fl oz	244	8.0	11.7	1.6	0.0	500	.10	0.2	.90	.79	381	235	.12		
lowfat 1% fat, pro fortified	119	218.3	2.9	0.1	10	145	3	.47	.12	15	143	349	39	1.11	
8 fl oz	246	9.7	13.6	1.8	0.0	500	.11	0.3	1.05	.92	444	273	.15		

[a] Lactose content is 3 g/cup.

	KCAL	H₂O (g)	FAT (g)	PUFA (g)	CHOL (mg)	A (RE)	C (mg)	B-2 (mg)	B-6 (mg)	FOL (mcg)	Na (mg)	Ca (mg)	Mg (mg)	Zn (mg)	Mn (mg)
	WT (g)	PRO (g)	CHO (g)	SFA (g)	DFIB (g)	A (IU)	B-1 (mg)	NIA (mg)	B-12 (mcg)	PANT (mg)	K (mg)	P (mg)	Fe (mg)	Cu (mg)	
lowfat 1% fat w/ nfdm	104	220.0	2.4	0.1	10	145	2	.42	.11	13	128	313	35	.98	
8 fl oz	245	8.5	12.2	1.5	0.0	500	.10	0.2	.94	.82	397	245	.12		
lowfat 2% fat	121	217.7	4.7	0.2	18	140	2	.40	.11	12	122	297	33	.95	
8 fl oz	244	8.1	11.7	2.9	0.0	500	.10	0.2	.89	.78	377	232	.12		
lowfat 2% fat, pro fortified	137	215.8	4.9	0.2	19	140	3	.48	.13	15	145	352	40	1.11	
8 fl oz	246	9.7	13.5	3.0	0.0	500	.11	0.3	1.05	.93	447	276	.15		
lowfat 2% fat w/ nfdm	125	217.7	4.7	0.2	18	140	2	.42	.11	13	128	313	35	.98	
8 fl oz	245	8.5	12.2	2.9	0.0	500	.10	0.2	.94	.82	397	245	.12		
skim	86	222.5	0.4	0.0	4	149	2	.34	.10	13	126	302	28	.98	
8 fl oz	245	8.4	11.9	0.3	0.0	500	.09	0.2	.93	.81	406	247	.10		
skim, dry	109	1.0	0.2	0.0	6	2	2	.47	.11	15	161	377	33	1.22	
1/4 cup	30	10.9	15.6	0.2	0.0	11	.12	.29	1.21	1.07	538	290	.10		
skim, dry, Ca reduced	100	1.4	0.1	0.0	1	1		.47	.08		646	79	17		
1 oz	28	10.1	14.7	0.0	0.0	2	.05	0.2	1.13	.94	193	287			
skim, dry, inst	326	3.6	0.7	0.0	17	646	5	1.59	.31	45	499	1120	107	4.01	
1 1/3 cups (3.2 oz envelope)[a]	91	31.9	47.5	0.4	0.0	2157	.38	0.8	3.63	2.94	1552	896	.28		
skim pro fortified	100	219.8	0.6	0.0	5	149	3	.48	.12	15	144	352	40	1.11	
8 fl oz	246	9.7	13.7	0.4	0.0	500	.11	0.3	1.05	.93	446	275	.15		
skim w/ nfdm	90	221.4	0.6	0.0	5	149	2	.43	.11	13	130	316	36	1.00	
8 fl oz	245	8.8	12.3	0.4	0.0	500	.10	0.2	.95	.83	418	255	.12		
whole, 3.3% fat	150	214.7	8.2	0.3	33	76	2	.40	.10	12	120	291	33	.93	
8 fl oz	244	8.0	11.4	5.1	0.0	307	.09	0.2	.87	.77	370	228	.12		
whole, 3.5% fat	150	213.3	8.0	0.2	34		5	.42	.02	37	122	288	24	1.00	
8 fl oz	244	8.0	11.0	4.9	0.0		.10	0.2	1.34	.85	351	227			
whole, 3.7% fat	157	214.0	8.9	0.3	35	83	4	.39	.10	12	119	290	33	.93	
8 fl oz	244	8.0	11.4	5.6	0.0	337	.09	0.2	.87	.76	368	227	.12		
whole, dry	159	0.8	8.6	0.2	31	90	3	.39	.10	12	119	292	27	1.07	
1/4 cup	32	8.4	12.3	5.4	0.0	295	.09	0.2	1.04	.73	426	248	.15		
whole, low Na	149	215.2	8.4	0.3	33	78		.26	.08		6	246	12		
8 fl oz	244	7.6	10.9	5.3	0.0	317	.05	0.1	.88	.74	617	209			

23.2. OTHER MILK

	KCAL	H₂O (g)	FAT (g)	PUFA (g)	CHOL (mg)	A (RE)	C (mg)	B-2 (mg)	B-6 (mg)	FOL (mcg)	Na (mg)	Ca (mg)	Mg (mg)	Zn (mg)	Mn (mg)
	WT (g)	PRO (g)	CHO (g)	SFA (g)	DFIB (g)	A (IU)	B-1 (mg)	NIA (mg)	B-12 (mcg)	PANT (mg)	K (mg)	P (mg)	Fe (mg)	Cu (mg)	
filled[b,c]	154	213.9	8.4	1.8	4	5	2	.30	.10	12	138	312	32	.88	
8 fl oz	244	8.1	11.6	1.9	0.0	17	.07	0.2	.83	.73	339	236	.12		
filled[b,d]	153	214.4	8.3	0.0	4	5	2	.30	.10	12	138	312	31	.88	
8 fl oz	244	8.1	11.6	7.6	0.0	17	.07	0.2	.83	.73	339	236	.12		
goat	168	212.4	10.1	0.4	28	137	3	.34	.11	1	122	326	34	.73	
8 fl oz	244	8.7	10.9	6.5	0.0	451	.12	0.7	.16	.76	499	270	.12		
human	21	27.0	1.4	0.2	4	20	1	.01	.00	2	5	10	1	.05	
1 fl oz	31	0.3	2.1	0.6	0.0	74	.00	0.1	.01	.07	16	4	.07		
imitation[c,e]	150	215.3	8.3	1.2	0	0	0	.22	.00	0	191	79	16	2.88	
8 fl oz	244	4.3	15.0	1.9	0.0	0	.03	0.0	.00	.00	279	181	.95		
imitation[d,e]	150	215.2	8.3	0.0	0	0	0	.22	.00	0	191	79	16	2.88	
8 fl oz	244	4.3	15.0	7.4	0.0	0	.03	0.0	.00	.00	279	181	.95		
indian buffalo	236	203.5	16.8	0.4	46	129	6	.33	.06	14	127	412	76	.54	
8 fl oz	244	9.2	12.6	11.2	0.0	434	.13	0.2	.89	.47	434	286	.29		
sheep	264	197.7	17.2	0.8		103	10	.87			108	474	45		
8 fl oz	245	14.7	13.1	11.3	0.0	360	.16	1.0	1.74	1.00	334	387	.24		
soy	79	223.9	4.6	2.0	0	8	0	.17	.10	4	30	10	45	.54	.408
1 cup	240	6.6	4.3	0.5	2.6	77	.39	0.4	.00	.12	338	117	1.38	.288	

[a] Reconstitutes to 1 qt fluid skim milk.
[b] Filled milk contains fats or oils other than milk fat; contains milk solids (milk, cream, or skim milk).
[c] Contains blend of hydrogenated soybean, cottonseed &/or safflower oils.
[d] Contains lauric acid oils which include modified coconut oil, hydrogenated coconut oil &/or palm kernel oil.
[e] Imitation milk contains fats or oils other than milk fat; contains food solids, excluding milk solids.

	KCAL / WT (g)	H₂O (g) / PRO (g)	FAT (g) / CHO (g)	PUFA (g) / SFA (g)	CHOL (mg) / DFIB (g)	A (RE) / A (IU)	C (mg) / B-1 (mg)	B-2 (mg) / NIA (mg)	B-6 (mg) / B-12 (mcg)	FOL (mcg) / PANT (mg)	Na (mg) / K (mg)	Ca (mg) / P (mg)	Mg (mg) / Fe (mg)	Zn (mg) / Cu (mg)	Mn (mg)

23.3. YOGURT (FROM COW MILK)

	KCAL / WT (g)	H₂O / PRO	FAT / CHO	PUFA / SFA	CHOL / DFIB	A(RE) / A(IU)	C / B-1	B-2 / NIA	B-6 / B-12	FOL / PANT	Na / K	Ca / P	Mg / Fe	Zn / Cu	Mn
custard style	160		4.0								140				
8 fl oz	227	12.0	18.0								560				
flavored[a]	187		4.0								97				
6 fl oz	170	7.0	31.4								291				
lowfat, breakfast, Yoplait[b]	228		3.7								91				
6 fl oz	170	8.0	40.6								354				
lowfat w/ nfdm	144	193.1	3.5	0.1	14	36	2	.49	.11	25	159	415	40	2.02	
8 fl oz	227	11.9	16.0	2.3	0.0	150	.10	0.3	1.28	1.34	531	326	.18		
coffee/van flavor	194	179.3	2.8	0.0	11	30	2	.46	.10	24	149	389	37	1.88	
8 fl oz	227	11.2	31.3	1.8	0.0	123	.10	0.2	1.20	1.25	498	306	.16		
fruit flavor	225	170.9	2.6	0.0	10	27	1	.37	.08	19	121	314	30	1.52	
8 fl oz	227	9.0	42.3	1.7		111	.08	0.2	.97	1.01	402	247	.14		
skim w/ nfdm	127	193.5	0.4	0.0	4	5	2	.53	.12	28	174	452	43	2.20	
8 fl oz	227	13.0	17.4	0.3	0.0	16	.11	0.3	1.39	1.46	579	355	.20		
whole	139	199.5	7.4	0.2	29	68	1	.32	.07	17	105	274	26	1.34	
8 fl oz	227	7.9	10.6	4.8	0.0	279	.07	0.2	.84	.88	351	215	.11		
fruit flavor, Yoplait[b]	190		4.0								105				
6 fl oz	170	7.0	32.0								350				
yo creme, Yoplait[c]	236		9.2								83				
5 fl oz		6.4	31.6								320				

23.4. MILK BEVERAGES

	KCAL / WT (g)	H₂O / PRO	FAT / CHO	PUFA / SFA	CHOL / DFIB	A(RE) / A(IU)	C / B-1	B-2 / NIA	B-6 / B-12	FOL / PANT	Na / K	Ca / P	Mg / Fe	Zn / Cu	Mn
Alba choc flavor nonfat dry milk in water	80		1.0				1	.29	.06		120	310			
8 fl oz		6.0	13.0			378	.07	0.2				169	.24		
carob flavor mix in whole milk	195	215.0	8.2	0.3	33	76	2	.39	.12	12	132	291	33	.93	.005
3 t powder in 8 fl oz milk	256	8.2	22.6	5.1		307	.10	0.3	.87	.77	370	228	.67	.023	
choc flavor mix in whole milk	226	215.2	8.8	0.3	33	76	3	.43	.10	12	165	300	54	1.26	.157
2-3 hp t powder in 8 fl oz milk	266	8.8	30.9	5.5		312	.10	0.3	.87	.77	498	256	.80	.176	
choc malted milk flavor mix															
in whole milk	229	215.0	8.9	0.4	34	80	3	.44	.14	16	172	304	47	1.09	.005
3 hp t powder in 8 fl oz milk	265	9.1	29.8	5.5		326	.13	0.6	.91	.77	499	265	.60	.066	
w/ added nutrients in whole milk	225	215.2	8.9	0.4	33	900	34	1.26	1.02	32	244	384	53	1.15	.135
4-5 hp t powder in 8 fl oz milk	265	9.1	29.1	5.5		3058	.73	10.9	.87	.91	620	312	3.77	.156	
choc milk															
1% fat milk	158	211.3	2.5	0.1	7	148	2	.42	.10	12	152	287	33	1.02	
8 fl oz	250	8.1	26.1	1.5		500	.10	0.3	.86	.76	426	256	.60		
2% fat milk	179	209.0	5.0	0.2	17	142	2	.41	.10	12	150	284	33	1.02	
8 fl oz	250	8.0	26.0	3.1		500	.09	0.3	.85	.75	422	254	.60		
whole milk	208	205.8	8.5	0.3	30	72	2	.41	.10	12	149	280	33	1.02	.120
8 fl oz	250	7.9	25.9	5.3		302	.09	0.3	.84	.74	417	251	.60	.146	
whole milk w/ malt	233	215.1	9.1	0.4	34	80	2	.43	.13	17	168	304	48	1.11	.095
8 fl oz	265	9.4	29.2	5.5		326	.14	0.7	.92	.77	500	265	.50	.066	
choc syrup in whole milk	232	229.0	8.5	0.3	33	77	2	.42	.10	14	156	297	57	1.20	.149
2 T syrup in 8 fl oz milk	282	8.8	33.5	5.3		319	.10	0.3	.87	.77	455	277	.91	.217	
choc syrup w/ added nutrients in whole	196	220.4	8.4	0.3	33	320	3	.55	.11	12	148	292	33	.93	.005
milk—1 T syrup in 8 fl oz milk	263	8.4	23.8	5.2		1125	.10	6.5	.87	.77	460	228	2.67	.024	
cocoa/hot chocolate															
prep w/ whole milk	218	204.0	9.1	0.3	33	85	2	.44	.11	12	123	298	56	1.22	
8 fl oz	250	9.1	25.8	5.6		318	.10	0.4	.87	.81	480	270	.78		

[a] Values are averages for 7 flavors.
[b] Values are averages for 9 flavors.
[c] Values are averages for 5 flavors.

	KCAL / WT (g)	H₂O (g) / PRO (g)	FAT (g) / CHO (g)	PUFA (g) / SFA (g)	CHOL (mg) / DFIB (g)	A (RE) / A (IU)	C (mg) / B-1 (mg)	B-2 (mg) / NIA (mg)	B-6 (mg) / B-12 (mcg)	FOL (mcg) / PANT (mg)	Na (mg) / K (mg)	Ca (mg) / P (mg)	Mg (mg) / Fe (mg)	Zn (mg) / Cu (mg)	Mn (mg)
prep w/ milk, sugar & high-fat plain	218	204.1	9.1	0.3	33	85	2	.44	.11	12	123	298	56	1.22	
cocoa—8 fl oz	250	9.1	25.8	5.6		318	.10	0.4	.87	.81	480	270	.78		
prep w/ water from mix	103	177.9	1.1	0.0		1	1	.16	.03	0	149	96	25	.46	.078
3/4 hp t powder in 6 fl oz water	206	3.1	22.5	0.7		4	.03	0.2	.37	.25	203	89	.35	.093	
sweetened w/ aspartame	48	177.5	0.4	0.0			0	.21	.05	2	173	90	33	.56	.100
.53 oz pkt in 6 fl oz water	192	3.8	8.5	0.3			.04	0.2		.57	405	134	.75	.121	
w/ added nutrients	120	178.1	3.0	0.1		150	6	.17			207	104	24	.27	.092
1 pkt in 6 fl oz water	209	1.9	24.1	1.8		501	.15	2.0			405	111	1.81	.111	
dairy drink mix w/ aspartame,	64	185.3	0.5	0.0		73	0	.41	0.2	9	172	192	47	.81	.157
choc, in water—3/4 oz pkt in water[a]	204	5.3	10.7	0.4	0.2	245	.02	0.3	.51	.46	479		1.65	.182	
eggnog, nonalcoholic	342	188.9	19.0	0.9	149	203	4	.48	.13	2	138	330	47	1.17	
8 fl oz	254	9.7	34.4	11.3	0.0	894	.09	0.3	1.14	1.06	420	278	.51		
mix in whole milk	260	214.5	8.4	0.3	33	76	2	.39	.10	12	163	291	33	.92	.005
2 hp t in 8 fl oz milk	272	8.1	39.0	5.1		307	.09	0.2	.87	.76	369	228	.38	.024	
mix in whole milk, Delmark	290		9.0			35		11	.65	.16	0	220	470	59	1.14	.010
1.23 oz mix in 8 fl oz milk	257	15.0	37.0	5.7	0.0	1090	.29	4.2	1.90	1.30	600	370	3.70	.070	
instant breakfast in whole milk	280		8.5					29	.60	.50	112	280	431	113	3.93
1 pkt in 8 fl oz milk[b]	265	15.0	34.5			2000	.39	5.2	1.47	2.70	670	376	4.60	.530	
choc, Delmark	290		9.0		40		20	.71	.66	130	270	470	132	4.90	.010
1.23 oz mix in 8 fl oz milk	279	15.0	37.0	0.0		1650	.50	6.6	2.00	3.30	720	420	5.90	.660	
van/strawberry, Delmark	290		9.0		40		20	.71	.66	130	220	470	132	4.90	.010
1.23 oz mix in 8 fl oz milk	279	15.0	37.0	0.0		1650	.50	6.6	2.00	3.30	630	420	5.90	.660	
instant breakfast, van/strawberry, liquid	250		8.0		20		15	.42	.50	100	190	300	60	3.70	1.000
pack, Delmark—8 fl oz	279	12.0	33.0			1250	.37	5.0	1.50	2.50	480	300	4.50	.500	
malted milk	236	215.2	9.9	0.6	37	93	2	.54	.18	22	215	347	52	1.14	.095
8 fl oz	265	10.8	26.6	6.0		376	.20	1.3	1.04	.77	529	307	.29	.066	
malted milk flavor mix															
in whole milk	237	215.1	9.8	0.6	37	94	3	.59	.19	22	223	354	52	1.13	.005
3 hp t in 8 fl oz milk	265	10.4	27.3	6.0		369	.20	1.3	1.03	.77	529	303	.27	.066	
w/ added nutrients in whole milk	230	215.3	8.7	0.4	33	743	29	1.15	.87		205	370	47	1.08	.093
4-5 hp t in 8 fl oz milk	265	9.9	28.4	5.4		2530	.71	10.4		.77	573	307	3.61	.082	
milk shake															
mix in whole milk, Delmark															
choc	230		9.0		35		2	.41	.10	0	240	300	49	.93	.120
.667 oz mix in 8 fl oz milk	263	9.0	28.0			340	.08	0.3	1.30	.83	450	270	.47	.034	
strawberry	230		9.0		35		2	.40	.10	0	160	290	39	.93	
.667 oz mix in 8 fl oz milk	263	8.0	29.0			340	.07	0.2	1.30	.80	360	230	.18	.040	
van	230		9.0		35		2	.40	.10	0	160	290	39	.93	
.667 oz mix in 8 fl oz milk	263	8.0	30.0			340	.07	0.2	1.30	.80	360	230	.17	.039	
thick															
choc	356	216.6	8.1	0.3	32		0	.67	.08	15	333	396	48	1.44	
1 avg	300	9.2	63.5	5.0		258	.14	0.4	.95	1.09	672	378	.93		
van	350	233.0	9.5	0.4	37	88	0	.61	.13	21	299	457	37	1.22	
1 avg	313	12.1	55.6	5.9		357	.09	0.5	1.63		572	361	.31		
strawberry flavored mix in whole milk	234	215.1	8.2	0.3	33	76	2	.42	.10	12	128	292	33	.93	.005
2-3 hp t powder in 8 fl oz milk	266	8.1	32.8	5.1		308	.09	0.2	.87	.77	370	228	.22	.024	

23.5. MILK BEVERAGE MIXES

	KCAL / WT (g)	H₂O (g) / PRO (g)	FAT (g) / CHO (g)	PUFA (g) / SFA (g)	CHOL (mg) / DFIB (g)	A (RE) / A (IU)	C (mg) / B-1 (mg)	B-2 (mg) / NIA (mg)	B-6 (mg) / B-12 (mcg)	FOL (mcg) / PANT (mg)	Na (mg) / K (mg)	Ca (mg) / P (mg)	Mg (mg) / Fe (mg)	Zn (mg) / Cu (mg)	Mn (mg)
carob flavor mix, powder	45	0.3	0.0	0.0	0		0		.02						
3 t	12	0.2	11.2	0.0			.00	0.1				12	.55		
choc flavor mix, powder	75	0.2	0.7	0.0	0	0	0	.03	.00		45	8	21	.33	.153
2-3 hp t	22	0.7	19.5	0.4		4	.01	0.1	.00		128	28	.68	.152	
choc malted milk flavor mix, powder	79	0.3	0.8	0.1	1	4	0	.04	.03	4	53	13	15	.17	
3/4 oz (3 hp t)	21	1.1	18.4	0.5		19	.04	0.4	.04		130	37	.48	.042	
choc malted milk flavor mix w/ added	75	0.5	0.7	0.1		824	32	.86	.92	20	125	93	20	.22	.132
nutrients, powder—3/4 oz (4-5 hp t)	21	1.0	17.7	0.4		2751	.64	10.7		.14	251	84	3.65	.133	
choc syrup	82	13.9	0.3	0.0	0	1	0	.02	.00	2	36	5	24	.27	.143
2 T (1 fl oz)	38	0.7	22.1	0.2		1	.00	0.1	.00	.01	84	48	.79	.192	

[a] 4 fl oz water w/ 3 ice cubes.

[b] Averages for various flavors of instant breakfast.

| | KCAL | H₂O (g) | FAT (g) | PUFA (g) | CHOL (mg) | A (RE) | C (mg) | B-2 (mg) | B-6 (mg) | FOL (mcg) | Na (mg) | Ca (mg) | Mg (mg) | Zn (mg) | Mn (mg) |
	WT (g)	PRO (g)	CHO (g)	SFA (g)	DFIB (g)	A (IU)	B-1 (mg)	NIA (mg)	B-12 (mcg)	PANT (mg)	K (mg)	P (mg)	Fe (mg)	Cu (mg)	
choc syrup w/ added nutrients	46	5.5	0.2	0.0	0	245	0	.16	.00		29				
1 T	19	0.3	12.4	0.1		817	.00	6.3			90		2.55		
cocoa mix, powder	102	0.4	1.1	0.0		1	1	.16	.03	0	143	93	24	.41	.076
1 oz pkt (3-4 hp t)	28	3.1	22.5	0.7		4	.03	0.2	.38	.25	202	89	.34	.081	
Alba	62		0.5		4		1	.29			134	323			
19.14 g pkt	19	5.4	10.2			283	.06	0.2			425	238	.07		
aspartame sweetened	48	0.5	0.4	0.0			0	.21	.05	2	168	86	31	.52	.100
.53 oz pkt	15	3.8	8.5	0.3			.04	0.2		.57	405	134	.74	.110	
choc & marshmallow	110	0.5	1.0				1	.15	.01	1	120	49	1	.12	.000
1 oz	28	2.0	23.0			1	.02	0.1	.05	.04	179	76	.23	.003	
milk choc	110	0.7	1.0				0	.17	.03	4	120	89	7	.25	.001
1 oz	28	3.0	23.0			3	.03	0.1	.30	.20	174	96	.21	.020	
rich choc	110	0.5	1.0				0	.15	.01	1	120	49	1	.12	.002
1 oz	28	2.0	23.0			1	.02	0.1	.06	.04	179	76	.21	.003	
rich choc, 70 cal	70	0.4	0.0				1	.16	.04	5	125	89	16	.40	.07
1 pkt (3 hp t)	21	3.0	15.0			3	.01	0.1	.37	.30	196	85	.46	.100	
sugar-free	50	0.6	0.0				1	.23	.04	6	160	123	20	.50	.09
1 pkt (4 hp t)	15	4.0	8.2			4	.03	0.2	.46	.34	290	131	.87	.120	
Hershey	76		0.4					.01			50	6	23	.38	.168
3 t (3/4 oz)	21	1.0	16.8			4	.01	0.1			89	17	.69	.168	
w/ added nutrients	120	0.5	3.0	0.1		150	6	.17			201	100	22	.22	.090
1.1 oz pkt	31	1.9	24.0	1.8		500	.15	2.0			404	111	1.80	.100	
dairy drink mix, choc powder, aspartame	63	2.7	0.5	0.0		73	0	.41	.02	9	166	187	45	.77	.155
sweetened—3/4 oz pkt	21	5.3	10.7	0.4	0.3	245	.02	0.3	.51	.46	477		1.64	.170	
dairy light, Alba	10		0.0				0	.06			15	40			
2.8 g pkt	2.8	1.0	1.0			0	.01	0.0					.08		
eggnog mix, powder, nonalcoholic	111	0.1	0.3	0.0		0			.00		44				
2 hp t	28	0.1	27.7	0.1			.00	0.0					.26		
fit 'n frosty, Alba															
choc	70		1.0				1	.30			210	310			
21.26 g pkt	21	6.0	11.0			304	.07	0.4				176	.43		
choc marshmallow	70		1.0				1	.29			250	310			
21.26 g pkt	21	6.0	11.0			305	.07	0.5				177	.39		
double fudge	70		1.0				1	.28			170	310			
21.26 g pkt	21	6.0	11.0			305	.07	0.5				173	.42		
strawberry	70		0.0				1	.36			150	310			
21.26 g pkt	21	5.0	12.0			265	.09	0.6				187	.10		
van	70		0.0				1	.30			150	310			
21.26 g pkt	21	5.0	12.0			218	.08	1.9				182	.10		
instant breakfast															
choc, Carnation	130	1.0	0.0				27	.24	.40	100	135	165	80	3.00	
1.25 oz pkt	35	7.0	24.0			1750	.30	5.0	.60	2.00	290	160	4.50	.500	
choc, Pillsbury	137	1.0	0.5				24	.37			149	97			
1 pouch	37	5.9	27.0			1777	.48	6.1			270	94	5.92		
choc malt, Carnation	130	1.0	1.0				27	.12	.40	100	170	80	80	3.00	
1.24 oz pkt	35	7.0	23.0			1750	.30	5.0	.60	2.00	200	115	4.50	.500	
choc malt, Pillsbury	134	1.1	0.5				24	.44			126	124			
1 pouch	37	5.9	26.6			1780	.48	6.1			278	132	5.92		
coffee, Carnation	130	1.0	0.0				27	.30	.40	100	130	200	80	3.00	
1.26 oz pkt	36	7.0	24.0			1750	.30	5.0	.60	2.00	400	180	4.50	.500	
eggnog, Carnation	130	1.0	0.0				27	.18	.40	100	180	135	80	3.00	
1.20 oz pkt	34	7.0	24.0			1750	.30	5.0	.60	2.00	220	125	4.50	.500	
strawberry, Carnation	130	1.0	0.0	.01	3		27	.30	.40	100	190	200	80	3.00	
1.25 oz pkt	35	7.0	24.0	.08		1750	.30	5.0	.60	2.00	360	180	4.50	.500	
strawberry, Pillsbury	137	0.9	0.3				24	.33			138	96			
1 pouch	37	5.7	27.9			1777	.48	6.1			265	81	5.70		
van, Carnation	130	1.0	0.0				27	.30	.40	100	135	200	80	3.00	
1.23 oz pkt	35	7.0	24.0			1750	.30	5.0	.60	2.00	360	180	4.50	.500	

	KCAL / WT (g)	H₂O (g) / PRO (g)	FAT (g) / CHO (g)	PUFA (g) / SFA (g)	CHOL (mg) / DFIB (g)	A (RE) / A (IU)	C (mg) / B-1 (mg)	B-2 (mg) / NIA (mg)	B-6 (mg) / B-12 (mcg)	FOL (mcg) / PANT (mg)	Na (mg) / K (mg)	Ca (mg) / P (mg)	Mg (mg) / Fe (mg)	Zn (mg) / Cu (mg)	Mn (mg)
van, Pillsbury	140	0.9	0.2					24	.26			210	84			
1 pouch	37	5.6	28.9			2073	.33	5.0			242	117	5.40			
instant breakfast, no sugar added, Carnation															
choc	70	0.9	1.0				27	.17	.40	100	140	100	80	3.00	
.69 oz pkt	20	7.0	8.0			1750	.30	5.0	.60	2.00	268	100	4.50	.500	
choc malted	70	0.9	1.0				27	.14	.40	100	150	60	80	3.00	
.71 oz pkt	20	7.0	9.0			1750	.30	5.0	.60	2.00	181	80	4.50	.500	
strawberry	70	0.8	0.0				27	.26	.40	100	120	150	80	3.00	
.68 oz pkt	19	7.0	9.0			1750	.30	5.0	.60	2.00	296	100	4.50	.500	
van	70	0.8	0.0				27	.26	.40	100	120	150	80	3.00	
.67 oz pkt	19	7.0	9.0			1750	.30	5.0	.60	2.00	296	100	4.50	.500	
malted milk flavor mix, powder	87	0.4	1.7	0.3	4	19	1	.19	.09	10	103	63	20	.21	
3/4 oz (3 hp t)	21	2.3	15.9	0.9		61	.11	1.1	.16		159	75	.15	.042	
malted milk flavor mix w/ added nutrients, powder—*3/4 oz (4-5 hp t)*	80	0.6	0.6	0.1		668	27	.75	.76		85	79	14	.15	.089
	21	1.8	17.1	0.3		222	.62	10.2			203	79	3.49	.059	
strawberry flavor mix, powder	84	0.1	0.0		0		0	.02	.00		8				
2-3 hp t	22	0.0	21.4				.00	0.0	.00				.09		

24. NUTS, NUT PRODUCTS & SEEDS

	KCAL / WT (g)	H₂O (g) / PRO (g)	FAT (g) / CHO (g)	PUFA (g) / SFA (g)	CHOL (mg) / DFIB (g)	A (RE) / A (IU)	C (mg) / B-1 (mg)	B-2 (mg) / NIA (mg)	B-6 (mg) / B-12 (mcg)	FOL (mcg) / PANT (mg)	Na (mg) / K (mg)	Ca (mg) / P (mg)	Mg (mg) / Fe (mg)	Zn (mg) / Cu (mg)	Mn (mg)
acorn flour, full-fat	142	1.7	8.6	1.7	0							12	31	.18	
1 oz	28	2.1	15.5	1.1				.00			202	29	.34	.174	
acorns															
raw	105	7.9	6.8	1.3	0	0	.03				0	12	18	.14	
1 oz	28	1.8	11.6	0.9			.03	0.5	.00		153	12	.22	.176	
dried	145	1.4	8.9	1.7	0	0	.04				0	15	23	.19	
1 oz	28	2.3	15.2	1.2			.04	0.7	.00		201	29	.30	.232	
almond butter	101	0.2	9.5	2.0	0	0	0	.10	.01	10	2ª	43	48	.49	.377
1 T	16	2.4	3.4	0.9		0	.02	0.5	.00	.04	121	84	.59	.144	
almond butter w/ honey & cinn	96	0.3	8.4	1.8	0	0	0	.10	.01	10	2ᵇ	43	48	.48	.373
1 T	16	2.5	4.3	0.8		0	.02	0.5	.00	.04	120	83	.59	.155	
almond meal, partially defatted	116	2.0	5.2	1.1	0	0		.48			2ᶜ	120			
1 oz	28	11.2	8.2	0.5		0	.09	1.8	.00		398	260	2.41		
almond paste	127	4.4	7.7	1.6	0	0	0	.21	.03	16	3	65	73	.73	.571
1 oz	28	3.4	12.4	0.7		0	.06	0.8	.00	.12	184	127	.90	.237	
almonds															
barbeque, Blue Diamond	166	0.4	14.9					.33			172	92			
1 oz	28	6.1	4.8				.03	1.0			116		1.07		
blanched, Blue Diamond	174	1.4	14.4	3.3	0			.20	.03	10	1	73	86	.94	
1 oz	28	6.0	5.2	1.2			.04	0.9			173	150	1.09	.310	
blanched, slivered, Blue Diamond	176	1.3	14.6	3.3	0			.19	.03	10	0	72	87	.98	
1 oz	28	6.0	5.1	1.2			.05	0.9			179	151	1.04	.330	
cheese, Blue Diamond	156	0.4	12.5					.34			95	102			
1 oz	28	6.1	7.4				.03	0.8			126		1.04		
chopped, Blue Diamond	175	1.3	14.6	3.3	0			.20	.03	10	0	77	85	.92	
1 oz	28	6.0	5.0	1.2			.06	0.9			174	151	1.15	.300	
dried	167	1.3	14.8	3.1	0	0	0	.22	.03	17	3	75	84	.83	.646
1 oz (24 nuts)	28	5.7	5.8	1.4	1.3	0	.06	1.0	.00	.13	208	148	1.04	.268	
dry roasted	167	0.9	14.7	3.1	0	0	0	.17	.02	18	3ᵈ	80	86	1.39	.561
1 oz	28	4.6	6.9	1.4		0	.04	0.8	.00	.07	219	156	1.08	.348	
dry roasted, hickory smoke, Blue Diamond—*1 oz*	166	0.7	15.0					.42			121	117			
	28	5.9	4.9				.08	1.4			109		1.31		
dry roasted, unsalted, Blue Diamond	168	0.5	15.1					.41			1	92			
1 oz	28	6.0	5.1				.01	0.6			175		1.35		
oil roasted	176	0.9	16.4	3.4	0	0	0	.28	.02	18	3ᵈ	66	86	1.39	.561
1 oz (22 nuts)	28	5.8	4.5	1.6		0	.04	1.0	.00	.07	194	155	1.09	.348	

ª If salt is added, Na is 126 mg.
ᵇ If salt is added, Na is 48 mg.
ᶜ If salt is added, Na is 209 mg.
ᵈ If salt is added, Na is 218 mg.

	KCAL	H₂O (g)	FAT (g)	PUFA (g)	CHOL (mg)	A (RE)	C (mg)	B-2 (mg)	B-6 (mg)	FOL (mcg)	Na (mg)	Ca (mg)	Mg (mg)	Zn (mg)	Mn (mg)
	WT (g)	PRO (g)	CHO (g)	SFA (g)	DFIB (g)	A (IU)	B-1 (mg)	NIA (mg)	B-12 (mcg)	PANT (mg)	K (mg)	P (mg)	Fe (mg)	Cu (mg)	
oil roasted, blanched, salted, Blue Diamond—*1 oz*	174	0.7	16.3					.40			70	65			
	28	5.9	4.0				.02	1.2			165		1.40		
oil roasted, salted, Blue Diamond	174	0.4	16.4					.31			95	87			
1 oz	28	6.3	3.4				.03	1.0			132		1.07		
onion garlic, Blue Diamond	160	0.5	13.6					.32			168	125			
1 oz	28	6.1	6.3				.02	0.9			112		1.12		
sliced, Blue Diamond	175	1.3	14.6	3.3	0			.20	.03	10	0	77	85	.92	
1 oz	28	6.0	5.0	1.2			.06	0.9			174	151	1.15	.300	
smokehouse, Blue Diamond	171	0.4	15.8					.34			123	101			
1 oz	28	5.9	4.3				.03	1.0			129		1.11		
toasted	167	0.7	14.4	3.0	0	0	0	.17	.02	18	3	80	87	1.40	.569
1 oz	28	5.8	6.5	1.4		0	.04	0.8	.00	.07	220	156	1.40	.349	
whole, Blue Diamond	173	1.4	14.2	3.3	0		0	.30	.03	10	0	79	86	.95	.510
1 oz	28	5.9	5.4	1.2	3.5	0	.05	0.9	.00	.00	179	149	1.30	.300	
beechnuts, dried	164	1.9	14.2	5.7	0							0	0		
1 oz	28	1.8	9.5	1.6				.00				0			
brazilnuts, dried	186	1.0	18.8	6.9	0		0	.04	.07	1	0	50	64	1.30	.220
1 oz (8 med nuts)	28	4.1	3.6	4.6			.28	0.5	.00	.07	170	170	1.0	.503	
butternuts, dried	174	1.0	16.2	12.1	0						0	15	67	.89	1.863
1 oz	28	7.1	3.4	0.4				.00			119	127	1.14	.128	
cashew butter	167	0.8	14.0	2.4	0	0	0	.05	.07	19	4ᵃ	12	73	1.47	
1 oz	28	5.0	7.8	2.8			.09	0.5	.00	.34	155	130	1.43	.622	
cashews															
dry roasted	163	0.5	13.2	2.2	0	0	0	.06	.07	20	4ᵇ	13	74	1.59	
1 oz	28	4.4	9.3	2.6		0	.06	0.4	.00	.35	160	139	1.70	.630	
honey roast, Planters	170		12.0	2							170				
1 oz	28	4.0	11.0	2							135				
oil roasted	163	1.1	13.7	2.3	0	0	0	.05	.07	19	5ᶜ	12	72	1.35	.229
1 oz (18 med nuts)	28	4.6	8.1	2.7		0	.12	0.5	.00	.34	151	121	1.16	.616	
cashew & peanuts, honey roast, Planters	170		12.0	3							170				
1 oz	28	5.0	9.0	2							150				
chestnuts, chinese															
raw	64	12.5	0.3	0.1	0	6	10	.05			1	5	24	.25	.455
1 oz	28	1.2	13.9	0.0		57	.05	0.2	.00		127	27	.40	.103	
boiled & steamed	44	17.5	0.2	0.1	0						1	3	16	.17	.312
1 oz	28	0.8	9.6	0.0				.00			87	19	.27	.071	
dried	103	2.5	0.5	0.1							2	8	39	.40	.739
1 oz	28	1.9	22.7	0.1				.00			206	44	.65	.168	
roasted	68	11.4	0.3	0.1	0	0		.03			1	5	26	.26	
1 oz	28	1.3	14.9	0.1		1	.04	0.4	.00		135	29	.43	.110	
chestnuts, european															
raw	60	13.8	0.6	0.3	0	1	12	.05	.11	18	1	8	9	.15	.270
1 oz (2 1/2 nuts)	28	0.7	12.9	0.1	2.8	8	.07	0.3	.00	.15	147	26	.29	.127	
boiled & steamed	37	19.4	0.4	0.2	0						8	13	15	.07	.243
1 oz	28	0.6	7.9	0.1				.00			203	28	.49	.134	
dried	106	2.7	1.3	0.5	0	0	4	.10	.19	31	11	19	21	.10	.369
1 oz	28	1.8	22.0	0.2		0	.08	0.2	.00	.26	280	50	.68	.185	
roasted	70	11.5	0.6	0.2	0	1	7	.05	.14	20	1	8	9	.16	.335
1 oz (3 1/2 nuts)	28	0.9	15.0	0.1	3.3	7	.07	0.4	.00	.16	168	30	.26	.144	
chestnuts, japanese															
raw	44	17.4	0.2	0.0	0	1	8	.05			4	9	14	.31	.452
1 oz	28	0.6	9.9	0.0		10	.10	0.4	.00	.06	94	21	.41	.160	
boiled & steamed	16	24.4	0.1	0.0	0	0	3	.02			1	3	5	.11	.164
1 oz	28	0.2	3.6	0.0		4	.04	0.2	.00	.02	34	7	.15	.058	
dried	102	2.8	0.4	0.1	0	2	17	.11			10	20	33	.73	1.054
1 oz	28	1.5	23.1	0.1		24	.23	1.0	.00	.14	218	48	.96	.373	
roasted	57	14.2	0.2	0.1	0	2	8	.07				10	18	.41	.586
1 oz	28	0.8	12.8	0.0		21	.13	0.2	.00			26	.60	.207	

ᵃ If salt is added, Na is 172 mg.
ᵇ If salt is added, Na is 179 mg.
ᶜ If salt is added, Na is 175 mg.

	KCAL	H₂O (g)	FAT (g)	PUFA (g)	CHOL (mg)	A (RE)	C (mg)	B-2 (mg)	B-6 (mg)	FOL (mcg)	Na (mg)	Ca (mg)	Mg (mg)	Zn (mg)	Mn (mg)
	WT (g)	PRO (g)	CHO (g)	SFA (g)	DFIB (g)	A (IU)	B-1 (mg)	NIA (mg)	B-12 (mcg)	PANT (mg)	K (mg)	P (mg)	Fe (mg)	Cu (mg)	
coconut															
raw	159	21.2	15.1	0.2	0	0	2	.01	.02	12	9	6	14	.50	.675
1 piece (2″x2″x1/2″)	45	1.5	6.9	13.4		0	.03	0.2	.00	.14	160	51	1.1	.196	
dried	187	0.9	18.3	0.2	0	0	0	.03	.09	3	11	7	26	.57	.780
1 oz	28	2.0	6.7	16.3		0	.02	0.2	.00	.23	154	59	.94	.226	
dried, sweetened, flaked, cnd	505	26.5	36.1	0.4	0	0	0	.02			23	16	56	1.81	2.475
4 oz	114	3.8	46.6	32.0		0	.03	0.3	.00		369	117	2.10	.351	
dried, sweetened, flaked, cnd, Angel Flake—*1/3 cup loosely packed*	114	5.5	8.7	0.1	0	0	0	.01	.04	1	6	3	11	.25	
	25	0.8	9.8	7.7		0	.01	0.1	.00	.10	67	25	.42	.099	
dried, sweetened, flaked, packaged	351	11.6	23.8	0.3	0	0	0	.02			189	10	36	1.30	1.767
1 cup	74	2.4	35.2	21.1		0	.02	0.2	.00		234	74	1.33	.223	
dried, sweetened, flaked, packaged, Angel Flake—*1/3 cup loosely pckd*	116	4.5	8.1	0.1	0	0	0	.01	.04	1	74	3	10	.24	
	25	0.8	10.4	7.2		0	.01	0.1	.00	.10	63	24	.39	.093	
dried, sweetened, shredded	466	11.7	33.0	0.4	0	0	1	.02			244	14	47	1.69	2.302
1 cup	93	2.7	44.3	29.3		0	.03	0.4	.00		313	99	1.78	.291	
dried, sweetened, shredded, Bakers	133	4.5	9.3	0.1	0	0	0	.01	.04	1	83	3	12	.27	
1/3 cup loosely packed	28	0.9	12.1	8.3		0	.01	0.1	.00	.11	72	27	.45	.107	
dried, toasted	168	0.3	13.4	0.1	0						11	8	26	.58	.795
1 oz	28	1.5	12.6	11.8					.00		157	60	.96	.231	
coconut cream, raw[a]	792	129.4	83.2	0.9	0	0	7	.00			10	26		2.30	3.130
1 cup	240	8.7	16.0	73.8		0	.07	2.1	.00		781	293	5.5	.907	
coconut cream, sweetened, cnd[a]	568	210.7	52.4	0.6	0			.12			149	4		1.77	2.412
1 cup	296	8.0	24.7	46.5			.07	.11	.00		299	66	1.50	.699	
coconut milk, raw[b]	552	162.3	57.2	0.6	0	0	7	.00			37	39	89	1.61	2.198
1 cup	240	5.5	13.3	50.7		0	.06	1.8	.00		630	240	3.94	.638	
coconut milk, cnd[b]	445	164.7	48.2	0.5	0		2				29	40	104	1.27	1.736
1 cup	226	4.6	6.3	42.7					.00		497	217	7.46	.504	
coconut water[c]	46	228.0	0.5	0.0	0	0	6	.14	.08		252	58	60	.25	
1 cup	240	1.7	8.9	0.4		0	.07	0.2	.00	.10	600	49	.69	.096	
filberts (hazelnuts)															
chopped/sliced, Blue Diamond	161	1.6	14.4				1	.07			0	38			
1 oz	28	5.5	5.1			0	.22	0.5			211		1.17		
dried	179	1.5	17.8	1.7	0	2	0	.03	.17	20	1	53	81	.68	.573
1 oz	28	3.7	4.4	1.3		19	.14	0.3	.00	.33	126	89	.93	.429	
dry roasted	188	0.5	18.8	1.8	0						1[d]	55	84	.71	.594
1 oz	28	2.8	5.1	1.4					.00		131	92	.96	.444	
oil roasted	187	0.3	18.1	1.7	0						1[e]	56	85	.71	.598
1 oz	28	4.1	5.4	1.3					.00		132	92	.97	.448	
roasted, salted, Blue Diamond	180	0.6	17.7				1	.06	.08	70	39	51	46	.65	
1 oz	28	4.6	3.9			0	.08	0.3	.00	.19	157	79	1.09	.350	
whole, Blue Diamond	166	1.7	15.4	2.8	0		0	.06			1	40			
1 oz	28	4.9	5.0	1.1		0	.17	0.3			167		1.18		
fruit 'n nut mix, Planters	150		9.0	2.0	0						90				
1 oz	28	5.0	13.0	2.0							200				
ginko nuts															
raw	52	15.7	0.5	0.2	0	16	4	.03			2	1	8	.10	.032
1 oz	28	1.2	10.7	0.1		158	.06	1.7	.00	.05	145	35	.28	.078	
cnd	32	20.7	0.5	0.2	0						87	1	5	.06	.019
1 oz (14 med nuts)	28	0.7	6.3	0.1					.00		51	15	.08	.047	
dried	99	3.5	0.6	0.2	0	31	8	.05			4	6	15	.19	.062
1 oz	28	2.9	20.6	0.1		310	.12	3.3	.00	.38	283	76	.45	.152	
hickorynuts, dried	187	0.8	18.3	6.2	0						0	17	49	1.22	
1 oz	28	3.6	5.2	2.0					.00		124	95	.60	.210	
macadamia nuts															
dried	199	0.8	20.9	0.4	0	0		.03			1	20	33	.49	
1 oz	28	2.4	3.9	3.1		0	.10	0.6	.00		104	39	.68	.084	
dry roasted, Blue Diamond	193	0.4	21.2	1.0	0		0	.03	.05	70	117	21	28	.60	
1 oz	28	2.4	2.5	3.1		79	.09	0.7	.00	.08	76	122	.70	.400	

[a] Liquid expressed from grated coconut.
[b] Liquid expressed from grated coconut & water.
[c] Liquid from coconuts.
[d] If salt is added, Na is 218 mg.
[e] If salt is added, Na is 220 mg.

	KCAL / WT (g)	H₂O (g) / PRO (g)	FAT (g) / CHO (g)	PUFA (g) / SFA (g)	CHOL (mg) / DFIB (g)	A (RE) / A (IU)	C (mg) / B-1 (mg)	B-2 (mg) / NIA (mg)	B-6 (mg) / B-12 (mcg)	FOL (mcg) / PANT (mg)	Na (mg) / K (mg)	Ca (mg) / P (mg)	Mg (mg) / Fe (mg)	Zn (mg) / Cu (mg)	Mn (mg)
oil roasted	204	0.5	21.7	0.4	0	0	0	.03			2ᵃ	13	33	.31	
1 oz (10-12 nuts)	28	2.1	3.7	3.3		3	.06	0.6	.00		94	57	.51	.085	
mixed nuts															
dry roastedᵇ	169	0.5	14.6	3.1	0	0	0	.06	.08	14	3ᶜ	20	64	1.08	.550
1 oz	28	4.9	7.2	2.0		4	.06	1.3	.00	.34	169	124	1.05	.363	
oil roastedᵈ	175	0.6	16.0	3.8	0	1	0	.06	.07	24	3ᵉ	31	67	1.44	.537
1 oz	28	4.8	6.1	2.5		6	.14	1.4	.00	.35	165	132	.91	.472	
w/o peanuts, oil roastedᶠ	175	0.9	16.0	3.3	0	1	0	.14	.05	16	3ᵍ	30	71	1.32	.439
1 oz	28	4.4	6.3	2.6		6	.14	0.6	.00	.27	154	127	.73	.510	
nuts, wheat-base formulated															
flavoredʰ	184	0.6	17.7	6.9	0			.09			26	6			
1 oz	28	3.7	5.9	2.7			.11	0.4	.00		91	104	.74		
macadamia flavoredⁱ	176	0.9	16.1	6.2	0			.06			13	6			
1 oz	28	3.2	7.9	2.4			.06	0.3	.00		74	85	.57		
unflavoredʲ	177	0.7	16.4	6.5	0			.09			143	7			
1 oz	28	3.9	6.7	2.5			.09	0.4	.00		90	105	.68		
nut topping, Planters	180		16.0	5.0	0						0				
1 oz	28	5.0	9.0	2.0							170				
peanut butter, creamy/smooth	95	0.2	8.2	2.5	0	0	0	.02	.06	13	3ᵏ	5	28	.47	.257
1 T	16	4.6	2.5	1.4		0	.02	2.2	.00	.15	110	60	.29	.094	
Jif	186	0.4	15.7	4.8				.03			157	13	48	.74	
2 T	32	9.0	5.4	2.3			.02	3.8			234	99	.58	.192	
Skippy	95	0.2	8.2	2.5	0			.01			75	5	30	.50	
1 T	16	4.6	2.2	1.7			.01	1.9		.20	105	60	.30		
peanut butter, chunk style/crunchy	188	0.4	16.0	4.6	0	0	0	.04	.14	29	156ˡ	13	51	.89	.597
2 T	32	7.7	6.9	3.1	0.9	0	.04	4.4	.00	.31	239	101	.61	.165	
Jif	186	0.3	15.7					.03			131	12			
2 T	32	9.0	5.4				.02	3.8					.58		
Skippy	95	0.2	8.4	2.5	0			.01			65	5	25	.50	
1 T	16	4.5	2.1	1.7			.01	1.9		.20	100	60	.30		
peanut flour															
defatted	13	0.3	0.0	0.0	0			.02			1ᵐ	6	15	.20	.196
1 T	4	2.1	1.4	0.0			.03	1.1			52	30	.08	.072	
low-fat	120	2.2	6.1	1.9							0	36	13	1.68	1.185
1 oz	28	9.5	8.8	0.9							380		1.33	.571	
peanuts															
boiled	102	13.4	7.0	2.2	0	0	0	.02	.05	24	240	18	33	.59	.327
1/2 cup	32	4.3	6.8	1.0		0	.08	1.7	.00	.26	58	63	.32	.160	
dried	161	1.9	14.0	4.4	0	0	0	.04	.08	29	5	17	51	.93	.316
1 oz	28	7.3	4.6	1.9		0	.19	4.0	.00	.79	204	109	.92	.285	
dry roasted	164	0.4	13.9	4.4	0	0	0	.03	.07	41	228ⁿ	15	49	.93	.583
1 oz	28	6.6	6.0	1.9		0	.12	3.8	.00	.39	184	100	.63	.188	
dry roasted, lite, Planters	90		6.0	2.0	0						180				
2/3 oz	19	6.0	5.0	1.0							160				
honey roast, Planters	170		13.0	4.0							180				
1 oz	28	6.0	8.0	2.0							160				

ᵃ If salt is added, Na is 73 mg.
ᵇ Cashews, almonds, peanuts, filberts & pecans.
ᶜ If salt is added, Na is 187 mg.
ᵈ Cashews, peanuts, brazilnuts, filberts, almonds & pecans.
ᵉ If salt is added, Na is 183 mg.
ᶠ Cashews, almonds, brazilnuts, pecans & filberts.
ᵍ If salt is added, Na is 196 mg.
ʰ Hydrogenated soybean oil, wheat germ, sugar, sodium caseinate, soy protein, natural & artificial flavors, & artificial color.

ⁱ Hydrogenated soybean oil, wheat germ, sugar, wheat starch, sodium caseinate, soy protein, natural & artificial flavor.
ʲ Hydrogenated soybean oil, wheat germ, fructose, wheat starch, sodium caseinate, soy protein, & salt.
ᵏ If salt is added, Na is 131 mg.
ˡ If no salt is added, sodium is 5 mg.
ᵐ If salt is added, Na is 50 mg.
ⁿ If no salt is added, sodium is 2 mg.

	KCAL	H₂O (g)	FAT (g)	PUFA (g)	CHOL (mg)	A (RE)	C (mg)	B-2 (mg)	B-6 (mg)	FOL (mcg)	Na (mg)	Ca (mg)	Mg (mg)	Zn (mg)	Mn (mg)
	WT (g)	PRO (g)	CHO (g)	SFA (g)	DFIB (g)	A (IU)	B-1 (mg)	NIA (mg)	B-12 (mcg)	PANT (mg)	K (mg)	P (mg)	Fe (mg)	Cu (mg)	
oil roasted	165	0.6	14.0	4.4	0	0	0	.03	.11	30	4ᵃ	24	53	1.88	.351
1 oz	28	7.6	5.3	1.9		0	.08	4.2	.00	.59	200	144	.54	.362	
peanuts, spanish, oil-roasted	162	0.5	13.7	4.7	0	0	0	.02	.07	35	121ᵇ	28	47	.56	.659
1 oz	28	7.8	4.9	2.1		0	.09	4.2	.00	.39	217	108	.64	.185	
peanuts, valencia, oil-roasted	165	0.6	14.4	5.0	0	0	0	.04	.07	35	216ᵇ	15	45	.86	.482
1 oz	28	7.6	4.6	2.2		0	.03	4.0	.00	.39	171	89	.46	.235	
peanuts, virginia, oil-roasted	161	0.6	13.6	4.1	0	0	0	.03	.07	35	121ᵇ	24	53	1.85	.562
1 oz	28	8.3	4.5	1.8		0	.08	4.1	.00	.39	183	142	.47	.356	
pecans															
dried	190	1.4	19.2	4.8	0	4	1	.04	.05	11	0	10	36	1.55	1.280
1 oz (31 large nuts)	28	2.2	5.2	1.5		36	.24	0.3	.00	.49	111	83	.60	.336	
dry roasted	187	0.3	18.4	4.5	0			.03		12	0ᶜ	10	38	1.61	1.337
1 oz	28	2.3	6.3	1.5			.09		.00		105	86	.62	.351	
oil roasted	195	1.2	20.2	5.0	0						0ᵈ	10	37	1.56	1.294
1 oz (15 halves)	28	2.0	4.6	1.6					.00		102	84	.60	.340	
pilinuts, dried	204	0.8	22.6	2.2	0	1		.03			1	41			
1 oz (15 nuts)	28	3.1	1.1	8.9		12	.26	0.1	.00		144	163	1.00		
pine nuts, pignolia, dried	146	1.9	14.4	6.1	0			.05			1	7		1.21	
1 oz	28	6.8	4.0	2.2			.23	1.0	.00		170	144	2.61	.291	
pine nuts, pinyon, dried	161	1.7	17.3	7.3	0	1	1	.06			20	2	67	1.22	
1 oz	28	3.3	5.5	2.7		8	.35	1.2	.00		178	10	.87	.294	
pistachio nuts															
dried	164	1.1	13.7	2.1	0	7		.05		17	2	38	45	.38	.093
1 oz (47 nuts)	28	5.8	7.1	1.7		66	.23	0.3	.00		310	143	1.92	.338	
dry roasted	172	0.6	15.0	2.3	0			.07			2ᶜ	20	37	.39	.095
1 oz	28	4.2	7.8	1.9			.12	0.4	.00		275	135	.90	.344	
dry roasted, Blue Diamond	162	0.8	13.9	4.5	0		1	.09	.04	120	173	41	33	.31	
1 oz	28	5.8	6.0	2.3		0	.16	0.3	.00	.06	230	51	1.20	.100	
dry roasted, red	158	0.7	13.4		0		0	.09			254	29			
1 oz	28	5.7	6.5			0	.16	0.3			254		1.20		
pumpkin & squash seeds															
dried	154	2.0	13.0	5.9	0	11		.09			5	12	152	2.12	
1 oz (142 seeds)	28	7.0	5.1	2.5		108	.06	0.5	.00		229	333	4.25	.394	
roasted	148	2.0	12.0	5.5	0			.03			5ᵉ	12	152	2.11	
1 oz	28	9.4	3.8	2.3					.00		229	333	4.24	.393	
sesame butter (tahini)	89	0.5	8.1	3.5	0		0	.07			17	64	14	.69	
1 T	15	2.6	3.2	1.1			.18	0.8	.00		62	110	1.34	.242	
sesame nut mix															
dry roasted, Planters	160		12.0	4.0							330				
1 oz	28	5.0	8.0	3.0							140				
oil roasted, Planters	160		13.0	2.0							220				
1 oz	28	5.0	8.0	2.0							110				
sesame seeds															
kernels, driedᶠ	47	0.4	4.4	1.9	0	1		.01			3	10	28	.82	
1 T	8	2.1	0.8	0.6		5	.06	0.4	.00	.05	33	62	.62		
kernels, toastedᶠ	161	1.4	13.6	6.0	0			.13			11	37	98	2.90	
1 oz	28	4.8	7.4	1.9			.34	1.5	.00		115	220	2.21		
whole, dried	52	0.4	4.5	2.0	0	0	0	.02	.07	9	1	88	32	.70	.221
1 T	9	1.6	2.1	0.6		1	.07	0.4	.00	.01	42	57	1.31	.367	
whole, toasted & roasted	161	0.9	13.6	6.0	0						3	281	101	2.03	.709
1 oz	28	4.8	7.3	1.9					.00		135	181	4.19	.701	
soybean nuts															
dry roasted	387	0.7	18.6	10.5	0	2	4	.65	.19	176	2	232	196	4.10	1.878
1/2 cup	86	34.0	28.1	2.7		20	.37	0.9	.00	.41	1173	558	3.40	.928	
roasted	405	1.7	21.8	12.3	0	17	2	.13	.18	182	140	119	125	2.70	1.856
1/2 cup	86	30.3	28.9	3.2		172	.09	1.2	.00	.39	1264	312	3.35	.712	

ᵃ If salt is added, Na is 121 mg.
ᵇ If no salt is added, sodium is 2 mg.
ᶜ If salt is added, Na is 218 mg.
ᵈ If salt is added, Na is 212 mg.
ᵉ If salt is added, Na is 161 mg.
ᶠ Decorticated seeds.

	KCAL	H₂O (g)	FAT (g)	PUFA (g)	CHOL (mg)	A (RE)	C (mg)	B-2 (mg)	B-6 (mg)	FOL (mcg)	Na (mg)	Ca (mg)	Mg (mg)	Zn (mg)	Mn (mg)
	WT (g)	PRO (g)	CHO (g)	SFA (g)	DFIB (g)	A (IU)	B-1 (mg)	NIA (mg)	B-12 (mcg)	PANT (mg)	K (mg)	P (mg)	Fe (mg)	Cu (mg)	
roasted & toasted	129	1.2	6.8	3.6	0	6	.1	.04	.09	64	1ᵃ	39	49	1.03	
1 oz (95 nuts)	28	10.5	8.7	0.9		57	.03	0.5	.00	.13	417	103	1.26	.302	
sunflower seeds															
dried	162	1.5	14.1	9.3	0	1		.07			1	33	100	1.44	.574
1 oz	28	6.5	5.3	1.5		.14	.65	1.3	.00		196	200	1.92	.498	
dry roasted	165	0.3	14.1	9.3	0			.07			1ᵇ	20	37	1.50	.599
1 oz	28	5.5	6.8	1.5			.03	2.0	.00		241	328	1.1	.520	
oil roasted	175	0.7	16.3	10.8	0		0	.08		67	1ᶜ	16	36	1.48	.591
1 oz	28	6.1	4.2	1.7			.09	1.2	.00		137	323	1.90	.512	
tavern nuts, Planters	170		15.0	4.0							65				
1 oz	28	7.0	6.0	2.0							190				
walnuts, black, dried	172	1.2	16.1	10.6	0	8		.03			0	16	57	.97	1.213
1 oz	28	6.9	3.4	1.0		84	.06	0.2	.00		149	132	.87	.290	
walnuts, english/persian, dried	182	1.0	17.6	11.1	0	4	1	.04	.16	19	3	27	48	.78	.823
1 oz (14 halves)	28	4.1	5.2	1.6		35	.11	0.3	.00	.18	142	90	.69	.394	
watermelon seeds, dried	158	1.4	13.5	8.0	0	0	0	.04		16	28	15	146		
1 oz (95 large seeds)	28	8.1	4.4	2.8		0	.05	1.0	.00		184	215	2.1		

25. POULTRYᵈ
25.1. CHICKEN
25.1.1. BROILER/FRYER MEAT

	KCAL	H₂O (g)	FAT (g)	PUFA (g)	CHOL (mg)	A (RE)	C (mg)	B-2 (mg)	B-6 (mg)	FOL (mcg)	Na (mg)	Ca (mg)	Mg (mg)	Zn (mg)	Mn (mg)
dark meat w/ skin, fried	285	50.8	16.9	3.9	92	31	0	.24	.32	8	89	17	24	2.60	.039
3.5 oz	100	27.2	4.1	4.6		104	.10	6.8	.30	1.16	230	176	1.50	.088	
dark meat w/ skin, roasted	253	58.6	15.8	3.5	91	58	0	.21	.31	7	87	15	22	2.49	.021
3.5 oz	100	26.0	0.0	4.4	0.0	201	.07	6.4	.29	1.11	220	168	1.36	.077	
dark meat w/ skin, stewed	233	63.0	14.7	3.2	82	54	0	.18	.17	6	70	14	18	2.26	.019
3.5 oz	100	23.5	0.0	4.1	0.0	186	.05	4.5	.20	.77	166	133	1.31	.068	
dark meat w/o skin, fried	239	55.7	11.6	2.8	96	24	0	.25	.37	9	97	18	25	2.91	.033
3.5 oz	100	29.0	2.6	3.1		79	.09	7.1	.33	1.26	253	187	1.49	.089	
dark meat w/o skin, roasted	205	63.1	9.7	2.3	93	22	0	.23	.36	8	93	15	23	2.80	.017
3.5 oz	100	27.4	0.0	2.7	0.0	72	.07	6.6	.32	1.21	240	179	1.33	.080	
dark meat w/o skin, stewed	192	65.8	9.0	2.1	88	21	0	.20	.21	7	74	14	20	2.66	.021
3.5 oz	100	26.0	0.0	2.5	0.0	69	.06	4.7	.22	.89	181	143	1.36	.075	
light & dark meat w/ skin, fried, batter dipped—3.5 oz	289	49.4	17.4	4.1	87	28	0	.19	.31	8	292	21	21	1.67	.057
	100	22.5	9.4	4.6		93	.12	7.0	.28	.89	185	155	1.37	.072	
light & dark meat w/ skin, fried, flour coated—3.5 oz	269	52.4	14.9	3.4	90	27	0	.19	.41	6	84	17	25	2.04	.034
	100	28.6	3.2	4.1		89	.09	9.0	.31	1.08	234	191	1.38	.075	
light & dark meat w/ skin, roasted	239	59.5	13.6	3.0	88	47	0	.17	.40	5	82	15	23	1.94	.020
3.5 oz	100	27.3	0.0	3.8	0.0	161	.06	8.5	.30	1.03	223	182	1.26	.066	
light & dark meat w/ skin, stewed	219	63.9	12.6	2.7	78	42	0	.15	.22	5	67	13	19	1.76	.019
3.5 oz	100	24.7	0.0	3.5	0.0	146	.05	5.6	.20	.67	166	139	1.16	.057	
light & dark meat w/o skin, fried	219	57.3	9.1	2.2	94	18	0	.20	.48	7	91	17	27	2.24	.028
3.5 oz	100	30.6	1.7	2.5		59	.09	9.7	.34	1.17	257	205	1.35	.075	
light & dark meat w/o skin, roasted	190	63.8	7.4	1.7	89	16	0	.18	.47	6	86	15	25	2.10	.019
3.5 oz	100	28.9	0.0	2.0	0.0	53	.07	9.2	.33	1.10	243	195	1.21	.067	
light & dark meat w/o skin, stewed	177	66.8	6.7	1.5	83	15	0	.16	.26	6	70	14	21	1.99	.019
3.5 oz	100	27.3	0.0	1.8	0.0	50	.05	6.1	.22	.75	180	150	1.17	.061	
light meat w/ skin, fried	246	54.7	12.1	2.7	87	20	0	.13	.54	4	77	16	27	1.26	.026
3.5 oz	100	30.5	1.8	3.3		68	.08	12.0	.33	.97	239	213	1.21	.058	
light meat w/ skin, roasted	222	60.5	10.9	2.3	84	32	0	.12	.52	3	75	15	25	1.23	.021
3.5 oz	100	29.0	0.0	3.1	0.0	110	.06	11.1	.32	.93	227	200	1.14	.053	
light meat w/ skin, stewed	201	65.1	10.0	2.1	74	28	0	.11	.27	3	63	13	20	1.14	.018
3.5 oz	100	26.1	0.0	2.8	0.0	96	.04	6.9	.20	.54	167	146	.98	.044	
light meat w/o skin, fried	192	60.1	5.5	1.3	90	9	0	.13	.63	4	81	16	29	1.27	.020
3.5 oz	100	32.8	0.4	1.5		30	.07	13.4	.36	1.03	263	231	1.14	.054	
light meat w/o skin, roasted	173	64.8	4.5	1.0	85	9	0	.12	.60	4	77	15	27	1.23	.017
3.5 oz	100	30.9	0.0	1.3	0.0	29	.07	12.4	34	.97	247	216	1.06	.050	
light meat w/o skin, stewed	159	68.0	4.0	0.9	77	8	0	.12	.33	3	65	13	22	1.19	.018
3.5 oz	100	28.9	0.0	1.1	0.0	27	.04	7.8	.23	.57	180	159	.93	.044	

ᵃ If salt is added, Na is 46 mg.
ᵇ If salt is added, Na is 218 mg.
ᶜ If salt is added, Na is 169 mg.
ᵈ Gram weights are edible portions.

	KCAL	H₂O (g)	FAT (g)	PUFA (g)	CHOL (mg)	A (RE)	C (mg)	B-2 (mg)	B-6 (mg)	FOL (mcg)	Na (mg)	Ca (mg)	Mg (mg)	Zn (mg)	Mn (mg)
	WT (g)	PRO (g)	CHO (g)	SFA (g)	DFIB (g)	A (IU)	B-1 (mg)	NIA (mg)	B-12 (mcg)	PANT (mg)	K (mg)	P (mg)	Fe (mg)	Cu (mg)	

25.1.2. BROILER/FRYER PARTS

	KCAL	H₂O	FAT	PUFA	CHOL	A	C	B-2	B-6	FOL	Na	Ca	Mg	Zn	Mn
back w/ skin, fried	238	31.7	14.9	3.5	64	27	0	.17	.22	6	65	17	17	1.78	.036
1/2 back	72	20.0	6.7	4.0		88	.08	5.3	.20	.79	163	119	1.17	.066	
breast w/ skin, fried	218	55.5	8.7	1.9	88	15	0	.13	.57	4	75	16	29	1.07	.015
1/2 breast	98	31.2	1.6	2.4		49	.08	13.5	.34	.98	253	228	1.17	.056	
breast w/ skin, roasted	193	61.2	7.6	1.6	83	26	0	.12	.54	3	69	14	27	1.00	.018
1/2 breast	98	29.2	0.0	2.2	0.0	91	.07	12.5	.32	.92	240	210	1.04	.049	
breast w/ skin, stewed	202	72.8	8.2	1.7	83	26	0	.13	.32	3	68	14	24	1.06	.020
1/2 breast	110	30.1	0.0	2.3	0.0	90	.05	8.6	.23	.60	195	172	1.01	.048	
breast w/o skin, fried	161	51.8	4.1	0.9	78	6	0	.11	.55	4	68	14	27	.93	.018
1/2 breast	86	28.8	0.4	1.1		20	.07	12.7	.31	.89	237	212	.98	.046	
breast w/o skin, roasted	142	56.1	3.1	0.7	73	5	0	.10	.51	3	63	13	25	.86	.015
1/2 breast	86	26.7	0.0	0.9	0.0	18	.06	11.8	.29	.83	220	196	.89	.042	
breast w/o skin, stewed	144	64.9	2.9	0.6	73	6	0	.11	.32	3	59	12	22	.92	.017
1/2 breast	95	27.5	0.0	0.8	0.0	18	.04	8.0	.22	.54	178	157	.84	.041	
drumstick w/ skin, fried	120	27.8	6.7	1.6	44	12	0	.11	.17	4	44	6	11	1.42	.014
1 drumstick	49	13.2	0.8	1.8		41	.04	3.0	.16	.60	112	86	.66	.039	
drumstick w/ skin, roasted	112	32.6	5.8	1.3	48	15	0	.11	.18	4	47	6	12	1.49	.011
1 drumstick	52	14.1	0.0	1.6	0.0	52	.04	3.1	.17	.63	119	91	.69	.040	
drumstick w/ skin, stewed	116	37.1	6.1	1.4	48	15	0	.11	.11	4	43	7	11	1.51	.011
1 drumstick	57	14.4	0.0	1.7	0.0	52	.03	2.4	.12	.49	105	80	.76	.040	
drumstick w/o skin, stewed	76	29.4	2.5	0.6	41	8	0	.10	.17	4	42	5	11	1.40	.009
1 drumstick	44	12.5	0.0	0.7	0.0	26	.03	2.7	.15	.57	108	81	.57	.035	
leg w/ skin, fried	285	61.9	16.2	3.7	105	31	0	.26	.38	9	99	15	27	3.00	.036
1 leg	112	30.1	2.8	4.4		103	.10	7.3	.35	1.34	261	204	1.60	.095	
leg w/ skin, roasted	265	69.5	15.4	3.4	105	45	0	.24	.37	8	99	14	26	2.96	.024
1 leg	114	29.6	0.0	4.2	0.0	154	.08	7.1	.35	1.32	256	199	1.52	.088	
leg w/ skin, stewed	275	80.0	16.2	3.6	105	46	0	.24	.23	8	92	14	24	3.04	.024
1 leg	125	30.2	0.0	4.5	0.0	156	.07	5.7	.25	1.02	220	174	1.69	.089	
leg w/o skin, stewed	187	67.1	8.1	1.9	90	18	0	.22	.22	8	78	11	21	2.81	.019
1 leg	101	26.5	0.0	2.2	0.0	60	.06	4.8	.23	.92	192	151	1.41	.078	
neck w/ skin, fried	119	17.1	8.5	2.0	34	21	0	.09	.09	2	29	11	7	1.11	.019
1 neck	36	8.6	1.5	2.3		69	.03	1.9	.09	.35	65	48	.87	.047	
neck w/ skin, simmered	94	23.5	6.9	1.5	27	18	0	.09	.04	1	20	10	5	1.03	.017
1 neck	38	7.5	0.0	1.9	0.0	61	.02	1.3	.05	.20	41	46	.87	.037	
neck w/o skin, simmered	32	12.1	1.5	0.4	14	7	0	.05	.03	1	12	8	3	.68	.009
1 neck	18	4.4	0.0	0.4	0.0	22	.01	0.7	.03	.12	25	23	.47	.023	
thigh w/ skin, fried	162	33.6	9.3	2.1	60	18	0	.15	.21	5	55	8	5	1.56	.022
1 thigh	62	16.6	2.0	2.5		61	.06	4.3	.19	.74	147	116	.93	.055	
thigh w/ skin, roasted	153	36.8	9.6	2.1	58	30	0	.13	.19	4	52	8	14	1.46	.013
1 thigh	62	15.5	0.0	2.7	0.0	102	.04	3.9	.18	.69	137	108	.83	.048	
thigh w/ skin, stewed	158	42.9	10.0	2.2	57	30	0	.13	.12	4	49	8	13	1.53	.013
1 thigh	68	15.8	0.0	2.8	0.0	103	.04	3.3	.13	.53	115	94	.93	.048	
thigh w/o skin, roasted	109	32.7	5.7	1.3	49	10	0	.12	.18	4	46	6	12	1.34	.010
1 thigh	52	13.5	0.0	1.6	0.0	34	.04	3.4	.16	.62	124	95	.68	.042	
wing w/ skin, fried	103	15.6	7.1	1.6	26	12	0	.04	.13	1	25	5	6	.56	.009
1 wing	32	8.4	0.8	1.9		40	.02	2.1	.09	.28	57	48	.40	.020	
wing w/ skin, roasted	99	18.7	6.6	1.4	29	16	0	.04	.14	1	28	5	7	.62	.006
1 wing	34	9.1	0.0	1.9	0.0	54	.01	2.3	.10	.31	62	51	.43	.019	
wing w/ skin, stewed	100	24.9	6.7	1.4	28	16	0	.04	.09	1	27	5	6	.65	.007
1 wing	40	9.1	0.0	1.9	0.0	53	.02	1.8	.07	.20	56	48	.45	.018	

25.1.3. CAPON

	KCAL	H₂O	FAT	PUFA	CHOL	A	C	B-2	B-6	FOL	Na	Ca	Mg	Zn	Mn
meat w/ skin, roasted	229	58.7	11.7	2.5	86	20	0	.17	.43	6	49	14	24	1.74	.021
3.5 oz	100	29.0	0.0	3.3	0.0	68	.07	8.9	.33	1.10	255	246	1.49	.069	

25.1.4. ROASTER

	KCAL	H₂O	FAT	PUFA	CHOL	A	C	B-2	B-6	FOL	Na	Ca	Mg	Zn	Mn
dark meat w/o skin, roasted	178	67.1	8.8	2.0	75	16	0	.19	.31	7	95	11	20	2.13	.019
3.5 oz	100	23.3	0.0	2.4	0.0	54	.06	5.7	.27	1.03	224	171	1.33	.070	

	KCAL	H₂O (g)	FAT (g)	PUFA (g)	CHOL (mg)	A (RE)	C (mg)	B-2 (mg)	B-6 (mg)	FOL (mcg)	Na (mg)	Ca (mg)	Mg (mg)	Zn (mg)	Mn (mg)
	WT (g)	PRO (g)	CHO (g)	SFA (g)	DFIB (g)	A (IU)	B-1 (mg)	NIA (mg)	B-12 (mcg)	PANT (mg)	K (mg)	P (mg)	Fe (mg)	Cu (mg)	
light & dark meat w/ skin, roasted	223	62.1	13.4	2.9	76	25	0	.14	.35	5	73	12	20	1.45	.018
3.5 oz	100	24.0	0.0	3.7	0.0	83	.06	7.4	.27	.92	211	179	1.26	.058	
light & dark meat w/o skin, roasted	167	67.4	6.6	1.5	75	12	0	.15	.41	5	75	12	21	1.52	.017
3.5 oz	100	25.0	0.0	1.8	0.0	41	.06	7.9	.29	.97	229	192	1.21	.057	
light meat w/o skin, roasted	153	67.9	4.1	0.9	75	8	0	.09	.54	3	51	13	23	.78	.015
3.5 oz	100	27.1	0.0	1.1	0.0	25	.06	10.5	.31	.91	236	217	1.08	.042	

25.1.5. STEWER

	KCAL	H₂O (g)	FAT (g)	PUFA (g)	CHOL (mg)	A (RE)	C (mg)	B-2 (mg)	B-6 (mg)	FOL (mcg)	Na (mg)	Ca (mg)	Mg (mg)	Zn (mg)	Mn (mg)
dark meat w/o skin, stewed	258	55.1	15.3	3.7	95	43	0	.35	.24	8	95	12	22	3.12	.023
3.5 oz	100	28.1	0.0	4.1	0.0	145	.13	4.6	.25	1.01	204	187	1.64	.144	
light & dark meat w/ skin, stewed	285	53.1	18.9	4.2	79	39	0	.24	.25	5	73	13	20	1.77	.021
3.5 oz	100	26.9	0.0	5.1	0.0	191	.09	5.8	.23	.75	182	180	1.37	.100	
light & dark meat w/o skin, stewed	237	56.4	11.9	2.8	83	33	0	.28	.31	6	78	13	22	2.06	.022
3.5 oz	100	30.4	0.0	3.1	0.0	112	.11	6.4	.26	.86	202	204	1.43	.116	
light meat w/o skin, stewed	213	57.8	8.0	1.9	70	22	0	.20	.39	4	58	14	23	.83	.020
3.5 oz	100	33.0	0.0	2.0	0.0	73	.09	8.5	.27	.69	199	225	1.19	.085	

25.1.6. UNSPECIFIED TYPE

	KCAL	H₂O (g)	FAT (g)	PUFA (g)	CHOL (mg)	A (RE)	C (mg)	B-2 (mg)	B-6 (mg)	FOL (mcg)	Na (mg)	Ca (mg)	Mg (mg)	Zn (mg)	Mn (mg)
	WT (g)	PRO (g)	CHO (g)	SFA (g)	DFIB (g)	A (IU)	B-1 (mg)	NIA (mg)	B-12 (mcg)	PANT (mg)	K (mg)	P (mg)	Fe (mg)	Cu (mg)	
breast cutlets, breaded & fried, Perdue	240		11.0		45						513				
4 oz cutlet	119	19.0	17.0												
breast, fried, frzn, Am Hosp Co	241		14.7		85			.15			525	20			
3.5 oz	100	16.5	10.8			67	.12	10.5			201		1.23		
breast, fried, frzn, Banquet	238		12.0				0	.07			772	16			
1/5 of 22 oz pkg	125	17.0	14.0			46	.06	6.6			141		1.00		
breast, oven roasted, Louis Rich	39	20.0	1.8	0.4	14						165	10	6		.17
1 oz slice	28	5.4	0.2	0.7							59	70	.13		
cnd	309	101.7	18.3				6	.19				33			
5.5 oz can	156	33.9	0.0	5.8		360	.06	6.9			215	385	2.30		
cnd, chunks o' chicken, Swanson	104		3.0				0	.12			230	7			
2.5 oz	71	19.2	0.0			0	.02	4.9			132		.90		
cnd, chunk, white & dark, Swanson	108		5.6				0	.09			263	9			
2.5 oz	71	14.5	0.1			0	.01	3.5			108		.90		
cnd, chunk white, Swanson	97		3.7				0	.07			236	7			
2.5 oz	71	15.7	0.2			0	.01	4.7			124		.60		
cnd, chunk white in water, Swanson	82		1.3				0	.04			239	5			
2.5 oz	71	16.8	0.7			0	.01	5.0			89		.50		
cnd, mixin chicken, Swanson	124		7.6				0	.08			225	38			
2.5 oz	71	13.9	0.1			0	.01	2.9			105		1.10		
drum snackers, breaded & fried, frzn, Banquet—3 oz	202		10.0				1	.08			532	23			
	85	14.0	14.0			46	.05	4.9			157		.90		
fried, frzn, Banquet	325		19.0				0	.10			1201	21			
1/5 of 32 oz pkg	181	18.0	20.0			85	.08	4.8			144		1.00		
fried, hot & spicy, frzn, Banquet	325		19.0				0	.10			1201	21			
1/5 of 32 oz pkg	181	18.0	20.0			85	.08	4.8			144		1.00		
nuggets															
breaded & fried, frzn, Banquet	233		14.0				0	.04			573	9			
3 oz	85	14.0	14.0			0	.08	3.0			150		.90		
breaded & fried, Perdue	160		8.0		40						312				
4 nuggets	74	12.0	8.0												
hot & spicy, fried, frzn, Banquet	235		14.0												
3 oz	85	13.0	13.0												
w/ cheddar, fried, frzn, Banquet	275		19.0												
3 oz	85	14.0	11.0												
patties, breaded & fried, frzn, Banquet	225		14.0				0	.05			513	8			
3 oz	85	13.0	13.0			0	.08	3.2			150		.90		
roasted, Perdue	220		16.0		88						396				
4 oz	113	24.0	0.0												
roll, light meat	159	68.6	7.4	1.6	50			.13			584	43	19	.72	
3.5 oz	100	19.5	2.5	2.0			.07	5.3			228	157	.97	.041	

	KCAL	H₂O (g)	FAT (g)	PUFA (g)	CHOL (mg)	A (RE)	C (mg)	B-2 (mg)	B-6 (mg)	FOL (mcg)	Na (mg)	Ca (mg)	Mg (mg)	Zn (mg)	Mn (mg)
	WT (g)	PRO (g)	CHO (g)	SFA (g)	DFIB (g)	A (IU)	B-1 (mg)	NIA (mg)	B-12 (mcg)	PANT (mg)	K (mg)	P (mg)	Fe (mg)	Cu (mg)	
sticks, breaded & fried, frzn, Banquet	228		13.0				0	.04			566	8			
3 oz	85	14.0	15.0			0	.08	3.3			150		.90		
tenders, breaded & fried, Perdue	212		12.0		68						192				
4 oz	113	20.0	8.0												
thighs & drumsticks, fried, frzn, Am Hosp	190		9.3		74			.18			360	15			
Co—*3 oz*	85	13.6	12.9			74	.10	4.4			159		1.15		
thighs & drumsticks, fried, frzn, Banquet	277		16.0				0	.13			892	16			
5 oz	142	16.0	16.0			69	.06	6.5			136		1.00		
winglets, breaded & fried, frzn, Banquet	202		10.0				1	.08			532	23			
3 oz	85	14.0	14.0			46	.05	4.9			157		.90		

25.2. TURKEY

	KCAL	H₂O (g)	FAT (g)	PUFA (g)	CHOL (mg)	A (RE)	C (mg)	B-2 (mg)	B-6 (mg)	FOL (mcg)	Na (mg)	Ca (mg)	Mg (mg)	Zn (mg)	Mn (mg)
	WT (g)	PRO (g)	CHO (g)	SFA (g)	DFIB (g)	A (IU)	B-1 (mg)	NIA (mg)	B-12 (mcg)	PANT (mg)	K (mg)	P (mg)	Fe (mg)	Cu (mg)	
breast															
barbequed, Louis Rich	39	19.1	1.3								316	2	5	.31	
1 oz	28	6.2	0.5								82	76	.23		
ckd, Louis Rich	51	17.6	1.9	0.4	12						21	2	7	.63	
1 oz	28	8.5	0.0	0.6							87	61	.33		
fillet w/ cheese, Land O' Lakes	304		15.6	3.7	34			.31			834	125			
5 oz	142	25.3	15.6	4.5			.10	8.7			315	392	1.36		
hickory smoked, Louis Rich	35	19.8	1.1	0.3	13						227	1	7	.30	
1 oz	28	6.2	0.2	0.3							69	80	.20		
oven roasted, Louis Rich	30	20.8	0.7	0.1	10						222	1	7	.31	
1 oz	28	5.8	0.1	0.3							63	76	.13		
sliced, ckd, Louis Rich	44	18.1	0.9	0.1	13						29	1	8	.46	
1 oz	28	8.9	0.0	0.2							115	72	.29		
smoked, Louis Rich	34	20.0	1.0	0.3	11						260	2	7	.32	
1 oz	28	5.8	0.3	0.4							69	82	.25		
smoked, sliced, Louis Rich	21	15.4	0.3	0.1	7						194	1	5	.24	
1 slice	21	4.5	0.1	0.1							58	61	.10		
tenderloins, ckd, Louis Rich	41	18.5	0.5	0.1	9						22	1	8	.46	
1 oz	28	9.0	0.0	0.2							115	75	.33		
ckd, Louis Rich	57	17.5	2.9	0.6	18						20	3	6	.61	
1 oz slice	28	7.7	0.0	0.7							81	55	.29		
cnd	315	101.2	19.5					.22				16			
5.5 oz can	156	32.6	0.0	5.7		200	.03	7.3					2.20		
dark meat w/ skin, roasted	221	60.2	11.5	3.1	89	0	0	.24	.32	9	76	33	23	4.16	.023
3.5 oz	100	27.5	0.0	3.5	0.0	0	.06	3.5	.36	1.16	274	196	2.27	.150	
dark meat w/o skin, roasted	187	63.1	7.2	2.2	85	0	0	.25	.36	9	79	32	24	4.46	.023
3.5 oz	100	28.6	0.0	2.4	0.0	0	.06	3.6	.37	1.29	290	204	2.33	.160	
drumstick, ckd, Louis Rich	55	17.4	2.4	0.6	33						21	3	6	1.49	
1 oz	28	8.2	0.0	0.9							49	43	.67		
ground, ckd, Louis Rich	61	17.1	3.5	0.9	24						30	7	6	.97	
1 oz	28	7.3	0.0	1.2							76	56	.48		
ham, cured thigh meat	128	71.4	5.1	1.5	62			.25	.28		996	10	21	3.06	
3.5 oz	100	18.9	0.4	1.7			.05	3.5	2.27		325	191	2.76	.100	
chopped, Louis Rich	42	19.5	2.3	0.6	17						254	2	5	.73	
1 oz	28	5.1	0.2	0.7							70	56	.34		
Louis Rich	34	20.1	1.2	0.3	18						278	2	6	.76	
1 oz	28	5.5	0.3	0.5							78	81	.36		
w/ added water, Louis Rich	34	20.5	1.4	0.4	21						281	1	5	.68	
1 oz	28	5.0	0.2	0.5							76	83	.34		
light & dark meat, w/ skin, roasted	208	61.7	9.7	2.5	82	0	0	.18	.41	7	68	26	25	2.96	.021
3.5 oz	100	28.1	0.0	2.8	0.0	0	.06	5.1	.35	.86	280	203	1.79	.093	
light & dark meat, w/o skin, roasted	170	64.9	5.0	1.4	76	0	0	.18	.46	7	70	25	26	3.10	.021
3.5 oz	100	29.3	0.0	1.6	0.0	0	.06	5.4	.37	.94	298	213	1.78	.094	
light meat, w/ skin, roasted	197	62.8	8.3	2.0	76	0	0	.13	.47	6	63	21	26	2.04	.020
3.5 oz	100	28.6	0.0	2.3	0.0	0	.06	6.3	.35	.63	285	208	1.41	.048	
light meat, w/o skin, roasted	157	66.3	3.2	0.9	69	0	0	.13	.54	6	64	19	28	2.04	.020
3.5 oz	100	29.9	0.0	1.0	0.0	0	.06	6.8	.37	.68	305	219	1.35	.042	

	KCAL	H₂O (g)	FAT (g)	PUFA (g)	CHOL (mg)	A (RE)	C (mg)	B-2 (mg)	B-6 (mg)	FOL (mcg)	Na (mg)	Ca (mg)	Mg (mg)	Zn (mg)	Mn (mg)
	WT (g)	PRO (g)	CHO (g)	SFA (g)	DFIB (g)	A (IU)	B-1 (mg)	NIA (mg)	B-12 (mcg)	PANT (mg)	K (mg)	P (mg)	Fe (mg)	Cu (mg)	
loaf, breast meat	110	71.9	1.6	0.3	41	0	0	.11	.36		1431	7	20	1.13	
3.5 oz	100	22.5	0.0	0.5		0	.04	8.3	2.02	.59	278	229	.40	.053	
patties, breaded & fried	266	46.7	16.9					.18			752	13			
1 patty	94	13.2	14.8				.09	2.2			259	254	2.07		
roll															
light & dark meat	149	70.2	7.0	1.8	55			.28			586	32	18	2.00	
3.5 oz	100	18.1	2.1	2.0			.09	4.8			270	168	1.35	.077	
light meat	147	71.6	7.2	1.7	43			.23			489	40	16	1.56	
3.5 oz	100	18.7	0.5	2.0			.09	7.0			251	183	1.28	.044	
sausage															
breakfast, ckd, Louis Rich	60	17.2	3.9	1.2	23						200	6	6	.93	
1 oz	28	6.2	0.0	1.5							80	53	.53		
smoked, Louis Rich	55	18.3	3.8	1.1	19						230	5	5	.71	
1 oz	28	4.6	0.4	1.4							60	37	.40		
smoked, Louis Rich	33	20.3	1.1	0.3	12						279	2	6	.52	
1 oz	28	5.5	0.2	0.4							76	73	.14		
sticks, breaded & fried	357	63.2	21.6					.23			1073	18			
2 sticks	128	18.2	21.8				.13	2.7			333	300	2.82		
thigh, ckd, Louis Rich	64	16.6	3.7	0.6	28						21	3	6	1.22	
1 oz	28	7.7	0.0	0.9							55	52	.52		
wing drumettes, ckd, Louis Rich	52	17.8	2.2	0.4	20						20	3	5	.85	
1 oz	28	8.1	0.0	0.7							43	40	.27		
wings, ckd, Louis Rich	54	18.0	2.6	0.6	26						18	3	6	.93	
1 oz	28	7.5	0.0	0.8							38	34	.28		

25.3. OTHER POULTRY

	KCAL	H₂O	FAT	PUFA	CHOL	A (RE)	C	B-2	B-6	FOL	Na	Ca	Mg	Zn	Mn
duck w/ skin, roasted	337	51.8	28.4	3.7	84	63	0	.27	.18	6	59	11	16	1.86	
3.5 oz	100	19.0	0.0	9.7	0.0	210	.17	4.8	.30	1.10	204	156	2.70	.227	
duck w/o skin, roasted	201	64.2	11.2	1.4	89	23	0	.47	.25	10	65	12	20	2.60	
3.5 oz	100	23.5	0.0	4.2	0.0	77	.26	5.1	.40	1.50	252	203	2.70	.231	
goose w/ skin, roasted	305	52.0	21.9	2.5	91	21	0	.32	.37	2	70	13	22		
3.5 oz	100	25.2	0.0	6.9	0.0	70	.08	4.2			329	270	2.83	.264	
goose w/o skin, roasted	238	57.2	12.7	1.5	96		0	.39	.47		76	14	25		
3.5 oz	100	29.0	0.0	4.6	0.0		.09	4.1			388	309	2.87	.276	
guinea hen w/o skin, raw	110	74.4	2.5		63										
3.5 oz	100	20.6	0.0		0.0										
pheasant w/ skin, raw	152	68.9	5.2												
3.5 oz	100	24.7	0.0												
pheasant w/o skin, raw	133	72.8	3.6	0.6		49	6	.15	.74		37	13	20	.97	.017
3.5 oz	100	23.6	0.0	1.2	0.0	165	.08	6.8	.84	.96	262	230	1.15	.069	
quail w/o skin, raw	134	70.0	4.5	1.2		17	7	.29			51	13			
3.5 oz	100	21.8	0.0	1.3	0.0	57	.28	8.2			237	307	4.50	.594	
squab (pigeon) w/o skin, raw	142	72.8	7.5	1.6	90										
3.5 oz	100	17.5	0.0	2.0	0.0			6.9							

25.4. INTERNAL ORGANS

25.4.1. GIBLETS

	KCAL	H₂O	FAT	PUFA	CHOL	A (RE)	C	B-2	B-6	FOL	Na	Ca	Mg	Zn	Mn
chicken, fried	277	47.9	13.5	3.4	446	3579	9	1.52	.61	379	113	18	25	6.27	.222
3.5 oz	100	32.5	4.4	3.8		11929	.10	11.0	13.31	4.45	330	286	10.32	.422	
chicken, simmered	157	67.6	4.8	1.1	393	2229	8	.95	.34	376	58	12	20	4.57	.170
3.5 oz	100	25.9	1.0	1.5	0.0	7431	.09	4.1	10.14	2.96	158	229	6.44	.255	
pheasant, raw	139	71.4	4.9												
3.5 oz	100	20.8	1.6												
squab (pigeon), raw	154	69.8	7.2												
3.5 oz	100	19.8	1.2												
turkey, simmered	167	65.4	5.1	1.1	418	1795	2	.90	.33	345	59	13	17	3.68	.175
3.5 oz	100	26.6	2.1	1.5	0.0	6036	.05	4.5	24.03	3.46	200	204	6.71	.391	

	KCAL	H₂O (g)	FAT (g)	PUFA (g)	CHOL (mg)	A (RE)	C (mg)	B-2 (mg)	B-6 (mg)	FOL (mcg)	Na (mg)	Ca (mg)	Mg (mg)	Zn (mg)	Mn (mg)
	WT (g)	PRO (g)	CHO (g)	SFA (g)	DFIB (g)	A (IU)	B-1 (mg)	NIA (mg)	B-12 (mcg)	PANT (mg)	K (mg)	P (mg)	Fe (mg)	Cu (mg)	

25.4.2. GIZZARD

	KCAL	H₂O	FAT	PUFA	CHOL	A(RE)	C	B-2	B-6	FOL	Na	Ca	Mg	Zn	Mn
chicken, simmered	153	67.3	3.7	1.1	194	56	2	.24	.12	53	67	10	20	4.38	.062
3.5 oz	100	27.2	1.1	1.0	0.0	188	.03	4.0	1.94	.71	179	155	4.15	.110	
goose, raw	139	73.0	5.3												
3.5 oz	100	21.4	0.0												
turkey, simmered	163	65.4	3.9	1.1	232	55	2	.33	.12	52	54	15	19	4.16	.098
3.5 oz	100	29.4	0.6	1.1	0.0	185	.03	3.1	1.90	.85	211	128	5.44	.173	

25.4.3. HEART

	KCAL	H₂O	FAT	PUFA	CHOL	A(RE)	C	B-2	B-6	FOL	Na	Ca	Mg	Zn	Mn
chicken, simmered	185	64.9	7.9	2.3	242	9	2	.74	.32	80	48	19	20	7.30	.107
3.5 oz	100	26.5	0.1	2.3	0.0	28	.07	2.8	7.29	2.65	132	199	9.03	.502	
turkey, simmered	177	64.2	6.1	1.8	226	8	2	.88	.32	79	55	13	22	5.27	.092
3.5 oz	100	26.8	2.1	1.8	0.0	28	.07	3.3	7.15	2.72	183	205	6.89	.627	

25.4.4. LIVER

	KCAL	H₂O	FAT	PUFA	CHOL	A(RE)	C	B-2	B-6	FOL	Na	Ca	Mg	Zn	Mn
chicken, simmered	157	68.3	5.5	0.9	631	4913	16	1.75	.58	770	51	14	21	4.34	.297
3.5 oz	100	24.4	0.9	1.8	0.0	16375	.15	4.5	19.39	5.41	140	312	8.47	.370	
duck, raw	136	71.8	4.6	0.6		11946						11			
3.5 oz	100	18.7	3.5	1.4	0.0	39907			54.00			269	30.53	5.962	
goose, raw	125	67.5	4.0	0.2		8728		.84	.72		132	40	23		
3.3 oz	94	15.4	5.9	1.5	0.0	29138	.53	6.1			216	245		7.701	
turkey, simmered	169	65.6	6.0	1.1	626	3741	2	1.42	.52	666	64	11	15	3.09	.250
3.5 oz	100	24.0	3.4	1.9	0.0	12581	.05	5.9	47.50	5.96	194	272	7.80	.560	

26. SALAD DRESSINGS
26.1. LOW CALORIE SALAD DRESSINGS

	KCAL	H₂O	FAT	PUFA	CHOL	A(RE)	C	B-2	B-6	FOL	Na	Ca	Mg	Zn	Mn
french	22	11.3	0.9	0.5	1						128	2		.03	
1 T	16	0.0	3.5	0.1							13	2	.10		
italian	16	12.3	1.5	0.9	1						118	0			
1 T	15	0.0	0.7	0.2							2	1	.00		
italian, lite, from mix, Good Seasons	26	10.9	2.5	1.0	0		0	.00	.00		175	1	0	.01	
1 T	15	0.1	0.8	0.4		7	.00	0.0	.00	.00	6	1	.03	.001	
italian, lite, zesty, from mix, Good Seasons—*1 T*	26	11.1	2.6	1.0	0		0	.00	.00		132	2	0	.01	
	15	0.1	0.8	0.4		8	.00	0.0	.00	.00	7	2	.03	.002	
italian, no oil, from mix, Good Seasons	7	15.1	0.0	0.0	0		0	.00	.00		31	1	0	.01	
1 T	17	0.1	1.8	0.0		11	.00	0.0	.00	.00	6	1	.02	.002	
russian	23	10.6	0.7	0.4	1						141	3			
1 T	16	0.1	4.5	0.1							26	6	.10		
thousand island	24	10.6	1.6	1.0	2	49					153	2			
1 T	15	0.1	2.5	0.2							17	3	.10		

26.2. REGULAR SALAD DRESSINGS

	KCAL	H₂O	FAT	PUFA	CHOL	A(RE)	C	B-2	B-6	FOL	Na	Ca	Mg	Zn	Mn
bleu cheese & herbs, from mix, Good Seasons—*1 T*	85	5.5	9.2	3.4	0		0	.00	.00	0	156	3	0	.02	
	16	0.1	0.7	1.4		9	.00	0.0	.01	.01	4	3	.02	.002	
blue (bleu) cheese	77	4.9	8.0	4.3		32	0	.00				12			
1 T	15	0.7	1.1	1.5			.00	.00				11	.00		
buttermilk farm style, from mix, Good Seasons—*1 T*	58	7.9	5.8	2.9	5		0	.02	.01	1	138	16	2	.05	
	16	0.5	1.2	1.0		31	.01	0.0	.04	.05	24	14	.05	.001	
cheese garlic, from mix, Good Seasons	85	5.5	9.2	3.4	0		0	.00	.00	0	173	4	0	.02	
1 T	16	0.2	0.5	1.4		3	.00	0.0	.01	.01	5	3	.03	.002	
cheese italian, from mix, Good Seasons	84	5.5	9.1	3.4	0		0	.00	.00		134	6	0	.01	
1 T	16	0.1	0.6	1.4		2	.00	0.0	.00	.01	7	4	.02	.001	
ckd, homemade	25	11.1	1.5	0.3		66	0	.02			117	13			
1 T	16	0.7	2.4	0.5			.01	0.0			19	14	.10		
classic herb, from mix, Good Seasons	85	5.7	9.3	3.5	0		0	.00	.00		150	2	1	.01	
1 T	16	0.1	0.3	1.4		2	.00	0.0	.00		6	2	.05	.000	
french	67	5.9	6.4	3.4							214	2		.01	
1 T	16	0.1	2.7	1.5							12	2	.10		
french, Catalina, Kraft	65	5.2	5.5	3.3	0		1	.00	.01	5	185	2	0	.05	
1 T	15	0.1	3.7	0.8		53	.00	0.1	.00	.01	15	1	.10	.010	

	KCAL	H₂O (g)	FAT (g)	PUFA (g)	CHOL (mg)	A (RE)	C (mg)	B-2 (mg)	B-6 (mg)	FOL (mcg)	Na (mg)	Ca (mg)	Mg (mg)	Zn (mg)	Mn (mg)
	WT (g)	PRO (g)	CHO (g)	SFA (g)	DFIB (g)	A (IU)	B-1 (mg)	NIA (mg)	B-12 (mcg)	PANT (mg)	K (mg)	P (mg)	Fe (mg)	Cu (mg)	
french, homemade	88	3.4	9.8	4.7		72	0	.00			92	1			
1 T	14	0.0	0.5	1.8			.00	0.0			3	0	.00		
garlic & herbs, from mix, Good Seasons	84	5.5	9.1	3.4	0		8	.00			186	2	0	.01	
1 T	16	0.1	0.7	1.3		0	.00	0.0	.00		5	2	.03	.000	
italian	69	5.6	7.1	4.1							116	1		.02	
1 T	15	0.1	1.5	1.0							2	1	.00		
italian, from mix, Good Seasons	84	5.6	9.2	3.5	0			.00	.00		150	1	0	.01	
1 T	16	0.1	0.5	1.4		6	.00	0.0	.00		5	1	.03	.001	
italian, mild, from mix, Good Seasons	88	5.7	9.4	3.5	0			.00			198	2	0	.01	
1 T	17	0.0	1.2	1.4		10	.00	0.0	.00		5	1	.03	.001	
italian, zesty, from mix, Good Seasons	85	5.6	9.2	3.5	0		0	.00	.00		123	1	0	.01	
1 T	16	0.1	0.6	1.4		0	.00	0.0	.00	.00	6	2	.03	.002	
lemon & herbs, from mix, Good Seasons	85	5.6	9.2	3.5	0		0	.00		0	145	2	0	.00	
1 T	16	0.0	0.6	1.4		7	.00	0.0	.00	.00	3	1	.04	.000	
mayonnaise type	57	5.9	4.9	2.6	4	32					105	2	0		
1 T	15	0.1	3.5	0.7	0.0						1	4	.00		
russian	76	5.3	7.8	4.5		106	1	.01			133	3		.07	
1 T	15	0.2	1.6	1.1			.01	0.1			24	6	.10		
sesame seed	68	6.0	6.9	3.8	0						153				
1 T	15	0.5	1.3	0.9											
thousand island	59	7.2	5.6	3.1		50					109	2		.02	
1 T	16	0.1	2.4	0.9							18	3	.10		
vinegar & oil, homemade	72	7.6	8.0	3.9							0				
1 T	16	0.0	0.4	1.5							1				

27. SAUCES, CONDIMENTS & GRAVIES
27.1. SAUCES & CONDIMENTS

	KCAL	H₂O (g)	FAT (g)	PUFA (g)	CHOL (mg)	A (RE)	C (mg)	B-2 (mg)	B-6 (mg)	FOL (mcg)	Na (mg)	Ca (mg)	Mg (mg)	Zn (mg)	Mn (mg)
alfredo fetticini sce, cnd, Prego	251		12.6				1	.17			369	79			
5 oz	142	7.7	26.6			244	.21	1.3			39		1.90		
alfredo sce, refrig, Fresh Chef	298		26.8				1	.20			731	202			
4 oz	113	11.2	3.0			147	.02	0.1			32		1.30		
barbeque sce	12	12.6	0.3	0.1	0	14	1	.00	.01	0	127	3	1	.012	
1 T	16	0.3	2.0	0.0		136	.01	0.1	0	.00	27	3	.14	.006	
Am Hosp Co	18		0.6				4	.01			22	3			
1 oz	28	0.3	2.9			265	.02	0.3			73		.30		
Heinz chunky/reg	20		0.0								230				
1 T		0.0	5.0												
Heinz hickory smoke/hot/mushroom	20		0.0								220				
1 T		0.0	5.0												
Heinz onion	20		0.0								200				
1 T		0.0	5.0												
Kraft	42	21.5	0.6	0.3	0		0	.01	.01	0	508	8	3	.07	
2 T	33	0.3	9.2	0.1		323	.01	0.2	.00	.01	49	6	.30	.018	
Open Pit	22	11.5	0.2	0.1	0		0	.00	.00		249	1	1	.01	
1 T	18	0.1	5.4	0.0		12	.00	0.0	.00	.00	5	1	.03	.007	
Open Pit hickory smoke flavor	22	11.5	0.2	0.1	0		0	.00	.00		230	1	1	.01	
1 T	18	0.1	5.4	0.0		12	.00	0.0	.00	.00	5	1	.03	.007	
Open Pit hickory thick & tangy	23	11.3	0.2	0.1	0		0	.00	.00		231	1	1	.02	
1 T	18	0.1	5.6	0.0		23	.00	0.0	.00	.00	8	1	.05	.012	
Open Pit hot 'n tangy flavor	22	11.6	0.2	0.1	0		0	.00	.00		208	2	1	.02	
1 T	18	0.1	5.4	0.0		17	.00	0.0	.00	.00	7	2	.10	.006	
Open Pit mesquite 'n tangy	23	11.2	0.2	0.1	0		0	.00	.00		250	1	1	.02	
1 T	18	0.1	5.5	0.0		19	.00	0.0	.00	.00	7	1	.04	.010	
Open Pit sweet 'n tangy flavor	24	10.9	0.2	0.1	0		0	.00	.01		188	3	1	.01	
1 T	18	0.1	6.0	0.0		27	.00	0.0	.00	.01	18	2	.11	.026	
Open Pit w/ onions	23	11.3	0.2	0.1	0		0	.00	.00		249	1	1	.01	
1 T	18	0.1	5.6	0.0		12	.00	0.0	.00	.00	9	2	.04	.007	
bearnaise sce, from mix	263	58.5	25.6	1.1	71						474				
1/4 pkt prep	96	3.1	6.6	15.7											

	KCAL	H₂O (g)	FAT (g)	PUFA (g)	CHOL (mg)	Vitamins A (RE)	C (mg)	B-2 (mg)	B-6 (mg)	FOL (mcg)	Minerals Na (mg)	Ca (mg)	Mg (mg)	Zn (mg)	Mn (mg)
	WT (g)	PRO (g)	CHO (g)	SFA (g)	DFIB (g)	A (IU)	B-1 (mg)	NIA (mg)	B-12 (mcg)	PANT (mg)	K (mg)	P (mg)	Fe (mg)	Cu (mg)	
bolognese sce, refrig, Fresh Chef	127		6.2				10	.13			585	44			
4 oz	113	5.6	12.2			591	.08	1.6			444		1.10		
brown sce, frzn, Am Hosp Co	14		0.1		0		0	.00			28	0			
1 oz	28	1.1	2.2			0	.00	1.1			82		.00		
burgundy wine sce, frzn, Am Hosp Co	10		0.3		0		1	.00			5	1			
1 oz	28	0.3	1.4			53	.01				5		.10		
catsup (ketchup), tomato	16	10.3	0.1				2	.01			156	3			
1 T	15	0.3	3.8			210	.01	0.2			54	8	.10		
Campbell's	17		0.0				2	.01			172	3			
1 T	14	0.2	4.0			129	.01	0.2			56		.10		
Heinz	18		0.0								180	7			
1 T		0.2	4.0										.10		
Heinz lite	8		0.0								110[a]	7			
1 T		0.2	2.0										.10		
chili hot dog sce, Wolf Brand	44	25.6	2.3					.10	.05	9	199	16	15	.00	.000
1/6 cup	35	1.5	4.4			410	.06	0.4	.00	.00	102	41	.57	.070	
chili sce, Heinz	17		0.0								191				
1 T		0.2	3.8												
clam sce, red, refrig, Fresh Chef	81		3.2				9	.07			556	32			
4 oz	113	3.6	9.3			641	.06	0.9			274		1.20		
clam sce, white, refrig, Fresh Chef	121		9.6				2	.03			639	28			
4 oz	113	4.5	4.2			112	.02	0.4			139		1.30		
curry sce, from mix	84	67.5	4.6	0.9	11						399	152			
1/4 pkt prep	85	3.4	8.1	1.9								88			
Heinz 57 sce	15		0.2								265				
1 T		0.4	2.7												
hollandaise sce															
w/ butterfat, from mix	47	42.7	3.9	0.2	10			.04			308	24			
1/4 pkt prep	51	0.9	2.7	2.3				0.0			25		.18		
w/ veg oil, from mix	264	58.5	25.6	1.1	71						425				
1/4 pkt prep	96	3.1	6.7	15.7											
horseradish, prep	6	13.1	0.0	0.0							14	9			
1 T	15	0.2	1.4	0.0							44	5	.10		
horseradish sce, Sauceworks	54	6.4	4.9	2.9	7		0	.00	.00	1	108	1	0	.02
1 T	14	0.1	2.3	0.8		9	.00	0.0	.02	.02	1	2	.03	.002	
madeira sce, frzn, Am Hosp Co	10		0.2		0		1	.01			14	2			
1 oz	28	0.2	2.0			52	.01	0.0			21		.11		
marinara (tomato based) sce	171	206.3	8.4	2.3	0	240	32	.15			1572	44	59	.67	
1 cup	250	4.0	25.4	1.2		2403	.11	4.0	.00		1061	88	2.00	.353	
marinara (tomato based) sce, Prego	91		4.4				18	.05			623	32			
4 oz	113	1.8	11.1			848	.08	1.3			435		.80		
mushroom sce, from mix	71	67.4	3.2	0.4	11						479				
1/4 pkt prep	83	3.5	7.4	1.7											
mustard															
brown	5	3.9	0.3								65	6			
1 t	5	0.3	0.3								7	7	.10		
mild, Heinz	5		0.2								71	4			
1 t		0.3	0.5										.10		
pourable, Heinz	5		0.2								71	4			
1 t		0.3	0.5										.10		
yellow	4	4.0	0.2								63	4			
1 t	5	0.2	0.3								7	4	.10		
pesto sce, refrig, Fresh Chef	155		14.6				0	.07			244	98			
1 oz	28	2.8	3.0			494	.01	0.2			27		.30		
pizza sce, cnd															
Contadina	40	50.0	2.0	0.7	0		2	.03			350	12			
1/4 cup	60	0.0	5.0	0.2		363	.01	0.7			230		.43		

[a] Low sodium lite ketchup contains 90 mg sodium per tablespoon.

	KCAL / WT (g)	H₂O (g) / PRO (g)	FAT (g) / CHO (g)	PUFA (g) / SFA (g)	CHOL (mg) / DFIB (g)	A (RE) / A (IU)	C (mg) / B-1 (mg)	B-2 (mg) / NIA (mg)	B-6 (mg) / B-12 (mcg)	FOL (mcg) / PANT (mg)	Na (mg) / K (mg)	Ca (mg) / P (mg)	Mg (mg) / Fe (mg)	Zn (mg) / Cu (mg)	Mn (mg)
w/ cheese, Contadina	40	48.4	2.0	0.7	0		2	.03			350	19			
1/4 cup	60	1.0	6.0	0.7		458	.01	0.7			302		.37		
w/ pepperoni, Contadina	45	48.2	2.0	0.7	0		2	.05			340	12			
1/4 cup	60	1.0	6.0	0.4		301	.01	0.8			300		.43		
w/ tomato chunks, Contadina	25	52.3	1.0				2	.01			280	19			
1/4 cup	60	0.0	5.0			292	.01	0.6			176		.38		
seafood cocktail sce, Heinz	20										160				
1 T															
sour cream sce, from mix	64	27.0	3.8	0.4	12			.09			126	68		.17	
1/4 pkt prep	39	2.4	5.7	2.0				0.1			92		.08	.010	
soy sce															
made w/ hydrolyzed veg protein	24	43.9	0.1	0.0	0	0	0	.06	.08	8	3300	3	3	.18	
1/4 cup	58	1.4	4.5	0.0		0	.02	1.6	.00	.16	88	54	.86	.056	
made w/ soy & wheat (shoyu)	30	41.2	0.1	0.0	0	0	0	.08	.10	9	3314[a]	10	20	.21	
1/4 cup	58	3.0	4.9	0.0		0	.03	1.9	.00	.19	104	64	1.17	.067	
made w/ soy (tamari)	35	38.3	0.1	0.0	0	0	0	.09	.12	11	3240	12	23	.25	
1/4 cup	58	6.1	3.2	0.0		0	.03	2.3	.00	.22	123	75	1.38	.078	
spaghetti sce, cnd	272	187.7	11.9	3.3	0	306	28	.15			1236	70	60	.53	
1 cup	249	4.5	39.7	1.7		3055	.14	3.7	.00		957	90	1.62	.284	
no salt added, Prego	100		5.7				19	.06			21	43			
4 oz	113	2.0	10.3			1642	.07	1.7			528		1.10		
Prego	136		5.6				18	.05			623	36			
4 oz	113	1.9	19.6			1085	.04	1.2			386	52	1.20		
w/ meat flavor, Prego	142		5.8				16	.06			625	34			
4 oz	113	2.4	20.2			1059	.04	1.5			419	51	1.20		
w/ mushrooms & chunk tomatoes, Prego—4 oz	121		4.7				15	.07			406	41			
	113	2.1	17.4			916	.04	0.9			325		1.20		
w/ mushrooms, Prego	133		5.2				17	.07			632	40			
4 oz	113	1.8	19.8			1002	.05	1.5			410	54	1.20		
spaghetti sce, frzn, w/ ground beef, Am Hosp Co—3 oz	116		4.6		15		18	.08			489	23			
	85	5.3	13.3			1061	.09	1.9			393		2.01		
stroganoff sce, from mix	73	62.4	2.9	0.1	10			.21			493	141		.30	
1/4 pkt prep	80	3.2	9.1	1.8				.23	0.2		181	81	.36	.023	
sweet & sour sce, from mix	55	44.7	0.0	0.0	0			.02			146	8		.02	
1/4 pkt prep	59	0.2	13.6	0.0				0			12		.30	.005	
sweet & sour sce, frzn, Am Hosp Co	19		0.0		0		0	.00			6	1			
1 oz	28	0.1	4.5			75	.00	0.1			27		.04		
tartar sce, Best Foods/Hellmann's	70	5.1	8.1	4.5	5						220				
1 T	14	0.2	0.1	1.2											
teriyaki sce															
bottled	15	12.2	0.0	0.0	0	0	0	.01	.02	4	690	4	11	.02	.000
1 T	18	1.1	2.9	0.0		0	.01	0.2	0	.04	41	28	.31	.018	
from mix	8	14.8	0.1	0.0	0						299	7			
1 T	18	0.3	1.7	0.0				0.1	0		14		.17		
tomato chili sce	16	10.2	0.0				2	.01			201	3			
1 T	15	0.4	3.7			210	.01	0.2			56	8	.10		
tomato sce, cnd	37	108.7	0.2	0.1	0	119	16	.07			738	17	23	.30	
1/2 cup	122	1.6	8.8	0.0		1195	.08	1.4	.00	.377	452	39	.94	.239	
al fresco, Prego	94		5.1				19	.05			610	36			
4 oz	113	1.7	10.3			1106	.07	1.4			426		.90		
Contadina	45	112.0	1.0	0.0	0		15	.05			520	15	19		
1/2 cup	124	2.0	9.0	0.0		750	.05	1.2			467	35	.92		
italian style, Contadina	40	113.0	1.0	0.0	0		1	.04			760	16			
1/2 cup	124	1.0	8.0	0.0		688	.02	1.3				468	.57		
no salt added, Heinz	32	91.3	0.1				5	.04			17	14	14	.17	
3.5 oz	100	1.0	6.9			618	.05	0.9			299	20	.50	.150	

[a] Low sodium types contain 1933 mg sodium.

	KCAL	H₂O (g)	FAT (g)	PUFA (g)	CHOL (mg)	A (RE)	C (mg)	Vitamins B-2 (mg)	B-6 (mg)	FOL (mcg)	Na (mg)	Ca (mg)	Minerals Mg (mg)	Zn (mg)	Mn (mg)
	WT (g)	PRO (g)	CHO (g)	SFA (g)	DFIB (g)	A (IU)	B-1 (mg)	NIA (mg)	B-12 (mcg)	PANT (mg)	K (mg)	P (mg)	Fe (mg)	Cu (mg)	
refrig, Fresh Chef	151		10.3				17	.09			686	73			
4 oz	113	3.1	11.6			681	.08	1.1			468		1.10		
spanish style	40	108.7	0.3	0.1	0	120	11	.08			576	20			
1/2 cup	122	1.8	8.8	0.0		1202	.09	1.6	.00				4.25		
w/ herbs & cheese	72	101.8	2.4	1.0		120	12	.15				45			
1/2 cup	122	2.6	12.5	0.8		1205	.09	1.5	.00				1.06		
w/ mushrooms	42	107.3	0.2	0.1	0	116	15	.13			552	16	23	.25	
1/2 cup	122	1.8	8.0	0.0		1165	.09	1.5	.00		464	39	1.1	.243	
w/ mushrooms al fresco, Prego	95		4.9				18	.06			545	25			
4 oz	113	1.6	11.2			802	.06	1.1			438		.50		
w/ red peppers, al fresco, Prego	92		5.0				21	.06			531	22			
4 oz	113	1.5	10.2			968	.07	1.1			423		.80		
w/ onions	52	105.0	0.2	0.1	0	104	16	.16			672	20	23	.28	
1/2 cup	122	1.9	12.1	0.0		1038	.09	1.5	.00		504	48	1.13	.221	
w/ onions, green peppers & celery	50	107.7	0.9	0.4	0	99	16	.15			672		16		
1/2 cup	122	1.2	10.7	0.2		988	.08	1.3	.00				.92		
w/ tomato tidbits	39	108.7	0.5	0.2	0	98	26	.12			18	13	24	.23	.268
1/2 cup	122	1.6	8.7	0.1		977	.09	1.4	.00		455	51	.83	.017	
tomato sce, frzn, Am Hosp Co	14		0.4		1		4	.01			19	2			
1 oz	28	0.6	1.9			248	.02	1.1			90		.20		
veloutee sce, frzn, Am Hosp Co[a]	19		0.5		1		0	.01			23	32			
1 oz	28	1.0	2.8			15	.01	0.0			11		.05		
white cream (bechamel) sce, cnd,	59		4.5				0	.02			265	14			
Campbell's—*2 oz*	57	0.7	3.8			0	.01	0.1			32		.20		
white sce, from mix	151	134.4	8.4	1.1	22			.28	.04		498	265	165	.34	
1/4 pkt prep	165	6.4	13.4	4.0			.05	0.3			277	160	.17	.028	
worcestershire sce, Heinz	11										234				
1 T															

27.2. GRAVIES

	KCAL	H₂O (g)	FAT (g)	PUFA (g)	CHOL (mg)	A (RE)	C (mg)	B-2 (mg)	B-6 (mg)	FOL (mcg)	Na (mg)	Ca (mg)	Mg (mg)	Zn (mg)	Mn (mg)
au jus															
cnd	24		0.3	0.0	0	0	1	.09			839	6			
1/2 can	149	1.8	3.7	0.1		0	.03	1.3			115	47	.89		
cnd, Franco-American	9		0.1				0	.01			325	3			
2 oz	57	0.6	1.5			0	.00	0.6			42		.20		
from mix	9	120.0	0.4	0.0	0.4						289	5		.02	
1/2 cup	123	0.3	1.2	0.2									.005		
beef															
cnd	77	127.3	3.4	0.1	4	0	0	.05	.01	0	73	8		1.45	.291
1/2 can	145	5.4	7.0	1.7		0	.04	0.9			118	43	1.02	.145	
cnd, Franco-American	25		1.0				0	.02			303	3			
2 oz	57	1.2	2.8			2	.01	0.5			37		.20		
brown															
cnd, Campbell's	47		3.2				0	.01			302	3			
2 oz	57	1.2	3.3			0	.00	0.2			26		.20		
cnd, homestyle, Heinz	29		2.3				1	.02			130	0			
2 oz	57	0.7	1.4			0	.00	0.1			80		.05		
from mix	4	136.8	0.1	0.0	0			.00			66	4	0	.01	.005
1/2 cup	138	0.1	0.8	0.0			.00	0.0			3	2	.01	.000	
from mix, Pillsbury	16		0.2				0	.00			328	4			
1/4 cup		0.4	3.2			0	.01	0.0			5	3	.04		
w/ onions, cnd, Franco-American	25		0.9				0	.00			332	6			
2 oz	57	0.4	3.7			0	.01	0.1			25		.20		
chicken															
cnd	118	127.1	8.5	2.2	3		0	.06	.01		859	30		1.19	.300
1/2 can	149	2.8	8.1	2.1		550	.02	0.6			162	43	.70	.149	

[a] Chicken-based sauce.

	KCAL / WT (g)	H₂O (g) / PRO (g)	FAT (g) / CHO (g)	PUFA (g) / SFA (g)	CHOL (mg) / DFIB (g)	A (RE) / A (IU)	C (mg) / B-1 (mg)	B-2 (mg) / NIA (mg)	B-6 (mg) / B-12 (mcg)	FOL (mcg) / PANT (mg)	Na (mg) / K (mg)	Ca (mg) / P (mg)	Mg (mg) / Fe (mg)	Zn (mg) / Cu (mg)	Mn (mg)
cnd, Campbell's	41		2.4				0	.02			245	18			
2 oz	57	1.3	3.5			149	.01	0.3			31		.20		
cnd, Franco-American	42		3.2				0	.01			305	11			
2 oz	57	0.8	2.6			217	.01	0.2			26		.20		
cnd, homestyle, Heinz	35		2.3				1	.02			110	0			
2 oz	57	0.7	2.9			0	.00	0.1			50		.04		
from mix	41	118.7	0.9	0.2	1			.07			566	19		.16	
1/2 cup	130	1.3	7.1	0.2										.013	
from mix, Pillsbury	25		1.0								230				
1/4 cup		1.0	4.0								20				
chicken giblet, cnd, Franco-American	29		1.5				0	.02			301	4			
2 oz	57	1.1	2.7			0	.00	0.2			15		.30		
homestyle, from mix, Pillsbury	16		0.1				0	.00			339	3			
1/4 cup		0.5	3.1			0	.01	0.0			5	3	.03		
mushroom															
cnd	75	132.6	4.0	1.5	0	0	0	.09	.03	0	849	11		1.04	.445
1/2 can	149	1.9	8.1	0.6		0	.05	1.0			158	22	.98	.149	
cnd, Franco-American	21		0.8				0	.03			287	3			
2 oz	57	0.6	2.7			0	.01	0.3			31		.20		
cnd, homestyle, Heinz	21		0.9				1	.02			200	0			
2 oz	57	0.4	2.7			0	.00	0.1			70		.10		
from mix	35	118.6	0.4	0.0	0						701	24		.16	
1/2 cup	129	1.0	6.9	0.2										.057	
onion															
cnd, homestyle, Heinz	24		0.8				1	.02			150	0			
2 oz	57	0.4	3.6			0	.00	0.1			80		.11		
from mix	40	118.8	0.3	0.0	0						518	34		.10	
1/2 cup	130	1.1	8.4	0.2										.021	
pork															
cnd, Franco-American	38		2.7				0	.01			346	3			
2 oz	57	0.7	2.8			17	.01	0.2			28		.20		
cnd, homestyle, Heinz	31		1.7				1	.02			130	0			
2 oz	57	1.8	2.1			0	.00	0.1			80		.06		
from mix	38	118.7	0.9	0.1	1			.03			617	16			
1/2 cup	129	0.9	6.7	0.4											
turkey															
cnd	76	132.0	3.1	0.7	3	0	0	.12			868	6			
1/2 can	149	3.9	7.6	0.9		0	.03	1.9			154	60	1.04		
cnd, Franco-American	26		1.2				0	.02			280	4			
2 oz	57	1.0	3.0			8	.01	0.5			36		.20		
cnd, homestyle, Heinz	29		2.0				1	.02			140	0			
2 oz	57	0.8	2.1			0	.00	0.0			90		.07		
from mix	43	118.8	0.9	0.2	1			.05			749	25			
1/2 cup	130	1.4	7.5	0.3											

28. SOUPS
28.1. CANNED, CONDENSED

	KCAL / WT (g)	H₂O (g) / PRO (g)	FAT (g) / CHO (g)	PUFA (g) / SFA (g)	CHOL (mg) / DFIB (g)	A (RE) / A (IU)	C (mg) / B-1 (mg)	B-2 (mg) / NIA (mg)	B-6 (mg) / B-12 (mcg)	FOL (mcg) / PANT (mg)	Na (mg) / K (mg)	Ca (mg) / P (mg)	Mg (mg) / Fe (mg)	Zn (mg) / Cu (mg)	Mn (mg)
celery, crm of	219	259.1	13.6	6.1	34	75	1	.12	.03	6	2308	98	16	.37	.610
1 can	305	4.0	21.5	3.4		745	.07	0.8			299	91	1.52	.345	
cheese	311	198.3	21.0	0.6	59	218	0	.27	.05		1920	284	8	1.29	.514
1 can	257	10.9	21.1	13.3		2177	.03	0.8	0		308	272	1.49	.257	
chicken, crm of	283	249.3	17.9	3.6	24	136	0	.15	.04	4	2397	83	6	1.53	.915
1 can	305	8.3	22.5	5.1		1362	.07	2.0			212	92	1.47	.305	
mushroom, crm of	313	247.7	23.1	10.8	3	0	3	.20	.03		2469	78	11	1.44	.610
1 can	305	4.9	22.6	6.3		0	.07	2.0			203	102	1.28	.305	
tomato	208	247.8	4.7	2.3	0	169	162	.12	.28	36	2120	32	18	.59	.610
1 can	305	5.0	40.3	0.9		1693	.21	3.4	.00		641	82	4.27	.610	

	KCAL / WT (g)	H₂O (g) / PRO (g)	FAT (g) / CHO (g)	PUFA (g) / SFA (g)	CHOL (mg) / DFIB (g)	A (RE) / A (IU)	C (mg) / B-1 (mg)	B-2 (mg) / NIA (mg)	B-6 (mg) / B-12 (mcg)	FOL (mcg) / PANT (mg)	Na (mg) / K (mg)	Ca (mg) / P (mg)	Mg (mg) / Fe (mg)	Zn (mg) / Cu (mg)	Mn (mg)

28.2. CANNED, CONDENSED, PREPARED WITH MILK

	KCAL/WT	H₂O/PRO	FAT/CHO	PUFA/SFA	CHOL/DFIB	A(RE)/A(IU)	C/B-1	B-2/NIA	B-6/B-12	FOL/PANT	Na/K	Ca/P	Mg/Fe	Zn/Cu	Mn
asparagus, crm of	161	213.3	8.2	2.2	22	83	4	.28	.06		1041	175	20	.92	.379
1 cup	248	6.3	16.4	3.3		599	.10	0.9			359	153	.87	.139	
celery, crm of	165	214.4	9.7	2.7	32	68	1	.25	.06	9	1010	186	22	.20	.253
1 cup	248	5.7	14.5	4.0		461	.07	0.4			309	151	.69	.154	
cheese	230	206.9	14.6	0.4	48	147	1	.33	.08		1020	288	20	.69	.259
1 cup	251	9.5	16.2	9.1		1243	.06	0.5	.44		340	250	.81	.141	
chicken, crm of	191	210.4	11.5	1.6	27	94	1	.26	.07	8	1046	180	18	.68	.379
1 cup	248	7.5	15.0	4.6		715	.07	0.9			273	152	.67	.139	
clam chowder, new england	163	211.4	6.6	1.1	22	40	4	.24	.13	10	992	187	23	.80	.253
1 cup	248	9.5	16.6	3.0		164	.07	1.0	10.25		300	157	1.48	1.39	
mushroom, crm of	203	209.7	13.6	4.6	20	38	2	.28	.06		1076	178	20	.64	.253
1 cup	248	6.1	15.0	5.1		154	.08	0.9			270	156	.59	.139	
onion, crm of	[a]		9.4	1.6	32	68	2	.27				180			
1 cup	248	6.8	[a]	4.0		451	.10	0.6					.69		
oyster stew	134	217.9	7.9	0.3	32	45	4	.23	.06		1040	167	21	10.33	.370
1 cup	245	6.1	9.8	5.1		225	.07	0.3	2.63		235	162	1.04	1.605	
pea, green	239	197.9	7.0	0.5	18	58	3	.27	.10	8	1048	173	55	1.76	.660
1 cup	254	12.6	32.2	4.0		356	.16	1.3	.44		377	238	2.01	.391	
potato, crm of[b]	148	215.0	6.5	0.6	22	67	1	.24	.09	9	1060	166	17	.68	.379
1 cup	248	5.8	17.2	3.8		443	.08	0.6			323	160	.54	.263	
shrimp, crm of	165	214.3	9.3	0.4	35	54	1	.23			1036	164		.80	
1 cup	248	6.8	13.9	5.8		313	.06	0.5					.59		
tomato	160	210.0	6.0	1.1	17	108	68	.25	.16	21	932	159	23	.29	.251
1 cup	248	6.1	22.3	2.9		849	.13	1.5	.44		450	148	1.82	.263	
tomato bisque	198	204.6	6.6	1.2	22	110	7	.27	.14		1108	186	25	.63	.259
1 cup	251	6.3	29.4	3.1		879	.11	1.3	.44		604	174	.88	.141	

28.3. CANNED, CONDENSED, PREPARED WITH WATER

	KCAL/WT	H₂O/PRO	FAT/CHO	PUFA/SFA	CHOL/DFIB	A(RE)/A(IU)	C/B-1	B-2/NIA	B-6/B-12	FOL/PANT	Na/K	Ca/P	Mg/Fe	Zn/Cu	Mn
asparagus, crm of	87	224.0	4.1	1.8	5	44	3	.08	.01		981	29	4	.88	.376
1 cup	244	2.3	10.7	1.0		445	.05	0.8			173	39	.80	.124	
bean w/ bacon	173	212.9	5.9	1.8	3	89	2	.03	.04	32	952	81	44	1.03	.670
1 cup	253	7.9	22.8	1.5		889	.09	0.6			403	132	2.05	.402	
bean w/ franks	187	207.6	7.0	1.6	12	87	1	.07	.13		1092	86	49	1.18	.788
1 cup	250	10.0	22.0	2.1		869	.11	1.0			477	166	2.34	.395	
beef broth & barley, Campbell's	59		1.0				0	.01			871	12			
1 cup	233	2.5	10.0			1568	.02	0.5			87		.60		
beef, Campbell's	71		1.5				2	.05			837	13			
1 cup	233	4.9	9.6			1443	.03	1.0			123		.80		
beef mushroom	[a]		3.0	0.1	7	0	5	.06				5			
1 cup	244	5.8	[a]	1.5		0	.04	0.9					.88		
beef noodle	84	224.5	3.1	0.5	5	63	0	.06	.04	4	952	15	6	1.54	.273
1 cup	244	4.8	9.0	1.2		629	.07	1.1	.20		99	46	1.10	.139	
beef noodle homestyle, Campbell's	80		2.9				2	.10			798	9			
1 cup	233	6.1	7.3			104	.08	1.7			138		1.00		
black bean	116	215.6	1.5	0.5	0	49	1	.05	.09	25	1198	45	42	1.41	.642
1 cup	247	5.6	19.8	0.4		506	.08	0.5	.02	.20	273	107	2.16	.385	
broccoli, creamy, Campbell's	69		2.7				11	.05			667	25			
1 cup	233	1.5	9.5			493	.02	0.5			45		.40		
cauliflower, creamy, Campbell's	97		5.9				6	.06			825	19			
1 cup	233	1.5	9.6			0	.03	0.6			63		.50		
celery, crm of	90	225.1	5.6	2.5	15	31	0	.05	.01	2	949	40	6	.15	.251
1 cup	244	1.7	8.8	1.4		306	.03	0.3			123	37	.62	.142	
cheese	155	217.7	10.5	0.3	30	109	0	.14	.03		959	142	4	.64	.257
1 cup	247	5.4	10.5	6.7		1088	.02	0.4	.00		154	136	.75	.128	
chicken alphabet, Campbell's	71		2.0				1	.08			802	13			
1 cup	233	3.6	9.6			1109	.06	1.4			89		.80		
chicken & dumplings	97	221.2	5.5	1.3	34	52	0	.07	.04		861	15	4	.37	.489
1 cup	241	5.6	6.0	1.3		518	.02	1.8	.16		116	61	.62	.123	

[a] Kcal & cho not available. [b] Includes vichyssoise.

Food	KCAL / WT (g)	H₂O (g) / PRO (g)	FAT (g) / CHO (g)	PUFA (g) / SFA (g)	CHOL (mg) / DFIB (g)	A (RE) / A (IU)	C (mg) / B-1 (mg)	B-2 (mg) / NIA (mg)	B-6 (mg) / B-12 (mcg)	FOL (mcg) / PANT (mg)	Na (mg) / K (mg)	Ca (mg) / P (mg)	Mg (mg) / Fe (mg)	Zn (mg) / Cu (mg)	Mn (mg)
chicken & stars, Campbell's	55		1.8				0	.06			875	13			
1 cup	233	3.1	6.5			409	.06	1.1			64		.60		
chicken & wild rice, Golden Classic, Campbell's—1 cup	76		2.3				2	.08			738	19			
	233	3.5	10.3			921	.04	1.2			119		.40		
chicken barley, Campbell's	70		1.8				0	.03			846	12			
1 cup	233	3.6	9.8			1170	.01	1.0			86		.70		
chicken broth	39	234.1	1.4	0.3	1	0	0	.07	.02		776	9	2	.25	.249
1 cup	244	4.9	0.9	0.4		0	.01	3.3	.24		210	73	.51	.124	
chicken broth & noodles, Campbell's	58		1.8				0	.08			864	11			
1 cup	233	2.4	7.9			903	.10	1.4			72		.80		
chicken broth & rice, Campbell's	44		1.0				0	.02			857	7			
1 cup	233	1.4	7.4			482	.01	0.7			50		.30		
chicken, crm of	116	221.1	7.4	1.5	10	56	0	.06	.02	2	986	34	3	.63	.376
1 cup	244	3.4	9.3	2.1		560	.03	0.8			87	38	.61	.124	
chicken, crm of, Special Request (1/3 less salt)—1 cup	106		6.6				0	.04			519	25			
	233	2.9	8.7			667	.02	0.6			69		.40		
chicken creamy mushroom, Campbell's	111		7.2				0	.07			925	21			
1 cup	233	3.4	8.2			790	.01	1.0			92		.50		
chicken gumbo	56	229.0	1.4	0.4	5	14	5	.05	.06		955	24	4	.38	.251
1 cup	244	2.6	8.4	0.3		136	.02	0.7			75	25	.89	.124	
chicken mushroom	ᵃ		9.2	2.3	10	112	0	.11				29			
1 cup	244	4.4	ᵃ	2.4		1135	.02	1.6					.88		
chicken noodle	75	221.7	2.5	0.6	7	72	0	.06	.03	2	1107	17	5	.40	.289
1 cup	241	4.0	9.4	0.7		711	.05	1.4			55	36	.78	.195	
chicken noodle, Special Request, (1/3 less salt)—1 cup	61		2.0				0	.07			578	11			
	233	3.1	7.6			332	.09	1.2			54		.80		
chicken noodle-o's, Campbell's	66		2.0				0	.07			816	13			
1 cup	233	3.3	8.6			458	.07	1.2			61		.80		
chicken rice	60	226.1	1.9	0.4	7	66	0	.02	.02	1	814	17	1	.26	.366
1 cup	241	3.5	7.2	0.5		660	.02	1.1			100	21	.75	.118	
chicken veg	74	223.3	2.8	0.6	10	266	1	.06	.05		944	18	6	.40	.366
1 cup	241	3.6	8.6	0.9		2656	.04	1.2			154	41	.87	.123	
chili beef	169	211.7	6.6	0.3	12	151	4	.08	.16		1035	43	30	1.40	1.050
1 cup	250	6.7	21.5	3.3		1510	.06	1.1	.32		525	148	2.13	.395	
clam chowder, manhattan	78	218.4	2.3	1.3	2	93	3	.05	.08	10	1808	34	10	.93	.376
1 cup	244	4.2	12.2	0.4		920	.06	1.3	2.19	.12	262	57	1.89	.148	
clam chowder, new england	95	220.9	2.9	1.1	5	1	2	.04	.08	4	914	43	7	.75	.251
1 cup	244	4.8	12.4	0.4		8	.02	1.0	8.01	.32	146	54	1.48	.124	
consomme w/ gelatin	29	231.9	0.0	0.0	0	0	1	.03	.02	3	637	8	0	.37	.366
1 cup	241	5.4	1.8	0.0		0	.02	0.7	.00		153	32	.53	.246	
corn potage, Campbell's	93		2.4				0	.05			898	27			
1 cup	233	1.8	16.1			584	.01	0.5			95		.30		
curly noodle w/ chicken, Campbell's	67		2.3				0	.07			955	13			
1 cup	233	3.2	8.3			1406	.07	1.2			72		.80		
garden veg, Campbell's	63		1.8				3	.04			801	18			
1 cup	233	2.2	9.7			3465	.03	0.8			167		.60		
gazpacho, Campbell's	41		0.2				6	.03			585	27			
1 cup	233	0.9	9.0			2089	.03	0.5			219		.70		
meatball alphabet, Campbell's	98		3.8				2	.07			909	16			
1 cup	233	4.8	11.0			1856	.06	1.4			163		1.00		
minestrone	83	220.1	2.5	1.1		234	1	.04	.10	16	911	34	7	.74	.366
1 cup	241	4.3	11.2	0.5		2337	.05	0.9	.00		312	56	.92	.123	
mushroom barley	ᵃ		2.3	0.7	0	20	0	.09		0		13			
1 cup	244	1.9	ᵃ	0.4		198	.02	0.9					.50		
mushroom, crm of	129	220.4	9.0	4.2	2	0	1	.09	.06		1031	46	5	.59	.251
1 cup	244	2.3	9.3	2.4		0	.05	0.7	.05	.29	101	50	.51	.124	
mushroom, golden, Campbell's	73		3.0				1	.09			865	8			
1 cup	233	2.6	9.3			832	.03	1.1			128		.50		
mushroom w/ beef stock	85	224.7	4.0	0.8	7	126	1	.10	.04	9	970	10	9	1.38	.376
1 cup	244	3.2	9.3	1.6		1255	.03	1.2	.00		158	36	.84	.251	
nacho cheese, Campbell's	105		7.2				5	.06			754	78			
1 cup	233	3.8	6.4			1354	.01	0.3			56		.60		

ᵃ Kcal & cho not available.

	KCAL / WT (g)	H₂O (g) / PRO (g)	FAT (g) / CHO (g)	PUFA (g) / SFA (g)	CHOL (mg) / DFIB (g)	A (RE) / A (IU)	C (mg) / B-1 (mg)	B-2 (mg) / NIA (mg)	B-6 (mg) / B-12 (mcg)	FOL (mcg) / PANT (mg)	Na (mg) / K (mg)	Ca (mg) / P (mg)	Mg (mg) / Fe (mg)	Zn (mg) / Cu (mg)	Mn (mg)
noodles & ground beef, Campbell's	90		3.5				0	.10			825	18			
1 cup	233	4.7	9.9			1242	.10	1.6			115		1.10		
onion	57	224.3	1.7	0.7	0	0	1	.02	.05	15	1053	26	2	.61	.246
1 cup	241	3.8	8.2	0.3		0	.03	0.6	.00		69	11	.67	.123	
onion, crm of	ᵃ		5.3	1.5	15	30	1	.08				34			
1 cup	244	2.8	ᵃ	1.5		296	.05	0.5					.63		
oyster stew	59	228.6	3.8	0.2	14	7	3	.04	.01		980	22	5	10.29	.366
1 cup	241	2.1	4.1	2.5		71	.02	0.2	2.19		49	48	.98	1.593	
pea, green	164	208.7	2.9	0.4	0	20	2	.07	.05	2	987	27	39	1.71	.658
1 cup	250	8.6	26.5	1.4		202	.11	1.2	.00		190	124	1.95	.378	
pea, split w/ ham	189	206.9	4.4	0.6	8	44	1	.08	.07	3	1008	22	48	1.32	.670
1 cup	253	10.3	28.0	1.8		444	.15	1.5			399	213	2.28	.369	
pea, split w/ ham & bacon, Campbell's	159		3.2				1	.06			793	18			
1 cup	233	8.8	24.0			491	.13	1.2			319		1.80		
pepperpot	103	217.3	4.6	0.4	10	87	1	.05	.06		970	23	5	1.22	.612
1 cup	241	6.4	9.4	2.1		865	.05	1.2	.17		152	42	.89	.123	
potato, crm ofᵇ	73	225.7	2.4	0.4	5	29	0	.04	.04	3	1000	20	1	.63	.379
1 cup	244	1.7	11.5	1.2		288	.03	0.5			137	46	.48	.251	
scotch broth	80	221.1	2.6	0.6	5	218	1	.05	.07		1012	15	4	1.59	.366
1 cup	241	5.0	9.5	1.1		2180	.02	1.2	.27		159	55	.83	.246	
shrimp, crm of	90	225.0	5.2	0.2	17	16	0	.03			976	18		.75	
1 cup	244	2.8	8.2	3.2		158	.02	0.4					.53		
spinach, creamy, Campbell's	88		5.5				2	.07			679	33			
1 cup	233	1.4	8.3			1639	.01	0.3			61		.60		
stockpot	100	223.7	3.9	1.8	5	398	2	.05	.09		1048	22	4	1.16	.257
1 cup	247	4.9	11.5	0.9		3980	.04	1.2	.00		238	54	.87	.128	
tomato	86	220.5	1.9	1.0	0	69	67	.05	.11	15	872	13	8	.24	.253
1 cup	244	2.1	16.6	0.4		688	.09	1.4	.00		263	34	1.76	.251	
tomato beef w/ noodle	140	211.5	4.3	0.7	5	53	0	.09	.09		917	18	8	.75	.251
1 cup	244	4.5	21.2	1.6		533	.08	1.9	.19		221	56	1.12	.124	
tomato bisque	123	215.3	2.5	1.1	4	72	6	.07	.09		1048	40	9	.59	.257
1 cup	247	2.3	23.7	0.5		721	.07	1.1	.00		417	60	.82	.128	
tomato, crm of, Campbell's	104		2.2				25	.04			823	14			
1 cup	233	1.3	19.9			596	.04	.09			249		1.10		
tomato garden, crispy, Campbell's	78		0.3				12	.05			492	34			
1 cup	233	1.4	17.4			3431	.05	1.2			406		.90		
tomato rice	120	217.6	2.7	1.4	2	76	15	.05	.08		815	23	5	.51	.385
1 cup	247	2.1	21.9	0.5		755	.06	1.1	.00		330	33	.79	.128	
tomato zesty, Campbell's	89		0.4				23	.06			769	36			
1 cup	233	2.4	19.0			1353	.04	1.2			269		.90		
tortellini & veg, Golden Classic, Campbell's—*1 cup*	79		2.1				4	.05			861	41			
	233	2.8	12.0			1736	.42	2.2			152		1.80		
turkey noodle	69	226.9	2.0	0.5	5	29	0	.06	.04		815	12	5	.58	.251
1 cup	244	3.9	8.6	0.6		292	.07	1.4			75	48	.94	.124	
turkey veg	74	223.9	3.0	0.7	2	244	0	.04	.05		905	17	4	.61	.246
1 cup	241	3.1	8.6	0.9		2445	.03	1.0	.17		175	40	.76	.123	
veg, Campbell's	79		1.5				5	.04			796	16			
1 cup	233	3.1	13.3			3186	.06	1.0			130		.70		
veg, Special Request (1/3 less salt)	79		1.5				4	.04			523	16			
1 cup	233	3.1	13.2			3091	.06	1.0			132		.70		
veg, old fashioned, Campbell's	60		1.6				3	.04			887	19			
1 cup	233	2.4	9.0			3094	.03	0.8			150		.60		
veg vegetarian	72	222.5	1.9	0.7	0	300	1	.05	.06	11	823	21	7	.46	.460
1 cup	241	2.1	12.0	0.3		3005	.05	0.9	.00		209	35	1.08	.123	
veg w/ beefᶜ	79	223.5	1.9	0.1	5	189	2	.05	.08	11	957	17	6	1.55	.315
1 cup	244	5.6	10.2	0.9		1891	.04	1.0	.31		173	40	1.11	.183	
veg w/ beef broth	81	220.5	1.9	0.8	2	209	2	.05	.06		810	18	7	.80	.337
1 cup	241	3.0	13.1	0.4		2091	.05	1.0	.00		192	39	.97	.154	
won ton, Campbell's	41		1.0				0	.06			878	10			
1 cup	233	3.4	4.5			25	.04	1.4			76		.60		

ᵃ Kcal & cho not available.
ᵇ Includes vichyssoise.
ᶜ Includes beef; beef veg; barley & veg beef.

	KCAL	H₂O (g)	FAT (g)	PUFA (g)	CHOL (mg)	A (RE)	C (mg)	B-2 (mg)	B-6 (mg)	FOL (mcg)	Na (mg)	Ca (mg)	Mg (mg)	Zn (mg)	Mn (mg)
	WT (g)	PRO (g)	CHO (g)	SFA (g)	DFIB (g)	A (IU)	B-1 (mg)	NIA (mg)	B-12 (mcg)	PANT (mg)	K (mg)	P (mg)	Fe (mg)	Cu (mg)	

28.4. CANNED, READY-TO-SERVE

	KCAL	H₂O	FAT	PUFA	CHOL	A(RE)	C	B-2	B-6	FOL	Na	Ca	Mg	Zn	Mn
bean, Campbell's	143		3.9				0	.02			822	72			
7.5 oz	213	6.3	20.7			830	.10	0.6			274		2.20		
bean w/ ham, chunky	231	191.1	8.5	1.0	22	395	4	.15			972	79			
1 cup	243	12.6	27.1	3.3		3950	.15	1.7					3.23		
beef barley, Progresso	200		9.0								1420				
10.5 oz	298	8.0	16.0								230				
beef broth/ bouillon	16	234.1	0.5	0.0	0	0	0	.05			782	15			
1 cup	240	2.7	0.1	0.3		0	.01	1.9			130	31	.41		
beef, chunky	171	200.0	5.1	0.2	14	261	7	.15	.13	13	867	31		2.64	2.40
1 cup	240	11.7	19.6	2.6		2611	.06	2.7	.61		336	120	2.32	.240	
beef minestrone, Progresso	150		4.0								1040				
9 oz	269	11.0	19.0								440				
beef noodle, Campbell's	73		2.6				0	.08			806	13			
7.27 oz	206	4.5	7.9			350	.10	1.3			79		1.10		
beef noodle, Home Cookin, Campbell's	144		3.8				8	.19			1129	34			
10.76 oz	305	16.9	10.4			2076	.07	4.0			351		3.00		
beef ravioli romano, chunky, Campbell's	231		8.3				13	.14			1047	64			
9.49 oz	269	8.1	30.9			6470	.11	3.1			412		3.10		
beef, Progresso	160		4.0								1390				
9.5 oz	269	12.0	19.0								416				
beef veg, Progresso	150		2.0								1140				
9.5 oz	269	12.0	19.0								450				
beef w/ mushroom, chunky, low Na, Campbell's—10.76 oz	204		6.6				11	.35			65	41			
	305	13.4	22.9			5949	.13	4.3			480		2.50		
beef w/ noodles, stroganoff style, chunky, Campbell's—10.76 oz	300		14.1				0	.31			1222	76			
	305	15.4	27.9			2985	.14	3.2			330		2.90		
chickarina w/ tiny meatballs, Progresso	90		6.0								1060				
9.5 oz	269	8.0	8.0								190				
chicken broth, low Na, Campbell's	19		0.6				1	.08			49	10			
7.27 oz	206	2.4	1.1			0	.02	2.3			99		.80		
chicken broth, Swanson	25		1.2				0	.07			895	9			
7.27 oz	206	2.6	0.8			0	.01	3.0			145		.60		
chicken, chunky	178	211.1	6.6	1.4	30	130	1	.17	.05	5	887	24		1.00	.251
1 cup	251	12.7	17.3	2.0		1299	.09	4.4	.25		176	113	1.73	.251	
chicken, crm of, Campbell's	85		5.5				0	.03			833	29			
7.27 oz	206	2.3	6.5			484	.02	0.5			65		.60		
chicken, hearty, Progresso	170		8.0								1010				
9.5 oz	269	10.0	12.0								180				
chicken, homestyle, Progresso	90		2.0								1190				
9.5 oz	269	9.0	8.0												
chicken minestrone, Progresso	150		6.0								1220				
9.5 oz	269	11.0	14.0								290				
chicken mushroom, creamy, chunky, Campbell's—9.38 oz	272		21.5				1	.22			1259	23			
	266	10.4	9.3			1212	.07	2.6			84		1.80		
chicken noodle, Campbell's	60		2.0				0	.07			870	14			
7.27 oz	206	3.3	7.1			340	.10	1.3			65		1.00		
chicken noodle, chunky	ᵃ		6.0	1.5	18	122	0	.17				24			
1 cup	240	12.7	ᵃ	1.4		1222	.07	4.3					1.44		
chicken noodle, Progresso	130		4.0								990				
9.5 oz	269	11.0	10.0								202				
chicken noodle w/ meatballs	99	225.0	3.6	0.8	10	233	8	.12			1039	30			
1 cup	248	8.1	8.4	1.1		2326	.12	2.15					1.74		
chicken rice, chunky	127	208.3	3.2	0.7	12	586	4	.10		4	888	35			
1 cup	240	12.3	13.0	1.0		5858	.02	4.1					1.87		
chicken rice, Campbell's	52		2.1				0	.01			810	12			
7.27 oz	206	2.9	5.5			556	.01	0.8			52		.50		
chicken rice w/ veg, Progresso	140		3.0								940				
9.5 oz	269	8.0	20.0								240				

ᵃ Kcal & cho not available.

	KCAL	H₂O (g)	FAT (g)	PUFA (g)	CHOL (mg)	A (RE)	C (mg)	B-2 (mg)	B-6 (mg)	FOL (mcg)	Na (mg)	Ca (mg)	Mg (mg)	Zn (mg)	Mn (mg)
	WT (g)	PRO (g)	CHO (g)	SFA (g)	DFIB (g)	A (IU)	B-1 (mg)	NIA (mg)	B-12 (mcg)	PANT (mg)	K (mg)	P (mg)	Fe (mg)	Cu (mg)	
chicken veg, chunky	167	200.3	4.8	1.0	17	599	6	.17			1068	25			
1 cup	240	12.3	18.9	1.4		5991	.04	3.3					1.47		
chicken veg, chunky, low Na, Campbell's	236		10.9				12	.41			91	44			
10.76 oz	305	15.4	19.0			7930	.20	5.6			449		1.70		
chicken veg, Progresso	130		5.0								760				
9.5 oz	269	9.0	20.0								450				
chicken veg w/ rice, Home Cookin, Campbell's—*10.76 oz*	131		2.7				9	.07			991	46			
	305	13.0	13.7			3752	.09	3.3			280		2.20		
chicken w/ noodles & mushrooms, chunky, Campbell's—*10.76 oz*	195		6.6				0	.24			1142	34			
	305	13.9	20.1			1406	.12	4.7			238		2.10		
chicken w/ noodles, Home Cookin, Campbell's—*10.76 oz*	129		3.4				4	.15			1126	38			
	305	13.8	10.9			3846	.08	4.2			183		2.00		
chicken w/ noodles, low Na, Campbell's	160		5.0				4	.35			82	32			
10.76 oz	305	13.7	15.1			2274	.17	5.5			254		2.10		
chicken w/ rice, chunky, Campbell's	142		4.3				5	.10			1072	37			
9.49 oz	269	11.1	14.8			6111	.01	3.2			229		1.30		
chili beef, chunky, Campbell's	256		5.7				6	.13			984	61			
9.74 oz	276	18.8	32.4			1433	.11	2.3			553		4.10		
clam chowder															
manhattan, Campbell's	67		1.8				4	.02			800	25			
7.27 oz	206	2.0	10.7			1756	.02	0.7			221		1.00		
manhattan, chunky	133	206.5	3.4	0.1	14	329	12	.06	.26	9	1000	67		1.68	.240
1 cup	240	7.3	18.8	2.1		3292	.06	1.8	7.92		384	84	2.64	.240	
manhattan, Progresso	140		3.0								1240				
9.5 oz	269	7.0	19.0								430				
new england, chunky, Campbell's	162		4.0				11	.11			1102	63			
10.76 oz	305	7.3	24.1			7084	.03	1.7			494		2.40		
crab	76	223.3	1.5	0.4	10	50	0	.07	.12		1234	65			
1 cup	244	5.5	10.3	0.4		505	.20	1.3	.20	.29	326	88	1.22		
escarole	27	240.3	1.8	0.4	2	217	5	.05			3865	32			
1 cup	248	1.5	1.8	0.5		2170	.08	2.3					.74		
escarole in chicken broth, Progresso	40		2.0								1050				
9.5 oz	269	2.0	3.0								250				
fisherman chowder, chunky, Campbell's	229		12.2				3	.04			1130	67			
9.49 oz	269	9.9	19.8			0	.02	1.3			240		1.60		
french onion, Progresso	120		8.0								1270				
9.5 oz	269	5.0	8.0								180				
gazpacho	57	228.8	2.2	1.3	0	20	3	.02	.15		1183	24			
1 cup	244	8.7	0.8	0.3		200	.05	0.9	.00	.17	224	37	.98		
ham & beans, Progresso	180		2.0								1050				
9.5 oz	269	11.0	30.0								590				
ham n' butter bean, chunky, Campbell's	273		9.8				6	.11			1180	41			
10.76 oz	305	12.7	33.6			3031	.14	2.1			617		2.80		
lentil, Home Cookin, Campbell's	164		1.0				5	.10			938	54			
10.76 oz	305	10.3	28.5			3900	.18	1.4			305		4.00		
lentil, Prego	165		1.5				6	.10			935	41			
9.49 oz	269	10.5	27.4			713	.08	1.4			386		4.20		
lentil, Progresso	170		2.0								1020				
9.5 oz	269	10.0	25.0								450				
lentil w/ ham	140	212.7	2.8	0.3	7	36	4	.11	.22	50	1318	42			
1 cup	248	9.3	20.2	1.1		360	.17	1.4	.30	.35	356	184	2.64		
macaroni & bean, Progresso	180		3.0								1290				
9.5 oz	269	6.0	30.0								536				
mediterranean veg, chunky, Campbell's	150		4.3				9	.13			1007	71			
9.49 oz	269	4.5	23.3			6019	.08	1.8			367		2.10		
minestrone, Campbell's	70		2.1				2	.04			724	25			
7.27 oz	206	2.9	9.8			3171	.06	0.9			165		1.00		
minestrone, chunky	127	208.1	2.8	0.3	5	435	5	.12			864	61			
1 cup	240	5.1	20.7	1.5		4352	.06	1.2					1.77		
minestrone, Progresso	150		3.0								850				
9.5 oz	269	6.0	22.0								410				

	KCAL	H₂O (g)	FAT (g)	PUFA (g)	CHOL (mg)	A (RE)	C (mg)	B-2 (mg)	B-6 (mg)	FOL (mcg)	Na (mg)	Ca (mg)	Mg (mg)	Zn (mg)	Mn (mg)
	WT (g)	PRO (g)	CHO (g)	SFA (g)	DFIB (g)	A (IU)	B-1 (mg)	NIA (mg)	B-12 (mcg)	PANT (mg)	K (mg)	P (mg)	Fe (mg)	Cu (mg)	
minestrone w/ italian sausage, Prego	169		8.4				6	.25			1170	48			
9.49 oz	269	6.5	16.9			5452	.14	1.7			322		1.50		
mushroom, crm of, Campbell's	99		6.7				0	.06			909	34			
7.27 oz	206	1.9	7.8			0	.02	0.5			85		.70		
mushroom, crm of, low Na, Campbell's	141		9.9				1	.18			33	50			
7.27 oz	206	2.3	10.8			0	.02	0.9			93		.80		
mushroom, creamy, chunky, Campbell's	226		18.5				2	.30			1134	26			
9.38 oz	266	4.4	10.6			0	.07	2.8			90				
onion, low Na, Campbell's	48		2.2				7	.14			32	22			
7.27 oz	206	1.5	5.3			0	.07	1.9			126		1.20		
pea, green, Campbell's	136		2.6				0	.03			782	19			
7.5 oz	213	6.8	21.3			46	.13	0.9			272		1.60		
pea, split															
low Na, Campbell's	159		2.9				4	.11			16	28			
7.27 oz	206	7.9	25.2			956	.13	1.6			322		1.90		
Progresso	190		2.0								1050				
9.5 oz	269	11.0	30.0								430				
w/ ham, chunky	184	194.3	4.0	0.6	7	487	7	.09		5	965	33			
1 cup	240	11.1	26.8	1.6		4871	.12	2.5					2.14		
w/ ham, Home Cookin, Campbell's	201		3.4				9	.13			1233	45			
10.76 oz	305	13.8	28.6			2332	.16	2.1			321		2.20		
w/ ham, Progresso	170		4.0								1030				
9.5 oz	269	10.0	26.0								380				
sirloin burger, chunky, Campbell's	191		7.3				9	.16			1105	34			
9.49 oz	269	11.0	20.3			5566	.05	3.0			412		2.30		
steak & potato, chunky, Campbell's	169		3.8				7	.19			990	15			
9.49 oz	269	12.7	21.0			57	.05	2.9			414		2.20		
tomato, Campbell's	106		1.7				30	.05			786	33			
7.27 oz	206	2.2	20.5			684	.05	1.0			283		.90		
tomato w/ macaroni shells, Progresso	130		2.0								1270				
10.5 oz	298	4.0	24.0												
tomato w/ tomato pieces, low Na,	124		3.5				30	.11			23	32			
Campbell's—*7.27 oz*	206	2.6	20.5			942	.10	2.1			429		1.20		
tomato w/ veg & macaroni, Progresso	120		2.0								1260				
9.5 oz	269	3.0	24.0								380				
tortellini, Progresso	80		3.0								1080				
9.5 oz	269	4.0	11.0								280				
tortilla, chunky, Campbell's	293		9.6				12	.11			1034	101			
10.76 oz	305	17.1	34.4			669	.05	2.6			390		3.50		
turkey, chunky	136	203.8	4.4	1.1	9	716	6	.11	.31	11	923	50		2.12	.236
1 cup	236	10.2	14.1	1.2		7156	.04	3.6	2.12		361	104	1.91	.236	
veg															
Campbell's	68		1.4				3	.04			724	19			
7.27 oz	206	2.4	11.3			3053	.04	0.9			141		1.00		
chunky	122	210.2	3.7	1.4	0	588	6	.07	.19	17	1010	56		3.12	.480
1 cup	240	3.5	19.0	0.6		5877	.07	1.2	.00		396	72	1.63	.240	
Progresso	130		5.0								1150				
9.5 oz	269	4.0	15.0								260				
veg w/ beef															
Campbell's	138		2.6				10	.13			1139	49			
10.76 oz	305	13.5	15.2			5113	.09	2.5			337		2.90		
chunky, Campbell's	153		4.1				8	.15			972	50			
9.49 oz	269	11.6	17.4			5705	.06	2.5			414		2.40		
chunky, low Na, Campbell's	156		4.8				13	.28			60	55			
10.76 oz	305	13.2	15.0			5603	.18	4.0			390		2.40		
won ton, imperial, chunky, Campbell's	121		3.1				9	.11			1062	39			
9.49 oz	269	7.3	15.7			633	.07	3.1			245		2.70		

	KCAL / WT (g)	H₂O (g) / PRO (g)	FAT (g) / CHO (g)	PUFA (g) / SFA (g)	CHOL (mg) / DFIB (g)	A (RE) / A (IU)	C (mg) / B-1 (mg)	B-2 (mg) / NIA (mg)	B-6 (mg) / B-12 (mcg)	FOL (mcg) / PANT (mg)	Na (mg) / K (mg)	Ca (mg) / P (mg)	Mg (mg) / Fe (mg)	Zn (mg) / Cu (mg)	Mn (mg)

28.5. DEHYDRATED

	KCAL / WT	H₂O / PRO	FAT / CHO	PUFA / SFA	CHOL / DFIB	A(RE) / A(IU)	C / B-1	B-2 / NIA	B-6 / B-12	FOL / PANT	Na / K	Ca / P	Mg / Fe	Zn / Cu	Mn
beef broth cube	6	0.1	0.1	0.0	0			.01			864		2	.01	.014
1 cube	4	0.6	0.6	0.1			.01	0.1			15	8	.08		
chicken broth cube	9	0.1	0.2	0.1	1			.02			1152		3	.01	.018
1 cube	5	0.7	1.1	0.1			.01	0.2			18	9	.09		
chicken noodle w/ meat, Campbell's	99		1.7				0	.12			805	9			
0.92 oz	26	5.7	15.0			0	.15	3.2			89		.90		
chicken rice, Campbell's	90		1.3				0	.18			800	7			
0.85 oz	24	3.7	15.8			0	.04	1.4			77		.10		
onion soup mix	115	1.43	2.3	0.3	2	1	1	.24		6	3493	55	25	.23	.248
1 pkt	39	4.5	20.9	0.5		8	.11	2.0			260	126	.58	.064	

28.6. DEHYDRATED, RECONSTITUTED SOUPS

	KCAL / WT	H₂O / PRO	FAT / CHO	PUFA / SFA	CHOL / DFIB	A(RE) / A(IU)	C / B-1	B-2 / NIA	B-6 / B-12	FOL / PANT	Na / K	Ca / P	Mg / Fe	Zn / Cu	Mn
asparagus, crm of	59	235.5	1.7	0.7	0						801				
1 cup	251	2.2	9.0	0.3											
bean w/ bacon	105	237.7	2.2	0.2	3						928				
1 cup	265	5.5	16.4	1.0							326				
beef broth/ bouillon	19	236.3	0.7	0.0	1			.02			1358	5	4		.037
1 cup	244	1.3	1.9	0.3		4	.01	0.4			36	26			
beef noodle[a]	41	239.2	0.8	0.2	2	1	1	.06	.04	2	1041	5	9	.10	
1 cup	251	2.2	6.0	0.3		9	.12	0.7			81	40	.33		
cauliflower	68	238.0	1.7	0.7	0						843				
1 cup	256	2.9	10.7	0.3											
celery, crm of	63	237.4	1.6	0.6	1						839				
1 cup	254	2.6	9.8	0.2											
chicken broth/bouillon[b]	21	236.2	1.1	0.4	1		0	.03			1484	15	4	.01	
1 cup	244	1.3	1.4	0.3		40	.01	0.2			25	13	.08		
chicken, crm of	107	237.5	5.3	0.4	3			.20			1184	76			
1 cup	261	1.8	13.4	3.4							215	96			
chicken noodle[c]	53	237.6	1.2	0.3	3	6	0	.06	.01	1	1284	32	7	.20	.024
1 cup	252	2.9	7.4	0.3		63	.07	0.9			31	32	.50	.035	
chicken rice	60	237.2	1.4	0.4	3						980	8			
1 cup	253	2.4	9.3	0.3				0.4			10	10			
chicken veg	49	237.2	0.8	0.2	3	1	1	.05	.09		808		21	.21	
1 cup	251	2.7	7.8	0.2		14	.07	0.7			68	32	.59	.025	
clam chowder															
manhattan	65		1.6	0.5	0						1336				
1 cup	250	2.1	10.9	0.3											
new england	95		3.7	1.2	1			.16			745	76			
1 cup	250	2.8	12.9	0.6							205	100			
consomme w/ gelatin	17	236.5	0.0	0.0	0						3299				
1 cup	249	2.2	2.1	0.0											
leek	71	235.6	2.1	0.1	3						966			.24	
1 cup	254	2.1	11.4	1.0										.036	
minestrone	79	232.8	1.7	0.1	3						1026				
1 cup	254	4.4	11.9	0.8											
mushroom	96	231.9	4.9	1.5	1	1		.11			1019	67		.09	
1 cup	253	2.2	11.1	0.8		7	.28	0.5		.24	199	77		.030	
onion[d]	28	236.9	0.6	0.1	0	0	0	.06		2	848	13	6	.06	.060
1 cup	246	1.1	5.1	0.1		2	.03	0.5			63	31	.14	.015	
oxtail	71	235.3	2.6	0.1	3						1210				
1 cup	253	2.8	9.0	1.3											
pea, green/split[e]	133	235.2	1.6	0.3	3	5	0	.15	.05	15	1220	22	46	.59	.268
1 cup	271	7.7	22.7	0.4		49	.22	1.3		.26	238	134	1.01	.192	
tomato[f]	102	237.7	2.4	0.2	1	82	5	.05	.10	7	943	54	15	.21	
1 cup	265	2.5	19.4	1.1		832	.06	0.8			295	66	.42	.093	
tomato veg[g]	55	236.6	0.9	0.1	0	20	6	.05			1146	8	20	.17	
1 cup	253	2.0	10.2	0.4		190	.06	0.8		.14	103	29	.63	.033	

[a] Includes beef & macaroni & beef-flavored noodle.
[b] Includes chicken consomme.
[c] Includes chicken broth w/ noodles.
[d] Includes french onion.
[e] Includes pea w/ ham.
[f] Includes crm of tomato.
[g] Includes italian veg & spring veg.

	KCAL	H_2O (g)	FAT (g)	PUFA (g)	CHOL (mg)	A (RE)	C (mg)	B-2 (mg)	B-6 (mg)	FOL (mcg)	Na (mg)	Ca (mg)	Mg (mg)	Zn (mg)	Mn (mg)
	WT (g)	PRO (g)	CHO (g)	SFA (g)	DFIB (g)	A (IU)	B-1 (mg)	NIA (mg)	B-12 (mcg)	PANT (mg)	K (mg)	P (mg)	Fe (mg)	Cu (mg)	
veg beef	53	237.5	1.1	0.1	1	24		.04	.05		1000			.27	
1 cup	253	2.9	8.0	0.6		238	.03	0.5				37	.85	.030	
veg, crm of	105	237.2	5.7	1.5	0	4	4	.11			1171				
1 cup	260	1.9	12.3	1.4		35	1.22	0.5			96	54			

29. SPECIAL DIETARY FOODS

	KCAL	H_2O (g)	FAT (g)	PUFA (g)	CHOL (mg)	A (RE)	C (mg)	B-2 (mg)	B-6 (mg)	FOL (mcg)	Na (mg)	Ca (mg)	Mg (mg)	Zn (mg)	Mn (mg)
	WT (g)	PRO (g)	CHO (g)	SFA (g)	DFIB (g)	A (IU)	B-1 (mg)	NIA (mg)	B-12 (mcg)	PANT (mg)	K (mg)	P (mg)	Fe (mg)	Cu (mg)	
breakfast bar, Carnation															
choc chip	200	1.9	11.0				28	.03	.40	100	180	20	60	3.00	
1.44 oz bar	41	6.0	20.0			1750	.30	5.0	.63	2.00	80	60	4.50	.500	
choc crunch	190	1.7	10.0				28	.03	.40	100	140	20	60	3.00	
1.34 oz bar	38	6.0	20.0			1750	.30	5.0	.63	2.00	103	60	4.50	.500	
peanut butter crunch	190	1.5	11.0				28	.03	.40	100	170	20	60	3.00	
1.35 oz bar	38	7.0	20.0			1750	.30	5.0	.63	2.00	85	60	4.50	.500	
peanut butter w/ choc chips	200	1.3	11.0				28	.03	.40	100	160	20	60	3.00	
1.39 oz bar	39	7.0	20.0			1750	.30	5.0	.63	2.00	87	60	4.50	.500	
figurine diet bar, Pillsbury															
choc	98	0.7	5.2				10	.13	.23	63	46	55	36	1.47	
1 bar	21	2.2	10.9			584	.23	2.6	.46		105	59	2.37	.252	
choc caramel	98	0.7	5.5				10	.13	.23	63	55	58	36	1.47	
1 bar	21	2.2	10.4			584	.23	2.6	.46		88	55	2.25	.252	
choc peanut butter	98	0.7	5.5				10	.13	.25	63	46	56	38	1.47	
1 bar	21	2.9	9.7			584	.23	3.0	.46		102	59	2.31	.256	
s'mores	99	0.7	5.3				10	.13	.25	63	45	55	36	1.47	
1 bar	21	2.1	11.0			584	.23	2.6	.46		83	50	2.31	.252	
van	99	0.7	5.2				10	.08	.25	63	46	58	36	1.47	
1 bar	21	2.1	11.1			584	.18	2.6	.46		86	54	2.25	.252	
fortified pudding, from mix, Delmark															
choc	250		8.0		20		11	.37	.36	70	140	270	72	2.70	.010
1/2 cup	164	9.0	35.0			900	.27	3.6	1.10	1.80	755	270	3.20	.360	
van/lemon	250		8.0		20		11	.37	.36	70	140	270	72	2.70	.010
1/2 cup	164	9.0	35.0			900	.27	3.6	1.10	1.80	675	270	3.20	.360	
gluten-free, wheat-free foods, Ener-G Foods															
baking mix															
rice, brown	653		23.0				0	.25			34	122	119		
1 cup	153	13.0	98.8			199	.29	6.0			252		3.52	.459	
rice, white	111		2.9				0	.01			93	8			
1 oz	28	1.7	19.5			0	.02	1.3			38		.72		
bread															
rice, brown	175	30.7	1.5				4	.11			16	66			
2 slices	73	4.2	36.1			0	.01	5.0			99		3.80		
rice, white	121	21.1	0.8				0	.03			230	45			
2 slices	51	2.8	25.8			0	.23	0.9					2.04		
rice, white, yeast-free	384	12.5	3.6				0				48	5			
2 slices	105	3.9	84.2			0	.81	1.2					1.05		
cookies															
almond butter	239	2.7	12.2				1	.25			118	5			
3 cookies	48	4.1	28.2			0	.08	0.2			66		.24		
choc	199	6.0	4.5				0	.04			167	88			
3 cookies	51	2.9	36.6			17	.04	1.7					2.04		
choc chip potato	243	8.5	10.1				1	.06			80	28	27	2.32	
3 cookies	57	3.1	34.6			57	.06	0.2			116	80	1.48	.274	
peanut butter	255	10.0	7.2				0	.05			211	104			
3 cookies	66	5.9	41.7			0	.06	2.3					2.64		

	KCAL / WT (g)	H₂O (g) / PRO (g)	FAT (g) / CHO (g)	PUFA (g) / SFA (g)	CHOL (mg) / DFIB (g)	A (RE) / A (IU)	C (mg) / B-1 (mg)	B-2 (mg) / NIA (mg)	B-6 (mg) / B-12 (mcg)	FOL (mcg) / PANT (mg)	Na (mg) / K (mg)	Ca (mg) / P (mg)	Mg (mg) / Fe (mg)	Zn (mg) / Cu (mg)	Mn (mg)
doughnut, rice, plain	186	19.6	10.7				0	.13			285	143			
1 doughnut	54	3.3	19.1			0	.04	0.5					.92		
granola	204	1.0	5.1				1	.14			1	35			
1/2 cup	50	5.0	34.5			0	.02	1.4			598		2.00		
nutquik milk powder	164	1.5	14.7				0	.29	.04	0	4	67	80	1.09	
1 oz	28	6.9	4.2			1	.06	1.1	.00		193	123	1.34	.336	
pizza shell, rice, low protein	191	20.4	6.8				0	.02			88	85	5	1.80	
2 oz	60	0.2	32.2			0	.16	0.1			14	25	.60		
rice mix	98		0.6				0	.01				90			
1 oz	28	1.8	21.4			0	.11	1.4					.36		
soup, mushroom, dry mix	66	2.4	0.0				0	.20			160	17			
2 3/4 T	20	2.4	13.9			0	.07	1.0			273		.66		
soup, tomato, dry mix	51	1.9	0.0				8	.06			23	13			
1 2/3 T	15	0.7	11.9			240	.06	0.8			35		.60		
soyquix milk powder	50	0.9	0.1				1	.06			39	36	43	.28	
2 T	14	7.3	4.9			0	.14	0.8			294	91	1.82	.140	
high protein foods, Delmark															
broth, beef, from mix	120		0.4		1	0	0	.02	.01	2	370	10	3	.07	.030
6 fl oz	183	7.0	22.0			0	.07	0.1	.11	.02	125	110	.08	.020	
broth, chicken, from mix	120		0.4		0	0	0	.02	.01	2	370	10	3	.07	.030
6 fl oz	183	7.0	22.0			0	.03	0.1	.11	.02	125	110	.08	.020	
gelatin dessert, all flavors, from mix	140		0.2		0	0	0	.44		15	250	10	17	.12	.009
5 fl oz	175	17.0	18.0			0	.02	0.2	.04	.12	210	20	.21	.090	
sego lite, liquid diet meal, cnd, Pet															
choc/dutch choc	150		3.0								475				
10 fl oz	a	11.0	20.0								690				
choc malt/double choc/choc jamocha	150		3.0								475				
almond—*10 fl oz*	a	11.0	20.0								690				
french van/strawberry	150		4.0								390				
10 fl oz	a	11.0	17.0								600				
van	150		4.0								490				
10 fl oz	a	11.0	17.0								600				
sego, liquid diet meal, cnd, Pet															
very choc/choc malt/dutch choc	225		1.0								445				
10 fl oz	a	11.0	43.0								690				
very van/strawberry	225		5.0								360				
10 fl oz	a	11.0	34.0								600				
slender bars, Carnation															
choc	270	1.8	14.0				15	.43	.42	100	285	250	100	3.75	
2 bars	55	11.0	26.0			1250	.38	5.0	1.50	2.50	345	250	4.50	.500	
choc chip	270	1.6	14.0				15	.43	.42	100	315	250	100	3.75	
2 bars	55	11.0	26.0			1250	.38	5.0	1.50	2.50	355	250	4.50	.500	
choc peanut butter	270	1.5	15.0				15	.43	.42	100	285	250	100	3.75	
2 bars	55	11.0	24.0			1250	.38	5.0	1.50	2.50	369	250	4.50	.500	
van	270	1.6	15.0				15	.43	.42	100	320	250	100	3.75	
2 bars	55	11.0	24.0			1250	.38	5.0	1.50	2.50	349	250	4.50	.500	
slender, liquid diet meal, cnd, Carnation															
banana	220	258.0	4.0				15	.43	.50	100	430	250	100	3.75	
10 fl oz	313	11.0	34.0			1250	.38	5.0	1.50	2.50	448	250	4.50	.500	
choc	220	257.0	4.0				15	.43	.50	100	515	250	100	3.75	
10 fl oz	313	11.0	34.0			1250	.38	5.0	1.50	2.50	555	250	4.50	.500	
choc fudge	220	255.0	4.0				15	.43	.50	100	550	250	100	3.75	
10 fl oz	313	11.0	34.0			1250	.38	5.0	1.50	2.50	585	250	4.50	.500	
choc malt	220	257.0	4.0				15	.43	.50	100	530	250	100	3.75	
10 fl oz	313	11.0	34.0			1250	.38	5.0	1.50	2.50	495	250	4.50	.500	
milk choc	220	257.0	4.0				15	.43	.50	100	520	250	100	3.75	
10 fl oz	313	11.0	34.0			1250	.38	5.0	1.50	2.50	560	250	4.50	.500	

ᵃ Gram wt not available.

	KCAL	H₂O (g)	FAT (g)	PUFA (g)	CHOL (mg)		A (RE)	C (mg)	Vitamins B-2 (mg)	B-6 (mg)	FOL (mcg)		Na (mg)	Ca (mg)	Minerals Mg (mg)	Zn (mg)	Mn (mg)
	WT (g)	PRO (g)	CHO (g)	SFA (g)	DFIB (g)		A (IU)	B-1 (mg)	NIA (mg)	B-12 (mcg)	PANT (mg)		K (mg)	P (mg)	Fe (mg)	Cu (mg)	
peach	220	258.0	4.0					15	.43				430	250			
10 fl oz	313	11.0	34.0				1250	.38	5.0				448				
strawberry	220	258.0	4.0					15	.43	.50	100		430	250	100	3.75	
10 fl oz	313	11.0	34.0				1250	.38	5.0	1.50	2.50		448	250	4.50	.500	
van	220	258.0	4.0						.43	.50	100		550	250	100	3.75	
10 fl oz	313	11.0	34.0				1250	.38	5.0	1.50	2.50		585	250	4.50	.500	

	KCAL / WT (g) / — / TRY (mg)	H₂O (g) / PRO (g) / — / THR (mg)	FAT (g) / CHO (g) / — / ISO (mg)	PUFA (g) / SFA (g) / — / LEU (mg)	CHOL (mg) / DFIB (g) / INOS (mg) / LYS (mg)	A (RE) / A (IU) / D (IU) / MET (mg)	C (mg) / B-1 (mg) / E (IU) / CYS (mg)	B-2 (mg) / NIA (mg) / K (mcg) / PHE (mg)	B-6 (mg) / B-12 (mcg) / BIO (mcg) / TYR (mg)	FOL (mcg) / PANT (mg) / CHLN (mg) / VAL (mg)	Na (mg) / K (mg) / Cl (mg) / ARG (mg)	Ca (mg) / P (mg) / I (mcg) / HIS (mg)	Mg (mg) / Fe (mg) / Mo (mg)	Zn (mg) / Cu (mg)	Mn (mg)

30. SPECIAL DIETARY FORMULAS, COMMERCIAL & HOSPITAL

amin-acid inst drink powder, Am McGaw[a]—5.2 oz pkt

	KCAL/WT	H₂O/PRO	FAT/CHO	PUFA/SFA	CHOL/DFIB/INOS	A	C	B-2	B-6	FOL	Na	Ca	Mg	Zn	Mn
	665		15.7												
	147	6.6	124.3												

casec protein supplement, Mead Johnson[b]—100 g dry powder

	370	5.5	2.0								150	1600			
	100	88.0	0.0								10	800			
										10					
	1120	3600	5000	8800	7500	2500	260	4800	5000	6300	3500	2800			

compleat, modified formula, Sandoznutrition[c]—250 ml can

	267		9.2			250	15	.43	.50	70	170	170	67	2.50	.670
		10.7	35.0			833	.37	3.3	1.00	1.70	350	230	3.00	.330	
						67	d	17	50	50	120	25	.050	.017	
	134	475	640	984	901	296	90	518	496	699	599	345			

compleat, regular formula, Sandoznutrition[e]—150 ml can/bottle

	267		10.7			250	15	.43	.50	70	320	170	67	2.50	.670
		10.7	32.0			833	.37	3.3	1.00	1.70	350	330	3.00	.330	
						67	d	17	50	50	220	25	.050	.017	
	139	484	622	965	885	262	116	479	455	663	538	326			

criticare HN, Mead Johnson[f] 8 fl oz

	250		0.8	0.6	0		38	.54	.62	50	150	125	50	2.50	.620
	260	9.0	52.5	0.1		625	.48	6.2	1.88	3.10	310	125	2.20	.250	
						50	9.4	31	38	62	250	19			
	145	440	550	925	775	405	35	433	198	675	360	270			

enrich, Ross Labs[g] 8 fl oz

	260	196	8.8	4.8	5		51	.44	.51	102	200	170	68	3.83	.850
	255	9.4	38.3	1.3	3.4	850	.39	5.1	1.60	2.55	370	170	3.06	.340	
						68	7.7	12	77	102	340	26			
	103	385	442	837	677	226	56	470	414	555	395	254			

ensure HN, Ross Labs[h] 8 fl oz

	250	199	8.4	4.6	4	268	54	.46	.54	108	220	179	72	4.02	.890
	253	10.5	33.4	1.2	0.0	893	.41	5.4	1.70	2.68	370	179	3.22	.360	
		8.2				72	8.1	13	81	107	340	27			
	126	431	473	914	725	273	53	504	515	525	389	252			

ensure plus, Ross Labs[i] 8 fl oz

	355	182	12.6	6.9	5		50	.57	.67	134	270	167	67	3.75	.830
	259	13.0	47.3	1.9	0.0	834	.50	6.7	2.00	3.34	500	167	3.00	.340	
		10.2				67	7.5	12	100	100	470	25			
	156	533	585	1131	897	338	65	624	637	650	481	312			

ensure plus HN, Ross Labs[j] 8 fl oz

	355	181	11.8	6.5	5		75	.85	1.00	200	280	250	100	5.63	1.250
	260	14.8	47.3	1.7	0.0	1250	.75	10.0	3.00	5.00	430	250	4.50	.500	
						100	11.3	18	150	150	380	38			
	178	607	666	1288	1021	385	74	710	725	740	548	355			

ensure pudding, van, Ross Labs[k] 5 oz

	250	91.0	9.7	0.8	5		15	.29	.34	68	240	200	68	3.83	.850
	142	6.8	34.0	2.3		850	.26	3.4	1.10	1.70	330	200	3.06	.340	
						68	7.7	12	51	25	220	60			
	102	299	340	673	530	184	54	320	306	435	224	163			

[a] Oral or tube feeding for uremic patients; available in orange, lemon-lime, strawberry and berry flavors; requires 250 g water to reconstitute.

[b] Used to increase the protein content of the diets of children & adults.

[c] Ready-to-use liquid tube feeding formulated from natural foods providing complete, balanced nutrition in a lactose-free, low sodium, isotonic formula; for short and long term tube-fed patients with normal gastrointestinal tracts.

[d] Alpha-tocopherol is 5.0 mg.

[e] Ready-to-use liquid tube feeding formulated from natural foods for short and long term tube-fed patients with normal gastrointestinal tracts; complete, balanced nutrition.

[f] Ready-to-use high nitrogen elemental diet.

[g] Fiber-containing, nutritionally complete liquid formula suitable for patients who do not require a low-residue diet; may be fed orally or by tube.

[h] Nutrient-dense, high-nitrogen, low-residue liquid food for patients w/ energy requirements less than 2000 kcal/day (elderly, comatose) or those with elevated nitrogen requirements; may be fed orally or by tube.

[i] Nutritionally complete, high-calorie liquid food providing 1.5 kcal/ml; useful when extra energy and nutrients, but normal protein concentration are needed in a limited volume; oral or tube feeding.

[j] Nutritionally complete, high-calorie, high-nitrogen liquid food, designed to meet the needs of stressed patients w/ higher energy and protein needs or for those with limited volume tolerance.

[k] Calorically dense supplement for patients in need of increased nutrient intake.

					Vitamins					Minerals				
KCAL	H₂O (g)	FAT (g)	PUFA (g)	CHOL (mg)	A (RE)	C (mg)	B-2 (mg)	B-6 (mg)	FOL (mcg)	Na (mg)	Ca (mg)	Mg (mg)	Zn (mg)	Mn (mg)
WT (g)	PRO (g)	CHO (g)	SFA (g)	DFIB (g)	A (IU)	B-1 (mg)	NIA (mg)	B-12 (mcg)	PANT (mg)	K (mg)	P (mg)	Fe (mg)	Cu (mg)	
				INOS (mg)	D (IU)	E (IU)	K (mcg)	BIO (mcg)	CHLN (mg)	Cl (mg)	I (mcg)	Mo (mg)		
TRY (mg)	THR (mg)	ISO (mg)	LEU (mg)	LYS (mg)	MET (mg)	CYS (mg)	PHE (mg)	TYR (mg)	VAL (mg)	ARG (mg)	HIS (mg)			

ensure, Ross Labs[a] — *8 fl oz*

250	200	8.8	4.8	4	188	38	.43	.50	100	200	125	50	2.82	.620
253	8.8	34.3	1.3	0.0	625	.38	5.0	1.50	2.50	370	125	2.25	.250	
					50	5.63	9	75	75	340	19			
106	361	396	766	607	229	44	422	431	440	326	211			

hepatic-aid II inst drink powder, Am McGaw[b] — *3.1 oz pkt*

400		12.3												
88	15.0	57.3												

isocal HCN, Mead Johnson[c] — *8 fl oz*

473		24.2	9.9	7		71	1.02	1.18	95	190	240	95	7.10	.780
260	17.7	47.3	2.5		1180	.90	11.8	3.50	5.90	400	240	4.30	.710	
					95	17.7	59	71.0	118.0	280	35			
227	738	1041	1797	1514	511	52	993	1041	1277	724	563			

isocal, Mead Johnson[d] — *8 fl oz*

250		10.5	5.3	0		38	.54	.62	50	125	150	50	2.50	.620
250	8.1	31.5	1.3		625	.48	6.2	1.88	3.10	310	125	2.20	.250	
					50	9.4	31	38.0	62.0	250	19			
145	320	425	775	575	193	40	400	375	550	305	215			

lonalac, Mead Johnson[e] — *8 fl oz (47 g powder in 6.25 fl oz water)*

240	203.4	13.2	0.2				.65			10	423	34		
250	12.7	17.9	9.6		360	.15	0.3			470	375			
										188				
163	528	720	1272	1080	362	38	696	720	912	504	398			

lytren, Mead Johnson[f] — *100 ml*

10		0.0								177				
	0.0	2.5								58				
										173				

MCT oil (medium chain triglycerides), Mead Johnson[g] — *1 T*

115	0.0	13.9												
14	0.0	0.0												

moducal, Mead Johnson[h] — *100 g powder*

380	5.0	0.0	0.0							70				
100	0.0	95.0	0.0							5				
										170				

osmolite HN, Ross Labs[i] — *8 fl oz*

250	199	8.7	2.8	5	268	54	.46	.54	108	220	179	72	4.02	.890
253	10.5	33.4	5.4	0.0	893	.41	5.4	1.70	2.68	370	179	3.22	.360	
				0	72	8.1	13	81	42	340	27			
126	431	473	914	725	273	53	504	515	525	389	252			

osmolite, Ross Labs[j] — *8 fl oz*

250	199	9.1	3.0	5	188	38	.43	.50	100	150	125	50	2.82	.620
253	8.8	34.3	4.4	0.0	625	.38	5.0	1.50	2.50	240	125	2.25	.250	
				0	50	5.6	9	75	75	200	19			
106	361	396	766	607	229	44	422	431	440	326	211			

phenyl-free, Mead Johnson[k] — *4 fl oz*

100	95.0	1.7				13	.25	.22	31	100	125	38	1.75	.250
120	5.0	16.2			300	.15	2.0	.62	.75	338	125	3.00	.150	
				7.5	38	2.5	25	7.5	21.0	230	11			
70	230	270	420	460	155	85	5	230	310		115			

[a] Oral/tube, liquid, low-residue feeding; complete, balanced formula.

[b] Oral or tube feeding for patients with chronic liver disease; available in choc, eggnog & custard flavors; requires 280 g water to reconstitute.

[c] For tube feedings that are hypermetabolic, fluid-restricted or volume-limited.

[d] Nutritionally complete formula to meet the dietary needs of most tube-fed patients.

[e] Low sodium, high protein beverage mix.

[f] Electrolytes plus carbohydrate for oral administration; contains dextrose, potassium citrate, sodium chloride, sodium citrate & citric acid.

[g] Substitute or supplemental source of fat calories for patients with poor digestion, absorption or utilization of conventional food fats. Lipid fraction of coconut oil contains < 6% fatty acids shorter than C8, 67% fatty acids with C8 (caprylic), 23% fatty acids with C10 (capric) & < 4% fatty acids longer than C10.

[h] Refined readily digestible carbohydrate which can be added to foods to increase caloric content for patients with increased caloric requirements; 100% maltodextrin from hydrolysis of corn starch.

[i] Nutrient-dense, high nitrogen, isotonic, low-residue liquid food for patients with energy requirements less than 2000 kcal/day or for patients with elevated nitrogen requirements who require tube feeding; may also be used for oral feeding.

[j] Isotonic, low-residue, liquid, complete, balanced formula for hospitalized patients who require tube feeding (nasogastric, nasoduodenal, jejunal); may also be used for oral feeding.

[k] Phenylalanine-free food used in the management of phenylketonuria.

	KCAL / WT (g) / / TRY (mg)	H₂O (g) / PRO (g) / / THR (mg)	FAT (g) / CHO (g) / / ISO (mg)	PUFA (g) / SFA (g) / / LEU (mg)	CHOL (mg) / DFIB (g) / INOS (mg) / LYS (mg)	A (RE) / A (IU) / D (IU) / MET (mg)	C (mg) / B-1 (mg) / E (IU) / CYS (mg)	B-2 (mg) / NIA (mg) / K (mcg) / PHE (mg)	B-6 (mg) / B-12 (mcg) / BIO (mcg) / TYR (mg)	FOL (mcg) / PANT (mg) / CHLN (mg) / VAL (mg)	Na (mg) / K (mg) / Cl (mg) / ARG (mg)	Ca (mg) / P (mg) / I (mcg) / HIS (mg)	Mg (mg) / Fe (mg) / Mo (mg)	Zn (mg) / Cu (mg)	Mn (mg)
polycose, Ross Labs[a] *100 ml*	200	70.0	0.0	0.0	0						70	20			
	120	0.0	50.0	0.0	0.0						6	3			
											140				
	0	0	0	0	0	0	0	0	0	0	0	0			
portagen, 30 cal/fl oz, Mead Johnson[b] *8 fl oz*	240		11.5	0.9	2		20	.45	.50	38	113	225	50	2.25	.300
	240	8.5	27.5	0.3		1875	.38	5.0	1.50	2.50	300	169	4.50	.375	
					11.3	188	7.5	38	18.8	31.3	206	18			
	108	346	480	840	720	240	24	461	480	600	338	264			
portagen, 60 cal/fl oz, Mead Johnson[b] *8 fl oz*	480		22.9	1.9	4		39	.90	1.00	75	225	450	100	4.50	.600
	240	16.8	55.3	0.7		3750	.75	10.0	3.00	5.00	600	338	9.00	.750	
					22.5	375	15	75	37.5	62.5	413	35			
	216	692	960	1680	1440	480	48	922	960	1200	676	528			
pro mod, Ross Labs[c] *1 scoop*	28	0.6	0.6								13	23			
	5	5.0	0.7								65	22			
	95	365	310	540	465	110	130	180	170	300	130	95			
pulmocare, Ross Labs[d] *8 fl oz*	355	186	21.8	12.0	6		75	.85	1.00	200	310	250	100	5.63	1.250
	250	14.8	25.0	3.2	0.0	1250	.75	10.0	3.00	5.00	450	250	4.50	.500	
						100	11.3	18	150	150	400	38			
	163	592	755	1362	1110	444	59	710	710	947	533	414			
ross SLD (surgical liquid diet), Ross Labs[e]—*200 ml (1.3 oz pkt in water)*	140	164.0	0.1	0.0	0		15	.43	.50	100	167	167	67	3.80	.840
	214	7.5	27.3	0.0		834	.39	5.0	1.50	2.50	167	167	3.00	.330	
						67	7.5	6	75	75	200	25			
	83	383	413	690	563	248	188	533	338	570	525	203			
sustacal HC, Mead Johnson[f] *8 fl oz*	360		13.6	5.4	4		18	.51	.60	120	200	200	80	3.00	.600
	260	14.4	45.0	2.1		1000	.45	6.0	1.80	3.00	350	200	3.60	.400	
					50	80	6	50	90	50	300	30			
	184	594	828	1440	1224	410	43	792	828	1044	580	450			
sustacal, Mead Johnson[g] *8 fl oz*	240		5.5	2.1	4		13	.40	.47	89	220	240	90	3.30	.670
	260	14.5	33.1	0.9		1110	.33	4.7	1.33	2.30	490	220	4.00	.470	
						89	6.7	56	67	56	370	33			
	158	552	768	1320	1056	346	58	744	696	912	624	389			
sustacal powder, vanilla, Mead Johnson *1 pkt*	200		0.3				20	.20	.55	133	190	290	105	4.00	1.000
	54	12.5	36.0			1290	.40	6.8	1.00	2.70	560	250	6.00	.700	
						33	10		93		220	40			
sustacal powder, vanilla w/ 8 fl oz whole milk, Mead Johnson—*8 fl oz*	360		9.2				20	.60	.70	133	330	580	135	5.00	1.000
	290	20.5	48.0			1670	.50	7.0	2.00	3.50	910	480	6.00	.700	
						133	10		100		480	50			
sustacal pudding, Mead Johnson[h] *5 fl oz*	240	92.0	9.5	3.7	5		9	.26	.30	60	120	220	60	2.20	.670
	142	6.8	32.0	1.5		750	.23	3.0	.90	1.50	320	220	2.70	.300	
						60	4.5		50		200	22			
	94	271	362	672	442	151	34	310	355	408	158	144			

[a] Glucose polymers for energy support to be mixed with food or beverage; may be used as enteral carbohydrate supplement.

[b] Nutritionally complete formula for children & adults who do not efficiently digest conventional fat or absorb the resulting long chain fatty acids.

[c] Protein supplement to be used enterally or orally.

[d] High-fat, low-carbohydrate enteral formula designed to reduce carbon dioxide production, thereby minimizing carbon dioxide retention resulting from chronic obstructive pulmonary disease or acute respiratory failure. Can be used for enteral tube feeding or oral supplementation and is appropriate for ambulatory or ventilation-dependent patients.

[e] Highly fortified, low-residue, low-fat formula for patients restricted to clear liquid diets.

[f] Complete or supplemental oral formula for patients who are fluid restricted and volume sensitive.

[g] Complete or supplemental oral formula for patients who have difficulty meeting nutritional requirements through normal diets.

[h] Nutritionally complete dietary supplement.

	KCAL / WT (g) / / TRY (mg)	H₂O (g) / PRO (g) / / THR (mg)	FAT (g) / CHO (g) / / ISO (mg)	PUFA (g) / SFA (g) / / LEU (mg)	CHOL (mg) / DFIB (g) / INOS (mg) / LYS (mg)	A (RE) / A (IU) / D (IU) / MET (mg)	C (mg) / B-1 (mg) / E (IU) / CYS (mg)	B-2 (mg) / NIA (mg) / K (mcg) / PHE (mg)	B-6 (mg) / B-12 (mcg) / BIO (mcg) / TYR (mg)	FOL (mcg) / PANT (mg) / CHLN (mg) / VAL (mg)	Na (mg) / K (mg) / Cl (mg) / ARG (mg)	Ca (mg) / P (mg) / I (mcg) / HIS (mg)	Mg (mg) / Fe (mg) / Mo (mg)	Zn (mg) / Cu (mg)	Mn (mg)
sustagen, Mead Johnson[a]	438		4.0		14		75	1.08	1.25	100	250	800	100	5.00	1.250
8 fl oz	260	26.8	75.0			1250	.95	12.5	3.75	6.25	800	600	4.50	.500	
					210	100	11	63	75	125	675	38			
	447	1051	1402	2540		657	184	1270	1270	1752	837	731			
traum-aid HBC powder, Am McGaw[b]	500		6.2				14	.24	.28	56	265		56	2.81	.546
4.2 oz pkt	119	28.0	83.0			700	.21	2.8	.84	1.39	585	168	2.52	.281	
						56	8.4	21	42	59	344	21	.021	.021	.007
traumaCal, Mead Johnson[c]	355	184.9	16.2	7.5	1		35	.51	.59	47	280	177	47	3.50	.590
8 fl oz	260	19.5	33.7	1.9		590	.45	5.9	1.77	3.00	330	117	2.10	.350	
						47	8.9	30	35	59	380	18			
	249	817	1101	1953	1669	554	57	1065	1101	1385	781	614			
two cal HN, Ross Labs	478		21.5				75	.68	.80	160	250	250	100	5.63	1.250
8 fl oz		19.8	51.4			1250	.60	8.0	2.40	4.00	550	250	4.50	.500	
						100	11.3	18	120	150	370	38			
	218	792	1010	1822	1485	594	79	950	950	1267	713	554			
vital high nitrogen, Ross Labs[d]	300	255	3.3	1.2	6		60	.68	.80	160	140	200	80	4.50	1.000
300 ml (79 g pkt w/ 255 ml water)	326	12.5	55.4	1.4	0.0	1000	.60	8.0	2.40	4.00	400	200	3.60	.400	
					0	80	9	14	120	120	270	30	.025	.006	.026
	138	475	575	1000	663	275	113	563	475	700	688	325			
vivonex flavor packets, Norwich Eaton Pharmaceuticals[e]**—1 pkt**	[f]		0.0												
	2.5	0.0	[g]												
vivonex high nitrogen, Norwich Eaton Pharmaceuticals[h]**—1000 ml**	1000	6.7	0.9	0.8	0		20	.57	.67	133	529	333	133	5.00	.940
	267	44.3	203.7	0.1	0.0	1667	.50	6.7	2.00	3.30	1173	333	6.00	.670	
					0	133	[i]	22	100	25		150	.050	.050	.017
	567	1833	1838	2900	2200	2029	0	3133	372	2029	186	50			
vivonex, standard, Norwich Eaton Pharmaceuticals[h]**—1000 ml**	1000	6.7	1.5	1.3	0		33	.94	1.11	220	468	556	222	8.33	1.560
	267	20.6	226.3	0.1	0.0	2778	.83	11.1	3.33	5.55	1172	556	10.00	1.110	
					0	222	[j]	37	170	41		83	.083	.083	.028
	290	937	937	1483	1114	960	0	1067	194	1034	1827	455			
vivonex, TEN, Norwich Eaton Pharmaceuticals[h]**—1000 ml**	1000	6.7	2.8	2.6	0		60	1.70	2.00	400	460	500	200	10.00	.940
	268	38.2	205.6	0.3	0.0	2500	1.50	20.0	6.00	10.00	782	500	9.00	1.000	
					0	200	[k]	22	300	74		75	.050	.050	.017
	489	1528	3159	6326	1948	1398	0	1971	321	3159	2918	901			

[a] High calorie, high protein nutritional supplement.

[b] Oral or tube feeding for catabolic patients; available in lemon-creme flavor; requires 420 g water to reconstitute.

[c] Oral or tube nutritionally complete feeding for patients with multiple trauma and major burns having elevated caloric requirements, exceptional nitrogen needs and volume restrictions.

[d] Nutritionally complete elemental diet for patients with impaired gastrointestinal function.

[e] Used to enhance the flavor of orally administered vivonex diets. They contain 20 mg of saccharin per pkt. Available in lemon-lime, orange-pineapple, strawberry & vanilla.

[f] 6 kcal/pkt for lemon-lime & orange-pineapple; 8 kcal/pkt for strawberry; 9 kcal/pkt for vanilla.

[g] Cho/pkt is 0.8 for lemon-lime, 0.9 for orange-pineapple, 1.3 for strawberry & 2.2 for vanilla.

[h] Elemental chemically defined diet.

[i] Alpha-tocopherol is 6.7 mg.

[j] Alpha-tocopherol is 11.19 mg.

[k] Alpha-tocopherol is 10.0 mg.

31. SPICES, HERBS, & FLAVORINGS

	KCAL	H_2O (g)	FAT (g)	PUFA (g)	CHOL (mg)	A (RE)	C (mg)	B-2 (mg)	B-6 (mg)	FOL (mcg)	Na (mg)	Ca (mg)	Mg (mg)	Zn (mg)	Mn (mg)
	WT (g)	PRO (g)	CHO (g)	SFA (g)	DFIB (g)	A (IU)	B-1 (mg)	NIA (mg)	B-12 (mcg)	PANT (mg)	K (mg)	P (mg)	Fe (mg)	Cu (mg)	
allspice, ground	5	0.2	0.2	0.0	0	1	1	.00			1	13	3	.02	
1 t	2	0.1	1.4	0.1		10	.00	0.0	.00		20	2	.13		
anise seed	7	0.2	0.3	0.1	0						0	14	4	.11	
1 t	2	0.4	1.1						.00		30	9	.78		
basil, ground	4	0.1	0.1		0	13	1	.00			0	30	6	.08	
1 t	1	0.2	0.9			131	.00	0.1	.00		48	7	.59		
bay leaf, crumbled	2	0.0	0.1	0.0	0	4	0	.00			0	5	1	.02	
1 t	1	0.1	0.5			37	.00	0.0	.00		3	1	.26		
caraway seed	7	0.2	0.3	0.1	0	1		.01			0	14	5	.12	
1 t	2	0.4	1.1	0.0		8	.01	0.1	.00		28	12	.34		
cardamom, ground	6	0.2	0.1	0.0	0			.00			0	8	5	.15	
1 t	2	0.2	1.4	0.0			.00	0.0	.00		22	4	.28		
celery seed	8	0.1	0.5	0.1	0	0	0				3	35	9	.14	
1 t	2	0.4	0.8	0.0		1		0.0			28	11	.90		
chervil, dried	1	0.0	0.0		0				.01		0	8	1	.05	
1 t	1	0.1	0.3						.00		28	3	.19		
chili powder[a]	8	0.2	0.4		0	91	2	.02			26	7	4	.07	
1 t	3	0.3	1.4			908	.01	0.2	.00		50	8	.37		
cinn, ground	6	0.2	0.1	0.0	0	1	1	.00			1	28	1	.05	
1 t	2	0.1	1.8	0.0		6	.00	0.0	.00		11	1	.88		
cloves, ground	7	0.1	0.4		0	1	2	.01			5	14	6	.02	
1 t	2	0.1	1.3	0.1		11	.00	0.0	.00		23	2	.18		
coriander leaf, dried	2	0.0	0.0		0		3	.01			1	7	4		
1 t	1	0.1	0.3				.01	0.1	.00		27	3	.25		
coriander seed	5	0.2	0.3	0.0	0			.01			1	13	6	.08	
1 t	2	0.2	1.0	0.0			.00	0.0	.00		23	7	.29		
cumin seed	8	0.2	0.5		0	3	0	.01			4	20	8	.10	
1 t	2	0.4	0.9			27	.01	0.1	.00		38	10	1.39		
curry powder	6	0.2	0.3		0	2	0	.01			1	10	5	.08	
1 t	2	0.3	1.2			20	.01	0.1	.00		31	7	.59		
dill seed	6	0.2	0.3	0.0	0	0		.01			0	32	5	.11	
1 t	2	0.3	1.2	0.0		1	.01	0.1	.00		25	6	.34		
dill weed, dried	3	0.1	0.0		0			.00	.02		2	18	5	.03	
1 t	1	0.2	0.6				.00	0.0	.00		33	5	.49		
fennel seed	7	0.2	0.2	0.0	0	0		.01			2	24	8	.07	
1 t	2	0.3	1.1	0.0		3	.01	0.1	.00		34	10	.37		
fenugreek seed	12	0.3	0.2		0	0		.01		2	2	6	7	.09	
1 t	4	0.9	2.2				.01	0.1	.00		28	11	1.24		
garlic powder	9	0.2	0.0		0			.00			1	2	2	.07	
1 t	3	0.5	2.0				.01	0.0	.00		31	12	.08		
ginger, ground	6	0.2	0.1	0.0	0	0		.00			1	2	3	.08	
1 t	2	0.2	1.3	0.0		3	.00	0.1	.00		24	3	.21		
mace, ground	8	0.1	0.6	0.1	0	1		.01			1	4	3	.04	
1 t	2	0.1	0.9	0.2		14	.01	0.0	.00		8	2	.24		
marjoram, dried	2	0.0			0	5	0	.00			0	12	2	.02	
1 t	1	0.4				48	.00	0.0	.00		9	2	.50		
mustard seed, yellow	15	0.2	1.0	0.2	0	0		.01			0	17	10	.19	
1 t	3	0.8	1.2	0.1		2	.02	0.3	.00		23	28	.33		
nutmeg, ground	12	0.1	0.8	0.0	0	0		.00			0	4	4	.05	
1 t	2	0.1	1.1	0.6		2	.01	0.0	.00		8	5	.07		
onion powder	7	0.1	0.0		0			.00	.00		1	8	3	.05	
1 t	2	0.2	1.7				.01	0.0	.00		20	7	.05		
oregano, ground	5	0.1	0.2	0.1	0	10					0	24	4	.07	
1 t	2	0.2	1.0	0.0		104	.01	0.1	.00		25	3	.66		
paprika	6	0.2	0.3	0.2	0	127	1	.04			1	4	4	.08	
1 t	2	0.3	1.2	0.0		1273	.01	0.3	.00		49	7	.50		
parsley, dried	4	0.1	0.1		0	30	2	.02	.01		6	19	3	.06	
1 t	1	0.3	0.7			303	.00	0.1	.00		49	5	1.27		

[a] Contains red pepper, cumin, oregano, salt & garlic powder.

	KCAL	H₂O (g)	FAT (g)	PUFA (g)	CHOL (mg)	A (RE)	C (mg)	B-2 (mg)	B-6 (mg)	FOL (mcg)	Na (mg)	Ca (mg)	Mg (mg)	Zn (mg)	Mn (mg)
	WT (g)	PRO (g)	CHO (g)	SFA (g)	DFIB (g)	A (IU)	B-1 (mg)	NIA (mg)	B-12 (mcg)	PANT (mg)	K (mg)	P (mg)	Fe (mg)	Cu (mg)	
pepper, black	5	0.2	0.1	0.0	0	0		.01			1	9	4		.03
1 t	2	0.2	1.4	0.0		4	.00	0.0	.00		26	4	.61		
pepper, red/cayenne	6	0.1	0.3	0.2	0	75	1	.02			1	3	3		.05
1 t	2	0.2	1.0	0.1		749	.01	0.2	.00		36	5	.14		
pepper, white	7	0.3	0.1		0			.00			0	6	2		.03
1 t	2	0.3	1.7				.00	0.0	.00		2	4	.34		
poppy seed	15	0.2	1.3	0.9	0			.01	.01		1	41	9		.29
1 t	3	0.5	0.7	0.1			.02	0.0	.00		20	24	.26		
poultry seasoning[a]	5	0.1	0.1		0	4	0	.00			0	15	3		.05
1 t	2	0.1	1.0			39	.00	0.0	.00		10	3	.53		
pumpkin pie spice[b]	6	0.1	0.2		0	0	0	.00			1	12	2		.04
1 t	2	0.1	1.2			4	.00	0.0	.00		11	2	.34		
rosemary, dried	4	0.1	0.2		0	4	1				1	15	3		.04
1 t	2	0.1	0.8			38	.01	0.0	.00		11	1	.35		
saffron	2	0.1	0.0		0						1	1			
1 t	1	0.1	0.5						.00		12	2	.08		
sage, ground	2	0.1	0.1	0.0	0	4	0	.00			0	12	3		.03
1 t	1	0.1	0.4	0.1		41	.01	0.0	.00		7	1	.20		
salt[c]	0		0.0	0.0							2300	0			
1 t	6	0.0	0.0	0.0							0				
salt, kosher, Morton	0		0.0								1880				
1 t	5	0.0	0.0												
salt substitutes															
Morton lite salt[d]	0		0.0	0.0	0						1100	0	4		
1 t	6	0.0	0.0	0.0	0.0						1500				
Morton salt substitute	0		0.0	0.0	0						0	30	0		
1 t	6	0.0	0.1	0.0	0.0						2800	28			
Morton seasoned salt substitute	2		0.0								0				
1 t	5	0.0	0.5								2100				
No Salt salt alternative					0						5				
		0.0									2500				
No Salt salt alternative, seasoned	4		0.0		0						2				
1 t	4.5	0.0	0.9								1330				
Nu-Salt											0	0	0		
1 t	1										1				
savorex, Loma Linda[e]	16	2.0	1.0		0			.42	.25	90	320	4	14		.35
1 t	7	3.0	1.0				.14	3.5		.42	210		.35		
savory, ground	4	0.1	0.1		0	7					0	30	5		.06
1 t	1	0.1	1.0			72	.01	0.1	.00		15	2	.53		
seasoning blend, Nature's Seasons, Morton—*1 t*	3		0.5								1300				
	4	0.0	0.0								10				
tarragon, ground	5	0.1	0.1		0	7		.00			1	18	6		.06
1 t	2	0.4	0.8			67	.00	0.1	.00		48	5	.52		
tumeric, ground	8	0.3	0.2		0		1	.01			1	4	4		.10
1 t	2	0.2	1.4				.00	0.1	.00		56	6	.91		

32. SPREADS (BUTTER, MARGARINE, MAYONNAISE, ETC.)

	KCAL	H₂O (g)	FAT (g)	PUFA (g)	CHOL (mg)	A (RE)	C (mg)	B-2 (mg)	B-6 (mg)	FOL (mcg)	Na (mg)	Ca (mg)	Mg (mg)	Zn (mg)	Mn (mg)
	WT (g)	PRO (g)	CHO (g)	SFA (g)	DFIB (g)	A (IU)	B-1 (mg)	NIA (mg)	B-12 (mcg)	PANT (mg)	K (mg)	P (mg)	Fe (mg)	Cu (mg)	
butter	36	0.8	4.1	0.2	11	38	0	.00	.00	0	41	1	0		.00
1 t	5	0.0	0.0	2.5	0.0	153	.00	0.0			1	1	.01		
	108	2.4	12.2	0.5	33	114	0	.01	.00	0	123	3	0		.00
1 T	15	0.1	0.0	7.6	0.0	459	.00	0.0			3	3	.03		
sweet (unsalted)	36	0.8	4.1	0.2	11	38	0	.00	.00	0	1	1	0		.00
1 t	5	0.0	0.0	2.5	0.0	165	.00	0.0			1	1	.01		
sweet (unsalted)	108	2.4	12.2	0.5	33	114	0	.01	.00	0	2	3	0		.00
1 T	15	0.1	0.0	7.6	0.0	495	.00	0.0			3	3	.03		

[a] Contains white pepper, sage, thyme, marjoram, savory, ginger, allspice & nutmeg.
[b] Contains cinn, ginger, nutmeg, allspice & cloves.
[c] Contains 3.6 g Cl & 400 mcg I if iodized; contains less than 100 mcg if not iodized.
[d] Contains 400 mcg I.
[e] Contains yeast extract, veg flavors & salt.

	KCAL / WT (g)	H₂O (g) / PRO (g)	FAT (g) / CHO (g)	PUFA (g) / SFA (g)	CHOL (mg) / DFIB (g)	A (RE) / A (IU)	C (mg) / B-1 (mg)	B-2 (mg) / NIA (mg)	B-6 (mg) / B-12 (mcg)	FOL (mcg) / PANT (mg)	Na (mg) / K (mg)	Ca (mg) / P (mg)	Mg (mg) / Fe (mg)	Zn (mg) / Cu (mg)	Mn (mg)
whipped	27	0.6	3.1	0.1	8	29	0	.00	.00	0	31	1	0	.00	
1 t	4	0.0	0.0	1.9	0.0	116	.00	0.0			1	1	.01		
whipped	81	1.8	9.2	0.3	24	87	0	.00	.00	0	93	3	0	.00	
1 T	11	0.1	0.0	5.7	0.0	319	.00	0.0			3	3	.03		
butter blend, stick/soft, Blue Bonnet	90		11.0	5.0	5						95[a]				
1 T		0.0	0.0	2.0							10[a]				
butter buds	12		0.1		0						170				
2 t dry[b]	4	0.0	2.9			7					3				
buttermatch blend, Shedd's	90		10.0												
1 T	14	0.0	0.0												
buttermatch blend, whipped, Shedd's	60		7.0												
1 T	10	0.0	0.0												
buttery blend, liquid, Mrs. Filberts	120		14.0	4.0	0						60				
1 T	14	0.0	0.0	3.0											
margarine by brand															
Blue Bonnet soft	100		11.0	4.0	0						95				
1 T		0.0	0.0	2.0							5				
Blue Bonnet soft, whipped	70		7.0	2.0	0						70				
1 T		0.0	0.0	2.0							5				
Blue Bonnet, stick	100		11.0	3.0	0						95				
1 T		0.0	0.0	2.0							5				
Blue Bonnet, stick, whipped	70		7.0	2.0	0						70				
1 T		0.0	0.0	2.0							5				
Fleischmann's soft	100		11.0	5.0	0						95[c]				
1 T		0.0	0.0	2.0							5				
Fleischmann's squeeze	100		11.0	6.0	0						85				
1 T		0.0	0.0	2.0							5				
Fleischmann's stick	100		11.0	4.0	0						95[d]				
1 T		0.0	0.0	2.0							5				
Fleischmann's whipped, lightly salted	70		7.0	3.0	0						60[e]				
1 T		0.0	0.0	2.0							5				
Mazola	100	2.2	11.3	4.0	0						100				
1 T	14	0.0	0.2	1.9		500									
Mazola unsalted	100	2.7	11.3	4.0	0						0				
1 T	14	0.0	0.0	1.9		500									
Nucoa	100	2.3	11.3	3.3	0						160				
1 T	14	0.0	0.0	2.4		500									
Nucoa soft	90	2.0	10.5	4.2	0						150				
1 T	13	0.0	0.1	2.0		520									
Shedd's liquid	100		11.0	6.0	0						100				
1 T	14	0.0	0.0	2.0											
margarine by form & type of oil															
liquid, soybean & cottonseed	34	0.7	3.8	1.7		47	0	.00	.00	0	37	3	0		
1 t	5	0.1	0.0	0.6		155	.00	0.0	.01	.01	4	2			
stick, corn	34	0.7	3.8	0.8		47	0	.00	.00	0	44[f]	1	0		
1 t	5	0.0	0.0	0.6		155	.00	0.0	.00	.00	2	1			
stick, safflower & soybean	34	0.7	3.8	1.5		47	0	.00	.00	0	44[f]	1	0		
1 t	5	0.0	0.0	0.6		155	.00	0.0	.00	.00	2	1			
stick, soybean	34	0.7	3.8	1.0		47	0	.00	.00	0	44[f]	1	0		
1 t	5	0.0	0.0	0.8		155	.00	0.0	.00	.00	2	1			
stick, soybean & cottonseed	34	0.7	3.8	0.9		47	0	.00	.00	0	44[f]	1	0		
1 t	5	0.0	0.0	0.8		155	.00	0.0	.00	.00	2	1			
tub, corn	34	0.8	3.8	1.5		47	0	.00	.00	0	51[g]	1	0		
1 t	5	0.0	0.0	0.7		158	.00	0.0	.00	.00	2	1			
tub, safflower	34	0.8	3.8	2.1		47	0	.00	.00	0	51[g]	1	0		
1 t	5	0.0	0.0	0.4		155	.00	0.0	.00	.00	2	1			
tub, soybean	34	0.8	3.8	1.3		47	0	.00	.00	0	51[g]	1	0		
1 t	5	0.0	0.0	0.6		155	.00	0.0	.00	.00	2	1			
tub, soybean & cottonseed	34	0.8	3.8	1.4		47	0	.00	.00	0	51[g]	1	0		
1 t	5	0.0	0.0	0.8		155	.00	0.0	.00	.00	2	1			

[a] Unsalted butter blend contains 0 mg sodium & 0 mg potassium.
[b] Corresponds to 2 T liquid (28 g).
[c] Unsalted soft margarine contains 0 mg sodium.
[d] Unsalted stick margarine contains 0 mg sodium.
[e] Unsalted whipped margarine contains 0 mg sodium.
[f] Unsalted stick margaring contains 1 mg sodium.
[g] Unsalted tub margarine contains 2 mg sodium.

	KCAL / WT (g)	H₂O (g) / PRO (g)	FAT (g) / CHO (g)	PUFA (g) / SFA (g)	CHOL (mg) / DFIB (g)	A (RE) / A (IU)	C (mg) / B-1 (mg)	B-2 (mg) / NIA (mg)	B-6 (mg) / B-12 (mcg)	FOL (mcg) / PANT (mg)	Na (mg) / K (mg)	Ca (mg) / P (mg)	Mg (mg) / Fe (mg)	Zn (mg) / Cu (mg)	Mn (mg)
margarine, imitation (diet) by brand															
Blue Bonnet	50		6.0	2.0	0						100				
1 T		0.0	0.0	1.0							5				
Fleischmann's	50		6.0	2.0	0						100				
1 T		0.0	0.0	1.0							5				
Fleischmann's w/ lite salt	50		6.0	3.0	0						50				
1 T		0.0	0.0	2.0							5				
Mazola diet	50	8.1	5.5	2.4	0						130				
1 T	14	0.0	0.0	0.9		500									
Parkay, diet soft	50	8.0	5.6	1.8	0		0	.00	.00	0	90	1	0	.00	
1 T	14	0.0	0.0	1.0		526	.00	0.0	.00	.00	4	1	.00	.007	
margarine, imitation (diet) by form & type of oil															
tub, corn	17	2.8	1.9	0.8		48	0	.00	.00	0	46	1	0		
1 t	5	0.0	0.0	0.3		159	.00	0.0	.00	.00	1	1			
tub, soybean & cottonseed	17	2.8	1.9	0.5		48	0	.00	.00	0	46	1	0		
1 t	5	0.0	0.0	0.3		159	.00	0.0	.00	.00	1	1			
mayonnaise															
Best Foods/Hellmann's	100	2.3	11.2	6.1	7						80				
1 T	14	0.2	0.1	1.7											
safflower & soybean	99	2.1	11.0	7.6		39	0				78	2		.02	
1 T	14	0.2	0.4	1.2	0.0		.00				5	4	.10		
soybean	99	2.1	11.0	5.7	8	39	1				78	2		.02	
1 T	14	0.2	0.4	1.6	0.0		.00				5	4	.10		
mayonnaise, imitation, soybean	35	9.4	2.9	1.6	4						75			.02	
1 T	15	0.0	2.4	0.5	0.0										
miracle whip, Kraft	69	5.1	6.9	4.1	6		0	.00	.00	5	84	1	0	.05	
1 T	14	0.1	1.7	1.1		14	.00	0.0	.02	.02	2	2	.05	.005	
miracle whip, light, Kraft	44	7.6	3.9	2.2	3		0	.00	.00	5	96	1	0	.05	
1 T	14	0.1	2.2	0.6		14	.00	0.0	.04	.01	1	3	.05	.005	
sandwich spread	60	6.2	5.2	3.1	12										
1 T	15	0.1	3.4	0.8											
sandwich spread, Best Foods/Hellmann's	55	7.1	4.9	2.6	5						175				
1 T	15	0.1	2.4	0.8											
shedd's spread, Country Crock															
corn oil	50		6.0												
1 T	10	0.0	0.0												
soybean oil	50		5.0	2.0	0						70				
1 T	10	0.0	0.0	1.0											
vegetable oil spread by brand															
Blue Bonnet, 52% veg oil	80		8.0	3.0	0						110				
1 T		0.0	0.0	2.0							15				
Blue Bonnet, light tasty, 52% veg oil	60		7.0	2.0	0						100				
1 T		0.0	0.0	1.0							10				
Blue Bonnet, stick, 60% fat	50		6.0	2.0	0						55				
1 T		0.0	0.0	1.0							5				
Blue Bonnet, stick, 70% fat	90		10.0	3.0	0						95				
1 T		0.0	0.0	2.0							10				
Blue Bonnet, stick, 75% fat	90		11.0	3.0	0						95				
1 T		0.0	0.0	2.0							10				
Fleischmann's soft/stick, light corn oil	80		8.0	3.0	0						70				
1 T		0.0	0.0	2.0							10				
vegetable oil spread by form & type of oil															
stick, soybean & palm	26	1.8	2.9	0.9		48	0	.00	.00	0	48	1	0		
1 t	5	0.0	0.0	0.7		159	.00	0.0	.00	.00	1	1			
stick, soybean & cottonseed	26	1.8	2.9	0.3		48	0	.00	.00	0	48	1	0		
1 t	5	0.0	0.0	0.6		159	.00	0.0	.00	.00	1	1			

	KCAL / WT (g)	H₂O (g) / PRO (g)	FAT (g) / CHO (g)	PUFA (g) / SFA (g)	CHOL (mg) / DFIB (g)	A (RE) / A (IU)	C (mg) / B-1 (mg)	B-2 (mg) / NIA (mg)	B-6 (mg) / B-12 (mcg)	FOL (mcg) / PANT (mg)	Na (mg) / K (mg)	Ca (mg) / P (mg)	Mg (mg) / Fe (mg)	Zn (mg) / Cu (mg)	Mn (mg)

33. SUGARS, SYRUPS & OTHER SWEETENERS

Item	KCAL / WT	H₂O / PRO	FAT / CHO	PUFA / SFA	CHOL / DFIB	A(RE) / A(IU)	C / B-1	B-2 / NIA	B-6 / B-12	FOL / PANT	Na / K	Ca / P	Mg / Fe	Zn / Cu	Mn
honey	64	3.6	0.0				0	.01			1	1			
1 T	21	0.1	17.3			0	.00	0.1			11	1	.10		
jams/preserves	54	5.8	0.0				0ᵃ	.01			2	4			
1 T	20	0.1	14.0			0	.00	0.0			18	2	.20		
jelly	49	5.2	0.0				1ᵇ	.01			3	4			
1 T	18	0.0	12.7			0	.00	0.0			14	1	.30		
jelly, all flavors, Kraft	16	2.3	0.0	0.0	0		0	.00	.00	5	2	1	0		
1 t	7	0.0	4.2	0.0		0	.00	0.0	.00	.00	4	0	.05		
jelly, imitation, sweetened w/ saccharin,	4		0.0		0						4				
all flavors, Estee—*1/2 oz*	14	0.0	1.0								20				
marmalade, citrus	51	5.8	0.0	0.0			1	.00			3	7			
1 T	20	0.1	14.0	0.0			.00	0.0			7	12	.10		
molasses															
barbados	54	4.8	0.0					.04				49			
1 T	20	0.0	14.0				.01					10			
blackstrap	43	4.8	0.0	0.0				.04			19	137			
1 T	20	0.0	11.0				.02	0.4			585	17	3.20		
light	50	4.8	0.0	0.0				.01			3	33			
1 T	20	0.0	13.0	0.0			.01	0.0			183	9	.90		
med	46	4.8	0.0	0.0				.02			7	58			
1 T	20	0.0	12.0					0.2			213	14	1.20		
preserves, strawberry, reduced calorie,	6	3.8	0.0	0.0	0	0	.00		.00	1	7	1	0	.00	.010
Kraft—*1 t*	6	0.0	1.7	0.0		0	.00	0.0	.00	.00	9	8	.02	.001	
sugar															
brown	541	3.0	0.0	0.0			0	.04			44	123			
1 cup not packed	145	0.0	139.8	0.0		0	.01	0.3			499	28	4.90		
powdered	462	0.6	0.0	0.0			0	.00			1	0			
1 cup	120	0.0	119.4	0.0		0	.00	0.0			4	0	.10		
white, granulated	15	0.0	0.0	0.0			0	.00			0	0			
1 t	4	0.0	4.0	0.0		0	.00	0.0			0	0	.00		
white, granulated	46	0.1	0.0	0.0			0	.00			0	0			
1 T	12	0.0	11.9	0.0		0	.00	0.0			0	0	.00		
white, granulated	770	1.0	0.0	0.0			0	.00			2	0			
1 cup	200	0.0	199.0	0.0		0	.00	0.0			6	0	.20		
sugar substitute															
equal powderᶜ	4		0.0								0				
1 g pkt	1	0.0	1.0												
equal tabletᵈ	0		0.0								0				
1 tablet		0.0	0.7												
sprinkle sweet	2	0.2	0.0				0	.00			1	0			
1 t	0.7	0.0	0.5			0	.00	0.0			0	0	.00		
sweet 'n low	4		0.0		0						4				
1 pkt	1	0.0	0.9								3				
sweet 10	0		0.0								0				
1/8 t	0.8	0.0	0.0			0	.00	0.0			0	0	.00		
syrup															
cane & maple	50	6.5	0.0				0	.00			0	3			
1 T	20	0.0	12.8			0	.00	0.0			5	0	.00		
corn	59	4.9	0.0				0	.00			14	9			
1 T	21	0.0	15.4			0	.00	0.0			1	3	.80		

ᵃ For cherry/strawberry jams/preserves Vit C = 3 mg.
ᵇ For guava jelly, Vit C = 7 mg.
ᶜ Contains dextrose w/ dried corn syrup, aspartame, silicon dioxide, cellulose, tribasic calcium phosphate & cellulose derivatives.
ᵈ Contains lactose, aspartame, maltodextrin, leucine, cellulose & cellulose derivatives.

	KCAL	H₂O (g)	FAT (g)	PUFA (g)	CHOL (mg)	A (RE)	C (mg)	B-2 (mg)	B-6 (mg)	FOL (mcg)	Na (mg)	Ca (mg)	Mg (mg)	Zn (mg)	Mn (mg)
	WT (g)	PRO (g)	CHO (g)	SFA (g)	DFIB (g)	A (IU)	B-1 (mg)	NIA (mg)	B-12 (mcg)	PANT (mg)	K (mg)	P (mg)	Fe (mg)	Cu (mg)	
corn, dark, Karo	60	5.4	0.0								40				
1 T	21	0.0	15.0ᵃ												
corn, light, Karo	60	5.4	0.0								30				
1 T	21	0.0	14.9ᵇ												
maple	50	6.5	0.0				0				2	20			
1 T	20	0.0	12.8								35	2	.20		
maple honey, Log Cabin	106	10.7	0.0	0.0	0		0	.00	.00		7	7			
1 fl oz	39	0.0	28.1	0.0	0.0	0	.00	0.0	.00	.00	6	2	.43	.047	
pancake & waffle															
Aunt Jemima	103		0.0					.00	.00	0	21	0	0	.00	.000
1 fl oz	40	0.0	25.8			0	.00	0.0	.00	.00	3	0	.00	.000	
Aunt Jemima butter lite	52		0.0					.00	.00	0	67	0	0	.00	.000
1 fl oz	34	0.0	12.9			0	.00	0.0	.00	.00	0	0	.00	.000	
Aunt Jemima lite	60		0.0					.00	.00	0	66	0	0	.00	.000
1 fl oz	34	0.0	15.1			0	.00	0.0	.00	.00	1	0	.00	.000	
Golden Griddle	57	5.7	0.0								15				
1 T	20	0.0	14.2ᶜ												
Karo	60	5.5	0.0								35				
1 T	21	0.0	14.9ᵈ												
Log Cabin	99	12.7	0.0	0.0	0		0	.00	.00	0	38	2	1	.04	
1 fl oz	39	0.0	26.1	0.0	0.0	0	.00	0.0	.00	.00	3	1	.06	.003	
Log Cabin buttered	105	11.9	0.6	0.0	2		0	.00	.00	0	74	5	0	.00	
1 fl oz	39	0.0	26.1	0.4	0.0	24	.00	0.0	.00	.00	1	2	.41	.036	
Log Cabin country kitchen	101	12.7	0.0	0.0	0		0	.00	.00	0	19	11			
1 fl oz	39	0.0	26.0	0.0	0.0	0	.00	0.0	.00	.00	1	4	.95	.083	
Log Cabin lite	61	19.8	0.0	0.0	0		0	.00	.00	0	84	0	0	.00	
1 fl oz	36	0.0	15.8	0.0	0.0	0	.00	0.0	.00	.00	1	15	.01	.001	
sorghum	53	4.7	0.0				.02					35			
1 T	21	0.0	14.0				0.0					5	2.60		

34. VEGETABLES, VEGETABLE PRODUCTS & VEGETABLE SALADSᵉ

	KCAL	H₂O (g)	FAT (g)	PUFA (g)	CHOL (mg)	A (RE)	C (mg)	B-2 (mg)	B-6 (mg)	FOL (mcg)	Na (mg)	Ca (mg)	Mg (mg)	Zn (mg)	Mn (mg)
	WT (g)	PRO (g)	CHO (g)	SFA (g)	DFIB (g)	A (IU)	B-1 (mg)	NIA (mg)	B-12 (mcg)	PANT (mg)	K (mg)	P (mg)	Fe (mg)	Cu (mg)	
adzuki beans															
boiled	294	152.5	0.2		0	1	0	.15			18	63	120	4.06	1.318
1 cup	230	17.3	57.0			13	.27	1.6	.00		1224	385	4.60	.685	
cnd, sweetened	702	120.1	0.1		0						646	66	91		
1 cup	296	11.3	162.8						.00		353	220	3.34		
w/ sugar (yokan)	112	15.2	0.1		0						36	12	8		
three 1/4" thick slices	43	1.4	26.1						.00		19	17	.50		
alfalfa seeds, sprouted, raw	10	30.1	0.2	0.1	0	5	3	.04	.01	12	2	10	9	.30	.062
1 cup	33	1.3	1.3	0.0	0.7	51	.03	0.2	.00	.19	26	23	.32	.052	
amaranth, boiled	14	60.4	0.1	0.1	0	183	27	.09			14	138	36		
1/2 cup	66	1.4	2.7	0.0		1828	.01	0.4	.00		423	47	1.49		
arrowhead, boiled	9	9.3	0.0			0	0	.01			2	1	6		
1 med corm (1" diameter)	12	0.5	1.9			0	.02	0.1	.00		106	24	.15		
artichoke															
boiled	53	103.8	0.2	0.1	0	17	9	.06	.10	54	79	47	47	.43	.328
1 med	300ᶠ	2.8	12.4	0.0		172	.07	0.7	.00	.24	316	72	1.62	.083	
frzn, boiled	36	69.2	0.4	0.2	0	13	4	.13	.07	95	42	17	25	.29	.218
2.8 oz	80	2.5	7.3	0.1		131	.05	0.7	.00	.16	211	49	.45	.049	
artichoke hearts, boiled	37	72.7	0.1	0.1	0	12	6	.04	.07	37	55	33	33	.30	.229
1/2 cup	84	1.9	8.7	0.0		121	.05	0.5	.00	.17	221	50	1.13	.051	
asparagus															
boiled	22	82.8	0.3	0.1	0	75	18	.11	.13	88	4	22	17	.43	.187
1/2 cup (6 spears)	90	2.3	4.0	0.1		746	.09	0.9		.15	279	54	.59	.090	

ᵃ 0.3 g fructose, 3.1 g glucose, 0.6 g sucrose, 2.0 g maltose, 1.6 g trisaccharides & 7.4 g other polysaccharides.

ᵇ 0.4 g fructose, 3.9 g glucose, 1.9 g maltose, 1.6 g trisaccharides & 7.1 g other polysaccharides.

ᶜ 0.9 g fructose, 3.2 g glucose, 2.2 g disaccharides, 1.6 g trisaccharides & 6.3 g other polysaccharides.

ᵈ 0.8 g fructose, 3.6 g glucose, 1.9 g maltose, 1.5 g trisaccharides & 7.1 g other polysaccharides.

ᵉ Beans are listed by type unless the product is specified simply as "beans".

ᶠ Weight w/ refuse.

	KCAL / WT (g)	H₂O (g) / PRO (g)	FAT (g) / CHO (g)	PUFA (g) / SFA (g)	CHOL (mg) / DFIB (g)	A (RE) / A (IU)	C (mg) / B-1 (mg)	B-2 (mg) / NIA (mg)	B-6 (mg) / B-12 (mcg)	FOL (mcg) / PANT (mg)	Na (mg) / K (mg)	Ca (mg) / P (mg)	Mg (mg) / Fe (mg)	Zn (mg) / Cu (mg)	Mn (mg)
cnd	24	113.7	0.8	0.3	0					116				.49	
1/2 cup	121	2.6	3.0	0.2					.00				2.22	.116	
frzn, boiled	17	54.7	0.3	0.1	0	49	15	.06	.01	81	2	14	8	.33	.111
4 spears	60	1.8	2.9	0.1		491	.04	0.6	.00	.10	131	33	.38	.103	
avocado, raw															
calif	306	125.5	30.0	3.5	0	106	14	.21	.48	113	21	19	70	.73	.422
1 med	173	3.6	12.0	4.5	4.7	1059	.19	3.3	.00	1.68	1097	73	2.04	.460	
florida	339	242.4	27.0	4.5	0	186	24	.37	.85	162	14	33	104	1.28	.517
1 med	304	4.8	27.1	5.3		1860	.33	5.8	.00	2.95	1484	119	1.60	.763	
bamboo shoots															
raw	21	69.2	0.2	0.1	0	2	3	.05			3	10	2		
1/2 cup	76	2.0	4.0	0.1	2.0	15	.11	0.5	.00		405	45	.38		
boiled	15	115.1	0.3	0.1	0	0	0	.06			5	14	3		
1 cup	120	1.8	2.3	0.1		0	.02	0.4	.00		640	24	.29		
cnd	25	123.6	0.5	0.2	0	1	1	.03			9	10	6		
1 cup	131	2.3	4.2	0.1		11	.03	.18	.00		104	33	.42		
bean salad, cnd															
3 bean, Joan of Arc	90		0.3				2	.05			920	44			
1/2 cup	127	2.2	19.8			239	.05	0.9			188	39	1.02		
3 bean, Pillsbury	80	91.5	0.2				3	.06			522	31			
1/2 cup	113	2.3	17.4			305	.03	0.4			116	42	.34		
4 bean, Joan of Arc	120		0.2				3	.04			894	46			
1/2 cup	138	2.8	26.9			282	.06	0.7			185	48	1.10		
beans, cnd															
baked															
B&M	330		8.0								770				
7/8 cup	227	16.0	49.0								840				
Friend's	360		5.0								1270				
1 cup	255	17.0	62.0												
B&M	290		4.0								1050				
7/8 cup	227	15.0	49.0								1010				
Campbell's	248		4.0				5	.07			1037	95			
7.9 oz	223	10.8	42.3			501	.10	1.1			632		3.00		
Special Recipe	217		3.5				5	.06			907	83			
6.9 oz	195	9.4	37.0			438	.09	1.0			552		2.60		
Joan of Arc	100		1.1				0	.10			691	35			
1/2 cup	119	6.1	16.7			495	.12	0.8			224	113	2.14		
mexican, Van Camp's	210	171.2	2.4					.14			718	75			
1 cup	227	11.4	35.8 ·			1595	.12	0.9			640		4.42		
baked, garden style, Special Recipe	191		1.5				3	.08			807	128			
6.9 oz	195	7.2	37.2			638	.09	0.5			421		2.90		
baked, homestyle, Campbell's	265		3.1				6	.07			1105	133			
8 oz	227	11.5	47.8			343	.09	1.1			584		4.00		
baked, honey, B&M	280		2.0								940				
7/8 cup	227	15.0	50.0								930				
baked, maple, Friend's	280		2.0								1000				
7/8 cup	227	15.0	50.0												
baked, new england style, Special Recipe—*6.9 oz*	256		4.4				5	.11			755	172			
	195	10.3	43.9			0	.06	0.4			350		4.20		
baked, old fashioned, Special Recipe	225		3.8				5	.08			817	126			
6.9 oz	195	7.7	39.9			0	.09	0.6			307		2.90		
baked, ranchero, Campbell's	216		4.6				6	.08			841	90			
7.7 oz	219	10.3	33.4			601	.10	0.9			668		3.80		
baked, tomato, B&M	270		3.0								1010				
7/8 cup	227	12.0	48.0								920				
baked, vegetarian	235	184.5	1.1	0.5	0	43		.15	.34	61	1008	128	82	3.55	.879
1 cup	254	12.2	52.1	0.3	6.6	434	.39	1.1	.00	.24	752	264	.74	.523	

	KCAL	H₂O (g)	FAT (g)	PUFA (g)	CHOL (mg)	A (RE)	C (mg)	B-2 (mg)	B-6 (mg)	FOL (mcg)	Na (mg)	Ca (mg)	Mg (mg)	Zn (mg)	Mn (mg)
	WT (g)	PRO (g)	CHO (g)	SFA (g)	DFIB (g)	A (IU)	B-1 (mg)	NIA (mg)	B-12 (mcg)	PANT (mg)	K (mg)	P (mg)	Fe (mg)	Cu (mg)	
baked, western, Van Camp's	207	173.7	3.8					.14			1006	73			
1 cup	227	11.1	32.0			1962	.11	0.9			589		2.71		
baked w/ brown sugar, Van Camp's	284	155.9	5.1					.12			692	121			
1 cup	227	11.6	47.8			0	.09	0.9			535		3.83		
baked w/ molasses & brown sugar, Campbell's—8 oz	265		2.9				9	.06			1017	128			
1 cup	227	11.3	48.5			116	.06	0.8			600				
baked w/ pork & sweet sce	282	178.9	3.7	0.5	17	29	8	.15	.22	95	849	155	87	3.80	.939
1 cup	253	13.4	53.1	1.4		289	.12	0.9	.06	.26	673	266	4.20	.253	
baked w/ pork & tomato sce	247	183.9	2.6	0.3	17	31	8	.12	.18	57	1113	141	88	14.83	1.240
1 cup	253	13.1	49.1	1.0	5.6	313	.13	1.3	.03	1.34	759	297	8.30	.643	
Campbell's	237		2.9				5	.05			730	102			
8 oz	227	9.9	42.9			330	.13	0.6			400		3.60		
Joan of Arc	130		0.7				1	.07			688	60			
1/2 cup	133	6.5	23.9			86	.07	1.0			245	82	3.86		
Van Camp's	216	168.9	1.9					.12			1011	97			
1 cup	227	10.9	38.8			164	.12	1.0			563		4.29		
beans, baked, homemade[a]	382	164.9	13.0	1.9	13	0	3	.12	.23	122	1068	155	110	1.84	.645
1 cup	253	14.0	54.1	4.9		1	.34	1.0	.03	.39	907	275	5.04	.402	
beans, refried, cnd	270	182.9	2.7	0.3			15	.14			1071	118	99	3.45	
1 cup	253	15.8	46.8	1.0			.12	1.2			994	214	4.47	1.040	
beet greens, boiled	20	64.2	0.1	0.1	0	367	18	.21	.10		173	82	49	.36	
1/2 cup	72	1.9	3.9	0.0		3672	.08	.36	.00	.24	654	29	1.37	.181	
beets															
boiled	26	77.3	0.0	0.0	0	1	5	.01	.03	45	42	9	31	.21	.204
1/2 cup slices	85	0.9	5.7	0.0		11	.03	0.2	.00	.08	266	26	.53	.048	
cnd	27	77.3	0.1	0.0	0							2		.18	.244
1/2 cup slices	85	0.8	6.1	0.0					.00				1.55	.050	
cnd, harvard	89	98.6	0.1	0.0	0		3	.06			199	13	24		
1/2 cup slices	123	1.0	22.4	0.0				0.1	.00		201	21	.44		
cnd, pickled	75	93.3	0.1	0.0	0	1	3	.06			301	13	18	.30	
1/2 cup slices	114	0.9	18.6	0.0		7	.03	0.3	.00		169	20	.47	.132	
black beans, boiled	227	113.1	0.9	0.4	0	1	0	.10	.12	256	1	47	121	1.92	.764
1 cup	172	15.2	40.8	0.2	7.2	10	.42	0.9	.00	.42	611	241	3.60	.359	
black turtle beans															
boiled	241	121.6	0.6	0.3	0	1	0	.10	.14	158	6	103	91	1.41	.605
1 cup	185	15.1	45.1	0.2		11	.42	1.0	.00	.48	801	282	5.27	.498	
cnd	218	181.5	0.7	0.3	0	1	6	.29	.13	146	922	84	84	1.30	.559
1 cup	240	14.5	39.7	0.2		10	.34	1.5	.00	.44	739	260	4.56	.461	
broadbeans															
boiled	186	121.6	0.7	0.3	0	3	1	.15	.12	177	8	62	73	1.72	.716
1 cup	170	12.9	33.4	0.1	8.7	26	.17	1.2	.00	.27	456	212	2.54	.440	
cnd	183	205.6	0.6	0.2	0	3	5	.13	.12	84	1161	67	82	1.60	.737
1 cup	256	14.0	31.8	0.1		26	.05	2.5	.00	.31	620	202	2.56	.279	
falafel[b]	170	17.7	9.1	2.1	0	1	1	.09	.06	40	150	27	42	.77	.352
3 patties w/ 2 1/4" diameter	51	6.8	16.2	1.2		7	.07	0.5	.00	.15	298	98	1.74	.132	
broccoli															
raw	12	39.9	0.2	0.1	0	68	41	.05	.07	31	12	21	11	.18	.101
1/2 cup chopped	44	1.3	2.3	0.0	0.6	678	.03	0.3	.00	.24	143	29	.39	.020	
boiled	23	70.4	0.2	0.1	0	110	49	.16	.15	54	8	89	47	.12	.191
1/2 cup	78	2.3	4.3	0.0		1099	.06	0.6	.00	.23	127	37	.89	.054	
frzn															
chopped, boiled	25	83.5	0.1	0.1	0	174	37	.08	.12	52	22	47	19	.28	.299
1/2 cup	92	2.9	4.9	0.0	2.0	1741	.05	0.4	.00	.25	166	51	.56	.040	
spears, boiled	25	83.5	0.1	0.1	0	174	37	.08	.12	79	22	47	19	.28	.299
1/2 cup	92	2.9	4.9	0.0	2.0	1741	.05	0.4	.00	.25	166	51	.56	.040	
spears w/ butter sce, Pillsbury	51	94.3	2.0				31	.07			387	38			
1/2 cup	106	2.1	5.9			245	.02	0.4			166	47	.39		

[a] White beans, onions, salt pork, molasses, brown sugar, vinegar, salt, pepper & dry mustard.   [b] Broadbeans, soybean oil, onions, flour, salt, garlic, coriander & cumin.

	KCAL	H₂O (g)	FAT (g)	PUFA (g)	CHOL (mg)	A (RE)	C (mg)	B-2 (mg)	B-6 (mg)	FOL (mcg)	Na (mg)	Ca (mg)	Mg (mg)	Zn (mg)	Mn (mg)
	WT (g)	PRO (g)	CHO (g)	SFA (g)	DFIB (g)	A (IU)	B-1 (mg)	NIA (mg)	B-12 (mcg)	PANT (mg)	K (mg)	P (mg)	Fe (mg)	Cu (mg)	
w/ cheddar cheese sce, Pillsbury	60	95.9	2.5				38	.10			436	70			
1/2 cup	109	3.1	6.2			308	.03	0.7			158	71	.33		
w/ cheese sce, Birds Eye	116	117.4	6.2	0.9	6		43	.19	.13	68	488	109	20	.47	
1/2 cup	142	4.8	11.6	1.9		3490	.06	0.5	.30	.42	249	103	.76	.040	
w/ creamy italian cheese sce, Birds Eye—*1/2 cup*	90	109.1	5.6	0.4	14		47	.15	.13	74	391	93	17	.43	
	128	5.0	6.6	3.0		1941	.06	0.5	.25	.32	225	86	.91	.036	
brussels sprouts															
boiled	30	68.1	0.4	0.2	0	56	48	.06	.14	47	17	28	16	.25	.177
1/2 cup (4 sprouts)	78	2.0	6.8	0.1	1.1	561	.08	0.5	.00	.20	247	44	.94	.065	
frzn															
boiled	33	67.7	0.3	0.2	0	46	36	.09	.23	79	18	19	19	.28	.250
1/2 cup	78	2.8	6.5	0.1	1.4	459	.08	0.4	.00	.27	254	42	.58	.065	
w/ butter sce, Pillsbury	60	81.9	1.0				59	.08			296	28			
1/2 cup	102	2.9	9.8			785	.09	0.7			298	49	.45		
w/ cheese sce, Birds Eye	113	103.0	5.6	0.9	5		56	.19	.13	90	417	77	21	.40	
1/2 cup	128	5.0	12.5	1.7		2622	.09	0.6	.26	.38	359	96	.84	.028	
w/ cheese sce, Pillsbury	79	97.9	1.5				59	.18			493	67			
1/2 cup	118	3.1	13.0			610	.06	1.2			325	185	.52		
burdock root, boiled	110	94.6	0.2		0	0		.07			5	62	49		
1 cup	125	2.6	26.4			0	.05	0.4	.00		450	116	.96		
butter beans															
cnd, Joan of Arc	100		0.4				0	.05			434	40			
1/2 cup	122	5.7	18.3			59	.06	0.6			254	93	1.83		
cnd, Van Camp's	162	180.9	0.5					.08			752	41			
1 cup	227	11.0	28.4			0	.09	0.5			639		3.78		
butterbur (fuki), boiled	8	96.7	0.0		0	3	19	.01			4	59	8		
3.5 oz	100	0.2	2.2			27	.01	0.1	.00		354	7	.10		
cabbage, chinese (pak-choi)															
raw	5	33.4	0.1	0.0	0	105	16	.03			23	37	7		
1/2 cup shredded	35	0.5	0.8	0.0		1050	.01	0.2	.00		83	13	.28		
boiled	10	81.2	0.1	0.1	0	218	22	.05			29	79	9		
1/2 cup shredded	85	1.3	1.5	0.0		2183	.03	0.4	.00		315	25	.88		
cabbage, green															
raw	8	32.4	0.1	0.0	0	4	17	.01	.03	20	6	16	5	.06	.056
1/2 cup shredded	35	0.4	1.9	0.0	0.4	44	.02	0.1	.00	.05	86	8	.20	.008	
boiled	16	70.2	0.1	0.1	0	6	18	.04	.05	15	14	25	11	.12	.097
1/2 cup shredded	75	0.7	3.6	0.0		64	.04	0.2	.00	.05	154	18	.29	.021	
cabbage, red															
raw	10	32.0	0.1	0.0	0	1	20	.01	.07	7	4	18	5	.07	.063
1/2 cup shredded	35	0.5	2.1	0.0	0.4	14	.02	0.1	.00	.11	72	15	.17	.034	
boiled	16	70.2	0.2	0.1	0	2	26	.02	.11	9	6	28	8	.11	.097
1/2 cup shredded	75	0.8	3.5	0.0		20	.03	0.2	.00	.17	105	21	.27	.052	
cabbage, savoy															
raw	10	31.9	0.0	0.0	0	35	11	.01	.07		10	12	10		
1/2 cup shredded	35	0.7	2.1	0.0		350	.03	0.1	.00		81	15	.14		
boiled	18	67.2	0.1	0.0	0	65	12	.02	.11		17	22	17		
1/2 cup shredded	73	1.3	4.0	0.0		649	.04	0.0	.00		134	24	.28		
cannellini beans (white kidney beans), cnd, Progresso—*8 oz*	180		1.0								717				
	227	12.0	30.0												
cardoon, boiled	22	93.5	0.1	0.0	0	12	2	.03			176	72	43		
3.5 oz	100	0.8	5.3	0.0		118	.02	0.3	.00		392	23	.73		
carrots															
raw	31	63.2	0.1	0.1	0	2025	7	.04	.11	10	25	19	11	.14	.102
1 med	72	0.7	7.3	0.0	1.1	20253	.07	0.7	.00	.14	233	32	.36	.034	

	KCAL	H₂O (g)	FAT (g)	PUFA (g)	CHOL (mg)	A (RE)	C (mg)	B-2 (mg)	B-6 (mg)	FOL (mcg)	Na (mg)	Ca (mg)	Mg (mg)	Zn (mg)	Mn (mg)
	WT (g)	PRO (g)	CHO (g)	SFA (g)	DFIB (g)	A (IU)	B-1 (mg)	NIA (mg)	B-12 (mcg)	PANT (mg)	K (mg)	P (mg)	Fe (mg)	Cu (mg)	
boiled	35	68.2	0.1	0.1	0	1915	2	.04	.19	11	52	24	10	.23	.587
1/2 cup slices	78	0.9	8.2	0.0	1.5	19152	.03	0.4	.00	.24	177	24	.48	.105	
cnd	17	67.9	0.1	0.1	0	1006	2	.02	.08	7	176[a]	19	6	.19	.329
1/2 cup slices	73	0.5	4.0	0.0	0.9	10050	.01	0.4	.00	.10	131	17	.47	.076	
frzn, boiled	26	65.6	0.1	0.0	0	1292	2	.03	.09	8	43	21	7	.18	.296
1/2 cup slices	73	0.9	6.0	0.0	1.3	12922	.02	0.3	.00	.12	115	19	.35	.053	
cassava, raw	120	68.5	0.4	0.1	0	1	48	.10			8	91	66		
3.5 oz	100	3.1	26.9	0.1		10	.23	1.4			764	70	3.60		
cauliflower															
raw	12	46.1	0.1	0.0	0	1	36	.03	.12	33	7	14	7	.09	.102
1/2 cup pieces	50	1.0	2.5	0.0		8	.04	0.3	.00	.07	178	23	.29	.016	
boiled	15	57.4	0.1	0.1	0	1	34	.03	.13	32	4	17	7	.15	.110
1/2 cup pieces	62	1.2	2.9	0.0	1.0	9	.04	0.3	.00	.08	200	22	.26	.056	
frzn															
boiled	17	84.6	0.2	0.1	0	2	28	.05	.08	37	16	15	8	.12	.135
1/2 cup pieces	90	1.5	3.4	0.0	1.5	20	.03	0.3	.00	.09	125	22	.37	.022	
w/ butter sce, Pillsbury	30	80.1	1.2				24	.04			319	21			
1/2 cup	88	1.1	3.5			64	.02	0.4			123	24	.35		
w/ cheddar cheese sce, Pillsbury	70	104.1	3.5				52	.12			446	69			
1/2 cup	121	3.4	7.1			70	.08	0.5			247	75	.79		
w/ cheese sce, Birds Eye	114	118.2	6.1	0.9	6		39	.16	.12	42	483	83	17	.35	
1/2 cup	142	4.0	11.7	1.9		2298	.06	0.4	.30	.33	240	88	.50	.026	
w/ cheese sce, Pillsbury	60	92.4	1.7				30	.13			427	56			
1/2 cup	107	1.5	9.2			703	.03	0.1			229	156	.37		
celeriac, raw	25	92.0	0.2		0	0	4	.04	.01		61	26	12		
3.5 oz	100	1.0	5.9			0	.03	0.4	.00		173	66	.43		
celery															
raw	6	37.9	0.1	0.0	0	5	3	.01	.01	4	35	14	5	.07	.054
1 stalk (7.5" long)	40	0.3	1.5	0.0	0.4	51	.01	0.1	.00	.07	114	10	.19	.014	
boiled	11	71.3	0.1	0.0	0	8	4	.02	.02	5	48	27	9	.12	.092
1/2 cup diced	75	0.4	2.6	0.0		81	.02	0.2	.00	.11	266	18	.10	.023	
celtuce, raw	22	94.5	0.3		0	350	20	.07			1	39	28		
12 leaves	100	0.9	3.7			3500	.06	0.6	.00		330	39	.55		
chard, swiss, boiled	18	81.5	0.1		0	276	16	.08			158	51	76		
1/2 cup chopped	88	1.7	3.6			2762	.03	0.3	.00	.14	483	29	1.99		
chayote, boiled	19	74.7	0.4		0	4	6	.03			1	10	9		
1/2 cup pieces	80	0.5	4.1			37	.02	0.3	.00	.33	138	23	.18		
chickpeas (garbanzo beans)															
boiled	269	98.8	4.3	1.9	0	4	2	.10	.23	282	11	80	78	2.51	1.689
1 cup	164	14.5	45.0	0.4	5.7	44	.19	0.9	.00	.47	477	275	4.74	.577	
cnd	285	167.3	2.7	1.2	0	6	9	.08	1.14	160	718	78	70	2.53	1.450
1 cup	240	11.9	54.3	0.3		58	.07	0.3	.00	.72	413	216	3.23	.418	
hummus[b]	420	159.7	20.8	7.8	0	6	19	.13	.98	146	599	124	71	2.70	1.400
1 cup	246	12.1	49.6	3.1		61	.23	1.0	.00	.71	427	275	3.87	.561	
chickory greens, raw	21	82.8	0.3	0.1	0	360	22	.09			41	90	27		
1/2 cup chopped	90	1.5	4.2	0.1		3600	.05	0.5	.00		378	42	.81		
chickory witloof, raw	7	42.8	0.1	0.0	0	0	5	.06	.02		3	6			
1/2 cup	45	0.5	1.4	0.0		0	.03	0.2	.00		82	9	.23		
chives															
raw	1	2.8	0.0	0.0	0	19	2	.01	.01		0	2	2		
1 T chopped	3	0.1	0.1	0.0		192	.00	0.0	.00	.01	8	2	.05		
freeze-dried	2	0.0	0.0	0.0	0	55	5	.01				7			
1/4 cup	0.8	0.2	0.5	0.0		546	.01	0.0	.00		24	4	.16		
coleslaw															
Fresh Chef	199		14.6				77	.03			214	38			
4 oz	113	0.9	16.0			337	.05	0.2			122		1.40		

[a] The special dietary pack contains 31 mg Na. [b] Chickpeas, lemon jce, tahini, olive oil & garlic.

	KCAL / WT (g)	H₂O (g) / PRO (g)	FAT (g) / CHO (g)	PUFA (g) / SFA (g)	CHOL (mg) / DFIB (g)	A (RE) / A (IU)	C (mg) / B-1 (mg)	B-2 (mg) / NIA (mg)	B-6 (mg) / B-12 (mcg)	FOL (mcg) / PANT (mg)	Na (mg) / K (mg)	Ca (mg) / P (mg)	Mg (mg) / Fe (mg)	Zn (mg) / Cu (mg)	Mn (mg)
homemade[a]	42	48.9	1.6	0.8	5	49	20	.04	.08	16	14	27	6	.12	.058
1/2 cup	60	0.8	7.5	0.2		381	.04	0.2	.02	.08	109	19	.35	.014	
collards															
boiled	27	181.9	0.3		0	422	19	.08	.08	12	36	148	21	1.22	.467
1 cup chopped	190	2.1	5.0			4218	.03	0.4	.00	.08	177	19	.78	.285	
frzn, boiled	31	75.2	0.4		0	508	23	.10	.10	65	42	179	26	.23	.564
1/2 cup chopped	85	2.5	6.1			5084	.04	0.5	.00	.10	214	23	.95	.047	
coriander, raw	1	3.7	0.0		0	11	0	.01			1	4	1		
1/4 cup	4	0.1	0.1			111	.00	0.0	.00		22	1	.08		
corn, white															
cnd, Pillsbury	90	70.5	0.4				9	.07			259	5			
1/2 cup	94	2.2	19.5			300	.06	1.4			198	60	.32		
frzn, Pillsbury	80	50.8	0.8				5	.04			2	6			
1/2 cup	70	2.0	16.0			13	.03	0.8			159	51	.25		
frzn w/ butter sce, Pillsbury	110	84.4	2.0				7	.06			301	7			
1/2 cup	111	2.6	20.5			31	.03	1.0			202	63	.26		
corn, yellow															
boiled	89	57.0	1.1	0.5	0	18[b]	5	.06	.05	38	14	2	26	.39	.159
1/2 cup	82	2.7	20.6	0.2		178[b]	.18	1.3	.00	.72	204	84	.50	.043	
cnd	66	63.1	0.8	0.4	0	13[b]								.32	.142
1/2 cup	82	2.2	15.2	0.1	1.1	128[b]			.00				.70	.048	
cream style	93	100.8	0.3	0.3	0	12[b]	6	.07	.08	57	365[c]	4	22	.68	.050
1/2 cup	128	2.2	23.2	0.1		124[b]	.03	1.2	.00	.23	172	65	.49	.067	
vacuum pack	83	80.4	0.5	0.2	0	25[b]	9	.08	.06	52	286[d]	5	24	.48	.070
1/2 cup	105	2.5	20.4	0.1		253[b]	.04	1.2	.00	.71	195	67	.44	.050	
w/ red & green peppers	86	88.4	0.6	0.3	0	26	10	.09			396	5	29	.42	
1/2 cup	114	2.7	20.7	0.1		265	.03	1.1	.00		174	71	.90	.068	
frzn															
boiled	67	62.1	0.1	0.0	0	20[b]	2	.06	.08	19	4	2	15	.28	.148
1/2 cup	82	2.5	16.8	0.0	1.7	204[b]	.06	1.1	.00	.18	114	39	.25	.027	
cream style, Pillsbury	120	97.5	0.6				5	.05			320	9			
1/2 cup	128	2.9	25.6			166	.09	0.6			183	65	.26		
on-the-cob, Birds Eye	120	90.5	1.0	0.5	0		8	.11	.22	49	4	5	40	.87	
1 ear	125	4.1	28.7	0.1		261	.15	2.1	.00	.37	378	113	.86	.063	
on-the-cob, Pillsbury	150	95.8	1.2				12	.20			20	4			
1 ear	133	3.9	31.0			258	.12	2.4			358	109	.65		
w/ butter sce, Niblets	100	87.7	2.1				6	.06			275	8			
1/2 cup	111	2.4	17.9			78	.08	1.5			176	57	.36		
w/ peppers, Mexicorn	80	70.8	0.3				6	.07			316	6			
1/2 cup	92	1.8	17.3			513	.05	1.0			151	51	.37		
corn pudding[e]	271	190.8	13.3	1.7	230	89	7	.32	.30	63	138	100	37	1.26	
1 cup	250	11.0	31.9	6.3		616	1.03	2.5	.23	.62	402	143	1.39	.108	
cornsalad, raw[f]	6	26.0	0.1		0										
1/2 cup	28	0.6	1.0						.00						
cowpeas, (blackeye peas)															
boiled	198	119.8	0.9	0.4	0	3	1	.09	.17	356	6	42	91	2.20	.812
1 cup	171	13.2	35.5	0.2	4.4	26	.35	0.8	.00	.70	476	266	4.29	.458	
cnd	184	191.1	1.3	0.6	0	3	7	.18	.11	123	718	48	66	1.68	.679
1 cup	240	11.4	32.7	0.3		32	.18	0.9	.00	.46	413	167	2.34	.281	
frzn, boiled	112	56.2	0.6	0.1	0	6	2	.05	.08	120	5	20	42	1.21	.672
1/2 cup	85	7.2	20.2	0.2		64	.22	0.6	.00	.18	319	104	1.80	.156	
young pods w/ seeds, boiled	16	42.1	0.1	0.0	0	66	8	.04			1	26			
1/2 cup	47	1.2	3.3	0.0		658	.04	0.4	.00		92	23	.33		
cowpeas, catjang, boiled	200	119.2	1.2	0.5	0	2	1	.08	.16	242	32	44	165	3.20	.809
1 cup	171	13.9	34.7	0.3		17	.28	1.2	.00	.66	641	242	5.22	.463	

[a] Cabbage, celery, table cream, sugar, green pepper, lemon jce, onion, pimiento, vinegar, salt, dry mustard & white pepper.
[b] For yellow varieties: white varieties contain only a trace of vitamin A.
[c] Special dietary pack contains 4 mg Na.
[d] Special dietary pack contains 3 mg Na.
[e] Yellow corn, whole milk, egg, sugar, butter, salt & pepper.
[f] European herb widely cultivated as a salad plant & potherb.

	KCAL / WT (g)	H₂O (g) / PRO (g)	FAT (g) / CHO (g)	PUFA (g) / SFA (g)	CHOL (mg) / DFIB (g)	A (RE) / A (IU)	C (mg) / B-1 (mg)	B-2 (mg) / NIA (mg)	B-6 (mg) / B-12 (mcg)	FOL (mcg) / PANT (mg)	Na (mg) / K (mg)	Ca (mg) / P (mg)	Mg (mg) / Fe (mg)	Zn (mg) / Cu (mg)	Mn (mg)
cranberry beans															
boiled	240	114.4	0.8	0.4	0	0	0	.12	.14	366	1	89	89	2.01	.655
1 cup	177	16.5	43.3	0.2	6.0	0	.37	0.9	.00	.43	685	240	3.70	.409	
cnd	216	201.6	0.7	0.3	0	0	2	.10	.14	201	863	87	83	2.18	.520
1 cup	260	14.4	39.3	0.2		0	.10	1.3	.00	.37	675	223	4.02	.369	
cucumber, raw	7	49.9	0.1	0.0	0	2	2	.01	.03	7	1	7	6	.12	.032
1/2 cup slices (1/6 cucumber)	52	0.3	1.5	0.0	0.3	23	.02	0.2	.00	.13	78	9	.14	.021	
dandelion greens															
raw	13	24.0	0.2		0	392	10	.07			21	52	10		
1/2 cup chopped	28	0.8	2.6			3920	.05		.00		111	18	.87		
boiled	17	46.7	0.3		0	608	9	.09			23	73			
1/2 cup	52	1.0	3.3			6084	.07		.00		121	22	.94		
dock															
raw	15	62.3	0.5		0	268	32	.07			3	29	69		
1/2 cup chopped	67	1.3	2.1			2680	.03	0.3	.00		261	42	1.61		
boiled	20	93.6	0.6		0	347	26	.09			3	38	89		
3.5 oz	100	1.8	2.9			3474	.03	0.4	.00		321	52	2.08		
eggplant															
raw	11	37.7	0.0	0.0	0	3	1	.01	.04	7	1	15	5	.06	.057
1/2 cup pieces	41	0.5	2.6	0.0	0.6	29	.04	0.2	.00	.03	90	13	.22	.046	
boiled	13	44.1	0.0	0.0	0	3	1	.01	.04	7	2	3	6	.07	.065
1/2 cup	48	0.4	3.2	0.0		31	.04	0.3	.00	.04	119	11	.17	.052	
endive, raw	4	23.5	0.1	0.0	0	51	2	.02	.01	36	6	13	4	.20	.105
1/2 cup chopped	25	0.3	0.8	0.0		513	.02	0.1	.00	.23	79	7	.21	.025	
fava beans, cnd, Progresso	180		1.0												
8 oz	227	14.0	31.0												
french beans, boiled	228	117.8	1.3	0.8	0	0	2	.11	.19	132	11	111	99	1.13	.676
1 cup	177	12.5	42.5	0.1		5	.23	1.0	.00	.39	655	181	1.92	.204	
garden cress															
raw	8	22.4	0.2	0.1	0	23	17	.07	.06			20			
1/2 cup	25	0.7	1.4	0.0		2325	.02	0.3	.00				.33		
boiled	16	62.9	0.4	0.1	0	524	16	.11			5	41			
1/2 cup	68	1.3	2.6	0.0		5236	.04	.54	.00		240	33	.54		
garden salad, cnd, Joan of Arc	80		0.3				3	.04			636	48			
1/2 cup	136	1.5	18.1			392	.08	0.5			194	45	.82		
garlic, raw	13	5.3	0.1	0.0	0	0	3	.01		0	2	16	2		
3 cloves	9	0.6	3.0	0.0		0	.02	0.1	.00		36	14	.15		
ginger root, raw	17	19.6	0.2	0.0	0	0	1	.01	.04		3	4	10		
1/4 cup slices	24	0.4	3.6	0.0		0	.01	0.2	.00	.05	100	7	.12		
gourd, calabash (white-flowered), boiled	11	69.6	0.0	0.0	0	0	6	.02			1	18	8		
1/2 cup cubes	73	0.4	2.7	0.0		0	.02	0.3	.00		124	9	.18		
gourd, dishcloth (towelgourd), boiled	50	75.0	0.3	0.1	0	23	5	.04			18	8	18		
1/2 cup slices	89	0.6	12.8	0.0	0.4	231	.04	0.2	.00		403	28	.32		
great northern beans															
boiled	210	122.1	0.8	0.3	0	0	2	.10	.21	181	4	121	88	1.55	.917
1 cup	177	14.8	37.3	0.2	6.0	2	.28	1.2	.00	.47	692	293	3.77	.437	
cnd	300	183.1	1.0	0.4	0	0	3	.16	.28	213	11	139	134	1.70	1.069
1 cup	262	19.3	55.1	0.3		3	.38	1.2	.00	.73	919	355	4.11	.419	
cnd, Joan of Arc	90		0.1				0	.06			421	65			
1/2 cup	118	5.3	16.9				.08	0.4			168	99	1.42		
green beans (snap beans)															
boiled[a]	22	55.3	0.2	0.1	0	41	6	.06	.04	21	2	29	16	.23	.182
1/2 cup	62	1.2	4.9	0.0	1.1	413[b]	.05	0.4	.00	.05	185	24	.79	.064	

[a] Includes Italian, green & yellow varieties.   [b] For green varieties; yellow varieties contain 50 IU.

	KCAL / WT (g)	H₂O (g) / PRO (g)	FAT (g) / CHO (g)	PUFA (g) / SFA (g)	CHOL (mg) / DFIB (g)	A (RE) / A (IU)	C (mg) / B-1 (mg)	B-2 (mg) / NIA (mg)	B-6 (mg) / B-12 (mcg)	FOL (mcg) / PANT (mg)	Na (mg) / K (mg)	Ca (mg) / P (mg)	Mg (mg) / Fe (mg)	Zn (mg) / Cu (mg)	Mn (mg)
cnd	13	63.4	0.1	0.0	0	24	3	.04		22	170[b]	18	9	.20	.136
1/2 cup[a]	68	0.8	3.1	0.0	0.9	237[c]	.01	0.1	.00		74	13	.61	.026	
cut, no salt added, Heinz	16	95.7	0.1				4	.01			1	25	16	.20	
3.5 oz	100	0.9	3.1			471	.02	0.3			78	19	.60	.060	
cut, Pillsbury	18	90.3	0.4				3	.05			259	22			
1/2 cup	95	0.8	2.7			202	.03	0.4			92	14	.67		
french style, no salt added, Heinz	16	95.8	0.1				4	.00			1	25	15	.19	
3.5 oz	100	0.7	3.2			500	.02	0.2			64	14	.60	.070	
french style, Pillsbury	18	103.6	0.1				3	.03			261	21			
1/2 cup	109	0.9	3.3			238	.02	0.1			106	12	.76		
w/ onions, red peppers & garlic	18	107.5	0.2	0.1	0	60	4	.06			425	25	15	.16	
1/2 cup	114	1.0	4.0	0.1		599	.03	0.3	.00		106	18	.54	.068	
frzn															
boiled[a]	18	62.5	0.1	0.0	0	36	6	.05	.04		9	31	15	.42	.252
1/2 cup	68	0.9	4.2	0.0	1.1	359[d]	.03	0.3	.00	.04	76	16	.56	.046	
cut, Birds Eye	25	77.4	0.1	0.1	0		9	.07	.05	11	3	35	17	.17	
1/2 cup	85	1.4	5.8	0.0	2.2	455	.05	0.3	.00	.06	131	23	.73	.039	
cut w/ butter sce, Pillsbury	35	80.0	1.4				2	.05			243	0			
1/2 cup	88	0.8	4.8			111	.03	0.0			108	17	.53		
french style, Birds Eye	26	77.1	0.2	0.1	0		8	.07	.04	11	3	39	18	.22	
1/2 cup	85	1.4	5.9	0.0	2.2	390	.05	0.3	.00	.09	145	21	.80	.044	
french style w/ butter sce, Pillsbury	40	90.7	1.4				7	.07			349	31			
1/2 cup	100	1.3	5.6			450	.04	0.3			120	24	.48		
french style w/ toasted almonds, Birds Eye—*1/2 cup*	52	71.2	1.6	0.5	0		8	.06	.05	11	335	36	25	.28	
	85	2.5	8.4	0.2		372	.05	0.4	.00	.11	167	32	.52	.070	
italian, Birds Eye	31	75.2	0.1	0.1	0		14	.10	.04	14	3	35	19	.25	
1/2 cup	85	1.8	7.2	0.0	2.7	374	.07	0.5	.00	.22	184	31	.74	.050	
whole, Birds Eye	30	103.9	0.2	0.1	0		12	.10	.05	13	2	44	23	.27	
4 oz	113	1.7	6.7	0.1	2.5	776	.07	0.3	.00	.18	200	28	.87	.047	
green bean salad, german style, cnd, Joan	90		2.0				3	.04			770	44			
of Arc—*1/2 cup*	132	1.8	14.9			384	.04	0.5			152	48	1.60		
hominy, cnd															
white, Van Camp's	138	213.4	0.7					.06			708	10			
1 cup	246	2.0	30.0			0	.02	0.1			37		1.44		
yellow, Van Camp's	128	211.0	0.6					.06			701	9			
1 cup	242	2.7	27.9			277	.01	0.1			33		2.27		
yellow w/ red & green peppers, Van Camp's—*1 cup*	129	210.6	0.5					.07			685	10			
	242	2.6	28.5			408	.01	0.2			46		1.46		
hyacinth beans, boiled	228	134.1	1.1		0		0	.07			13	77	159	5.53	
1 cup	194	15.8	40.2				.52	0.8	.00		653	233	8.88	.662	
jerusalem artichoke, raw	57	58.5	0.0	0.0	0	2	3	.05				10	13		.045
1/2 cup slices	75	1.5	13.1	0.0		15	.15	1.0				58	2.55		
jew's ear (pepeao)															
raw	25	91.7	0.0		0	0	1	.20			9	16	25		
1 cup slices	99	0.5	6.7			0	.08	0.1	.00		42	14	.55		
dried	36	1.3	0.1		0	0	0	.04			8	14	17		
1/2 cup	12	0.6	9.7			0		0.4	.00		85	14	.74		
jute, potherb, boiled	16	37.5	0.1	0.0	0	223	14	.08			5	91	27		
1/2 cup	43	1.6	3.1	0.0		2230	.04	0.4	.00		237	31	1.35		
kale															
boiled	21	59.3	0.3	0.1	0	481	27	.05	.09	9	15	47	12	.15	.270
1/2 cup chopped	65	1.2	3.7	0.0		4810	.03	0.3	.00	.03	148	18	.59	.101	
frzn, boiled	20	58.8	0.3	0.2	0	413	16	.07	.06	9	10	90	12	.12	.292
1/2 cup chopped	65	1.9	3.4	0.0		4130	.03	0.4	.00	.03	209	18	.61	.031	
kale, scotch, boiled	18	59.3	0.3	0.1	0	130	34	.03	.09	9	29	86	37	.15	.271
1/2 cup chopped	65	1.2	3.7	0.0		1296	.03	0.5	.00	.03	178	25	1.25	.101	

[a] Includes Italian, green & yellow varieties.
[b] Special dietary pack contains 1 mg Na.
[c] For green varieties; yellow varieties contain 71 IU.
[d] For green varieties; yellow varieties contain 76 IU.

	KCAL / WT (g)	H₂O (g) / PRO (g)	FAT (g) / CHO (g)	PUFA (g) / SFA (g)	CHOL (mg) / DFIB (g)	Vitamins A (RE) / A (IU)	C (mg) / B-1 (mg)	B-2 (mg) / NIA (mg)	B-6 (mg) / B-12 (mcg)	FOL (mcg) / PANT (mg)	Minerals Na (mg) / K (mg)	Ca (mg) / P (mg)	Mg (mg) / Fe (mg)	Zn (mg) / Cu (mg)	Mn (mg)
kidney beans															
boiled	225	118.5	0.9	0.5	0	0	2	.10	.21	229	4	50	80	1.89	.844
1 cup	177	15.4	40.4	0.1	6.4	0	.28	1.0	.00	.39	713	252	5.20	.428	
cnd	208	199.6	0.8	0.4	0	0	3	.18	.18	126	889	69	79	1.41	.556
1 cup	256	13.3	38.1	0.1		0	.28	1.3	.00	.37	658	269	3.14	.384	
kidney beans, red															
boiled	225	118.5	0.9	0.5	0	0	2	.10	.21	229	4	50	80	1.89	.844
1 cup	177	15.4	40.4	0.1	6.4	0	.28	1.0	.00	.39	713	252	5.20	.428	
cnd	216	198.0	0.9	0.5	0	0	3	.23	.06	129	873	62	73	1.41	.620
1 cup	256	13.4	39.9	0.1		0	.27	1.2	.00	.38	658	240	3.22	.384	
kidney beans, red, california, boiled	219	118.5	0.2	0.1	0	0	2	.11	.18	131	7	116	85	1.52	.563
1 cup	177	16.2	39.7	0.0		5	.23	1.0	.00	.39	741	242	5.27	.512	
kidney beans, royal, red, boiled	218	118.6	0.3	0.2	0	0	2	.12	.18	130	8	78	74	1.58	.451
1 cup	177	16.8	38.7	0.0		5	.17	1.0	.00	.39	669	251	4.90	.464	
kohlrabi, boiled	24	74.1	0.1	0.0	0	2	44	.02			17	20	16		
1/2 cup slices	82	1.5	5.5	0.0		29	.03	0.3	.00		279	37	.33		
lambsquarters, boiled	29	80.0	0.6	0.3	0	873	33	.23				232			
1/2 cup chopped	90	2.9	4.5	0.0		8730	.09	0.8	.00			41	.63		
leeks															
raw	16	21.6	0.1	0.0	0	2	3	.01		17	5	15	7		
1/4 cup chopped	26	0.4	3.7	0.0	0.3	25	.02	0.1	.00		47	9	.55		
boiled	8	23.6	0.1	0.0	0	1	1	.01		6	3	8	4		
1/4 cup chopped	26	0.2	2.0	0.0		12	.01	0.1	.00		23	4	.29		
freeze-dried	3	0.0	0.0	0.0	0	0	1	.00		3	0	3	1		
1/4 cup	0.8	0.1	0.6	0.0		2	.01	0.0	.00		19	3	.06		
lentils, boiled	231	137.9	0.7	0.3	0	2	3	.15	.35	358	4	37	71	2.50	.978
1 cup	198	17.9	39.9	0.1	7.9	15	.34	2.1	.00	1.26	731	356	6.59	.497	
lentils, sprouted, stir-fried	101	68.7	0.5	0.2	0	4	13	.09				14	35	1.60	.502
3.5 oz	100	8.8	21.3	0.1		41	.22	1.2	.00	.57	284	153	3.10	.337	
lettuce															
butterhead, rawᵃ	2	14.3	0.0	0.0	0	15	1	.01		11	1			.03	.020
2 leaves	15	0.2	0.4	0.0	0.1	146	.01	0.0	.00		39		.04	.003	
cos/romaine, raw	4	26.6	0.1	0.0	0	73	7	.03		38	2	10	2		
1/2 cup shredded	28	0.5	0.7	0.0		728	.03	0.1	.00		81	13	.31		
iceberg, rawᵇ	3	19.2	0.0	0.0	0	7	1	.01	.01	11	2	4	2	.04	.030
1 leaf	20	0.2	0.4	0.1	0.2	66	.01	0.0	.00	.01	32	4	.10	.006	
looseleaf, raw	5	26.3	0.1	0.0	0	53	5	.02	.02		3	19	3		
1/2 cup shredded	28	0.4	1.0	0.0		532	.01	0.1	.00	.06	74	7	.39		
lima beans															
boiled	217	131.2	0.7	0.3	0	0	0	.10	.30	156	4	32	82	1.79	.970
1 cup	188	14.7	39.3	0.2	6.2	0	.30	0.8	.00	.79	955	208	4.50	.442	
cnd	191	185.8	0.4	0.2	0	0	0	.08	.22	121	809	50	94	1.57	.875
1 cup	241	11.9	35.9	0.1		0	.13	0.6	.00	.62	531	178	4.35	.434	
lima beans, baby															
boiled	229	122.2	0.7	0.3	0	0	0	.10	.14	273	5	52	97	1.87	1.065
1 cup	182	14.6	42.4	0.2	7.8	0	.29	1.2	.00	.86	729	231	4.36	.391	
frzn, boiled	94	65.1	0.3	0.1	0	15	5	.05	.10	80	26	25	50	.50	.732
1/2 cup	90	6.0	17.5	0.1		150	.06	0.7	.00	.16	370	101	1.76	.177	
frzn w/ butter sce, Pillsbury	110	73.7	1.7				14	.04			399	32			
1/2 cup	101	5.4	18.4			222	.06	1.0			340	77	1.31		
lima beans, fordhook, frzn, boiled	85	62.5	0.3	0.1	0	16	11	.05	.10	74	45	19	29	.37	.264
1/2 cup	85	5.2	16.0	0.1		162	.06	0.9	.0	.14	347	54	1.16	.047	
lotus root															
raw	45	64.1	0.1	0.0	0	0	36	.18			33	36	18		
10 slices	81	2.1	14.0	0.0		0	.13	0.3	.00		450	81	.94		

ᵃ Includes Boston & Bibb types.   ᵇ Includes crisphead types.

	KCAL / WT (g)	H₂O / PRO (g)	FAT / CHO (g)	PUFA / SFA (g)	CHOL / DFIB (mg/g)	A (RE) / A (IU)	C (mg) / B-1 (mg)	B-2 (mg) / NIA (mg)	B-6 (mg) / B-12 (mcg)	FOL (mcg) / PANT (mg)	Na (mg) / K (mg)	Ca (mg) / P (mg)	Mg (mg) / Fe (mg)	Zn (mg) / Cu (mg)	Mn (mg)
boiled	59	72.5	0.1	0.0	0	0	24	.01			40	23	20		
10 slices	89	1.4	14.3	0.0		0	.11	0.3	.00		323	69	.80		
lupins, boiled	197	118.0	4.8	1.2	0			.09			7	85	90	2.29	
1 cup	166	25.8	16.4	0.6			.22	0.8	.00		407	212	1.99	.383	
mixed veg															
cnd[a]	39	71.4	0.2	0.1	0	995	4	.04	.07	19	122	22	13	.34	
1/2 cup	82	2.1	7.6	0.0	2.0	9551	.04	0.5	.00	.12	239	34	.86	.060	
frzn[b]	54	75.7	0.1	0.1	0	389	3	.11	.07	17	32	22	20	.45	.345
1/2 cup	91	2.6	11.9	0.0	2.1	3892	.07	0.8	.00	.14	154	46	.75	.076	
Birds Eye[c]	59	78.5	0.4	0.2	0		9	.07	.13	27	41	22	18	.40	
1/2 cup	95	2.6	12.9	0.1	2.5	6507	.10	1.2	.00	.15	187	52	.81	.063	
Pillsbury	50	54.3	0.3				5	.05			27	16			
1/2 cup	67	2.1	9.6			1627	.05	0.3			155	41	.54		
w/ butter sce, Pillsbury	80	94.6	1.7				9	.07			374	23			
1/2 cup	114	2.9	13.3			2453	.06	0.9			190	56	.68		
w/ onion sce, Birds Eye[d]	97	55.6	5.2	1.0	1		5	.11	.10	15	340	43	15	.26	
1/3 cup	76	2.2	11.6	1.1		4155	.06	0.6	.19	.24	182	59	.53	.041	
mothbeans, boiled	207	122.5	1.0	0.5	0	2	2	.04			17	6	184	1.04	.933
1 cup	177	13.8	37.1	0.2		17	.22	1.2	.00		538	265	5.56	.290	
mountain yam, hawaii, steamed	59	55.5	0.1	0.0	0	0	0	.01			9	5	7		
1/2 cup cubes	72	1.2	14.4	0.0		0	.06	0.1	.00		356	29	.31		
mung beans, boiled	213	146.8	0.8	0.3	0	5	2	.12	.14	321	4	55	97	1.70	.602
1 cup	202	14.2	38.7	0.2	5.1	48	.33	1.2	.00	.83	536	201	2.83	.315	
mung beans, sprouted[e]															
raw	16	47.0	0.1	0.0	0	1	7	.06	.05	32	3	7	11	.21	.098
1/2 cup	52	1.6	3.1	0.0	0.6	11	.04	0.4	.00	.20	77	28	.47	.085	
cnd	8	59.6	0.0	0.0	0	1	0	.04		6	9	5			.045
1/2 cup	62	0.9	1.3	0.0		14	.02	0.1	.00		17	20	.27	.097	
ckd	13	57.9	0.1	0.0	0	1	7	.06			6	7	9	.29	.087
1/2 cup	62	1.3	2.6	0.0		8	.03	0.5	.00	.15	63	17	.40	.076	
stir-fried	31	52.3	0.1	0.0	0		10	.11					8	.56	
1/2 cup	62	2.7	6.6	0.0			.09	0.7	.00				1.18		
mungo beans, boiled	190	130.5	1.0	0.6	0	6	2	.14	.10	170	13	95	113	1.50	.742
1 cup	180	13.6	33.0	0.1		56	.27	2.7	.00	.78	416	280	3.14	.250	
mushroom marinara salad, Fresh Chef	75		5.8				1	.10			392	35			
3.2 oz	92	1.7	4.1			210	.03	1.0			153		.50		
mushrooms															
raw[f]	9	32.1	0.2	0.1	0	0	1	.16	.03	7	1	2	4	.17	.000
1/2 cup pieces	35	0.7	1.6	0.0		0	.04	1.4	.00	.77	130	36	.43	.039	
boiled	21	71.0	0.4	0.1	0	0	3	.23	.07	14	2	4	10	.68	.090
1/2 cup pieces	78	1.7	4.0	0.0		0	.06	3.5	.00	1.69	277	68	1.36	.393	
cnd	19	71.0	0.2	0.1	0	0				10				.56	.067
1/2 cup pieces	78	1.5	3.9	0.0		0			.00				.62	.183	
cnd w/ butter sce, Pillsbury	25	98.6	0.4				5	.17			559	8			
2 oz	106	2.3	3.0				.00	1.6			162	47	.64		
mushrooms, shitake															
ckd	40	60.1	0.2	0.0	0	0	0	.12			3	2	10		
4 mushrooms	72	1.1	10.3	0.0		0	.03	1.1	.00		85	21	.32		
cnd, Campbell's	45		0.3				0	.11			74	10			
4 oz	113	3.1	7.4			0	.02	2.9			146		1.20		
dried	44	1.4	0.2	0.0	0	0	1	.19			2	2	20		
4 mushrooms	15	1.4	11.3	0.0		0	.05	2.1	.00		230	44	.26		
mushrooms, sovereign, cnd, Campbell's	39		0.4				2	.41			73	6			
4 oz	113	3.7	5.1			0	.08	1.9			220		1.50		

[a] Carrots, green peas, snap beans & lima beans.
[b] Corn, lima beans, snapp beans, green peas & carrots.
[c] Corn, carrots, peas & green beans.
[d] Carrots, corn, green beans, peas & pearl onions.
[e] Seeds attached to sprouts.
[f] Agaricus bisporus.

	KCAL / WT (g)	H₂O (g) / PRO (g)	FAT (g) / CHO (g)	PUFA (g) / SFA (g)	CHOL (mg) / DFIB (g)	A (RE) / A (IU)	C (mg) / B-1 (mg)	B-2 (mg) / NIA (mg)	B-6 (mg) / B-12 (mcg)	FOL (mcg) / PANT (mg)	Na (mg) / K (mg)	Ca (mg) / P (mg)	Mg (mg) / Fe (mg)	Zn (mg) / Cu (mg)	Mn (mg)
mustard greens															
boiled	11	66.1	0.2	0.0	0	212	18	.04			11	52	10		
1/2 cup chopped	70	1.6	1.5	0.0		2122	.03	0.3	.00	.08	141	29	.49		
frzn, boiled	14	70.4	0.2	0.0	0	335	10	.04	.08		19	75	10	.15	.221
1/2 cup chopped	75	1.7	2.3	0.0		3352	.03	0.2	.00	.01	104	18	.84	.044	
mustard spinach (tendergreen)															
raw	17	69.2	0.2		0	743	98					158			
1/2 cup chopped	75	1.7	2.9			7425			.00			21	1.13		
boiled	14	85.1	0.2		0	738	59					142			
1/2 cup chopped	90	1.5	2.5			7380			.00			16	.72		
navy beans															
boiled	259	115.0	1.0	0.4	0	0	2	.11	.30	255	2	128	107	1.93	1.012
1 cup	182	15.8	47.9	0.3	6.6	3	.37	1.0	.00	.46	669	285	4.51	.537	
cnd	296	184.6	1.1	0.5	0	0	2	.14	.27	163	1173	123	122	2.02	.983
1 cup	262	19.7	53.6	0.3		4	.37	1.3	.00	.45	755	352	4.84	.545	
new zealand spinach															
raw	4	26.3	0.1	0.0	0	123	8	.04			36	16	11		
1/2 cup chopped	28	0.4	0.7	0.0		1232	.01	0.1	.00	.09	36	8	.22		
boiled	11	85.3	0.2	0.1	0	326	14	.10			97	43	29		
1/2 cup chopped	90	1.2	2.0	0.0		3260	.03	0.4	.00	.23	92	20	.59		
okra															
boiled	25	71.9	0.1	0.0	0	46	13	.04	.15	37	4	50	46	.44	.729
1/2 cup slices	80	1.5	5.8	0.0		460	.11	0.7	.00	.17	257	45	.36	.069	
frzn, boiled	34	83.8	0.3	0.1	0	47	11	.11	.04	134	3	88	47	.57	.939
1/2 cup slices	92	1.9	7.5	0.1		473	.09	0.7	.00	.22	215	42	.62	.089	
onion rings, frzn, heated[a]	285	20.0	18.7	3.6	0	16	1	.10	.05	9	263	21	14	.29	.294
7 rings	70	3.7	26.7	6.0		158	.20	2.5	.00	.16	90	57	1.18	.056	
onions															
raw	27	72.7	0.2	0.1	0	0	7	.01	.13	16	2	20	8	.14	.106
1/2 cup chopped	80	0.9	5.9	0.0	0.6	0	.05	0.1	.00	.11	124	23	.29	.032	
boiled	29	96.9	0.2	0.1	0	0	6	.01	.19	13	8	29	11	.19	.118
1/2 cup chopped	105	1.0	6.6	0.0		0	.04	0.1	.00	.13	159	24	.21	.042	
cnd	21	105.4	0.1	0.0	0						416	51	6	.32	.114
1/2 cup chopped	112	1.0	4.5	0.0	1.2				.00		124	31	.15	.062	
dehydrated flakes	45	0.6	0.1	0.0	0	0	11	.01	.22	23	3	36	13	.26	.194
1/4 cup	14	1.3	11.7	0.0		0	.07	0.1	.00	.19	227	42	.22	.058	
frzn															
boiled	28	92.2	0.1	0.0	0	2	5	.02	.07	13	8	27	8	.09	.040
3.5 oz	100	0.7	6.7	0.0		21	.02	0.1	.00	.08	101	2	.34	.024	
small, whole, Birds Eye	40	101.9	0.1	0.0	0		9	.03	.10	24	10	40	12	.13	
1/2 cup	113	1.0	9.6	0.0		19	.03	0.2	.00	.11	156	26	.52	.057	
small w/ cream sce, Birds Eye	100	65.4	5.9	1.1	1		6	.10	.08	15	345	55	12	.12	
1/2 cup	85	1.6	10.8	1.2		251	.03	0.2	.22	.24	168	55	.36	.038	
onions, spring, raw[b]	13	46.0	0.1	0.0	0	250	23	.07		7	2	30	10	.22	
1/2 cup chopped	50	0.9	2.8	0.0		2500	.04	0.1	.00	.07	128	16	.94	.030	
onions, welsh, raw	34	90.5	0.4	0.2	0		27	.09				18			
3.5 oz	100	1.9	6.5	0.1			.05	0.4	.00			49			
parsley															
raw	10	26.5	0.1		0	156	27	.03	.05	55	12	39	13	.22	.048
1/2 cup chopped	30	0.7	2.1			1560	.02	0.2	.00	.09	161	12	1.86	.017	
freeze-dried	4	0.0	0.1		0	89	2	.03	.02	22	5	2	5	.09	.019
1/4 cup	1.4	0.4	0.6			885	.02	0.1	.00	.04	88	8	.75	.006	
parsnips, boiled	63	60.6	0.2	0.0	0	0	10	.04	.07	45	8	29	23	.20	.229
1/2 cup slices	78	1.0	15.2	0.0	2.1	0	.07	.6	.00	.46	287	54	.45	.108	

[a] Breaded & parboiled in veg oil.   [b] Includes tops & bulbs.

	KCAL / WT (g)	H₂O (g) / PRO (g)	FAT (g) / CHO (g)	PUFA (g) / SFA (g)	CHOL (mg) / DFIB (g)	A (RE) / A (IU)	C (mg) / B-1 (mg)	B-2 (mg) / NIA (mg)	B-6 (mg) / B-12 (mcg)	FOL (mcg) / PANT (mg)	Na (mg) / K (mg)	Ca (mg) / P (mg)	Mg (mg) / Fe (mg)	Zn (mg) / Cu (mg)	Mn (mg)
peas, green															
raw	63	61.5	0.3	0.1	0	50	31	.10	.13	51	4	19	26	.97	.320
1/2 cup	78	4.2	11.3	0.1	2.7	499	.21	1.6	.00	.08	190	84	1.15	.137	
boiled	67	62.3	0.2	0.1	0	48	11	.12	.17	51	2	22	31	.95	.420
1/2 cup	80	4.3	12.5	0.0	3.0	478	.21	1.6	.00	.12	217	94	1.24	.138	
cnd	59	69.4	0.3	0.1	0	65	8	.07	.05	38	186[a]	17	15	.60	.258
1/2 cup	85	3.8	10.7	0.0	3.5	653	.10	0.6	.00	.11	147	57	.81	.070	
w/ onions, red peppers & garlic	57	98.6	0.3	0.1	0	49	13	.08			290	18	17	.74	
1/2 cup	114	3.5	10.6	0.1		494	.11	0.8	.00		139	61	1.37	.113	
frzn															
boiled	63	63.6	0.2	0.1	0	53	8	.08	.09	47	70	19	23	.75	.331
1/2 cup	80	4.1	11.4	0.0	3.0	534	.23	1.2	.00	.11	134	72	1.26	.111	
w/ butter sce, Le Sueur	90	93.2	2.0				17	.09			542	19			
1/2 cup	115	4.9	13.6			626	.30	1.8			138	83	1.84		
w/ cream sce, Birds Eye	118	51.5	5.6	1.0	1		6	.13	.08	37	368	42	23	.63	
1/2 cup	76	4.1	13.4	1.1	2.7	637	.19	1.0	.20	.24	171	91	1.02	.089	
w/ cream sce, Pillsbury	100	92.9	4.2				11	.11			317	57			
1/2 cup	114	3.8	11.5			412	.46	2.1			164	25	1.03		
peas, green, early june															
cnd, Pillsbury	60	101.5	0.2				8	.08			367	21			
1/2 cup	118	3.4	11.1			485	.13	0.4			100	63	1.18		
frzn, Le Sueur	60	60.8	0.2				11	.30			169	21			
1/2 cup	76	4.7	9.1			406	.18	1.4			116	74	1.14		
peas, mature seeds, sprouted, boiled[b]	118	74.4	0.5	0.2	0	11	7	.29	.13	36	3	26	41	.78	.325
3.5 oz	100	7.1	21.9	0.1	3.3	107	.22	1.1	.00	.68	268	24	1.67	.020	
peas, split, boiled	231	136.2	0.8	0.3	0	1	1	.11	.09	127	4	26	71	1.96	.776
1 cup	196	16.4	41.4	0.1		14	.37	1.7	.00	1.17	710	195	2.52	.355	
peppers, hot chili															
raw[c]	18	39.5	0.1	0.0	0	35	109	.04	.13	11	3	8	11	.14	.107
1 pepper	45	0.9	4.3	0.0		346[d]	.04	0.4	.00	.03	153	20	.54	.078	
cnd[c]	17	62.9	0.1	0.0	0	42	46	.03				5			
1/2 cup chopped	68	0.6	4.2	0.0		415[e]	.01	0.5	.00			12	.34		
cnd, Del Monte	28	86.8	0.1				10	.06			1882	49	17	.16	
3.5 oz	100	1.1	6.9			272	.03	0.5			138	26	.70	.180	
peppers, jalapeno, cnd	17	61.1	0.4	0.2	0	116	9	.03			995	18	8	.13	
1/2 cup chopped	68	0.5	3.3	0.0		1156	.02	0.3	.00	.73	92	12	1.90	.095	
peppers, sweet															
raw[c]	12	46.4	0.2	0.1	0	26	64[f]	.03	.08	8	2	3	7	.09	.070
1/2 cup chopped	50	0.4	2.7	0.0	0.6	265[g]	.04	0.3	.00	.02	98	11	.63	.052	
boiled[c]	12	64.4	0.2	0.1	0	26	76	.02	.07	7	1	3	7	.08	.066
1/2 cup chopped	68	0.4	2.7	0.0		264[h]	.04	0.2	.00	.02	88	10	.60	.048	
cnd[c]	13	63.9	0.2	0.1	0	11	33	.02			958	28[i]	8	.12	
1/2 cup halves	70	0.6	2.7	0.0		109[j]	.02	0.4	.00		102	14	.56	.091	
cherry, cnd, Del Monte	42	82.4	0.1				69	.12			1970	18	20	.24	
3.5 oz	100	1.8	10.4			2977	.05	1.2			217	39	.80	.180	
freeze-dried[c]	5	0.0	0.1	0.0	0	10	30	.02	.04	4	3	2	3	.04	.030
1/4 cup	1.6	0.3	1.1	0.0		100[k]	.02	0.1	.00	.01	51	5	.17	.022	
frzn, boiled[c]	18	94.7	0.2	0.1	0	29	41	.03	.11	10	4	8	7	.05	.097
3.5 oz	100	1.0	3.9	0.0		290[l]	.05	1.1	.00	.02	72	13	.52	.044	
pigeonpeas, boiled	204	115.2	0.6	0.3	0	0	0	.10	.08	186	9	72	77	1.51	.842
1 cup	168	11.4	39.1	0.1	7.9	4	.25	1.3	.00	.54	644	201	1.86	.452	

[a] Special dietary pack contains 2 mg Na.
[b] Seeds attached to sprouts.
[c] Includes green & red varieties.
[d] For green varieties; red varieties contain 484 IU.
[e] For green varieties; red varieties contain 8087 IU.
[f] For green varieties; red varieties contain 95 mg.
[g] For green varieties; red varieties contain 2850 IU.
[h] For green varieties; red varieties contain 2557 IU.
[i] Ca added as firming agent.
[j] For green varieties; red varieties contain 364 IU.
[k] For green varieties; red varieties contain 1236 IU.
[l] For green varieties; red varieties contain 3343 IU.

	KCAL	H₂O (g)	FAT (g)	PUFA (g)	CHOL (mg)	Vitamins A (RE)	C (mg)	B-2 (mg)	B-6 (mg)	FOL (mcg)	Minerals Na (mg)	Ca (mg)	Mg (mg)	Zn (mg)	Mn (mg)
	WT (g)	PRO (g)	CHO (g)	SFA (g)	DFIB (g)	A (IU)	B-1 (mg)	NIA (mg)	B-12 (mcg)	PANT (mg)	K (mg)	P (mg)	Fe (mg)	Cu (mg)	
pimientos, cnd, Dromedary	10		0.0								5				
1 oz	28	0.0	2.0								45				
pink beans, boiled	252	103.4	0.8	0.4	0	0	0	.11	.30	284	3	88	110	1.63	.926
1 cup	169	15.3	47.2	0.2	7.4	0	.43	1.0	.00	.51	858	279	3.89	.458	
pinto beans															
boiled	235	109.9	0.9	0.3	0	0	4	.16	.27	294	3	82	95	1.85	.951
1 cup	171	14.0	43.9	0.2	6.8	3	.32	0.7	.00	.49	800	273	4.47	.439	
cnd	186	189.1	0.8	0.3	0	0	2	.15	.18	145	998	89	64	1.66	.550
1 cup	240	11.0	34.9	0.2		3	.24	0.7	.00	.33	723	220	3.85	.336	
poi	134	86.0	0.2	0.1	0	2	5	.05			14	19	29		
1/2 cup	120	0.5	32.7	0.0		24	.16	1.3	.00		220	47	1.06		
pokeberry shoots, boiled	16	76.2	0.3		0	713	67	.21				43			
1/2 cup	82	1.9	2.5			7134	.06	0.9	.00			27	.98		
potato															
raw w/o skin	88	88.4	0.1	0.0	0		22	.04	.29	14	7	8	24	.44	.295
1 potato	112	2.3	20.1	0.0			.10	1.7	.00	.43	608	52	.85	.290	
baked w/ skin	220	143.8	0.2	0.1	0		26	.07	.70	22	16	20	55	.65	.463
1 potato	202	4.7	51.0	0.1			.22	3.3	.00	1.12	844	115	2.75	.616	
baked w/o skin	145	117.7	0.2	0.1	0		20	.03	.47	14	8	8	39	.45	.251
1 potato	156	3.1	33.6	0.0			.16	2.2	.00	.87	610	78	.55	.335	
baked w/ cheese topping, frzn, Pillsbury	200	96.6	5.5				7	.07			521	42			
5 oz	142	4.5	32.9			339	.06	2.4			625	112	.85		
baked w/ sour cream & chives, frzn, Pillsbury—*5 oz*	231	93.7	9.9				5	.09			579	50			
	142	4.7	30.7			439	.07	2.4			575	111	.99		
boiled w/o skin	116	104.6	0.1	0.1	0		10	.03	.36	12	7	10	26	.37	.189
1 potato	135	2.3	27.0	0.0			.13	1.8	.00	.69	443	54	.42	.225	
cnd w/o skin	54	75.9	0.2	0.1	0		5	.01	.17	6		5	12	.25	.087
1/2 cup	90	1.3	12.3	0.0			.06	0.8	.00	.32	206	25	1.13	.051	
cottage fries, frzn, heated[a]	109	26.5	4.1	1.7	0		5	.02	.12	8	23	5	11	.21	.152
10 strips	50	1.7	17.0	1.9			.06	1.2	.00	.35	240	33	.75	.100	
french fried, from restaurant[b]	158	19.0	8.3	3.8	0		5	.01	.12	15	108	10	17	.19	.096
10 pieces	50	2.0	20.0	2.5			.09	1.6	.00	.33	366	47	.38	.069	
french fried, frzn, heated[a]	111	26.5	4.4	0.3	0		6	.02	.12	8	15	4	11	.21	.148
10 pieces	50	1.7	17.0	2.1			.06	1.2	.00	.33	229	43	.67	.082	
frzn, boiled w/o skin	65	82.8	0.1	0.1	0		9	.03	.20	8	20	7	11	.25	.185
3.5 oz	100	2.0	14.5	0.0			.10	1.3	.00	.28	287	26	.84	.078	
hash brown, frzn, prep[c]	170	43.8	9.0	1.0			5	.02	.10		27	12	13	.25	.174
1/2 cup	78	2.5	21.9	3.5			.09	1.9	.00	.35	340	56	1.17	.119	
hash brown, homemade[c]	163	48.0	10.9	1.3			5	.02	.22	6	19	6	16	.23	.119
1/2 cup	78	1.9	16.6	4.2			.06	1.6	.00	.39	251	32	.63	.140	
hash brown w/ butter sce, frzn, prep	178	63.7	8.9	1.8	23	16	4	.03			101	33			
3.5 oz	100	2.0	24.1	3.4		111	.05	1.4			327		.99		
mashed, from flakes[d]	118	80.1	5.9	0.3	15	22	10	.05	.01	8	349	52	19	.18	
1/2 cup	105	2.0	15.8	3.6		189	.12	0.7	.08	.13	245	59	.23	.017	
mashed, from granules[d]	137	77.7	6.5	1.4	18	192	1	.06	.44	7	358	57	19	.09	
1/2 cup	105	2.0	17.6	1.2		778	.04	0.7	.46	1.57	223	50	4.27		
mashed, from mix, French's	130		6.0								340				
1/2 cup	105	2.0	16.0								270				
mashed, from mix, Hungry Jack	140		7.0								380				
1/2 cup	105	3.0	17.0								320				
mashed, homemade[e]	111	80.1	4.4	1.3	2[f]	21	6	.04	.24	8	309	27	19	.29	.120
1/2 cup	105	2.0	17.5	1.1		177	.09	1.1	.05	.60	303	49	.28	.144	
microwaved w/ skin	212	145.5	0.2	0.1	0		31	.07	.70	24	16	22	54	.73	.590
1 potato	202	4.9	48.7	0.1			.24	3.5	.00		903	212	2.50	.675	
microwaved w/o skin	156	114.7	0.2	0.1	0		24	.04	.50	19	11	8	39	.51	.265
1 potato	156	3.3	36.3	0.0			.20	2.5	.00	.93	641	170	.64	.370	

[a] Pan-fried in veg oil.
[b] Fried in veg oil; french fries fried in animal fat & veg oil contain 0.5 g PUFA, 3.4 g SFA & 6 mg cholestrol.
[c] Prepared w/ veg oil.
[d] Whole milk & butter added.
[e] Whole milk, margarine & salt added.
[f] If butter is used in place of margarine, cholestrol is 13 mg.

	KCAL	H₂O (g)	FAT (g)	PUFA (g)	CHOL (mg)	A (RE)	C (mg)	B-2 (mg)	B-6 (mg)	FOL (mcg)	Na (mg)	Ca (mg)	Mg (mg)	Zn (mg)	Mn (mg)
	WT (g)	PRO (g)	CHO (g)	SFA (g)	DFIB (g)	A (IU)	B-1 (mg)	NIA (mg)	B-12 (mcg)	PANT (mg)	K (mg)	P (mg)	Fe (mg)	Cu (mg)	
potato pancakes															
from mix, French's	80		1.0								410				
1/2 cup	140	3.0	16.0								140				
homemade[a]	495	30.3	12.6	2.5	93	27	0	.10	.29	22	388	21	24	.68	.296
1 pancake	76	4.6	26.4	3.4		89	.10	1.6	.22	.71	538	78	1.21	.273	
potato puffs, frzn, prep[b]	138	32.8	6.7	0.5	0	1	4	.05	.14	10	462	19	12	.19	.168
1/2 cup	62	2.1	18.9	3.2		10	.12	1.3	.00	.41	236	30	.97	.037	
potato salad, commercial															
Fresh Chef	210		14.0				3	.11			320	21			
4 oz	113	2.1	18.9			0	.04	0.8			244		1.30		
german style, cnd, Joan of Arc	120		1.3				13	.07			854	6			
1/2 cup	130	2.0	24.6			172	.04	0.9			216	27	.65		
home style, cnd, Joan of Arc	160		8.6				5	.08			728	16			
1/2 cup	120	2.0	18.2			187	.10	1.4			281	54	.72		
potato salad, homemade[c]	179	95.0	10.3	4.7	86	41	13	.08	.18	8	661	24	19	.39	.126
1/2 cup	125	3.4	14.0	1.8		261	.10	1.1	.19	.67	317	65	.81	.148	
potatoes, au gratin															
from mix[d]	127	108.2	5.6	0.2			4	.11	.06		601	114	21	.33	
1/6 of 5.5 oz pkg	137	3.2	17.6	3.5			.03	1.3		.33	300	130	.44	.063	
from mix, French's	140		6.0								470				
1/2 cup	220	4.0	19.0								220				
frzn, Am Hosp Co	164		8.8		8		8	.10			147	94			
4 oz	113	5.6	15.6			133	.07				456		.32		
frzn, Banquet	98		2.0		15		0	.08			472	41			
4 oz	113	2.0	17.0			19	.04	0.3			127		.40		
homemade[e]	160	90.3	9.3	0.3	29[f]	47	12	.14	.21	10	528	146	24	.84	.196
1/2 cup	122	6.2	13.7	5.8		322	.08	1.2	.25	.47	483	138	.78	.195	
potatoes, creamy ital style w/ parmesan sce, from mix, French's—*1/2 cup*	130		4.0								430				
	190	3.0	21.0								190				
potatoes, creamy stroganoff, from mix, French's—*1/2 cup*	130		4.0								520				
	210	3.0	20.0								210				
potatoes, o'brien															
frzn, prep	204	62.0	13.2	3.5		19	10	.14			43	20			
3.5 oz	100	2.2	21.9	3.3		188	.05	1.4					.96		
homemade[g]	157	154.4	2.5	0.1	7	111	32	.11	.41	16	421	70	35	.59	.235
1 cup	194	4.6	30.0	1.5		934	.15	2.0	.16	.85	516	97	.91	.250	
potatoes, scalloped															
from mix[h]	127	108.5	5.9	0.3			5	.08	.06	2	467	49	19	.34	
1/6 of 5.5 oz pkg	137	2.9	17.5	3.6			.03	1.4		.45	278	77	.52	.067	
from mix, crispy top w/ savory onion, French's—*1/2 cup*	140		5.0								420				
	200	3.0	20.0								200				
from mix w/ cheese, French's	140		5.0								370				
1/2 cup	200	3.0	20.0								200				
homemade[i]	105	98.8	4.5	0.2	14[j]	23	13	.11	.22	11	409	70	23	.49	.203
1/2 cup	122	3.5	13.2	2.8		165	.08	1.3	.17	.63	461	77	.70	.199	
potatoes w/ sour cream & chives, from mix, French's—*1/2 cup*	150		6.0								560				
	220	3.0	20.0								220				
pumpkin															
boiled	24	114.3	0.1	0.0	0	132	6	.10			2	18	11		
1/2 cup mashed	122	0.9	6.0	0.0		1320	.04	0.5	.00		281	37	.70		
cnd	41	109.8	0.3	0.0	0	2691	5	.07	.07	15	6	32	28	.21	
1/2 cup	122	1.3	9.9	0.2		26908	.03	0.4	.00	.49	251	42	1.70	.131	

[a] Potatoes, eggs, onion, margarine, flour & salt.
[b] Pan-fried in veg oil.
[c] Potatoes, egg, mayonnaise, celery, sweet pickle relish, onion, green pepper, pimiento, salt & dry mustard.
[d] Water, whole milk & margarine added.
[e] Potatoes, whole milk, cheddar cheese, butter, flour & salt.
[f] If margarine is used in place of butter, cholesterol is 18 mg.
[g] Potatoes, whole milk, onions, green pepper, bread crumbs, salt, butter & black pepper.
[h] Water, whole milk & butter added.
[i] Potatoes, whole milk, butter, flour & salt.
[j] If margarine is used in place of butter, cholesterol is 7 mg.

	KCAL	H₂O (g)	FAT (g)	PUFA (g)	CHOL (mg)	A (RE)	C (mg)	B-2 (mg)	B-6 (mg)	FOL (mcg)	Na (mg)	Ca (mg)	Mg (mg)	Zn (mg)	Mn (mg)
	WT (g)	PRO (g)	CHO (g)	SFA (g)	DFIB (g)	A (IU)	B-1 (mg)	NIA (mg)	B-12 (mcg)	PANT (mg)	K (mg)	P (mg)	Fe (mg)	Cu (mg)	
pumpkin pie mix, cnd	141	96.5	0.2	0.0	0	1120	5	.16			280	49	22	.36	
1/2 cup	135	1.5	35.6	0.1		11203	.02	0.5	.00		186	60	1.43	.092	
purslane, boiled	10	54.2	0.1		0	107	6	.05			26	45	39		
1/2 cup	58	0.9	2.1			1074	.02	0.3	.00		283	22	.45		
radish, raw	7	42.7	0.2	0.0	0	0	10	.02	.03	12	11	9	4	.13	.032
10 radishes	45	0.3	1.6	0.0		3	.00	0.1	.00	.04	104	8	.13	.018	
radish, oriental[a]															
raw	8	41.6	0.0	0.0	0	0	10	.01		0	9	12	7		
1/2 cup slices	44	0.3	1.8	0.0		0	.01	0.1			100	10	.18		
boiled	13	70.3	0.2	0.1	0	0	11	.02			10	12	7		
1/2 cup slices	74	0.5	2.5	0.1		0	.00	0.1	.00		211	18	.11		
dried	157	11.4	0.4	0.2	0	0	0	.39			161	365	99		
1/2 cup	58	4.6	36.8	0.1		0	.16	2.0	.00		2027	118	3.90		
red beans															
Joan of Arc	100		0.4			0		.11			445	39			
1/2 cup	118	5.3	18.9			28	.08	1.0			152	94	1.77		
Van Camp's	194	173.0	0.6					.10			928	84			
1 cup	227	11.5	35.6			0	.10	0.6			546		3.65		
rhubarb															
frzn, raw	29	128.1	0.2		0	15	7	.04	.03	11	2	266	25	.14	.133
1 cup	137	0.8	7.0			147	.04	0.3	.00	.09	148	17	.39	.032	
frzn, ckd, sweetened	139	81.4	0.1			8	4	.03	.02	6	2	174	15	.10	.088
1/2 cup	120	0.5	37.4			83	.02	0.2	.00	.06	115	10	.25	.032	
rice, brown, ckd	232	137.1	1.2		0	0	0	.04			550[b]	23			
1 cup	195	4.9	49.7			0	.18	2.7	.00		137	142	1.00		
rice, white															
enr, ckd	223	148.8	0.2			0		.02			767[b]	21			
1 cup	205	4.1	49.6			0	.23	2.1			57	57	1.80		
enr, inst, ckd	180	120.3	0.0			0					450[b]	5			
1 cup	165	3.6	39.9			0	.21	1.7				31	1.30		
enr, parboiled, ckd	186	128.5	0.2			0		.02			627[b]	33			
1 cup	175	3.7	40.8			0	.19	2.1			75	100	1.40		
extra long grain, enr, ckd	85	57.0	0.0	0.0	0	0	0	.00	.00	0	0[c]	22			
1/2 cup	80	2.0	19.0	0.0	0.1	0	.00	1.0	.00		22	8	.70		
extra long grain, enr, parboiled, ckd, Chef-Way—*1/2 cup*	83	57.0	0.1	0.0	0	0	0	.00		0	0[c]	15			
	80	2.0	18.0	0.0	0.0	0	.00	0.9	.00		34	45	.60		
from mix, Minute	141	92.0	2.0	0.1	5		0	.01	.01	1	21	5	6	.37	
2/3 cup	124	2.6	27.4	1.2		72	.15	1.2	.00	.10	1	34	1.08	.041	
for rib roast	152	82.6	4.1	0.2	10		0	.02	.02	2	724	10	10	.36	
1/2 cup	117	3.0	25.2	2.4		146	.16	1.4	.01	.08	25	38	1.21	.041	
fried	159	62.1	4.8	1.8	0		0	.01	.02	2	551	10	9	.37	
1/2 cup	97	3.2	25.2	0.7		1	.16	1.4	.00	.08	47	37	1.30	.043	
w/ chicken flavor	153	72.7	4.2	0.3	10		0	.01	.01	2	688	11	12	.39	
1/2 cup	107	3.1	25.1	2.5		148	.17	1.4	.01	.07	22	40	1.25	.039	
frzn, w/ herb butter sce, Rice Originals	150	86.5	5.0				0	.04			406	18			
1/2 cup	119	2.9	23.3			206	.24	2.5			49	40	1.79		
rice, long grain & wild															
from mix, Minute	148	81.9	4.1	0.2	10		1	.02	.02	3	571	14	10	.29	
1/2 cup	115	2.8	24.7	2.4		153	.20	2.4	.01	.13	40	25	1.45	.046	
frzn, Rice Originals	120	80.7	2.1				3	.03			530	14			
1/2 cup	109	2.6	22.5			89	.21	1.3			39	40	1.42		
rice pilaf, frzn, Rice Originals	121	87.8	1.6				1	.04			537	12			
1/2 cup	117	2.8	23.4			35	.26	2.6			88	35	1.52		
rutabaga, boiled	29	76.7	0.2	0.1	0	0	19	.03	.08	13	15	36	18	.26	.130
1/2 cup cubes	85	0.9	6.6	0.0		0	.06	0.5	.00	.12	244	42	.40	.031	

[a] Includes daikon (Japanese) & Chinese.

[b] Salt added as specified on pkg; if no salt is added, Na is negligible.

[c] No salt added during cooking.

	KCAL / WT (g)	H₂O (g) / PRO (g)	FAT (g) / CHO (g)	PUFA (g) / SFA (g)	CHOL (mg) / DFIB (g)	A (RE) / A (IU)	C (mg) / B-1 (mg)	B-2 (mg) / NIA (mg)	B-6 (mg) / B-12 (mcg)	FOL (mcg) / PANT (mg)	Na (mg) / K (mg)	Ca (mg) / P (mg)	Mg (mg) / Fe (mg)	Zn (mg) / Cu (mg)	Mn (mg)
salad, Fresh Chef															
chicken w/ pineapple tidbits	346		29.7				3	.07			622	29			
4.66 oz	132	13.2	6.7			82	.05	1.1			201		.90		
ham & cheddar cheese	322		28.1				4	.16			1061	79			
4.65 oz	132	11.6	5.5			149	.32	1.5			256		1.10		
salsify, boiled	46	55.1	0.1		0	0	3	.12			11	32	12		
1/2 cup slices	68	1.9	10.5			0	.04	0.3	.00		192	38	.37		
sauerkraut															
bottled, Claussen	6	25.9	0.1				3	.00	.03	1	189	9	3	.04	
1 oz	28	0.2	1.2				.01	0.0		.03	43	5	.16	.010	
cnd	22	109.2	0.2	0.1	0	2	17	.03	.15		780	36	15	.22	
1/2 cup	118	1.1	5.1	0.0		21	.03	0.2	.00	.11	201	23	1.73	.113	
seaweed															
agar, raw	26	91.3	0.0	0.0	0	0	0	.02			9	54	67		.373
3.5 oz	100	0.5	6.8	0.0		0	.01	0.1	.00		226	5	1.86		
agar, dried	306	8.7	0.3	0.1	0	0	0	.22			102	625	770		4.300
3.5 oz	100	6.2	80.9	0.1		0	.01	0.2	.00		1125	52	21.40		
irishmoss, raw	49	81.3	0.2	0.1	0		0	.59			67	72		1.95	.370
3.5 oz	100	1.5	12.3	0.0			.47	0.2	.00		63	157	8.90	.149	
kelp (kombu/tangle), raw	43	81.6	0.6	0.0	0	12		.15		180	233	168	121	1.23	.200
3.5 oz	100	1.7	9.6	0.2		116	.05	0.5	.00		89	42	2.85	.130	
laver (nori), raw	35	85.0	0.3	0.1	0	520	39	.45	.16		48	70	2	1.05	.988
3.5 oz	100	5.8	5.1	0.1		5202	.10	1.5	.00		356	58	1.80	.264	
spirulina, raw	26	90.7	0.4	0.1	0	1		.34	.03		98				
3.5 oz	100	5.9	2.4	0.1			.22	1.2	.00	.33	127	.11			
spirulina, dried	290	4.7	7.7	2.1	0	10		3.67	.36		1048		195		
3.5 oz	100	57.5	23.9	2.7			2.38	12.8	.00	3.48	1363	118	28.50		
wakame, raw	45	80.0	0.6	0.2	0	36	3	.23			872	150	107	.38	1.400
3.5 oz	100	3.0	9.1	0.1		360	.06	1.6	.00		50	80	2.18	.284	
shallots															
raw	7	8.0	0.0	0.0	0		1	.00			1	4			
1 T chopped	10	0.3	1.7	0.0			.01	0.0	.00		33	6	.12		
freeze-dried	13	0.1	0.0	0.0	0		1	.00	.06	4	2	7	4	.07	.051
1/4 cup	4	0.4	2.9	0.0			.01	0.0	.00	.05	59	11	.22	.015	
shellie beans, cnd	37	110.6	0.2	0.1	0	28	4	.07			408	36			
1/2 cup	122	2.1	7.6	0.0		278	.04	0.3	.00		133		1.21		
soybean products															
miso	284	57.2	8.4	4.7	0	12	0	.35	.30	46	5032	92	58	4.58	1.185
1/2 cup	138	16.3	38.6	1.2	3.9	120	.13	1.2	.29	.36	226	211	3.78	.603	
natto	187	48.4	9.7	5.5	0	0	11	.17			6	191	101	2.67	1.345
1/2 cup	88	15.6	12.6	1.4		0	.14	0.0			642	153	7.57	.587	
tempeh	165	45.6	6.4	3.6	0	57	0	.09	.25	43	5	77	58	1.50	1.187
1/2 cup	83	15.7	14.1	0.9		569	.11	3.8	.70	.30	305	171	1.88	.556	
tofu, dried-frozen (koyadufu)	82	1.0	5.2	2.9	0	9	0	.05	.05	16	1	62	10	.83	.627
1 piece	17	8.2	2.5	0.7		88	.08	0.2	.00	.07	3	82	1.65	.200	
tofu, fried	35	6.6	2.6	1.5	0	0	0	.01	.01	4	2	48	8	.26	.194
1 piece	13	2.2	1.4	0.4		0	.02	0.0	.00	.02	19	37	.63	.052	
tofu, okara	47	49.8	1.1	0.5	0	0	0	.01			6	49	16		
1/2 cup	61	2.0	7.7	0.1		0	.01	0.1	.00		130	37	.79		
tofu, raw	94	104.8	5.9	3.3	0	11	0	.06	.06	19	9	130	127	1.00	.750
1/2 cup	124	10.0	2.3	0.9	0.6	105	.10	0.2	.00	.08	150	120	6.65	.239	
tofu, raw, firm	183	88.0	11.0	6.2	0	21	0	.13	.12	37	17	258	118	1.98	1.488
1/2 cup	126	19.9	5.4	1.6		209	.20	0.5	.00	.17	298	239	13.19	.476	
tofu, salted & fermented (fuyu)	13	7.7	0.9	0.5	0						316	5	6		
1 block	11	0.9	0.6	0.1					.00		8	8	.22		
soybeans, green, boiled	127	61.7	5.8	3.2	0	14	15	.14				131			
1/2 cup	90	11.1	10.0	0.6		140	.23	1.1	.00			142	2.25		

	KCAL	H₂O (g)	FAT (g)	PUFA (g)	CHOL (mg)	A (RE)	C (mg)	B-2 (mg)	B-6 (mg)	FOL (mcg)	Na (mg)	Ca (mg)	Mg (mg)	Zn (mg)	Mn (mg)
	WT (g)	PRO (g)	CHO (g)	SFA (g)	DFIB (g)	A (IU)	B-1 (mg)	NIA (mg)	B-12 (mcg)	PANT (mg)	K (mg)	P (mg)	Fe (mg)	Cu (mg)	
soybeans, mature, boiled	298	107.6	15.4	8.7	0	2	3	.49	.40	93	1	175	148	1.98	1.417
1 cup	172	28.6	17.1	2.2		15	.27	0.7	.00	.31	886	421	8.84	.700	
soybeans, mature, sprouted[a]															
steamed	38	37.3	2.1	1.2	0	1	4	.03			5	28	28	.49	.334
1/2 cup	47	4.0	3.1	0.2		5	.10	0.5	.00	.35	167	64	.62	.155	
stir-fried[b]	125	67.2	7.1		0	2	12	.19				82	96	2.10	1.133
3.5 oz	100	13.1	9.4			17	.42	1.1	.00	1.19	567	216	.40	.527	
spinach															
raw	6	25.6	0.1	0.0	0	188	8	.05	.06	54	22	28	22	.15	.251
1/2 cup chopped	28	0.8	1.0	0.0	0.9	1880	.02	0.2	.00	.02	156	14	.76	.036	
boiled	21	82.1	0.2	0.1	0	737	9	.21	.22	131	63	122	79	.69	.842
1/2 cup	90	2.7	3.4	0.0	1.7	7371	.09	.44	.00	.13	419	50	3.21	.157	
cnd	25	98.2	0.5	0.2	0	939	15	.15	.11	105	29[c]	135	81	.49	.639
1/2 cup	107	3.0	3.6	0.1	3.0	9391	.02	0.4	.00	.05	370	47	2.46	.193	
frzn															
boiled	27	85.5	0.2	0.1	0	739	12	.16	.14	102	82	139	65	.66	.895
1/2 cup	95	3.0	5.1	0.0	2.0	7395	.06	0.4	.00	.08	283	46	1.44	.134	
creamed, Birds Eye	59	72.7	3.8	0.7	1		11	.12	.06	16	312	72	27	.16	
1/3 cup	85	2.1	5.1	0.8	1.5	4231	.05	0.2	.13	.14	199	43	1.08	.042	
creamed, Pillsbury	80	93.0	2.9				3	.22			475	130			
1/2 cup	112	3.6	9.7			1848	.67	0.1			385	84	.90		
w/ butter sce, Pillsbury	60	130.5	1.5				26	.26			579	207			
1/2 cup	150	4.5	6.6			2907	.06	1.1			530	36	2.25		
squash, summer															
all varieties, raw	13	60.9	0.1	0.1	0	13	10	.02	.07	17	1	13	15	.17	.102
1/2 cup slices	65	0.8	2.8	0.0	0.7	127	.04	0.4	.00	.07	126	23	.30	.049	
all varieties, boiled	18	84.3	0.3	0.1	0	26	5	.04	.06	18	1	24	22	.35	.192
1/2 cup slices	90	0.8	3.9	0.1	1.0	259	.04	0.5	.00	.12	173	35	.32	.093	
crookneck, raw[d]	12	61.2	0.2	0.1	0	22	5	.03	.07	15	1	14	14	.19	.102
1/2 cup slices	65	0.6	2.6	0.0	0.7	220	.03	0.3	.00	.07	138	21	.31	.066	
crookneck, boiled[d]	18	84.3	0.3	0.1	0	26	5	.04	.09	18	1	24	22	.35	.192
1/2 cup slices	90	0.8	3.9	0.1	1.0	259	.04	0.5	.00	.12	173	35	.32	.093	
crookneck, cnd[d]	14	103.7	0.1	0.0	0	13	3	.03	.05	11	5[c]	13	14	.32	.105
1/2 cup slices	108	0.7	3.2	0.0	1.1	130	.02	0.5	.00	.05	104	22	.77	.086	
crookneck, frzn, boiled[d]	24	88.6	0.2	0.1	0	19	7	.05	.10	12	6	19	26	.32	.252
1/2 cup slices	96	1.2	5.3	0.0	1.2	187	.04	0.4	.00	.10	243	40	.49	.070	
scallop, raw	12	61.2	0.1	0.1	0	7	12	.02	.07	20	1	12	15	.19	.102
1/2 cup slices	65	0.8	2.5	0.0		72	.05	0.4	.00	.07	118	23	.26	.066	
scallop, boiled	14	85.5	0.2	0.1	0	8	10	.02	.08	19	1	14	17	.22	.115
1/2 cup slices	90	0.9	3.0	0.0		77	.05	0.4	.00	.07	126	25	.29	.075	
zucchini, raw	9	61.9	0.1	0.0	0	22	6	.02	.06	14	2	10	14	.13	.083
1/2 cup slices	65	0.8	1.9	0.0	0.3	221	.05	0.3	.00	.05	161	21	.28	.037	
zucchini, boiled	14	85.3	0.1	0.0	0	22	4	.04	.07	15	2	12	19	.16	.160
1/2 cup slices	90	0.6	3.5	0.0		216	.04	0.4	.00	.10	228	36	.32	.077	
zucchini, frzn, boiled	19	106.1	0.2	0.1	0	48	4	.05	.05	9	2	19	14	.22	.258
1/2 cup slices	112	1.3	4.0	0.0		483	.05	0.4	.00	.30	218	28	.54	.053	
zucchini, italian style, cnd[e]	33	103.3	0.1	0.1	0	61	3	.05			427	19	16	.29	
1/2 cup slices	114	1.2	7.8	0.0		615	.05	0.6	.00		312	33	.78	.112	
squash, winter															
all varieties, baked	39	90.8	0.6	0.3	0	363	10	.02	.07	29	1	14	8	.27	.215
1/2 cup cubes	102	0.9	8.9	0.1	1.2	3628	.09	0.7	.00	.36	445	20	.33	.097	
acorn, baked	57	84.6	0.1	0.1	0	44	11	.01	.20	19	4	45	43	.18	
1/2 cup cubes	102	1.1	14.9	0.0		437	.17	0.9	.00	.51	446	46	.95	.088	
acorn, boiled, mashed	41	109.4	0.1	0.0	0	31	8	.01	.14	14	3	32	31	.13	
1/2 cup	122	0.8	10.7	0.0		315	.12	0.6	.00	.37	321	33	.68	.063	
butternut, boiled	41	89.6	0.1	0.0	0	714	15	.02	.13	20	4	42	30	.13	
1/2 cup cubes	102	0.9	10.7	0.0		7141	.07	1.0	.00	.37	290	27	.61	.066	

[a] Seeds attached to sprouts.
[b] Stir-fried in veg oil.
[c] No salt used in processing.
[d] Includes straightneck varieties.
[e] Packed in tomato jce.

	KCAL	H_2O (g)	FAT (g)	PUFA (g)	CHOL (mg)	A (RE)	C (mg)	B-2 (mg)	B-6 (mg)	FOL (mcg)	Na (mg)	Ca (mg)	Mg (mg)	Zn (mg)	Mn (mg)
	WT (g)	PRO (g)	CHO (g)	SFA (g)	DFIB (g)	A (IU)	B-1 (mg)	NIA (mg)	B-12 (mcg)	PANT (mg)	K (mg)	P (mg)	Fe (mg)	Cu (mg)	
butternut, frzn, boiled	47	105.4	0.1	0.0	0	401	4	.05	.08		2	23	11	.14	
1/2 cup mashed	120	1.5	12.1	0.0		4007	.06	0.6	.00	.19	160	17	.70	.043	
hubbard, baked	51	86.8	0.6	0.3	0	616	10	.05	.18	17	8	17	22	.15	
1/2 cup cubes	102	2.5	11.0	0.1		6156	.08	0.6	.00	.46	365	23	.48	.046	
hubbard, boiled, mashed	35	107.5	0.4	0.2	0	473	8	.03	.12	12	6	12	16	.11	
1/2 cup	118	1.8	7.6	0.1		4726	.05	0.4	.00	.35	252	17	.33	.055	
spaghetti, boiled/baked	23	72.0	0.2	0.1	0	9	3	.02	.08	6	14	17	8	.16	
1/2 cup	78	0.5	5.0	0.0		86	.03	0.6	.00	.28	91	11	.26	.027	
succotash															
boiled	111	65.6	0.8	0.4	0	28	8	.09	.11		16	16	51	.61	
1/2 cup	96	4.9	23.4	0.1		282	.16	1.3	.00	.54	393	112	1.46	.172	
cnd	81	104.9	0.6	0.3	0	19	6	.07	.06	41	283	14	24	.64	.468
1/2 cup	128	3.3	17.9	0.1		187	.04	0.8	.00	.40	209	71	.68	.140	
cnd w/ cream style corn	102	104.0	0.7	0.3	0	19	9	.09	.17	59	325	15	1		.858
1/2 cup	133	3.5	23.4	0.1		187	.04	0.8	.00	.29	243	78	.83	.237	
frzn, boiled	79	63.0	0.8	0.4	0	20	5	.06	.08	28	38	13	19	.38	.238
1/2 cup	85	3.7	17.0	0.1		196	.06	1.1	.00	.20	225	57	.76	.051	
swamp cabbage															
raw	11	51.8	0.1		0	353	31	.06			63	43	40		
1 cup chopped	56	1.5	1.8			3528	.02	0.5	.00		174	22	.94		
boiled	10	45.5	0.1		0	255	8	.04			60	26	15		
1/2 cup chopped	49	1.0	1.8			2548	.01	0.2	.00		139	21	.65		
sweet potato															
baked	118	83.0	0.1	0.1	0	2488	28	.15	.28	26	12	32	23	.33	.638
1 sweet potato	114	2.0	27.7	0.0	2.1	24877	.08	0.7	.00	.74	397	62	.52	.237	
boiled	172	119.5	0.5	0.2	0	2797	28	.23	.40	18	21	35	16	.43	.553
1/2 cup mashed	164	2.7	39.8	0.1		27968	.09	1.1	.00	.87	301	44	.92	.264	
candied[a]	144	70.3	3.4	0.2	8[b]	440	7	.04	.04	12	73	27	12	.16	
1 piece 2 1/2" long & 2" diameter	105	0.9	29.3	1.4		4399	.02	0.4	.03		198	27	1.19	.107	
cnd	183	152.1	0.4	0.2	0	1597	53	.11	.38	33	107	44	45	.36	.910
1 cup pieces	200	3.3	42.3	0.1		15965	.07	1.5	.00	1.05	625	98	1.77	.278	
mashed, Joan of Arc	130		0.2				3	.09			59	24			
1/2 cup	118	1.9	29.5			24411	.07	1.3			312	54	.83		
syrup pack	106	71.0	0.3	0.1	0	701	11	.04	.06		38	16	12	.16	.603
1/2 cup	98	1.3	24.9	0.1		7014	.03	0.3	.00	.39	189	25	.93	.164	
w/ orange pineapple sce, Joan of Arc	180		0.2				6	.03			57	23			
1/2 cup	114	1.3	43.3			16509	.06	0.6			302	29	.80		
frzn, baked	88	64.9	0.1	0.0	0	1444	8	.05	.16	20	7	31	18	.26	.585
1/2 cup cubes	88	1.5	20.6	0.0	1.3	14441	.06	0.5	.00	.49	332	39	.47	.161	
taro, ckd	94	42.1	0.1	0.0	0	0	3	.02			10	12	20		
1/2 cup slices	66	0.3	22.8	0.0		0	.07	0.3	.00		319	50	.48		
taro, tahitian, ckd	30	58.8	0.5	0.2	0	120	26	.14			37	101	34		
1/2 cup slices	68	2.8	4.7	0.1		1200	.03	0.3	.00		423	45	1.06		
tomato, green, raw	30	114.4	0.3	0.1	0	79	29	.05			16	16	13	.09	.123
1 tomato	123	1.5	6.3	0.0		789	.07	0.6	.00	.62	251	35	.63	.111	
tomato, red															
raw	24	115.6	0.3	0.1	0	139	22	.06	.06	12	10	8	14	.13	.150
1 tomato	123	1.1	5.3	0.0	1.0	1394	.07	0.7	.00	.30	254	29	.59	.095	
boiled	30	110.9	0.3	0.1	0	162	25	.07	.04	11	13	10	17	.16	.184
1/2 cup	120	1.3	6.8	0.0		1623	.09	0.9	.00	.35	312	35	.72	.116	
bottled, kosher, Claussen	5	25.8	0.0		0		0	.00	.02	1	324	4	2	.02	
1 oz	28	0.2	1.0				.01	0.1		.03	41	3	.08	.020	
cnd															
crushed in tomato puree, Contadina	35	108.0	1.0	0.0	0		8	.04			340	33			
1/2 cup	119	1.0	6.0	0.0		615	.02	1.3			352		.51		

[a] Canned sweet potato & syrup, brown sugar, butter & salt.   [b] If margarine is used in place of butter, cholesterol is 0 mg.

	KCAL	H₂O (g)	FAT (g)	PUFA (g)	CHOL (mg)	A (RE)	C (mg)	B-2 (mg)	B-6 (mg)	FOL (mcg)	Na (mg)	Ca (mg)	Mg (mg)	Zn (mg)	Mn (mg)
	WT (g)	PRO (g)	CHO (g)	SFA (g)	DFIB (g)	A (IU)	B-1 (mg)	NIA (mg)	B-12 (mcg)	PANT (mg)	K (mg)	P (mg)	Fe (mg)	Cu (mg)	
italian style (pear), Contadina	25	111.9	1.0	0.1	0		12	.03			320	20	12		
1/2 cup	120	1.0	5.0	0.0		500	.03	0.8			247	20	.36		
sliced, Contadina	35		1.0				5	.03			370	17			
1/2 cup	120	1.0	9.0			239	.02	0.8			374		.48		
stewed	34	116.9	0.2	0.1	0	71	17	.05			325	42ᵃ	15	.21	
1/2 cup	128	1.2	8.3	0.0		710	.06	0.9	.00		307	25	.93	.143	
stewed, Contadina	35		1.0	0.0			15	.34			330	40	16		
1/2 cup	123	1.0	9.0	0.0		500	.06	0.8			265	20	.36		
wedges in tomato jce	34	120.2	0.2	0.1	0	76	19	.04			285	34ᵃ	15	.21	
1/2 cup	131	1.0	8.3	0.0		757	.07	0.9	.00		329	31	.61	.136	
whole, peeled	24	112.4	0.3	0.1	0	72	18	.04	.11		195ᵇ	32ᵃ	14	.19	
1/2 cup	120	1.1	5.2	0.0	0.8	725	.05	0.9	.00	.20	265	23	.73	.132	
whole, peeled, Contadina	25	111.9	1.0	0.1	0		18	.03			360	40	16		
1/2 cup	120	1.0	6.0	0.0		500	.06	0.8			256	20	.36		
w/ green chili	18	113.1	0.1	0.0	0	47	8	.02			481	24ᵃ	13	.16	
1/2 cup	120	0.8	4.3	0.0		468	.04	0.7	.00		129	17	.31	.108	
stewedᶜ	59	85.0	2.2	0.4	0	102	15	.06	.03	10	374	19	13	.17	.109
1 cup	101	1.8	10.4	0.4		1017	.07	0.8	.00	.21	170	32	.78	.069	
tomato paste, cnd	110	97.0	1.2	0.5	0	323	55	.25	.50		86ᵈ	46	67	1.05	
1/2 cup	131	5.0	24.7	0.2		3234	.20	4.2	.00	.99	1221	104	3.91	.776	
Contadina	38	31.5	0.3				13	.04			10	12	21		
1/4 cup	43	1.5	8.8			875	.05	1.4			497	33	.72		
italian, Contadina	70	37.9	1.0	0.6	0		1	.08			710	24			
1/4 cup	57	2.0	13.0	0.2		855	.03	1.6			474		.81		
no salt added, Heinz	87	73.3	0.1				27	.18			65	47	55	.56	
3.5 oz	100	3.6	20.4			2108	.18	3.0			967	75	1.30	.570	
tomato puree	102	218.2	0.3	0.1	0	340	88	.14	.38		49ᵉ	37	60	.54	
1 cup	250	4.2	25.1	0.0		3402	.18	4.3	.00	1.10	1051	99	2.32	.408	
Contadina	50	107.0	1.0				15	.34			80	16	16		
1/2 cup	119	2.0	11.0			1250	.60	1.2			577	40	.72		
turnip															
boiled	14	73.0	0.1	0.0	0	0	9	.02	.05	7	39	18	6		
1/2 cup cubes	78	0.6	3.8	0.0		0	.02	0.2	.00	.11	106	15	.17		
frzn, boiled	23	93.6	0.2	0.1	0	2	4	.03			36	32	14		
3.5 oz	100	1.5	4.4	0.0		25	.04	0.6	.00		182	26	.98		
turnip greens															
raw	7	25.5	0.1	0.0	0	213	17	.03	.07	54	11	53	9	.05	.130
1/2 cup chopped	28	0.4	1.6	0.0		2128	.02	0.2	.00	.11	83	12	.31	.098	
boiled	15	67.1	0.2	0.1	0	396	20	.05	.13	85	21	99	16	.10	.243
1/2 cup chopped	72	0.8	3.1	0.0		3959	.03	0.3	.00	.20	146	21	.57	.182	
cnd	17	110.8	0.4	0.1	0	420	18	.07	.04	48	325	138	24	.27	.324
1/2 cup	117	1.6	2.8	0.1		4196	.01	0.4	.00	.05	165	24	1.77	.097	
frzn, boiled	24	74.1	0.4	0.1	0	654	18	.06	.06	32	12	125	21	.34	.390
1/2 cup	82	2.8	4.1	0.1		6540	.04	0.4	.00	.06	184	27	1.59	.123	
turnip greens & turnips, frzn, boiled	17	94.2	0.2	0.1	0	516	9	.07			15	91	12		
3.5 oz	100	2.1	2.9	0.0		5161	.03	0.3	.00		62	17	1.33		
vegetable combinations, cnd															
peas & carrots	48	112.8	0.4	0.2	0	739	8	.07	.11	24	332ᶠ	29	18	.74	.457
1/2 cup	128	2.8	10.9	0.1		7386	.10	0.7	.00	.15	128	58	.97	.132	
peas & onions	30	51.8	0.2	0.1	0	10	2	.04			265	10	10	.35	
1/2 cup	60	2.0	5.1	0.0		96	.06	0.8	.00		57	30	.52	.060	
peas & onions, Pillsbury	60	103.2	0.3				10	.08			530	20			
1/2 cup	120	3.5	10.8			558	.11	1.7			115	62	.96		

ᵃ Ca added as firming agent.
ᵇ Special dietary pack contains 16 mg Na.
ᶜ Tomatoes, bread crumbs, margarine, sugar, onions, salt & pepper.
ᵈ If salt is added, Na is 1035 mg.
ᵉ If salt is added, Na is 998 mg.
ᶠ Special dietary pack contains 5 mg Na.

	KCAL	H₂O (g)	FAT (g)	PUFA (g)	CHOL (mg)	A (RE)	C (mg)	B-2 (mg)	B-6 (mg)	FOL (mcg)	Na (mg)	Ca (mg)	Mg (mg)	Zn (mg)	Mn (mg)
	WT (g)	PRO (g)	CHO (g)	SFA (g)	DFIB (g)	A (IU)	B-1 (mg)	NIA (mg)	B-12 (mcg)	PANT (mg)	K (mg)	P (mg)	Fe (mg)	Cu (mg)	
vegetable combinations, frzn															
am style, Pillsbury	90	87.5	2.1				23	.04			322	17			
1/2 cup	108	2.4	15.1			819	.06	1.1			164	52	.43		
broccoli & carrots, Pillsbury	20	66.6	0.1				23	.01			27	21			
1/2 cup	74	1.3	3.1			1654	.03	0.5			158	25	.37		
broccoli & cauliflower															
Pillsbury	60	96.0	1.2				84	.06			460	23			
1/2 cup	111	2.2	9.9			1995	.06	0.9			248	42	.47		
w/ creamy italian cheese sce, Birds	89	109.6	5.5	0.4	14		45	.14	.12	60	388	79	15	.36	
Eye—*1/2 cup*	128	4.6	6.7	3.0		1084	.05	0.4	.25	.28	220	78	.77	.029	
broccoli, carrots & pasta w/ lightly	87	76.6	4.4	0.9	0		22	.07	.12	35	269	35	13	.27	
seasoned sce, Birds Eye—*2/3 cup*	95	2.2	10.6	0.8		8850	.06	0.6	.00	.16	164	40	.76	.053	
broccoli, carrots & water chestnuts,	45	99.8	0.3	0.1	0		42	.11	.20	67	34	49	18	.37	
Birds Eye—*3/4 cup*	113	2.7	9.4	0.0	3.5	6032	.06	0.7	.00	.30	278	54	.83	.087	
broccoli, cauliflower & carrots															
Birds Eye	33	102.8	0.3	0.1	0		45	.08	.17	60	33	42	15	.30	
3/4 cup	113	2.4	6.8	0.0	2.7	7821	.06	0.6	.00	.22	234	48	.82	.048	
w/ butter sce, Pillsbury	30	72.1	1.2				21	.05			318	24			
1/2 cup	81	1.2	3.6			1750	.02	0.6			132	26	.41		
w/ cheese sce, Birds Eye	99	119.9	4.7	0.7	4		39	.15	.15	53	381	86	18	.40	
1/2 cup	142	4.0	11.6	1.5		6158	.06	0.6	.23	.35	257	84	.73	.042	
w/ cheese sce, Pillsbury	70	101.9	2.1				12	.18			516	68			
1/2 cup	119	2.9	10.0			1587	.04	0.4			255	174	.36		
broccoli, corn & red peppers, Birds Eye	59	95.2	0.5	0.2	0		39	.09	.16	66	16	33	20	.40	
2/3 cup	113	3.2	13.3	0.1	2.9	1682	.07	1.2	.00	.23	225	66	.77	.047	
broccoli fanfare, Pillsbury	80	91.8	1.8				35	.06			417	32			
3/4 cup	111	3.0	12.8			1254	.10	0.8			141	51	.78		
broccoli, green beans, pearl onions &	32	103.1	0.2	0.1	0		42	.09	.12	58	18	51	18	.29	
red peppers, Birds Eye—*3/4 cup*	113	2.4	6.6	0.0	2.9	1848	.06	0.6	.00	.19	199	44	.91	.052	
broccoli, red peppers, bamboo shoots &	29	103.6	0.3	0.1	0		51	.10	.12	66	21	43	14	.25	
straw mushrm, Birds Eye—*3/4 cup*	113	2.8	5.6	0.0	3.0	2233	.07	0.8	.00	.19	253	54	.81	.037	
brussel sprouts, cauliflower & carrots,	40	100.6	0.3	0.2	0		58	.09	.21	87	26	31	18	.28	
Birds Eye—*3/4 cup*	113	3.0	8.2	0.1	3.7	4047	.08	0.7	.00	.29	308	53	.81	.039	
carrots, peas & pearl onions, Birds Eye	48	81.8	0.2	0.1	0		9	.05	.16	22	61	29	14	.35	
1/2 cup	95	2.1	10.1	0.0	2.0	9344	.10	1.0	.00	.17	181	37	.86	.067	
cauliflower & carrots, Pillsbury	60	97.7	2.9				27	.07			287	16			
1/2 cup	111	2.2	7.0			1485	.11	0.7			161	39	.67		
cauliflower, carrots & snow peas, Birds	38	101.9	0.2	0.1	0		33	.07	.18	33	33	33	15	.26	
Eye—*2/3 cup*	113	2.0	8.1	0.0	2.9	7670	.05	0.7	.00	.26	228	36	.88	.039	
chinese style															
Birds Eye[a]	68	80.3	3.9	0.7	0		16	.05	.08	32	299	32	24	.20	
1/2 cup	95	2.2	7.7	0.7		1669	.04	0.4	.00	.15	148	32	.77	.038	
stir fry, Birds Eye[b]	36	82.8	0.2	0.1	0		16	.11	.08	23	540	44	24	.27	
1/2 cup	95	2.0	7.9	0.0	1.9	2153	.07	0.6	.00	.20	234	36	.98	.087	
chow mein style, Birds Eye[c]	89	75.5	4.2	0.8	0		6	.09	.09	20	368	25	16	.26	
1/2 cup	95	1.6	12.4	0.8		2745	.06	0.6	.00	.20	202	36	.63	.097	
corn & broccoli, Pillsbury	45	68.9	0.6				24	.02			14	17			
1/2 cup	82	2.1	7.9			284	.02	1.0			188	44	.49		
corn, green beans & pasta curls w/ light	108	71.1	4.9	0.9	1		8	.07	.07	17	283	53	18	.37	
cream sce, Birds Eye—*1/2 cup*	95	3.0	14.9	1.1		630	.07	0.9	.01	.10	141	68	.63	.040	
green beans, bavarian style & spaetzle	98	74.8	5.3	1.0	14		4	.06	.05	13	357	40	14	.25	
(noodles), Birds Eye—*1/2 cup*	95	2.4	11.5	1.0		572	.04	0.3	.10	.10	73	36	.72	.046	
italian style															
Birds Eye[d]	102	74.7	5.5	1.0	0		27	.08	.08	16	489	38	15	.17	
1/2 cup	95	2.1	11.1	1.0		736	.04	0.4	.01	.11	102	33	.62	.057	
Pillsbury	50	97.9	2.0				32	.08			301	52			
1/2 cup	110	2.9	5.0			1415	.03	0.4			134	51	.55		

[a] Bean sprouts, cabbage, pea pods, spinach, red peppers & water chestnuts.
[b] Snow peas, spinach, bean sprouts, celery, water chestnuts & red peppers.
[c] Bean sprouts, french green beans, water chestnuts, carrots, celery, straw mushrooms & sce.
[d] Italian green beans, chickpeas, red peppers, onions & ripe olives.

	KCAL / WT (g)	H₂O (g) / PRO (g)	FAT (g) / CHO (g)	PUFA (g) / SFA (g)	CHOL (mg) / DFIB (g)	A (RE) / A (IU)	C (mg) / B-1 (mg)	B-2 (mg) / NIA (mg)	B-6 (mg) / B-12 (mcg)	FOL (mcg) / PANT (mg)	Na (mg) / K (mg)	Ca (mg) / P (mg)	Mg (mg) / Fe (mg)	Zn (mg) / Cu (mg)	Mn (mg)
japanese style															
Birds Eye[a]	89	76.6	5.0	1.0	0		18	.06	.08	26	426	31	16	.23	
1/2 cup	95	2.0	10.0	0.9		774	.04	0.2	.01	.11	134	37	.57	.048	
Pillsbury	45		0.9				23	.06			405	33			
1/2 cup	110	2.0	7.2			1235	.03	0.4			173	34	.77		
stir fry, Birds Eye[b]	30	84.9	0.2	0.1	0		24	.09	.09	29	516	28	12	.20	
1/2 cup	95	1.6	6.6	0.0		917	.05	0.7	.00	.28	157	35	.70	.074	
le sueur style, Pillsbury	90	93.2	2.3				11	.07			346	26			
1/2 cup	115	4.5	12.8			461	.28	0.9			89	63	1.73		
mandarin style, Birds Eye[c]	87	76.0	4.1	0.8	0		20	.07	.13	25	390	41	15	.21	
1/2 cup	95	1.4	11.9	0.8		4053	.04	0.6	.00	.20	223	32	.64	.052	
mexican style, Pillsbury	151	105.7	5.2				24	.11			529	74			
1/2 cup	140	4.6	22.4			1812	.08	1.5			256	136	.50		
new england style, Birds Eye[d]	125	70.4	6.2	1.6	0		10	.08	.11	29	357	26	21	.24	
1/2 cup	95	3.1	14.2	1.2		885	.09	0.6	.01	.25	157	42	.61	.048	
pasta primavera style, Birds Eye[e]	122	69.0	5.2	0.9	3		17	.13	.10	29	338	95	20	.53	
1/2 cup	95	5.0	14.3	1.4		4279	.14	1.1	.13	.23	172	97	.95	.067	
peas & carrots	38	68.6	0.3	0.2	0	621	7	.05	.07	21	55	18	13	.36	.162
1/2 cup	80	2.5	8.1	0.1		6209	.18	0.9	.00	.13	127	39	.75	.061	
peas & onions, boiled	40	79.4	0.2	0.1	0	31	6	.06				13			
1/2 cup	90	2.3	7.8	0.0		313	.14	0.9	.00				.84		
peas & pearl onions															
Birds Eye	71	75.6	0.2	0.1	0		19	.08	.13	44	442	15	18	.48	
1/2 cup	95	4.6	13.5	0.0	3.5	636	.24	1.8	.00	.17	172	59	.85	.083	
w/ cheese sce, Birds Eye	137	111.6	4.9	0.8	4		15	.16	.13	43	446	76	25	.72	
1/2 cup	142	5.7	18.1	1.5		2145	.22	1.5	.23	.29	217	109	1.22	.094	
peas & potatoes w/ cream sce, Birds Eye—*1/2 cup*	127	50.0	6.2	1.1	1		6	.13	.12	26	390	42	21	.48	
	76	3.4	14.8	1.3		520	.15	0.9	.23	.33	208	84	.74	.093	
peas, onions & carrots w/ butter sce, Le Sueur—*1/2 cup*	80		3.0								470				
		4.0	11.0								115				
peas, pea pods & water chestnuts w/ butter sce, Le Sueur—*1/2 cup*	80	99.1	2.5				192	.07			426	31			
	118	4.1	10.4			861	.25	1.1			124	68	1.32		
rice & broccoli w/ cheese sce, Rice Originals—*1/2 cup*	120	82.6	4.1				6	.04			487	47			
	109	3.4	17.3			952	.22	1.5			69	72	1.42		
rice & peas w/ mushrooms, Birds Eye	108	38.0	0.2	0.1	0		10	.14	.07	18	321	8	8	.33	
2/3 cup	66	3.5	23.0	0.0		424	.16	1.8	.00	.04	68	39	.84	.049	
rice & spinach w/ cheese sce, Rice Originals—*1/2 cup*	170	79.4	6.6				0	.08			406	67			
	115	4.1	23.5			601	.22	2.0			101	60	2.19		
rice, french style, Birds Eye[f]	107	67.3	0.4	0.1	0		1	.02	.03	6	615	11	8	.37	
1/2 cup	95	2.7	22.9	0.1		48	.05	0.5	.00	.07	33	41	.95	.109	
rice, italian style, Birds Eye[g]	120	64.4	0.7	0.1	0		8	.02	.05	8	349	13	8	.35	
1/2 cup	95	2.9	26.0	0.1		319	.06	0.7	.00	.11	51	41	1.02	.100	
rice jubilee, Rice Originals	150	80.6	5.7				0	.03			333	15			
1/2 cup	112	2.4	22.2			1025	.19	2.4			54	34	1.57		
rice medley, Rice Originals	150	105.6	3.4				0	.04			326	17			
1/2 cup	140	3.6	26.3			545	.41	2.9			66	53	2.24		
rice, spanish style, Birds Eye[h]	111	66.5	0.5	0.2	0		33	.02	.07	11	547	8	8	.32	
1/2 cup	95	2.5	23.8	0.1		422	.08	0.8	.00	.09	57	38	.71	.081	
san francisco style, Birds Eye[i]	90	77.0	4.5	0.9	0		10	.07	.07	25	334	21	15	.25	
1/2 cup	95	2.2	10.4	0.9		489	.06	0.5	.00	.15	154	38	.53	.077	
sweet peas & cauliflower, Pillsbury	30	62.8	0.2				21	.05			60	19			
1/2 cup	73	2.5	4.5			968	.10	0.9			131	42	.66		
veg & rice w/ butter, Banquet	183		9.0		25		0	.04			650	20			
4 oz	113	3.0	23.0			227	.04	0.4			30		.60		
vinespinach, raw	19	93.1	0.3		0	800	102					109			
3.5 oz	100	1.8	3.4			8000	.05	0.5	.00			52	1.20		

[a] French green beans, broccoli, pearl onions, mushrooms & red peppers.
[b] Green beans, broccoli, green peppers, mushrooms & red peppers.
[c] Cabbage, carrots, bok choy, water chestnuts, snow pea pods, straw mushrooms & sce.
[d] French style green beans, corn, broccoli & red peppers.
[e] Peas, broccoli, carrots, red peppers, macaroni & sce.
[f] W/ french style green beans, onions, wild rice & mushrooms.
[g] W/ italian green beans, ripe olives & red peppers.
[h] W/ onions, peas, green peppers & red peppers.
[i] French style green beans, bean sprouts, celery, mushrooms & red peppers.

	KCAL	H₂O (g)	FAT (g)	PUFA (g)	CHOL (mg)	A (RE)	C (mg)	B-2 (mg)	B-6 (mg)	FOL (mcg)	Na (mg)	Ca (mg)	Mg (mg)	Zn (mg)	Mn (mg)
	WT (g)	PRO (g)	CHO (g)	SFA (g)	DFIB (g)	A (IU)	B-1 (mg)	NIA (mg)	B-12 (mcg)	PANT (mg)	K (mg)	P (mg)	Fe (mg)	Cu (mg)	
water chestnuts, chinese															
raw	66	45.5	0.1		0	0	3	.12			9	7	14		
1/2 cup slices	62	0.9	14.8				.09	0.6	.00		362	39	.37		
cnd	35	60.5	0.0		0	0	1	.02			6	3	3	.27	
1/2 cup slices	70	0.6	8.7			3	.01	0.3	.00		82	14	.61	.070	
watercress, raw	2	16.2	0.0	0.0	0	80	7	.02	.02		7	20	4		
1/2 cup chopped	17	0.4	0.2	0.0		799	.02	0.0	.00	.05	56	10	.03		
wax beans, cut, cnd, Joan of Arc	25		0.2				6	.04			321	87			
1/2 cup	119	1.3	4.5			81	.04	0.3			261	17	3.09		
waxgourd, boiled[a]	11	83.6	0.2	0.1	0	0	9	.00			93	16			
1/2 cup cubes	87	0.4	2.6	0.0		0	.03	0.3	.00		5	15	.33		
white beans															
boiled	249	112.9	0.6	0.3	0	0	0	.08	.17	145	11	161	113	2.46	1.138
1 cup	179	17.4	44.9	0.2		0	.21	0.3	.00	.41	1003	202	6.61	.514	
cnd	306	183.7	0.8	0.3	0	0	0	.10	.20	171	13	191	134	2.92	1.349
1 cup	262	19.0	57.5	0.2		0	.25	0.3	.00	.49	1189	239	7.84	.608	
white beans, small, boiled	253	113.2	1.2	0.5	0	0	0	.11	.23	245	4	131	122	1.96	.913
1 cup	179	16.1	46.2	0.3	7.9	0	.42	0.5	.00	.45	828	302	5.09	.267	
wild rice, raw	565	13.6	1.1			0	0	1.01			11	30			
1 cup	160	22.6	120.5			0	.72	9.9			352	542	6.70		
winged beans, boiled	252	115.6	10.1	2.7	0	0	0	.22	.08	18	22	244	94	2.48	2.062
1 cup	172	18.3	25.7	1.4		0	.51	1.4	.00		481	264	7.45	1.330	
yam, boiled/baked[b]	79	47.7	0.1	0.0	0	0	8	.02	.16	11	6	9	12	.13	
1/2 cup cubes	68	1.0	18.8	0.0		0	.07	0.4	.00	.21	455	33	.35	.103	
yambean, boiled	46	87.9	0.1		0	0	16	.03			6	16	17		
3.5 oz	100	1.2	10.4			0	.04	0.3	.00		181	23	.29		
yardlong bean, boiled	202	117.7	0.8	0.3	0	3	1	.11	.16	249	9	72	167	1.84	.833
1 cup	171	14.2	36.1	0.2		27	.36	0.9	.00	.68	539	309	4.51	.385	
yellow beans															
baked, cnd, B&M	326		7.0								770				
7/8 cup	227	15.0	50.0								760				
baked, cnd, Friend's	360		6.0								1040				
1 cup	255	18.0	60.0												
boiled	254	111.5	1.9	0.8	0	0	3	.18	.23	143	8	110	131	1.87	.805
1 cup	177	16.2	44.7	0.5		4	.33	1.3	.00	.41	576	324	4.39	.329	

35. MISCELLANEOUS

	KCAL	H₂O (g)	FAT (g)	PUFA (g)	CHOL (mg)	A (RE)	C (mg)	B-2 (mg)	B-6 (mg)	FOL (mcg)	Na (mg)	Ca (mg)	Mg (mg)	Zn (mg)	Mn (mg)
	WT (g)	PRO (g)	CHO (g)	SFA (g)	DFIB (g)	A (IU)	B-1 (mg)	NIA (mg)	B-12 (mcg)	PANT (mg)	K (mg)	P (mg)	Fe (mg)	Cu (mg)	
baking choc, unsweetened, Bakers	139	0.4	14.6	0.4	0		0	.07	.01	3	1	23	85		
1 oz	28	3.1	8.4	8.7		17	.01	0.4	.00	.05	242	112	1.96	.779	
baking choc, unsweetened, Hershey	185		15.8					.13			3	20	84	1.12	.560
1 oz	28	4.0	6.7				6	.02	0.3			224	123	2.04	.560
baking powder															
Calumet	3	0.2	0.0	0.0	0		0	.00	.00	0	426	241	0	.00	
1 t	4	0.0	0.7	0.0		0	.00	0.0	.00	.00	0	83	.00	.000	
cream of tartar	7	1.0	0.0				0	.00			694	0			
1 T	10	0.0	1.8			0	.00	0.0			361	0	.00		
sodium aluminum silicate	14	0.2	0.0				0	.00			1205	213			
1 T	11	0.0	3.4			0	.00	0.0			17	319			
bubble gum															
blueberry/grape/raspberry/strawberry,	23		0.0	0.0	0	0	0	.00	.00	0	0	0	0	.00	.000
Hubba Bubba—*1 piece*	8	0.0	5.8	0.0		0	.00	0.0	.00	.00	0	0	.000		
Bubble Yum[c]	25		0.0								0				
1 piece		0.0	7.0								0				

[a] Chinese preserving melon.
[b] This is the true yam of tropical areas (Dioscorea species), not the american yam which is really a sweet potato.
[c] Values are for 9 flavors.

	KCAL	H₂O (g)	FAT (g)	PUFA (g)	CHOL (mg)	A (RE)	C (mg)	B-2 (mg)	B-6 (mg)	FOL (mcg)	Na (mg)	Ca (mg)	Mg (mg)	Zn (mg)	Mn (mg)
	WT (g)	PRO (g)	CHO (g)	SFA (g)	DFIB (g)	A (IU)	B-1 (mg)	NIA (mg)	B-12 (mcg)	PANT (mg)	K (mg)	P (mg)	Fe (mg)	Cu (mg)	
Bubble Yum, sugarless[a]	20	.	0.0								0				
1 piece		0.0	5.0								0				
Carefree, sugarless[b]	10	0.0	0.0								0				
1 piece		0.0	2.0								0				
fruit/original, Hubba Bubba	23		0.0	0.0	0	0	0	.00	.00	0	0	30	0	.00	.000
1 piece	8	0.0	5.8	0.0		0	.00	0.0	.00	.00	0	0		.00	.000
fruit stripe, cherry/fruit/grape/lemon	10		0.0								0				
1 piece		0.0	2.0								0				
chewing gum															
big red/doublemint/freedent/juicy fruit/ spearmint, Wrigley's—*1 piece*	10		0.0	0.0	0	0	0	.00	.00	0	0	3	0	.00	.000
	3.2	0.0	2.3	0.0		0	.00	0.0	.00	.00	0	0		.00	.000
candy coated Beechies, fruit/pepper-mint/pepsin/spearmint—*1 piece*	6		0.0								0				
		0.0	2.0								0				
Carefree, sugarless[a]	8		0.0								0				
1 piece		0.0	2.0								0				
cinn/fruit/peppermint/spearmint, Beech-Nut—*1 piece*	10		0.0								0				
		0.0	2.0								0				
fruit stripe, cherry/lemon/lime/orange	9		0.0								0				
1 piece		0.0	2.0								0				
peppermint/spearmint, extra sugarfree, Wrigley's[c]—*1 piece*	8	0.0	0.0	0.0	0	0	0	.00	.00	0	0	5	0	.00	.000
	2.7	0.0	0.0	0.0		0	.00	0.0	.00	.00	0	0		.00	.000
cocoa															
dry powder	14	0.2	1.0				0	.02			0	7	21		
1 T	5	0.9	2.8	0.6		0	.01	0.1			82	35	.60		
dry powder, processed w/ alkali	14	0.2	1.0				0	.02			39	7	21		
1 T	5	0.9	2.6	0.6		0	.01	0.1			35	35	.60		
Hershey	115		3.6					.15			11	41	154	2.13	1.064
1/3 cup (1 oz)	28	7.6	12.8			22	.02	0.6			476	221	4.54	1.008	
cornstarch	30	0.8	0.0				0	.00			0	0			
1 T	8	0.0	7.2			0	.00	0.0			0	0	.00		
fruit pectin															
Certo	2	13.1	0.0	0.0	0		0	.00	.00	0	1	0	0	.00	
1 T	14	0.0	0.4	0.0		0	.00	0.0	.00	.00	33		.00		
light, sweetened, Sure-Jell	32	0.9	0.0	0.0	0		0	.00	.00	0	1	0	0	.00	
1/4 pkt	12	0.0	9.6	0.0		0	.00	0.0	.00	.00	0	0	.00	.000	
sweetened, Sure-Jell	37	1.0	0.0	0.0	0		0	.00	.00	0	1	0	0	.00	
1/4 pkt	12	0.0	10.3	0.0		0	.00	0.0	.00	.00	0	0	.00	.000	
gelatin, dry	23	0.9	0.0												
1 pkt	7	6.0	0.0												
meat/poultry coatings															
oven fry															
extra crispy recipe for chicken	112	1.3	2.0	0.8	0		0	.03	.02	15	804	31	9	.20	
1/4 pouch	29	3.3	20.1	0.3		190	.02	0.4	.00	.14	49	30	1.08	.068	
extra crispy recipe for pork	116	1.2	2.5	1.0	0		0	.03	.01	11	689	10	11	.22	
1/4 pouch	29	2.7	20.6	0.4		372	.02	0.3	.00	.09	49	34	.59	.052	
light crispy homestyle recipe	82	1.3	2.0	0.3	0		0	.04	.02	2	942	14	12	.11	
1/4 pouch	22	1.3	14.7	1.0		250	.05	0.5	.00	.03	27	27	.43	.022	
shake & bake															
barbeque for chicken	93	0.7	2.1	0.1	0		1	.01	.00	2	848	17	7	.09	
1/4 pouch	25	0.8	18.5	0.9		239	.01	0.1	.00	.03	88	16	.38	.027	
barbeque for pork	75	0.6	1.7	0.1	0		2	.02	.01	4	690	15	6	.09	
1/4 pouch	20	0.9	14.5	0.8		313	.02	0.2	.00	.05	75	15	.33	.027	
country mild recipe	77	1.0	3.8	0.2	0		1	.02	.01	4	506	5	3	.08	
1/4 pouch	17	0.8	10.0	2.1		450	.01	0.2	.00	.04	26	11	.27	.017	
for chicken	75	1.1	1.7	0.3	0		0	.07	.07	18	440	6	6	.16	
1/4 pouch	19	1.6	13.3	0.6		307	.06	0.8	.19	.05	38	22	.33	.041	

[a] Values are for 4 flavors.
[b] Values are for 3 flavors.
[c] Contains 2.0 g polyols which are metabolized as carbohydrates, but more slowly.

	KCAL	H₂O (g)	FAT (g)	PUFA (g)	CHOL (mg)	A (RE)	C (mg)	B-2 (mg)	B-6 (mg)	FOL (mcg)	Na (mg)	Ca (mg)	Mg (mg)	Zn (mg)	Mn (mg)
	WT (g)	PRO (g)	CHO (g)	SFA (g)	DFIB (g)	A (IU)	B-1 (mg)	NIA (mg)	B-12 (mcg)	PANT (mg)	K (mg)	P (mg)	Fe (mg)	Cu (mg)	
for fish	74	1.0	1.3	0.5	0		0	.01	.01	1	414	4	4	.14	
1/4 pouch	19	1.3	14.2	0.2		149	.01	0.1	.00		25	12	.39	.018	
for pork	81	0.7	1.1	0.4	0		0	.00			595		9	4	.20
1/4 pouch	21	1.3	16.2	0.3		162	.00	0.0	.00		34	16	.72	.015	
italian herb recipe	75	1.1	1.1	0.4	0		0	.01	.01	6	626	14	5	.15	
1/4 pouch	20	1.6	14.4	0.2		107	.01	0.2	.01	.05	30	18	.50	.036	
olives, pickled, green	45	36.0	4.9								926	24			
10 large	46	0.5	0.5	0.5		120					21	7	.60		
olives, pickled, ripe, greek style (salt cured, oil coated)—*10 med*	65	10.5	6.9								631				
	24	0.4	1.7	0.8								6			
olives, pickled, ripe, manzanillo/mission	4	2.6	0.4	0.0			0	.00	.00		29	3			
1 small	3.3	0.0	0.2	0.0	0.1	13	.00	0.0			0		.11	.008	
	5	3.2	0.4	0.0			0	.00	.00		35	4			
1 med	4.0	0.0	0.3	0.1	0.1	16	.00	0.0			0		.13	.010	
	5	3.7	0.5	0.0			0	.00	.00		40	4			
1 large	4.6	0.0	0.3	0.1	0.1	19	.00	0.0			0		.15	.012	
	7	4.8	0.6	0.0			0	.00	.00		52	5			
1 extra large	6.0	0.1	0.4	0.1	0.2	24	.00	0.0			0		.20	.015	
olives, pickled, ripe, sevillano/ascolano	7	7.7	0.6	0.1			0	.00	.00		82	9			
1 jumbo	9.1	0.1	0.5	0.1	0.3	31	.00	0.0			1		.30	.021	
	9	9.5	0.8	0.1			0	.00	.00		101	11			
1 colossal	11.3	0.1	0.6	0.1	0.3	39	.00	0.0			1		.38	.026	
	13	13.7	1.1	0.1			0	.00	.00		145	15			
1 super colossal	16.2	0.2	0.9	0.2	0.4	56	.00	0.0			1		.54	.037	
pickle relish															
hamburger, Del Monte	38	17.0	0.2				1	.01			308	2	2	.03	
1 oz	28	0.2	10.2			76	.01	0.2			22	5	.34	.023	
hot dog, Del Monte	25	20.5	0.1				0	.01			328	5	5	.06	
1 oz	28	0.5	6.4			47	.01	0.1			22	11	.40	.023	
picalilli, Heinz	30		0.0								140				
1 oz	28	0.0	7.0												
sweet	21	9.5	0.1								107	3			
1 T	15	0.1	5.1									2	.10		
sweet, Del Monte	38	17.4	0.1				0	.01			235	0	1	.04	
1 oz	28	0.1	10.1			44		0.1			7	4	.28	.023	
pickles															
bread & butter	11	11.8	0.0				1	.00			101	5			
2 slices	15	0.1	2.7			20	.00	0.0				4	.30		
Claussen	20	22.5	0.1		0		1	.00	.01		170	22	3	.11	
1 oz	28	0.2	4.7				.00	0.0			33	9	.11	.030	
Heinz	25		0.0								170				
1 oz	28	0.0	6.0												
chow chow															
sour[a]	35	105.1	1.5								1605	38			
1/2 cup	120	1.7	4.9									63	3.10		
sweet[a]	142	84.1	1.1								645	28			
1/2 cup	122	1.8	33.1									27	1.80		
dill															
2 slices	1	12.1	0.0				1	.00			186	3			
	13	0.1	0.3			10	.00	0.0			26	3	.10		
1 med (3 3/4" long, 1 1/4" diameter)	7	60.6	0.1				4	.01			928	17			
	65	0.5	1.4			70	.00	0.0			130	14	.70		
halves, del Monte	3	26.6	0.1				0	.00			440	4	3	.04	
1 oz	28	0.1	0.5			62		0.0			26	4	.23	.020	

[a] Pickled cucumber w/ cauliflower, onions & mustard.

	KCAL / WT (g)	H₂O (g) / PRO (g)	FAT (g) / CHO (g)	PUFA (g) / SFA (g)	CHOL (mg) / DFIB (g)	A (RE) / A (IU)	C (mg) / B-1 (mg)	B-2 (mg) / NIA (mg)	B-6 (mg) / B-12 (mcg)	FOL (mcg) / PANT (mg)	Na (mg) / K (mg)	Ca (mg) / P (mg)	Mg (mg) / Fe (mg)	Zn (mg) / Cu (mg)	Mn (mg)
hamburger chips, Del Monte	2	26.4	0.1				0	.00			494	1	1	.02	
1 oz	28	0.1	0.5			15					10	3	.28	.028	
hot garlic, Del Monte	6	25.7	0.1				0	.01			383	5	5	.08	
1 oz	28	0.2	1.2			73	.01	0.0			49	7	.34	.037	
kosher, halves, Claussen	4	26.3	0.1		0		1	.00	.01	1	325	4	2	.03	
1 oz	28	0.1	0.6				.00	0.0		.05	28	4	.07	.010	
kosher, slices, Claussen	3	26.4	0.1				0	.01	.01		315	5	2	.03	
1 oz	28	0.1	0.4				.00	0.0			30	4	.08	.010	
kosher, whole, Claussen	3	26.3	0.1				1	.01	.01	1	323	4	2	.03	.017
1 oz	28	0.2	0.5				.00	0.1		.06	33	5	.09	.030	
kosher, whole, Del Monte	2	26.5	0.1				0	.00			464	1	1	.01	
1 oz	28	0.1	0.5			50					11	3	.14	.017	
no garlic, Claussen	6	25.8	0.1								297	10	2	.03	
1 oz	28	0.2	1.0								54	5	.08	.010	
polish style, Heinz	4		0.0								285				
1 oz	28	0.0	1.0												
sweet halves, Del Monte	29	19.2	0.1				0	.01			430	1	1	.03	
1 oz	28	0.1	8.0			72	.01	0.1			8	3	.17	.028	
whole genuine, Del Monte	2	26.6	0.1				0	.00			418	4	3	.05	
1 oz	28	0.1	0.5			59		0.0			27	5	.23	.014	
polskie ogorki, Del Monte	6	26.1	0.1				1	.01			273	4	4	.04	
1 oz	28	0.2	1.3			133	.01				42	7	.11	.023	
sour															
1 med (3 3/4" long, 1 1/2"	7	61.6	0.1				5	.01			879	11			
diameter)	65	0.3	1.3			70	.00	0.0				10	2.10		
whole, Del Monte	3	26.7	0.1				0	.00			342	1	1	.01	
1 oz	28	0.1	0.6			41					6	4	.11	.023	
sweet															
1 small (2 1/3" long, 3/4"	22	9.1	0.1				1	.00				2			
diameter)	15	0.1	5.5			10	.00	0.0				2	.20		
chips, Del Monte	30	19.4	0.1				0	.01			322	0	1	.01	
1 oz	28	0.1	8.2			33	.00	0.0			5	3	.11	.028	
Claussen	50	15.5	0.4								167	13	1	.05	
1 oz	28	0.1	11.6								34	3	.10	.020	
Del Monte	38	17.1	0.1				0	.01			268	0	1	.02	
1 oz	28	0.1	10.4			30		0.1			5	3	.23	.028	
mild, Del Monte	31	19.0	0.1				0	.01			262	0	1	.01	
1 oz	28	0.1	8.6			48	.00	0.0			3	3	.11	.026	
sweet mixed															
Del Monte	37	17.3	0.1				0	.01			345	1	1	.02	
1 oz	28	0.1	10.1			43	.00	0.1			6	3	.23	.026	
Heinz	40		0.0								200				
1 oz	28	0.0	9.0												
sweet salad cubes, Heinz	30		0.0								270				
1 oz	28	0.0	7.0												
rennin (salt, starch, rennin enzyme)	12	1.0	0.1				0	.00			2453	386			
1 pkt	11	0.0	2.7			0	.00	0.0				22			
soybean protein concentrate	92	1.6	0.1	0.1	0	0	0	.04	.04	95	1	102	88	1.23	1.173
1 oz	28	17.8	7.1	0.0		0	.09	0.2	.00	.02	617	235	3.02	.273	
soybean protein isolate	94	1.4	1.0	0.5	0	0	0	.03		49	281	50	11	1.13	.418
1 oz	28	24.7	0.0	0.1		0	.05	0.4	.00	.02	23	217	4.06	.448	
tapioca, Minute	32	1.1	0.0	0.0	0	0	0	.00	.00	0	0	1	0		
1 T	9	0.1	7.8	0.0		0	.00	0.0	.00	.00	2	1	.04	.005	
tapioca, quick cooking, pearl/granulated	30	1.1	0.0	0.0			0	.00			0	1			
1 T	8	0.1	7.3	0.0			.00	0.0			2	2	.00		
tomato powder	302	3.1	0.4	0.2	0	1725	117	.76	.46	120	134	166	178	1.71	1.951
3 1/2 oz	100	12.9	74.7	0.1		17247	.91	9.1	.00	3.76	1927	295	4.56	1.241	

	KCAL / WT (g)	H₂O (g) / PRO (g)	FAT (g) / CHO (g)	PUFA (g) / SFA (g)	CHOL (mg) / DFIB (g)	A (RE) / A (IU)	C (mg) / B-1 (mg)	B-2 (mg) / NIA (mg)	B-6 (mg) / B-12 (mcg)	FOL (mcg) / PANT (mg)	Na (mg) / K (mg)	Ca (mg) / P (mg)	Mg (mg) / Fe (mg)	Zn (mg) / Cu (mg)	Mn (mg)
vinegar															
cider	2	14.1	0.0								0	1			
1 T	15	0.0	0.9								15	1	.10		
distilled	2	14.3	0.0								0				
1 T	15	0.0	0.8								2				
whey															
acid, dry	10	0.1	0.0	0.0		0	0	.06	.02	1	28	59	6	.18	
1 T	3	0.3	2.1	0.0	0.0	2	.02	0.0	.07	.16	66	39	.04		
acid, fluid	59	229.8	0.2	0.0		2	0	.34	.10	5	118	253	24	1.06	
1 cup	246	1.9	12.6	0.1	0.0	17	.10	1.2	.44	.94	352	191	.20		
sweet, dry	26	0.2	0.1	0.0	0	1	0	.17	.04	1	80	59	13	.15	
1 T	8	1.0	5.6	0.1	0.0	3	.04	0.1	.18	.42	155	70	.07		
sweet, fluid	66	229.1	0.9	0.0	5	10	0	.39	.08	2	132	115	20	.32	
1 cup	246	2.1	12.6	0.6	0.0	39	.09	0.2	.68	.94	396	112	.15		
yeast															
active dry, Fleischmann's	20		0.0								10				
1 pkt		3.0	3.0								150				
bakers, compressed	24	19.9	1.0				0	.47			5	4			
1 oz	28	3.4	3.1			0	.20	3.2			173	112	1.40		
bakers, dry	80	1.4	0.5				0	1.53			15	12			
1 oz	28	10.5	11.0			0	.66	10.4			566	366	4.6		
brewers	80	1.4	0.3				0	1.21			34	60			
1 oz	28	11.0	10.9			0	4.43	10.7			537	497	4.90		
compressed, Fleischmann's	15		0.0								5				
1 cube		2.0	2.0								100				
rapid rise, Fleischmann's	20		0.0								10				
1 pkt		3.0	3.0								150				
torula	79	1.7	0.3				0	1.43			4	120			
1 oz	28	10.9	10.5			0	3.97	12.6			580	486	5.50		
torula, dried, Lake States	348	4.8	4.4						4.45	225	33	350	175	12.90	2.500
3.5 oz	100	47.5	29.0						.59	7.75	2100	1375	22.50	.295	

Supplementary Tables

Amino Acids[a] (mg)

	TRY	THR	ISO	LEU	LYS	MET	CYS	PHE	TYR	VAL	ARG	HIS
1. Beverages												
1.1. Alcoholic Beverages												
1.1.1. Ales, Beers & Malt Liquors												
beer—*12 fl oz (356 g)*	11	18	18	21	25	4	11	21	53	32	32	18
beer, light—*12 fl oz (354 g)*	11	14	14	18	18	4	7	18	42	25	25	14
1.1.2. Cocktails & Cocktail Mixes												
bourbon & soda—*4 fl oz cocktail (116 g)*	0	0	0	0	0	0	0	0	0	0	0	0
screwdriver (orange jce & vodka)—*7 fl oz cocktail (213 g)*	2	13	13	21	15	6	9	15	6	19	79	4
1.1.3. Distilled Spirits												
gin, 90 proof—*1.5 fl oz jigger (42 g)*	0	0	0	0	0	0	0	0	0	0	0	0
gin/rum/vodka/whiskey, 94 proof—*1.5 fl oz jigger (42 g)*	0	0	0	0	0	0	0	0	0	0	0	0
rum, 80 proof—*1.5 fl oz jigger (42 g)*	0	0	0	0	0	0	0	0	0	0	0	0
vodka, 80 proof—*1.5 fl oz jigger (42 g)*	0	0	0	0	0	0	0	0	0	0	0	0
whiskey, 86 proof—*1.5 fl oz jigger (42 g)*	0	0	0	0	0	0	0	0	0	0	0	0
1.1.4. Liqueurs												
coffee w/ cream, 34 proof—*1.5 fl oz (47 g)*	19	60	80	129	105	33	12	64	64	88	48	36
creme de menthe—*1.5 fl oz (50 g)*	0	0	0	0	0	0	0	0	0	0	0	0
1.2. Carbonated Beverages												
cream soda—*12 fl oz (371 g)*	0	0	0	0	0	0	0	0	0	0	0	0
grape soda—*12 fl oz (372 g)*	0	0	0	0	0	0	0	0	0	0	0	0
lemon-lime soda—*12 fl oz (368 g)*	0	0	0	0	0	0	0	0	0	0	0	0
orange soda—*12 fl oz (372 g)*	0	0	0	0	0	0	0	0	0	0	0	0
pepper type soda—*12 fl oz (368 g)*	0	0	0	0	0	0	0	0	0	0	0	0
tonic water/quinine water—*12 fl oz (366 g)*	0	0	0	0	0	0	0	0	0	0	0	0
1.3. Carbonated Beverages, Low Calorie												
club soda—*12 fl oz (355 g)*	0	0	0	0	0	0	0	0	0	0	0	0
diet cherry coke, Coca-Cola—*12 fl oz (354 g)*								104[b]				
diet coke, Coca-Cola—*12 fl oz (354 g)*								104[b]				
diet Sprite—*12 fl oz (354 g)*								96[b]				
Fresca—*12 fl oz (354 g)*								98[b]				
Tab—*12 fl oz (354 g)*								16[b]				
1.4. Cereal Grain Beverages												
powder—*1 t (2.3 g)*	2	4	5	9	5	2	3	6	4	6	6	3
prep from powder												
w/ water—*6 fl oz water & 1 t powder (180 g)*	2	4	5	9	5	2	2	5	4	7	7	2
w/ whole milk—*6 fl oz milk & 1 t powder (185 g)*	85	278	368	598	481	154	57	296	294	409	224	165
1.5. Coffee												
brewed—*6 fl oz (177 g)*	0	2	4	9	2	0	4	5	4	5	2	4
inst powder—*1 rd t (1.8 g)*	1	3	3	9	2	0	4	5	3	5	1	3
cappuccino flavor, sugar sweetened—*2 rd t (14 g)*	1	5	6	16	3	1	7	9	5	9	2	5
decaffeinated—*1 rd t (1.8 g)*	1	2	3	8	2	0	3	5	3	5	1	3
french flavor, sugar sweetened—*2 rd t (11.5 g)*	1	6	7	19	4	1	8	10	6	11	2	6
mocha flavor, sugar sweetened—*2 rd t (11.5 g)*	5	14	14	24	17	3	6	18	14	21	18	7

[a] Amino acids for infant formulas and special dietary formulas are in the main table.

[b] Phenylalanine is from aspartame.

	TRY	THR	ISO	LEU	LYS	MET	CYS	PHE	TYR	VAL	ARG	HIS
w/ chicory—*1 rd t (1.8 g)*	0	2	2	6	1	0	3	3	2	4	1	2
prep from inst powder—*6 fl oz water & 1 rd t powder (179 g)*	0	2	4	9	2	0	4	5	4	5	2	4
cappuccino flavor, sugar sweetened—*6 fl oz water & 2 rd t powder (192 g)*	2	4	6	15	4	0	6	8	6	10	2	6
decaffeinated—*6 fl oz water & 1 rd t powder (179 g)*	0	2	4	9	2	0	4	5	4	5	2	4
french flavor, sugar sweetened—*6 fl oz water & 2 rd t powder (189 g)*	2	6	8	19	4	0	8	9	6	11	2	6
mocha flavor, sugar sweetened—*6 fl oz water & 2 rd t powder (188 g)*	6	13	13	24	17	4	6	17	13	21	19	8
w/ chicory—*6 fl oz water & 1 rd t powder (179 g)*	0	2	2	5	2	0	2	4	2	4	0	2
1.6. Fruit Juice Drinks & Fruit Flavored Beverages												
lemonade, from powder—*8 fl oz (240 g)*	0	0	0	0	0	0	0	0	0	0	0	0
orange breakfast drink, from frozn conc—*6 fl oz (188 g)*	0	2	2	4	4	2	2	4	2	4	13	2
orange flavor gelatin drink, prep from powder—*1 pkt in 4 fl oz water (136 g)*	0	122	94	201	261	48		144	26	169	521	54
thirst quencher, bottled—*8 fl oz (241 g)*	0	0	0	0	0	0	0	0	0	0	0	0
1.8. Tea, hot/iced												
brewed 3 min—*6 fl oz (178 g)*	0	0	0	0	0	0	0	0	0	0	0	0
inst powder—*1 t (0.7 g)*	1	0	0	0	0	0	0	0	0	0	0	0
w/ lemon flavor—*1 rd t (1.4 g)*	1	0	0	0	0	0	1	0	0	0	1	0
w/ sugar & lemon flavor—*3 rd t (23 g)*	1	1	1	1	1	0	1	0	1	1	1	0
w/ sodium saccharin & lemon flavor—*2 t (1.6 g)*	1	0	0	0	0	0	0	0	0	0	0	0
prep from inst powder—*1 t powder in 8 fl oz water (237 g)*	0	0	0	0	0	0	0	0	0	0	0	0
w/ lemon flavor—*1 rd t powder in 8 fl oz water (238 g)*	0	0	0	0	0	0	0	0	0	0	0	0
w/ sugar & lemon flavor—*3 rd t powder in 8 fl oz water (259 g)*	0	0	0	0	0	0	0	0	0	0	0	0
w/ sodium saccharin & lemon flavor—*2 t powder in 8 fl oz water (238 g)*	0	0	0	0	0	0	0	0	0	0	0	0
1.9. Water												
bottled, Perrier—*6.5 fl oz bottled (192 g)*	0	0	0	0	0	0	0	0	0	0	0	0
bottled, Poland Spring—*8 fl oz (237 g)*	0	0	0	0	0	0	0	0	0	0	0	0
municipal—*8 fl oz (237 g)*	0	0	0	0	0	0	0	0	0	0	0	0
2. Candy												
milk choc, Cadbury—*1 oz (28 g)*	120	300	400	680	480	160	80	410	200	440	270	220
3. Cereals, Cooked or to-be-Cooked												
corn grits, reg/quick, enr, ckd—*1 cup (242 g)*	19	121	136	520	68	82	77	177	148	177	114	94
cream of rice, ckd—*¾ cup (183 g)*	24	81	27	134	68	48	27	68	90	104	132	48
cream of wheat, inst ckd—*¾ cup (181 g)*	45	105	145	252	85	62	74	179	105	161	143	76
mix & eat—*1 pkt prep (142 g)*	38	86	120	208	72	51	62	149	88	135	122	62
mix & eat, flavored[a]—*1 pkt prep (150 g)*	34	78	106	186	64	46	55	132	78	118	109	57
cream of wheat, quick, ckd—*¾ cup (179 g)*	38	86	120	206	70	50	61	147	86	132	118	63
cream of wheat, reg, ckd—*¾ cup (188 g)*	39	90	126	216	73	53	64	154	90	137	124	66
farina, ckd, enr—*¾ cup (175 g)*	35	77	110	191	58	46	56	137	79	121	102	56

[a] Apple w/cinn, banana & spice, or maple & brown sugar.

	TRY	THR	ISO	LEU	LYS	MET	CYS	PHE	TYR	VAL	ARG	HIS
oatmeal, quick/reg, ckd—¾ cup (from ⅓ cup or 1 oz dry) (175 g)	61	158	207	345	184	75	112	247	161	264	338	102
4. Cereals, Ready-to-Eat												
alpha-bits, Post—*1 cup (1 oz) (28 g)*	26	76	97	200	79	39	53	117	81	124	143	51
bran, 100%—*½ cup (1 oz) (28 g)*	55	115	113	210	137	50	69	136	101	166	248	97
bran flakes,—*⅔ cup (1 oz) (28 g)*	56	103	120	214	107	50	65	147	96	156	190	82
cap'n crunch, Quaker—¾ cup (1 oz) (28 g)	12	51	61	184	39	31	34	76	59	78	67	37
cap'n crunch's crunchberries, Quaker—¾ cup (1 oz) (28 g)	13	52	63	183	42	32	33	77	60	79	67	38
peanut butter, Quaker—¾ cup (1 oz) (28 g)	20	68	85	210	65	35	39	108	86	105	164	54
cheerios, General Mills—1¼ cups (1 oz) (28 g)	58	149	196	327	175	71	107	233	153	251	320	97
cocoa pebbles, Post—⅞ cup (1 oz) (28 g)	18	65	69	108	56	38	22	58	73	86	106	39
corn flakes, Post Toasties, Post—1 cup (1 oz) (28 g)	14	80	89	343	45	54	51	117	97	117	75	62
crispy wheats & raisins, General Mills— ¾ cup (1 oz) (28 g)	33	58	75	129	53	31	38	92	55	90	90	43
C.W. Post—¼ cup (1 oz) (28 g)	37	95	120	199	103	48	60	135	98	151	187	62
C.W. Post w/ raisins—¼ cup (1 oz) (28 g)	34	89	112	185	96	45	56	125	91	140	173	57
fortified oat flakes, Post—⅔ cup (1 oz) (28 g)	92	252	295	490	344	99	131	278	200	318	342	130
frosted rice krinkles—⅞ cup (1 oz) (28 g)	20	69	73	114	58	41	24	59	77	89	113	41
fruity pebbles, Post—⅞ cup (1 oz) (28 g)	17	57	60	95	48	34	20	48	64	74	93	34
golden grahams, General Mills—¾ cup (1 oz) (28 g)	18	56	67	193	46	35	33	82	64	84	63	42
granola homemade—¼ cup (1 oz) (28 g)	42	127	153	249	151	60	71	168	110	189	311	87
Nature Valley[a]—⅓ cup (1 oz) (28 g)	39	101	134	222	118	49	72	158	103	170	218	66
grape-nut flakes, Post—⅞ cup (1 oz) (28 g)	55	97	124	215	93	51	60	154	93	152	155	73
grape-nuts, Post—¼ cup (1 oz) (28 g)	60	106	136	235	101	56	66	168	102	166	169	80
honeycomb, Post—1⅓ cups (1 oz) (28 g)	14	57	67	207	43	35	38	85	66	87	75	42
honey nut cheerios, General Mills— ¾ cup (1 oz) (28 g)	79	114	136	367	145	54	75	160	108	177	240	76
king vitamin, Quaker—1¼ cups (1 oz) (28 g)	12	52	62	191	39	33	34	77	61	79	67	38
kix, General Mills—1½ cups (1 oz) (28 g)	20	88	103	325	65	55	58	131	102	134	112	65
life, cinn, Quaker—⅔ cup (1 oz) (28 g)	75	210	277	446	297	96	107	284	206	313	368	133
life, Quaker—⅔ cup (1 oz) (28 g)	75	210	277	446	297	96	107	284	206	313	368	133
lucky charms, General Mills—1 cup (1 oz) (28 g)	35	90	119	198	105	43	64	141	92	151	193	58
puffed rice—1 cup (½ oz) (14 g)	13	45	48	75	38	27	15	38	50	58	74	27
puffed wheat—1 cup (½ oz) (14 g)	32	64	89	151	58	36	41	108	63	100	101	54
Quaker 100% natural ¼ cup (1 oz) (28 g)	46	123	167	273	160	62	63	174	125	197	227	82
w/ apples & cinn—¼ cup (1 oz) (28 g)	42	111	150	244	149	58	57	154	113	177	182	72
w/ raisins & dates—¼ cup (1 oz) (28 g)	40	106	143	234	138	53	53	149	106	169	192	71
quisp, Quaker—1 cup (1 oz) (28 g)	12	50	60	179	38	31	33	75	57	76	66	36
raisin bran, Post—½ cup (1 oz) (28 g)	39	74	86	153	77	37	46	105	68	112	133	58
shredded wheat 1 oz (28 g)	60	104	123	215	96	54	60	150	91	154	162	72
1 biscuit (24 g)	49	86	101	177	79	45	49	124	75	127	133	60
sugar sparkled flakes, Post—¾ cup (1 oz) (28 g)	9	52	58	223	29	35	33	76	63	76	48	40

[a] Cinn & raisin, coconut & honey, fruit & nut or toasted oat.

	TRY	THR	ISO	LEU	LYS	MET	CYS	PHE	TYR	VAL	ARG	HIS
super sugar crisp—⅞ *cup (1 oz) (28 g)*	28	56	78	133	50	32	36	95	56	88	89	47
team flakes—*1 cup (1 oz) (28 g)*	26	77	87	156	69	44	36	87	84	109	127	50
total, General Mills—*1 cup (1 oz) (28 g)*	54	94	111	195	87	49	54	136	83	139	147	66
trix, General Mills—*1 cup (1 oz) (28 g)*	11	55	63	214	37	35	36	82	65	82	65	41
wheaties, General Mills—*1 cup (1 oz) (28 g)*	51	90	106	186	83	47	51	130	79	133	140	62

5. Cheese & Cheese Products
5.1. Cheese

	TRY	THR	ISO	LEU	LYS	MET	CYS	PHE	TYR	VAL	ARG	HIS
american, processed—*1 oz (28 g)*	92	204	290	555	623	162	40	319	344	376	263	256
blue—*1 oz (28 g)*	89	223	319	545	526	166	30	309	368	442	202	215
brick—*1 oz (28 g)*	92	250	322	636	602	160	37	349	316	417	248	233
brie—*1 oz (28 g)*	91	213	288	547	525	168	32	328	340	380	208	203
camembert—*1 oz (28 g)*	87	203	275	522	501	160	31	313	325	362	199	194
cheddar—*1 oz (28 g)*	91	251	438	676	588	185	35	372	341	471	267	248
cheddar—*3.5 oz (100 g)*	320	886	1546	2385	2072	652	125	1311	1202	1663	941	874
cheddar, grated—*1 cup, not packed (113 g)*	362	1001	1746	2695	2342	737	141	1482	1358	1879	1063	988
cheshire—*1 oz (28 g)*	85	236	411	635	551	173	33	349	320	442	250	233
colby—*1 oz (28 g)*	87	240	418	645	561	176	34	355	325	450	254	236
cottage cheese, creamed—*1 rd T (28 g)*	39	155	206	360	283	105	32	188	186	216	160	116
cottage cheese, creamed—*4 oz (113 g)*	157	626	830	1451	1141	425	131	761	752	874	644	469
cottage cheese, creamed—*1 cup, not packed (210 g)*	292	1163	1542	2697	2121	789	243	1414	1398	1624	1196	872
cottage cheese, creamed w/ fruit—*1 cup (226 g)*	249	992	1315	2301	1810	673	207	1206	1192	1385	1021	743
cottage cheese, dry curd—*1 cup, not packed (145 g)*	279	1111	1472	2575	2025	754	232	1350	1334	1550	1142	832
cottage cheese, low fat, 1% fat—*1 cup (226 g)*	312	1242	1646	2879	2265	843	259	1510	1492	1734	1277	930
cottage cheese, low fat, 2% fat—*1 cup (226 g)*	346	1377	1825	3193	2511	934	287	1674	1655	1923	1416	1032
cream cheese—*1 oz (2T) (28 g)*	19	91	113	207	192	51	19	119	102	125	81	77
edam—*1 oz (28 g)*		264	371	728	754	204		406	413	513	273	293
gjetost—*1 oz (28 g)*	38	111	147	281	231	90	16	153	154	217	93	83
gouda—*1 oz (28 g)*		264	370	727	752	204		406	412	512	273	293
gruyere—*1 oz (28 g)*	119	309	457	880	768	233	86	494	503	636	276	317
limburger—*1 oz (28 g)*	82	209	346	593	475	176		316	339	408	198	164
monterey—*1 oz (28 g)*	89	247	431	665	578	182	35	365	335	463	262	244
mozzarella—*1 oz (28 g)*		210	264	537	559	154	33	287	318	344	236	207
mozzarella, low moisture—*1 oz (28 g)*		233	294	597	622	171	36	320	354	383	263	230
mozzarella, part skim—*1 oz (28 g)*		262	330	671	699	192	41	359	398	430	295	259
mozzarella, part skim, low moisture—*1 oz (28 g)*		184	232	471	491	135	29	252	280	302	208	182
muenster—*1 oz (28 g)*	93	252	325	641	606	161	37	352	318	420	250	235
neufchatel—*1 oz (28 g)*	25	120	149	274	253	68	25	157	135	166	107	101
parmesan, grated—*1 T (5 g)*	28	77	110	201	192	56	14	112	116	143	77	80
parmesan, hard—*1 oz (28 g)*	137	373	537	979	937	272	67	545	566	696	373	392
pimento, processed—*1 oz (28 g)*	91	204	290	555	622	162	40	319	343	376	263	256
port du salut—*1 oz (28 g)*	97	248	410	704	563	208		375	403	484	235	194
provolone—*1 oz (28 g)*		278	309	651	750	194	33	365	431	465	290	316
ricotta, part skim—*½ cup (124 g)*		649	739	1532	1678	352	124	697	739	868	793	576
ricotta, whole milk—*½ cup (124 g)*		641	731	1514	1659	348	123	689	731	858	783	569
swiss—*1 oz (28 g)*	114	294	436	839	733	222	82	471	480	606	263	302
swiss, processed—*1 oz (28 g)*	102	227	324	620	696	181	45	356	384	420	293	286
tilsit, whole milk—*1 oz (28 g)*	100	255	421	722	578	214		385	413	497	241	200

5.2. Cheese Products
cheese spread

	TRY	THR	ISO	LEU	LYS	MET	CYS	PHE	TYR	VAL	ARG	HIS
american—*1 T (16 g)*		101	135	288	243	87		150	144	221	88	82
american—*1 oz (28 g)*		178	236	505	427	152		264	252	387	155	144

	TRY	THR	ISO	LEU	LYS	MET	CYS	PHE	TYR	VAL	ARG	HIS
7. Creams & Cream Substitutes												
creamers												
liquid/frzn[a]—*½ fl oz (15 g)*	2	7	8	13	10	2	3	8	6	8	12	4
liquid/frzn[b]—*½ fl oz (15 g)*	2	6	9	15	12	5	1	8	9	11	6	4
powdered[b]—*1 t (2 g)*	1	4	6	9	8	3	0	5	5	7	4	3
half & half cream—*1 T (15 g)*	6	20	27	43	35	11	4	21	21	30	16	12
light (coffee/table) cream—*1 T (15 g)*	6	18	25	40	32	10	4	20	20	27	15	11
medium (25% fat) cream—*1 T (15 g)*	5	17	22	36	29	9	3	18	18	25	13	10
sour cream, cultured—*1 T (12 g)*	4	15	20	32	25	8		15		22		
whipped cream, pressurized—*1 T (3 g)*	1	4	6	9	8	2	1	5	5	6	3	3
whipping cream												
heavy, fluid—*1 T (15 g)*	4	14	19	30	24	8	3	15	15	21	11	8
light, fluid—*1 T (15 g)*	5	15	20	32	26	8	3	16	16	22	12	9
whipped topping												
from mix, prep w/ whole milk—*1 T (4 g)*	2	6	9	14	11	4	1	7	7	10	5	4
frzn—*1 T (4 g)*	1	2	3	5	4	2	0	3	3	4	2	1
pressurized—*1 T (4 g)*	1	2	2	4	3	1	0	2	2	3	2	1
8. Desserts												
8.7. Frozen Desserts												
ice cream												
french van, soft serve—*1 cup (173 g)*	100	325	423	680	549	177	72	337	335	466	281	188
van, reg (10% fat)—*1 cup (133 g)*	68	217	290	470	381	120	44	232	232	321	174	130
van, rich (16% fat)—*1 cup (148 g)*	58	186	250	405	327	104	38	199	199	276	150	112
ice milk												
van—*1 cup (131 g)*	73	233	312	506	409	129	48	249	249	345	187	140
van, soft serve—*1 cup (175 g)*	113	363	486	787	637	201	74	388	388	538	291	218
sherbet, orange—*1 cup (193 g)*	30	98	131	212	171	54	20	104	104	145	78	59
sorbet, Dole												
mandarin orange—*½ cup (104 g)*								3				
peach—*½ cup (104 g)*								3				
pineapple—*½ cup (104 g)*								3				
raspberry—*½ cup (104 g)*								3				
strawberry—*½ cup (104 g)*								3				
9. Eggs, Egg Dishes & Egg Substitutes												
9.1. Eggs, Chicken												
boiled, hard/soft—*1 large (50 g)*	97	298	380	533	410	196	145	343	253	437	388	147
fried—*1 large (46 g)*	86	264	336	472	363	174	128	303	224	387	344	130
omelet, plain—*1 large egg (64 g)*	94	290	372	565	410	188	133	332	252	426	364	146
poached—*1 large (50 g)*	97	297	378	531	408	195	144	341	251	435	387	146
scrambled w/ milk & fat—*1 large egg (64 g)*	94	290	372	565	410	188	133	332	252	426	364	146
white, fresh/frzn—*white of 1 large egg (33 g)*	51	149	204	291	206	130	83	210	134	251	195	76
whole, dried, stabilized (glucose reduced)—*1 T (5 g)*	39	118	151	212	163	78	57	136	100	173	154	58
whole, fresh/frzn—*1 large (50 g)*	97	298	380	533	410	196	145	343	253	437	388	147
yolk, fresh—*yolk of 1 large egg (17 g)*	41	151	160	237	189	71	50	121	120	170	193	67
9.2. Eggs, Other												
duck, whole—*1 egg (70 g)*	182	515	419	768	666	403	199	588	429	620	535	224
9.3. Egg (Chicken) Dishes												
souffle, spinach[c]—*1 cup (136 g)*	166	458	650	994	782	298	143	574	487	719	533	313
10. Entrees & Meals												
10.2. Canned Entrees												
beans, baked												
w/ beef—*1 cup (266 g)*	205	713	742	1357	1202	277	181	875	484	878	1067	484

[a] Contains hydrogenated veg oil & soy protein; veg oils are usually soybean, cottonseed, safflower, or blends thereof.
[b] Contains lauric acid oils and Na caseinate; lauric oils include modified coconut oil, hydrogenated coconut oil, and/or palm kernel oil.
[c] Contains whole milk, spinach, egg white, cheddar cheese, egg yolk, butter, flour, salt & pepper.

	TRY	THR	ISO	LEU	LYS	MET	CYS	PHE	TYR	VAL	ARG	HIS
w/ franks—*1 cup (257 g)*	198	717	763	1372	1213	270	190	897	488	887	1097	488
w/ pork—*1 cup (253 g)*	157	559	587	1063	913	202	144	721	374	696	825	372
chili w/ beans—*1 cup (255 g)*	176	612	638	1163	1043	242	156	745	418	750	918	418
cowpeas, common w/ pork—*1 cup (240 g)*	82	250	269	504	446	94	72	384	214	314	456	204
11. Fast Foods												
shake												
choc—*10 fl oz (283 g)*	136	436	583	945	764	241	88	464	464	645	348	263
strawberry—*10 fl oz (283 g)*	133	430	577	931	753	238	88	458	458	637	342	260
van—*10 fl oz (283 g)*	139	447	597	965	781	246	91	475	475	659	354	269
12. Fats												
beef suet, raw—*1 oz (28 g)*		12	8	24	21	5		14	9	15	27	6
pork separable fat—*1 oz (28 g)*	6	61	49	130	153	27	16	69	30	88	191	21
salt pork, raw—*1 oz (28 g)*	5	47	38	100	119	21	12	54	23	69	148	16
13. Fish, Shellfish & Crustacea												
abalone												
raw—*3 oz (85 g)*	163	626	632	1023	1086	328	190	521	465	635	1061	279
fried[a]—*3 oz (85 g)*	190	712	726	1178	1218	375	222	608	533	731	1199	321
alewife												
raw—*3.5 oz (100 g)*	194	834	989	1474	1707	563		718		1028		
cnd—*3.5 oz (100 g)*	162	697	826	1231	1426	470		599		859		
anchovy												
raw—*3 oz (85 g)*	194	758	797	1406	1589	512	185	675	584	891	1034	509
cnd in olive oil—*5 anchovies (20 g)*	65	253	266	470	531	171	62	226	195	298	346	170
paste—*1 t (7 g)*	14	60	71	106	123	41		52		74		
pickled—*1 oz (28 g)*	59	252	299	445	515	170		216		311		
bass, black												
raw—*3.5 oz (100 g)*	192	826	979	1459	1690	557		710		1018		
baked—*4 oz (113 g)*	212	1062	1227	1864	2384	661		897		1298		
stuffed, baked[b]—*3.5 oz (100 g)*	162	687	826	1231	1426	470		599		859		
bass, freshwater, raw—*3 oz (85 g)*	179	703	739	1303	1472	474	172	626	541	825	959	472
bass, striped, raw—*3 oz (85 g)*	169	660	694	1225	1384	446	162	588	509	777	902	444
bluefish, raw—*3 oz (85 g)*	190	746	785	1385	1564	504	183	665	575	877	1019	502
bullhead, black, raw—*3.5 oz (100 g)*	163	701	831	1239	1435	473		603		864		
burbot, raw—*3 oz (85 g)*	184	720	757	1335	1508	486	176	641	554	846	983	484
butterfish, raw—*3 oz (85 g)*	165	644	677	1194	1349	435	157	574	496	757	879	433
carp												
raw—*3 oz (85 g)*	170	665	699	1232	1392	449	162	592	512	781	907	446
ckd by dry heat—*3 oz (85 g)*	218	852	896	1579	1785	575	208	759	656	1001	1163	572
catfish, channel												
raw—*3 oz (85 g)*	173	677	712	1255	1419	457	166	604	522	796	925	455
breaded & fried[c]—*3 oz (85 g)*	170	672	711	1282	1357	448	174	617	529	802	915	451
caviar, black & red, granular—*1 T (16 g)*	52	202	166	341	293	103	72	171	155	202	254	104
cisco												
raw—*3 oz (85 g)*	181	707	744	1312	1482	478	173	630	545	831	966	475
smoked—*3 oz (85 g)*	156	609	641	1131	1278	411	149	543	469	717	832	410
clam liquid, cnd—*1 cup (240 g)*		0	0	0	0			0	0	0	0	0
clams												
raw—*3 oz (4 large or 9 small) (85 g)*	122	468	473	764	811	245	143	389	348	474	792	208
breaded & fried[d]—*3 oz (9 small) (85 g)*	143	512	541	870	843	275	173	467	394	553	837	238
cnd—*3 oz (85 g)*	243	934	945	1528	1623	490	285	778	694	949	1584	417
ckd by moist heat—*3 oz (9 small) (85 g)*	243	934	945	1528	1623	490	285	778	694	949	1584	417
cod, atlantic												
raw—*3 oz (85 g)*	169	664	698	1230	1390	448	162	591	511	779	906	445

[a] Dipped in flour & salt before frying.
[b] Stuffed w/ bacon, butter, celery, onion & bread cubes.
[c] Breading consists of cornmeal, egg, milk & salt.
[d] Prepared w/ bread crumbs, egg, milk & salt.

	TRY	THR	ISO	LEU	LYS	MET	CYS	PHE	TYR	VAL	ARG	HIS
cnd—*3 oz (85 g)*	217	848	892	1573	1777	573	207	756	653	997	1158	570
ckd by dry heat—*3 oz (85 g)*	218	851	894	1578	1782	575	208	757	655	1000	1161	571
dried & salted—*3 oz (85 g)*	598	2341	2461	4340	4904	1580	572	2084	1803	2751	3195	1572
cod, pacific, raw—*3 oz (85 g)*	170	667	701	1237	1397	451	163	594	513	784	910	448
crab, alaska king												
raw—*3 oz (85 g)*	217	630	754	1234	1353	438	174	657	518	732	1358	316
ckd by moist heat—*3 oz (85 g)*	229	666	797	1306	1431	463	184	694	547	774	1437	334
imitation, made from surimi—*3 oz (85 g)*	62	494	478	809	933	347	110	400	412	519	678	235
crab, blue												
raw—*3 oz (85 g)*	213	621	744	1218	1336	432	172	649	511	722	1340	312
cnd—*3 oz (85 g)*	243	706	846	1384	1518	491	196	737	581	820	1524	354
ckd by moist heat—*3 oz (85 g)*	239	695	832	1363	1494	484	192	725	571	808	1500	349
crab cakes[a]—*1 cake (60 g)*	169	497	598	970	1040	341	143	522	409	588	1038	248
crab, dungeness, raw—*3 oz (85 g)*	206	599	717	1174	1288	417	166	625	492	696	1293	301
crab, queen, raw—*3 oz (85 g)*	219	637	762	1248	1369	443	176	664	524	740	1374	320
crayfish												
raw—*3 oz (85 g)*	221	642	768	1259	1380	446	178	670	528	746	1386	322
ckd by moist heat—*3 oz (85 g)*	283	823	986	1613	1770	573	228	859	677	956	1777	413
croaker, atlantic												
raw—*3 oz (85 g)*	169	663	696	1228	1388	447	162	590	510	779	904	445
breaded & fried[b]—*3 oz (85 g)*	177	670	720	1258	1335	446	184	631	528	805	902	446
cusk, raw—*3 oz (85 g)*	181	708	744	1312	1482	478	173	631	545	832	966	475
cuttlefish, raw—*3 oz (85 g)*	155	594	601	972	1031	311	181	495	442	603	1007	265
dolphinfish, raw—*3 oz (85 g)*	176	689	724	1278	1444	466	168	614	531	810	941	463
drum, freshwater, raw—*3 oz (85 g)*	167	654	687	1211	1309	441	160	582	503	768	892	439
fish pieces, frzn, reheated[c]—*1 piece (4″x2″x½″) (57 g)*	111	368	425	722	640	242	129	398	307	476	483	239
fish sticks, frzn[c]—*1 stick (4″x2″x½″) (28 g)*	54	181	209	354	314	119	64	196	151	234	237	117
flatfish												
raw—*3 oz (85 g)*	179	702	738	1302	1471	474	172	626	541	825	959	472
ckd by dry heat—*3 oz (85 g)*	230	900	946	1669	1886	608	220	802	694	1058	1229	604
flounder/sole												
raw—*3.5 oz (100 g)*	148	646	756	1125	1306	434		553		794		
baked—*3.5 oz (100 g)*	300	1290	1530	2250	2640	870		1110		1590		
gefiltefish w/ broth, sweet—*1 piece (42 g)*	36	205	204	340	354	107	47	207	160	230	250	110
grouper												
raw—*3 oz (85 g)*	184	722	759	1339	1512	488	177	643	556	848	985	485
ckd by dry heat—*3 oz (85 g)*	236	926	973	1716	1940	625	226	825	713	1088	1264	621
haddock												
raw—*3 oz (85 g)*	180	705	740	1306	1476	476	173	627	542	828	961	473
ckd by dry heat—*3 oz (85 g)*	231	904	949	1675	1893	610	221	804	695	1062	1233	607
smoked—*3 oz (85 g)*	241	940	988	1743	1969	635	230	836	724	1105	1283	632
halibut, atlantic & pacific												
raw—*3 oz (85 g)*	198	775	815	1438	1624	524	190	691	598	911	1058	521
ckd by dry heat—*3 oz (85 g)*	254	995	1046	1844	2083	672	243	886	766	1169	1357	668
halibut, greenland, raw—*3 oz (85 g)*	137	536	563	993	1122	361	131	477	412	629	731	360
herring, atlantic												
raw—*3 oz (85 g)*	171	669	704	1241	1403	452	164	596	515	786	914	450
ckd by dry heat—*3 oz (85 g)*	219	859	902	1591	1798	580	210	764	661	1009	1171	576
kippered—*1 piece (4⅜″x1¾″x ¼″) (40 g)*	110	431	453	799	903	291	106	384	332	506	588	290
pickled—*1 piece (1¾″x⅞″x½″) (15 g)*	24	93	98	173	195	63	23	83	72	110	127	63
herring, pacific, raw—*3 oz (85 g)*	156	611	642	1132	1280	412	150	544	470	718	834	411
ling, raw—*3 oz (85 g)*	181	707	744	1312	1482	478	173	630	545	831	966	475
lingcod, raw—*3 oz (85 g)*	168	658	692	1220	1379	445	161	586	507	774	898	442

[a] Prepared w/ crab meat, egg, onion & margarine.
[b] Breading consists of bread crumbs, egg, milk & salt.
[c] Prepared from walleye pollock, bread crumbs, egg, milk & salt.

	TRY	THR	ISO	LEU	LYS	MET	CYS	PHE	TYR	VAL	ARG	HIS
lobster, northern												
raw—*3 oz (85 g)*	223	647	774	1268	1391	450	179	675	532	751	1396	325
ckd by moist heat—*3 oz (85 g)*	242	706	845	1383	1516	490	196	736	580	819	1522	354
mackerel, atlantic												
raw—*3 oz (85 g)*	177	693	728	1285	1452	468	169	617	534	814	946	466
ckd by dry heat—*3 oz (85 g)*	227	888	934	1647	1822	600	218	791	684	1044	1213	597
mackerel, jack, cnd—*1 cup (190 g)*	494	1932	2029	3582	4047	1303	473	1720	1488	2271	2637	1298
mackerel, king, raw—*3 oz (85 g)*	193	756	795	1401	1584	510	184	673	582	888	1032	507
mackerel, pacific & jack, raw—*3 oz (85 g)*	191	748	785	1386	1567	505	183	666	575	879	1021	502
mackerel, spanish												
raw—*3 oz (85 g)*	184	719	756	1333	1505	485	176	640	553	845	981	483
ckd by dry heat—*3 oz (85 g)*	224	879	924	1629	1841	593	215	783	677	1033	1199	590
milkfish, raw—*3 oz (85 g)*	196	765	804	1419	1603	517	187	682	589	899	1045	513
monkfish, raw—*3 oz (85 g)*	138	540	567	1000	1131	365	132	480	416	634	737	362
mullet, striped												
raw—*3 oz (85 g)*	184	721	758	1337	1510	487	176	642	555	847	984	485
ckd by dry heat—*3 oz (85 g)*	236	925	972	1714	1936	624	226	823	711	1086	1261	621
mussels, blue												
raw—*3 oz (85 g)*	113	435	440	712	756	228	133	362	324	442	738	194
ckd by moist heat—*3 oz (85 g)*	227	871	881	1425	1512	456	265	725	648	884	1476	388
ocean perch, atlantic												
raw—*3 oz (85 g)*	178	694	729	1287	1454	468	170	618	535	816	948	466
ckd by dry heat—*3 oz (85 g)*	227	890	935	1650	1864	601	218	792	685	1046	1215	598
oysters, eastern												
raw—*6 med (84 g)*	66	255	258	417	444	134	78	213	190	259	433	114
breaded & fried[a]—*3 oz (about 6 med) (85 g)*	89	310	337	542	495	169	111	299	247	348	497	149
cnd—*3 oz (85 g)*	67	258	261	422	449	135	79	215	192	262	438	116
ckd by moist heat—*3 oz (12 med) (85 g)*	134	517	522	845	897	271	157	430	384	524	876	230
oysters, pacific, raw—*3 oz (85 g)*	53	204	206	333	353	107	62	170	151	207	345	91
perch												
raw—*3 oz (85 g)*	184	723	759	1340	1514	488	177	643	557	849	986	485
ckd by dry heat—*3 oz (85 g)*	236	927	973	1717	1941	626	226	825	713	1089	1264	622
pike, northern												
raw—*3 oz (85 g)*	184	717	754	1330	1503	485	175	639	553	843	979	482
ckd by dry heat—*3 oz (85 g)*	235	920	967	1706	1927	621	225	819	708	1081	1255	618
pike, pickerel, raw—*3.5 oz (100 g)*	187	804	954	1402	1646	542		692		991		
pike, walleye, raw—*3 oz (85 g)*	182	713	750	1322	1494	481	174	635	549	838	973	479
pollock, atlantic, raw—*3 oz (85 g)*	185	724	762	1343	1518	490	177	645	558	852	989	486
pollock, walleye												
raw—*3 oz (85 g)*	163	640	673	1187	1341	433	156	570	493	752	874	430
ckd by dry heat—*3 oz (85 g)*	224	876	921	1624	1835	592	214	780	675	1029	1196	588
pompano, florida												
raw—*3 oz (85 g)*	176	689	723	1277	1442	465	168	613	530	809	940	462
ckd by dry heat—*3 oz (85 g)*	225	882	928	1636	1849	596	216	786	680	1037	1204	592
porgy/scup												
raw—*3.5 oz (100 g)*	190	817	969	1425	1672	551		703		1007		
fried—*3.3 oz (93 g)*	227	976	1158	1702	1998	658		840		1203		
pout, ocean, raw—*3 oz (85 g)*	158	620	652	1149	1299	418	151	553	478	728	847	417
rockfish, pacific												
raw—*3 oz (85 g)*	179	699	734	1295	1464	472	171	622	538	821	954	469
ckd by dry heat—*3 oz (85 g)*	229	896	942	1661	1877	605	219	797	690	1052	1222	602
roe, mixed species, raw—*1 oz (28 g)*	82	285	320	548	476	155	109	306	314	366	358	170
roughy, orange, raw—*3 oz (85 g)*	140	547	575	1016	1148	370	134	488	422	643	748	368
sablefish												
raw—*3 oz (85 g)*	128	500	525	927	1047	337	122	445	385	587	683	336
smoked—*3 oz (85 g)*	168	658	692	1220	1379	445	161	586	507	774	898	442
salmon, atlantic, raw—*3 oz (85 g)*	189	740	777	1371	1549	499	181	659	570	869	1009	496

[a] Prepared w/ bread crumbs, egg, milk & salt.

	TRY	THR	ISO	LEU	LYS	MET	CYS	PHE	TYR	VAL	ARG	HIS
salmon, chinook												
raw—*3 oz (85 g)*	191	747	785	1386	1566	505	183	666	575	878	1020	502
smoked—*3 oz (85 g)*	174	681	716	1263	1427	460	167	607	524	801	930	457
salmon, chum, raw												
raw—*3 oz (85 g)*	192	751	789	1391	1572	507	184	668	578	881	1024	504
cnd w/ bone—*3 oz (85 g)*	204	799	840	1481	1673	539	196	711	615	938	1090	536
salmon, coho												
raw—*3 oz (85 g)*	206	805	847	1493	1687	544	197	717	621	946	1090	541
ckd by moist heat—*3 oz (85 g)*	260	1020	1072	1890	2136	689	249	908	785	1199	1391	685
salmon, pink												
raw—*3 oz (85 g)*	190	743	781	1378	1556	502	182	661	572	873	1014	499
cnd w/ bone—*3 oz (85 g)*	189	737	775	1367	1544	498	180	656	568	866	1006	495
salmon, sockeye												
raw—*3 oz (85 g)*	203	794	835	1471	1663	536	194	707	611	932	1084	533
cnd w/ bone—*3 oz (85 g)*	195	762	802	1414	1598	515	186	679	587	897	1041	513
ckd by dry heat—*3 oz (85 g)*	260	1017	1069	1886	2132	687	249	906	784	1196	1389	683
sardines												
atlantic, cnd in soybean oil—*2 sardines (3″x1″x½″) (24 g)*	66	259	272	480	542	175	63	231	199	304	354	174
pacific, cnd in tomato sce—*1 sardine (4¾″x1⅛″x⅜″) (38 g)*	60	302	302	516	532	181	53	301	238	352	399	258
scallops												
raw—*3 oz (6 large or 14 small) (85 g)*	160	614	621	1004	1066	322	187	511	456	623	1040	274
breaded & fried[a]—*2 large (31 g)*	65	238	248	401	397	126	78	213	182	254	392	110
imitation, made from surimi—*3 oz (85 g)*	66	524	507	859	992	368	116	426	438	551	721	250
scup, raw—*3 oz (85 g)*	179	703	740	1304	1473	475	172	626	541	826	960	473
sea bass												
raw—*3 oz (85 g)*	175	687	722	1273	1439	464	168	612	529	808	938	462
ckd by dry heat—*3 oz (85 g)*	225	881	926	1633	1845	594	215	785	678	1034	1202	592
seatrout, raw—*3 oz (85 g)*	160	624	656	1157	1307	422	152	556	480	734	851	419
shad, american, raw—*3 oz (85 g)*	162	631	663	1170	1322	426	155	562	486	741	861	423
sheepshead												
raw—*3 oz (85 g)*	192	753	791	1397	1578	508	184	671	580	885	1029	506
ckd by dry heat—*3 oz (85 g)*	247	970	1019	1798	2032	655	237	864	746	1139	1323	651
shrimp												
raw—*3 oz (12 large) (85 g)*	241	699	837	1370	1503	486	194	729	575	813	1509	351
breaded & fried[a]—*3 oz (11 large) (85 g)*	254	731	886	1443	1499	502	219	789	611	871	1516	371
cnd—*3 oz (85 g)*	273	794	951	1557	1708	553	220	829	653	923	1714	399
ckd by moist heat—*3 oz (15½ large) (85 g)*	247	719	862	1410	1547	501	199	751	592	836	1552	361
imitation, made from surimi—*3 oz (85 g)*	64	509	492	834	962	357	113	413	424	534	700	242
smelt, atlantic												
raw—*4-5 med (100 g)*	186	800	949	1395	1637	539		688		986		
cnd—*4-5 med (100 g)*	184	791	938	1380	1619	534		681		975		
smelt, rainbow												
raw—*3 oz (85 g)*	167	657	690	1218	1376	444	161	585	506	772	897	441
ckd by dry heat—*3 oz (85 g)*	215	842	885	1561	1765	569	206	750	649	989	1149	565
snapper												
raw—*3 oz (85 g)*	196	764	803	1417	1601	516	187	681	588	898	1043	513
ckd by dry heat—*3 oz (85 g)*	250	980	1030	1816	2053	661	240	873	755	1152	1337	658
sole												
raw—*3.5 oz (100 g)*	148	646	756	1125	1306	434		553		794		
spiny lobster, raw—*3 oz (85 g)*	244	709	848	1389	1523	493	196	740	582	824	1529	355
spot, raw—*3 oz (85 g)*	176	690	725	1279	1445	466	168	615	531	811	942	463
sturgeon												
raw—*3 oz (85 g)*	154	602	632	1115	1261	406	147	536	463	707	821	404
ckd by dry heat—*3 oz (85 g)*	197	771	811	1430	1616	521	189	687	594	906	1052	518
smoked—*3 oz (85 g)*	297	1163	1222	2156	2439	785	284	1035	895	1366	1387	781

[a] Prepared w/ bread crumbs, egg, milk & salt.

	TRY	THR	ISO	LEU	LYS	MET	CYS	PHE	TYR	VAL	ARG	HIS
sucker, white, raw—*3 oz (85 g)*	160	625	656	1158	1308	422	153	556	481	734	853	419
sunfish, pumpkinseed, raw—*3 oz (85 g)*	184	723	760	1340	1515	488	177	643	557	849	987	485
surimi[a]—*3 oz (85 g)*	78	624	603	1022	1179	438	139	506	520	655	857	298
swordfish												
raw—*3 oz (85 g)*	189	738	775	1368	1545	498	180	657	568	867	1007	496
ckd by dry heat—*3 oz (85 g)*	241	946	995	1754	1982	638	231	842	728	1112	1291	635
tilefish												
raw—*3 oz (85 g)*	167	652	685	1209	1366	440	160	581	502	767	890	438
ckd by dry heat—*3 oz (85 g)*	233	913	959	1692	1912	616	224	813	703	1073	1245	613
tomcod, atlantic, raw—*3.5 oz (100 g)*	172	740	877	1290	1514	499		636		912		
trout, dolly varden—*3.5 oz (100 g)*	199	856	1015	1512	1751	577		736		1055		
trout, mixed species, raw—*3 oz (85 g)*	198	774	813	1435	1621	523	190	689	596	910	1056	519
trout, rainbow												
raw—*3 oz (85 g)*	196	766	805	1420	1604	517	187	682	590	899	1045	514
ckd by dry heat—*3 oz (85 g)*	251	982	1032	1820	2056	663	240	874	756	1153	1340	659
tuna												
cnd in oil												
light—*3 oz (85 g)*	277	1085	1141	2013	2274	733	265	966	836	1276	1482	729
white—*3 oz (85 g)*	252	989	1040	1833	2071	667	241	881	762	1162	1350	664
cnd in spring water												
light—*3 oz (85 g)*	281	1102	1159	2043	2309	745	269	982	849	1295	1505	740
white—*3 oz (85 g)*	254	994	1045	1842	2082	671	243	885	765	1168	1356	667
salad[b]—*½ cup (205 g)*	369	1437	1515	2651	2987	964	353	1283	1105	1689	1982	957
tuna, bluefin												
raw—*3 oz (85 g)*	222	870	914	1612	1821	587	213	774	669	1022	1187	584
ckd in dry heat—*3 oz (85 g)*	285	1114	1171	2066	2335	752	273	993	859	1310	1522	748
tuna, skipjack, raw—*3 oz (85 g)*	209	819	862	1520	1717	553	201	730	632	963	119	551
tuna, yellowfin, raw—*3 oz (85 g)*	223	871	915	1615	1825	588	213	776	671	1023	1189	585
turbot, european, raw—*3 oz (85 g)*	153	598	629	1109	1253	404	146	533	461	703	816	402
weakfish (sea trout)												
raw—*3.5 oz (100 g)*	165	710	842	1238	1452	478		610		874		
broiled—*3.5 oz (100 g)*	246	1058	1255	1845	2165	713		910		1304		
whelk												
raw—*3 oz (85 g)*	263	908	704	1618	1245	513	159	700	645	881	2098	415
ckd by moist heat—*3 oz (85 g)*	525	1816	1407	3236	2491	1024	318	1401	1290	1764	4196	830
whitefish												
raw—*3 oz (85 g)*	182	711	748	1318	1490	480	174	633	547	836	971	478
baked, stuffed[c]—*3.5 oz (100 g)*	152	654	775	1140	1338	441		562		806		
smoked—*3 oz (85 g)*	223	872	916	1617	1827	589	213	777	672	1025	1190	586
whiting												
raw—*3 oz (85 g)*	174	683	717	1265	1430	461	167	608	525	802	932	458
ckd by dry heat—*3 oz (85 g)*	224	875	920	1623	1833	591	214	779	674	1029	1194	587
wolffish, atlantic, raw—*3 oz (85 g)*	167	652	685	1209	1366	440	160	581	502	767	890	438
yellowtail, raw—*3 oz (85 g)*	220	863	906	1599	1807	582	211	768	664	1013	1177	579
14. Fruit & Vegetable Juices												
beef broth & tomato jce—*5.5 fl oz can (168 g)*	2	18	15	30	37	7	2	22	5	25	69	10
clam & tomato jce, cnd—*5.5 fl oz can (166 g)*	7	25	20	30	30	5	7	22	15	22	22	17
grape jce, cnd/bottled—*8 fl oz (253 g)*		40	18	30	25	3		30	8	25	119	18
grape jce, from frzn conc, sweetened—*8 fl oz (250 g)*		13	5	10	8			10	3	8	40	5
orange jce, fresh—*8 fl oz (248 g)*	5	20	20	32	22	7	12	22	10	27	117	7
orange jce, cnd—*8 fl oz (249 g)*	5	17	15	27	20	7	10	17	7	22	100	7
orange jce, from frzn conc—*8 fl oz (249 g)*	5	20	17	32	22	7	12	20	10	27	112	7
tangerine jce, fresh—*8 fl oz (247 g)*	2	15	12	25	17	5	10	15	7	20	84	5

[a] Prepared from walleye pollock.
[b] Prepared w/ light tuna cnd in oil, pickle relish, salad dressing, onions & celery.
[c] Stuffed w/ bacon, onion, celery & bread crumbs.

	TRY	THR	ISO	LEU	LYS	MET	CYS	PHE	TYR	VAL	ARG	HIS
tangerine jce, cnd, sweetened—*8 fl oz (249 g)*	2	15	12	25	17	5	10	15	7	20	85	5
tangerine jce, from frzn conc, sweetened—*8 fl oz (241 g)*	2	12	12	19	14	5	7	12	5	17	70	5
tomato jce—*6 fl oz (182 g)*	9	31	27	38	40	7	7	29	18	27	27	22

15. Fruits

	TRY	THR	ISO	LEU	LYS	MET	CYS	PHE	TYR	VAL	ARG	HIS
apple												
raw, w/ skin—*1 med (138 g)*	3	10	11	17	17	3	4	7	6	12	8	4
raw, w/o skin—*1 med (128 g)*	1	6	8	12	12	3	3	5	4	9	6	3
boiled, w/o skin—*1 cup (171 g)*	3	15	17	27	27	5	7	12	9	21	14	7
cnd, sliced, sweetened—*½ cup (102 g)*	2	6	7	11	11	2	2	5	3	8	6	3
dried, sulfured—*10 rings (64 g)*	6	21	24	36	37	6	8	17	11	28	19	10
micro ckd w/o skin—*1 cup (170 g)*	5	17	19	29	31	5	7	14	9	22	15	7
applesce, cnd, sweetened—*½ cup (128 g)*	3	9	9	14	14	3	3	6	4	10	8	4
applesce, cnd, unsweetened—*½ cup (122 g)*	2	7	7	12	12	2	2	6	4	10	6	4
apricots												
raw—*3 med (106 g)*	16	50	43	82	103	6	3	55	31	50	48	29
cnd, heavy syrup—*4 halves (90 g)*	8	16	14	27	32	3	2	19	11	17	18	7
cnd, jce pack—*3 halves (84 g)*	9	19	16	31	37	3	2	22	13	19	20	8
cnd, light syrup—*3 halves (85 g)*	8	16	14	26	31	3	2	19	10	16	18	8
cnd, water pack—*4 halves (90 g)*	11	23	19	36	43	3	2	25	14	23	24	10
dried, sulfured—*10 halves (35 g)*	23	46	39	75	89	6	4	53	30	47	49	21
frzn, sweetened—*½ cup (121 g)*	10	29	24	47	59	4	2	31	18	29	27	16
banana, raw—*1 med (114 g)*	14	39	38	81	55	13	19	43	27	54	54	92
blueberries												
raw—*1 cup (145 g)*	4	26	30	58	17	16	10	35	12	41	49	15
cnd, heavy syrup—*½ cup (128 g)*	4	23	26	50	15	13	9	29	10	35	42	13
frzn, sweetened—*1 cup (230 g)*	5	25	28	53	16	14	9	32	12	39	46	14
breadfruit, raw—*¼ small (96 g)*		50	61	62	36	10	9	25	18	45		
carambola, raw—*1 med (127 g)*	5	29	29	51	51	14		24	29	33	14	5
crabapples, raw—*1 cup slices (110 g)*	4	15	18	28	28	4	6	12	9	21	14	7
custard apple, raw—*3.5 oz (100 g)*	7				37.	4						
dates, dried—*10 dates (83 g)*	42	43	39	73	50	18	37	46	25	55	55	25
elderberries, raw—*1 cup (145 g)*	19	39	39	87	38	20	22	58	74	48	68	22
figs, raw—*1 med (50 g)*	3	12	12	17	15	3	6	9	16	14	9	6
figs, cnd, heavy syrup—*3 figs (85 g)*	3	10	10	14	13	3	5	8	14	12	8	4
figs, dried—*10 figs (187 g)*	49	187	174	249	228	47	94	138	247	215	131	80
grapefruit												
raw, pink & red—*½ med (123 g)*	2				17	2						
raw, white—*½ med (118 g)*	2				21	2						
cnd, jce pack—*½ cup (124 g)*	2				22	2						
cnd, light syrup—*½ cup (127 g)*	3				18	3						
grapes												
american (slip skin), raw—*1 cup (92 g)*	3	16	5	12	13	19	9	12	10	16	42	21
european (adherent skin), raw—*1 cup (160 g)*	5	29	8	22	24	35	18	22	19	29	78	38
thompson seedless, cnd, heavy syrup—*½ cup (128 g)*	3	17	5	13	14	20	10	13	10	17	45	23
guava, raw—*1 med (90 g)*	6	28	27	50	21	5		2	9	25	19	6
guava, strawberry, raw—*1 cup (244 g)*	12	54	51	95	39	10		2	17	49	37	12
lime, raw—*1 med (67 g)*	2				9	1						
longans												
raw—*31 fruits (100 g)*		34	26	54	46	13		30	25	58	35	12
dried—*3.5 oz (100 g)*		128	97	202	172	49		112	94	217	131	45
loquats, raw—*10 med (100 g)*	5	15	15	26	23	4	6	14	13	21	14	7
lychees												
raw—*10 med (100 g)*	7				41	9						
dried—*3.5 oz (100 g)*	33				187	42						

	TRY	THR	ISO	LEU	LYS	MET	CYS	PHE	TYR	VAL	ARG	HIS
mammy apple, raw—⅛ med (100 g)	5				37	6						
mandarin oranges, cnd												
jce pack—½ cup (124 g)	7	12	21	19	38	16	7	25	12	32	53	14
light syrup—½ cup (126 g)	5	9	15	14	29	13	6	19	10	24	39	10
mango, raw—1 med (207 g)	17	39	37	64	85	10		35	21	54	39	25
orange												
navel, raw—1 med (140 g)	14	24	39	36	73	31	15	48	24	62	101	27
valencia, raw—1 med (121 g)	12	21	34	31	64	27	13	41	22	53	88	24
papaya, raw—1 med (304 g)	24	33	24	49	76	·6		27	15	30	30	15
peach												
raw—1 med (87 g)	2	23	17	35	20	15	5	19	16	33	16	11
cnd, heavy syrup—1 cup (256 g)	3	46	33	67	38	28	10	36	31	64	31	20
cnd, heavy syrup, spiced—1 med (88 g)	1	14	11	20	11	9	3	11	10	20	9	7
cnd, jce pack—1 cup (248 g)	5	62	45	89	50	37	12	50	40	84	40	30
cnd, light syrup—1 cup (251 g)	3	45	33	63	35	28	10	35	30	63	28	20
cnd, water pack—1 cup (244 g)	2	41	32	61	34	27	10	34	27	59	27	20
dried, sulfured—10 halves (130 g)	13	183	135	265	151	113	38	148	122	256	120	87
frzn, sweetened—1 cup (250 g)	5	60	45	88	50	38	13	50	40	85	40	30
pear												
raw—1 med (166 g)	17	18	33	23	8	7	17	5	23	12	7	22
cnd, heavy pack—1 cup (255 g)	13	15	26	18	5	5	13	5	18	8	5	18
cnd, jce pack—1 cup (248 g)	22	25	42	30	10	7	22	7	30	15	10	27
cnd, light syrup—1 cup (251 g)	13	15	25	18	5	5	13	5	18	8	5	15
cnd, water pack—1 cup (244 g)	12	12	22	17	5	5	12	5	17	7	5	15
dried, sulfured—10 halves (175 g)	86	95	165	116	39	32	86	28	116	56	35	109
persimmon, raw—1 med (25 g)	4	10	9	15	11	2	5	9	6	11	9	4
persimmon, japanese												
raw—1 med (168 g)	17	50	42	71	55	8	22	44	27	50	42	20
dried—1 med (34 g)	8	24	20	34	27	4	10	21	13	24	20	9
pineapple												
raw—1 cup pieces (155 g)	8	19	20	29	39	17	3	19	19	25	28	14
cnd, heavy syrup—1 cup pieces (255 g)	13	23	23	33	41	23	3	23	20	28	31	10
cnd, jce pack—1 cup pieces (250 g)	13	25	25	40	48	28	3	25	25	33	35	23
plantain, ckd—1 cup slices (154 g)	14	32	34	55	57	15	18	42	31	43	102	60
plum												
raw—1 med (66 g)		11	11	14	11	4	3	11	4	13	9	9
heavy syrup—3 plums (133 g)		11	9	13	11	4	3	11	4	12	8	8
cnd, jce pack—3 plums (95 g)		10	10	13	10	4	3	10	4	11	9	8
sapodilla, raw—1 med (170 g)	9	20	26	41	66	5		22	24	27	29	27
sapote, raw—1 med (225 g)	52	131	104	189	216	36		119	124	173	124	95
soursop, raw—1 cup (225 g)	25				135	16						
strawberries												
raw—1 cup (149 g)	10	28	21	46	37	1	7	27	31	27	39	18
frzn, sweetened—1 cup (255 g)	15	41	31	69	56	3	13	38	46	38	59	26
frzn, unsweetened—1 cup (149 g)	7	19	15	33	25	1	6	18	21	18	27	12
sugar apple, raw—1 med (155 g)	16				85	11						
tamarind, raw—1 cup (120 g)	22				167	17						
tangerine, raw—1 med (84 g)	5	8	14	13	27	11	6	18	9	23	37	10
watermelon, raw—1 cup (160 g)	11	43	30	29	99	'10	3	24	19	26	94	10

16. Grain Fractions

	TRY	THR	ISO	LEU	LYS	MET	CYS	PHE	TYR	VAL	ARG	HIS
corn germ, Ener-G Foods—1 cup (100 g)		675	597	1849	1076	269		527	354	974	1565	604
potato flour—½ cup (90 g)	120	292	311	443	431	112	73	329	234	372	390	173
rice polish, Ener-G Foods—½ cup (56 g)		16	14	27	2	8		17	16	20	38	13
soybean flour												
defatted—1 cup (100 g)	683	2042	2281	3828	3129	634	757	2453	1778	2346	3647	1268
full fat—1 cup (85 g)	427	1275	1424	2390	1953	396	473	1532	1110	1465	2277	791

	TRY	THR	ISO	LEU	LYS	MET	CYS	PHE	TYR	VAL	ARG	HIS
full fat, roasted—*1 cup (85 g)*	430	1284	1435	2409	1969	399	477	1544	1119	1476	2295	797
low fat—*1 cup (88 g)*	595	1778	1986	3334	2725	552	660	2137	1549	2043	3177	1104
soy meal, defatted—*1 cup (122 g)*	797	2381	2660	4465	3649	739	883	2862	2074	2736	4254	1479

17. Grain Products
17.7. Pasta

	TRY	THR	ISO	LEU	LYS	MET	CYS	PHE	TYR	VAL	ARG	HIS
macaroni, enr, ckd—*1 cup (140 g)*	58	186	238	308	152	72	90	248	158		272	112
noodles, enr, ckd—*1 cup (160 g)*	74	276	324	436	220	112	128	316	164	388		156
spaghetti, enr, ckd—*1 cup (140 g)*	58	186	238	308	152	72	90	248	158		272	112

19. Infant, Junior & Toddler Foods
19.1. Baked Products

	TRY	THR	ISO	LEU	LYS	MET	CYS	PHE	TYR	VAL	ARG	HIS
teething biscuit—*1 biscuit (11 g)*	21	62	93	161	39	30	17	58	73	97	66	44

19.2. Cereals

	TRY	THR	ISO	LEU	LYS	MET	CYS	PHE	TYR	VAL	ARG	HIS
barley												
dry—*1 T (2.4 g)*	3	9	10	19	9	5	6	16	10	14	14	6
prep w/ whole milk—*1 oz (28 g)*	17	53	67	114	79	29	19	68	57	80	54	33
cereal & egg yolks												
jr—*1 jar (213 g)*	58	175	213	366	271	109	60	196	179	260	222	96
str—*1 jar (128 g)*	35	105	128	220	163	65	36	118	108	156	133	58
cereal, egg yolks & bacon, str—*1 jar (128 g)*	29	120	127	241	173	46		125	123	145	175	76
grits & egg yolks, str—*1 jar (128 g)*	28	92	111	234	142	69	35	125	114	136	108	72
high protein												
dry—*1 T (2.4 g)*	13	35	42	71	57	15	17	45	34	45	67	24
prep w/ whole milk—*1 oz (28 g)*	37	103	129	213	172	48	41	126	103	138	158	68
w/ apple & orange, dry—*1 T (2.4 g)*	8	28	3	50	32	15		32	26	36	45	15
mixed												
dry—*1 T (2.4 g)*	4	10	11	25	10	6	8	16	12	16	17	7
prep w/ whole milk—*1 oz (28 g)*	18	54	69	125	81	31	23	69	60	82	61	34
w/ applesce & bananas, jr—*1 jar (220 g)*	29	75	95	189	77	53	57	136	95	128	143	59
w/ applesce & bananas, str—*1 jar (135 g)*	18	47	59	119	49	34	36	85	59	80	90	38
w/ bananas, dry—*1 T (2.4 g)*	3	9	10	23	11	5	5	13	11	14	12	8
w/ bananas, prep w/ whole milk—*1 oz (28 g)*	17	52	67	121	83	30	16	63	58	80	52	36
oatmeal												
dry—*1 T (2.4 g)*	4	11	13	26	14	6	12	13	13	19	25	6
prep w/ whole milk—*1 oz (28 g)*	19	57	73	126	88	32	29	63	63	88	76	32
w/ applesce & bananas, jr—*1 jar (220 g)*	37	101	108	220	119	68	66	147	114	152	207	75
w/ applesce & bananas, str—*1 jar (135 g)*	23	62	66	134	72	41	39	89	70	92	126	46
w/ bananas, dry—*1 T (2.4 g)*	4	10	12	23	13	6	7	15	12	17	17	9
w/ bananas, prep w/ whole milk—*1 oz (28 g)*	18	55	70	122	88	32	20	67	61	84	61	38
rice												
dry—*1 T (2.4 g)*	2	8	7	13	7	4	4	9	8	11	16	5
prep w/ whole milk—*1 oz (28 g)*	15	50	60	102	76	28	15	54	53	73	58	30
w/ applesce & bananas, str—*1 jar (135 g)*	16	54	65	157	92	27	19	78	76	92	54	43
w/ bananas, dry—*1 T (2.4 g)*	4	11	11	20	11	5	4	10	9	13	14	7
w/ bananas, prep w/ whole milk—*1 oz (28 g)*	18	56	69	115	83	30	16	56	55	77	56	34
w/ mixed fruit, jr—*1 jar (220 g)*	24	81	110	213	139	42	35	119	108	139	143	64

19.3. Desserts

	TRY	THR	ISO	LEU	LYS	MET	CYS	PHE	TYR	VAL	ARG	HIS
banana pudding, str, Heinz—*1 jar (128 g)*						38						
choc custard pudding												
jr—*1 jar (220 g)*		196	229	411	304	103		191	154	268	185	106
str—*1 jar (135 g)*		111	129	233	173	59		109	88	152	105	60

	TRY	THR	ISO	LEU	LYS	MET	CYS	PHE	TYR	VAL	ARG	HIS
cottage cheese w/ pineapple												
jr—*1 jar (220 g)*		163	216	398	312	57		233	178	262	132	117
str—*1 jar (135 g)*		99	132	244	190	35		143	109	159	81	72
orange pudding, str—*1 jar (135 g)*		61	77	153	122	20		57	61	95	115	42
tutti frutti												
jr, Heinz—*1 jar (213 g)*								34				
str, Heinz—*1 jar (128 g)*								8				
van custard pudding												
jr—*1 jar (220 g)*		130	167	130	251	101		163	130	196	139	90
str—*1 jar (128 g)*		76	96	76	146	59		93	76	13	79	52

19.4. Dinners

	TRY	THR	ISO	LEU	LYS	MET	CYS	PHE	TYR	VAL	ARG	HIS
beef & egg noodles												
jr—*1 jar (213 g)*	58	211	294	443	394	113	62	249	198	281	334	158
str—*1 jar (128 g)*	32	114	159	238	212	60	33	134	106	151	180	84
beef & egg noodles w/ veg, toddler, Gerber—*1 jar (170 g)*	68	308	265	529	517	68		292	187	403	541	226
beef & egg rice, toddler—*1 jar (177 g)*	81	354	434	694	666	218	96	349	285	487	598	227
beef lasagna, toddler—*1 jar (177 g)*	83	289	365	573	526	145	85	322	234	409	448	181
beef stew, toddler—*1 jar (177 g)*	94	372	435	687	697	251	94	356	281	481	625	227
chicken & noodles												
jr—*1 jar (213 g)*	47	173	213	351	300	83	51	190	153	243	266	100
str—*1 jar (128 g)*	31	113	141	230	198	54	35	125	101	160	175	67
chicken soup, crm of, str—*1 jar (128 g)*	38	136	156	264	236	65	35	137	119	183	197	76
chicken stew, toddler—*1 jar (170 g)*	97	376	437	689	697	190	88	362	292	496	563	218
macaroni alphabets w/ tomato sce & cheese, toddler, Gerber—*1 jar (177 g)*	35	106	142	258	133	48		156	104	159	122	80
macaroni & cheese												
jr—*1 jar (213 g)*	68	166	277	498	315	192	68	281	262	313	217	132
str—*1 jar (128 g)*	41	100	166	300	189	115	41	169	157	188	131	79
macaroni, tomato & beef												
jr—*1 jar (213 g)*	60	192	251	422	317	83	72	239	175	271	283	136
str—*1 jar (128 g)*	32	104	137	229	173	46	40	129	95	147	154	74
noodles & chicken w/ carrots & peas, toddler, Gerber—*1 jar (170 g)*	75	313	323	575	525	138		303	216	388	551	235
potatoes & ham, toddler, Gerber—*1 jar (170 g)*	56	213	223	398	386	92		204	173	264	337	162
spaghetti, tomato & meat, toddler—*1 jar (177 g)*	115	349	474	731	554	184	117	427	326	506	503	250
spaghetti, tomato sce & beef, toddler, Gerber—*1 jar (177 g)*	58	237	276	490	365	106		193	193	304	370	173
split peas & ham, jr—*1 jar (213 g)*	72	262	285	515	479	115	60	326	245	328	624	192
turkey & rice												
jr—*1 jar (213 g)*	40	158	192	307	309	104	40	162	138	217	281	92
str—*1 jar (128 g)*	26	100	122	195	196	67	26	102	87	137	178	58
veg & bacon												
jr—*1 jar (213 g)*	36	130	175	273	232	79	55	164	121	207	258	85
str—*1 jar (128 g)*	19	69	84	133	109	41	29	79	70	109	142	44
veg & beef												
jr—*1 jar (213 g)*	49	181	213	345	358	81	49	175	130	256	302	113
str—*1 jar (128 g)*	26	91	106	173	179	41	24	87	65	128	151	56
toddler, Gerber—*1 jar (177 g)*	57	303	219	474	522	41		235	161	359	520	214
veg & chicken												
jr—*1 jar (213 g)*	43	143	187	283	251	75	58	162	124	215	245	79
str—*1 jar (128 g)*	27	88	116	174	155	46	35	100	76	133	151	49
toddler, Gerber—*1 jar (177 g)*	46	253	198	418	480	35		289	235	310	462	234
veg & ham												
jr—*1 jar (213 g)*	55	194	232	381	354	109	55	194	158	271	343	141
str—*1 jar (128 g)*	23	84	100	165	152	47	23	84	68	118	148	60
toddler—*1 jar (177 g)*	101	301	370	575	540	143	83	320	251	405	471	202
veg & lamb												
jr—*1 jar (213 g)*	51	164	202	330	328	66	38	177	141	217	300	102
str—*1 jar (128 g)*	29	93	115	188	187	38	22	101	79	124	170	59

	TRY	THR	ISO	LEU	LYS	MET	CYS	PHE	TYR	VAL	ARG	HIS
veg & liver												
jr—*1 jar (213 g)*	58	158	183	339	290	89	51	183	134	243	232	94
str—*1 jar (128 g)*	41	114	133	244	209	64	37	133	97	175	168	68
veg & turkey												
jr—*1 jar (213 g)*	38	145	175	281	266	81	38	136	124	204	228	77
str—*1 jar (128 g)*	23	84	102	165	156	47	23	79	73	120	134	45
toddler—*1 jar (177 g)*	106	329	441	689	586	156	92	354	308	492	522	202

19.5. Dinners, High Meat/Cheese

	TRY	THR	ISO	LEU	LYS	MET	CYS	PHE	TYR	VAL	ARG	HIS
beef w/ veg												
jr—*1 jar (128 g)*	64	307	338	599	594	225	87	303	230	398	495	242
str—*1 jar (128 g)*	59	279	307	544	539	204	78	275	209	361	449	219
chicken w/ veg												
jr—*1 jar (128 g)*	93	375	417	705	694	237	78	348	261	442	588	261
str—*1 jar (128 g)*	83	332	370	625	614	209	69	308	230	392	520	230
cottage cheese w/ pineapple, str—*1 jar (135 g)*	128	293	374	752	585	267	61	394	405	474	288	238
ham w/ veg												
jr—*1 jar (128 g)*	79	317	360	608	636	201	93	288	227	378	502	291
str—*1 jar (128 g)*	78	312	353	599	626	198	92	284	224	371	494	287
turkey w/ veg												
jr—*1 jar (128 g)*	82	320	371	604	599	197	78	311	243	385	499	261
str—*1 jar (128 g)*	77	303	352	571	567	187	74	294	230	364	472	247
veal w/ veg												
jr—*1 jar (128 g)*	79	312	351	608	620	178	93	301	227	385	529	250
str—*1 jar (128 g)*	78	307	343	596	608	174	91	294	221	378	520	224

19.6. Fruit Juices

	TRY	THR	ISO	LEU	LYS	MET	CYS	PHE	TYR	VAL	ARG	HIS
apple-apricot, str, Heinz—*1 jar (130 g)*								4				
apple-pineapple, str, Heinz—*1 jar (130 g)*								7				

19.7. Fruits

	TRY	THR	ISO	LEU	LYS	MET	CYS	PHE	TYR	VAL	ARG	HIS
apples & apricots												
jr, Heinz—*1 jar (213 g)*								26				
str, Heinz—*1 jar (128 g)*								15				
apples & cranberries w/ tapioca												
jr, Heinz—*1 jar (213 g)*								4				
str, Heinz—*1 jar (128 g)*								3				
apples & pears												
jr, Heinz—*1 jar (213 g)*								11				
str, Heinz—*1 jar (128 g)*								6				

19.8. Meat/Egg Yolks

	TRY	THR	ISO	LEU	LYS	MET	CYS	PHE	TYR	VAL	ARG	HIS
beef												
jr—*1 jar (99 g)*	145	629	652	1150	1194	441	167	555	477	726	978	487
str—*1 jar (99 g)*	136	591	613	1080	1122	414	157	522	448	681	919	457
beef w/ beef heart, str—*1 jar (99 g)*	125	506	613	1000	1040	326	129	518	361	683	824	336
chicken												
jr—*1 jar (99 g)*	165	653	686	1125	1216	390	191	593	466	733	1018	441
str—*1 jar (99 g)*	154	609	639	1047	1133	363	178	552	435	682	947	411
chicken sticks, jr—*1 jar (71 g)*	83	406	518	811	824	229	90	478	353	542	712	328
egg yolks, str—*1 jar (94 g)*	101	433	532	795	737	255	161	389	389	603	658	199
ham												
jr—*1 jar (99 g)*	148	649	711	1196	1270	382	184	570	501	771	1012	509
str—*1 jar (99 g)*	137	597	654	1100	1168	351	169	525	461	709	931	467
lamb												
jr—*1 jar (99 g)*	149	686	707	1188	1336	471	207	592	527	762	986	380
str—*1 jar (99 g)*	139	636	655	1101	1237	437	192	548	488	707	914	352
liver, beef, str—*1 jar (99 g)*	220	676	694	1307	926	372	213	678	569	898	831	363
meat sticks, jr—*1 jar (71 g)*	65	413	474	740	734	219	52	431	370	491	618	327
pork, str—*1 jar (99 g)*	135	611	673	1116	1146	394	152	564	505	695	939	443
turkey												
jr—*1 jar (99 g)*	158	674	764	1213	1261	472	185	628	536	774	969	389
str—*1 jar (99 g)*	148	627	710	1127	1172	439	172	584	498	720	900	362

	TRY	THR	ISO	LEU	LYS	MET	CYS	PHE	TYR	VAL	ARG	HIS
turkey sticks, jr—*1 jar (71 g)*	72	388	447	760	835	216	87	433	343	462	627	261
veal												
jr—*1 jar (99 g)*	174	632	682	1168	1215	334	196	585	484	736	1004	464
str—*1 jar (99 g)*	145	558	604	1034	1074	295	173	518	428	650	888	411
19.9. Vegetables												
beans, green												
jr—*1 jar (206 g)*	29	105	111	161	122	37	21	101	87	134	136	68
str—*1 jar (128 g)*	19	70	74	109	83	24	14	68	59	91	92	46
beans, green, buttered												
jr—*1 jar (206 g)*	31	113	119	175	132	39	23	109	95	144	146	74
str—*1 jar (128 g)*	18	65	69	100	76	23	13	63	54	83	84	42
beans, green, creamed—*1 jar (213 g)*	32	81	102	166	92	47	23	96	92	124	102	49
beets, str—*1 jar (128 g)*	15	41	49	59	44	13	9	23	45	58	38	27
carrots												
jr—*1 jar (213 g)*	23	49	51	70	45	19	13	53	43	66	111	28
str—*1 jar (128 g)*	14	28	31	41	26	12	8	31	24	38	64	15
carrots, buttered												
jr—*1 jar (213 g)*	23	49	51	70	45	19	13	53	43	66	111	28
str—*1 jar (128 g)*	15	31	32	44	27	12	8	33	26	41	68	17
corn, creamed												
jr—*1 jar (213 g)*	32	111	138	300	175	85	38	104	145	166	128	98
str—*1 jar (128 g)*	19	67	83	179	104	51	23	63	86	99	77	59
garden veg, str—*1 jar (128 g)*	35	93	110	183	148	55	26	113	122	127	251	58
mixed veg												
jr—*1 jar (213 g)*	32	96	119	194	100	45	60	117	111	147	198	66
str—*1 jar (128 g)*	17	49	60	99	51	23	31	60	56	74	101	33
peas, str—*1 jar (128 g)*	44	174	193	301	300	51	33	183	152	216	498	96
peas, buttered												
jr—*1 jar (206 g)*	70	284	315	490	488	84	56	299	249	352	814	157
str—*1 jar (128 g)*	46	183	204	316	315	54	36	192	160	228	525	101
spinach, creamed												
jr—*1 jar (213 g)*	94	258	288	564	379	141	79	247	294	386	388	162
str—*1 jar (128 g)*	46	129	143	283	189	70	40	123	147	193	195	82
squash												
jr—*1 jar (213 g)*	26	53	70	102	66	23	15	62	60	77	100	34
str—*1 jar (128 g)*	15	32	42	60	40	13	9	37	36	46	59	20
squash, buttered												
jr—*1 jar (213 g)*	21	45	60	87	55	19	13	53	51	66	85	28
str—*1 jar (128 g)*	12	24	32	46	29	10	8	28	28	36	45	15
sweet potatoes												
jr—*1 jar (220 g)*	46	117	110	167	95	51	33	134	88	156	117	55
str—*1 jar (135 g)*	30	73	69	105	59	32	20	84	55	99	73	35
sweet potatoes, buttered												
jr—*1 jar (220 g)*	33	84	79	121	68	37	24	97	64	114	84	40
str—*1 jar (135 g)*	24	61	57	86	49	27	18	70	46	82	61	28
20. Meat Analogues & Meat Analogue Entrees												
bacon, simulated meat product—*1 strip (8 g)*	13	36	45	73	58	12	14	49	32	47	70	24
meat extender, simulated meat product—*1 oz (28 g)*	161	452	559	914	727	146	176	611	400	592	874	299
sausage, simulated meat product—*1 link (25 g)*	70	196	243	397	316	63	76	265	174	257	380	130
sausage, simulated meat product—*1 patty (38 g)*	106	298	369	603	480	96	116	403	264	391	577	197
21. Meats												
21.1. Beef												
breakfast strips, ckd—*3 slices (34 g)*	97	402	460	782	816	247	136	383	347	468	657	339
brisket												
sep lean & fat, braised—*3.5 oz (100 g)*	258	1004	1034	1817	1913	589	258	898	773	1118	1453	787
sep lean, braised—*3.5 oz (100 g)*	329	1283	1321	2322	2445	752	329	1147	987	1429	1857	1006

	TRY	THR	ISO	LEU	LYS	MET	CYS	PHE	TYR	VAL	ARG	HIS
brisket, flat half												
sep lean & fat, braised—*3.5 oz (100 g)*	246	960	988	1736	1828	562	246	858	738	1069	1388	752
sep lean, braised—*3.5 oz (100 g)*	312	1225	1261	2217	2333	718	314	1095	942	1364	1772	960
brisket, point half												
sep lean & fat, braised—*3.5 oz (100 g)*	276	1077	1109	1949	2051	631	276	963	828	1199	1558	844
sep lean, braised—*3.5 oz (100 g)*	353	1377	1417	2492	2623	807	353	1231	1059	1533	1992	1079
chuck arm pot roast												
sep lean & fat, braised—*3.5 oz (100 g)*	303	1183	1218	2141	2254	693	303	1057	910	1318	1712	927
sep lean, braised—*3.5 oz (100 g)*	370	1442	1485	2610	2747	845	370	1289	1109	1606	2087	1131
chuck blade roast												
sep lean & fat, braised—*3.5 oz (100 g)*	285	1111	1144	2011	2117	651	285	993	855	1238	1608	871
sep lean, braised—*3.5 oz (100 g)*	348	1357	1396	2455	2584	795	348	1212	1043	1511	1963	1063
corned beef												
cured brisket, ckd—*3.5 oz (100 g)*	166	686	785	1334	1392	421	232	654	593	799	1122	578
cured, cnd—*3.5 oz (100 g)*	247	1023	1170	1990	2076	629	347	975	884	1192	1673	863
dried (chipped)—*1 oz (28 g)*	67	346	338	616	673	199	98	309	249	379	557	239
flank												
sep lean & fat, braised—*3.5 oz (100 g)*	308	1202	1237	2175	2289	704	308	1074	924	1338	1739	942
sep lean & fat, broiled—*3.5 oz (100 g)*	280	1094	1126	1979	2084	641	280	978	841	1218	1583	857
sep lean, braised—*3.5 oz (100 g)*	314	1224	1260	2215	2331	717	314	1094	941	1363	1771	959
sep lean, broiled—*3.5 oz (100 g)*	284	1109	1142	2008	2113	650	284	992	852	1235	1605	870
ground, extra lean												
baked, med—*3.5 oz (100 g)*	301	1026	1049	1961	2044	572	235	928	763	1186	1652	779
baked, well done—*3.5 oz (100 g)*	373	1270	1299	2429	2531	708	291	1149	945	1469	2046	965
broiled, med—*3.5 oz (100 g)*	313	1065	1089	2036	2121	593	244	963	792	1231	1715	809
broiled, well done—*3.5 oz (100 g)*	352	1198	1226	2291	2387	668	274	1084	892	1386	1930	910
pan fried, med—*3.5 oz (100 g)*	308	1046	1070	2001	2085	583	240	947	779	1210	1685	795
pan fried, well done—*3.5 oz (100 g)*	345	1173	1200	2243	2338	654	269	1061	873	1357	1890	891
ground, lean												
baked, med—*3.5 oz (100 g)*	295	1003	1026	1918	1999	559	230	907	747	1160	1616	762
baked, well done—*3.5 oz (100 g)*	365	1241	1269	2373	2472	691	284	1122	924	1435	1999	942
broiled, med—*3.5 oz (100 g)*	305	1036	1060	1981	2065	577	237	937	771	1198	1669	787
broiled, well done—*3.5 oz (100 g)*	347	1182	1209	2261	2355	659	271	1069	880	1367	1904	898
pan fried, med—*3.5 oz (100 g)*	299	1016	1039	1942	2024	566	233	919	756	1175	1636	772
pan fried, well done—*3.5 oz (100 g)*	340	1157	1183	2212	2305	645	265	1047	861	1338	1864	879
ground, regular												
baked, med—*3.5 oz (100 g)*	284	965	987	1845	1923	538	221	873	718	1116	1554	733
baked, well done—*3.5 oz (100 g)*	355	1207	1235	2309	2405	673	276	1092	899	1396	1945	917
broiled, med—*3.5 oz (100 g)*	297	1009	1032	1929	2010	562	231	913	751	1167	1625	766
broiled, well done—*3.5 oz (100 g)*	335	1140	1166	2180	2272	635	261	1031	849	1319	1837	866
pan fried, med—*3.5 oz (100 g)*	295	1003	1026	1917	1998	559	230	907	746	1160	1615	762
pan fried, well done—*3.5 oz (100 g)*	333	1132	1158	2164	2255	631	259	1024	842	1309	1823	860
rib eye, small end (rib 10-12)												
sep lean & fat, broiled—*3.5 oz (100 g)*	284	1108	1141	2006	2111	650	284	991	853	1234	1604	869
sep lean, broiled—*3.5 oz (100 g)*	314	1225	1261	2216	2333	718	314	1095	942	1364	1772	960
rib, large end (rib 6-9)												
sep lean & fat, broiled—*3.5 oz (100 g)*	225	879	905	1590	1674	515	225	785	676	979	1272	689
sep lean & fat, roasted—*3.5 oz (100 g)*	254	992	1021	1795	1890	581	254	887	763	1105	1435	778
sep lean, broiled—*3.5 oz (100 g)*	276	1076	1108	1947	2050	631	276	962	828	1198	1557	844
sep lean, roasted—*3.5 oz (100 g)*	308	1202	1238	2176	2290	705	308	1075	925	1339	1740	942
rib, shortribs												
sep lean & fat, braised—*3.5 oz (100 g)*	242	942	969	1704	1794	552	242	842	725	1049	1363	738
sep lean, braised—*3.5 oz (100 g)*	344	1343	1383	2431	2559	787	344	1201	1033	1496	1944	1053
rib, small end (ribs 10-12)												
sep lean & fat, broiled—*3.5 oz (100 g)*	269	1050	1081	1900	2000	615	269	938	808	1169	1519	823
sep lean & fat, roasted—*3.5 oz (100 g)*	249	972	1001	1760	1852	570	249	869	748	1083	1407	762
sep lean, broiled—*3.5 oz (100 g)*	314	1225	1261	2216	2333	718	314	1095	942	1364	1772	960
sep lean, roasted—*3.5 oz (100 g)*	299	1168	1202	2113	2225	684	299	1044	898	1301	1690	915
rib, whole (ribs 6-12)												
sep lean & fat, broiled—*3.5 oz (100 g)*	240	885	884	1625	1678	510	240	823	678	1005	1356	672

	TRY	THR	ISO	LEU	LYS	MET	CYS	PHE	TYR	VAL	ARG	HIS
sep lean & fat, roasted—*3.5 oz (100 g)*	245	957	985	1732	1824	561	245	856	736	1066	1385	750
sep lean, broiled—*3.5 oz (100 g)*	291	1136	1170	2057	2165	666	291	1016	874	1266	1644	891
sep lean, roasted—*3.5 oz (100 g)*	305	1188	1223	2150	2263	696	305	1062	914	1323	1719	931
round, bottom												
sep lean & fat, braised—*3.5 oz (100 g)*	334	1302	1340	2356	2480	763	334	1164	1001	1450	1884	1021
sep lean, braised—*3.5 oz (100 g)*	354	1380	1420	2497	2628	809	354	1233	1061	1536	1996	1082
round, eye of												
sep lean & fat, roasted—*3.5 oz (100 g)*	300	1169	1203	2115	2227	685	300	1045	899	1302	1691	916
sep lean, roasted—*3.5 oz (100 g)*	325	1266	1303	2291	2412	742	325	1132	974	1410	1832	993
round, full cut												
sep lean & fat, broiled—*3.5 oz (100 g)*	286	1115	1148	2018	2125	654	286	997	858	1242	1614	874
sep lean, broiled—*3.5 oz (100 g)*	319	1243	1279	2249	2367	728	319	1111	956	1384	1798	974
round, tip												
sep lean & fat, roasted—*3.5 oz (100 g)*	296	1156	1190	2092	2202	678	296	1033	889	1287	1673	906
sep lean, roasted—*3.5 oz (100 g)*	322	1254	1291	2269	2389	735	322	1121	965	1397	1815	983
round, top												
sep lean & fat, broiled—*3.5 oz (100 g)*	345	1346	1386	2436	2565	789	345	1203	1036	1499	1948	1055
sep lean & fat, pan fried—*3.5 oz (100 g)*	355	1384	1425	2505	2636	811	355	1237	1065	1541	2003	1085
sep lean, broiled—*3.5 oz (100 g)*	355	1384	1425	2505	2637	811	355	1237	1065	1542	2003	1085
sep lean, pan fried—*3.5 oz (100 g)*	393	1532	1576	2771	2917	898	393	1369	1178	1705	2216	1201
sausage, smoked, cnd												
—*1 oz (28 g)*	37	151	173	294	306	93	51	144	130	176	247	127
—*1 sausage (8 per 12 oz pkg) (43 g)*	55	229	262	445	465	141	78	218	198	267	375	193
shank, crosscuts												
sep lean & fat, simmered—*3.5 oz (100 g)*	354	1381	1421	2498	2630	809	354	1234	1062	1537	1998	1082
sep lean, simmered—*3.5 oz (100 g)*	377	1471	1514	2662	2802	862	377	1315	1132	1638	2129	1153
short loin porterhouse steak												
sep lean & fat, broiled—*3.5 oz (100 g)*	281	1096	1128	1983	2088	642	281	980	843	1221	1586	859
sep lean, broiled—*3.5 oz (100 g)*	315	1230	1266	2226	2343	721	315	1099	946	1370	1780	964
short loin T-bone steak												
sep lean & fat, broiled—*3.5 oz (100 g)*	269	1048	1078	1896	1996	614	269	936	806	1167	1516	821
sep lean, broiled—*3.5 oz (100 g)*	315	1229	1265	2223	2341	720	315	1098	945	1368	1778	963
short loin tenderloin												
sep lean & fat, broiled—*3.5 oz (100 g)*	291	1134	1167	2052	2160	665	291	1013	872	1263	1641	889
sep lean & fat, roasted—*3.5 oz (100 g)*	274	1069	1100	1934	2035	626	274	955	822	1190	1546	838
sep lean, broiled—*3.5 oz (100 g)*	316	1234	1270	2233	2350	723	316	1103	949	1374	1785	967
sep lean, roasted—*3.5 oz (100 g)*	308	1203	1238	2177	2291	705	308	1075	925	1339	1740	943
short loin, top loin												
sep lean & fat, broiled—*3.5 oz (100 g)*	288	1124	1157	2033	2140	659	288	1004	864	1251	1626	881
sep lean, broiled—*3.5 oz (100 g)*	321	1250	1287	2262	2381	733	321	1117	962	1392	1809	980
wedge-bone sirloin												
sep lean & fat, broiled—*3.5 oz (100 g)*	307	1196	1231	2165	2279	701	307	1069	920	1332	1731	938
sep lean & fat, pan fried—*3.5 oz (100 g)*	308	1199	1234	2170	2284	703	308	1072	923	1335	1735	940
sep lean, broiled—*3.5 oz (100 g)*	340	1327	1365	2400	2527	777	340	1186	1020	1477	1919	1040
sep lean, pan fried—*3.5 oz (100 g)*	364	1419	1460	2567	2702	832	364	1268	1091	1580	2053	1112

21.3. Pork

	TRY	THR	ISO	LEU	LYS	MET	CYS	PHE	TYR	VAL	ARG	HIS
arm picnic												
sep lean & fat, braised—*3.5 oz (100 g)*	346	1241	1264	2163	2620	647	341	1069	929	1428	1908	1304
sep lean & fat, roasted—*3.5 oz (100 g)*	286	1030	1047	1799	2178	536	283	889	769	1188	1598	1076
sep lean, braised—*3.5 oz (100 g)*	434	1518	1559	2622	3180	795	418	1291	1151	1729	2240	1636
sep lean, roasted—*3.5 oz (100 g)*	359	1255	1289	2169	2630	657	346	1067	952	1430	1853	1353
arm picnic, cured												
sep lean & fat, roasted—*3.5 oz (100 g)*	227	885	861	1602	1728	516	294	871	637	894	1403	683
sep lean, roasted—*3.5 oz (100 g)*	299	1109	1094	1980	2115	659	375	1078	818	1082	1620	894

	TRY	THR	ISO	LEU	LYS	MET	CYS	PHE	TYR	VAL	ARG	HIS
bacon, canadian style[a]												
unheated—*2 slices (57 g)*	116	470	442	824	921	318	146	380	354	466	638	425
grilled—*2 slices (47 g)*	112	452	425	793	887	306	140	366	341	449	615	409
bacon, cured												
broiled/pan fried—*3 med pieces (19 g)*	55	222	235	403	430	128	59	223	169	279	354	167
broiled/pan fried—*4.48 oz*[b] *(127 g)*	371	1485	1571	2691	2871	853	396	1491	1126	1862	2363	1114
raw—*3 med slices (68 g)*	56	226	239	409	437	130	61	227	171	284	360	169
blade roll, cured, sep lean & fat, roasted—*3.5 oz (100 g)*	207	768	757	1371	1465	456	260	746	567	749	1122	619
boston blade												
sep lean & fat, braised—*3.5 oz (100 g)*	342	1224	1247	2131	2582	638	337	1052	918	1407	1875	1289
sep lean & fat, broiled—*3.5 oz (100 g)*	282	1012	1031	1764	2137	527	278	872	757	1165	1558	1064
sep lean & fat, roasted—*3.5 oz (100 g)*	282	1011	1031	1759	2131	527	278	869	758	1161	1544	1067
sep lean, braised—*3.5 oz (100 g)*	419	1466	1505	2532	3071	768	404	1246	1112	1670	2163	1580
sep lean, broiled—*3.5 oz (100 g)*	338	1184	1216	2046	2481	620	326	1007	898	1349	1748	1277
sep lean, roasted—*3.5 oz (100 g)*	327	1146	1177	1980	2401	600	316	975	869	1306	1692	1236
breakfast strips, ckd—*3 slices (34 g)*	95	378	400	685	731	217	101	379	287	474	601	284
center loin												
sep lean & fat, braised—*3.5 oz (100 g)*	382	1366	1393	2375	2878	712	376	1173	1025	1569	2083	1442
sep lean & fat, broiled—*3.5 oz (100 g)*	357	1272	1299	2215	2683	663	350	1093	954	1462	1943	1344
sep lean & fat, pan fried—*3.5 oz (100 g)*	296	1071	1089	1873	2268	557	294	927	799	1238	1670	1116
sep lean & fat, roasted—*3.5 oz (100 g)*	332	1182	1208	2054	2490	617	325	1014	888	1356	1795	1253
sep lean, braised—*3.5 oz (100 g)*	467	1636	1680	2826	3427	857	451	1391	1241	1864	2414	1763
sep lean, broiled—*3.5 oz (100 g)*	431	1505	1547	2601	3154	788	415	1280	1141	1715	2222	1623
sep lean, pan fried—*3.5 oz (100 g)*	387	1353	1390	2338	2835	709	373	1151	1026	1542	1997	1459
sep lean, roasted—*3.5 oz (100 g)*	383	1340	1377	2315	2808	702	369	1140	1016	1527	1978	1445
center rib												
sep lean & fat, braised—*3.5 oz (100 g)*	370	1324	1350	2307	2794	691	364	1140	993	1523	2030	1395
sep lean & fat, broiled—*3.5 oz (100 g)*	317	1139	1161	1983	2403	593	312	980	853	1310	1749	1198
sep lean & fat, pan fried—*3.5 oz (100 g)*	271	989	1002	1735	2100	513	271	860	734	1147	1566	1020
sep lean & fat, roasted—*3.5 oz (100 g)*	321	1149	1172	1999	2422	599	316	987	863	1320	1753	1213
sep lean, braised—*3.5 oz (100 g)*	463	1620	1664	2799	3394	849	446	1378	1229	1846	2391	1747
sep lean, broiled—*3.5 oz (100 g)*	387	1356	1393	2342	2840	710	373	1153	1028	1545	2001	1462
sep lean, pan fried—*3.5 oz (100 g)*	376	1315	1351	2273	2756	689	362	1119	998	1499	1942	1418
sep lean, roasted—*3.5 oz (100 g)*	379	1327	1363	2293	2781	695	366	1129	1007	1512	1959	1431
ham, cured, lean												
cnd—*3.5 oz (100 g)*	210	826	796	1438	1589	482	219	713	607	829	1145	731
cnd, roasted—*3.5 oz (100 g)*	240	944	911	1645	1818	552	250	816	694	948	1310	836
roasted[c]—*3.5 oz (100 g)*	251	931	918	1661	1775	553	315	904	687	908	1360	750
unheated[c]—*3.5 oz (100 g)*	232	860	848	1535	1640	511	291	836	634	839	1257	693
ham, cured, regular												
center slice, sep lean & fat, unheated[c]—*3.5 oz (100 g)*	235	888	871	1593	1708	523	299	867	649	877	1337	705
cnd—*3.5 oz (100 g)*	193	758	731	1320	1458	443	201	655	557	760	1051	671
cnd, roasted—*3.5 oz (100 g)*	233	916	883	1596	1764	535	243	791	673	920	1271	811
roasted[c]—*3.5 oz (100 g)*	238	882	870	1574	1682	524	298	857	651	860	1289	711
unheated[c]—*3.5 oz (100 g)*	211	781	770	1394	1489	464	264	759	576	762	1141	629
ham patties[d]												
grilled—*1 patty (60 g)*	92	352	346	625	672	209	107	336	259	345	511	289
unheated—*1 patty (65 g)*	97	371	364	659	708	220	112	353	273	363	537	304
leg												
sep lean & fat, roasted—*3.5 oz (100 g)*	326	1163	1187	2023	2451	607	320	998	873	1335	1771	1230
sep lean, roasted—*3.5 oz (100 g)*	381	1332	1368	2302	2791	698	367	1133	1010	1518	1966	1436

[a] Bacon is fully cooked as purchased.
[b] Yield from 1 lb raw bacon.
[c] Ham is fully cooked as purchased.
[d] Fully cooked as purchased.

	TRY	THR	ISO	LEU	LYS	MET	CYS	PHE	TYR	VAL	ARG	HIS
loin												
sep lean & fat, braised—*3.5 oz (100 g)*	350	1257	1281	2191	2654	655	345	1082	941	1447	1934	1321
sep lean & fat, broiled—*3.5 oz (100 g)*	303	1090	1110	1901	2303	568	300	940	815	1256	1681	1144
sep lean & fat, roasted—*3.5 oz (100 g)*	304	1086	1108	1892	2292	566	299	934	814	1249	1665	1144
sep lean, braised—*3.5 oz (100 g)*	443	1551	1594	2681	3251	813	427	1319	1177	1768	2290	1673
sep lean, broiled—*3.5 oz (100 g)*	374	1309	1345	2263	2744	686	361	1114	993	1492	1933	1412
sep lean, roasted—*3.5 oz (100 g)*	362	1265	1300	2186	2651	663	349	1076	960	1442	1868	1364
loin blade												
sep lean & fat, braised—*3.5 oz (100 g)*	305	1103	1121	1928	2335	574	303	954	823	1274	1718	1150
sep lean & fat, broiled—*3.5 oz (100 g)*	262	950	965	1662	2012	494	261	823	708	1098	1486	989
sep lean & fat, pan fried—*3.5 oz (100 g)*	232	856	864	1505	1821	443	235	747	623	995	1374	873
sep lean & fat, roasted—*3.5 oz (100 g)*	270	971	988	1696	2053	505	267	838	725	1120	1505	1016
sep lean, braised—*3.5 oz (100 g)*	399	1397	1435	2414	2927	732	385	1188	1060	1592	2062	1506
sep lean, broiled—*3.5 oz (100 g)*	335	1172	1204	2025	2455	614	323	997	889	1335	1730	1264
sep lean, pan fried—*3.5 oz (100 g)*	327	1144	1175	1977	2397	599	315	973	868	1303	1689	1233
sep lean, roasted—*3.5 oz (100 g)*	332	1161	1192	2006	2432	608	320	987	880	1323	1713	1252
rump												
sep lean & fat, roasted—*3.5 oz (100 g)*	350	1242	1270	2155	2613	649	342	1063	936	1423	1873	1322
sep lean, roasted—*3.5 oz (100 g)*	392	1371	1408	2368	2872	718	378	1166	1040	1562	2023	1478
sausage, fresh												
ckd—*1 link (13 g)*	20	101	93	171	194	62	26	85	74	103	151	74
ckd—*1 patty (27 g)*	42	210	194	356	403	129	53	177	153	213	313	153
w/ beef, ckd—*1 link (13 g)*	17	72	69	127	141	43	18	62	53	77	111	54
shank												
sep lean & fat, roasted—*3.5 oz (100 g)*	314	1126	1148	1962	2376	587	310	969	844	1296	1727	1186
sep lean, roasted—*3.5 oz (100 g)*	379	1327	1363	2293	2780	695	366	1128	1006	1512	1959	1431
shoulder												
sep lean & fat, roasted—*3.5 oz (100 g)*	284	1019	1038	1776	2152	530	280	878	763	1173	1568	1071
sep lean, roasted—*3.5 oz (100 g)*	341	1193	1226	2062	2501	624	329	1015	905	1360	1762	1287
sirloin												
sep lean & fat, braised—*3.5 oz (100 g)*	362	1296	1322	2258	2735	675	356	1115	971	1491	1987	1365
sep lean & fat, broiled—*3.5 oz (100 g)*	311	1117	1138	1946	2357	582	307	961	836	1285	1717	1174
sep lean & fat, roasted—*3.5 oz (100 g)*	329	1166	1192	2026	2456	609	321	999	878	1338	1764	1239
sep lean, braised—*3.5 oz (100 g)*	450	1576	1619	2723	3302	825	434	1340	1195	1796	2326	1699
sep lean, broiled—*3.5 oz (100 g)*	380	1329	1365	2296	2784	696	366	1130	1008	1514	1962	1433
sep lean, roasted—*3.5 oz (100 g)*	370	1293	1328	2235	2710	677	356	1100	981	1474	1909	1394
spareribs, sep lean & fat, braised—*3.5 oz (100 g)*	391	1367	1404	2362	2864	716	377	1162	1037	1557	2018	1474
tenderloin, sep lean, roasted—*3.5 oz (100 g)*	387	1354	1391	2340	2837	709	373	1152	1027	1543	1999	1460
top loin												
sep lean & fat, braised—*3.5 oz (100 g)*	355	1278	1301	2229	2700	666	351	1102	956	1473	1974	1340
sep lean & fat, broiled—*3.5 oz (100 g)*	303	1095	1114	1912	2315	570	300	945	818	1263	1698	1145
sep lean & fat, pan fried—*3.5 oz (100 g)*	269	985	997	1728	2091	511	270	856	730	1142	1561	1014
sep lean & fat, roasted—*3.5 oz (100 g)*	312	1121	1142	1953	2366	584	308	965	840	1290	1721	1179
sep lean, braised—*3.5 oz (100 g)*	463	1620	1664	2799	3394	849	446	1378	1229	1846	2391	1747
sep lean, broiled—*3.5 oz (100 g)*	387	1356	1393	2342	2840	710	373	1153	1028	1545	2001	1462
sep lean, pan fried—*3.5 oz (100 g)*	376	1315	1351	2273	2756	689	362	1119	998	1499	1942	1418
sep lean, roasted—*3.5 oz (100 g)*	379	1327	1363	2293	2781	695	366	1129	1007	1512	1959	1431
21.5. Variety Cuts												
brains												
beef, pan fried—*3.5 oz (100 g)*	103	597	487	943	752	261	223	635	446	617	686	320
beef, simmered—*3.5 oz (100 g)*	90	526	429	831	662	230	197	560	393	544	604	282
pork, braised—*3.5 oz (100 g)*	155	567	561	1058	954	241		618	509	691	635	326

	TRY	THR	ISO	LEU	LYS	MET	CYS	PHE	TYR	VAL	ARG	HIS
chitterlings, pork, simmered—*3.5 oz (100 g)*	61	451	420	809	656	194		410	379	502	840	215
ears, pork, simmered—*1 ear (111 g)*	34	525	402	963	805	141		561	351	702	1404	210
feet, pork												
cured, pickled—*3.5 oz (100 g)*	27	365	230	595	582	149		392	216	338	1014	149
simmered—*3.5 oz (100 g)*	38	518	326	845	826	211		557	307	480	1440	211
heart												
beef, simmered—*3.5 oz (100 g)*	322	1359	1262	2547	2372	737	378	1303	1046	1502	1925	792
pork, braised—*1 heart (129 g)*	351	1335	1467	2748	2518	779	546	1344	1042	1613	2046	774
jowl, pork, raw—*3.5 oz (100 g)*	21	210	168	446	528	95	56	239	104	305	659	72
kidneys												
beef, simmered—*3.5 oz (100 g)*	347	1231	1040	2043	1696	530	200	1223	958	1590	1496	665
pork, braised—*3.5 oz (100 g)*	329	1053	1357	2280	1829	545	557	1199	914	1463	1561	610
liver												
beef, braised—*3.5 oz (100 g)*	351	1116	1116	2294	1693	616	374	1299	967	1506	1533	667
beef, pan fried—*3.5 oz (100 g)*	385	1222	1222	2513	1855	675	410	1423	1060	1650	1680	731
pork, braised—*3.5 oz (100 g)*	366	1107	1320	2319	2007	645	491	1274	887	1607	1603	708
lungs												
beef, braised—*3.5 oz (100 g)*	186	761	973	1498	1446	408	313	829	460	1005	1234	620
pork, braised—*3.5 oz (100 g)*	146	584	664	1288	1211	268		691		988	863	420
pancreas												
beef, braised—*3.5 oz (100 g)*	351	1257	1370	2116	1999	490		1127	1184	1453	1548	533
pork, braised—*3.5 oz (100 g)*	625	1281	1496	2130	1965	470		1222	1195	1537	1642	552
spleen												
beef, braised—*3.5 oz (100 g)*	261	988	968	2217	1815	462	727	1008	715	1510	1454	900
pork, braised—*3.5 oz (100 g)*	289	1128	1259	2306	2107	523		1205	790	1534	1539	672
stomach, pork, raw—*3.5 oz (100 g)*	98	510	560	990	874	288		525	412	692	924	246
tail, pork, simmered—*3.5 oz (100 g)*	102	595	391	952	1020	306		510	1173	510	1173	306
thymus, beef, braised—*3.5 oz (100 g)*	168	790	745	1458	1818	304		626		1947	1440	385
tongue												
beef, simmered—*3.5 oz (100 g)*	170	962	952	1652	1705	467	290	913	715	1058	1408	573
pork, braised—*3.5 oz (100 g)*	278	1018	1099	1932	1970	540		999		1253	1488	605
tripe, beef, raw—*3.5 oz (100 g)*	114	503	589	948	1044	315	168	471	396	613	995	363
21.6. Other Meats												
eel												
raw—*3 oz (85 g)*	176	688	723	1274	1440	464	168	612	530	808	938	462
ckd by dry heat—*3 oz (85 g)*	225	881	927	1634	1845	595	215	785	678	1035	1203	592
octopus, raw—*3 oz (85 g)*	142	546	552	892	947	286	167	454	405	553	925	243
shark												
raw—*3 oz (85 g)*	200	782	822	1449	1637	528	191	696	602	919	1067	525
batter-dipped & fried[a]—*3 oz (85 g)*	180	717	737	1288	1389	460	180	638	538	820	930	457
squid												
raw—*3 oz (85 g)*	148	570	576	932	989	298	173	474	423	578	966	254
fried[b]—*3 oz (85 g)*	172	649	663	1077	1114	343	206	558	490	668	1097	296
22. Meats, Luncheon												
beef												
chopped, smoked—*1 oz (28 g)*	47	240	234	428	467	138	68	214	173	263	386	166
jellied lunch meat—*1 oz (28 g)*	39	210	202	372	414	119	56	191	145	234	373	141
summer sausage—*1 slice (23 g)*	33	137	157	267	278	84	46	131	118	160	224	115
thin sliced lunch meat—*5 slices (21 g)*	48	247	242	441	482	143	70	221	179	271	399	171
berliner (pork & beef)—*1 slice (23 g)*	40	150	157	276	303	90	51	138	110	162	239	135
bockwurst (pork, veal, milk, etc.), raw—*1 link (65 g)*	88	358	378	610	681	208	96	320	278	396	528	259
bologna												
beef—*1 slice (23 g)*	26	106	122	207	216	65	36	101	92	124	174	90
beef & pork—*1 slice (23 g)*	24	118	117	207	203	64	31	106	83	143	161	73
pork—*1 slice (23 g)*	34	147	152	269	277	95	39	135	111	170	231	111
bratwurst, pork, ckd—*1 link (85 g)*	96	473	437	802	910	291	121	400	345	481	706	345
braunschweiger (pork liver sausage) —*1 slice (18 g)*	26	96	87	186	164	56	45	100	77	111	138	58

[a] Prepared w/ flour, oil, egg, milk, baking powder & salt.   [b] Dipped in flour & salt before frying.

	TRY	THR	ISO	LEU	LYS	MET	CYS	PHE	TYR	VAL	ARG	HIS
brotwurst (pork & beef w/ nfdm)—*1 link (70 g)*	92	419	424	756	797	258	114	379	310	473	662	305
chorizo (pork & beef)—*1 link (60 g)*	167	884	1324	1025	1448	282		689		548	1016	433
corned beef loaf, jellied—*1 oz slice (28 g)*	47	253	244	449	499	143	67	230	175	282	450	170
frankfurter												
beef—*1 frank (8 per 1 lb pkg) (57 g)*	63	259	296	504	526	159	88	247	224	302	424	218
beef—*1 frank (10 per 1 lb pkg) (45 g)*	50	204	234	398	415	126	69	195	177	238	334	172
beef & pork—*1 frank (45 g)*	37	183	218	369	407	103	58	162	141	212	382	158
ham, cured												
chopped, cnd—*1 slice (21 g)*	38	151	145	262	290	88	40	131	111	151	209	134
chopped, packaged—*1 slice (21 g)*	44	160	162	289	322	95	54	145	112	168	252	146
minced—*1 slice (21 g)*	33	154	147	264	286	96	40	135	113	157	215	127
sliced, lean (5% fat)—*1 slice (28 g)*	67	244	247	441	490	145	82	221	170	256	385	223
sliced, reg (11% fat)—*1 slice (28 g)*	61	221	225	400	445	132	75	201	155	233	350	202
ham & cheese loaf/roll—*1 slice (28 g)*	59	204	214	384	428	124	67	196	159	227	317	192
headcheese (pork)—*1 slice (28 g)*	24	126	154	286	274	75	63	172	132	187	325	84
honey loaf (pork & beef)—*1 slice (28 g)*	52	223	206	389	420	135	44	190	164	220	305	173
italian sausage, pork, ckd—*1 link (67 g)*	108	531	490	900	1020	326	135	449	387	539	793	387
kielbasa/kolbassy (pork & beef w/ nfdm)—*1 slice (26 g)*	36	112	166	227	263	72	59	130	127	166	245	82
knackwurst/knockwurst (pork & beef)—*1 slice (68 g)*	73	326	317	558	634	195	100	277	245	350	482	245
lebanon bologna (beef)—*1 slice (23 g)*	36	186	182	332	363	107	53	167	135	204	300	129
livercheese (pork liver)—*1 slice (38 g)*	78	247	240	506	448	130	125	272	177	306	318	149
liver pate, unspecified, cnd—*1 oz (28 g)*	45	161	157	298	238	81	48	165	129	218	254	84
liver sausage/liverwurst (pork)—*1 slice (18 g)*	28	122	119	207	210	52	27	112	66	156	147	81
luxury loaf (pork)—*1 slice (28 g)*	61	249	231	430	468	133	43	204	183	253	332	186
meat spread, ham salad—*1 oz (28 g)*	25	116	114	206	219	65	14	100	79	127	168	99
mortadella (beef & pork)—*1 slice (15 g)*	23	95	106	182	189	59	31	90	80	110	154	78
new england brand sausage (pork & beef)—*1 slice (23 g)*	44	174	175	313	348	103	57	157	123	185	276	150
old fashioned loaf (pork & beef)—*1 slice (28 g)*	42	175	158	305	305	91	30	145	126	175	227	117
olive loaf (pork)—*1 slice (28 g)*	29	134	119	246	229	85	41	119	109	144	165	83
pastrami, beef—*1 oz (28 g)*	45	185	211	359	375	113	63	176	160	215	302	156
peppered loaf (pork & beef)—*1 slice (28 g)*	56	216	222	393	428	127	69	197	156	234	332	184
pepperoni (pork & beef)—*1 slice (6 g)*	11	47	50	87	90	29	14	43	37	54	74	37
pickle & pimento loaf (pork)—*1 slice (28 g)*	33	148	139	269	254	72	33	124	111	158	189	97
picnic loaf (pork & beef)—*1 slice (28 g)*	41	186	165	325	338	107	46	152	130	182	252	124
polish sausage (pork)—*1 oz (28 g)*	39	168	173	305	315	107	45	153	126	192	262	126
pork & beef lunch meat—*1 slice (28 g)*	38	154	182	297	336	82	65	145	142	202	262	115
pork & beef luncheon sausage—*1 slice (23 g)*	37	153	181	294	333	81	65	143	141	200	260	114
pork lunch meat, cnd—*1 slice (21 g)*	26	104	121	203	198	71	45	104	81	139	184	76
salami												
beerwurst, beef—*1 slice (23 g)*	26	108	123	210	219	66	37	103	93	126	176	91
beerwurst, pork—*1 slice (23 g)*	26	130	112	217	237	82	25	104	93	117	186	96
ckd, beef—*1 slice (23 g)*	32	131	150	254	265	80	44	124	113	152	214	110
ckd, beef & pork—*1 slice (23 g)*	26	120	155	214	255	69	45	111	127	154	197	83
dry/hard pork—*1 slice (10 g)*	25	101	108	163	188	47	29	94	69	112	137	61
dry/hard pork & beef—*1 slice (10 g)*	21	96	97	173	182	59	26	87	71	108	152	70
smoked link sausage												
pork—*1 link (68 g)*	147	632	647	1151	1187	405	169	577	475	726	989	475
pork & beef—*1 link (68 g)*	73	316	330	551	614	248	71	275	262	316	520	269
pork & beef w/ american cheese—*1 link (43 g)*	64	233	266	472	498	154	65	243	222	298	360	203
pork & beef w/ flour & nfdm—*1 link (68 g)*	95	389	422	732	734	244	110	367	312	460	591	295
pork & beef w/ nfdm—*1 link (68 g)*	92	375	411	710	709	234	103	354	307	119	553	282

	TRY	THR	ISO	LEU	LYS	MET	CYS	PHE	TYR	VAL	ARG	HIS
thuringer (cervelat/summer sausage) beef—*1 slice (23 g)*	35	158	177	241	318	81	45	133	125	185	228	108
turkey breast meat—*1 slice (21 g)*	54	210	246	377	445	137	49	188	187	251	330	147
turkey ham (cured thigh meat)—*2 slices (57 g)*	122	477	558	855	1012	311	112	426	424	570	749	335
turkey pastrami—*2 slices (57 g)*	115	452	521	806	942	290	117	405	393	540	731	311
vienna sausage, cnd (beef & pork)— *1 sausage (16 g)*	17	57	89	128	127	42	28	68	55	92	113	44

23. Milk, Yogurt, Milk Beverages & Milk Beverage Mixes
23.1. Cow Milk

	TRY	THR	ISO	LEU	LYS	MET	CYS	PHE	TYR	VAL	ARG	HIS
buttermilk, cultured—*8 fl oz (245 g)*	88	386	500	807	679	198	76	427	339	596	309	233
buttermilk, dry—*1 T (7 g)*	31	101	135	218	177	56	21	108	108	149	81	60
condensed, sweetened, cnd—*1 fl oz (38 g)*	43	136	186	296	240	76	28	146	146	202	109	82
evaporated, lowfat, cnd, Carnation— *4 fl oz (127 g)*	117	323	447	687	644	201	74	402	421	481		185
evaporated, skim, cnd—*1 fl oz (32 g)*	34	109	146	236	191	60	22	116	116	161	87	65
evaporated, whole, cnd—*1 fl oz (32 g)*	30	97	130	210	170	54	20	104	104	144	78	58
evaporated, whole, cnd—*4 fl oz (126 g)*	121	387	519	841	681	215	79	414	414	574	311	233
lowfat 1% fat—*8 fl oz (244 g)*	113	362	486	786	637	201	74	388	388	537	291	218
lowfat 1% fat, pro fortified—*8 fl oz (246 g)*	136	436	585	947	767	242	89	467	467	647	350	262
lowfat 1% fat w/ nfdm—*8 fl oz (245 g)*	120	385	516	835	676	214	79	412	412	571	309	231
lowfat 2% fat—*8 fl oz (244 g)*	115	367	492	796	644	204	75	392	392	544	294	220
lowfat 2% fat, pro fortified—*8 fl oz (246 g)*	137	439	588	952	771	244	90	469	469	650	352	263
lowfat 2% fat w/ nfdm—*8 fl oz (245 g)*	120	385	516	835	676	214	79	412	412	571	309	231
skim—*8 fl oz (245 g)*	118	377	505	818	663	210	77	403	403	559	302	227
skim, dry—*¼ cup (30 g)*	153	490	656	1063	860	272	100	524	524	726	393	294
skim, dry, Ca reduced—*1 oz (28 g)*	142	454	609	986	798	252	93	486	486	674	364	273
skim, dry, inst—*1⅓ cups (3.2 oz envelope)[a] (91 g)*	451	1442	1932	3129	2533	801	295	1542	1542	2138	1156	866
skim pro fortified—*8 fl oz (246 g)*	137	440	589	954	773	244	93	470	470	652	353	264
skim w/ nfdm—*8 fl oz (245 g)*	123	395	529	857	694	219	81	422	422	585	317	237
whole, 3.3% fat—*8 fl oz (244 g)*	113	362	486	786	637	201	74	388	388	537	291	218
whole, 3.5% fat—*8 fl oz (244 g)*	119	391	544	842	663	204		408		586		176
whole, 3.7% fat—*8 fl oz (244 g)*	113	361	484	784	635	201	74	386	386	536	290	217
whole, dry—*¼ cup (32 g)*	119	380	510	825	668	211	78	407	407	564	305	228
whole, low Na—*8 fl oz (244 g)*	107	341	458	741	600	190	70	365	365	506	274	205

23.2. Other Milk

	TRY	THR	ISO	LEU	LYS	MET	CYS	PHE	TYR	VAL	ARG	HIS
filled[b,c]—*8 fl oz (244 g)*	115	367	492	796	644	204	75	392	392	544	294	220
filled[c,d]—*8 fl oz (244 g)*	115	367	492	796	644	204	75	392	392	544	294	220
goat—*8 fl oz (244 g)*	106	398	505	765	708	196	113	377	437	585	291	218
human—*1 fl oz (31 g)*	5	14	17	29	21	6	6	14	16	19	13	7
imitation[b,e]—*8 fl oz (244 g)*	59	181	262	422	344	130	18	229	244	305	169	126
imitation[d,e]—*8 fl oz (244 g)*	59	181	262	422	344	130	18	229	244	305	169	126
indian buffalo—*8 fl oz (244 g)*	131	445	496	892	683	237	116	394	447	534	278	189
sheep—*8 fl oz (245 g)*	207	657	829	1438	1256	379	85	696	689	1098	485	409
soy—*1 cup (240 g)*	103	271	346	578	430	96	113	362	269	338	514	170

23.3. Yogurt (From Cow Milk)
lowfat w/ nfdm

	TRY	THR	ISO	LEU	LYS	MET	CYS	PHE	TYR	VAL	ARG	HIS
—*8 fl oz (227 g)*	67	489	650	1201	1068	351		650	601	986	359	295
coffee/van flavor—*8 fl oz (227 g)*	63	460	610	1128	1003	330		610	565	926	337	277
fruit flavor—*8 fl oz (227 g)*	51	371	493	911	810	266		493	456	748	272	224

[a] Reconstitutes to 1 qt fluid skim milk.
[b] Contains blend of hydrogenated soybean, cottonseed &/or safflower oils.
[c] Filled milk contains fats or oils other than milk fat; contains milk solids (milk, cream, or skim milk).
[d] Contains lauric acid oils which include modified coconut oil, hydrogenated coconut oil &/or palm kernel oil.
[e] Imitation milk contains fats or oils other than milk fat; contains food solids, excluding milk solids.

	TRY	THR	ISO	LEU	LYS	MET	CYS	PHE	TYR	VAL	ARG	HIS
skim w/ nfdm—*8 fl oz (227 g)*	73	534	709	1311	1166	383		709	656	1076	391	322
whole—*8 fl oz (227 g)*	44	323	430	794	706	232		430	398	652	237	195

23.4. Milk Beverages

	TRY	THR	ISO	LEU	LYS	MET	CYS	PHE	TYR	VAL	ARG	HIS
carob flavor mix in whole milk—*3 t powder in 8 fl oz milk (256 g)*	113	364	486	786	637	202	74	389	389	538	289	218
choc flavor mix in whole milk—*2-3 hp t powder in 8 fl oz milk (266 g)*	122	388	511	825	668	207	80	418	412	575	325	229
choc malted milk flavor mix in whole milk—*3 hp t powder in 8 fl oz milk (265 g)*	125	395	519	848	670	215	93	427	419	578	334	239
w/ added nutrients in whole milk—*4-5 hp t powder in 8 fl oz milk (265 g)*	111	363	485	784	636	201	74	387	387	538	292	217
choc milk												
1% fat milk—*8 fl oz (250 g)*	114	366	490	793	642	203	75	391	391	542	293	220
2% fat milk—*8 fl oz (250 g)*	113	362	486	786	636	201	74	387	387	537	291	218
whole milk—*8 fl oz (250 g)*	112	358	479	776	629	199	73	383	383	530	287	215
whole milk w/ malt—*8 fl oz (265 g)*	38	124	162	265	209	68	28	136	131	187		75
choc syrup in whole milk—*2 T syrup in 8 fl oz milk (282 g)*	121	389	510	823	668	209	82	417	412	572	324	228
choc syrup w/ added nutrients in whole milk—*1 T syrup in 8 fl oz milk (263 g)*	116	376	497	805	652	205	76	402	400	555	308	224
cocoa/hot chocolate												
prep w/ whole milk—*8 fl oz (250 g)*	128	411	551	891	722	228	84	439	439	609	329	247
prep w/ milk, sugar & high-fat plain cocoa—*8 fl oz (250 g)*	128	411	551	891	722	228	84	439	439	609	329	247
prep w/ water from mix—*¾ hp t powder in 6 fl oz water (206 g)*	43	130	163	262	212	62	29	142	134	192	119	72
prep w/ water from mix w/ added nutrients—*1 pkt in 6 fl oz water (209 g)*	27	79	100	161	132	38	19	88	84	117	73	46
eggnog, nonalcoholic—*8 fl oz (254 g)*	137	444	583	937	758	222	97	463	462	643	378	240
mix in whole milk—*2 hp t in 8 fl oz milk (272 g)*	114	370	492	794	645	204	76	392	392	544	299	220
mix in whole milk, Delmark—*1.23 oz mix in 8 fl oz milk (257 g)*	210	700	970	1480	1160	390	150	750	760	1050	580	390
instant breakfast in whole milk—*1 pkt in 8 fl oz milk*[a] *(265 g)*	196	703	927	1490	1210	387	142	746	757	1030		424
choc, Delmark—*1.23 oz mix in 8 fl oz milk (279 g)*	260	850	1150	1790	1400	440	190	920	930	1260	710	480
van/strawberry, Delmark—*1.23 oz mix in 8 fl oz milk (279 g)*	380	1250	1730	2670	2110	660	240	1310	1380	1870	990	710
malted milk—*8 fl oz (265 g)*	143	433	567	944	712	239	129	479	464	630		272
malted milk flavor mix in whole milk—*3 hp t in 8 fl oz milk (265 g)*	138	427	559	925	702	236	122	469	456	620	382	265
milk shake, thick												
choc—*1 avg (300 g)*	129	413	554	896	726	229	85	442	442	612	331	248
van—*1 avg (313 g)*	170	545	731	1184	958	303	112	583	583	809	437	328
strawberry flavored mix in whole milk—*2-3 hp t powder in 8 fl oz milk (266 g)*	112	364	487	787	638	202	74	388	388	537	290	218

23.5. Milk Beverage Mixes

	TRY	THR	ISO	LEU	LYS	MET	CYS	PHE	TYR	VAL	ARG	HIS
choc flavor mix, powder—*2-3 hp t (22 g)*	10	24	23	37	31	5	8	30	23	36	34	10
choc malted milk flavor mix, powder—*¾ oz (3 hp t) (21 g)*	12	30	34	61	34	13	20	39	32	41	43	21
choc syrup—*2 T (1 fl oz) (38 g)*	10	24	24	37	31	6	8	30	24	36	34	11
choc syrup w/ added nutrients—*1 T (19 g)*	5	11	11	18	15	3	4	14	11	17	16	5
cocoa mix, powder—*1 oz pkt (3-4 hp t) (28 g)*	43	129	162	262	213	62	29	142	135	191	119	73

[a] Averages for various flavors ſof instant breakfast.

	TRY	THR	ISO	LEU	LYS	MET	CYS	PHE	TYR	VAL	ARG	HIS
Carnation												
choc & marshmallow—*1 oz (28 g)*	32	97	110	175	140	43	26	81	77	117		42
milk choc—*1 oz (28 g)*	31	100	138	212	168	53	19	105	108	149		57
rich choc—*1 oz (28 g)*	32	98	110	174	140	43	26	81	77	117		43
rich choc, 70 cal—*1 pkt (3 hp t) (21 g)*	40	130	170	260	210	70	20	130	140	180		70
sugar-free—*1 pkt (4 hp t) (15 g)*	59	191	252	384	307	97	39	236	196	264		104
w/ added nutrients—*1.1 oz pkt (31 g)*	26	79	100	161	131	38	18	87	83	117	73	45
eggnog mix, powder, nonalcoholic—*2 hp t (28 g)*	2	6	7	10	8	3	2	5	5	7	8	3
instant breakfast												
choc, Carnation—*1.25 oz pkt (35 g)*	80	331	418	673	545	173	70	347	351	471		194
choc malt, Carnation—*1.24 oz pkt (35 g)*	86	289	381	612	469	160	71	337	331	424		182
coffee, Carnation—*1.26 oz pkt (36 g)*	83	367	468	752	621	196	70	369	385	522		218
eggnog, Carnation—*1.20 oz pkt (34 g)*	85	330	447	705	573	176	55	356	578	498		206
strawberry, Carnation—*1.25 oz pkt (35 g)*	83	367	468	752	621	196	70	369	385	522		218
van, Carnation—*1.23 oz pkt (35 g)*	83	367	467	751	621	195	70	369	385	521		218
instant breakfast, no sugar added, Carnation												
choc—*.69 oz pkt (20 g)*	102	320	423	659	503	205	102	410	302	491		167
choc malted—*.71 oz pkt (20 g)*	96	288	394	611	447	203	102	399	312	462		171
strawberry—*.68 oz pkt (19 g)*	111	346	470	730	571	224	99	434	320	534		180
van—*.67 oz pkt (19 g)*	111	346	470	730	571	224	99	434	320	534		180
malted milk flavor mix, powder—*¾ oz (3 hp t) (21 g)*	26	63	73	140	67	34	49	82	68	83	90	49
24. Nuts, Nut Products & Seeds												
acorn flour, full-fat—*1 oz (28 g)*	26	82	99	169	133	36	38	93	64	120	164	59
acorns												
raw—*1 oz (28 g)*	21	67	81	139	109	29	31	76	53	98	134	48
dried—*1 oz (28 g)*	28	89	107	183	143	39	41	101	70	129	177	64
almond butter—*1 T (16 g)*	43	89	105	188	81	28	43	135	85	124	302	68
almond butter w/ honey & cinn—*1 T (16 g)*	45	94	110	197	85	29	45	141	89	131	317	71
almond meal, partially defatted—*1 oz (28 g)*	201	416	487	873	375	128	201	626	396	578	1403	314
almond paste—*1 oz (28 g)*	60	125	146	262	112	38	60	188	119	174	421	94
almonds												
blanched, Blue Diamond—*1 oz (28 g)*	48	184	223	416	192	45	133	320	164	266	708	144
blanched, slivered, Blue Diamond—*1 oz (28 g)*	51	204	238	447	229	34	133	374	192	283	822	164
chopped, Blue Diamond—*1 oz (28 g)*	48	187	227	419	192	39	138	326	167	266	722	144
dried—*1 oz (24 nuts) (28 g)*	102	210	246	441	189	64	102	316	200	282	709	158
dry roasted—*1 oz (28 g)*	83	172	201	361	155	53	83	259	164	239	580	130
oil roasted—*1 oz (22 nuts) (28 g)*	104	215	252	450	193	66	104	323	204	298	724	162
sliced, Blue Diamond—*1 oz (28 g)*	48	187	227	419	192	39	138	326	167	266	722	144
toasted—*1 oz (28 g)*	104	215	251	450	193	66	104	323	204	298	724	162
whole, Blue Diamond—*1 oz (28 g)*	48	184	227	416	192	36	124	326	167	269	720	144
brazilnuts, dried—*1 oz (8 med nuts) (28 g)*	74	131	171	337	154	288	99	212	130	259	679	114
butternuts, dried—*1 oz (28 g)*	104	267	335	625	219	174	137	410	277	438	1381	229
cashew butter—*1 oz (28 g)*	77	193	238	419	266	89	92	258	160	339	567	130
cashews												
dry roasted—*1 oz (28 g)*	67	168	208	365	232	78	80	225	139	295	494	113
oil roasted—*1 oz (18 med nuts) (28 g)*	71	178	219	385	245	82	85	237	147	312	522	119
chestnuts, chinese												
raw—*1 oz (28 g)*	14	47	45	74	65	29	31	54	36	62	122	34
boiled & steamed—*1 oz (28 g)*	10	33	31	51	44	20	21	37	24	43	84	24
dried—*1 oz (28 g)*	23	77	72	120	105	47	51	88	58	102	199	56
roasted—*1 oz (28 g)*	15	51	47	78	69	31	33	58	38	67	130	37

	TRY	THR	ISO	LEU	LYS	MET	CYS	PHE	TYR	VAL	ARG	HIS
chestnuts, european												
raw—*1 oz (2½ nuts) (28 g)*	8	24	27	41	41	16	22	29	19	38	49	19
boiled & steamed—*1 oz (28 g)*	6	20	22	34	34	13	18	24	16	32	41	16
dried—*1 oz (28 g)*	20	65	72	107	107	43	57	77	50	101	130	50
roasted—*1 oz (3½ nuts) (28 g)*	10	32	36	53	53	21	29	38	25	51	64	25
chestnuts, japanese												
raw—*1 oz (28 g)*	9	26	32	39	42	15	18	25	18	38	42	16
boiled & steamed—*1 oz (28 g)*	3	9	11	14	15	6	7	9	7	14	16	6
dried—*1 oz (28 g)*	21	60	73	92	97	36	43	58	43	89	98	37
roasted—*1 oz (28 g)*	12	34	41	52	55	20	24	33	24	50	55	21
coconut												
raw—*1 piece (2"x2"x½") (45 g)*	18	54	59	111	66	28	30	76	46	91	246	35
dried—*1 oz (28 g)*	23	71	77	145	86	37	39	99	60	118	321	45
dried, sweetened, flaked, cnd—*4 oz (114 g)*	44	139	150	284	169	72	75	194	119	231	628	88
dried, sweetened, flaked, packaged—*1 cup (74 g)*	28	88	95	180	107	45	48	123	75	147	398	56
dried, sweetened, shredded—*1 cup (93 g)*	32	98	105	199	118	50	53	136	83	163	440	61
dried, toasted—*1 oz (28 g)*	18	55	59	112	66	28	30	76	47	91	247	35
coconut cream, raw[a]—*1 cup (240 g)*	101	317	341	646	384	163	173	442	269	528	1428	199
coconut cream, sweetened, cnd[a]—*1 cup (296 g)*	92	290	314	592	352	148	157	406	246	482	1308	184
coconut milk, raw[b]—*1 cup (240 g)*	65	199	216	408	242	103	108	278	170	334	902	127
coconut milk, cnd[b]—*1 cup (226 g)*	54	167	179	339	201	86	90	231	140	276	748	104
coconut water[c]—*1 cup (240 g)*	19	62	67	127	77	31	34	89	53	106	283	41
filberts (hazelnuts)												
dried—*1 oz (28 g)*	61	127	161	312	113	46	65	195	129	188	612	93
dry roasted—*1 oz (28 g)*	47	98	124	240	87	36	50	150	99	144	470	71
oil roasted—*1 oz (28 g)*	67	139	176	341	124	50	71	213	141	205	669	102
roasted, salted, Blue Diamond—*1 oz (28 g)*	36	164	184	345	121	90	85	176	130	263	710	124
whole, Blue Diamond—*1 oz (28 g)*	167	147	181	320	136	51	65	218	138	223	691	104
ginko nuts												
raw—*1 oz (28 g)*	20	76	59	90	59	16	7	49	17	80	119	29
cnd—*1 oz (14 med nuts) (28 g)*	11	40	32	48	31	8	3	26	9	43	63	15
dried—*1 oz (28 g)*	48	182	142	214	140	38	16	116	41	192	285	69
hickorynuts, dried—*1 oz (28 g)*	39	120	164	292	141	85	77	202	129	207	592	110
macadamia nuts												
dried—*1 oz (28 g)*		75	69	131	92	26	27	74	96	91	255	48
dry roasted, Blue Diamond—*1 oz (28 g)*	17	85	82	161	93	82	62	90	138	96	371	59
oil roasted—*1 oz (10-12 nuts) (28 g)*		65	61	115	81	23	24	64	84	80	223	42
mixed nuts												
dry roasted[d]—*1 oz (28 g)*	75	170	211	389	202	65	82	271	192	265	637	136
oil roasted[e]—*1 oz (28 g)*	70	162	206	382	187	96	83	261	186	266	575	134
w/o peanuts, oil roasted[f]—*1 oz (28 g)*	72	161	200	359	194	98	86	233	145	270	565	118
nuts, wheat-base formulated												
flavored[g]—*1 oz (28 g)*	49	157	157	276	252	78	64	166	138	202	279	108
macadamia flavored[h]—*1 oz (28 g)*	39	135	138	242	218	68	53	145	122	176	232	92
unflavored[i]—*1 oz (28 g)*	47	164	158	281	258	79	72	171	137	207	302	112
peanut butter, creamy/smooth—*1 T (16 g)*	55	132	177	342	176	47	58	260	219	206	613	133
peanut butter, chunk style/crunchy—*2 T (32 g)*	75	263	270	499	276	94	99	399	313	323	920	195

[a] Liquid expressed from grated coconut.
[b] Liquid expressed from grated coconut & water.
[c] Liquid from coconuts.
[d] Cashews, almonds, peanuts, filberts & pecans.
[e] Cashews, peanuts, brazilnuts, filberts, almonds & pecans.
[f] Cashews, almonds, brazilnuts, pecans & filberts.

[g] Hydrogenated soybean oil, wheat germ, sugar, sodium caseinate, soy protein, natural & artificial flavors, & artificial color.
[h] Hydrogenated soybean oil, wheat germ, sugar, wheat starch, sodium caseinate, soy protein, natural & artificial flavor.
[i] Hydrogenated soybean oil, wheat germ, fructose, wheat starch, sodium caseinate, soy protein, & salt.

	TRY	THR	ISO	LEU	LYS	MET	CYS	PHE	TYR	VAL	ARG	HIS
peanut flour												
defatted—*1 T (4 g)*	25	60	81	157	81	21	27	119	100	94	281	61
low-fat—*1 oz (28 g)*	92	324	333	613	340	116	121	491	385	397	1132	239
peanuts												
boiled—*½ cup (32 g)*	42	148	152	280	155	53	55	224	176	181	517	109
dried—*1 oz (28 g)*	88	211	283	548	282	75	93	417	350	330	982	212
dry roasted—*1 oz (28 g)*	64	227	233	430	238	81	85	344	270	278	793	168
oil roasted—*1 oz (28 g)*	92	220	295	571	294	78	97	435	365	344	1024	222
peanuts, spanish, oil-roasted—*1 oz (28 g)*	76	269	276	508	281	96	101	407	319	329	938	198
peanuts, valencia, oil-roasted—*1 oz (28 g)*	74	259	266	491	272	93	97	393	308	318	906	192
peanuts, virginia, oil-roasted—*1 oz (28 g)*	70	248	255	470	260	89	93	375	295	304	866	183
pecans												
dried—*1 oz (31 large nuts) (28 g)*	57	72	91	148	83	53	59	116	81	110	314	64
dry roasted—*1 oz (28 g)*	58	74	94	152	85	54	61	120	83	113	323	66
oil roasted—*1 oz (15 halves) (28 g)*	51	64	82	133	74	47	53	104	72	98	282	58
pilinuts, dried—*1 oz (15 nuts) (28 g)*		116	137	253	105	112		141	108	199	431	72
pine nuts, pignolia, dried—*1 oz (28 g)*	86	216	265	491	256	122	124	261	249	352	1326	163
pine nuts, pinyon, dried—*1 oz (28 g)*	41	104	128	237	123	59	60	126	120	170	639	79
pistachio nuts												
dried—*1 oz (47 nuts) (28 g)*	80	205	277	476	363	108	146	336	203	400	621	152
dry roasted—*1 oz (28 g)*	59	149	201	346	263	78	106	244	147	291	450	110
dry roasted, Blue Diamond—*1 oz (28 g)*	70	249	292	493	345	144	124	306	170	385	637	161
pumpkin & squash seeds												
dried—*1 oz (142 seeds) (28 g)*	122	256	359	590	521	156	85	347	289	560	1145	193
roasted—*1 oz (28 g)*	164	344	482	793	699	210	115	466	389	753	1539	260
sesame butter (tahini)—*1 T (15 g)*	56	106	110	195	82	84	51	135	107	143	378	75
sesame seeds												
kernels, dried[a]—*1 T (8 g)*	38	94	103	172	66	72	42	122	90	118	266	54
kernels, toasted[a]—*1 oz (28 g)*	105	200	207	369	154	159	97	255	202	269	714	142
whole, dried—*1 T (9 g)*	35	66	69	122	51	53	32	85	67	89	237	47
whole, toasted & roasted—*1 oz (28 g)*	105	200	207	369	154	159	97	255	202	269	714	142
soybean nuts												
dry roasted—*½ cup (86 g)*	495	1478	1651	2772	2265	459	549	1777	1287	1699	2641	918
roasted—*½ cup (86 g)*	440	1316	1470	2466	2016	409	488	1581	1146	1512	2350	817
roasted & toasted—*1 oz (95 nuts) (28 g)*	160	456	535	856	734	142	178	548	417	525	855	283
sunflower seeds												
dried—*1 oz (28 g)*	99	264	323	471	266	140	128	332	189	376	682	179
dry roasted—*1 oz (28 g)*	84	224	275	400	226	119	109	282	160	317	579	152
oil roasted—*1 oz (28 g)*	93	247	303	442	250	131	120	311	178	350	640	168
walnuts, black, dried—*1 oz (28 g)*	91	207	278	484	205	134	133	314	213	365	1040	193
walnuts, english/persian, dried—*1 oz (14 halves) (28 g)*	54	127	161	282	110	80	98	178	125	205	597	102
watermelon seeds, dried—*1 oz (95 large seeds) (28 g)*	111	316	381	610	252	237	124	577	289	442	1391	220

25. Poultry[b]
25.1. Chicken
25.1.1. Broiler/Fryer Meat

	TRY	THR	ISO	LEU	LYS	MET	CYS	PHE	TYR	VAL	ARG	HIS
dark meat w/ skin, fried—*3.5 oz (100 g)*	308	1123	1370	1994	2202	725	365	1071	884	1326	1681	804
dark meat w/ skin, roasted—*3.5 oz (100 g)*	289	1071	1288	1883	2105	688	348	1007	832	1258	1634	759
dark meat w/ skin, stewed—*3.5 oz (100 g)*	261	969	1167	1705	1906	623	314	911	754	1139	1477	687
dark meat w/o skin, fried—*3.5 oz (100 g)*	340	1221	1528	2176	2441	799	375	1156	978	1438	1742	897
dark meat w/o skin, roasted—*3.5 oz (100 g)*	320	1156	1445	2053	2325	757	350	1086	924	1357	1651	849

[a] Decorticated seeds.   [b] Gram weights are edible portions.

	TRY	THR	ISO	LEU	LYS	MET	CYS	PHE	TYR	VAL	ARG	HIS
dark meat w/o skin, stewed—*3.5 oz (100 g)*	303	1097	1371	1949	2206	719	322	1030	877	1288	1566	806
light & dark meat w/ skin, fried, batter dipped—*3.5 oz (100 g)*	257	922	1125	1653	1765	591	311	899	732	1100	1378	655
light & dark meat w/ skin, fried, flour coated—*3.5 oz (100 g)*	323	1181	1439	2092	2320	762	382	1121	928	1392	1766	845
light & dark meat w/ skin, roasted—*3.5 oz (100 g)*	305	1128	1362	1986	2223	726	364	1061	879	1325	1711	802
light & dark meat w/ skin, stewed—*3.5 oz (100 g)*	276	1020	1233	1797	2011	657	329	959	796	1199	1545	726
light & dark meat w/o skin, fried—*3.5 oz (100 g)*	358	1289	1612	2294	2583	844	393	1217	1031	1516	1839	947
light & dark meat w/o skin, roasted—*3.5 oz (100 g)*	338	1222	1528	2171	2458	801	370	1148	977	1435	1745	898
light & dark meat w/o skin, stewed—*3.5 oz (100 g)*	319	1153	1441	2048	2318	755	349	1083	921	1353	1646	847
light meat w/ skin, fried—*3.5 oz (100 g)*	344	1262	1537	2231	2487	815	405	1192	991	1485	1888	904
light meat w/ skin, roasted—*3.5 oz (100 g)*	326	1202	1458	2119	2374	776	385	1130	940	1412	1811	858
light meat w/ skin, stewed—*3.5 oz (100 g)*	294	1084	1316	1910	2142	699	347	1019	848	1273	1629	774
light meat w/o skin, fried—*3.5 oz (100 g)*	383	1386	1732	2463	2784	908	420	1303	1108	1628	1978	1018
light meat w/o skin, roasted—*3.5 oz (100 g)*	361	1305	1632	2319	2626	855	396	1226	1043	1533	1864	959
light meat w/o skin, stewed—*3.5 oz (100 g)*	337	1220	1525	2167	2454	799	370	1146	975	1433	1742	896
25.1.2. Broiler/Fryer Parts												
back w/ skin, fried—*½ back (72 g)*	266	822	1001	1463	1599	529	272	790	647	973	1235	587
breast w/ skin, fried—*½ breast (98 g)*	357	1300	1599	2306	2581	845	410	1228	1028	1531	1915	940
breast w/ skin, roasted—*½ breast (98 g)*	333	1219	1495	2154	2424	791	382	1146	960	1432	1800	879
breast w/ skin, stewed—*½ breast (110 g)*	343	1257	1541	2221	2499	815	395	1181	990	1476	1858	906
breast w/o skin, fried—*½ breast (86 g)*	335	1214	1518	2158	2439	796	368	1142	970	1427	1733	892
breast w/o skin, roasted—*½ breast (86 g)*	311	1127	1409	2002	2266	739	341	1059	900	1324	1609	828
breast w/o skin, stewed—*½ breast (95 g)*	322	1163	1454	2066	2339	762	352	1093	929	1395	1661	855
drumstick w/ skin, fried—*1 drumstick (49 g)*	150	549	672	972	1085	355	174	518	432	646	815	395
drumstick w/ skin, roasted—*1 drumstick (52 g)*	158	583	708	1028	1152	376	186	548	456	684	876	417
drumstick w/ skin, stewed—*1 drumstick (57 g)*	163	599	730	1057	1187	388	191	563	470	704	897	429
leg w/ skin, fried—*1 leg (112 g)*	340	1245	1520	2205	2452	804	400	1180	980	1467	1858	893
leg w/ skin, roasted—*1 leg (114 g)*	332	1226	1487	2160	2421	791	393	1153	959	1440	1847	876
leg w/ skin, stewed—*1 leg (125 g)*	339	1253	1520	2208	2474	808	401	1178	980	1471	1884	894
leg w/o skin, stewed—*1 leg (101 g)*	310	1120	1401	1991	2253	734	339	1052	896	1316	1600	823
neck w/ skin, fried—*1 neck (36 g)*	92	364	403	608	658	218	122	331	264	410	559	237
neck w/ skin, simmered—*1 neck (38 g)*	76	295	331	510	557	183	107	278	218	347	500	196
neck w/o skin, simmered—*1 neck (18 g)*	52	187	233	332	375	122	57	175	149	219	267	137
thigh w/ skin, fried—*1 thigh (62 g)*	187	685	835	1214	1345	442	222	652	539	808	1027	490
thigh w/ skin, roasted—*1 thigh (62 g)*	174	643	779	1133	1269	415	206	605	502	755	971	458
thigh w/ skin, stewed—*1 thigh (68 g)*	177	654	792	1153	1291	422	211	615	511	768	990	466
thigh w/o skin, roasted—*1 thigh (52 g)*	158	570	712	1012	1146	373	173	535	456	669	814	419
wing w/ skin, fried—*1 wing (32 g)*	90	338	396	592	650	214	116	321	258	398	538	233
wing w/ skin, roasted—*1 wing (34 g)*	98	370	432	645	714	234	126	348	281	435	592	255
wing w/ skin, stewed—*1 wing (40 g)*	98	370	434	646	716	235	125	348	282	435	588	256
25.1.3. Capon												
meat w/ skin, roasted—*3.5 oz (100 g)*	325	1200	1455	2115	2370	774	384	1128	938	1409	1806	857
25.1.4. Roaster												
dark meat w/o skin, roasted—*3.5 oz (100 g)*	272	982	1228	1745	1975	644	298	923	785	1153	1402	722

	TRY	THR	ISO	LEU	LYS	MET	CYS	PHE	TYR	VAL	ARG	HIS
light & dark meat w/ skin, roasted—3.5 oz (100 g)	266	989	1191	1739	1945	636	320	929	769	1162	1506	701
light & dark meat w/o skin, roasted—3.5 oz (100 g)	292	1056	1321	1877	2125	693	320	993	844	1240	1509	776
light meat w/o skin, roasted—3.5 oz (100 g)	317	1146	1433	2036	2305	751	347	1077	916	1346	1637	842

25.1.5. Stewer

	TRY	THR	ISO	LEU	LYS	MET	CYS	PHE	TYR	VAL	ARG	HIS
dark meat w/o skin, stewed—3.5 oz (100 g)	329	1189	1486	2112	2391	779	360	1117	950	1396	1698	874
light & dark meat w/ skin, stewed—3.5 oz (100 g)	301	1113	1348	1961	2197	717	357	1046	869	1307	1680	793
light & dark meat w/o skin, stewed—3.5 oz (100 g)	356	1285	1606	2283	2585	842	389	1207	1027	1509	1835	944
light meat w/o skin, stewed—3.5 oz (100 g)	386	1396	1744	2479	2807	914	423	1311	1115	1639	1993	1025

25.2. Turkey

	TRY	THR	ISO	LEU	LYS	MET	CYS	PHE	TYR	VAL	ARG	HIS
dark meat w/ skin, roasted—3.5 oz (100 g)	304	1200	1379	2138	2503	774	302	1076	1044	1432	1936	828
dark meat w/o skin, roasted—3.5 oz (100 g)	325	1271	1485	2276	2693	828	297	1134	1129	1518	1993	893
ham, cured thigh meat —3.5 oz (100 g)	215	842	984	1508	1784	548	197	751	748	1006	1321	591
light & dark meat, w/ skin, roasted—3.5 oz (100 g)	311	1227	1409	2184	2557	790	308	1100	1066	1464	1979	845
light & dark meat, w/o skin, roasted—3.5 oz (100 g)	333	1304	1525	2336	2763	849	305	1164	1159	1557	2045	915
light meat, w/ skin, roasted—3.5 oz (100 g)	315	1247	1432	2220	2599	803	314	1117	1084	1487	2013	859
light meat, w/o skin, roasted—3.5 oz (100 g)	340	1330	1555	2383	2818	866	311	1187	1182	1588	2086	933
loaf, breast meat—3.5 oz (100 g)	256	1001	1170	1793	2120	652	234	893	889	1195	1570	702

25.3. Other Poultry

	TRY	THR	ISO	LEU	LYS	MET	CYS	PHE	TYR	VAL	ARG	HIS
duck w/ skin, roasted—3.5 oz (100 g)	232	773	872	1465	1486	475	299	752	640	938	1284	462
duck w/o skin, roasted—3.5 oz (100 g)	327	1003	1206	1983	2009	635	361	984	894	1228	1499	620
goose w/ skin, roasted—3.5 oz (100 g)		1123	1183	2109	1988	608		1055	805	1232	1566	700
pheasant w/o skin, raw—3.5 oz (100 g)	328	1180	1324	1995	2157	686	309	920	773	1305	1433	939
quail w/o skin, raw—3.5 oz (100 g)	341	1090	1187	1866	1905	689	380	944	1010	1180	1379	825

25.4. Internal Organs
25.4.1. Giblets

	TRY	THR	ISO	LEU	LYS	MET	CYS	PHE	TYR	VAL	ARG	HIS
chicken, fried—3.5 oz (100 g)	373	1467	1630	2601	2348	810	437	1480	1067	1737	2159	759
chicken, simmered—3.5 oz (100 g)	295	1172	1297	2066	1886	646	343	1170	848	1379	1727	602
turkey, simmered—3.5 oz (100 g)	307	1204	1339	2140	1951	662	354	1207	877	1430	1768	625

25.4.2. Gizzard

	TRY	THR	ISO	LEU	LYS	MET	CYS	PHE	TYR	VAL	ARG	HIS
chicken, simmered—3.5 oz (100 g)	243	1251	1281	1907	1877	712	356	1129	825	1216	1950	547
turkey, simmered—3.5 oz (100 g)	264	1356	1389	2067	2034	772	386	1224	895	1318	2114	593

25.4.3. Heart

	TRY	THR	ISO	LEU	LYS	MET	CYS	PHE	TYR	VAL	ARG	HIS
chicken, simmered—3.5 oz (100 g)	338	1196	1415	2303	2214	638	359	1183	946	1496	1694	693
turkey, simmered—3.5 oz (100 g)	343	1212	1435	2333	2243	646	364	1199	959	1516	1717	702

25.4.4. Liver

	TRY	THR	ISO	LEU	LYS	MET	CYS	PHE	TYR	VAL	ARG	HIS
chicken, simmered—3.5 oz (100 g)	343	1083	1294	2198	1843	577	327	1212	857	1535	1493	647
duck, raw—3.5 oz (100 g)	264	833	995	1691	1418	444	252	932	660	1181	1148	498
goose, raw—3.3 oz (94 g)	216	684	818	1388	1165	365	207	766	541	970	943	409
turkey, simmered—3.5 oz (100 g)	338	1066	1273	2163	1814	568	322	1193	844	1511	1469	637

27. Sauces

	TRY	THR	ISO	LEU	LYS	MET	CYS	PHE	TYR	VAL	ARG	HIS
marinara (tomato based) sce—1 cup (250 g)	28	93	78	113	115	20	23	85	55	83	80	63
soy sce												
made w/ soy & wheat (shoyu)—¼ cup (58 g)	43	121	142	240	171	44	53	158	109	148	207	78
made w/ soy (tamari)—¼ cup (58 g)	105	236	282	426	424	97	62	310	198	304	235	125

	TRY	THR	ISO	LEU	LYS	MET	CYS	PHE	TYR	VAL	ARG	HIS
spaghetti sce, cnd—*1 cup (249 g)*	32	105	87	127	129	22	27	95	60	92	90	72
tomato sce, cnd—*½ cup (122 g)*	11	37	32	45	46	9	10	34	22	33	33	26
spanish style—*½ cup (122 g)*	12	40	34	49	50	9	11	37	23	35	35	28
w/ herbs & cheese—*½ cup (122 g)*	26	72	85	138	135	34	20	85	73	100	109	61
w/ mushrooms—*½ cup (122 g)*	21	52	45	67	90	17	9	48	28	50	51	34
w/ onions—*½ cup (122 g)*	20	45	51	60	72	12	22	44	35	41	139	31
w/ onions, green peppers & celery—*½ cup (122 g)*	10	27	26	34	37	6	9	26	17	24	45	18
w/ tomato tidbits—*½ cup (122 g)*	11	37	32	45	46	9	10	34	22	33	33	26

28. Soups
28.1. Canned, Condensed

	TRY	THR	ISO	LEU	LYS	MET	CYS	PHE	TYR	VAL	ARG	HIS
celery, crm of—*1 can (305 g)*	46	143	189	302	180	73	46	189	131	217	143	95
cheese—*1 can (257 g)*	144	380	645	1025	761	234	90	558	499	761	324	293
chicken, crm of—*1 can (305 g)*	104	317	415	641	522	195	122	372	287	421	406	223
mushroom, crm of—*1 can (305 g)*	70	189	235	384	265	95	61	226	183	262	204	113
tomato—*1 can (305 g)*	49	125	143	241	122	55	67	171	104	162	146	88

28.2. Canned, Condensed, Prep w/ Milk

	TRY	THR	ISO	LEU	LYS	MET	CYS	PHE	TYR	VAL	ARG	HIS
asparagus, crm of—*1 cup (248 g)*	84	260	340	558	432	144	67	290	270	384	231	159
celery, crm of—*1 cup (248 g)*	74	241	322	518	394	131	55	273	248	360	206	149
cheese—*1 cup (251 g)*	128	371	567	906	700	218	83	474	444	650	309	256
chicken, crm of—*1 cup (248 g)*	99	312	414	657	533	181	87	347	312	444	312	201
clam chowder, new england—*1 cup (248 g)*	117	350	451	719	605	203	102	374	337	489	407	265
mushroom, crm of—*1 cup (248 g)*	84	260	340	553	429	141	62	288	270	377	231	156
pea, green—*1 cup (254 g)*	130	485	541	1016	831	206	117	572	447	711	853	279
potato, crm of[a]—*1 cup (248 g)*	82	243	320	513	402	131	64	278	255	362	221	149
tomato—*1 cup (248 g)*	77	233	303	494	370	124	64	265	238	335	206	146
tomato bisque—*1 cup (251 g)*	80	248	321	520	409	131	60	271	254	356	211	153

28.3. Canned, Condensed, Prep w/ Water

	TRY	THR	ISO	LEU	LYS	MET	CYS	PHE	TYR	VAL	ARG	HIS
asparagus, crm of—*1 cup (244 g)*	29	78	98	163	112	41	29	95	76	115	85	49
bean w/ bacon—*1 cup (253 g)*	83	326	385	650	536	99	89	440	235	435	415	205
bean w/ franks—*1 cup (250 g)*	413	488	823	680	125	110	558	298	550	525	260	498
beef noodle—*1 cup (244 g)*	46	154	188	315	261	90	59	195	124	207	198	112
black bean—*1 cup (247 g)*	64	249	287	422	415	62	59	311	173	284	331	163
celery, crm of—*1 cup (244 g)*	20	59	78	124	73	29	20	78	54	90	59	39
cheese—*1 cup (247 g)*	72	190	321	511	380	116	44	279	249	380	163	146
chicken & dumplings—*1 cup (241 g)*	53	193	243	407	378	108	72	224	142	277	292	137
chicken, crm of—*1 cup (244 g)*	41	129	171	264	215	81	51	154	117	173	166	93
chicken gumbo—*1 cup (244 g)*	22	83	100	168	161	46	17	98	68	117	122	59
chicken noodle[b]—*1 cup (241 g)*	39	128	159	265	219	77	46	164	106	176	166	94
chicken rice—*1 cup (241 g)*	41	142	178	270	251	92	51	152	123	188	234	101
chicken veg—*1 cup (241 g)*	31	113	135	231	222	60	24	133	94	159	169	80
chili beef—*1 cup (250 g)*	70	275	328	553	455	85	73	373	200	368	350	175
clam chowder, new england—*1 cup (244 g)*	54	149	183	288	251	90	56	159	127	195	229	137
minestrone—*1 cup (241 g)*	31	104	130	236	183	43	34	154	84	178	198	72
mushroom, crm of—*1 cup (244 g)*	34	90	112	181	127	44	27	107	88	122	95	54
pea, green—*1 cup (250 g)*	73	303	298	623	510	105	80	378	250	443	708	170
pea, split w/ ham—*1 cup (253 g)*	101	364	435	711	696	139	134	455	319	491	703	215
pepperpot—*1 cup (241 g)*	41	195	234	402	311	92	60	234	157	299	494	92
potato, crm of[a]—*1 cup (244 g)*	24	61	76	117	83	29	27	83	61	93	76	39
scotch broth—*1 cup (241 g)*	43	154	186	316	304	84	34	181	128	217	231	108
stockpot—*1 cup (247 g)*	42	151	183	311	299	82	35	178	126	215	227	109
tomato—*1 cup (244 g)*	20	51	59	100	51	22	27	71	41	66	61	37

[a] Includes vichyssoise.

[b] Includes chicken noodle-o's; chicken w/ stars; curly noodle w/ chicken & chicken alphabet

	TRY	THR	ISO	LEU	LYS	MET	CYS	PHE	TYR	VAL	ARG	HIS
tomato beef w/ noodle—*1 cup (244 g)*	41	144	171	290	242	83	54	181	115	193	183	102
tomato bisque—*1 cup (247 g)*	22	67	77	126	89	30	25	77	59	86	67	44
turkey noodle—*1 cup (244 g)*	37	124	151	254	212	73	46	156	102	168	159	90
turkey veg—*1 cup (241 g)*	27	96	116	198	190	53	22	113	82	135	145	67
veg vegetarian—*1 cup (241 g)*	14	75	99	147	99	24	24	99	48	99	99	48
veg w/ beef[a]—*1 cup (244 g)*	49	173	210	359	344	95	39	205	146	246	261	122
veg w/ beef broth—*1 cup (241 g)*	22	72	92	164	125	31	24	106	58	125	137	51
28.4. Canned, Ready-to-Serve												
beef, chunky—*1 cup (240 g)*	465	592	898	929	248	123	481	353	636	609	275	725
chicken, chunky—*1 cup (251 g)*	123	437	552	924	853	243	163	505	326	628	660	311
chicken noodle, chunky—*1 cup (240 g)*		437	552	924	854	245	163	504	326	629	662	312
pea, split w/ ham, chunky—*1 cup (240 g)*	110	391	468	766	749	149	144	490	343	526	758	233
turkey, chunky—*1 cup (236 g)*	99	404	514	781	809	215	106	418	307	552	531	238
veg, chunky—*1 cup (240 g)*	26	108	161	271	190	26	26	161	82	190	190	82
29. Special Dietary Foods												
breakfast bar, Carnation												
choc chip—*1.44 oz bar (41 g)*	76	201	260	405	331	89	50	255	240	282		129
choc crunch—*1.34 oz bar (38 g)*	61	170	224	348	281	74	46	222	205	242		112
peanut butter crunch—*1.35 oz bar (38 g)*	72	202	265	408	332	93	49	257	248	294		134
peanut butter w/ choc chips—*1.39 oz bar (39 g)*	78	207	270	419	339	93	53	267	251	298		136
gluten-free, wheat-free foods, Ener-G Foods												
bread												
rice, brown—*2 slices (73 g)*		112	110	229	140	0		158	50	176	267	68
rice, white—*2 slices (51 g)*		72	63	135	101	0		79	27	105	189	36
nutquik milk powder—*1 oz (28 g)*	144	244	230	414	161	69	96	311	180	280	725	142
pizza shell, rice, low protein—*2 oz (60 g)*		10	10	17	13	0		10	3	13	9	5
soup, tomato, dry mix—*1⅔ T(15 g)*		14	12	20	15	4		15	7	16	16	9
soyquix milk powder—*2 T (14 g)*		1	1	1	1			1	1	1	1	
slender bars, Carnation												
choc—*2 bars (55 g)*	175	533	649	997	807	345	138	591	552	742		297
choc chip—*2 bars (55 g)*	181	527	631	980	812	333	122	569	550	717		289
choc peanut butter—*2 bars (55 g)*	153	467	583	910	726	289	111	543	524	679		288
van—*2 bars (55 g)*	186	559	684	1059	864	346	139	631	595	776		320
slender, liquid diet meal, cnd, Carnation												
banana—*10 fl oz (313 g)*	165	569	761	1167	961	308	88	590	651	847		333
choc—*10 fl oz (313 g)*	113	466	498	1008	804	282	69	538	541	663		291
choc fudge—*10 fl oz (313 g)*	116	485	545	1052	834	294	66	563	570	692		304
choc malt—*10 fl oz (313 g)*	116	485	545	1052	834	294	66	563	570	692		304
milk choc—*10 fl oz (313 g)*	116	485	541	1048	836	294	78	560	570	692		304
peach—*10 fl oz (313 g)*	165	569	761	1167	961	308	88	590	651	847		333
strawberry—*10 fl oz (313 g)*	165	569	761	1167	961	308	88	590	651	847		333
van—*10 fl oz (313 g)*	119	432	516	1014	814	250	88	516	504	635		263
31. Spices, Herbs, & Flavorings												
basil, ground—*1 t (1 g)*	3	8	8	15	9	3	2	10	6	10	9	4
dill seed—*1 t (2 g)*		12	16	19	22	3		14		24	27	7
fennel seed—*1 t (2 g)*	5	12	14	20	15	6	4	13	8	18	14	7
fenugreek seed—*1 t (4 g)*	14	33	46	65	62	13	14	40	28	41	91	25
garlic powder—*1 t (3 g)*	6	13	18	29	16	9	5	14	6	20	47	9
ginger, ground—*1 t (2 g)*	1	3	5	7	5	1	1	4	2	7	4	3
mustard seed, yellow—*1 t (3 g)*	17	36	36	59	50	16	19	35	25	44	58	25
onion powder—*1 t (2 g)*	3	4	6	7	10	2	4	5	5	5	28	3
poppy seed—*1 t (3 g)*	7	25	25	42	31	13	13	25	19	36	56	15

[a] Includes beef; beef veg; barley & veg beef.

	TRY	THR	ISO	LEU	LYS	MET	CYS	PHE	TYR	VAL	ARG	HIS
32. Spreads (butter, margarine, mayonnaise, etc.)												
butter												
—*1 t (5 g)*	1	2	3	4	3	1	0	2	2	3	2	1
—*1 T (15 g)*	3	6	9	12	9	3	0	6	6	9	6	3
whipped—*1 t (4 g)*	0	1	2	3	3	1	0	2	2	2	1	1
whipped—*1 T (11 g)*	0	3	6	9	9	3	0	6	6	6	3	3
margarine by form & type of oil												
liquid, soybean & cottonseed—*1 t (5 g)*	1	4	5	9	7	2	1	4	4	6	3	2
stick, corn—*1 t (5 g)*	1	2	2	4	3	1	0	2	2	3	1	1
stick, safflower & soybean—*1 t (5 g)*	1	2	2	4	3	1	0	2	2	3	1	1
stick, soybean—*1 t (5 g)*	1	2	2	4	3	1	0	2	2	3	1	1
stick, soybean & cottonseed—*1 t (5 g)*	1	2	2	4	3	1	0	2	2	3	1	1
tub, corn—*1 t (5 g)*	1	2	2	3	3	1	0	2	2	2	1	1
tub, safflower—*1 t (5 g)*	1	2	2	3	3	1	0	2	2	2	1	1
tub, soybean—*1 t (5 g)*	1	2	2	3	3	1	0	2	2	2	1	1
tub, soybean & cottonseed—*1 t (5 g)*	1	2	2	3	3	1	0	2	2	2	1	1
margarine, imitation (diet) by form & type of oil												
tub, corn—*1 t (5 g)*	0	1	1	2	2	1	0	1	1	2	1	1
tub, soybean & cottonseed—*1 t (5 g)*	0	1	1	2	2	1	0	1	1	2	1	1
mayonnaise												
safflower & soybean—*1 T (14 g)*	2	8	9	13	10	5	3	8	6	10	10	4
soybean—*1 T (14 g)*	2	8	9	13	10	5	3	8	6	10	10	4
vegetable oil spread by form & type of oil												
stick, soybean & palm—*1 t (5 g)*	0	1	2	3	2	1	0	1	1	2	1	1
stick, soybean & cottonseed—*1 t (5 g)*	0	1	2	3	2	1	0	1	1	2	1	1
33. Sugars, Syrups & Other Sweeteners												
sugar substitute												
equal powder[a]—*1 g pkt (1 g)*								19				
equal tablet[b]—*1 tablet (g)*				3				19				
34. Vegetables, Vegetable Products & Vegetable Salads												
adzuki beans												
boiled—*1 cup (230 g)*	166	587	690	1454	1304	182	161	915	515	890	1118	455
cnd, sweetened—*1 cup (296 g)*	107	382	447	944	847	118	104	595	334	580	728	296
w/ sugar (yokan)—*three ¼" thick slices (43 g)*	14	48	56	119	107	15	13	75	42	73	92	37
alfalfa seeds, sprouted, raw—*1 cup (33 g)*		44	47	88	71					48		
amaranth, boiled—*½ cup (66 g)*	18	56	67	110	72	20	17	75	45	78	69	29
aspargus												
boiled—*½ cup (6 spears) (90 g)*	23	65	86	101	111	23	28	55	37	90	109	36
cnd—*½ cup (121 g)*	25	73	96	113	122	25	30	62	41	99	121	40
frzn, boiled—*4 spears (60 g)*	17	49	65	77	84	17	21	42	28	68	83	28
avocado, raw												
calif—*1 med (173 g)*	38	121	130	227	173	67	38	125	90	178	109	52
florida—*1 med (304 g)*	52	161	173	301	228	88	52	164	119	237	143	70
bamboo shoots												
raw—*½ cup (76 g)*	21	65	67	106	102	23	17	68		81	74	32
boiled—*1 cup (120 g)*	19	60	61	98	95	20	16	64		74	68	30
cnd—*1 cup (131 g)*	24	75	76	122	117	26	18	79		93	84	37
beans, cnd												
baked, vegetarian—*1 cup (254 g)*	145	513	538	973	836	183	132	658	343	638	754	338
baked w/ pork & sweet sce—*1 cup (253 g)*	159	564	592	1073	921	205	147	726	377	703	832	374
baked w/ pork & tomato sce—*1 cup (253 g)*	154	549	577	1042	896	197	142	706	367	683	807	364

[a] Contains dextrose w/ dried corn syrup, aspartame, silicon dioxide, cellulose, tribasic calcium phosphate & cellulose derivatives.

[b] Contains lactose, aspartame, maltodextrin, leucine, cellulose & cellulose derivatives.

	TRY	THR	ISO	LEU	LYS	MET	CYS	PHE	TYR	VAL	ARG	HIS
beans, baked, homemade[a]—*1 cup (253 g)*	170	577	612	1083	959	218	157	726	392	713	901	387
beans, refried, cnd—*1 cup (253 g)*	187	663	696	1260	1083	238	172	853	443	825	977	438
beet greens, boiled—*½ cup (72 g)*	29	55	38	83	54	15	17	49	44	55	53	28
beets												
boiled—*½ cup slices (85 g)*	10	26	27	38	32	10	11	26	21	31	24	12
cnd—*½ cup slices (85 g)*	9	23	23	33	28	9	9	22	19	27	20	10
cnd, harvard—*½ cup slices (123 g)*	12	31	31	43	37	12	12	30	25	36	27	14
cnd, pickled—*½ cup slices (114 g)*	10	27	27	39	33	10	11	26	22	32	24	13
black beans, boiled—*1 cup (172 g)*	181	642	673	1218	1046	229	165	824	430	798	944	425
black turtle beans												
boiled—*1 cup (185 g)*	179	636	668	1208	1040	228	165	818	426	792	938	422
cnd—*1 cup (240 g)*	170	610	638	1154	994	218	158	782	408	756	895	403
broadbeans												
boiled—*1 cup (170 g)*	122	459	520	972	826	105	165	546	410	575	1193	328
cnd—*1 cup (256 g)*	133	497	566	1052	896	115	179	591	443	622	1293	356
falafel[b]—*3 patties w/ 2¼" diameter (51 g)*	68	251	289	481	437	95	93	361	173	287	653	186
broccoli												
raw—*½ cup chopped (44 g)*	13	40	48	58	62	15	9	37	28	56	64	22
boiled—*½ cup (78 g)*	24	75	90	108	117	28	16	70	52	106	120	41
frzn												
chopped, boiled—*½ cup (92 g)*	29	93	111	133	144	34	20	86	64	131	148	51
spears, boiled—*½ cup (92 g)*	29	93	111	133	144	34	20	86	64	131	148	51
brussels sprouts												
boiled—*½ cup (4 sprouts) (78 g)*	22	71	78	89	90	19	12	58		91	119	44
frzn, boiled—*½ cup (78 g)*	31	101	112	128	129	27	18	83		130	170	64
burdock root, boiled—*1 cup (125 g)*	10	44	51	55	115	15	10	56	30	58	180	53
cabbage, chinese (pak-choi)												
raw—*½ cup shredded (35 g)*	5	17	30	31	31	3	6	15	10	23	29	9
boiled—*½ cup shredded (85 g)*	13	43	76	77	79	8	14	39	26	59	74	23
cabbage, green												
raw—*½ cup shredded (35 g)*	4	15	21	22	20	4	4	14	7	18	24	9
boiled—*½ cup shredded (75 g)*	8	25	36	37	34	8	6	23	12	31	41	14
cabbage, red												
raw—*½ cup shredded (35 g)*	5	17	25	25	23	5	4	15	8	21	28	10
boiled—*½ cup shredded (75 g)*	8	27	40	41	38	8	7	26	14	34	45	16
cabbage, savoy												
raw—*½ cup shredded (35 g)*	7	24	35	36	33	7	6	22	12	30	40	14
boiled—*½ cup shredded (73 g)*	13	45	66	68	62	13	11	42	23	56	74	27
carrots												
raw—*1 med (72 g)*	8	27	30	31	29	5	6	23	14	32	31	12
boiled—*½ cup slices (78 g)*	9	31	34	36	34	5	7	27	16	36	35	13
cnd—*½ cup slices (73 g)*	5	17	18	20	18	3	4	15	9	20	19	7
frzn, boiled—*½ cup slices (73 g)*	9	31	34	36	34	6	7	27	17	37	36	13
cassava, raw—*3.5 oz (100 g)*	43	65	61	90	100	26	65	60	40	79	314	45
cauliflower												
raw—*½ cup pieces (50 g)*	13	36	38	58	54	14	12	36	22	50	48	20
boiled—*½ cup pieces (62 g)*	16	42	44	68	62	16	14	42	25	58	56	24
frzn, boiled—*½ cup pieces (90 g)*	19	53	55	85	77	21	17	52	32	73	70	30
celery												
raw—*1 stalk (7.5" long) (40 g)*	4	8	8	12	10	2	2	8	4	10	8	4
boiled—*½ cup diced (75 g)*	5	11	12	18	15	3	2	11	5	15	11	6
celtuce, raw—*12 leaves (100 g)*	6	39	55	52	55	10	10	36	21	46	46	15
chard, swiss, boiled—*½ cup chopped (88 g)*	16	76	136	119	91	18		100		100	107	33
chayote, boiled—*½ cup pieces (80 g)*	6	25	26	46	24	1		29	19	38	21	9
chickpeas (garbanzo beans)												
boiled—*1 cup (164 g)*	139	540	623	1035	973	190	195	779	361	610	1369	400

[a] White beans, onions, salt pork, molasses, brown sugar, vinegar, salt, pepper & dry mustard.

[b] Broadbeans, soybean oil, onions, flour, salt, garlic, coriander & cumin.

	TRY	THR	ISO	LEU	LYS	MET	CYS	PHE	TYR	VAL	ARG	HIS
cnd—*1 cup (240 g)*	115	442	509	845	794	156	161	636	295	499	1118	326
hummus[a]—*1 cup (246 g)*	116	426	487	905	768	101	155	509	381	536	1109	308
chickory greens, raw—*½ cup chopped (90 g)*	28	42	91	67	60	9		37		69	112	26
chickory witloof, raw—*½ cup (45 g)*	8	13	27	20	18	3		1		20	33	8
chives												
raw—*1 T chopped (3 g)*	1	3	4	5	4	1		3	2	4	6	1
freeze-dried—*¼ cup (0.8 g)*	2	7	7	10	8	2		5	5	7	12	3
coleslaw, homemade[b]—*½ cup (60 g)*	10	29	37	49	43	11	9	28	20	37	42	17
collards												
boiled—*1 cup chopped (190 g)*	27	74	86	129	101	29	21	74	57	103	106	40
frzn, boiled—*½ cup chopped (85 g)*	32	89	103	156	120	34	26	89	68	123	129	48
corn, yellow												
boiled—*½ cup (82 g)*	19	109	109	294	116	57	22	127	103	157	111	75
cnd—*½ cup (82 g)*	15	86	86	233	92	45	17	100	82	124	88	59
cream style—*½ cup (128 g)*	15	90	90	241	95	46	18	104	84	128	91	61
vacuum pack—*½ cup (105 g)*	18	102	102	273	107	53	21	118	97	145	103	69
w/ red & green peppers—*½ cup (114 g)*	19	106	106	280	113	55	23	122	99	150	108	72
frzn, boiled—*½ cup (82 g)*	17	99	99	267	105	52	21	116	94	142	101	68
corn pudding[c]—*1 cup (250 g)*	133	495	588	1075	658	293	163	560	455	718	543	288
cornsalad, raw[d]—*½ cup (28 g)*	7	21	28	37	28	7	6	25	10	28	25	10
cowpeas, (blackeye peas)												
boiled—*1 cup (171 g)*	162	503	537	1012	894	188	145	771	428	629	915	410
cnd—*1 cup (240 g)*	139	432	463	871	770	161	125	665	367	542	787	353
frzn, boiled—*½ cup (85 g)*	83	269	387	515	474	103	107	396	296	418	506	233
cowpeas, catjang, boiled—*1 cup (171 g)*	171	528	564	1065	941	198	154	812	450	662	963	431
cranberry beans												
boiled—*1 cup (177 g)*	196	696	729	1320	1135	248	181	894	466	866	1023	460
cnd—*1 cup (260 g)*	172	606	637	1152	991	216	156	780	406	754	892	400
cucumber, raw—*½ cup slices (1/6 cucumber) (52 g)*	2	8	9	12	11	2	2	8	5	9	18	4
dock												
raw—*½ cup chopped (67 g)*	63	68	112	77	23		76	56	89	72	36	88
boiled—*3.5 oz (100 g)*		86	93	152	105	32		104	75	121	98	49
eggplant												
raw—*½ cup pieces (41 g)*	4	16	20	28	21	5	2	19	12	23	25	10
boiled—*½ cup (48 g)*	4	14	17	25	19	4	2	17	11	21	22	9
endive, raw—*½ cup chopped (25 g)*	1	13	18	25	16	4	3	13	10	16	16	6
french beans, boiled—*1 cup (177 g)*	147	526	550	997	857	188	136	674	352	653	773	347
garlic, raw—*3 cloves (9 g)*	6	14	20	28	25	7	6	16	7	26	57	10
ginger root, raw—*¼ cup slices (24 g)*	3	9	12	18	14	3	2	11	5	18	10	7
gourd, calabash (white-flowered), boiled—*½ cup cubes (73 g)*	2	12	23	26	15	3		10		19	10	3
great northern beans												
boiled—*1 cup (177 g)*	175	621	651	1177	1012	221	161	798	416	772	913	411
cnd—*1 cup (262 g)*	228	812	852	1541	1326	291	210	1045	545	1011	1195	537
green beans (snap beans)[e]												
boiled—*½ cup (62 g)*	12	51	43	72	56	14	11	43	27	58	47	22
cnd—*½ cup (68 g)*	8	34	29	48	37	10	7	29	18	39	31	15
w/ onions, red peppers & garlic—*½ cup (114 g)*	10	41	34	58	46	11	9	34	22	47	39	18
frzn, boiled—*½ cup (68 g)*	10	40	34	57	45	12	9	34	22	46	37	18
hyacinth beans, boiled—*1 cup (194 g)*	132	611	757	1341	1079	126	184	795	565	819	1160	452
jute, potherb, boiled—*½ cup (43 g)*	10	56	75	132	74	22	14	72	50	84	84	37
kale												
boiled—*½ cup chopped (65 g)*	15	55	74	86	74	12	16	63	44	68	69	26
frzn, boiled—*½ cup chopped (65 g)*	23	83	110	129	110	18	25	95	66	101	103	39

[a] Chickpeas, lemon jce, tahini, olive oil & garlic.
[b] Cabbage, celery, table cream, sugar, green pepper, lemon jce, onion, pimiento, vinegar, salt, dry mustard & white pepper.

[c] Yellow corn, whole milk, egg, sugar, butter, salt & pepper.
[d] European herb widely cultivated as a salad plant & potherb.
[e] Includes Italian, green & yellow varieties.

	TRY	THR	ISO	LEU	LYS	MET	CYS	PHE	TYR	VAL	ARG	HIS
kale, scotch, boiled—½ cup chopped (65 g)	15	55	73	86	73	12	16	63	44	68	68	26
kidney beans												
boiled—1 cup (177 g)	182	646	678	1227	1053	230	166	830	432	804	950	428
cnd—1 cup (256 g)	156	561	586	1062	911	200	146	719	374	696	824	371
kidney beans, red												
boiled—1 cup (177 g)	182	646	678	1227	1053	230	166	830	432	804	950	428
cnd—1 cup (256 g)	159	566	594	1073	922	202	146	727	379	701	832	374
kidney beans, red, california, boiled—1 cup (177 g)	191	680	713	1290	1110	242	175	874	455	846	1000	450
kidney beans, royal, red, boiled—1 cup (177 g)	198	706	742	1340	1152	253	182	908	473	878	1039	467
kohlrabi, boiled—½ cup slices (82 g)	9	43	68	58	48	11	6	34		43	91	16
lambsquarters, boiled—½ cup chopped (90 g)	26	112	174	240	243	33	61	113	121	155	174	79
leeks												
raw—¼ cup chopped (26 g)	3	16	14	25	20	5	7	14	11	15	20	7
boiled—¼ cup chopped (26 g)	2	9	7	14	11	3	4	8	6	8	11	4
freeze-dried—¼ cup (0.8 g)	1	5	4	8	6	1	2	4	3	5	6	2
lentils, boiled—1 cup (198 g)	160	640	772	1295	1247	152	234	881	477	887	1380	503
lentils, sprouted, stir-fried—3.5 oz (100 g)		322	320	617	698	103	328	434	248	391	600	252
lettuce												
butterhead, raw[a]—2 leaves (15 g)	1	9	12	12	13	2	2	8	5	10	11	3
cos/romaine, raw—½ cup shredded (28 g)	3	21	29	27	29	6	5	19	11	24	25	8
iceberg, raw[b]—1 leaf (20 g)	2	11	15	14	15	3	3	10	6	12	13	4
looseleaf, raw—½ cup shredded (28 g)	3	17	24	22	24	4	4	15	9	20	20	6
lima beans												
boiled—1 cup (188 g)	173	634	773	1265	983	186	162	844	519	882	899	447
cnd—1 cup (241 g)	140	513	624	1024	795	149	130	684	419	713	728	364
lima beans, baby												
boiled—1 cup (182 g)	173	632	770	1263	981	186	162	843	517	881	897	448
frzn, boiled—½ cup (90 g)	78	254	385	470	395	97	73	295	193	374	400	203
lima beans, fordhook, frzn, boiled—½ cup (85 g)	68	218	332	405	341	51	63	254	166	322	345	175
lotus root												
raw—10 slices (81 g)	16	41	44	56	76	18	18	36	23	45	71	31
boiled—10 slices (89 g)	11	28	29	37	51	12	12	25	15	30	47	20
lupins, boiled—1 cup (166 g)	208	951	1154	1960	1381	183	319	1026	971	1079	2771	735
mixed veg												
cnd[c]—½ cup (82 g)	21	85	103	141	126	25	20	89	55	111	143	54
frzn[d]—½ cup (91 g)	26	105	126	173	155	31	24	109	67	136	176	66
mothbeans, boiled—1 cup (177 g)	89	687	929	752	133	71	620		443		466	
mountain yam, hawaii, steamed—½ cup cubes (72 g)	10	44	42	78	48	17	15	58	33	50	104	27
mung beans, boiled—1 cup (202 g)	154	465	600	1099	990	170	125	859	424	735	994	414
mung beans, sprouted[e]												
raw—½ cup (52 g)	19	41	69	91	86	18	9	61	27	68	102	36
cnd—½ cup (62 g)	12	25	42	55	53	11	6	37	16	42	63	22
ckd—½ cup (62 g)	17	36	61	81	76	16	7	53	24	60	91	32
stir-fried—½ cup (62 g)	36	76	128	171	162	33	16	113	50	126	192	68
mungo beans, boiled—1 cup (180 g)	140	472	693	1125	900	198	126	792	421	761	884	380
mushrooms												
raw[f]—½ cup pieces (35 g)	16	33	29	45	74	14	2	28	16	34	36	20
boiled—½ cup pieces (78 g)	40	79	69	107	177	34	5	69	37	80	87	47
cnd—½ cup pieces (78 g)	34	69	60	92	153	29	4	59	33	69	75	41
mushrooms, shitake												
ckd—4 mushrooms (72 g)	3	49	40	67	34	18	19	48	32	48	64	16
dried—4 mushrooms (15 g)	5	75	61	102	51	27	29	73	48	73	97	24

[a] Includes Boston & Bibb types.
[b] Includes crisphead types.
[c] Carrots, green peas, snap beans & lima beans.
[d] Corn, lima beans, snap beans, green peas & carrots.
[e] Seeds attached to sprouts.
[f] Agaricus bisporus.

	TRY	THR	ISO	LEU	LYS	MET	CYS	PHE	TYR	VAL	ARG	HIS
mustard greens												
boiled—½ cup chopped (70 g)	18	42	57	48	72	15	24	42	83	62	116	28
frzn, boiled—½ cup chopped (75 g)	19	45	62	53	77	16	26	45	90	66	125	31
navy beans												
boiled—1 cup (182 g)	187	666	699	1265	1087	238	173	855	446	828	981	440
cnd—1 cup (262 g)	233	831	870	1575	1355	296	215	1066	555	1032	1221	548
okra												
boiled—½ cup slices (80 g)	13	49	52	78	60	16	14	49	65	68	62	23
frzn, boiled—½ cup slices (92 g)	16	63	66	100	77	20	18	63	84	87	80	30
onion rings, frzn, heated[a]—7 rings (70 g)	49	105	150	245	106	57	76	170	110	151	228	76
onions												
raw—½ cup chopped (80 g)	14	22	34	33	45	8	17	24	23	22	126	15
boiled—½ cup chopped (105 g)	14	23	34	34	45	7	17	24	23	22	127	16
cnd—½ cup chopped (112 g)	13	22	32	31	44	8	17	24	22	21	122	15
dehydrated flakes—¼ cup (14 g)	18	30	44	44	60	10	22	32	31	29	168	20
frzn, boiled—3.5 oz (100 g)	10	17	25	25	34	6	13	18	18	16	95	12
onions, spring, raw[b]—½ cup chopped (50 g)	10	34	37	52	44	10		28	25	39	63	15
onions, welsh, raw—3.5 oz (100 g)	21	74	81	113	95	21		61	55	84	137	33
parsley												
raw—½ cup chopped (30 g)	11				66	5						
freeze-dried—¼ cup (1.4 g)	7				44	3						
peas, green												
raw—½ cup (78 g)	29	158	152	252	247	64	25	156	88	183	334	83
boiled—½ cup (80 g)	30	161	154	256	251	65	26	158	90	186	338	84
cnd—½ cup (85 g)	26	140	135	224	220	57	22	139	79	163	297	74
w/ onions, red peppers & garlic—½ cup (114 g)	24	132	127	210	206	54	21	130	74	153	278	70
frzn, boiled—½ cup (80 g)	28	154	148	246	242	62	24	152	86	178	326	81
peas, mature seeds, sprouted, boiled[c]—3.5 oz (100 g)		240	221	473	497	89	200	325	164	285	627	217
peas, split, boiled—1 cup (196 g)	182	580	674	1172	1180	167	249	753	474	772	1458	398
peppers, hot chili[d]												
raw—1 pepper (45 g)	12	33	29	47	40	11	17	28	19	38	43	18
cnd—½ cup chopped (68 g)	8	22	20	32	27	7	12	19	13	26	29	12
peppers, jalapeno, cnd—½ cup chopped (68 g)	7	20	18	29	24	7	10	17	12	23	26	11
peppers, sweet[d]												
raw—½ cup chopped (50 g)	6	16	14	22	19	5	8	13	9	18	21	9
boiled—½ cup chopped (68 g)	5	16	14	22	19	5	8	13	9	18	20	9
cnd—½ cup halves (70 g)	7	20	18	29	25	7	11	18	12	24	27	11
freeze-dried—¼ cup (1.6 g)	4	11	9	15	13	3	6	9	6	12	14	6
frzn, boiled—3.5 oz (100 g)	12	35	31	49	42	11	18	29	20	40	45	19
pigeonpeas, boiled—1 cup (168 g)	111	402	412	811	796	128	131	973	282	491	680	405
pink beans, boiled—1 cup (169 g)	181	644	676	1222	1051	230	167	828	431	801	948	426
pinto beans												
boiled—1 cup (171 g)	166	592	621	1122	964	212	152	759	395	735	870	392
cnd—1 cup (240 g)	130	461	482	874	751	166	120	593	307	574	679	305
potato												
raw w/o skin—1 potato (112 g)	36	84	94	139	141	37	29	103	86	131	106	50
baked w/ skin—1 potato (202 g)	73	170	188	279	283	73	59	206	172	263	214	101
baked w/o skin—1 potato (156 g)	47	111	125	184	186	48	39	136	114	172	140	67
boiled w/o skin—1 potato (135 g)	36	84	95	139	140	36	30	103	86	130	107	51
cnd w/o skin—½ cup (90 g)	20	46	51	77	77	20	16	57	47	72	59	28
cottage fries, frzn, heated[e]—10 strips (50 g)	23	78	74	104	92	20	11	74	43	88	82	29
french fried, from restaurant[f]—10 pieces (50 g)	27	92	87	122	107	23	13	86	51	103	96	34

[a] Breaded & parboiled in veg oil.
[b] Includes tops & bulbs.
[c] Seeds attached to sprouts.
[d] Includes green & red varieties.
[e] Pan-fried in veg oil.
[f] Fried in veg oil.

	TRY	THR	ISO	LEU	LYS	MET	CYS	PHE	TYR	VAL	ARG	HIS
french fried, frzn, heated[a]—*10 pieces (50 g)*	24	79	75	105	92	20	11	74	44	88	82	29
frzn, boiled w/o skin—*3.5 oz (100 g)*	31	72	80	119	120	31	25	88	73	111	91	43
hash brown, frzn, prep[b]—*½ cup (78 g)*	33	112	106	148	131	27	16	105	62	126	116	41
hash brown, homemade[b]—*½ cup (78 g)*	26	86	81	114	100	21	12	80	48	96	90	32
mashed, from flakes[c]—*½ cup (105 g)*	20	89	102	158	140	34	23	95	86	125	88	48
mashed, from granules[c]—*½ cup (105 g)*	32	144	161	247	223	51	39	153	138	200	147	78
mashed, homemade[d]—*½ cup (105 g)*	30	77	91	139	130	37	24	90	80	118	86	46
microwaved w/ skin—*1 potato (202 g)*	77	180	200	297	301	79	63	220	184	279	228	109
microwaved w/o skin—*1 potato (156 g)*	52	119	133	197	200	51	42	145	122	184	151	72
potato pancakes, homemade[e]—*1 pancake (76 g)*	72	192	234	337	282	107	86	231	178	286	258	106
potato salad, homemade[f]—*½ cup (125 g)*	53	145	176	253	214	83	64	169	130	215	190	78
potatoes, au gratin, homemade[g]—*½ cup (122 g)*	85	234	346	540	465	143	54	310	281	397	248	184
potatoes, o'brien												
frzn, prep—*3.5 oz (100 g)*	34	80	89	131	132	34	30	95	79	120	111	48
homemade[h]—*1 cup (194 g)*	68	173	213	328	279	83	58	206	178	258	211	105
potatoes, scalloped, homemade[i]—*½ cup (122 g)*	51	140	176	275	234	71	41	165	148	212	144	85
pumpkin												
boiled—*½ cup mashed (122 g)*	11	26	28	41	48	10	2	28	37	31	48	13
cnd—*½ cup (122 g)*	16	39	41	62	73	15	4	43	56	46	72	21
pumpkin pie mix, cnd—*½ cup (135 g)*	18	42	46	68	80	16	4	47	61	51	78	23
purslane, boiled—*½ cup (58 g)*	9	29	31	53	38	8	6	34	14	42	33	13
radish, raw—*10 radishes (45 g)*	2	13	14	17	16	3	2	10	6	14	18	6
radish, oriental[j]												
raw—*½ cup slices (44 g)*	1	11	11	14	13	3	2	9	5	12	15	5
boiled—*½ cup slices (74 g)*	3	21	21	26	24	4	4	16	10	23	29	10
dried—*½ cup (58 g)*	25	189	200	240	228	44	36	152	87	212	264	86
rutabaga, boiled—*½ cup cubes (85 g)*	10	36	39	30	31	8	9	25	18	37	116	23
seaweed												
kelp (kombu/tangle), raw—*3.5 oz (100 g)*	48	55	76	83	82	25	98	43	26	72	65	24
laver (nori), raw—*3.5 oz (100 g)*	43	232	259	501	222	145	100	273	254	402	285	140
spirulina, raw—*3.5 oz (100 g)*	96	306	331	509	312	118	68	286	266	362	427	112
spirulina, dried—*3.5 oz (100 g)*	929	2970	3209	4947	3025	1149	662	2777	2584	3512	4147	1085
wakame, raw—*3.5 oz (100 g)*	35	165	87	257	112	63	28	112	49	209	92	15
shallots												
raw—*1 T chopped (10 g)*	3	10	11	15	13	3		8	7	11	18	4
freeze-dried—*¼ cup (4 g)*	5	17	19	26	22	5		14	13	20	32	8
soybean products												
miso—*½ cup (138 g)*	197	882	1119	1558	911	206	131	822	500	1024	1031	454
natto—*½ cup (88 g)*	196	715	819	1328	1008	183	194	828	489	896	800	451
tempeh—*½ cup (83 g)*	234	639	832	1358	934	220	265	840	608	813	1093	413
tofu, dried-frozen (koyadufu)—*1 piece (17 g)*	127	333	404	619	537	104	113	397	273	411	542	237
tofu, fried—*1 piece (13 g)*	35	91	111	170	147	29	31	109	75	113	149	65
tofu, okara—*½ cup (61 g)*	31	80	97	149	129	25	27	96	66	99	131	57
tofu, raw—*½ cup (124 g)*	156	409	496	761	660	128	139	487	335	506	667	291
tofu, raw, firm—*½ cup (126 g)*	310	811	985	1511	1309	255	275	968	665	1003	1323	578

[a] Pan-fried in veg oil.
[b] Prepared w/ veg oil.
[c] Whole milk & butter added.
[d] Whole milk, margarine & salt added.
[e] Potatoes, eggs, onion, margarine, flour & salt.
[f] Potatoes, egg, mayonnaise, celery, sweet pickle relish, onion, green pepper, pimiento, salt & dry mustard.

[g] Potatoes, whole milk, cheddar cheese, butter, flour & salt.
[h] Potatoes, whole milk, onions, green pepper, bread crumbs, salt, butter & black pepper.
[i] Potatoes, whole milk, butter, flour & salt.
[j] Includes daikon (Japanese) & Chinese.

	TRY	THR	ISO	LEU	LYS	MET	CYS	PHE	TYR	VAL	ARG	HIS
tofu, salted & fermented (fuyu)— *1 block (11 g)*	14	37	44	68	59	11	12	44	30	45	60	26
soybeans, green, boiled—*½ cup (90 g)*	135	443	489	795	665	135	102	503	399	494	895	299
soybeans, mature, boiled—*1 cup (172 g)*	416	1244	1388	2331	1906	385	461	1495	1084	1429	2221	772
soybeans, mature, sprouted[a]												
steamed—*½ cup (47 g)*	91	231	199	334	278	45	21	159	131	223	191	107
stir-fried[b]—*3.5 oz (100 g)*	300	759	654	1100	916	147	69	524	432	734	629	352
spinach												
raw—*½ cup chopped (28 g)*	11	34	41	62	49	15	10	36	30	45	45	18
boiled—*½ cup (90 g)*	36	114	137	208	164	50	32	121	102	151	151	59
cnd—*½ cup (107 g)*	41	128	154	234	184	56	36	136	113	169	170	66
frzn, boiled—*½ cup (95 g)*	40	127	152	232	182	55	36	135	113	168	169	66
squash, summer												
all varieties, raw—*½ cup slices (65 g)*	7	18	27	45	42	11	8	27	20	34	33	16
all varieties, boiled—*½ cup slices (90 g)*	7	20	30	48	45	12	9	29	22	37	34	18
crookneck, raw[c]—*½ cup slices (65 g)*	5	15	22	36	34	9	7	21	16	27	26	13
crookneck, boiled[c]—*½ cup slices (90 g)*	7	20	30	48	45	12	9	29	22	37	34	18
crookneck, cnd[c]—*½ cup slices (108 g)*	5	16	24	39	37	10	6	24	17	30	28	14
crookneck, frzn, boiled[c]—*½ cup slices (96 g)*	11	30	44	72	68	17	13	43	33	56	52	27
scallop, raw—*½ cup slices (65 g)*	7	19	28	46	44	11	8	27	21	35	33	17
scallop, boiled—*½ cup slices (90 g)*	8	23	33	55	51	14	10	32	25	42	40	20
zucchini, raw—*½ cup slices (65 g)*	7	18	27	44	42	11	8	27	20	34	32	16
zucchini, boiled—*½ cup slices (90 g)*	5	14	21	33	32	8	6	20	15	26	24	13
zucchini, frzn, boiled—*½ cup slices (112 g)*	11	31	46	75	72	18	13	45	35	58	54	28
zucchini, italian style, cnd[d]—*½ cup slices (114 g)*	10	29	42	68	65	17	13	41	31	52	49	25
squash, winter												
all varieties, baked—*½ cup cubes (102 g)*	13	28	36	51	34	11	8	36	31	39	50	17
acorn, baked—*½ cup cubes (102 g)*	16	34	45	65	42	14	10	45	39	49	64	21
acorn, boiled, mashed—*½ cup (122 g)*	12	24	32	46	31	10	7	32	28	35	45	16
butternut, boiled—*½ cup cubes (102 g)*	13	28	36	52	34	11	8	36	31	40	51	17
butternut, frzn, boiled—*½ cup mashed (120 g)*	20	44	58	84	54	18	13	58	49	64	82	28
hubbard, baked—*½ cup cubes (102 g)*	21	45	59	86	56	18	13	59	51	65	84	29
hubbard, boiled, mashed—*½ cup (118 g)*	25	52	68	99	65	21	15	68	59	76	97	33
spaghetti, boiled/baked—*½ cup (78 g)*	7	14	19	27	17	5	4	19	16	20	26	9
succotash												
boiled—*½ cup (96 g)*	55	203	275	428	285	65	53	235	166	296	284	155
cnd—*½ cup (128 g)*	37	138	188	293	195	45	36	161	114	202	195	106
cnd w/ cream style corn—*½ cup (133 g)*	40	146	198	309	205	47	39	169	120	213	205	112
frzn, boiled—*½ cup (85 g)*	41	152	207	322	214	49	40	177	125	223	213	116
swamp cabbage												
raw—*1 cup chopped (56 g)*		78	58	82	61	25	16	71	45	76	83	26
boiled—*½ cup chopped (49 g)*		55	41	57	43	17	11	50	31	53	58	18
sweet potato												
baked—*1 sweet potato (114 g)*	24	98	98	144	97	48	16	117	81	128	91	36
boiled—*½ cup mashed (164 g)*	33	134	134	198	133	67	21	62	112	177	126	51
candied[e]—*1 piece 2½" long & 2" diameter (105 g)*	12	45	46	68	46	22	7	55	38	60	42	18
cnd—*1 cup pieces (200 g)*	40	164	166	242	162	82	26	198	136	216	154	62
syrup pack—*½ cup (98 g)*	16	63	63	92	62	31	10	75	52	82	59	24
frzn, baked—*½ cup cubes (88 g)*	18	75	76	111	74	37	12	91	62	99	70	28

[a] Seeds attached to sprouts.
[b] Stir-fried in veg oil.
[c] Includes straightneck varieties.

[d] Packed in tomato jce.
[e] Canned sweet potato & syrup, brown sugar, butter & salt.

	TRY	THR	ISO	LEU	LYS	MET	CYS	PHE	TYR	VAL	ARG	HIS
taro, ckd—½ cup slices (66 g)	5	16	13	25	15	5	7	18	13	18	24	8
tomato, green, raw—1 tomato (123 g)	11	37	36	54	54	12	20	38	26	38	36	22
tomato, red												
raw—1 tomato (123 g)	9	27	26	41	41	10	15	28	18	28	27	16
boiled—½ cup (120 g)	10	34	32	49	49	12	17	35	23	35	32	20
cnd												
stewed—½ cup (128 g)	9	29	28	44	44	10	15	31	20	31	28	18
wedges in tomato jce—½ cup (131 g)	8	26	25	38	38	9	13	28	18	26	25	16
whole, peeled—½ cup (120 g)	8	29	26	41	41	10	14	29	19	29	28	17
stewed[a]—1 cup (101 g)	20	52	63	102	48	23	26	74	45	68	57	35
tomato paste, cnd—½ cup (131 g)	34	113	96	138	141	25	29	105	67	101	100	79
tomato puree—1 cup (250 g)	28	95	80	115	120	23	25	88	55	85	83	65
turnip												
boiled—½ cup cubes (78 g)	5	16	23	20	22	7	3	11	9	18	15	9
frzn, boiled—3.5 oz (100 g)	15	42	62	57	61	19	9	30	23	51	41	24
turnip greens												
raw—½ cup chopped (28 g)	7	23	22	38	27	10	5	26	16	29	26	10
boiled—½ cup chopped (72 g)	14	45	42	76	53	19	9	50	32	56	52	20
cnd—½ cup (117 g)	27	87	82	145	103	36	18	97	62	108	99	39
frzn, boiled—½ cup (82 g)	48	151	142	252	179	62	31	169	107	187	172	66
turnip greens & turnips, frzn, boiled—3.5 oz (100 g)	31	97	101	156	120	41	20	102	66	120	108	45
vegetable combinations, cnd												
peas & carrots—½ cup (128 g)	20	104	101	160	157	40	17	101	58	120	210	54
peas & onions—½ cup (60 g)	14	72	71	115	114	29	13	71	42	83	162	38
vegetable combinations, frzn												
peas & carrots—½ cup (80 g)	18	93	90	143	140	35	15	90	51	107	187	48
peas & onions, boiled—½ cup (90 g)	17	84	82	133	132	33	15	83	49	96	188	44
vinespinach, raw—3.5 oz (100 g)	28	55	53	101	86	19	27	85	48	65	70	39
watercress, raw—½ cup chopped (17 g)	5	23	16	28	23	3	1	19	11	23	26	7
waxgourd, boiled[b]—½ cup cubes (87 g)	2				8	3						
white beans												
boiled—1 cup (179 g)	206	732	768	1389	1196	261	190	942	490	911	1078	485
cnd—1 cup (262 g)	225	799	838	1517	1305	286	207	1027	534	996	1176	529
white beans, small, boiled—1 cup (179 g)	190	675	709	1282	1103	242	175	868	453	840	993	448
winged beans, boiled—1 cup (172 g)	401	619	771	1311	1121	187	286	750	765	803	991	415
yam, boiled/baked[c]—½ cup cubes (68 g)	8	35	34	64	39	14	12	47	27	41	84	22
yambean, boiled—3.5 oz (100 g)		30	26	40	42	11	10	27	19	35	61	31
yardlong bean, boiled—1 cup (171 g)	174	540	576	1086	959	202	156	828	458	675	982	439
yellow beans, boiled—1 cup (177 g)	191	683	717	1296	1113	244	177	878	457	848	1004	451
35. Miscellaneous												
soybean protein concentrate—1 oz (28 g)	234	693	824	1377	1100	228	248	918	644	858	1300	442
soybean protein isolate—1 oz (28 g)	312	878	1191	1899	1492	316	293	1286	902	1147	1868	645
tomato powder—3½ oz (100 g)	89	295	250	359	370	66	76	273	173	264	258	204
whey												
acid, dry—1 T (3 g)	7	17	17	32	29	6	6	11	9	17	9	7
acid, fluid—1 cup (246 g)	38	94	93	178	161	35	34	62	48	92	52	37
sweet, dry—1 T (8 g)	15	61	54	88	77	18	19	30	27	52	28	18
sweet, fluid—1 cup (246 g)	33	132	116	192	166	39	41	66	59	113	61	38
yeast, tortula, dried, Lake States—3.5 oz (100 g)	523	2565	2660	4133	4133	285	190	2613	1663	3088	2803	1093

[a] Tomatoes, bread crumbs, margarine, sugar, onions, salt & pepper.
[b] Chinese preserving melon.
[c] This is the true yam of tropical areas (Dioscorea species), not the american yam which is really a sweet potato.

Alcohol (Ethanol) Content of Alcoholic Beverages

	weight (g)	volume (%)		weight (g)	volume (%)
ALES, BEERS & MALT LIQUORS			**COCKTAILS & COCKTAIL MIXES**		
ale, Blatz Cream—*12 fl oz (360g)*	15.5		pina colada		
beer—*12 fl oz (356g)*	12.8	4.5	cnd—*6.8 fl oz can (222g)*	20.0	11.2
Anheuser-Busch—*12 fl oz (355g)*	13.5		pineapple jce, rum, sugar & coconut		
Black Label—*12 fl oz (360g)*	12.6		cream—*4.5 fl oz cocktail (141g)*	14.0	12.3
Blatz—*12 fl oz (360g)*	13.1	4.6	screwdriver (orange jce & vodka)—*7 fl oz*		
Coors Premium—*12 fl oz (360g)*	13.2		*cocktail (213g)*	14.1	8.2
Heileman's Old Style—*12 fl oz (360g)*	13.7		tequila sunrise		
Heileman's Special Export—*12 fl oz (360g)*	15.5		cnd—*6.8 fl oz can (211g)*	19.8	11.7
Heileman's Special Export Dark—*12 fl oz (360g)*	15.5		orange jce, tequila, lime jce & grenadine—*5.5 fl oz cocktail (172g)*	18.7	13.5
Herman Joseph's—*12 fl oz (360g)*	14.2	4.95	tom collins (club soda, gin, lemon jce & sugar)—*7.5 fl oz cocktail (222g)*	16.0	9.0
Killian's—*12 fl oz (360g)*	15.6	5.43	whiskey sour		
Rainier—*12 fl oz (360g)*	13.1		cnd—*6.8 fl oz cocktail (209g)*	19.9	11.8
Schmidt—*12 fl oz (360g)*	13.1		lemon jce, whiskey & sugar—*3 fl oz cocktail (90g)*	15.1	20.6
Stroh's American Lager—*12 fl oz (355g)*	12.0		prep from bottled mix—*2 fl oz mix & 1.5 fl oz whiskey (106g)*	14.9	17.4
beer, light—*12 fl oz (354g)*	11.3[a]	4.0	prep from powdered mix—*17 g pkt, 1.5 fl oz water & 1.5 fl oz whiskey (103g)*	15.0	18.0
Anheuser-Busch—*12 fl oz (355g)*	11.0				
Blatz—*12 fl oz (360g)*	11.9		**DISTILLED SPIRITS**		
Coors—*12 fl oz (360g)*	11.9	4.18	gin, 90 proof—*1.5 fl oz jigger (42g)*	15.9	45.0
Heileman's Old Style—*12 fl oz (360g)*	10.3		gin/rum/vodka/whiskey, 94 proof—*1.5 fl oz jigger (42g)*	16.7	47.0
Heileman's Special Export—*12 fl oz (360g)*	11.9		gin/rum/vodka/whiskey, 100 proof—*1.5 fl oz jigger (42g)*	17.9	50.0
beer, LA (low alcohol)			rum, 80 proof—*1.5 fl oz jigger (42g)*	14.0	40.0
Anheuser-Busch—*12 fl oz (355g)*	6.4		vodka, 80 proof—*1.5 fl oz jigger (42g)*	14.0	40.0
Blatz—*12 fl oz (360g)*	6.5		whiskey, 86 proof—*1.5 fl oz jigger (42g)*	15.1	43.0
Heileman's Old Style—*12 fl oz (360g)*	6.5				
malt beverages, nonalcoholic—*12 fl oz (360g)*	0.7	0.3	**LIQUEURS**		
Kingsbury—*12 fl oz (360g)*	1.0		coffee, 53 proof—*1.5 fl oz (52g)*	11.3	26.5
malt liquor			coffee, 63 proof—*1.5 fl oz (52g)*	13.5	31.5
Blatz Old Fashioned Private Stock—*12 fl oz (360g)*	13.7		coffee w/ cream, 34 proof—*1.5 fl oz (47g)*	6.5	17.0
Colt 45—*12 fl oz (360g)*	15.7		creme de menthe—*1.5 fl oz (50g)*	14.9	36.0
COCKTAILS & COCKTAIL MIXES			**WINES & WINE BEVERAGES**		
bloody mary (tomato jce, vodka & lemon jce)—*5 fl oz cocktail (148g)*	13.9	11.7	sparkling cooler, citrus, La Croix—*12 fl oz (360g)*	12.6	
bourbon & soda—*4 fl oz cocktail (116g)*	15.1	16.1	sparkling cooler, strawberry, La Croix—*12 fl oz (360g)*	12.6	
daiquiri			wine, dessert, dry—*2 fl oz (59g)*	9.0	18.8
cnd—*6.8 oz can (207g)*	19.9	11.9	wine, dessert, sweet—*2 fl oz (59g)*	9.0	18.8
rum, lime jce & sugar—*2 fl oz cocktail (60g)*	13.9	28.3	wine, table, all types—*3.5 fl oz (103g)*	9.6	11.5
gin & tonic (tonic water, gin & lime jce)—*7.5 fl oz cocktail (225g)*	16.0	8.8	wine, table, red—*3.5 fl oz (103g)*	9.6	11.5
manhattan (whiskey & vermouth)—*2 fl oz cocktail (57g)*	17.4	36.9	wine, table, rose—*3.5 fl oz (103g)*	9.6	11.5
martini (gin & vermouth)—*2.5 fl oz cocktail (70g)*	22.4	38.4	wine, table, white—*3.5 fl oz (103g)*	9.6	11.5

[a] Range from 7.8 to 15.6 mg.

Biotin (mcg)

CEREALS, COOKED OR TO-BE-COOKED
corn grits, inst, white hominy, enr

Quaker—*1 pkt (4/5 cup ckd) (23 g)*	1
w/ cheddar cheese flavor, Quaker—*1 oz pkt (28 g)*	1
w/ imit bacon bits, Quaker—*1 oz pkt (28 g)*	2
w/ imit ham bits, Quaker—*1 oz pkt (28 g)*	1

corn grits, quick, yellow hominy, enr, Quaker—*3 T (1 oz) (28 g)* — 1

corn grits, reg/quick, white hominy, enr, Quaker/Aunt Jemima—*3 T (1 oz) (28 g)* — 1

farina, dry, quick, creamy wheat, Quaker—*2½ T (1 oz) (28 g)* — 1

oatmeal, inst, Quaker—*1 pkt (¾ cup ckd) (28 g)*	7
w/ apples & cinn—*1 pkt (¾ cup ckd) (35 g)*	4
w/ bran & raisins—*1 pkt (¾ cup ckd) (43 g)*	7
w/ cinn & spice—*1 pkt (¾ cup ckd) (46 g)*	6
w/ honey & graham—*1 pkt (¾ cup ckd) (35 g)*	4
w/ maple & brown sugar—*1 pkt (¾ cup ckd) (43 g)*	7
w/ peaches & cream—*1 pkt (¾ cup ckd) (35 g)*	4
w/ raisins & spice—*1 pkt (¾ cup ckd) (43 g)*	6
w/ raisins, dates & walnuts—*1 pkt (¾ cup) (37 g)*	4
w/ strawberries & cream—*1 pkt (¾ cup) (35 g)*	4

oatmeal, quick/old fashioned, Quaker—*⅓ cup dry (⅔ cup ckd) (28 g)* — 4

oatmeal, quick/reg, Ralston Purina—*1 oz dry (28 g)* — 4.5

ralston, ckd—*¾ cup (190 g)* — 1.9

CEREALS, READY-TO-EAT

bran chex, Ralston Purina—*⅔ cup (1 oz) (28 g)*	3.3
cap'n crunch, Quaker—*¾ cup (1 oz) (28 g)*	1

cap'n crunch's

choco crunch, Quaker—*¾ cup (1 oz) (28 g)*	1
crunchberries, Quaker—*¾ cup (1 oz) (28 g)*	1
peanut butter, Quaker—*¾ cup (1 oz) (28 g)*	2

cookie crisp, choc chip, Ralston Purina—*1 cup (1 oz) (28 g)* — 0.5

cookie crisp, vanilla wafer, Ralston Purina—*1 cup (1 oz) (28 g)* — 0.8

corn bran, Quaker—*⅔ cup (1 oz) (28 g)*	3
corn chex, Ralston Purina—*1 cup (1 oz) (28 g)*	1.1
corn flakes, Ralston Purina—*1 cup (1 oz) (28 g)*	0.4

crispy oatmeal & raisin chex, Ralston Purina—*¾ cup (1.3 oz) (38 g)* — 1.2

halfsies, Quaker—*¾ cup (1 oz) (28 g)*	1
king vitamin, Quaker—*1¼ cups (1 oz) (28 g)*	1
life, cinn, Quaker—*⅔ cup (1 oz) (28 g)*	6
life, Quaker—*⅔ cup (1 oz) (28 g)*	6
life w/ raisins, Quaker—*⅔ cup (1 oz) (28 g)*	6
Mr. T, Quaker—*1 cup (1 oz) (28 g)*	1
puffed wheat, Quaker—*1 cup (½ oz) (14 g)*	1
Quaker 100% natural—*¼ cup (1 oz) (28 g)*	5
w/ apples & cinn—*¼ cup (1 oz) (28 g)*	3
w/ raisins & dates—*¼ cup (1 oz) (28 g)*	4
quisp, Quaker—*1 cup (1 oz) (28 g)*	1
rice chex, Ralston Purina—*1⅛ cups (1 oz) (28 g)*	0.4

shredded wheat

Quaker—*1 oz (28 g)*	4

CEREALS, READY-TO-EAT

sun flakes, corn & rice, Ralston Purina—*1 cup (1 oz) (28 g)* — 0.5

wheat & raisin chex, Ralston Purina—*¾ cup (1.3 oz) (37 g)* — 2.9

wheat chex, Ralston Purina—*⅔ cup (1 oz) (28 g)* — 1.7

CHEESE

cheddar, med, Kraft—*1 oz (28 g)*	0.6
cream cheese, light, Philadelphia Brand—*1 oz (28 g)*	0.7
cream cheese, Philadelphia Brand—*1 oz (28 g)*	0.5
cream cheese w/ herb & garlic, Cremerie—*1 oz (28 g)*	0.5
havarti, Casino—*1 oz (28 g)*	0.5

CHEESE PRODUCTS

Light n' Lively, singles—*1 oz (28 g)*	1.0
Velveeta sharp cheese spread—*1 oz (28 g)*	1.0

DESSERTS
Cakes

coffee cake mix, Aunt Jemima—*1.3 oz dry mix (37 g)* — 1

Frozen Desserts
bon bon ice cream nuggets, Carnation

bavarian mint—*5 nuggets (48 g)*	0.5
choc—*5 nuggets (49 g)*	0.4
choc peanut butter—*5 nuggets (49 g)*	0.4
van—*5 nuggets (48 g)*	0.5

Granola Desserts

caramel nut dipps, Quaker—*1.1 oz bar (31 g)*	2
caramel nut, Smores—*1 oz bar (28 g)*	3.5
choc chip, chewy, Quaker—*1 oz bar (28 g)*	3
choc chip dipps, Quaker—*1 oz bar (29 g)*	1
choc chip, Smores—*1 oz bar (28 g)*	2.1
choc, graham & marshmallow, chewy, Quaker—*1 oz bar (28 g)*	3
chunky nut & raisin, chewy, Quaker—*1 oz bar (28 g)*	4
honey & oats, chewy, Quaker—*1 oz bar (28 g)*	3
honey & oats dipps, Quaker—*1 oz bar (28 g)*	2
mint choc chip dipps, Quaker—*1 oz bar (28 g)*	2
peanut butter & choc chip, chewy, Quaker—*1 oz bar (28 g)*	5
peanut butter & choc chip, Smores—*1 oz bar (28 g)*	4.1
peanut butter, chewy, Quaker—*1 oz bar (28 g)*	6
peanut butter dipps, Quaker—*1 oz bar (28 g)*	5
peanut butter, Smores—*1 oz bar (28 g)*	4.0
raisin & almond dipps, Quaker—*1 oz bar (28 g)*	2
raisin & cinn, chewy, Quaker—*1 oz bar (28 g)*	3
rocky road dipps, Quaker—*1 oz bar (28 g)*	2

EGGS, CHICKEN

white, fresh/frzn—*white of 1 large egg (33 g)*	2.6
whole, fresh/frzn—*1 large (50 g)*	11.0
yolk, fresh—*yolk of 1 large egg (17 g)*	8.4

ENTREES
Box Mix Entrees

macaroni & cheese deluxe dinner, prep, Kraft—*¾ cup (147 g)* — 3.4

pasta shells & cheese, prep, Velvetta—*¾ cup (150 g)* — 2.1

ENTREES & MEALS
Canned Entrees
chili-mac, Wolf Brand—*scant cup (213 g)* — 4
chili w/ beans
    Wolf Brand—*1 cup (227 g)* — 9
    Wolf Brand, extra spicy—*scant cup (213 g)* — 8
chili w/o beans
    Wolf Brand—*1 cup (227 g)* — 5
    Wolf Brand, extra spicy—*scant cup (213 g)* — 5
tamales, Wolf Brand—*scant cup (213 g)* — 6
Frozen Entrees
pizza, canadian style bacon
    Celeste—*7.75 oz pizza (220 g)* — 9
    Celeste—*¼ of 19 oz pizza (135 g)* — 5
pizza, cheese
    Celeste—*6.5 oz pizza (184 g)* — 7
    Celeste—*¼ of 18 oz pizza (126 g)* — 5
pizza, deluxe
    Celeste—*8.25 oz pizza (234 g)* — 12
    Celeste—*¼ of 22 oz pizza (158 g)* — 6
pizza, pepperoni
    Celeste—*6.75 oz pizza (191 g)* — 10
    Celeste—*¼ of 19 oz pizza (135 g)* — 5
pizza, sausage
    Celeste—*7.5 oz pizza (213 g)* — 9
    Celeste—*¼ of 20 oz pizza (142 g)* — 6
pizza, sausage & mushroom
    Celeste—*8.5 oz pizza (241 g)* — 12
    Celeste—*¼ of 25 oz pizza (177 g)* — 6
pizza, suprema
    Celeste—*9 oz pizza (255 g)* — 13
    Celeste—*¼ of 23 oz pizza (163 g)* — 8

GRAIN FRACTIONS
barley, pearled
    med, Scotch Brand—*¼ cup (48 g)* — 3
    quick, Scotch Brand—*¼ cup (48 g)* — 3
corn meal mix, Aunt Jemima
    white, bolted—*1/6 cup (1 oz) (28 g)* — 2
    white, buttermilk, self-rising—*3 T (1 oz) (28 g)* — 1
    yellow, bolted—*1/6 cup (1 oz) (28 g)* — 1
corn meal, white, bolted, enr, self-rising, Aunt Jemima—
    1/6 cup (1 oz) (28 g) — 2
corn meal, white, degermed
    enr, Quaker/Aunt Jemima—*3 T (1 oz) (28 g)* — 1
    enr, self-rising, Aunt Jemima—*1/6 cup (1 oz) (28 g)* — 2
corn meal, yellow, degermed, enr, Aunt Jemima—*3 T
    (1 oz) (28 g)* — 1
masa harina de maiz, Quaker—*1/3 cup (37 g)* — 3
masa trigo, Quaker—*1/3 cup (37 g)* — 1
oat bran, Quaker—*1/3 cup (1 oz) (28 g)* — 10
wheat bran, unprocessed, Quaker—*2 T (7 g)* — 2
wheat flour, self-rising, enr, Aunt Jemima—*¼ cup (1 oz)
    (28 g)* — 1

GRAIN PRODUCTS
Breads & Bread Products
corn bread, mix, Aunt Jemima—*1.7 oz dry mix (48 g)* — 2
Crackers
ry-krisp—*¼ large square (14 g)* — 6.8
    seasoned—*¼ large square (14 g)* — 2.2
    sesame—*2 triple crackers (14 g)* — 0.7

GRAIN PRODUCTS
French Toast
frzn
    Aunt Jemima—*2 slices (85 g)* — 4
    cinn swirl, Aunt Jemima—*2 slices (85 g)* — 5
    raisin, Aunt Jemima—*2 slices (85 g)* — 4
Muffins
corn muffin mix, Flako—*1 oz dry mix (28 g)* — 1
Pancakes
from frzn batter, Aunt Jemima
    blueberry—*three 4" pancakes (113 g)* — 2
    buttermilk—*three 4" pancakes (113 g)* — 3
    plain—*three 4" pancakes (113 g)* — 3
frzn, Aunt Jemima
    blueberry—*three 4" pancakes (106 g)* — 3
    buttermilk—*three 4" pancakes (106 g)* — 3
    plain—*three 4" pancakes (106 g)* — 7
pancake/waffle mix, Aunt Jemima
    buckwheat—*¼ cup dry mix (31 g)* — 3
    buttermilk—*1/3 cup dry mix (51 g)* — 2
    buttermilk, complete—*½ cup dry mix (71 g)* — 4
    plain—*¼ cup dry mix (31 g)* — 2
    plain, complete—*½ cup dry mix (74 g)* — 4
    whole wheat—*1/3 cup dry mix (43 g)* — 5
Pasta
noodles, enr, ckd—*1 cup (160 g)* — 4
Rolls
popover mix, Flako—*1 oz dry mix (28 g)* — 1
Waffles
frzn, Aunt Jemima
    apple & cinn—*2 waffles (71 g)* — 1
    blueberry—*2 waffles (71 g)* — 1
    buttermilk—*2 waffles (71 g)* — 1
    plain—*2 waffles (71 g)* — 1
    raisin—*2 waffles (82 g)* — 3

MEATS, LUNCHEON
meat spread, Carnation
    chicken—*1.9 oz (53 g)* — 0.6
    ham—*1.9 oz (53 g)* — 0.6
    tuna—*1.9 oz (53 g)* — 0.6
    turkey—*1.9 oz (53 g)* — 0.6

MILK, COW
evaporated, cnd, Carnation—*4 fl oz (126 g)* — 7.2
evaporated, skim, cnd, Carnation—*4 fl oz (128 g)* — 7.6

MILK BEVERAGES
eggnog, nonalcoholic
    mix in whole milk, Delmark—*1.23 oz mix in 8 fl oz
    milk (257 g)* — 10
instant breakfast in whole milk
    choc, Delmark—*1.23 oz mix in 8 fl oz milk (279 g)* — 90
    van/strawberry, Delmark—*1.23 oz mix in 8 fl oz milk
    (279 g)* — 90
instant breakfast, van/strawberry, liquid pack, Delmark—
    8 fl oz (279 g) — 70

MILK BEVERAGE MIXES
cocoa mix, powder, Carnation
    choc & marshmallow—*1 oz (28 g)* — 0.4
    milk choc—*1 oz (28 g)* — 0.2
    rich choc—*1 oz (28 g)* — 0.4
    rich choc, 70 cal—*1 pkt (3 hp t) (21 g)* — 2
    sugar-free—*1 pkt (4 hp t) (15 g)* — 3.7

MILK BEVERAGE MIXES
instant breakfast, Carnation
    strawberry—*1.25 oz pkt (35 g)*   5.0
    van—*1.23 oz pkt (35 g)*   5.0

NUTS
almonds, whole, Blue Diamond—*1 oz (28 g)*   23

SAUCES
chili hot dog sce, Wolf Brand—*1/6 cup (35 g)*   2

SPECIAL DIETARY FOODS
fortified pudding, from mix, Delmark
    choc—*½ cup (164 g)*   60
    van/lemon—*½ cup (164 g)*   60
gluten-free, wheat-free foods, Ener-G Foods
    nutquik milk powder—*1 oz (28 g)*   22
    soyquix milk powder—*2 T (14 g)*   10
high protein foods, Delmark
    broth, beef, from mix—*6 fl oz (183 g)*   1

SPECIAL DIETARY FOODS
    broth, chicken, from mix—*6 fl oz (183 g)*   1
    gelatin dessert, all flavors, from mix—*5 fl oz (175 g)*   7
slender bars, Carnation
    choc—*2 bars (55 g)*   75
    choc chip—*2 bars (55 g)*   75
    choc peanut butter—*2 bars (55 g)*   75
    van—*2 bars (55 g)*   75
slender, liquid diet meal, cnd, Carnation
    banana—*10 fl oz (313 g)*   75
    choc—*10 fl oz (313 g)*   75
    choc fudge—*10 fl oz (313 g)*   75
    choc malt—*10 fl oz (313 g)*   75
    milk choc—*10 fl oz (313 g)*   75
    strawberry—*10 fl oz (313 g)*   75
    van—*10 fl oz (313 g)*   75

MISCELLANEOUS
yeast, torula, dried, Lake States—*3.5 oz (100 g)*   85

Caffeine (mg)

BEVERAGES
Carbonated Beverages[a]
cherry coke, Coca-Cola—*12 fl oz (370 g)*   46
cherry cola Slice—*12 fl oz (360 g)*   48
cherry RC—*12 fl oz (360 g)*   36
Coca-Cola—*12 fl oz (370 g)*   46
Coca-Cola Classic—*12 fl oz (369 g)*   46
cola—*12 fl oz (370 g)*   37
cola, RC—*12 fl oz (360 g)*   36
Mello Yello—*12 fl oz (372 g)*   52
Mr. Pibb—*12 fl oz (369 g)*   40
Mountain Dew—*12 fl oz (360 g)*   54
pepper type soda—*12 fl oz (368 g)*   37
Pepsi Cola—*12 fl oz (360 g)*   38
Carbonated Beverages, Low Calorie[a]
diet cherry coke, Coca-Cola—*12 fl oz (354 g)*   46[a]
diet cherry cola Slice—*12 fl oz (360 g)*   48
diet coke, Coca-Cola—*12 fl oz (354 g)*   46
diet cola, aspartame sweetened—*12 fl oz (355 g)*   50
diet Pepsi—*12 fl oz (360 g)*   36
diet RC—*12 fl oz (360 g)*   48
Pepsi Light—*12 fl oz (360 g)*   36
Tab—*12 fl oz (354 g)*   46
Coffee
brewed—*6 fl oz (177 g)*   103
inst powder—*1 rd t (1.8 g)*   57
    decaffeinated—*1 rd t (1.8 g)*   2
    w/ chicory—*1 rd t (1.8 g)*   37
prep from inst powder—*6 fl oz water & 1 rd t powder (179 g)*   57

BEVERAGES
amaretto, General Foods—*6 fl oz water & 11.5 g powder (189 g)*   60
amaretto, sugar-free, General Foods—*6 fl oz water & 7.7 g powder (185 g)*   60
decaffeinated—*6 fl oz water & 1 rd t powder (179 g)*   2
francais, General Foods—*6 fl oz water & 11.5 g powder (189 g)*   53
francais, sugar-free, General Foods—*6 fl oz water & 7.7 g powder (185 g)*   59
irish creme, General Foods—*6 fl oz water & 12.8 g powder (190 g)*   53
irish creme, sugar-free, General Foods—*6 fl oz water & 7.1 g powder (185 g)*   48
irish mocha mint, General Foods—*6 fl oz water & 11.5 g powder (189 g)*   27
irish mocha mint, sugar-free, General Foods—*6 fl oz water & 6.4 g powder (184 g)*   25
orange cappuccino, General Foods—*6 fl oz water & 14 g powder (191 g)*   73
orange cappuccino, sugar-free, General Foods—*6 fl oz water & 6.7 g powder (184 g)*   71
suisse mocha, General Foods—*6 fl oz water & 11.5 g powder (189 g)*   41
suisse mocha, sugar-free, General Foods—*6 fl oz water & 6.4 g powder (184 g)*   40
vienna, General Foods—*6 fl oz water & 14 g powder (191 g)*   56
vienna, sugar-free, General Foods—*6 fl oz water & 6.7 g powder (184 g)*   55
w/ chicory—*6 fl oz water & 1 rd t powder (179 g)*   38

[a] Caffeine-free carbonated beverages & most non-cola carbonated beverages contain no caffeine.

BEVERAGES
Tea, Hot/Iced

brewed 3 min—*6 fl oz (178 g)*	36
inst powder—*1 t (0.7 g)*	31
w/ lemon flavor—*1 rd t (1.4 g)*	25
w/ sugar & lemon flavor—*3 rd t (23 g)*	29
w/ sodium saccharin & lemon flavor—*2 t (1.6 g)*	36
prep from inst powder	
1 t powder in 8 fl oz water (237 g)	31
Crystal Light—*8 fl oz (238 g)*	11
w/ lemon flavor—*1 rd t powder in 8 fl oz water (238 g)*	26
w/ sugar & lemon flavor—*3 rd t powder in 8 fl oz water (259 g)*	29
w/ sodium saccharin & lemon flavor—*2 t powder in 8 fl oz water (238 g)*	36

CANDY
chocolate

german sweet, Bakers—*1 oz square (28 g)*	8
semi-sweet, Bakers—*1 oz square (28 g)*	13
choc chips	
Bakers—*¼ cup (43 g)*	12
german sweet, Bakers—*¼ cup (43 g)*	15
semi-sweet, Bakers—*¼ cup (43 g)*	14
milk choc, Cadbury—*1 oz (28 g)*	15

DESSERTS
Frozen Desserts
pudding pops, Jell-O

choc—*1 pop (47 g)*	2
choc caramel swirl—*1 pop (47 g)*	1
choc fudge—*1 pop (47 g)*	3
choc van swirl—*1 pop (47 g)*	1
choc w/ choc chips—*1 pop (48 g)*	3
choc w/ choc coating—*1 pop (49 g)*	3
double choc swirl—*1 pop (47 g)*	2
milk choc—*1 pop (47 g)*	2
van w/ choc chips—*1 pop (48 g)*	1
van w/ choc coating—*1 pop (49 g)*	1

Pies

choc mousse, from mix, Jell-O—*⅛ pie (95 g)*	6

DESSERTS
Puddings, from inst mix
choc

Jell-O—*½ cup (150 g)*	5
sugar-free, D-Zerta—*½ cup (130 g)*	4
sugar-free, Jell-O—*½ cup (133 g)*	4
choc fudge	
Jell-O—*½ cup (150 g)*	8
sugar-free, Jell-O—*½ cup (135 g)*	9
choc fudge mousse, Jell-O—*½ cup (86 g)*	12
choc mousse, Jell-O—*½ cup (86 g)*	9
choc tapioca, Jell-O—*½ cup (147 g)*	8
milk choc, Jell-O—*½ cup (150 g)*	5

MILK BEVERAGES

choc flavor mix in whole milk—*2-3 hp t powder in 8 fl oz milk (266 g)*	8
choc malted milk flavor powder	
in whole milk—*3 hp t powder in 8 fl oz milk (265 g)*	8
w/ added nutrients in whole milk—*4-5 hp t powder in 8 fl oz milk (265 g)*	5
choc syrup in whole milk—*2 T syrup in 8 fl oz milk (282 g)*	6
cocoa/hot chocolate, prep w/ water from mix—*¾ hp t powder in 6 fl oz water (206 g)*	4

MILK BEVERAGE MIXES

choc flavor mix, powder—*2-3 hp t (22 g)*	8
choc malted milk flavor mix, powder—*¾ oz (3 hp t) (21 g)*	8
choc malted milk flavor mix w/ added nutrients, powder—*¾ oz (4-5 hp t) (21 g)*	6
choc syrup—*2 T (1 fl oz) (38 g)*	5
cocoa mix powder—*1 oz pkt (3-4 hp t) (28 g)*	5

SPECIAL DIETARY FORMULAS, COMMERCIAL & HOSPITAL

ensure, choc, Ross Labs—*8 fl oz (253 g)*	8
ensure, coffee, ross Labs—*8 fl oz (253 g)*	8
ensure HN, choc, Ross Labs—*8 fl oz (253 g)*	8
ensure plus, choc, Ross Labs—*8 fl oz (259 g)*	10
ensure, plus, coffee, Ross Labs—*8 fl oz (259 g)*	12

MISCELLANEOUS

baking choc, unsweetened, Bakers—*1 oz (28 g)*	25

Choline (mg)

BEVERAGES
Alcoholic
   beer, Stroh's American Lager—*12 fl oz (355 g)* 80

EGGS, CHICKEN
white, fresh/frzn—*white of 1 large egg (33 g)* 0.46
whole, fresh/frzn—*1 large (50 g)* 238.4
yolk, fresh—*yolk of 1 large egg (17 g)* 238

MILK BEVERAGES
eggnog, nonalcoholic
   mix in whole milk, Delmark—*1.23 oz mix in 8 fl oz milk (257 g)* 48
instant breakfast in whole milk
   choc, Delmark—*1.23 oz mix in 8 fl oz milk (279 g)* 48
   van/strawberry, Delmark—*1.23 oz mix in 8 fl oz milk (279 g)* 50
instant breakfast, van/strawberry, liquid pack, Delmark—*8 fl oz (279 g)* 30

MILK BEVERAGES
milk shake mix in whole milk, Delmark
   choc—*.667 oz mix in 8 fl oz milk (263 g)* 30
   strawberry—*.667 oz mix in 8 fl oz milk (263 g)* 30
   van—*.667 oz mix in 8 fl oz milk (263 g)* 30

SPECIAL DIETARY FOODS
fortified pudding, from mix, Delmark
   choc—*½ cup (164 g)* 28
   van/lemon—*½ cup (164 g)* 48
high protein foods, Delmark
   broth, beef, from mix—*6 fl oz (183 g)* 0.97
   broth, chicken, from mix—*6 fl oz (183 g)* 1.1

MISCELLANEOUS
yeast, torula, dried, Lake States—*3.5 oz (100 g)* 275

Gluten-Containing & Gluten-Free Grains & Products[a]

GLUTEN-CONTAINING GRAINS
barley
buckwheat
oats
rye
wheat

GLUTEN-FREE GRAINS & PRODUCTS
corn flour
corn meal
cornstarch
gluten-free wheat starch
lima bean flour
potato flour
rice
rice flour
soy flour

[a] Nontropical sprue and other malabsorption symptoms may be relieved/improved by a restriction of gluten-containing grains.

Iodine (mcg)

CHEESE
cheddar, med, Kraft—*1 oz (28 g)*   12
cream cheese, light, Philadelphia Brand—*1 oz (28 g)*   11
cream cheese, Philadelphia Brand—*1 oz (28 g)*   9
havarti, Casino—*1 oz (28 g)*   9

CHEESE PRODUCTS
Light n' Lively, singles—*1 oz (28 g)*   25
Velveeta sharp cheese spread—*1 oz (28 g)*   17

DESSERTS, FROZEN
bon bon ice cream nuggets, Carnation
  bavarian mint—*5 nuggets (48 g)*   3
  choc—*5 nuggets (49 g)*   3
  choc peanut butter—*5 nuggets (49 g)*   3
  van—*5 nuggets (48 g)*   3

EGGS, CHICKEN
white, fresh/frzn—*white of 1 large egg (33 g)*   1
whole, fresh/frzn—*1 large (50 g)*   26

ENTREES, BOX MIX
pasta shells & cheese, prep, Velvetta—*¾ cup (150 g)*   52[a]

MEATS, LUNCHEON
meat spread, Carnation
  chicken—*1.9 oz (53 g)*   4
  ham—*1.9 oz (53 g)*   4
  tuna—*1.9 oz (53 g)*   4
  turkey—*1.9 oz (53 g)*   4

MILK, COW
evaporated, cnd, Carnation—*4 fl oz (126 g)*   52
evaporated, skim, cnd, Carnation—*4 fl oz (128 g)*   54

MILK BEVERAGES
eggnog, nonalcoholic
  mix in whole milk, Delmark—*1.23 oz mix in 8 fl oz milk
    (257 g)*   122
instant breakfast in whole milk—*1 pkt in 8 fl oz milk
    (265 g)*   48
  choc, Delmark—*1.23 oz mix in 8 fl oz milk (279 g)*   97
milk shake
  mix in whole milk, Delmark
    choc—*.667 oz mix in 8 fl oz milk (263 g)*   31
    strawberry—*.667 oz mix in 8 fl oz milk (263 g)*   51
    van—*.667 oz mix in 8 fl oz milk (263 g)*   51

MILK BEVERAGE MIXES
cocoa mix, powder, Carnation
  choc & marshmallow—*1 oz (28 g)*   2
  milk choc—*1 oz (28 g)*   11
  rich choc—*1 oz (28 g)*   2
  rich choc, 70 cal—*1 pkt (3 hp t) (21 g)*   14
  sugar-free—*1 pkt (4 hp t) (15 g)*   17

MILK BEVERAGE MIXES
instant breakfast, Carnation
  choc—*1.25 oz pkt (35 g)*   3
  choc malt—*1.24 oz pkt (35 g)*   3
  coffee—*1.26 oz pkt (36 g)*   3
  eggnog—*1.20 oz pkt (34 g)*   3
  strawberry—*1.25 oz pkt (35 g)*   3
  van—*1.23 oz pkt (35 g)*   3
instant breakfast, no sugar added, Carnation
  choc—*.69 oz pkt (20 g)*   3
  choc malted—*.71 oz pkt (20 g)*   3
  strawberry—*.68 oz pkt (19 g)*   3
  van—*.67 oz pkt (19 g)*   3

NUTS
almonds, whole, Blue Diamond—*1 oz (28 g)*   4

SPECIAL DIETARY FOODS
figurine diet bar, Pillsbury
  choc—*1 bar (21 g)*   21
  choc caramel—*1 bar (21 g)*   21
  choc peanut butter—*1 bar (21 g)*   21
  s'mores—*1 bar (21 g)*   21
  van—*1 bar (21 g)*   21
fortified pudding, from mix, van/lemon, Delmark—*½ cup
    (164 g)*   40
high protein foods, Delmark
  gelatin dessert, all flavors, from mix—*5 fl oz (175 g)*   6
slender bars, Carnation
  choc—*2 bars (55 g)*   38
  choc chip—*2 bars (55 g)*   38
  choc peanut butter—*2 bars (55 g)*   38
  van—*2 bars (55 g)*   38
slender, liquid diet meal, cnd, Carnation
  banana—*10 fl oz (313 g)*   38
  choc—*10 fl oz (313 g)*   38
  choc fudge—*10 fl oz (313 g)*   38
  choc malt—*10 fl oz (313 g)*   38
  milk choc—*10 fl oz (313 g)*   38
  strawberry—*10 fl oz (313 g)*   38
  van—*10 fl oz (313 g)*   38

SPICES
salt, iodized—*1 t (6 g)*   400
salt, noniodized—*1 t (6 g)*   <100
salt substitute, Morton lite salt—*1 t (6 g)*   400

MISCELLANEOUS
yeast, torula, dried, Lake States—*3.5 oz (100 g)*   110

[a] 25 mcg iodine if prepared w/ uniodized salt added to cooking water.

Pectin (g)

BEVERAGES	
alcoholic	
beer—*12 fl oz (356 g)*	0.712
whiskey sour, prep from bottled mix—*2 fl oz mix & 1.5 fl oz whiskey (106 g)*	0.106
whiskey sour mix, bottled—*2 fl oz cocktail (65 g)*	0.130
coffee	
inst powder—*1 rd t (1.8 g)*	0.004
inst powder, mocha flavor, sugar sweetened—*2 rd t (11.5 g)*	0.104
prep from inst powder, mocha flavor, sugar sweetened—*6 fl oz water & 2 rd t powder (188 g)*	0.188
fruit juice drinks & fruit flavored beverages	
lemonade, from frzn conc—*6 fl oz (240 g)*	0.248
orange breakfast drink, from frozn conc—*6 fl oz (188 g)*	0.188
tea, black, inst powder—*1 t (0.7 g)*	0.026
FRUITS	
apple	
raw, w/ skin—*1 med (138 g)*	1.48
raw, w/o skin—*1 med (128 g)*	0.49
boiled, w/o skin—*1 cup (171 g)*	0.46
cnd, sliced, sweetened—*½ cup (102 g)*	0.44
cnd, sliced, sweetened, heated—*½ cup (102 g)*	0.50
frzn, unsweetened—*½ cup (86 g)*	0.40
frzn, unsweetened, heated—*½ cup (103 g)*	0.40
micro ckd w/o skin—*1 cup (170 g)*	0.75
applesce, cnd, sweetened—*½ cup (128 g)*	0.38
kiwifruit, raw—*1 med (76 g)*	0.32

NUTS	
almonds, dried, unblanched—*1 oz (24 nuts) (28 g)*	0.38
chestnuts, european	
raw, unpeeled—*1 oz (2½ nuts) (28 g)*	0.34
roasted—*1 oz (3½ nuts) (28 g)*	0.34
VEGETABLES	
broccoli, frzn, boiled—*½ cup (92 g)*	0.92
broccoli spears, frzn, boiled—*½ cup (92 g)*	0.92
brussel sprouts, frzn, boiled—*½ cup (78 g)*	1.09
carrots, raw—*1 med (72 g)*	0.72
corn, sweet, yellow/white, frzn, boiled—*½ cup (82 g)*	0.33
green beans, frzn, boiled—*½ cup (68 g)*	0.61
peas, green, frzn—*½ cup (72 g)*	0.50
peas, green, frzn, boiled—*½ cup (80 g)*	0.48
spinach	
raw—*½ cup chopped (28 g)*	0.22
boiled—*½ cup (90 g)*	0.72
frzn, boiled—*½ cup (95 g)*	1.05
squash, summer, all varieties	
raw—*½ cup slices (65 g)*	0.39
boiled—*½ cup slices (90 g)*	0.45
squash, summer, crookneck	
raw—*½ cup slices (65 g)*	0.39
boiled—*½ cup slices (90 g)*	0.45
sweet potato	
baked—*1 sweet potato (114 g)*	0.91
frzn, baked—*½ cup cubes (88 g)*	0.70

Phytosterol (mg)

CREAM SUBSTITUTES	
powdered, coffee-mate, Carnation—*1 t (2 g)*	0.6
ENTREES, BOX MIX	
pizza, cheese, from Contadina Pizzeria Kit	
thick crust—*¼ pizza (128 g)*	3.1
thin crust—*¼ pizza (104 g)*	3.1
FATS, OILS & SHORTENINGS	
corn oil, Mazola—*1 T (14 g)*	140
veg oil spray, Mazola No Stick—*2.5 sec spray (0.7 g)*	6
FRUITS	
apple, raw, w/ skin—*1 med (138 g)*	17
apricots, raw—*3 med (106 g)*	19
banana, raw—*1 med (114 g)*	18
cantaloupe, raw—*1 cup pieces (160 g)*	16
cherries, sweet, raw—*10 cherries (68 g)*	8
figs, raw—*1 med (50 g)*	16
grapefruit, raw, white—*½ med (118 g)*	20
grapes, european (adherent skin), raw—*1 cup (160 g)*	6
lemon peel—*1 T (6 g)*	2
lemon, raw w/ peel—*1 med (108 g)*	13

FRUITS	
loquats, raw—*10 med (100 g)*	2
orange, navel, raw—*1 med (140 g)*	34
orange peel—*1 T (6 g)*	2
peach, raw—*1 med (87 g)*	9
pear, raw—*1 med (166 g)*	13
persimmon, japanese, raw—*1 med (168 g)*	7
pineapple, raw—*1 cup pieces (155 g)*	9
plum, raw—*1 med (66 g)*	5
pomegranate, raw—*1 med (154 g)*	26
strawberries, raw—*1 cup (149 g)*	18
watermelon, raw—*1 cup (160 g)*	3
MEATS, LUNCHEON	
meat spread	
chicken, Carnation—*1.9 oz (53 g)*	16.5
ham, Carnation—*1.9 oz (53 g)*	9.9
tuna, Carnation—*1.9 oz (53 g)*	16.0
turkey, Carnation—*1.9 oz (53 g)*	16.3
NUTS, NUT PRODUCTS & SEEDS	
almonds, dried—*1 oz (24 nuts) (28 g)*	41
cashews, dry roasted—*1 oz (28 g)*	45

NUTS, NUT PRODUCTS & SEEDS

chestnuts, european, raw—*1 oz (2½ nuts) (28 g)*	6
coconut, raw—*1 piece (2"x2"x½") (45 g)*	21
coconut milk, raw—*1 cup (240 g)*	2
peanut butter	
creamy/smooth, Skippy—*1 T (16 g)*	16
chunk style/crunchy—*2 T (32 g)*	33
chunk style/crunchy, Skippy—*1 T (16 g)*	16
pecans	
dried—*1 oz (31 large nuts) (28 g)*	31
pine nuts, pignolia, dried—*1 oz (28 g)*	40
pistachio nuts, dried—*1 oz (47 nuts) (28 g)*	31
sesame seeds, whole, dried—*1 T (9 g)*	64
sunflower seeds, dried—*1 oz (28 g)*	152
walnuts, english/persian, dried—*1 oz (14 halves) (28 g)*	31

SAUCES

pizza sce, cnd	
Contadina—*¼ cup (60 g)*	3.2
w/ cheese, Contadina—*¼ cup (60 g)*	3.2
w/ pepperoni, Contadina—*¼ cup (60 g)*	3.2
tomato sce, cnd	
Contadina—*½ cup (124 g)*	0.1
italian style, Contadina—*½ cup (124 g)*	5.3
refrig, Fresh Chef—*4 oz (113 g)*	10.6

SPREADS

margarine	
Mazola—*1 T (14 g)*	70
Mazola unsalted—*1 T (14 g)*	70
margarine, imitation (diet), Mazola diet—*1 T (14 g)*	58
mayonnaise, Best Foods/Hellmann's—*1 T (14 g)*	40
sandwich spread, Best Foods/Hellmann's—*1 T (15 g)*	20

VEGETABLES

aspargus, raw—*4 spears (58 g)*	14
bamboo shoots, raw—*½ cup (76 g)*	14
beet greens, raw—*½ cup (19 g)*	4
beets, raw—*½ cup (68 g)*	17

VEGETABLES

brussel sprouts, raw—*½ cup (44 g)*	11
cabbage, green, raw—*½ cup shredded (35 g)*	4
carrots, raw—*1 med (72 g)*	9
cauliflower, raw—*½ cup pieces (50 g)*	9
celery, raw—*1 stalk (7.5" long) (40 g)*	2
celtuce, raw—*12 leaves (100 g)*	11
chives, raw—*1 T chopped (3 g)*	0.3
cucumber, raw—*½ cup slices (1/6 cucumber) (52 g)*	7
eggplant, raw—*½ cup pieces (41 g)*	3
ginger root, raw—*¼ cup slices (24 g)*	4
green beans, frzn, boiled—*½ cup (68 g)*	2
lettuce, iceberg, raw—*1 leaf (20 g)*	2
lettuce, looseleaf, raw—*½ cup shredded (28 g)*	11
mung beans, sprouted, raw—*½ cup (52 g)*	8
okra, raw—*½ cup (50 g)*	12
onions, raw—*½ cup chopped (80 g)*	12
parsley, raw—*½ cup chopped (30 g)*	2
peppers, sweet, raw—*½ cup chopped (50 g)*	4
potato, raw w/o skin—*1 potato (112 g)*	6
pumpkin, raw—*½ cup (58 g)*	7
radish, raw—*10 radishes (45 g)*	3
shallots, raw—*1 T chopped (10 g)*	1
soybeans, green, raw—*½ cup (128 g)*	64
spinach, raw—*½ cup chopped (28 g)*	2
taro, raw—*½ cup slices (52 g)*	10
tomato, red	
raw—*1 tomato (123 g)*	9
cnd, Contadina	
crushed in tomato puree—*½ cup (119 g)*	3.1
italian style (pear)—*½ cup (120 g)*	8.4
stewed—*½ cup (123 g)*	0.2
whole, peeled—*½ cup (120 g)*	8.4
tomato paste, cnd, italian, Contadina—*¼ cup (57 g)*	5.7
turnip, raw—*½ cup cubes (65 g)*	5
turnip greens, raw—*½ cup chopped (28 g)*	3
yam, raw—*½ cup cubes (75 g)*	8

Purine-Yielding Foods[a]

FOODS HIGHEST IN PURINES (150–825 mg/100 g)
anchovies (363 mg/100 g)
brains
kidney (beef—200 mg/100 g)
game meats
gravies
herring
liver (calf/beef—233 mg/100 g)
mackerel
meat extracts (160–400 mg/100 g)
sardines (295 mg/100 g)
scallops
sweetbreads (825 mg/100 g)

FOODS HIGH IN PURINES (50–150 mg/100 g)
asparagus
breads & cereals, whole grain
cauliflower
eel
fish, fresh & saltwater
legumes, beans/lentils/peas
meat—beef/lamb/pork/veal
meat soups & broths

FOODS HIGH IN PURINES (50–150 mg/100 g)
mushrooms
oatmeal
peas, green
poultry—chicken/duck/turkey
shellfish—crab/lobster/oysters
spinach
wheat germ & bran

FOODS LOWEST IN PURINES (0–50 mg/100 g)
beverages—coffee/tea/sodas
breads & cereals except whole grain
cheese
eggs
fats
fish roe
fruits & fruit juices
gelatin
milk
nuts
sugars, syrups, sweets
vegetables (except those listed above)
vegetable & cream soups

[a] Purines are normally formed in the body during the metabolic breakdown of nucleoproteins. In certain genetic disorders, including gout, the relatively insoluble purine *uric acid* tends to accumulate and deposit in the toes and in other joints. Drug treatment is generally prescribed for patients with gout; however, dietary restriction of purine-yielding foods may also be advised.

Salicylates (mg/100 g)[a]

<.10 mg/100 g
FRUITS

apple (yellow), banana, paw paw, pear w/o skin, plum (green), pomegranate

VEGETABLES

bamboo shoots, bean sprouts, blackeyed peas, brown beans, brussels sprouts, cabbage (green/red), celery, chives, garbanzo beans, leek, lentils, lettuce, lima beans, mung beans, peas (green/split), potato w/o peel, shallots, soy beans, summer squash (chayote), swede

GRAINS & GRAIN PRODUCTS

arrowroot, barley, buckwheat, maize, millet, oats, rice (brown/white), rye, soy grits, wheat

NUTS & SEEDS

cashews, poppyseeds

ANIMAL PRODUCTS

beef, cheese, chicken, egg, kidney, lamb, liver, milk, oysters, pork, salmon, scallops, shrimp, tripe, tuna, yogurt

OTHER

carob powder, cocoa powder, coffee powder (decaffeinated), gin, maple syrup, Ovaltine powder, parsley leaves, saffron, soy sauce, sugar, tanodri powder, tea bag (camomile), vinegar (malt), vodka, whisky

.10–.49 mg/100 g
FRUITS

apple (red), apple juice, apricot nectar, cherries (sour), custard apple, figs (fresh/cnd), grapes (light, seedless), grape juice (light), grapefruit juice, kiwi fruit, lemon, loquat, lychee, mango, nectarine, orange juice, passion fruit, peach nectar, pear w/ skin, persimmon, pineapple juice, plum (red), tamarillo, watermelon

VEGETABLES

asparagus, beet, carrot, cauliflower, corn (reg/creamed), eggplant w/o peel, green french beans, horseradish, mushrooms (fresh), onion, parsnip, pimentos, potato w/ peel, pumpkin, rhubarb, spinach (frzn), squash (marrow), sweetpotato (yellow), tomato (fresh), tomato juice, turnip

NUTS & SEEDS

brazil nuts, coconut (dried), hazelnuts, peanut butter, pecans, sesame seeds, sunflower seeds, walnuts

OTHER

beer, brandy, caramels, cereal coffee powder, cider (hard), cola soda, coriander leaves, garlic, molasses, olives (black), sherry (dry), syrup (corn), tabasco sauce, tea bag (decaffeinated/fruit herbal/rose hip), wine (claret/rose/white), vermouth

.50–.99 mg/100 g
FRUITS

apple (cnd), apple (granny smith), avocado, cherries (sweet), figs (dried), grapes (red), grape juice (dark), grapefruit, mandarin orange, mulberries, peach, tangelo

VEGETABLES

alfalfa, broad beans, broccoli, chili peppers (green/yellow-green), cucumber w/o peel, eggplant w/peel, mushrooms (cnd), okra, spinach (fresh), squash, sweetpotato (white), tomato (canned), watercress

NUTS

macadamia nuts, pine nuts, pistachios

OTHER

coffee powder,[b] fennel powder, sherry (sweet), wine (cabernet/claret riesling/sauvignon)

1.00–4.99 mg/100 g
FRUITS

apricot, blackberries, blueberries, boysenberries, cantaloupe, cherries (cnd), cranberries, cranberry sauce, currants (black/red), dates (fresh/dried), guava, grapes (sultana), logenberries, orange, pineapple, plum (dark red), raspberries (frzn), strawberries, youngberries

VEGETABLES

chicory, chili peppers (red), endive, mushrooms (cnd), peppers (sweet green), radishes, tomato paste, tomato sauce, zucchini

NUTS

almonds, peanuts, waterchestnuts

OTHER

bay leaves, basil, caraway, champagne, chili flakes, chili powder, ginger root, mints, nutmeg, olives (green), pepper (white), peppermints,[b] pimento powder, port, rum, tea bags/leaves (regular/herbal/peppermint),[b] vanilla flavoring, vinegar (white)

5.00–10.00 mg/100 g
FRUITS

raisins, prunes (cnd), raspberries (fresh)

OTHER

allspice, cardamom, cloves, dill (fresh), licorice, liqueurs,[b] mint (fresh), paprika (sweet), pepper (black), pickles

>10.00 mg/100 g
OTHER

aniseed, canella powder, cayenne, celery powder, cinnamon, cumin, curry, dill powder, fenugreek powder, garam masala, honey,[b] mace, mustard powder, oregano, paprika (hot), rosemary, sage, tarragon, tumeric, thyme, worcestershire sauce

---

[a] Adapted from Swain et al, 1985.
[b] Salicylate ranges for these items (mg/100 g) are: coffee powder, .0–.96; honey, 2.5–11.24; liqueurs, .66–9.04; peppermints, .77–7.50; tea bags/leaves, 1.9–7.34.

Sugars (g)

BEVERAGES
Alcoholic Beverages

ale, Blatz Cream—*12 fl oz (360 g)*	3.1[a]
beer	
Black Label—*12 fl oz (360 g)*	3.7[a]
Blatz—*12 fl oz (360 g)*	3.3[a]
Heileman's Old Style—*12 fl oz (360 g)*	3.2[a]
Heileman's Special Export—*12 fl oz (360 g)*	3.1[a]
Heileman's Special Export Dark—*12 fl oz (360 g)*	3.1[a]
Rainier—*12 fl oz (360 g)*	3.5[a]
Schmidt—*12 fl oz (360 g)*	4.0[a]
Stroh's American Lager—*12 fl oz (355 g)*	13.4
beer, light	
Anheuser-Busch—*12 fl oz (355 g)*	0.1
Blatz—*12 fl oz (360 g)*	1.6[a]
Heileman's Old Style—*12 fl oz (360 g)*	1.7[a]
Heileman's Special Export—*12 fl oz (360 g)*	1.8[a]
beer, LA (low alcohol)	
Blatz—*12 fl oz (360 g)*	1.8[a]
Heileman's Old Style—*12 fl oz (360 g)*	1.8[a]
brandy, cherry—*2 fl oz (60 g)*	19.6
distilled spirits, rum/vodka—*1.5 fl oz (42 g)*	0.0
liqueurs	
coffee—*1.5 fl oz (52 g)*	20.4
orange—*1.5 fl oz (52 g)*	14.7
malt liquor	
Blatz Old Fashioned Private Stock—*12 fl oz (360 g)*	3.2[a]
Colt 45—*12 fl oz (360 g)*	3.9[a]
sherry, med—*2 fl oz (60 g)*	2.2
vermouth	
dry—*2 fl oz (60 g)*	3.3
sweet—*2 fl oz (60 g)*	9.5
wine	
rose—*3.5 oz (103 g)*	2.6
white—*3.5 oz (103 g)*	0.6

Carbonated Beverages

all flavors—*12 fl oz (360 g)*	37.8
diet, all flavors—*12 fl oz (360 g)*	0.0

Cereal Grain Beverages

prep from powder w/ water, Postum—*6 fl oz (181 g)*	0.8
prep from powder w/ water, coffee flavored, Postum—*6 fl oz (180 g)*	0.7

Coffee, Prep From Inst Powder

amaretto, General Foods—*6 fl oz water & 11.5 g powder (189 g)*	5.6
amaretto, sugar-free, General Foods—*6 fl oz water & 7.7 g powder (185 g)*	0.2
francais, General Foods—*6 fl oz water & 11.5 g powder (189 g)*	5.3
francais, sugar-free, General Foods—*6 fl oz water & 7.7 g powder (185 g)*	0.2
irish creme, General Foods—*6 fl oz water & 12.8 g powder (190 g)*	6.8
irish creme, sugar-free, General Foods—*6 fl oz water & 7.1 g powder (185 g)*	0.2
irish mocha mint, General Foods—*6 fl oz water & 11.5 g powder (189 g)*	6.7

BEVERAGES

irish mocha mint, sugar-free, General Foods—*6 fl oz water & 6.4 g powder (184 g)*	0.2
mocha flavor, sugar sweetened—*6 fl oz water & 2 rd t powder (188 g)*	4.9
orange cappuccino, General Foods—*6 fl oz water & 14 g powder (191 g)*	8.7
orange cappuccino, sugar-free, General Foods—*6 fl oz water & 6.7 g powder (184 g)*	0.2
suisse mocha, General Foods—*6 fl oz water & 11.5 g powder (189 g)*	6.6
suisse mocha, sugar-free, General Foods—*6 fl oz water & 6.4 g powder (184 g)*	0.2
vienna, General Foods—*6 fl oz water & 14 g powder (191 g)*	9.0
vienna, sugar-free, General Foods—*6 fl oz water & 6.7 g powder (184 g)*	0.2

Fruit Juice Drinks & Fruit Flavored Beverages

appleberry jce works, Campbell's—*6 fl oz (182 g)*	23.4
apple jce works, Campbell's—*6 fl oz (182 g)*	22.0
cherry, cnd—*8 fl oz (240 g)*	25.7
cherry jce works, Campbell's—*6 fl oz (182 g)*	21.1
fruit punch drink	
cnd—*8 fl oz (240 g)*	27.1
from frzn conc—*8 fl oz (240 g)*	24.5
from powder—*2 rd t in 8 fl oz water (240 g)*	27.8
grape jce works, Campbell's—*6 fl oz (182 g)*	22.1
kool-aid, from powder, all flavors—*8 fl oz (246 g)*	25.1
kool-aid, from sugar sweetened powder, all flavors—*8 fl oz (241 g)*	20.3
kool-aid kooler, from powder, all flavors—*8.5 fl oz (262 g)*	30.4
lemonade	
from frzn conc—*8 fl oz (240 g)*	22.1
from powder—*8 fl oz (240 g)*	13.2
from powder, Country Time—*8 fl oz (242 g)*	20.4
lemonade, pink, from powder, Country Time—*8 fl oz (242 g)*	20.4
lemon lime, from powder, Country Time—*8 fl oz (242 g)*	20.4
orange flavor breakfast drink	
from frzn conc—*8 fl oz (240 g)*	24.0
prep from powder—*8 fl oz (240 g)*	25.9
orange, cnd—*8 fl oz (240 g)*	17.3
orange jce works, Campbell's—*6 fl oz (182 g)*	20.4
strawberry jce works, Campbell's—*6 fl oz (182 g)*	22.6
tang, orange, from powder—*6 fl oz (185 g)*	21.7
thirst-quencher drink, cnd—*8 fl oz (240 g)*	14.2
malt beverages, nonalcoholic, Kingsbury—*12 fl oz (360 g)*	4.5[a]
tea, herb, brewed—*6 fl oz (178 g)*	0.0

CANDY
chocolate

dark, sweet—*1 oz (28 g)*	13.7
german sweet, Bakers—*1 oz square (28 g)*	14.3
milk—*1 oz (28 g)*	15.3
milk w/ almonds—*1 oz (28 g)*	13.7
milk w/ crisped rice—*1 oz (28 g)*	12.9

[a] Specified as maltose.

CANDY

semi-sweet—*1 oz (28 g)*	15.1
Bakers—*1 oz square (28 g)*	12.5
chocolate chips	
Bakers—*¼ cup (43 g)*	28.4
german sweet, Bakers—*¼ cup (43 g)*	22.9
semi-sweet, Bakers—*¼ cup (43 g)*	23.8
chocolate covered	
coconut center & almonds—*1 oz (28 g)*	5.6[a]
crunchy peanut butter—*1 oz (28 g)*	8.3[a]
fudge, caramel & peanuts—*1 oz (28 g)*	6.6[a]
malt nougat & caramel—*1 oz (28 g)*	7.9[a]
malted milk balls—*1 oz (28 g)*	4.0[a]
mint flavored fondant, discs—*1 oz (28 g)*	12.7[a]
nougat—*1 oz (28 g)*	15.3
nougat & caramel—*1 oz (28 g)*	7.5[a]
peanut butter nougat, caramel & peanuts—*1 oz (28 g)*	12.6
peanuts—*1 oz (28 g)*	9.6[a]
hard candy—*1 oz (28 g)*	18.7
jelly beans—*1 oz (28 g)*	16.5
licorice—*1 oz (28 g)*	5.5[a]
peanut, caramel & van fudge bar—*1 oz (28 g)*	9.3[a]
sugar-coated	
choc & peanut discs—*1 oz (28 g)*	13.2
choc disks—*1 oz (28 g)*	16.2
toffee—*1 oz (28 g)*	15.5

CEREALS, COOKED

farina, quick/inst—*¾ cup (175 g)*	0.2
oatmeal, quick/reg—*¾ cup (175 g)*	0.7
whole wheat, reg—*¾ cup (182 g)*	0.7

CEREALS, READY-TO-EAT

almond delight, Ralston Purina—*¾ cup (1 oz) (28 g)*	8.0
alpha-bits, Post—*1 cup (1 oz) (28 g)*	11.1
apple jacks, Kellogg's—*1 cup (1 oz) (28 g)*	14.9
body buddies, brown sugar & honey, General Mills—*1 cup (1 oz) (28 g)*	6.0
body buddies, natural fruit flavor, General Mills—*1 cup (1 oz) (28 g)*	6.0
booberry, General Mills—*1 cup (1 oz) (28 g)*	13.0
bran, Kellogg's all-bran—*⅓ cup (1 oz) (28 g)*	5.4
bran, 100%—*½ cup (1 oz) (28 g)*	6.3
bran chex, Ralston Purina—*⅔ cup (1 oz) (28 g)*	5.0
bran flakes—*1 oz (28 g)*	3.4
bran flakes, Post—*⅔ cup (1 oz) (28 g)*	5.0
bran flakes w/ raisins—*1 oz (28 g)*	7.4
bran muffin crisp, General Mills—*⅔ cup (1.2 oz) (35 g)*	12.0
cap'n crunch, Quaker—*¾ cup (1 oz) (28 g)*	11.4
cap'n crunch's	
crunchberries, Quaker—*¾ cup (1 oz) (28 g)*	12.5
peanut butter, Quaker—*¾ cup (1 oz) (28 g)*	9.1
cheerios, General Mills—*1¼ cups (1 oz) (28 g)*	1.0
cinnamon toast crunch, General Mills—*¾ cup (1 oz) (28 g)*	9.0
circus fun, General Mills—*1 cup (1 oz) (28 g)*	12.0
cocoa krispies, Kellogg's—*¾ cup (1 oz) (28 g)*	12.6
cocoa pebbles, Post—*⅞ cup (1 oz) (28 g)*	12.2
cocoa puffs, General Mills—*1 cup (1 oz) (28 g)*	11.0

CEREALS, READY-TO-EAT

cookie crisp, choc chip, Ralston Purina—*1 cup (1 oz) (28 g)*	13.0
cookie crisp, vanilla wafer, Ralston Purina—*1 cup (1 oz) (28 g)*	11.0
corn chex, Ralston Purina—*1 cup (1 oz) (28 g)*	3.0
corn flakes—*1 oz (28 g)*	1.9
country, General Mills—*1 cup (1 oz) (28 g)*	3.0
Kellogg's—*1¼ cups (1 oz) (28 g)*	2.0
Post Toasties, Post—*1 cup (1 oz) (28 g)*	1.9
Ralston Purina—*1 cup (1 oz) (28 g)*	2.0
corn total, General Mills—*1 cup (1 oz) (28 g)*	3.0
count chocula, General Mills—*1 cup (1 oz) (28 g)*	13.0
cracklin bran, Kellogg's—*⅓ cup (1 oz) (28 g)*	8.1
crispy oatmeal & raisin chex, Ralston Purina—*¾ cup (1.3 oz) (38 g)*	12.0
crispy wheats & raisins, General Mills—*¾ cup (1 oz) (28 g)*	10.0
fiber one, General Mills—*½ cup (1 oz) (28 g)*	2.0
fortified oat flakes, Post—*⅔ cup (1 oz) (28 g)*	5.5
frankenberry, General Mills—*1 cup (1 oz) (28 g)*	13.0
froot loops, Kellogg's—*1 cup (1 oz) (28 g)*	13.9
frosted mini-wheats, Kellogg's—*4 biscuits (1 oz) (28 g)*	7.4
frosted rice krinkles—*⅞ cup (1 oz) (28 g)*	12.0
fruit & fiber, Post	
harvest medley—*½ cup (1 oz) (28 g)*	6.9
mountain trail—*½ cup (1 oz) (28 g)*	6.8
tropical fruit—*½ cup (1 oz) (28 g)*	6.4
w/ dates, raisins & walnuts—*½ cup (1 oz) (28 g)*	7.6
fruity pebbles, Post—*⅞ cup (1 oz) (28 g)*	12.1
golden grahams, General Mills—*¾ cup (1 oz) (28 g)*	9.0
granola	
hearty, Post—*¼ cup (1 oz) (28 g)*	6.8
hearty w/ raisins, Post—*¼ cup (1 oz) (28 g)*	8.0
Kretschmer Sun Country	
w/ almonds—*¼ cup (1 oz) (28 g)*	9.0
w/ raisins—*¼ cup (1 oz) (28 g)*	7.7
grape-nut flakes, Post—*⅞ cup (1 oz) (28 g)*	4.7
grape-nuts, Post—*¼ cup (1 oz) (28 g)*	3.4
grape-nuts w/ raisins, Post—*¼ cup (1 oz) (28 g)*	6.1
heartland natural, w/ coconut—*¼ cup (1 oz) (28 g)*	6.3
heartland natural, w/ raisins—*¼ cup (1 oz) (28 g)*	7.4
honey buc wheat crisp, General Mills—*¾ cup (1 oz) (28 g)*	8.0
honeycomb, Post—*1⅓ cups (1 oz) (28 g)*	10.7
honey nut cheerios, General Mills—*¾ cup (1 oz) (28 g)*	10.0
ice cream cones, choc chip, General Mills—*¾ cup (1 oz) (28 g)*	11.0
ice cream cones, vanilla, General Mills—*¾ cup (1 oz) (28 g)*	10.0
kaboom, General Mills—*1 cup (1 oz) (28 g)*	6.0
kix, General Mills—*1½ cups (1 oz) (28 g)*	3.0
life, cinn, Quaker—*⅔ cup (1 oz) (28 g)*	5.1
life, Quaker—*⅔ cup (1 oz) (28 g)*	5.1
lucky charms, General Mills—*1 cup (1 oz) (28 g)*	11.0
oat cereal—*1 oz (28 g)*	0.8
pac man, General Mills—*1 cup (1 oz) (28 g)*	12.0
product 19, Kellogg's—*¾ cup (1 oz) (28 g)*	3.0
puffed rice—*1 cup (½ oz) (14 g)*	0.1
puffed wheat—*1 cup (½ oz) (14 g)*	0.7

[a] Specified as sucrose.

CEREALS, READY-TO-EAT

Quaker 100% natural—*¼ cup (1 oz) (28 g)*	6.1
w/ apples & cinn—*¼ cup (1 oz) (28 g)*	7.1
w/ raisins & dates—*¼ cup (1 oz) (28 g)*	8.0
quisp, Quaker—*1 cup (1 oz) (28 g)*	11.5
raisin bran	
Kellogg's—*¾ cup (1.3 oz) (37 g)*	10.9
Post—*½ cup (1 oz) (28 g)*	8.8
raisin nut bran, General Mills—*½ cup (1 oz) (28 g)*	8.0
rice chex, Ralston Purina—*1⅛ cups (1 oz) (28 g)*	2.0
rice crispy—*1 oz (28 g)*	2.5
rice krispies, Kellogg's—*1 cup (1 oz) (28 g)*	2.3
rice krispies, frosted, Kellogg's—*1 cup (1 oz) (28 g)*	10.8
rice toasties, Post—*¾ cup (1 oz) (28 g)*	2.7
rocky road, General Mills—*⅔ cup (1 oz) (28 g)*	12.0
shredded wheat—*1 oz (28 g)*	0.1
shredded wheat—*1 biscuit (24 g)*	0.1
s'mores crunch, General Mills—*¾ cup (1 oz) (28 g)*	10.0
special K, Kellogg's—*1⅓ cups (1 oz) (28 g)*	2.2
sugar corn pops, Kellogg's—*1 cup (1 oz) (28 g)*	13.2
sugar frosted flakes, Kellogg's—*¾ cup (1 oz) (28 g)*	11.1
sugar smacks, Kellogg's—*¾ cup (1 oz) (28 g)*	15.8
sugar sparkled flakes, Post—*¾ cup (1 oz) (28 g)*	11.3
super golden crisp, Post—*⅞ cup (1 oz) (28 g)*	13.1
super sugar crisp—*⅞ cup (1 oz) (28 g)*	12.8
team flakes—*1 cup (1 oz) (28 g)*	4.5
total, General Mills—*1 cup (1 oz) (28 g)*	3.0
trix, General Mills—*1 cup (1 oz) (28 g)*	12.0
wheat & malted barley nuggets—*1 oz (28 g)*	2.5
wheat & raisin chex, Ralston Purina—*¾ cup (1.3 oz) (37 g)*	10.0
wheat chex, Ralston Purina—*⅔ cup (1 oz) (28 g)*	2.0
wheat flakes—*1 oz (28 g)*	2.2
wheaties, General Mills—*1 cup (1 oz) (28 g)*	3.0

CHEESE

cheddar—*1 oz (28 g)*	0.5
cottage, creamed—*1 cup (210 g)*	1.3
cottage, lowfat—*1 cup (226 g)*	7.2
cream cheese—*1 oz (2T) (28 g)*	0.5
ricotta, whole/skim—*1 oz (28 g)*	0.4
swiss—*1 oz (28 g)*	0.2

CHEESE PRODUCTS

cheese food, american—*1 oz (28 g)*	2.7

CHIPS & SNACK FOODS

chex party mix—*⅔ cup (1 oz) (28 g)*	5.0
chex party mix, nacho—*⅔ cup (1 oz) (28 g)*	5.0
chex party mix, sweet & nutty—*¾ cup (1 oz) (28 g)*	6.0
popcorn, oil-popped—*1 cup (9 g)*	0.1
popcorn w/ caramel—*1 oz (28 g)*	11.0

CREAMS & CREAM SUBSTITUTES

cream, whipping, unwhipped—*1 T (15 g)*	0.4
whipped topping	
from mix, Dream Whip—*1 T (5 g)*	1.0
from mix, w/ nutrasweet, D-Zerta—*1 T (from 1.112 g dry mix) (5 g)*	0.3
frzn, Cool Whip extra creamy, dairy—*1 T (5 g)*	1.2
frzn, Cool Whip non-dairy—*1 T (4 g)*	0.7

DESSERTS

Cakes

cheesecake from mix, Jell-O—*⅛ cheesecake (99 g)*	28.0
fruitcake—*1 piece (43 g)*	18.5
sponge, jam-filled—*1 piece (66 g)*	31.5

Cookies

animal crackers—*10 pieces (26 g)*	5.9
choc wafers, Famous—*5 cookies (28 g)*	11.3
Doughnut, cake type—*1 doughnut (25 g)*	4.2

Frozen Desserts

fruit jce bar, all flavors, Jell-O—*1 bar (52 g)*	9.4
gelatin pops, all flavors, Jell-O—*1 pop (44 g)*	6.8
pudding pops, Jell-O	
choc—*1 pop (47 g)*	9.8
choc caramel swirl—*1 pop (47 g)*	10.1
choc fudge—*1 pop (47 g)*	9.9
choc van swirl—*1 pop (47 g)*	10.1
choc w/ choc chips—*1 pop (48 g)*	11.2
choc w/ choc coating—*1 pop (49 g)*	13.0
double choc swirl—*1 pop (47 g)*	10.1
milk choc—*1 pop (47 g)*	10.3
van—*1 pop (47 g)*	10.4
van w/ choc chips—*1 pop (48 g)*	11.3
van w/ choc coating—*1 pop (49 g)*	13.1

Gelatin Desserts, from mix

all flavors, Jell-O—*½ cup (140 g)*	18.7
orange—*½ cup (140 g)*	17.9
raspberry—*½ cup (140 g)*	8.7

Granola Desserts

caramel grand slam, New Trail—*1.8 oz bar (51 g)*	19.7
caramel nut, Smores—*1 oz bar (28 g)*	9.0
choc chip, New Trail—*1.3 oz bar (37 g)*	14.5
choc chip, Smores—*1 oz bar (28 g)*	10.0
choc chip raisin, Smores—*1 oz bar (28 g)*	10.0
choc fudge, Smores—*1 oz bar (28 g)*	12.0
cocoa, choc covered, New Trail—*1.3 oz bar (37 g)*	14.0
honey graham, choc covered, New Trail—*1.3 oz bar (37 g)*	12.4
peanut butter & choc chip, New Trail—*1.3 oz bar (37 g)*	10.5
peanut butter & choc chip, Smores—*1 oz bar (28 g)*	9.0
peanut butter, New Trail—*1.3 oz bar (37 g)*	10.0
peanut butter, Smores—*1 oz bar (28 g)*	10.0
peanut, choc covered, New Trail—*1.3 oz bar (37 g)*	12.3
plain—*1 oz bar (28 g)*	5.5

Pie, from mix

banana cream, Jell-O—*⅛ pie (92 g)*	17.7
choc mousse, Jell-O—*⅛ pie (95 g)*	17.4
coconut cream, Jell-O—*⅛ pie (94 g)*	20.7

Pies, Snack

apple, baked—*1 snack pie (90 g)*	27.8
apple, fried—*1 snack pie (90 g)*	10.4

Pudding, cnd

choc—*½ cup (150 g)*	20.0[a]
van—*½ cup (150 g)*	22.0[a]

Pudding, from inst mix

banana cream, Jell-O—*½ cup (147 g)*	24.0
banana cream, sugar-free, Jell-O—*½ cup (130 g)*	6.0
butter pecan, Jell-O—*½ cup (147 g)*	23.7
butterscotch, Jell-O—*½ cup (147 g)*	23.9
butterscotch, sugar-free, D-Zerta—*½ cup (130 g)*	6.0

[a] Specified as sucrose.

DESSERTS

butterscotch, sugar-free, Jell-O—½ cup (131 g)	6.1
choc, Jell-O—½ cup (150 g)	24.7
choc, sugar-free, D-Zerta—½ cup (130 g)	6.0
choc, sugar-free, Jell-O—½ cup (133 g)	6.0
choc fudge, Jell-O—½ cup (150 g)	23.7
choc fudge, sugar-free, Jell-O—½ cup (135 g)	6.1
choc fudge mousse, Jell-O—½ cup (86 g)	14.9
choc mousse, Jell-O—½ cup (86 g)	16.2
choc tapioca, Jell-O—½ cup (147 g)	21.0
coconut cream, Jell-O—½ cup (147 g)	22.1
french van, Jell-O—½ cup (147 g)	24.1
golden egg custard, Jell-O—½ cup (143 g)	23.0
lemon, Jell-O—½ cup (147 g)	24.4
milk choc, Jell-O—½ cup (150 g)	24.8
pineapple cream, Jell-O—½ cup (147 g)	24.2
pistachio, Jell-O—½ cup (147 g)	23.7
pistachio, sugar-free, Jell-O—½ cup (131 g)	6.0
rice, Jell-O—½ cup (149 g)	19.0
van, Jell-O—½ cup (147 g)	24.6
van, sugar-free, D-Zerta—½ cup (130 g)	6.1
van, sugar-free, Jell-O—½ cup (131 g)	6.0
vanilla tapioca, Jell-O—½ cup (145 g)	21.1

Dessert Toppings

choc syrup—2 T (38 g)	25.3
confectioner's coating	
carob—1 oz (28 g)	10.6
white choc—1 oz (28 g)	17.5
icing/frosting, ready-to-spread	
choc—amt for 1/12 cake (37 g)	20.6
other flavors—amt for 1/12 cake (37 g)	26.3

ENTREES
Canned Entrees

beans, baked	
w/ bacon, Special Recipe—6.9 oz (195 g)	9.4
w/ beef in bbq sce, Special Recipe—6.9 oz (195 g)	8.4
w/ chicken in bbq sce, Special Recipe—6.9 oz (195 g)	9.0
w/ franks, Campbell's—8 oz (227 g)	13.2
w/ franks in tomato sce, Campbell's—7.9 oz (223 g)	13.0
beef & veg stew, Bounty—7.6 oz (216 g)	1.9
beef & veg stew, Campbell's—7.5 oz (213 g)	1.6
beef ravioli, Franco-American—7.5 oz (213 g)	9.2
beef raviolio's, Franco-American—7.5 oz (213 g)	9.0
beef sirloin w/ onions, Prego Plus—4 oz (113 g)	11.5
chicken a la king, Swanson—5.3 oz (149 g)	1.4
chicken & dumplings, Swanson—7.5 oz (213 g)	0.8
chicken & veg stew, Bounty—7.5 oz (213 g)	1.6
chicken & veg stew, Campbell's—7.6 oz (216 g)	1.4
chicken w/ veg & pasta, Prego—9.49 oz (269 g)	4.9
chili-mac, Bounty—7.8 oz (220 g)	4.3
chili w/ beans, Bounty—7.8 oz (220 g)	2.6
chili w/ beans, hot, Bounty—7.8 oz (220 g)	2.5
dumplings & chicken, Bounty—7.5 oz (213 g)	1.2
linguini w/ white clam sce, Prego—11 oz (312 g)	6.0
macaroni & cheese, Franco-American—7.4 oz (209 g)	1.3
macaroni w/ meat sce, Hearty—7.5 oz (213 g)	8.0
noodles & chicken, Bounty—7.5 oz (213 g)	1.4
pasta twists w/ meat sce, Hearty—7.5 oz (213 g)	18.0
pasta w/ chicken cacciatore sce, Prego—12.2 oz (347 g)	7.7

ENTREES

pizzo's, Franco-American—7.5 oz (213 g)	12.1
potatoes & beef in gravy, Bounty—7.5 oz (213 g)	2.9
ravioli w/ meat sce, Franco-American—7.5 oz (213 g)	10.3
ravioli w/ meat sce, Hearty—7.5 oz (213 g)	10.3
sausage, italian & green peppers, Prego Plus—4 oz (113 g)	10.0
spaghetti 'n beef, Franco-American—7.5 oz (213 g)	7.5
spaghettio's, Franco-American	
w/ franks—7.4 oz (209 g)	8.0
w/ meatballs—7.4 oz (209 g)	7.7
w/ tomato & cheese sce—7.5 oz (213 g)	11.1
spaghetti w/ meatballs & tomato sce	
Franco-American—7.4 oz (209 g)	9.9
Prego—13 oz (369 g)	16.6
spaghetti w/ meat sce	
Franco-American—7.5 oz (213 g)	8.1
Hearty—7.4 oz (209 g)	12.8
Prego—12.5 oz (354 g)	8.7
spaghetti w/ tomato sce & cheese, Franco-American— 7.8 oz (220 g)	12.5
spaghetti w/ tomato sce, Franco-American—7.4 oz (209 g)	11.9
UFO's, Franco-American—7.5 oz (213 g)	12.5
UFO's w/ meteors, Franco-American—7.5 oz (213 g)	11.4
veal & mushrooms, Prego Plus—4 oz (113 g)	11.0

Frozen Entrees (1-2 items/entree)

beef marsala w/ noodles, Prego—11.3 oz (319 g)	10.2
chicken piccata w/ rice, Prego—11 oz (312 g)	5.3

FAST FOODS

cheeseburger—1 reg sandwich (112 g)	5.8
eggs, scrambled—1 serving (98 g)	0.9
english muffin w/ egg, cheese & canadian bacon— 1 sandwich (138 g)	2.9
fish sandwich—1 sandwich (143 g)	4.7
hamburger—1 reg sandwich (98 g)	4.6
ice milk, soft serve in cone—1 serving (115 g)	20.9
shake, all flavors—1 shake (290 g)	52.2
sundae	
caramel—1 sundae (165 g)	41.3
hot fudge—1 sundae (165 g)	41.7
strawberry—1 sundae (165 g)	44.6

FRUIT JUICES

apple jce, unsweetened, cnd—8 fl oz (248 g)	27.0
grape jce, from frzn conc—8 fl oz (253 g)	35.9
grapefruit jce, cnd—8 fl oz (247 g)	18.5
lemon jce, fresh—8 fl oz (244 g)	5.9
orange jce, fresh—8 fl oz (248 g)	25.3
orange jce, from frzn conc—8 fl oz (249 g)	26.4
pineapple jce, unsweetened, cnd—8 fl oz (250 g)	31.3
prune jce, bottled—8 fl oz (256 g)	34.3
tomato jce, cnd—6 fl oz (182 g)	5.3
tomato jce, from concentrate, Campbell's—6 fl oz (182 g)	5.3
V-8 vegetable cocktail, Campbell's—6 fl oz (182 g)	5.6
no salt added—6 fl oz (182 g)	6.2
spicy hot—6 fl oz (182 g)	6.0

FRUITS

apple, raw—1 med (138 g)	18.4
applesce, cnd, sweetened—½ cup (128 g)	21.1

FRUITS

apricots

raw—*3 med (106 g)*	9.0
dried, sulfured—*10 halves (35 g)*	13.6
banana, raw—*1 med (114 g)*	17.8
blueberries, raw—*1 cup (145 g)*	10.6
cantaloupe—*1 cup pieces (160 g)*	13.9
cherries, sour, raw—*1 cup (114 g)*	9.2
cherries, sweet, raw—*10 cherries (68 g)*	9.9
currants—*½ cup (56 g)*	4.5
dates, dried—*10 dates (83 g)*	53.3
figs, dried—*10 figs (187 g)*	124.4
fruit cocktail, cnd, jce pack—*½ cup (124 g)*	19.0

fruit salad, Fresh Chef

fruit cooler—*2/5 cup (99 g)*	8.1
tropical delight—*4/5 cup (198 g)*	25.2
grapefruit, raw—*½ med (118 g)*	7.3
grapes, american, raw—*1 cup (92 g)*	15.1
kiwifruit, raw—*1 med (76 g)*	8.0
lemon, raw—*1 med (58 g)*	1.5
lime, raw—*1 med (67 g)*	0.3
mixed fruit, frzn in syrup, Birds Eye—*½ cup (142 g)*	15.9
nectarine, raw—*1 med (136 g)*	11.6
orange, raw—*1 med (140 g)*	12.5

peach

raw—*1 med (87 g)*	7.6
cnd, jce pack—*1 cup (248 g)*	43.2
dried, sulfured—*10 halves (130 g)*	58.0

pear

raw—*1 med (166 g)*	17.4
cnd, heavy pack—*1 cup (255 g)*	38.8
cnd, jce pack—*1 cup (248 g)*	24.1
cnd, water pack—*1 cup (244 g)*	14.9

pineapple

raw—*1 cup pieces (155 g)*	18.4
cnd, jce pack—*1 cup pieces (250 g)*	35.5
plum, raw—*1 med (66 g)*	5.0
prunes, dried—*10 prunes (84 g)*	37.0
raisins, dried—*⅔ cup (100 g)*	65.0
raspberries, frzn in lite syrup, Birds Eye—*½ cup (142 g)*	19.6

strawberries

raw—*1 cup (149 g)*	8.6
frzn, in lite syrup, Birds Eye—*½ cup (142 g)*	20.6
tangelo, raw—*1 med (170 g)*	12.6
watermelon, raw—*1 cup (160 g)*	14.4

GRAIN FRACTIONS

wheat bran—*1 oz (28 g)*	0.6

wheat flour

white—*1 cup (113 g)*	1.9
whole wheat—*1 cup (120 g)*	2.4
wheat germ, crude/toasted—*¼ cup (1 oz) (28 g)*	3.4

GRAIN PRODUCTS

biscuits, from mix—*1 biscuit (28 g)*	1.2

bread

white—*1 slice (24 g)*	0.9
white, toasted—*1 slice (21 g)*	0.8

GRAIN PRODUCTS

whole wheat—*1 slice (25 g)*	1.0
whole wheat, toasted—*1 slice (21 g)*	0.9
crackers, rye—*¼ large square (14 g)*	0.4
muffin, english, buttered—*1 muffin (57 g)*	2.2
pasta salad, italian, Fresh Chef—*3.2 oz (92 g)*	1.2
pasta salad, seafood, Fresh Chef—*4.3 oz (123 g)*	1.9
roll, hamburger—*1 roll (40 g)*	3.0

stuffing, from mix, Stove Top

americana san francisco style—*½ cup (107 g)*	2.5
chicken flavored—*½ cup (107 g)*	2.6
chicken flavored w/ rice—*½ cup (109 g)*	2.5
chicken florentine—*½ cup (102 g)*	3.2
cornbread—*½ cup (107 g)*	2.5
for beef—*½ cup (108 g)*	3.2
for pork—*½ cup (107 g)*	2.7
garden herb—*½ cup (95 g)*	3.2
homestyle herb—*½ cup (96 g)*	2.5
long grain & wild rice flavor—*½ cup (109 g)*	2.3
savory herbs flavor—*½ cup (107 g)*	2.5
turkey flavor—*½ cup (107 g)*	2.6
veg & almond—*½ cup (95 g)*	3.4
wild rice & mushroom—*½ cup (95 g)*	3.1
tortilla, corn—*1 tortilla (30 g)*	0.2

INFANT FORMULA

isomil, 20 cal/fl oz, Ross Labs—*1 fl oz (30 g)*	1.0[a]
isomil, SF, 20 cal/fl oz, Ross Labs—*1 fl oz (30 g)*	1.0[a]
nursoy, 20 cal/fl oz, Wyeth—*1 fl oz (30 g)*	1.1[a]

MEATS

pork sausage, ckd—*3.5 oz (100 g)*	2.6

MEATS, LUNCHEON

beef & pork spiced loaf—*1 oz (28 g)*	0.9
bologna, beef—*1 slice (23 g)*	0.6
frankfurter, beef & pork—*1 frank (45 g)*	0.9
ham, smoked, ckd—*1 oz (28 g)*	0.3
meat spread, chicken, cnd, Swanson—*1.02 oz (29 g)*	0.3
pastrami, beef—*1 oz (28 g)*	0.2
salami, beef—*1 slice (23 g)*	0.3

MILK, COW

buttermilk, cultured—*8 fl oz (245 g)*	11.8
lowfat, dry—*¼ cup (30 g)*	15.1
skim—*8 fl oz (245 g)*	10.8
whole—*1 cup (244 g)*	12.2
whole, dry—*¼ cup (32 g)*	11.5

YOGURT

lowfat, plain—*1 cup (227 g)*	11.6
lowfat, strawberry—*1 cup (227 g)*	34.7

MILK BEVERAGES

choc malted flavor powder in whole milk—*3 hp t powder in 8 fl oz milk (265 g)*	18.0

MILK BEVERAGE MIXES

choc malted milk flavor mix, powder—*¾ oz (3 hp t) (21 g)*	11.0[a]

cocoa mix, powder, Carnation

choc & marshmallow—*1 oz (28 g)*	16.0[a]
milk choc—*1 oz (28 g)*	16.0[a]

[a] Specified as sucrose.

MILK BEVERAGE MIXES

rich choc—*1 oz (28 g)*	17.0[a]
rich choc, 70 cal—*1 pkt (3 hp t) (21 g)*	11.0[a]
instant breakfast, Carnation	
choc—*1.25 oz pkt (35 g)*	20.0[b]
choc malt—*1.24 oz pkt (35 g)*	19.0[b]
coffee—*1.26 oz pkt (36 g)*	21.0[b]
eggnog—*1.20 oz pkt (34 g)*	21.0[b]
strawberry—*1.25 oz pkt (35 g)*	22.0[b]
vanilla—*1.23 oz pkt (35 g)*	21.0[b]

NUTS, NUT PRODUCTS & SEEDS
almonds

blanched, Blue Diamond—*1 oz (28 g)*	1.6[a]
blanched, slivered, Blue Diamond—*1 oz (28 g)*	1.6[a]
chopped, Blue Diamond—*1 oz (28 g)*	1.6[a]
dried/roasted—*1 oz (28 g)*	1.7
sliced, Blue Diamond—*1 oz (28 g)*	1.6[a]
whole, Blue Diamond—*1 oz (28 g)*	1.9[a]
brazilnuts, oil-roasted—*1 oz (28 g)*	0.7
cashews, oil-roasted—*1 oz (28 g)*	1.4
chestnuts, european—*1 oz (2½ nuts) (28 g)*	3.0
coconut	
raw—*1 piece (2″x2″x½″) (45 g)*	1.6
dried, sweetened, flaked, cnd—*4 oz (114 g)*	39.2
dried, sweetened, flaked, cnd, Angel Flake—*⅓ cup loosely packed (25 g)*	7.7
dried, sweetened, flaked, packaged, Angel Flake—*⅓ cup loosely packed (25 g)*	8.5
dried, sweetened, shredded, Bakers—*⅓ cup loosely packed (28 g)*	9.8
macadamia nuts, oil-roasted—*1 oz (10-12 nuts) (28 g)*	1.7
mixed nuts, oil roasted—*1 oz (28 g)*	1.1
peanut butter, creamy/smooth—*1 T (16 g)*	1.2
peanuts	
dried—*1 oz (28 g)*	1.2
dry/oil-roasted—*1 oz (28 g)*	1.3
pecans, dried—*1 oz (31 large nuts) (28 g)*	1.2
pistachio nuts, dried—*1 oz (47 nuts) (28 g)*	1.8
pumpkin & squash seeds, dried—*1 oz (142 seeds) (28 g)*	0.3
sesame seeds, dried/roasted—*1 oz (28 g)*	0.3
sunflower seeds, dried—*1 oz (28 g)*	0.9
walnuts, english/persian, dried—*1 oz (14 halves) (28 g)*	0.6

POULTRY
chicken, cnd, Swanson

chunks o'chicken—*2.5 oz (71 g)*	0.2
chunk, white & dark—*2.5 oz (71 g)*	0.2
chunk white—*2.5 oz (71 g)*	0.2
chunk white in water—*2.5 oz (71 g)*	0.3
mixin chicken—*2.5 oz (71 g)*	0.2
turkey breast—*3.5 oz (100 g)*	0.2

SALAD DRESSINGS
Low Calorie, from mix, Good Seasons

italian, lite—*1 T (15 g)*	0.4
italian, lite zesty—*1 T (15 g)*	0.3
italian, no oil—*1 T (17 g)*	1.3

SALAD DRESSINGS
Regular, from mix, Good Seasons

bleu cheese & herbs—*1 T (16 g)*	0.3
buttermilk farm style—*1 T (16 g)*	0.8
cheese garlic—*1 T (16 g)*	0.2
cheese italian—*1 T (16 g)*	0.2
classic herb—*1 T (16 g)*	0.1
italian—*1 T (16 g)*	0.3
italian, mild—*1 T (17 g)*	1.0
italian, zesty—*1 T (16 g)*	0.3
lemon & herbs—*1 T (16 g)*	0.3

SAUCES, CONDIMENTS & GRAVIES
Sauces & Condiments

alfredo fetticini sce, cnd, Prego—*5 oz (142 g)*	2.4
alfredo sce, refrig, Fresh Chef—*4 oz (113 g)*	0.6
barbeque sce, Open Pit—*1 T (18 g)*	4.5
hickory smoke flavor—*1 T (18 g)*	4.5
hickory thick & tangy—*1 T (18 g)*	4.5
hot 'n tangy flavor—*1 T (18 g)*	4.4
mesquite 'n tangy—*1 T (18 g)*	4.5
sweet 'n tangy flavor—*1 T (18 g)*	5.1
w/ onions—*1 T (18 g)*	4.7
bolognese sce, refrig, Fresh Chef—*4 oz (113 g)*	9.2
catsup (ketchup), tomato, Campbell's—*1 T (14 g)*	2.4
clam sce, red, refrig, Fresh Chef—*4 oz (113 g)*	6.1
clam sce, white, refrig, Fresh Chef—*4 oz (113 g)*	2.0
marinara (tomato based) sce, Prego—*4 oz (113 g)*	6.9
pesto sce, refrig, Fresh Chef—*1 oz (28 g)*	0.5
spaghetti sce, cnd	
no salt added, Prego—*4 oz (113 g)*	6.9
Prego—*4 oz (113 g)*	12.4
w/ meat flavor, Prego—*4 oz (113 g)*	12.8
w/ mushrooms & chunk tomatoes, Prego—*4 oz (113 g)*	9.6
w/ mushrooms, Prego—*4 oz (113 g)*	12.2
tomato sce, cnd	
al fresco, Prego—*4 oz (113 g)*	7.1
refrig, Fresh Chef—*4 oz (113 g)*	10.6
w/ mushrooms al fresco, Prego—*4 oz (113 g)*	6.9
w/ red peppers, al fresco, Prego—*4 oz (113 g)*	6.5
white cream (bechamel) sce, cnd, Campbell's—*2 oz (57 g)*	0.3

Gravies, cnd

au jus, Franco-American—*2 oz (57 g)*	0.3
beef, Franco-American—*2 oz (57 g)*	0.2
brown, Campbell's—*2 oz (57 g)*	0.7
brown w/ onions, Franco-American—*2 oz (57 g)*	1.2
chicken, Campbell's—*2 oz (57 g)*	0.3
chicken, Franco-American—*2 oz (57 g)*	0.3
chicken giblet, Franco-American—*2 oz (57 g)*	0.1
mushroom, Franco-American—*2 oz (57 g)*	0.2
pork, Franco-American—*2 oz (57 g)*	0.3
turkey, Franco-American—*2 oz (57 g)*	0.4

SOUPS
Canned, Condensed, Prep w/ Water

beef broth & barley, Campbell's—*1 cup (233 g)*	0.6
beef, Campbell's—*1 cup (233 g)*	0.7
beef noodle homestyle, Campbell's—*1 cup (233 g)*	0.7

[a] Specified as sucrose.

[b] Sucrose is 12 g for choc, 10 g for choc malt, 8 g for coffee, 13 g for eggnog, 9 g for strawberry & 9 g for vanilla.

SOUPS

broccoli, creamy, Campbell's—*1 cup (233 g)*	0.8
cauliflower, creamy, Campbell's—*1 cup (233 g)*	1.6
chicken alphabet, Campbell's—*1 cup (233 g)*	0.7
chicken & stars, Campbell's—*1 cup (233 g)*	0.6
chicken & wild rice, Golden Classic, Campbell's—*1 cup (233 g)*	4.1
chicken barley, Campbell's—*1 cup (233 g)*	0.6
chicken broth & noodles, Campbell's—*1 cup (233 g)*	0.6
chicken broth & rice, Campbell's—*1 cup (233 g)*	0.2
chicken, crm of, Special Request (⅓ less salt)—*1 cup (233 g)*	0.6
chicken creamy mushroom, Campbell's—*1 cup (233 g)*	0.6
chicken noodle, Special Request, (⅓ less salt)—*1 cup (233 g)*	0.5
chicken noodle-o's, Campbell's—*1 cup (233 g)*	0.6
corn potage, Campbell's—*1 cup (233 g)*	5.8
curly noodle w/ chicken, Campbell's—*1 cup (233 g)*	0.6
garden veg, Campbell's—*1 cup (233 g)*	2.3
gazpacho, Campbell's—*1 cup (233 g)*	5.2
meatball alphabet, Campbell's—*1 cup (233 g)*	1.2
mushroom, golden, Campbell's—*1 cup (233 g)*	0.7
nacho cheese, Campbell's—*1 cup (233 g)*	1.8
noodles & ground beef, Campbell's—*1 cup (233 g)*	0.8
pea, split w/ ham & bacon, Campbell's—*1 cup (233 g)*	2.9
spinach, creamy, Campbell's—*1 cup (233 g)*	0.9
tomato, crm of, Campbell's—*1 cup (233 g)*	11.7
tomato garden, crispy, Campbell's—*1 cup (233 g)*	12.3
tomato zesty, Campbell's—*1 cup (233 g)*	14.0
tortellini & veg, Golden Classic, Campbell's—*1 cup (233 g)*	5.4
veg, Campbell's—*1 cup (233 g)*	4.3
veg, Special Request (⅓ less salt)—*1 cup (233 g)*	4.2
veg, old fashioned, Campbell's—*1 cup (233 g)*	1.6
won ton, Campbell's—*1 cup (233 g)*	0.4
Canned, Ready-to-Serve	
bean, Campbell's—*7.5 oz (213 g)*	2.3
beef noodle, Campbell's—*7.27 oz (206 g)*	0.6
beef noodle, Home Cookin, Campbell's—*10.76 oz (305 g)*	2.6
beef ravioli romano, chunky, Campbell's—*9.49 oz (269 g)*	12.4
beef w/ mushroom, chunky, low Na, Campbell's—*10.76 oz (305 g)*	3.3
beef w/ noodles, stroganoff style, chunky, Campbell's—*10.76 oz (305 g)*	3.6
chicken broth, low Na, Campbell's—*7.27 oz (206 g)*	0.8
chicken broth, Swanson—*7.27 oz (206 g)*	0.6
chicken, crm of, Campbell's—*7.27 oz (206 g)*	0.6
chicken mushroom, creamy, chunky, Campbell's—*9.38 oz (266 g)*	1.5
chicken noodle, Campbell's—*7.27 oz (206 g)*	0.6
chicken rice, Campbell's—*7.27 oz (206 g)*	0.4
chicken veg, chunky, low Na, Campbell's—*10.76 oz (305 g)*	3.0
chicken veg w/ rice, Home Cookin, Campbell's—*10.76 oz (305 g)*	3.6
chicken w/ noodles & mushrooms, chunky, Campbell's—*10.76 oz (305 g)*	1.0
chicken w/ noodles, Home Cookin, Campbell's—*10.76 oz (305 g)*	1.5
chicken w/ noodles, low Na, Campbell's—*10.76 oz (305 g)*	2.9
chicken w/ rice, chunky, Campbell's—*9.49 oz (269 g)*	2.6
chili beef, chunky, Campbell's—*9.74 oz (276 g)*	2.6

SOUPS

clam chowder	
manhattan, Campbell's—*7.27 oz (206 g)*	1.5
new england, chunky, Campbell's—*10.76 oz (305 g)*	3.1
fisherman chowder, chunky, Campbell's—*9.49 oz (269 g)*	1.9
ham n' butter bean, chunky, Campbell's—*10.76 oz (305 g)*	3.9
lentil, Home Cookin, Campbell's—*10.76 oz (305 g)*	4.9
lentil, Prego—*9.49 oz (269 g)*	5.8
mediterranean veg, chunky, Campbell's—*9.49 oz (269 g)*	6.6
minestrone, Campbell's—*7.27 oz (206 g)*	1.3
minestrone w/ italian sausage, Prego—*9.49 oz (269 g)*	5.8
mushroom, crm of, Campbell's—*7.27 oz (206 g)*	0.8
mushroom, crm of, low Na, Campbell's—*7.27 oz (206 g)*	3.4
mushroom, creamy, chunky, Campbell's—*9.38 oz (266 g)*	1.7
onion, low Na, Campbell's—*7.27 oz (206 g)*	4.2
pea, green, Campbell's—*7.5 oz (213 g)*	3.7
pea, split, low Na, Campbell's—*7.27 oz (206 g)*	2.9
pea, split w/ ham, Home Cookin, Campbell's—*10.76 oz (305 g)*	5.4
sirloin burger, chunky, Campbell's—*9.49 oz (269 g)*	2.5
steak & potato, chunky, Campbell's—*9.49 oz (269 g)*	0.9
tomato, Campbell's—*7.27 oz (206 g)*	12.6
tomato w/ tomato pieces, low Na, Campbell's—*7.27 oz (206 g)*	11.7
tortilla, chunky, Campbell's—*10.76 oz (305 g)*	6.9
veg, Campbell's—*7.27 oz (206 g)*	2.3
veg w/ beef	
Campbell's—*10.76 oz (305 g)*	3.9
chunky, Campbell's—*9.49 oz (269 g)*	3.0
chunky, low Na, Campbell's—*10.76 oz (305 g)*	5.4
won ton, imperial, chunky, Campbell's—*9.49 oz (269 g)*	3.9
Dehydrated	
chicken noodle w/ meat, Campbell's—*0.92 oz (26 g)*	2.9
chicken rice, Campbell's—*0.85 oz (24 g)*	0.7

SPECIAL DIETARY FOODS

breakfast bar, Carnation	
choc chip—*1.44 oz bar (41 g)*	10.0[a]
choc crunch—*1.34 oz bar (38 g)*	8.0[a]
peanut butter crunch—*1.35 oz bar (38 g)*	7.0[a]
peanut butter w/ choc chips—*1.39 oz bar (39 g)*	10.0[a]
gluten-free, wheat-free foods, Ener-G Foods	
corn mix—*1 cup (128 g)*	0.8[a]
nutquik milk powder—*1 oz (28 g)*	1.6[a]
rice mix—*1 oz (28 g)*	0.1[a]
soyquix milk powder—*2 T (14 g)*	0.6[a]
slender bars, Carnation	
choc—*2 bars (55 g)*	18.0[a]
choc chip—*2 bars (55 g)*	20.0[a]
choc peanut butter—*2 bars (55 g)*	18.0[a]
vanilla—*2 bars (55 g)*	17.0[a]
slender, liquid diet meal, cnd, Carnation	
banana—*10 fl oz (313 g)*	21.0[a]
choc—*10 fl oz (313 g)*	25.0[a]
choc fudge—*10 fl oz (313 g)*	25.0[a]
choc malt—*10 fl oz (313 g)*	23.0[a]
milk choc—*10 fl oz (313 g)*	23.0[a]
peach—*10 fl oz (313 g)*	21.0[a]
strawberry—*10 fl oz (313 g)*	21.0[a]
vanilla—*10 fl oz (313 g)*	21.0[a]

[a] Specified as sucrose.

SPECIAL DIETARY FORMULAS, COMMERCIAL & HOSPITAL

enrich, Ross Labs—*8 fl oz (255 g)*	12.2[a]
ensure HN, Ross Labs—*8 fl oz (253 g)*	13.4[a]
ensure plus, Ross Labs—*8 fl oz (259 g)*	10.9[a]
ensure plus HN, Ross Labs—*8 fl oz (260 g)*	10.4[a]
ensure, Ross Labs—*8 fl oz (253 g)*	10.3[a]
pulmocare, Ross Labs—*8 fl oz (250 g)*	13.5[a]
vital high nitrogen, Ross Labs—*300 ml (79 g pkt w/ 255 ml water) (326 g)*	9.4[a]

SUGARS, SYRUPS & OTHER SWEETENERS

honey—*1 T (21 g)*	17.2
molasses, reg—*1 T (20 g)*	12.0
sugar	
brown—*1 cup not packed (145 g)*	130.2
powdered—*1 cup (120 g)*	111.6
white, granulated—*1 cup (200 g)*	193.6
syrup	
corn, dark—*1 T (21 g)*	7.8
corn, high fructose—*1 T (21 g)*	15.6
corn, light—*1 T (21 g)*	10.7
maple—*1 T (20 g)*	12.7
maple honey, Log Cabin—*1 fl oz (39 g)*	24.6
pancake & waffle, Log Cabin	
1 fl oz (39 g)	21.3
buttered—*1 fl oz (39 g)*	22.7
country kitchen—*1 fl oz (39 g)*	19.5
lite—*1 fl oz (36 g)*	15.0

VEGETABLES

artichoke	
boiled—*1 med (300 g)*[b]	3.3
frzn, boiled—*2.8 oz (80 g)*[b]	4.7
asparagus	
boiled—*½ cup (6 spears) (90 g)*	1.4
frzn, boiled—*4 spears (60 g)*	1.7
avocado, raw—*1 med calif (173 g)*	1.6
beans, cnd, baked	
bbq, Campbell's—*7.9 oz (223 g)*	10.7
bbq, Special Recipe—*6.9 oz (195 g)*	9.4
garden style, Special Recipe—*6.9 oz (195 g)*	8.4
homestyle, Campbell's—*8 oz (227 g)*	16.0
new england style, Special Recipe—*6.9 oz (195 g)*	8.8
old fashioned, Special Recipe—*6.9 oz (195 g)*	5.5
ranchero, Campbell's—*7.7 oz (219 g)*	5.5
w/ molasses & brown sugar, Campbell's—*8 oz (227 g)*	18.1
w/ pork—*1 cup (253 g)*	21.0
w/ pork & tomato sce, Campbell's—*8 oz (227 g)*	11.5
w/ tomato sce—*1 cup (253 g)*	14.4
beans, common, ckd—*1 cup (253 g)*	5.6
beet, raw—*½ cup (68 g)*	4.4
broadbeans	
immature, boiled—*1 cup (170 g)*	0.9
mature, ckd—*1 cup (170 g)*	3.1
broccoli	
raw—*½ cup chopped (44 g)*	0.7
frzn, chopped, boiled—*½ cup (92 g)*	1.8

VEGETABLES

frzn, spears, boiled—*½ cup (92 g)*	1.7
frzn, w/ cheese sce, Birds Eye—*½ cup (142 g)*	4.4
frzn, w/ creamy italian sce, Birds Eye—*½ cup (128 g)*	2.8
brussel sprouts	
boiled—*½ cup (78 g)*	1.5
w/ cheese sce, Birds Eye—*½ cup (128 g)*	4.1
cabbage, chinese (pak-choi), raw—*½ cup shredded (35 g)*	0.5
cabbage, green, raw—*½ cup shredded (35 g)*	1.3
cabbage, red	
raw—*½ cup shredded (35 g)*	1.9
boiled—*½ cup shredded (75 g)*	2.1
cabbage, savoy, raw—*½ cup shredded (35 g)*	1.0
carrot	
raw—*1 med (72 g)*	4.8
boiled—*½ cup slices (78 g)*	3.4
cnd—*½ cup slices (73 g)*	2.4
frzn, boiled—*½ cup slices (73 g)*	4.2
cauliflower	
raw—*½ cup pieces (50 g)*	1.2
frzn	
boiled—*½ cup pieces (90 g)*	2.3
w/ cheese sce, Birds Eye—*½ cup (142 g)*	5.1
celery, raw—*1 stalk (7.5" long) (40 g)*	0.4
chickpeas (garbanzo beans), ckd—*1 cup (164 g)*	7.9
coleslaw, Fresh Chef—*4 oz (113 g)*	11.4
corn, yellow, sweet	
boiled—*½ cup (82 g)*	2.1
cnd—*½ cup (82 g)*	2.3
frzn, boiled—*½ cup (82 g)*	3.1
frzn, on-the-cob, Birds Eye—*1 ear (125 g)*	4.9
cucumber, raw—*½ cup slices (1/6 cucumber) (52 g)*	1.2
eggplant, raw—*½ cup pieces (41 g)*	1.4
endive, raw—*½ cup chopped (25 g)*	0.3
green beans (snap beans)	
raw/frzn, boiled—*½ cup (62 g)*	1.4
cnd—*½ cup (68 g)*	1.1
frzn, Birds Eye	
cut—*½ cup (85 g)*	2.8
french style—*½ cup (85 g)*	2.8
french style w/ toasted almonds—*½ cup (85 g)*	4.2
italian—*½ cup (85 g)*	3.6
whole—*4 oz (113 g)*	3.2
jerusalem artichoke, raw	
freshly harvested—*½ cup slices (75 g)*	1.9
stored—*½ cup slices (75 g)*	7.2
kale, raw—*3.5 oz (100 g)*	2.2
lentils, boiled—*1 cup (198 g)*	3.6
lettuce	
cos/romaine, raw—*½ cup shredded (28 g)*	0.6
iceberg, raw—*1 leaf (20 g)*	0.5
lima beans, boiled—*1 cup (188 g)*	5.5
lima beans, baby, frzn, boiled—*½ cup (90 g)*	4.5
lima beans, fordhook, frzn, boiled—*½ cup (85 g)*	2.7
mixed veg, frzn, Birds Eye—*½ cup (95 g)*	4.2
w/ onion sce—*⅓ cup (76 g)*	5.4
mung beans, sprouted, raw—*½ cup (52 g)*	1.1

[a] Specified as sucrose.

[b] Weight w/ refuse.

VEGETABLES

mushroom marinara salad, Fresh Chef—*3.2 oz (92 g)*	1.2
mushrooms	
shitake, cnd, Campbell's—*4 oz (113 g)*	1.8
sovereign, cnd, Campbell's—*4 oz (113 g)*	0.1
okra, raw—*1 cup (100 g)*	2.4
onion, raw	
mature—*½ cup chopped (80 g)*	5.0
spring—*½ cup chopped (80 g)*	1.6
onion, frzn, Birds Eye	
small, whole—*½ cup (113 g)*	6.0
small w/ cream sce—*½ cup (85 g)*	6.7
parsley, raw—*½ cup chopped (30 g)*	0.3
peas, green	
boiled—*½ cup (80 g)*	4.5
cnd—*½ cup (85 g)*	3.0
frzn	
boiled—*½ cup (80 g)*	2.8
w/ cream sce, Birds Eye—*½ cup (76 g)*	6.6
peas, split, boiled—*1 cup (196 g)*	1.6[a]
peppers, sweet, green, raw—*½ cup chopped (50 g)*	1.3
potato	
baked—*1 med (202 g)*	3.4
french fried—*10 pieces (50 g)*	0.3
hashed brown—*½ cup (78 g)*	0.2
potato salad, Fresh Chef—*4 oz (113 g)*	6.2
radish, raw—*10 radishes (45 g)*	1.2
rice, brown, ckd—*1 cup (195 g)*	0.6
rice, white, enr, ckd—*1 cup (205 g)*	0.2[a]
parboiled—*1 cup (175 g)*	0.4[a]
rice, white, from mix, Minute—*⅔ cup (124 g)*	0.3
for rib roast—*½ cup (117 g)*	1.7
fried—*½ cup (97 g)*	1.4
w/ chicken flavor—*½ cup (107 g)*	0.8
rice, long grain & wild, from mix, Minute—*½ cup (115 g)*	1.0
rutabaga, raw—*1 cup (140 g)*	7.8
salad, Fresh Chef	
chicken w/ pineapple tidbits—*4.66 oz (132 g)*	3.9
ham & cheddar cheese—*4.65 oz (132 g)*	5.0
soybean product, tofu, raw, firm—*½ cup (126 g)*	0.5
soybeans, mature, boiled—*1 cup (172 g)*	5.2
spinach	
raw—*½ cup chopped (28 g)*	0.1
frzn, boiled—*½ cup (95 g)*	1.1
frzn, creamed, Birds Eye—*⅓ cup (85 g)*	2.4
squash, summer, all varieties, raw—*½ cup slices (65 g)*	1.4
squash, winter, all varieties, raw—*¼ squash (122 g)*	4.6
sweet potato, raw—*1 sweet potato (180 g)*	10.3
tomato, red	
raw—*1 tomato (123 g)*	3.4
cnd—*½ cup (128 g)*	3.2
tomato paste, cnd—*½ cup (131 g)*	16.9
turnip, raw—*½ cup cubes (65 g)*	4.9
vegetable combinations, frzn, Birds Eye	
broccoli & cauliflower w/ creamy italian cheese sce—*½ cup (128 g)*	3.2

VEGETABLES

broccoli, carrots & pasta w/ lightly seasoned sce—*⅔ cup (95 g)*	2.7
broccoli, carrots & water chestnuts—*¾ cup (113 g)*	4.5
broccoli, cauliflower & carrots—*¾ cup (113 g)*	3.4
broccoli, cauliflower & carrots w/ cheese sce—*½ cup (142 g)*	5.1
broccoli, corn & red peppers—*⅔ cup (113 g)*	3.0
broccoli, green beans, pearl onions & red peppers—*¾ cup (113 g)*	3.0
broccoli, red peppers, bamboo shoots & straw mushrooms—*¾ cup (113 g)*	1.6
brussel sprouts, cauliflower & carrots—*¾ cup (113 g)*	3.3
carrots, peas & pearl onions—*½ cup (95 g)*	5.1
cauliflower, carrots & snow peas—*⅔ cup (113 g)*	4.3
chinese style—*½ cup (95 g)*	2.8
chinese style, stir-fry—*½ cup (95 g)*	3.4
chow mein style—*½ cup (95 g)*	3.7
corn, green beans & pasta curls w/ light cream sce—*½ cup (95 g)*	3.3
green beans, bavarian style & spaetzle (noodles)—*½ cup (95 g)*	2.5
italian style—*½ cup (95 g)*	3.8
japanese style—*½ cup (95 g)*	3.7
japanese style, stir-fry—*½ cup (95 g)*	3.2
mandarin style—*½ cup (95 g)*	3.6
new england style—*½ cup (95 g)*	2.5
pasta primavera style—*½ cup (95 g)*	3.9
peas & pearl onions—*½ cup (95 g)*	5.6
peas & pearl onions w/ cheese sce—*½ cup (142 g)*	6.9
peas & potatoes w/ cream sce—*½ cup (76 g)*	6.6
rice & peas w/ mushrooms—*⅔ cup (66 g)*	1.7
rice, french style—*½ cup (95 g)*	1.5
rice, italian style—*½ cup (95 g)*	1.3
rice, spanish style—*½ cup (95 g)*	2.0
san francisco style—*½ cup (95 g)*	2.8
yam, boiled/baked—*½ cup cubes (68 g)*	0.3

MISCELLANEOUS

baking choc, unsweetened, Bakers—*1 oz (28 g)*	4.0
chewing gum—*1 stick (3.2 g)*	2.2
chewing gum, sugarless—*1 stick (2.7 g)*	0.0
fruit pectin	
Certo—*1 T (14 g)*	0.2
light, sweetened, Sure-Jell—*¼ pkt (12 g)*	6.7
sweetened, Sure-Jell—*¼ pkt (12 g)*	9.3
oven fry, meat/poultry coatings	
extra crispy recipe for chicken—*¼ pouch (29 g)*	2.5
extra crispy recipe for pork—*¼ pouch (29 g)*	1.3
light crispy homestyle recipe—*¼ pouch (22 g)*	0.5
shake & bake, meat/poultry coatings	
barbeque for chicken—*¼ pouch (25 g)*	9.9
barbeque for pork—*¼ pouch (20 g)*	7.5
country mild recipe—*¼ pouch (17 g)*	0.3
for chicken—*¼ pouch (19 g)*	0.9
for fish—*¼ pouch (19 g)*	1.4
for pork—*¼ pouch (21 g)*	1.5
italian herb recipe—*¼ pouch (20 g)*	0.8

[a] Specified as sucrose.

Theobromine (mg)

BEVERAGES
tea, hot/iced

brewed 3 min—*6 fl oz (178 g)*	3.6
inst powder—*1 t (0.7 g)*	2.1
prep from inst powder—*1 t powder in 8 fl oz water (237 g)*	2.4

CANDY

milk choc, Cadbury—*1 oz (28 g)*	44.0

MILK BEVERAGES

choc flavor mix in whole milk—*2-3 hp t powder in 8 fl oz milk (266 g)*	120.0
choc malted milk flavor mix in whole milk—*3 hp t powder in 8 fl oz milk (265 g)*	106.0
choc malted milk flavor mix w/ added nutrients in whole milk—*4-5 hp t powder in 8 fl oz milk (265 g)*	217.0

MILK BEVERAGES

choc syrup in whole milk—*2 T syrup in 8 fl oz milk (282 g)*	90.0
cocoa/hot chocolate, prep w/ water from mix—*¾ hp t powder in 6 fl oz water (206 g)*	68.0

MILK BEVERAGE MIXES

choc flavor mix, powder—*2-3 hp t (22 g)*	123.0
choc malted milk flavor mix, powder—*¾ oz (3 hp t) (21 g)*	106.0
choc malted milk flavor mix w/ added nutrients, powder—*¾ oz (4-5 hp t) (21 g)*	217.0
choc syrup—*2 T (1 fl oz) (38 g)*	89.0
cocoa mix, powder—*1 oz pkt (3-4 hp t) (28 g)*	67.0

SPECIAL DIETARY FORMULAS, COMMERCIAL & HOSPITAL

ensure HN, choc, Ross Labs—*8 fl oz (253 g)*	45.5
ensure plus, choc, Ross Labs—*8 fl oz (259 g)*	62.1
ensure, choc, Ross Labs—*8 fl oz (253 g)*	45.5

Vitamin D (IU)

CEREALS, READY-TO-EAT

bran flakes, Post—*⅔ cup (1 oz) (28 g)*	49
cocoa pebbles, Post—*⅞ cup (1 oz) (28 g)*	49
corn flakes	
Post Toasties, Post—*1 cup (1 oz) (28 g)*	49
Ralston Purina—*1 cup (1 oz) (28 g)*	40
crispy oatmeal & raisin chex, Ralston Purina—*¾ cup (1.3 oz) (38 g)*	5
fortified oat flakes, Post—*⅔ cup (1 oz) (28 g)*	49
fruit & fiber, Post	
harvest medley—*½ cup (1 oz) (28 g)*	49
mountain trail—*½ cup (1 oz) (28 g)*	49
tropical fruit—*½ cup (1 oz) (28 g)*	49
w/ dates, raisins & walnuts—*½ cup (1 oz) (28 g)*	49
fruity pebbles, Post—*⅞ cup (1 oz) (28 g)*	49
granola, hearty, Post—*¼ cup (1 oz) (28 g)*	49
granola, hearty w/ raisins, Post—*¼ cup (1 oz) (28 g)*	49
grape-nut flakes, Post—*⅞ cup (1 oz) (28 g)*	49
grape-nuts, Post—*¼ cup (1 oz) (28 g)*	49
grape-nuts w/ raisins, Post—*¼ cup (1 oz) (28 g)*	49
honeycomb, Post—*1⅓ cups (1 oz) (28 g)*	49
raisin bran, Post—*½ cup (1 oz) (28 g)*	49
sugar sparkled flakes, Post—*¾ cup (1 oz) (28 g)*	49
sun flakes, Ralston Purina	
corn & rice—*1 cup (1 oz) (28 g)*	40
wheat & rice—*1 cup (1 oz) (28 g)*	40
super golden crisp, Post—*⅞ cup (1 oz) (28 g)*	49

CHEESE

cheddar, med, Kraft—*1 oz (28 g)*	3
cream cheese, Philadelphia Brand—*1 oz (28 g)*	2
cream cheese w/ herb & garlic, Cremerie—*1 oz (28 g)*	3
havarti, Casino—*1 oz (28 g)*	3

CHEESE PRODUCTS

cheese food, american, grated, Kraft—*1 oz (28 g)*	8
cheese products, Light n' Lively	
sharp cheddar—*1 oz (28 g)*	1
singles—*1 oz (28 g)*	2
swiss—*1 oz (28 g)*	1
cheese spread	
Cheez Whiz, Kraft—*1 oz (28 g)*	1
Velveeta sharp—*1 oz (28 g)*	3

CREAMS & CREAM SUBSTITUTES

whipped topping, from mix, Dream Whip—*1 T (5 g)*	2

DESSERTS
Cakes

cheesecake, from mix, Jell-O—*⅛ cheesecake (99 g)*	28

Frozen Desserts

bon bon ice cream nuggets, Carnation	
bavarian mint—*5 nuggets (48 g)*	8
choc—*5 nuggets (49 g)*	6
choc peanut butter—*5 nuggets (49 g)*	6
van—*5 nuggets (48 g)*	8

ª Prepared w/ vitamin D fortified milk.

DESSERTS

Pies

banana cream, from mix, Jell-O—*⅛ pie (92 g)*	29[a]
choc mousse, from mix, Jell-O—*⅛ pie (95 g)*	30[a]
coconut cream, from mix, Jell-O—*⅛ pie (94 g)*	30[a]

Puddings, from inst mix

banana cream	
Jell-O—*½ cup (147 g)*	50[a]
sugar-free, Jell-O—*½ cup (130 g)*	50[a]
butter pecan, Jell-O—*½ cup (147 g)*	50[a]
butterscotch	
Jell-O—*½ cup (147 g)*	50[a]
sugar-free, D-Zerta—*½ cup (130 g)*	50[a]
sugar-free, Jell-O—*½ cup (131 g)*	50[a]
choc	
Jell-O—*½ cup (150 g)*	50[a]
sugar-free, D-Zerta—*½ cup (130 g)*	50[a]
sugar-free, Featherweight—*½ cup (134 g)*	50[a]
sugar-free, Jell-O—*½ cup (133 g)*	50[a]
choc fudge	
Jell-O—*½ cup (150 g)*	50[a]
sugar-free, Jell-O—*½ cup (135 g)*	50[a]
choc fudge mousse, Jell-O—*½ cup (86 g)*	25[a]
choc mousse, Jell-O—*½ cup (86 g)*	25[a]
choc tapioca, Jell-O—*½ cup (147 g)*	50[a]
coconut cream, Jell-O—*½ cup (147 g)*	50[a]
french van, Jell-O—*½ cup (147 g)*	50[a]
golden egg custard, Jell-O—*½ cup (143 g)*	55[a]
lemon, Jell-O—*½ cup (147 g)*	50[a]
milk choc, Jell-O—*½ cup (150 g)*	50[a]
pineapple cream, Jell-O—*½ cup (147 g)*	50[a]
pistachio	
Jell-O—*½ cup (147 g)*	50[a]
sugar-free, Jell-O—*½ cup (131 g)*	50[a]
rice, Jell-O—*½ cup (149 g)*	50[a]
vanilla	
Jell-O—*½ cup (147 g)*	50[a]
sugar-free, D-Zerta—*½ cup (130 g)*	50[a]
sugar-free, Featherweight—*½ cup (134 g)*	50[a]
sugar-free, Jell-O—*½ cup (131 g)*	50[a]
vanilla tapioca, Jell-O—*½ cup (145 g)*	50[a]

EGGS, CHICKEN

whole, fresh/frzn—*1 large (50 g)*	27
yolk, fresh—*yolk of 1 large egg (17 g)*	27

ENTREES, BOX MIX

chili con carne, concentrate, Oscar Mayer—*1 oz (28 g)*	9

GRAIN PRODUCTS

Stuffing, from mix, Stove Top

americana san francisco style—*½ cup (107 g)*	9
chicken flavored—*½ cup (107 g)*	9
chicken flavored w/ rice—*½ cup (109 g)*	9
chicken florentine—*½ cup (102 g)*	13
cornbread—*½ cup (107 g)*	9
for beef—*½ cup (108 g)*	9
for pork—*½ cup (107 g)*	9
garden herb—*½ cup (95 g)*	13

GRAIN PRODUCTS

homestyle herb—*½ cup (96 g)*	7
long grain & wild rice flavor—*½ cup (109 g)*	9
savory herbs flavor—*½ cup (107 g)*	9
turkey flavor—*½ cup (107 g)*	9
veg & almond—*½ cup (95 g)*	13
wild rice & mushroom—*½ cup (95 g)*	13

MEATS

pork, bacon pieces, bacon bits, Oscar Mayer—*¼ cup (7 g)*	4

MEATS, LUNCHEON, OSCAR MAYER

bbq loaf (pork & beef)—*1 oz slice (28 g)*	10
bologna—*1 slice (23 g)*	9
bologna, beef—*1 slice (23 g)*	7
bologna, beef garlic—*1 slice (23 g)*	8
bologna, w/ cheese—*1 slice (23 g)*	10
braunschweiger (pork liver sausage)	
german brand—*1 oz (28 g)*	12
in tube—*1 oz (28 g)*	9
corned beef loaf, jellied—*1 oz slice (28 g)*	5
frankfurter—*1 frank (45 g)*	16
beef—*1 frank (10 per 1 lb pkg) (45 g)*	11
beef & pork, small, Oscar Mayer—*1 frank (9 g)*	5
ham & cheese loaf/roll—*1 slice (28 g)*	12
headcheese (pork)—*1 slice (28 g)*	13
honey loaf (pork & beef)—*1 slice (28 g)*	10
lebanon bologna (beef)—*1 slice (23 g)*	11
livercheese, pork fat wrapped, Oscar Mayer—*1 slice (38 g)*	19
luxury loaf (pork)—*1 slice (28 g)*	8
meat spread—*1 oz (28 g)*	10
ham & cheese—*1 oz (28 g)*	7
ham salad—*1 oz (28 g)*	8
old fashioned loaf (pork & beef)—*1 slice (28 g)*	11
olive loaf (pork)—*1 slice (28 g)*	11
peppered loaf (pork & beef)—*1 slice (28 g)*	9
pickle & pimento loaf (pork)—*1 slice (28 g)*	9
picnic loaf (pork & beef)—*1 slice (28 g)*	12
salami, hard—*1 slice (9 g)*	6
thuringer (cervelat/summer sausage)	
beef—*1 slice (23 g)*	9
beef & pork—*1 slice (23 g)*	11

MILK

evaporated, cnd, Carnation—*4 fl oz (126 g)*	100
evaporated, lowfat, cnd, Carnation—*4 fl oz (127 g)*	100
evaporated, skim, cnd, Carnation—*4 fl oz (128 g)*	100

MILK BEVERAGES

Alba choc flavor nonfat dry milk in water—*8 fl oz*	70
choc malted milk flavor mix w/ added nutrients in whole milk—*4-5 hp t powder in 8 fl oz milk (265 g)*	278[a]
eggnog, nonalcoholic mix in whole milk, Delmark—*1.23 oz mix in 8 fl oz milk (257 g)*	140
instant breakfast in whole milk—*1 pkt in 8 fl oz milk (265 g)*	100
choc, Delmark—*1.23 oz mix in 8 fl oz milk (279 g)*	132
van/strawberry, Delmark—*1.23 oz mix in 8 fl oz milk (279 g)*	100

[a] Prepared w/ vitamin D fortified milk.

MILK BEVERAGES

instant breakfast, van/strawberry, liquid pack, Delmark—
8 fl oz (279 g) — 100

malted milk flavor mix w/ added nutrients in whole
milk—*4-5 hp t in 8 fl oz milk (265 g)* — 278[a]

milk shake mix in whole milk, Delmark

choc—*.667 oz mix in 8 fl oz milk (263 g)* — 100

strawberry—*.667 oz mix in 8 fl oz milk (263 g)* — 100

van—*.667 oz mix in 8 fl oz milk (263 g)* — 100

MILK BEVERAGE MIXES

cocoa mix, powder, Alba—*19.14 g pkt (19 g)* — 56

dairy light, Alba—*2.8 g pkt (2.8 g)* — 16

fit 'n frosty, Alba

choc—*21.26 g pkt (21 g)* — 56

choc marshmallow—*21.26 g pkt (21 g)* — 56

double fudge—*21.26 g pkt (21 g)* — 56

strawberry—*21.26 g pkt (21 g)* — 49

van—*21.26 g pkt (21 g)* — 41

malted milk flavor mix w/ added nutrients, powder—¾
oz (4-5 hp t) (21 g) — 178

SALAD DRESSINGS

buttermilk farm style, from mix, Good Seasons—*1 T
(16 g)* — 3

SAUCES

horseradish sce, Sauceworks—*1 T (14 g)* — 1

SPECIAL DIETARY FOODS

figurine diet bar, Pillsbury

choc—*1 bar (21 g)* — 10

choc caramel—*1 bar (21 g)* — 10

choc peanut butter—*1 bar (21 g)* — 10

s'mores—*1 bar (21 g)* — 10

van—*1 bar (21 g)* — 10

fortified pudding, from mix, Delmark

choc—*½ cup (164 g)* — 72

van/lemon—*½ cup (164 g)* — 72

gluten-free, wheat-free foods, Ener-G Foods

nutquik milk powder—*1 oz (28 g)* — 3

slender bars, Carnation

choc—*2 bars (55 g)* — 100

choc chip—*2 bars (55 g)* — 100

SPECIAL DIETARY FOODS

choc peanut butter—*2 bars (55 g)* — 100

van—*2 bars (55 g)* — 100

slender, liquid diet meal, cnd, Carnation

banana—*10 fl oz (313 g)* — 100

choc—*10 fl oz (313 g)* — 100

choc fudge—*10 fl oz (313 g)* — 100

choc malt—*10 fl oz (313 g)* — 100

milk choc—*10 fl oz (313 g)* — 100

strawberry—*10 fl oz (313 g)* — 100

van—*10 fl oz (313 g)* — 100

SPREADS

margarine

Mazola—*1 T (14 g)* — 60

Mazola unsalted—*1 T (14 g)* — 60

Nucoa—*1 T (14 g)* — 60

Nucoa soft—*1 T (13 g)* — 60

margarine, imitation (diet), Mazola diet—*1 T (14 g)* — 60

miracle whip, light, Kraft—*1 T (14 g)* — 1

VEGETABLES

broccoli, frzn

w/ cheese sce, Birds Eye—*½ cup (142 g)* — 2

w/ creamy italian cheese sce, Birds Eye—*½ cup (128 g)* — 1

brussel sprouts, frzn, w/ cheese sce, Birds Eye—*½ cup
(128 g)* — 1

cauliflower, frzn, w/ cheese sce, Birds Eye—*½ cup (142 g)* — 2

rice, white, from mix, Minute—*⅔ cup (124 g)* — 2

for rib roast—*½ cup (117 g)* — 4

w/ chicken flavor—*½ cup (107 g)* — 4

rice, long grain & wild, from mix, Minute—*½ cup (115 g)* — 4

vegetable combinations, frzn, Birds Eye

broccoli & cauliflower, w/ creamy italian cheese sce—
½ cup (128 g) — 1

broccoli, cauliflower & carrots, w/ cheese sce—*½ cup
(142 g)* — 1

green beans, bavarian style & spaetzle (noodles)—
½ cup (95 g) — 2

pasta primavera style—*½ cup (95 g)* — 1

peas & pearl onions, w/ cheese sce—*½ cup (142 g)* — 1

MISCELLANEOUS

shake & bake, for chicken—*¼ pouch (19 g)* — 6

[a] Prepared w/ vitamin D fortified milk.

Vitamin E (IU)

BEVERAGES
coffee, prep from inst powder

francais, sugar-free, General Foods—*6 fl oz water & 7.7 g powder (185 g)*	.06
irish creme, sugar-free, General Foods—*6 fl oz water & 7.1 g powder (185 g)*	.06
irish mocha mint, sugar-free, General Foods—*6 fl oz water & 6.4 g powder (184 g)*	.07
orange cappuccino, sugar-free, General Foods—*6 fl oz water & 6.7 g powder (184 g)*	.06
suisse mocha, sugar-free, General Foods—*6 fl oz water & 6.4 g powder (184 g)*	.06
vienna, sugar-free, General Foods—*6 fl oz water & 6.7 g powder (184 g)*	.06

CANDY

chocolate, semi-sweet, Bakers—*1 oz square (28 g)*	.09
choc chips	
Bakers—*¼ cup (43 g)*	.13
german sweet, Bakers—*¼ cup (43 g)*	.19
semi-sweet, Bakers—*¼ cup (43 g)*	.19
choc fudgies, Kraft—*1 piece (8 g)*	.10

CEREALS, READY-TO-EAT

king vitamin, Quaker—*1¼ cups (1 oz) (28 g)*	9.00
Quaker 100% natural—*¼ cup (1 oz) (28 g)*	1.00
w/ raisins & dates—*¼ cup (1 oz) (28 g)*	1.00

CHEESE

cheddar, med, Kraft—*1 oz (28 g)*	.10
cream cheese, Philadelphia Brand—*1 oz (28 g)*	.10
cream cheese w/ herb & garlic, Cremerie—*1 oz (28 g)*	.10

CHEESE PRODUCTS

cheese food, american, grated, Kraft—*1 oz (28 g)*	.20
cheese products	
Light n' Lively, singles—*1 oz (28 g)*	.10
Light n' Lively, swiss—*1 oz (28 g)*	.10

CREAMS & CREAM SUBSTITUTES
whipped topping

from mix, Dream Whip—*1 T (5 g)*	.07
from mix, w/ nutrasweet, D-Zerta—*1 T (from 1.112 g dry mix) (5 g)*	.07

DESSERTS
Cakes

cheesecake, from mix, Jell-O—*⅛ cheesecake (99 g)*	.39
coffee cake, mix, Aunt Jemima—*1.3 oz dry mix (37 g)*	.20

Frozen Desserts
bon bon ice cream nuggets, Carnation

bavarian mint—*5 nuggets (48 g)*	.02
choc—*5 nuggets (49 g)*	.01
choc peanut butter—*5 nuggets (49 g)*	.01
van—*5 nuggets (48 g)*	.02
fruit juice bar, all flavors, Jell-O—*1 bar (52 g)*	.02
pudding pops, Jell-O	
choc w/ choc coating—*1 pop (49 g)*	.20
van w/ choc coating—*1 pop (49 g)*	.18

DESSERTS
Granola Desserts

choc chip, chewy, Quaker—*1 oz bar (28 g)*	1.00
choc, graham & marshmallow, chewy, Quaker—*1 oz bar (28 g)*	1.00
chunky nut & raisin, chewy, Quaker—*1 oz bar (28 g)*	1.00
honey & oats, chewy, Quaker—*1 oz bar (28 g)*	1.00
mint choc chip dipps, Quaker—*1 oz bar (28 g)*	1.00
peanut butter & choc chip, chewy, Quaker—*1 oz bar (28 g)*	1.00
peanut butter, chewy, Quaker—*1 oz bar (28 g)*	1.00
peanut butter dipps, Quaker—*1 oz bar (28 g)*	1.00
raisin & almond dipps, Quaker—*1 oz bar (28 g)*	1.00
raisin & cinn, chewy, Quaker—*1 oz bar (28 g)*	2.00

Pies, from mix, Jell-O

banana cream—*⅛ pie (92 g)*	1.07
choc mousse—*⅛ pie (95 g)*	.77
coconut cream—*⅛ pie (94 g)*	.70

Pie Crust

mix, Flako—*1.7 oz dry mix (48 g)*	.20

Puddings, from inst mix
banana cream

Jell-O—*½ cup (147 g)*	.22
sugar-free, Jell-O—*½ cup (130 g)*	.01
butter pecan, Jell-O—*½ cup (147 g)*	.29
butterscotch, Jell-O—*½ cup (147 g)*	.22
choc	
Jell-O—*½ cup (150 g)*	.21
sugar-free, D-Zerta—*½ cup (130 g)*	.03
choc fudge	
Jell-O—*½ cup (150 g)*	.23
sugar-free, Jell-O—*½ cup (135 g)*	.11
choc fudge mousse, Jell-O—*½ cup (86 g)*	.52
choc mousse, Jell-O—*½ cup (86 g)*	.50
choc tapioca, Jell-O—*½ cup (147 g)*	.19
coconut cream, Jell-O—*½ cup (147 g)*	.12
french van, Jell-O—*½ cup (147 g)*	.22
golden egg custard, Jell-O—*½ cup (143 g)*	.13
lemon, Jell-O—*½ cup (147 g)*	.13
milk choc, Jell-O—*½ cup (150 g)*	.23
pineapple cream, Jell-O—*½ cup (147 g)*	.22
pistachio	
Jell-O—*½ cup (147 g)*	.29
sugar-free, Jell-O—*½ cup (131 g)*	.08
vanilla, Jell-O—*½ cup (147 g)*	.21

Sauces, Syrups & Toppings for Desserts

fudge topping, Kraft—*1 T (20 g)*	.20

ENTREES
Canned Entrees

beef & veg stew, Wolf Brand—*scant cup (213 g)*	2.00
chili-mac, Wolf Brand—*scant cup (213 g)*	2.00
chili w/ beans	
Wolf Brand—*1 cup (227 g)*	2.00
Wolf Brand, extra spicy—*scant cup (213 g)*	2.00

ENTREES

chili w/o beans

Wolf Brand—*1 cup (227 g)*	2.00
Wolf Brand, extra spicy—*scant cup (213 g)*	2.00
tamales, Wolf Brand—*scant cup (213 g)*	2.00

Frozen Entrees

pizza, canadian style bacon

Celeste—*7.75 oz pizza (220 g)*	4.00
Celeste—*¼ of 19 oz pizza (135 g)*	3.00

pizza, cheese

Celeste—*6.5 oz pizza (184 g)*	2.00
Celeste—*¼ of 18 oz pizza (126 g)*	1.00

pizza, deluxe

Celeste—*8.25 oz pizza (234 g)*	2.00
Celeste—*¼ of 22 oz pizza (158 g)*	3.00

pizza, pepperoni

Celeste—*6.75 oz pizza (191 g)*	4.00
Celeste—*¼ of 19 oz pizza (135 g)*	3.00

pizza, sausage

Celeste—*7.5 oz pizza (213 g)*	4.00
Celeste—*¼ of 20 oz pizza (142 g)*	3.00

pizza, sausage & mushroom

Celeste—*8.5 oz pizza (241 g)*	5.00
Celeste—*¼ of 25 oz pizza (177 g)*	3.00

pizza, suprema

Celeste—*9 oz pizza (255 g)*	3.00
Celeste—*¼ of 23 oz pizza (163 g)*	3.00

GRAIN FRACTIONS

corn germ, Ener-G Foods—*1 cup (100 g)*	6.00

corn meal mix

white, bolted, Aunt Jemima—*1/6 cup (1 oz) (28 g)*	1.00
corn meal, white bolted, enr, self-rising, Aunt Jemima—*1/6 cup (1 oz) (28 g)*	1.00
corn meal, white, degermed, enr, Quaker/Aunt Jemima—*3 T (1 oz) (28 g)*	1.00
corn meal, yellow, degermed, enr, Quaker/Aunt Jemima—*3 T (1 oz) (28 g)*	1.00

wheat germ

toasted—*¼ cup (1 oz) (28 g)*	6.00
toasted w/ brown sugar & honey—*¼ cup (1 oz) (28 g)*	4.30

GRAIN PRODUCTS

Breads & Bread Products

corn bread, mix, Aunt Jemima—*1.7 oz dry mix (48 g)*	.60

French Toast. frzn

cinn swirl, Aunt Jemima—*2 slices (85 g)*	3.00
raisin, Aunt Jemima—*2 slices (85 g)*	4.00

Muffins

corn, muffin mix, Flako—*1 oz dry mix (28 g)*	.20

Pancakes

from frzn batter. Aunt Jemima

buttermilk—*three 4″ pancakes (113 g)*	2.00
plain—*three 4″ pancakes (113 g)*	1.00

frzn, Aunt Jemima

blueberry—*three 4″ pancakes (106 g)*	1.00
buttermilk—*three 4″ pancakes (106 g)*	1.00
plain—*three 4″ pancakes (106 g)*	1.00

pancake/waffle mix, Aunt Jemima

buckwheat—*¼ cup dry mix (31 g)*	.10
buttermilk—*⅓ cup dry mix (51 g)*	.20

GRAIN PRODUCTS

buttermilk, complete—*½ cup dry mix (71 g)*	.50
plain—*¼ cup dry mix (31 g)*	.20
plain, complete—*½ cup dry mix (74 g)*	.50
whole wheat—*⅓ cup dry mix (43 g)*	3.00

Rolls

popover mix, Flako—*1 oz dry mix (28 g)*	.10

Stuffing

from mix, Stove Top

americana san francisco style—*½ cup (107 g)*	.59
chicken flavored—*½ cup (107 g)*	.73
chicken flavored w/ rice—*½ cup (109 g)*	.71
chicken florentine—*½ cup (102 g)*	1.10
for beef—*½ cup (108 g)*	.73
for pork—*½ cup (107 g)*	.73
garden herb—*½ cup (95 g)*	.78
homestyle herb—*½ cup (96 g)*	2.11
long grain & wild rice flavor—*½ cup (109 g)*	.71
savory herbs flavor—*½ cup (107 g)*	.73
turkey flavor—*½ cup (107 g)*	.72
veg & almond—*½ cup (95 g)*	1.22
wild rice & mushroom—*½ cup (95 g)*	.82

MEATS, LUNCHEON

meat spread, Carnation

chicken—*1.9 oz (53 g)*	.72
ham—*1.9 oz (53 g)*	.43
tuna—*1.9 oz (53 g)*	.70
turkey—*1.9 oz (53 g)*	.71

MILK BEVERAGES

eggnog, nonalcoholic

mix in whole milk, Delmark—*1.23 oz mix in 8 fl oz milk (257 g)*	.10
instant breakfast in whole milk—*1 pkt in 8 fl oz milk (265 g)*	7.80
choc, Delmark—*1.23 oz mix in 8 fl oz milk (279 g)*	9.90
van/strawberry, Delmark—*1.23 oz mix in 8 fl oz milk (279 g)*	9.90
instant breakfast, van/strawberry, liquid pack, Delmark—*8 fl oz (279 g)*	7.50

milk shake

mix in whole milk, Delmark

choc—*.667 oz mix in 8 fl oz milk (263 g)*	.10
strawberry—*.667 oz mix in 8 fl oz milk (263 g)*	.10
van—*.667 oz mix in 8 fl oz milk (263 g)*	.10

MILK BEVERAGE MIXES

instant breakfast, Carnation

choc—*1.25 oz pkt (35 g)*	7.50
choc malt—*1.24 oz pkt (35 g)*	7.50
coffee—*1.26 oz pkt (36 g)*	7.50
eggnog—*1.20 oz pkt (34 g)*	7.50
strawberry—*1.25 oz pkt (35 g)*	7.50
van—*1.23 oz pkt (35 g)*	7.50

instant breakfast, no sugar added, Carnation

choc—*.69 oz pkt (20 g)*	7.50
choc malted—*.71 oz pkt (20 g)*	7.50
strawberry—*.68 oz pkt (19 g)*	7.50
van—*.67 oz pkt (19 g)*	7.50

NUTS

almonds

blanched, Blue Diamond—*1 oz (28 g)*	8.75
blanched, slivered, Blue Diamond—*1 oz (28 g)*	8.50
chopped, Blue Diamond—*1 oz (28 g)*	8.50
sliced, Blue Diamond—*1 oz (28 g)*	8.50
whole, Blue Diamond—*1 oz (28 g)*	10.10
filberts (hazelnuts), roasted, salted, Blue Diamond—*1 oz (28 g)*	4.40
macadamia nuts, dry roasted, Blue Diamond—*1 oz (28 g)*	.30
pistachio nuts, dry roasted, Blue Diamond—*1 oz (28 g)*	.60

SALAD DRESSINGS

french, Catalina, Kraft—*1 T (15 g)* 1.00

from mix, Good Seasons

bleu cheese & herbs—*1 T (16 g)*	.74
buttermilk farm style—*1 T (16 g)*	4.09
cheese garlic—*1 T (16 g)*	2.23
cheese italian—*1 T (16 g)*	7.49
classic herb—*1 T (16 g)*	2.29
garlic & herbs—*1 T (16 g)*	2.22
italian—*1 T (16 g)*	2.26
italian, lite—*1 T (15 g)*	2.09
italian, lite, zesty—*1 T (15 g)*	.63
italian, mild—*1 T (17 g)*	2.30
italian, zesty—*1 T (16 g)*	2.26
lemon & herbs—*1 T (16 g)*	2.26

SAUCES

barbeque sce, Open Pit—*1 T (18 g)* .... .12

hickory smoke flavor—*1 T (18 g)*	.12
hickory thick & tangy—*1 T (18 g)*	.12
hot 'n tangy flavor—*1 T (18 g)*	.12
mesquite 'n tangy—*1 T (18 g)*	.12
sweet 'n tangy flavor—*1 T (18 g)*	.10
w/ onions—*1 T (18 g)*	.12

chili hot dog sce, Wolf Brand—*1/6 cup (35 g)* 1.00

SPECIAL DIETARY FOODS

breakfast bar, Carnation

choc chip—*1.44 oz bar (41 g)*	7.50
choc crunch—*1.34 oz bar (38 g)*	7.50
peanut butter crunch—*1.35 oz bar (38 g)*	7.50
peanut butter w/ choc chips—*1.39 oz bar (39 g)*	7.50

figurine diet bar, Pillsbury

choc—*1 bar (21 g)*	4.18
choc caramel—*1 bar (21 g)*	4.18
choc peanut butter—*1 bar (21 g)*	4.18
s'mores—*1 bar (21 g)*	4.20
van—*1 bar (21 g)*	4.20

fortified pudding, from mix, Delmark

choc—*½ cup (164 g)*	5.40
van/lemon—*½ cup (164 g)*	5.40

gluten-free, wheat-free foods, Ener-G Foods

nutquik milk powder—*1 oz (28 g)* 4.28

slender bars, Carnation

choc—*2 bars (55 g)*	7.50
choc chip—*2 bars (55 g)*	7.50
choc peanut butter—*2 bars (55 g)*	7.50
van—*2 bars (55 g)*	7.50

slender, liquid diet meal, cnd, Carnation

banana—*10 fl oz (313 g)*	7.50
choc—*10 fl oz (313 g)*	7.50

SPECIAL DIETARY FOODS

choc fudge—*10 fl oz (313 g)*	7.50
choc malt—*10 fl oz (313 g)*	7.50
milk choc—*10 fl oz (313 g)*	7.50
strawberry—*10 fl oz (313 g)*	7.50
van—*10 fl oz (313 g)*	7.50

VEGETABLES

artichoke, frzn, boiled—*2.8 oz (80 g)* .... .22

broccoli, frzn

chopped, boiled—*½ cup (92 g)*	.65
spears, boiled—*½ cup (92 g)*	.65
w/ cheese sce, Birds Eye—*½ cup (142 g)*	1.86
w/ creamy italian cheese sce, Birds Eye—*½ cup (128 g)*	.97

brussel sprouts, frzn

boiled—*½ cup (78 g)*	1.02
w/ cheese sce, Birds Eye—*½ cup (128 g)*	1.16

carrots, frzn, boiled—*½ cup slices (73 g)* .... .50

cauliflower, frzn

boiled—*½ cup pieces (90 g)*	.05
w/ cheese sce, Birds Eye—*½ cup (142 g)*	1.36

corn, yellow, frzn, boiled—*½ cup (82 g)* .... .11

green beans (snap beans), frzn

boiled—*½ cup (68 g)*	.17
cut, Birds Eye—*½ cup (85 g)*	.21
french style, Birds Eye—*½ cup (85 g)*	.21
french style w/ toasted almonds, Birds Eye—*½ cup (85 g)*	.71
whole, Birds Eye—*4 oz (113 g)*	.28

mixed veg, frzn, Birds Eye—*½ cup (95 g)*	.35
onions, frzn, small, whole, Birds Eye—*½ cup (113 g)*	.21
peas, green, frzn, boiled—*½ cup (80 g)*	.33
rice, white, from mix, Minute—*⅔ cup (124 g)*	.05
for rib roast—*½ cup (117 g)*	.15
w/ chicken flavor—*½ cup (107 g)*	.16

rice, long grain & wild

from mix, Minute—*½ cup (115 g)* .... .15

spinach, frzn, boiled—*½ cup (95 g)* 2.70

vegetable combinations, frzn, Birds Eye

broccoli & cauliflower, w/ creamy italian cheese sce—*½ cup (128 g)*	.70
broccoli, carrots & pasta w/ lightly seasoned sce—*⅔ cup (95 g)*	1.76
broccoli, cauliflower & carrots—*¾ cup (113 g)*	.54
w/ cheese sce—*½ cup (142 g)*	1.45
broccoli, corn & red peppers—*⅔ cup (113 g)*	.53
broccoli, green beans, pearl onions & red peppers—*¾ cup (113 g)*	.61
brussel sprouts, cauliflower & carrots—*¾ cup (113 g)*	.88
carrots, peas & pearl onions—*½ cup (95 g)*	.51
chinese style—*½ cup (95 g)*	1.29
chinese style, stir fry—*½ cup (95 g)*	.67
corn, green beans & pasta curls w/ light cream sce—*½ cup (95 g)*	1.39
green beans, bavarian style & spaetzle (noodles)—*½ cup (95 g)*	1.60
italian style—*½ cup (95 g)*	1.80
japanese style—*½ cup (95 g)*	1.79
japanese style, stir fry—*½ cup (95 g)*	.42
new england style—*½ cup (95 g)*	2.78
peas & pearl onions—*½ cup (95 g)*	.32

VEGETABLES

peas & pearl onions w/ cheese sce—*½ cup (142 g)*	1.32
rice, italian style—*½ cup (95 g)*	.17
rice, spanish style—*½ cup (95 g)*	.15
san francisco style—*½ cup (95 g)*	1.17

MISCELLANEOUS

baking choc, unsweetened, Bakers—*1 oz (28 g)*	1.40
meat/poultry coatings	
oven fry	
extra crispy recipe for chicken—*¼ pouch (29 g)*	1.48

MISCELLANEOUS

extra crispy recipe for pork—*¼ pouch (29 g)*	1.57
light crispy homestyle recipe—*¼ pouch (22 g)*	.81
shake & bake	
barbeque for chicken—*¼ pouch (25 g)*	.85
barbeque for pork—*¼ pouch (20 g)*	.68
country mild recipe—*¼ pouch (17 g)*	2.03
for chicken—*¼ pouch (19 g)*	1.00
for fish—*¼ pouch (19 g)*	.95
for pork—*¼ pouch (21 g)*	.52
italian herb recipe—*¼ pouch (20 g)*	.63

Vitamin E as Alpha-Tocopherol (mg)

CHIPS & SNACKS

potato chips—*1 oz (28 g)*	1.20
potato sticks—*1 oz (28 g)*	2.23

EGGS, CHICKEN

whole, fresh/frzn—*1 large (50 g)*	0.88
yolk, fresh—*yolk of 1 large egg (17 g)*	0.87

ENTREES, BOX MIX

pizza, cheese, from Contadina Pizzeria Kit

thick crust—*¼ pizza (128 g)*	0.14
thin crust—*¼ pizza (104 g)*	0.14

FATS, OILS & SHORTENINGS
Animal Fats

beef tallow, raw—*1 T (13 g)*	0.30
pork fat (lard), raw—*1 T (13 g)*	0.20

Vegetable Oils

almond oil—*1 T (14 g)*	5.30
coconut oil—*1 T (14 g)*	0.10
corn oil—*1 T (14 g)*	1.90
corn oil, Mazola—*1 T (14 g)*	3.00
cottonseed oil—*1 T (14 g)*	4.80
olive oil—*1 T (14 g)*	1.60
palm oil—*1 T (14 g)*	2.60
peanut oil—*1 T (14 g)*	1.60
safflower oil—*1 T (14 g)*	4.60
sesame oil—*1 T (14 g)*	0.20
soybean oil—*1 T (14 g)*	1.50
soybean oil, hydrogenated—*1 T (14 g)*	1.10
sunflower oil—*1 T (14 g)*	6.10
veg oil spray, Mazola No Stick—*2.5 sec spray (0.7 g)*	0.51[a]
wheat germ oil—*1 T (14 g)*	20.30

FRUIT & VEGETABLE JUICES

apple jce, cnd/bottled—*8 fl oz (248 g)*	0.03
grapefruit jce, cnd—*8 fl oz (247 g)*	0.10
orange jce, fresh—*8 fl oz (248 g)*	0.10
tomato jce—*6 fl oz (182 g)*	0.40

FRUITS

apple

raw, w/ skin—*1 med (138 g)*	0.81
raw, w/o skin—*1 med (128 g)*	0.35
apricots, cnd, in heavy syrup—*4 halves (90 g)*	0.80
banana, raw—*1 med (114 g)*	0.31
blackberries, raw—*½ cup (72 g)*	0.35
cantaloupe, raw—*1 cup pieces (160 g)*	0.22
cherries, sour, raw—*½ cup (78 g)*	0.10
currants, european black, raw—*½ cup (56 g)*	0.56
currants, red & white, raw—*½ cup (56 g)*	0.06
gooseberries, raw—*1 cup (150 g)*	0.56
grapefruit, raw, red & white—*½ med (123 g)*	0.30
mango, raw—*1 med (207 g)*	2.32
mixed fruit, frzn, in syrup, Birds Eye—*½ cup (142 g)*	0.06
orange, navel or valencia, raw—*1 fruit (131 g)*	0.30
pear, raw—*1 med (166 g)*	0.83
pineapple, raw—*1 cup pieces (155 g)*	0.16
raspberries	
raw—*1 cup (123 g)*	0.37
frzn, in lite syrup, Birds Eye—*½ cup (142 g)*	0.27
strawberries	
raw—*1 cup (149 g)*	0.18
frzn, in lite syrup, Birds Eye—*½ cup (142 g)*	0.13
frzn, sweetened or unsweetened—*1 cup (149 g)*	0.31

GRAIN PRODUCTS
Pasta

macaroni, enr, ckd—*1 cup (140 g)*	1.03[a]
spaghetti, enr, ckd—*1 cup (140 g)*	1.03[a]

NUTS, NUT PRODUCTS & SEEDS

almonds

dried—*1 oz (24 nuts) (28 g)*	6.72
oil roasted—*1 oz (22 nuts) (28 g)*	1.55
toasted—*1 oz (28 g)*	1.41
whole, Blue Diamond—*1 oz (28 g)*	1.66
brazilnuts, dried—*1 oz (8 med nuts) (28 g)*	2.13
cashews, dry roasted—*1 oz (28 g)*	0.16

[a] Specified as tocopherols.

NUTS, NUT PRODUCTS & SEEDS

coconut, raw—*1 piece (2″x2″x½″) (45 g)*	0.33
filberts (hazelnuts), dried—*1 oz (28 g)*	6.70
peanut butter, creamy/smooth, Skippy—*1 T (16 g)*	3.00
peanut butter, chunk style/crunchy, Skippy—*1 T (16 g)*	3.00
peanuts	
dried—*1 oz (28 g)*	2.56
dry roasted—*1 oz (28 g)*	2.18
oil roasted—*1 oz (28 g)*	2.07
pecans, dried—*1 oz (31 large nuts) (28 g)*	0.87
pistachio nuts, dried—*1 oz (47 nuts) (28 g)*	1.46
sesame seeds, whole, dried—*1 T (9 g)*	0.20
walnuts, english/persian, dried—*1 oz (14 halves) (28 g)*	0.73

SPREADS

butter—*1 T (15 g)*	0.20
margarine by brand	
Mazola—*1 T (14 g)*	8.00
Mazola unsalted—*1 T (14 g)*	8.00
margarine by form & type of oil	
liquid, soybean & cottonseed—*1 t (5 g)*	0.20
stick, safflower & soybean—*1 t (5 g)*	0.80
stick, soybean—*1 t (5 g)*	0.10
stick, soybean & cottonseed—*1 t (5 g)*	0.30
tub, corn—*1 t (5 g)*	0.50
tub, safflower—*1 t (5 g)*	0.60
tub, soybean—*1 t (5 g)*	0.10
tub, soybean & cottonseed—*1 t (5 g)*	0.30
margarine, imitation (diet) by brand	
Mazola diet—*1 T (14 g)*	3.00
Parkay, diet soft—*1 T (14 g)*	0.40
margarine, imitation (diet) by form & type of oil	
tub, soybean & cottonseed—*1 t (5 g)*	0.40
mayonnaise	
Best Foods/Hellmann's—*1 T (14 g)*	11.00
soybean—*1 T (14 g)*	2.90
miracle whip, Kraft—*1 T (14 g)*	0.50
miracle whip, light, Kraft—*1 T (14 g)*	0.40
sandwich spread, Best Foods/Hellmann's—*1 T (15 g)*	5.00

VEGETABLES

asparagus	
cnd—*½ cup (121 g)*	0.46
frzn, boiled—*4 spears (60 g)*	0.81
raw—*4 spears (58 g)*	1.15
avocado, raw, calif—*1 med (173 g)*	2.32
beet greens, raw—*1 cup (38 g)*	0.57
beets, cnd, harvard—*½ cup slices (123 g)*	0.04
broccoli, raw—*½ cup chopped (44 g)*	0.20
brussels sprouts	
raw—*½ cup chopped (44 g)*	0.39
boiled—*½ cup (4 sprouts) (78 g)*	0.66

VEGETABLES

cabbage, chinese (pak-choi), raw—*½ cup shredded (35 g)*	0.05
cabbage, green, raw—*½ cup shredded (35 g)*	0.58
carrots	
raw—*1 med (72 g)*	0.32
boiled—*½ cup slices (78 g)*	0.33
cauliflower, raw—*½ cup pieces (50 g)*	0.02
celery, raw—*1 stalk (7.5″ long) (40 g)*	0.14
corn, sweet, yellow/white, cnd—*½ cup (128 g)*	0.05
corn, sweet, yellow/white, frzn—*½ cup (82 g)*	0.02
cucumber, raw—*½ cup slices (1/6 cucumber) (52 g)*	0.08
dandelion greens, raw—*½ cup chopped (28 g)*	0.70
eggplant, raw—*½ cup pieces (41 g)*	0.01
garden cress, raw—*½ cup (25 g)*	0.18
garlic, raw—*3 cloves (9 g)*	0.001
green beans (snap beans)	
raw—*½ cup (55 g)*	0.01
cnd—*½ cup (68 g)*	0.03
frzn—*½ cup (62 g)*	0.06
frzn, boiled—*½ cup (68 g)*	0.09
leeks, raw—*¼ cup chopped (26 g)*	0.24
lettuce, iceberg, raw—*¼ head (135 g)*	0.54
mushrooms, raw—*½ cup pieces (35 g)*	0.03
mustard greens, raw—*½ cup chopped (28 g)*	0.56
onion rings, frzn, heated—*7 rings (70 g)*	0.48
onions, raw—*½ cup chopped (80 g)*	0.25
parsley, raw—*½ cup chopped (30 g)*	0.52
parsnips, raw—*½ cup (67 g)*	0.67
peas, green	
raw—*½ cup (78 g)*	0.10
frzn—*½ cup (72 g)*	0.09
frzn, boiled—*½ cup (80 g)*	0.10
peppers, sweet, raw—*½ cup chopped (50 g)*	0.34
potato	
raw w/o skin—*1 potato (112 g)*	0.07
baked w/o skin—*1 potato (156 g)*	0.05
boiled w/o skin—*1 potato (135 g)*	0.05
french fried, frzn, heated—*10 pieces (50 g)*	0.10
pumpkin, raw—*½ cup (58 g)*	0.58
rutabaga, boiled—*½ cup cubes (85 g)*	0.13
seaweed, kelp (kombu/tangle), raw—*3.5 oz (100 g)*	0.87
spinach	
raw—*½ cup chopped (28 g)*	0.53
cnd—*½ cup (107 g)*	0.02
squash, winter, all varieties, baked—*½ cup cubes (102 g)*	0.12
sweet potato, raw—*1 med (130 g)*	5.93
tomato, red, raw—*1 tomato (123 g)*	0.42
turnip greens, raw—*½ cup chopped (28 g)*	0.63
watercress, raw—*½ cup chopped (17 g)*	0.17

Vitamin K (mcg/100 g)

BEVERAGES	
coffee, dry	38
tea, green, dry	712
EGGS	
whole, fresh/frzn	50
yolk, fresh	147
FATS & OILS	
Vegetable Oils	
coconut oil	10
corn oil	60
cottonseed oil	0
olive oil	0
palm oil	8
peanut oil	0
safflower, linoleic	0
safflower, oleic	3
soybean oil	540
FRUITS	
apple, raw w/ skin	3
orange, raw	5
strawberries, raw	14
GRAINS	
oats, dry	63
wheat bran	83
wheat flour, whole wheat	30
wheat germ	39
INFANT FORMULA[a]	
concentrate, reconstituted	13
powder, reconstituted	15
ready-to-feed	13
MEATS	
Beef	
ground, regular, raw	4
Variety Cuts	
heart, beef, raw	0
kidneys, beef, raw	0
liver, raw	
beef	104
chicken	80
lamb	0
pigeon	0
pork	88
rabbit	35
turkey	0
veal	27
MILK, COW	
non-fat dry,reg	10
skim	4
whole, 3.7% fat	4

MILK, OTHER	
human	2
SWEETENERS	
honey	25
VEGETABLES	
asparagus spears	
frzn, boiled	27
raw	39
beet, raw	5
broccoli	
raw	132
frzn	68
cabbage, green, raw	149
carrot, raw	13
cauliflower, raw	191
chickpeas (garbanzo beans)	
mature seeds, dry	264
sprouted seeds, raw	48
corn, yellow, raw	7
cucumber, raw	5
green beans (snap beans)	
frzn, boiled	32
raw	28
lentils, mature seeds, dry	223
lentils, sprouted seeds, raw	39
lettuce, iceberg, raw	112
mung beans	
mature seeds, dry	170
sprouted seeds, raw	33
mushrooms, raw	8
nettle leaves, raw	372
peas, mature seeds, dry	81
peas, sprouted seeds, raw	28
potato	
raw	16
baked	4
seaweed	
dulse, dried	1700
rockweed, dried	1700
seagrass	246
sealettuce	68
soybeans, mature, raw	190
spinach	
raw	266
frzn	138
tomato, green, raw	47
tomato, red, raw	23
turnip greens, raw	650
watercress, raw	57

[a] Vitamin K values are listed in the main table for specific infant formulas and for special dietary formulas, commercial and hospital.

Sources of Data for the Tables

Agriculture Handbook No. 8, Composition of Foods, Raw, Processed, Prepared, U.S. Department of Agriculture, Washington, D.C., 1963.

Agriculture Handbook No. 8, Revised, Composition of Foods, Raw, Processed, Prepared, U.S. Department of Agriculture, Washington, D.C.

8-1	Dairy and Egg Products, November 1976
8-2	Spices and Herbs, January 1977
8-3	Baby Foods, December 1978
8-4	Fats and Oils, June 1979
8-5	Poultry Products, August 1979
8-6	Soups, Sauces and Gravies, February 1980
8-7	Sausages and Luncheon Meats, September 1980
8-8	Breakfast Cereals, July 1982
8-9	Fruits and Fruit Juices, August 1982
8-10	Pork Products, August 1983
8-11	Vegetables and Vegetable Products, August 1984
8-12	Nut and Seed Products, September 1984
8-13	Beef Products, August 1986
8-14	Beverages, May 1986
8-15	Finfish and Shellfish Products, September 1987
8-16	Legumes and Legume Products, December 1986

Agriculture Handbook No. 456, Nutritive Value of American Foods in Common Units, U.S. Department of Agriculture, 1975.

Fletcher, D. C., Do clotting factors in vitamin K-rich vegetables hinder anticoagulant therapy? (Questions and Answers), JAMA, 237(17):1871, 1977.

Swain, A. R., S. P. Dutton and A. S. Truswell, Salicylates in foods, J. Am. Diet. Assoc. 85(8):950–960, 1985.

U.S. Department of Agriculture Provisional Tables, Washington, D.C.
Fatty Acid and Cholesterol Content of Selected Foods, March 1984
Nutrient Content of Bakery Foods & Related Items, August 1981
Nutrient Content of Beverages, September 1981
Nutrient Content of Fast Foods, June 1984
Sugar Content of Selected Foods, October 1986
Vitamin K Content of Foods, June 1986

Information provided upon request by the food industry (individual companies and trade associations)

Previous editions of Food Values of Portions Commonly Used

Other Sources of Food Composition Data

Additional references are provided in three sections:

- References by substance (references for one major nutrient/compound in a food or foods)

- References by food group (references for more than one substance in a food or group of similar foods)

- References that report various substances in various foods

Most of these references are relatively recent: 74% are from the 1980s, 22% are from the 1970s, and 4% were published prior to 1970.

REFERENCES BY SUBSTANCE

Biotin

Evans, R. J., J. A. Davidson, D. Bauer and H. A. Butts, The biotin content of fresh and stored shell eggs, Poultry Sci. 32:680–683, 1953.

Guilarte, T. R., Analysis of biotin levels in selected foods using a radiometric microbiological method, Nutr. Rep. Inter. 32(4):837–845, 1985.

Caffeine and Theobromine

Ashoor, S. H., G. J. Seperich, W. C. Monte and J. Welty, High performance liquid chromatographic determination of caffeine in decaffeinated coffee, tea, and beverage products, J. Assoc. Off. Anal. Chem. 66(3):606–609, 1983.

Blauch, J. L. and S. M. Tarka Jr., HPLC determination of caffeine and theobromine in coffee, tea, and instant hot cocoa mixes, J. Food Sci. 48:745–747, 750, 1983.

Bunker, M. L. and M. McWilliams, Caffeine content of common beverages, J. Am. Diet. Assoc. 74:28–32, 1979.

Burg, A. W., Effects of caffeine on the human system, Tea and Coffee Trade J. 147:40–42, 88, 1975.

Caffeine and pregnancy, Office of Public Affairs, Food and Drug Administration, Public Health Service, U.S. Department of Health and Human Services, HHS Publication No. (FDA) 81–1081, 1981.

Caffeine: how to consume less, Consumer Reports 46(10):597–599, 1981.

Craig, W. J. and T. T. Nguyen, Caffeine and theobromine levels in cocoa and carob products, J. Food Sci. 49(1):302–303, 305, 1984.

Dulitzky, M., E. de la Tega and H. F. Lewis, Determination of caffeine in tea by high-performance liquid chromatography and a modified digestion procedure, J. Chromat. 317:403–405, 1984.

Graham, D. M., Caffeine—its identity, dietary sources, intake and biological effects, Nutr. Rev. 36(4):97–102, 1978.

Groisser, D. S., A study of caffeine in tea, Am. J. Clin. Nutr. 31:1727, 1978.

Hallal, J. C., Caffeine: is it hazardous to your patients' health?, Am. J. Nurs. 86(4):422–425, 1986.

Hsu, T. J., C. C. Liao and M. Y. Chen, Liquid chromatography of adenine, caffeine, theobromine, theophylline, tannic acid, xanthine and its application to tea analysis, J. Chin. Chem. Soc. 25(3):153–160, 1978.

Hurst, W. J., K. P. Snyder and R. A. Martin Jr., Use of microbore high-performance liquid chromatography for the determination of caffeine, theobromine and theophylline in cocoa, J. Chromatogr. 318(2):408–411, 1985.

Institute of Food Technologists' Expert Panel on Food Safety and Nutrition, Caffeine, Food Tech. 37(4):87–91, 1983.

Kreiser, W. R. and R. A. Martin Jr., High pressure liquid chromatographic determination of theobromine and caffeine in cocoa and chocolate products: collaborative study, J. Assoc. Off. Anal. Chem. 63(3):591–594, 1980.

Lecos, C., The latest caffeine scorecard, FDA Consumer, March 1984 (HHS Publication No. [FDA]84–2184).

Rajasekharan, T., A. S. V. Rao, K. O. Abraham and M. L. Shankaranarayana, Analysis of coffee–chicory blends, Indian Coffee 50(1):19–24, 1986.

Shivashankar, S., C. Balachandran, Y. S. Lewis and C. F. Natarajan, Determination of theobromine in cocoa products, J. Food Sci. Tech., India 15(4):153–161, 1978.

Sjoeberg, A. M. and J. Rajama, Simple method for the determination of alkaloids in cocoa using paper chromatography and UV spectrometry, J. Chromatogr. 295(1):291–294, 1984.

Terada, H. and Y. Sakabe, High-performance liquid chromatographic determination of theobromine, theophylline and caffeine in food products, J. Chromatogr. 291:453–459, 1984.

Trugo, L. C., R. Macrae and J. Dick, Determination of purine alkaloids and trigonelline in instant coffee and other beverages using high-performance liquid chromatography, J. Sci. Food Agr. 34(3):300–306, 1983.

Tyler, T. A., Liquid chromatographic determination of sodium saccharin, caffeine, aspartame, and sodium benzoate in cola beverages, J. Assoc. Off. Anal. Chem. 67(4):745–747, 1984.

Weidner, G. and J. Istvan, Dietary sources of caffeine (letter), N. Engl. J. Med. 313(22):1421, 1985.

What's in soft drinks, National Soft Drink Association, Washington, D.C., September 1982.

Zoumas, B. L., W. R. Kreiser and R. A. Martin, Theobromine and caffeine content of chocolate products, J. Food Sci. 45(2):314–316, 1980.

Carotenoids

Balasubramanian, T., Studies on quality and nutritional aspects of tomato, J. Food Sci. Tech., India 21(6):419–421, 1984.

Beecher, G. R. and F. Khachik, Evaluation of vitamin A and carotenoid data in food composition tables, J. Nat. Cancer Inst. 73(6):1397–1404, 1984.

Bushway, R. J., Determination of alpha- and beta-carotene in some raw fruits and vegetables by high-performance liquid chromatography, J. Agr. Food Chem. 34(3):409–412, 1986.

Edwards, C. G. and C. Y. Lee, Measurement of provitamin A carotenoids in fresh and canned carrots and green peas, J. Food Sci. 51(2):534–535, 1986.

Ezeala, D. O., Nutrients, carotenoids and mineral compositions of the leaf vegetables *Amaranthus viridis L.* and *A. caudatus L.,* Trop. Agr. 62(2):95–96. 1985.

Hsieh, Y. P. C. and M. Karel, Rapid extraction and determination of alpha- and beta-carotenes in foods, J. Chromatogr. 259(3):515–518, 1983.

Johjima, T. and H. Ogura, Analysis of tomato carotenoids by thin-layer chromatography and a cis-form gamma-carotene newly identified in tangerine tomato, J. Jap. Soc. Hort. Sci. 52(2):200–209, 1983.

Johnson, C. D., R. R. Eitenmiller, D. A. Lillard and M. Rao, Vitamin A activity of selected fruits, J. Am. Diet. Assoc. 85(12):1627–1629, 1985.

Landen, W. O. Jr., D. M. Hines, T. W. Hamill, J. I. Martin, E. R. Young, R. R. Eitenmiller and A. G. Soliman, Vitamin A and vitamin E content of infant formulas produced in the United States, J. Assoc. Off. Anal. Chem. 68(3):509–511, 1985.

Picha, D. H., Crude protein, minerals, and total carotenoids in sweet potatoes, J. Food Sci. 50(6):1768–1769, 1985.

Premachandra, B. R., A simple TLC method for the determination of provitamin A content of fruits and vegetables, Int. J. Vit. Nutr. Res. 55(2):139–147, 1985.

Takahashi, T., Vitamin A in new food composition tables from Japan, J. Nat. Cancer Inst. 73(6):1405–1407, 1984.

Cholesterol

Feeley, R. M., P. E. Criner and B. K. Watt, Cholesterol content of foods, J. Am. Diet. Assoc. 61:134–149, 1972.

Choline

Koning, A. J. de, Phospholipids of marine origin. V. The crab—a comparative study of a marine species (*Cyclograpsus punctatus*) and a fresh water species (*Potamon*), J. Sci. Food Agr. 21(6):290–293, 1970.

Parsons, J. G., P. G. Leeney and S. Patton, Identification and quantitative analysis of phospholipids in cocoa beans, J. Food Sci. 34(6):497–499, 1969.

Saucerman, J. R., C. E. Winstead and T. M. Jones, Quantitative gas chromatographic headspace determination of choline in adult and infant formula products, J. Assoc. Off. Anal. Chem. 67(5):982–985, 1984.

Zeisel, S. H., D. Char and N. F. Sheard, Choline, phosphatidylcholine and sphingomyelin in human and bovine milk and infant formulas, J. Nutr. 116(1):50–58, 1986.

Chromium

Farre, R., M. J. Lagarda and R. Montoro, Atomic absorption spectrophotometric determination of chromium in foods. J. Assoc. Off. Anal. Chem. 69(5):876–879, 1986.

Smart, G. A. and J. C. Sherlock, Chromium in foods and the diet, Food Addit. Contam. 2(2):139–147, 1985.

Dietary Fiber

Cummings, J. H., H. N. Englyst and R. Wood, Determination of dietary fibre in cereals and cereal products—collaborative trials. I. Initial trials, J. Assoc. Pub. Anal. 23(1):1–35, 1985.

Holloway, W. D., Composition of fruit, vegetable and cereal dietary fibre, J. Sci. Food Agric. 34(11):1236–1240, 1983.

Holloway, W. D., J. A. Monro, J. C. Gurnsey, E. W. Pomare and N. H. Stace, Dietary fiber and other constituents of some Tongan foods, J. Food. Sci. 50:1756–1757, 1985.

Horvath-Mosonvi, M., J. Rigo and E. Hegedues-Voelgvesi, An investigation into the dietary fibre content of different bread and dietary bran samples, Develop. Food Sci. 5B:1115–1120, 1983.

Kunerth, W. H. and V. L. Youngs, Effect of variety and growing year on the constituents of durum bran fiber, Cereal Chem. 61(4):350–352, 1984.

Lanza, E. and R. R. Butrum, A critical review of food fiber analysis and data, J. Am. Diet. Assoc. 86(6):732–743, 1986.

Lund, E. D., J. M. Smoot and N. T. Hall, Dietary fiber content of eleven tropical fruits and vegetables, J. Agric. Food Chem. 31(5):1013–1016, 1983.

Marlett, J. A. and J. G. Chesters, Measuring dietary fiber in human foods, J. Food Sci. 50(2):410–414, 423, 1985.

McNutt, K. W., Perspective—Fiber, J. Nutr. Ed. 8(4):150–152, 1976.

Patrow, C. J. and J. A. Marlett, Variability in the dietary fiber content of wheat and mixed-grain commercial breads, J. Am. Diet. Assoc. 86(6):794–796, 1986.

Paul, A. A. and D. A. T. Southgate, McCance and Widdowson's The Composition of Foods: Dietary fibre in egg, meat and fish dishes, J. Hum. Nutr. 33:335–336, 1979.

Polizzoto, L. M., A. M. Tinsley, C. W. Weber and J. W. Berry, Dietary fibers in muffins, J. Food Sci. 48:111–113, 1983.

Prosky, L., N. G. Asp, I. Furda, J. W. Devries, T. F. Schweizer and B. F. Harland, Determination of total dietary fiber in foods and food products: Collaborative study, J. Assoc. Off. Anal. Chem. 68(4):677–679, 1985.

Prosky, L., N. G. Asp, I. Furda, J. W. Devries, T. F. Schweizer and B. F. Harland, Determination of total dietary fiber in foods, food products, and total diets: Interlaboratory study, J. Assoc. Off. Anal. Chem. 67(6):1044–1052, 1984.

Ross, J. K., C. English and C. A. Perlmutter, Dietary fiber constituents of selected fruits and vegetables, J. Am. Diet. Assoc. 85(9):1111–1116, 1985.

Saraswathi, G., S. Kanchana and K. S. Shurpalekar, Estimation of dietary fibre of some selected Indian foods by different methods, Indian J. Med. Res. 77:833–838, 1983.

Varo, P., R. Laine, K. Veijalainen, A. Espo, A. Wetterhoff and P. Koivistoinen, Dietary fibre and available carbohydrates in Finnish vegetables and fruits, J. Agr. Sci. in Finland 56(1):49–59, 1984.

Varo, P., K. Veijalainen and P. Koivistoinen, Effect of heat treatment on the dietary fibre contents of potato and tomato, J. Food Tech. 19(4):485–492, 1984.

Visser, F. R. and C. Gurnsey, Inconsistent differences between neutral detergent fiber and total dietary fiber values of fruits and vegetables, J. Assoc. Off. Anal. Chem. 69(4):565–567, 1986.

Wenlock, R. W., L. M. Sivell and I. B. Agater, Dietary fibre fractions in cereal and cereal-containing products in Britain, J. Sci. Food Agric. 36:113–121, 1985.

Zyren, J., E. R. Elkins, J. A. Dudek and R. E. Hagen, Fiber contents of selected raw and processed vegetables, fruits and fruit juices as served, J. Food Sci. 48:600–603, 1983.

Fatty Acids

Ahn, P. C., N. Kassim and P. V. J. Hegarty, Total fat and fatty acid composition of commercially available chocolate candies, J. Am. Diet. Assoc. 79(5):552–554, 1981.

Anderson, B. A., Comprehensive evaluation of fatty acids in foods, VII. Pork products, J. Am. Diet. Assoc. 69:44–49, 1976.

Anderson, B. A., Comprehensive evaluation of fatty acids in foods, XIII. Sausages and luncheon meats, J. Am. Diet. Assoc. 72(1):48–52, 1978.

Anderson, B. A., G. Fristrom and J. L. Weihrauch, Comprehensive evaluation of fatty acids in foods, X. Lamb and veal, J. Am. Diet. Assoc. 70:53–58, 1977.

Anderson, B. A., J. E. Kinsella and B. K. Watt, Comprehensive evaluation of fatty acids in foods, II. Beef products, J. Am. Diet. Assoc. 67:35–41, 1975.

Brignoli, C. A., J. E. Kinsella and J. L. Weihrauch, Comprehensive evaluation of fatty acids in foods, V. Unhydrogenated fats and oils, J. Am. Diet. Assoc. 68:224–229, 1976.

Chanmugam, P., M. Boudreau and D. H. Hwang, Differences in the omega-3 fatty acid contents in pond-reared and wild fish and shellfish, J. Food Sci. 51:1556–1557, 1986.

Dutta, A. K., P. K. Pal, A. Ghosh, S. Misra, S. Nandi and Y. A. Choudhur, Lipids and fatty acids of the gastropod mollusc *Cerethidea cingulata*, J. Am. Oil Chem. Soc. 63(2):223–225, 1986.

Exler, J., R. M. Avena and J. L. Weihrauch, Comprehensive evaluation of fatty acids in foods, XI. Leguminous seeds, J. Am. Diet. Assoc. 71(4):412–415, 1977.

Exler, J. and J. L. Weihrauch, Comprehensive evaluation of fatty acids in foods, XII. Shellfish, J. Am. Diet. Assoc. 71(5):518–521, 1977.

Exler, J. and J. L. Weihrauch, Provisional table on the content of omega-3 fatty acids and other fat components in selected foods, Nutrient Data Research Branch, Nutrition Monitoring Division, Human Nutrition Information Service, HNIS/PT-103, U.S. Department of Agriculture, May 1986.

Fogerty, A. C., A. J. Evans, G. L. Ford and B. H. Kennett, Distribution of omega-6 and omega-3 fatty acids in lipid classes in Australian fish, Nutr. Rep. Inter. 33(5):777–786, 1986.

Fristrom, G. A., B. C. Stewart, J. L. Weihrauch and L. P. Posati, Comprehensive evaluation of fatty acids in foods, IV. Nuts, peanuts, and soups, J. Am. Diet. Assoc. 67:351–355, 1975.

Garcia, V. V., J. K. Palmer and R. W. Young, Fatty acid composition of the oil of winged beans, *Psophocarpus tetragonolobus (L.) DC*. J. Am. Oil Chem. Soc. 56:931–932, 1979.

Gibson, R. A., Australian fish—an excellent source of both arachidonic acid and omega-3 polyunsaturated fatty acids, Lipids 18(11):743–775, 1983.

Guild, L., D. Deethardt and E. Rust, Fatty acids in foods served in a university food service, J. Am. Diet. Assoc. 61:149–151, 1972.

Hepburn, F. N., J. Exler and J. L. Weihrauch, Provisional table on the content of omega-3 fatty acids and other fat components of selected foods, J. Am. Diet. Assoc. 86(6):788–793, 1986.

Kinsella, J. E., J. L. Shimp, J. Mai and J. Weihrauch, Fatty acid content and composition of freshwater finfish, J. Am. Oil Chem. Soc. 54:424–429, 1977.

Miljanich, P. and R. Ostwald, Fatty acids in newer brands of margarine, J. Am. Diet. Assoc. 56(1):29–30, 1970.

Ostwald, R., Fatty acids in eleven brands of margarine, J. Am. Diet. Assoc. 39(4):313–316, 1961.

Posati, L. P., J. E. Kinsella and B. K. Watt, Comprehensive evaluation of fatty acids in foods, I. Dairy products, J. Am. Diet. Assoc. 66:482–487, 1975.

Posati, L. P., J. E. Kinsella and B. K. Watt, Comprehensive evaluation of fatty acids in foods, III. Eggs and egg products, J. Am. Diet. Assoc. 67:111–115, 1975.

Sheppard, A. J., J. L. Iverson and J. L. Weihrauch, Composition of selected dietary fats, oils, margarine, and butter, In Handbook of Lipid Research, vol. 1, edited by A. Kuksis, Plenum, New York, 1973.

van den Reek, M. M., M. C. Craig-Schmidt and A. J. Clark, Use of published analyses of food items to determine dietary trans octadecenoic acid, J. Am. Diet. Assoc. 86(10):1391–1394, 1986.

Weihrauch, J. L., C. A. Brignoli, J. B. Reeves III and J. L. Iverson, Fatty acid composition of margarines, processed fats, and oils: A new compilation of data for tables of food composition, Food Tech. 31(2):80–91, 1977.

Weihrauch, J. L., J. E. Kinsella and B. K. Watt, Comprehensive evaluation of fatty acids in foods, VI. Cereal products, J. Am. Diet. Assoc. 68:335–355, 1976.

Williams, G., B. C. Davidson, P. Stevens and M. A. Crawford, Comparative fatty acids of the dolphin and the herring, J. Am. Oil Chem. Soc. 54:328–330, 1977.

Worthington, R. E., The linolenic acid content of peanut oil, J. Am. Oil Chem. Soc. 54:167–169, 1977.

Fluoride

Marier, J. R. and D. Rose, The fluoride content of some foods and beverages—a brief survey using a modified Zr-SPANDS method, J. Dent. Res. 31:941–946, 1966.

Ophaug, R. H., L. Singer and B. F. Harland, Estimated fluoride intake of 6-month-old infants in four dietary regions of the United States, Am. J. Clin. Nutr. 33:324–327, 1980.

Read, J. I. and R. Collins, Estimation of total fluoride in foodstuffs, J. Assoc. Pub. Anal. 20(4);109–111, 115, 1982.

Moody, G. J., B. Ong, K. Quinlan, A. H. Riah and J. D. R. Thomas, The determination of fluorine in coffee and tea using a microprocessor coupled with a fluoride ion-selective electrode, J. Food Tech. 15(3):335–343, 1980.

Siebert, G. and K. Trautner, Fluoride content of selected human food, pet food and related materials, Z. Ernahrungswiss. 24(1):54–66, 1985.

Smith, G. E., Fluoride, the environment, and human health, Perspect. Biol. Med. 29(4):560–572, 1986.

Taves, D. R., Dietary intake of fluoride ashed (total fluoride) v. unashed (inorganic fluoride) analysis of individual foods, Br. J. Nutr. 49(3):295–301, 1983.

Waldbott, G. L., Fluoride in food, Am. J. Clin. Nutr. 12:455–462, 1963.

Walters, C. B., J. C. Sherlock, W. H. Evans and J. I. Read, Dietary intake of fluoride in the United Kingdom and fluoride content of some foodstuffs, J. Sci. Food Agric. 34(5):523–528, 1983.

Folate

Huq, R. S., J. A. Abalaka and W. L. Stafford, Folate content of various Nigerian foods, J. Sci. Food Agric. 34(4):404–406, 1983.

Perloff, B. P. and R. R. Butrum, Folacin in selected foods, J. Am. Diet. Assoc. 70:161–172, 1977.

Gluten

Olson, G. B. and G. R. Gallo, Gluten in pharmaceutical and nutritional products, Am. J. Hosp. Pharm. 40(1):121–122, 1983.

Iodine

Berg, J. N. and D. Padgitt, Iodine concentrations in milk from iodophor teat dips, J. Dairy Sci. 68(2):457–461, 1985.

Bruhn, J. C. and A. A. Franke, An indirect method for the estimation of the iodine content in raw milk, J. Dairy Sci. 61(11):1557–1560, 1978.

Bruhn, J. C., A. A. Franke, R. B. Bushnell, H. Weisheit, G. H. Hutton and G. C. Gurtle, Sources and content of iodine in California milk and dairy products, J. Food Prot. 46(1):41–46, 1983.

Galton, D. M., L. G. Petersson and H. N. Erb, Milk iodine residues in herds practicing iodophor premilking teat disinfection, J. Dairy Sci. 69(1):267–271, 1986.

Galton, D. M., L. G. Petersson, W. G. Merrill, D. K. Bandler and D. E. Shuster, Effects of premilking udder preparation on bacterial population, sediment, and iodine residue in milk, J. Dairy Sci. 67(11):2580–2589, 1984.

Gushurst, C. A., J. A. Mueller, J. A. Green and F. Sedor, Breast milk iodide: reassessment in the 1980s, Pediatrics 73(3):354–357, 1984.

van Ryssen, J. B., S. van Malsen and J. G. van Blerk, The iodine content of fresh milk samples in Natal and the effect of iodophor teat dips on milk iodine content, J. S. Afr. Vet. Assoc. 56(4):181–185, 1985.

Wenlock, R. W., D. H. Buss, R. E. Moxon and N. G. Bunton, Trace nutrients. 4. Iodine in British foods, Br. J. Nutr. 47:381–390, 1982.

Wheeler, S. M., G. H. Fleet and R. J. Ashley, Effect of processing upon concentration and distribution of natural and iodophor-derived iodine in milk, J. Dairy Sci. 66(2):187–195, 1983.

Inositol

Clements, R. S. Jr. and B. Darnell, *Myo*-inositol content of common foods: development of a high-*myo*-inositol diet, Am. J. Clin. Nutr. 33:1954–1967, 1980.

Olano, A., Presence of trehalose and sugar alcohols in sherry, Am. J. Enology Viticulture 34:148–151, 1983.

Sandberg, A. S. and R. Ahderinne, HPLC method for determination of inositol tri-, tetra-, penta-, and hexaphosphates in foods and intestinal contents, J. Food Sci. 51(3):547–550, 1986.

Nickel

Smart, G. A. and J. C. Sherlock, Nickel in foods and the diet, Food Addit. Contam. 4(1):61–71, 1987.

Solomons, N. W., F. Viteri, T. R. Schuler and F. H. Nielsen, Bioavailability of nickel in man: effects of foods and chemically defined dietary constituents on the absorption of inorganic nickel, J. Nutr. 112(1):39–50, 1982.

Veien, N. K. and M. R. Andersen, Nickel in Danish food, Acta. Derm. Venereol. (Stockholm) 66(6):502–509, 1986.

Nitrate and Nitrite

Crosby, N. T., J. K. Foreman, J. F. Palframan and R. Sawyer, Estimation of steam-volatile N-nitrosamines in foods at the 1 ug/kg level, Nature 238:342–343, 1972.

Eggers, N. J. and D. L. Cattle, High-performance liquid chromatographic method for the determination of nitrate and nitrite in cured meat, J. Chromatogr. 354:490–494, 1986.

Fiddler, W., R. C. Doerr, R. A. Gates and J. B. Fox Jr., Comparison of chemiluminescent and AOAC methods for determining nitrite in commercial cured meat products, J. Assoc. Off. Anal. Chem. 67(3):525–538, 1984.

Hotchkiss, J. H., Sources of N-nitrosamine contamination in foods, Adv. Exp. Med. Biol. 177:287–298, 1984.

Hunt, J. and D. J. Seymour, Method for measuring nitrate-nitrogen in vegetables using anion-exchange high-performance liquid chromatography, Analyst 110(2):131–133, 1985.

Jackson, P. E., P. R. Haddad and S. Dilli, Determination of nitrate and nitrite in cured meats using high-performance liquid chromatography, J. Chromatogr. 295(2):471–478, 1984.

Jancar, J. C., M. D. Constant and W. C. Herwig, Determination of inorganic anions in beer by ion chromatography, J. Am. Soc. Brew. Chem. 42(2):90–93, 1984.

Kikugawa, K., T. Kato and H. Hayatsu, Mutagenicity of smoked, dried bonito products, Mutat. Res. 158(1-2):35–44, 1985.

Kleijn, J. P. de and K. Hoven, Determination of nitrite and nitrate in meat products by high-performance liquid chromatography, Analyst 109(4):527–528, 1984.

Kok, S. H., K. A. Buckle and M. Wootton, Determination of nitrate and nitrite in water using high-performance liquid chromatography, J. Chromatogr. 260(1):189–192, 1983.

Revelle, R., Toxins in the human food system, Basic Life Sci. 34:11–26, 1985.

Ross, H. D. and J. H. Hotchkiss, Determination of nitrate in dried foods by gas chromatography–thermal energy analyzer, J. Assoc. Off. Anal. Chem. 68(1):41–43, 1985.

Tanaka, A., N. Nose, H. Masaki, Y. Kikuchi and H. Iwasaki, Gas–liquid chromatographic determination of trace amounts of nitrite in egg, egg white, and egg yolk, J. Assoc. Off. Anal. Chem. 66(2):260–263, 1983.

White, J. W. Jr., Relative significance of dietary sources of nitrate and nitrite, J. Agric. Food Chem. 23(5):886–891, 1975.

Wooton, M., S. H. Kok and K. A. Buckle, Determination of nitrite and nitrate levels in meat and vegetable products by high-performance liquid chromatography, J. Sci. Food Agr. 36(4):297–304, 1985.

Oxalic Acid

Awadalla, M. Z., N. S. Bassily and N. H. Saba, Oxalate content of some common Egyptian food stuffs, J. Egypt. Public Health Assoc. 60(3–4):41–51, 1985.

Bushway, R. J., J. L. Bureau and D. F. Mcgann, Determinations of organic acids in potatoes by high-performance liquid chromatography, J. Food Sci. 49:75–77, 1984.

Kozukue, E., N. Kozukue and T. Kurosaki, Organic acid, sugar and amino acid composition of bamboo shoots, J. Food Sci. 48:935–938, 1983.

Ohkawa, H., Gas chromatographic determination of oxalic acid in foods, J. Assoc. Off. Anal. Chem. 68(1):108–111, 1985.

Oxalic acid content of selected vegetables, Vegetables and Vegetable Products, Agriculture Handbook No. 8-11, Composition of Foods, Raw, Processed, Prepared, p. 11, U.S. Department of Agriculture, Washington, D.C., August 1984.

Pilac, L. M., I. C. Abdon and E. P. Mandap, Oxalic acid content and its relation to the calcium present in some Philippine plant foods, Philippine J. Nutr. 24(1):21–36, 1971.

Ramasastri, B. V., Calcium, iron and oxalate content of some condiments and spices, Qual. Plan.—Plant Foods for Human Nutr. 33(1):11–15, 1983.

Sachde, A. G., A. Y. Al-Bakir, M. A. A. Aziz and J. A. Faris, Oxalic acid content of some common Iraqi vegetables, Iraqi J. Agr. Sci. (Zanco) 2(3):33–38, 1984.

Phytate

Faridi, H. A., P. L. Finney and G. L. Rubenthaler, Effect of soda leavening on phytic acid content and physical characteristics of Middle Eastern breads, J. Food Sci. 48:1654–1658, 1983.

Harland, B. F. and D. Oberleas, Anion-exchange method for determination of phytate in foods: collaborative study, J. Assoc. Off. Anal. Chem. 69(4):667–670, 1986.

Harland, B. F. and D. Oberleas, Phytate and zinc contents of coffees, cocoas, and teas, J. Food Sci. 50:832–842, 1985.

Kim, W. J., N. M. Kim and H. S. Sung, Effect of germination on phytic acid and soluble minerals in soymilk, Korean J. Food Sci. Tech. 16(3):358–362, 1984.

Oberleas, D. and B. F. Harland, Phytate content of foods: effect on dietary zinc bioavailability, J. Am. Diet. Assoc. 79:433–435, 1981.

Ologhobo, A. D. and B. L. Fetuga, Investigations on the trypsin inhibitor, hemagglutinin, phytic and tannic acid contents of cowpea *Vigna unguiculata,* Food Chem. 12(4):249–254, 1983.

O'Neill, I. K., M. Sargent and M. L. Trimble, Determination of phytate in foods by phosphorus-31 fourier transform nuclear magnetic resonance spectrometry, Anal. Chem. 52:1288–1291, 1980.

Phillippy, B. Q. and M. R. Johnston, Determination of phytic acid in foods by ion chromatography with post-column derivatization, J. Food Sci. 50(2):541–542, 1985.

Reddy, N. R., S. K. Sathe and D. K. Salunkhe, Phytates in legumes and cereals, Adv. Food Res. 28:1–92, 1982.

Tan, N. H., Z. H. Rahim, H. T. Khor and K. C. Wong, Winged bean (*Psophocarpus tetragonolobus*) tannin level, phytate content, and hemagglutinating activity, J. Agric. Food Chem. 31(4):916–917, 1983.

Tangkongchitr, U., P. A. Seib and R. C. Hoseney, Phytic acid, II. Its fate during breadmaking, Cereal Chem. 58(3):229–234, 1981.

Teklenburg, E., M. E. Zabik, M. A. Uebersax, J. Deitz and E. W. Lusas, Mineral and phytic acid partitioning among air-classified bean flour fractions, J. Food Sci. 49(2):569–572, 584, 1984.

Vinh, L. T. and E. Dworschak, Phytate content of some foods from plant origin from Vietnam and Hungary, Nahrung 29(2):161–166, 1985.

Purines

Burns, B. G. and P. J. Ke, Liquid chromatographic determination of hypoxanthine content in fish tissue, J. Assoc. Off. Anal. Chem. 68(3):444–448, 1985.

Savaiano, D. A., L. M. Salati, W. D. Rosamond, J. Salinsky and B. W. Willis, Nucleic acid, purine and proximate analyses of mechanically separated beef and veal, J. Food Sci. 48:1356–1357, 1983.

Shinoda, T., Y. Aoyagi and T. Sugahara, Contents of purine bases in fishes and fish products, J. Jap. Soc. Food Nutr. 34(2):153–162, 1981.

Shinoda, T. E., Y. Aoyagi and T. Sugahara, Purine base contents in foods and effects of cooking methods (sautéing, broiling, boiling), J. Jap. Soc. Food Nutr. 35(2):103–109, 1982.

Tiemeyer, W., M. Stohrer and D. Giesecke, Metabolites of nucleic acids in bovine milk, J. Dairy Sci. 67(4):723–728, 1984.

Warthesen, J. J., P. T. Waletzko and F. F. Busta, High-pressure liquid chromatographic determination of hypoxanthine in refrigerated fish, J. Agric. Food Chem. 28(6):1308–1309, 1980.

Selenium

Amer, M. A. and G. J. Brisson, Selenium in human food stuffs collected at the Ste-Foy (Quebec) food market, J. Inst. Can. Sci. Technol. Aliment 6(3):184–187, 1973.

Arthur, D., Selenium content of Canadian foods, Can. Inst. Food Sci. Technol. J. 5(3):165–169, 1972.

Cappon, C. J. and J. C. Smith, Chemical form and distribution of mercury and selenium in edible seafood, J. Anal. Toxicol. 6:10–21, 1982.

Cappon, C. J. and J. C. Smith, Chemical form and distribution of mercury and selenium in canned tuna, J. Appl. Toxicol. 2(4):181–189, 1982.

Cappon, C. J. and J. C. Smith, Mercury and selenium content and chemical form in fish muscle, Arch. Environ. Contam. Toxicol. 10:305–319, 1981.

Ferretti, R. J. and O. A. Levander, Effect of milling and processing on the selenium content of grains and cereal products. J. Agric. Food Chem. 22(6):1049–1051, 1974.

Ferretti, R. J. and O. A. Levander, Selenium content of soybean foods, J. Agric. Food Chem. 24(1):54–56, 1976.

Hadjimarkos, D. M. and C. W. Bonhorst, The selenium content of eggs, milk, and water in relation to dental caries in children, J. Pediatr. 59(2):256–259, 1961.

Hahn, M. H., R. W. Kuennen, J. A. Caruso and F. L. Fricke, Determination of trace amounts of selenium in corn, J. Agric. Food Chem. 29:792–796, 1981.

Higgs, D. J., V. C. Morris and O. A. Levander, Effect of cooking on selenium content of foods, J. Agric. Food Chem. 20(3):678–680, 1972.

Koivistoinen, P. and J. K. Huttunen, Selenium in food and nutrition in Finland: An overview on research and action, Ann. Clin. Res. 18(1):13–17, 1986.

Landsberger, S. and E. Hoffman, Rapid determination of selenium in various marine species by instrumental neutron activation analysis, J. Radioanal. Nucl. Chem. 87(1):41–50, 1984.

Lane, H. W., B. J. Taylor, E. Stool, D. Servance and D. C. Warren, Selenium content of selected foods, J. Am. Diet. Assoc. 82:24–28, 1983.

Levander, O. A., A review: selenium and chromium in human nutrition, J. Am. Diet. Assoc. 66:338–344, 1975.

Luten, J. B., A. Ruiter, T. M. Ritskes, A. B. Rauchbaar and G. Riekwel-Booy, Mercury and selenium in marine- and freshwater fish, J. Food Sci. 45:416–419, 1980.

Martin, R. F., V. R. Young and M. Janghorbani, Selenium content of enteral formulas, J. Parenter. Enteral. Nutr. 10(2):213–215, 1986.

Morris, V. C. and O. A. Levander, Selenium content of foods, J. Nutr. 100:1383–1388, 1970.

Moxon, A. L. and D. L. Palmquist, Selenium content of foods grown or sold in Ohio, Ohio Report 65(1):13–14, 1980.

Olson, O. E. and I. S. Palmer, Selenium in foods purchased or produced in South Dakota, J. Food Sci. 49:446–452, 1984.

Olson, O. E., I. S. Palmer and M. Howe Sr., Selenium in foods consumed by South Dakotans, Proc. S. D. Acad. Sci. 57:113–121, 1978.

Robberecht, H., O. Van Schoor and H. Deelstra, Selenium and chronium content of European beers as determined by AAS, J. Food Sci. 49:300–301, 1984.

Schroeder, H. A., D. V. Frost and J. J. Balassa, Essential trace metals in man: Selenium, J. Chronic Dis. 23:227–243, 1970.

Schubert, A., J. M. Holden and W. R. Wolf, Selenium content of a core group of foods based on a critical evaluation of published analytical data, J. Am. Diet. Assoc. 87(3):285–299, 1987.

Shum, G. T. C., H. C. Freeman and J. E. Uthe, Flameless atomic absorption spectrophotometry of selenium in fish and food products, J. Assoc. Off. Anal. Chem. 60(5):1010–1014, 1977.

Snook, J. T., D. Kinsey, D. L. Palmquist, J. P. DeLany, V. M. Vivian and A. L. Moxon, Selenium content of foods purchased or produced in Ohio, J. Am. Diet. Assoc. 87:744–749, 1987.

Taussky, H. H., A. Washington, E. Zubillaga and A. T. Milhorat, Selenium content of fresh eggs from normal and dystrophic chickens, Nature 200(4912):1211, 1963.

Thompson, J. N., P. Erdody, P. Smith and D. C. Smith, Selenium content of food consumed by Canadians, J. Nutr. 105:274–277, 1975.

Thorn, J., J. Robertson and D. H. Buss, Trace nutrients: selenium in British food, Br. J. Nutr. 39:391–396, 1978.

Sugars

Kozukue, E., N. Kozukue and T. Kurosaki, Organic acid, sugar and amino acid composition of bamboo shoots, J. Food Sci. 48:935–938, 1983.

Li, B. W. and P. J. Schuhmann, Sugar analysis of fruit juices: Content and method, J. Food Sci. 48:633–635, 1983.

Li, B. W., P. J. Schuhmann and J. M. Holden, Determination of sugars in yogurt by gas-liquid chromatography, J. Agric. Food Chem. 31(5):985–989, 1983.

Southgate, D. A. T., A. A. Paul, A. C. Dean and A. A. Christie, Free sugars in foods, J. Hum. Nutr. 32:335–347, 1978.

Tin

Dabeka, R. W., A. D. McKenzie and R. H. Albert, Atomic absorption spectrophotometric determination of tin in canned foods, using nitric acid–hydrochloric acid digestion and nitrous oxide–acetylene flame: collaborative study, J. Assoc. Off. Anal. Chem. 68(2):209–213, 1985.

Sherlock, J. C. and G. A. Smart, Tin in foods and the diet, Food Addit. Contam. 1(3):277–282, 1984.

Tyramine

Buteau, C., C. L. Duitschaever and G. C. Ashton, High-performance liquid chromatographic detection and quantitation of amines in must and wine, J. Chromatogr. 284(1):201–210, 1984.

Chang, S. F., J. W. Ayres and W. E. Sandine, Analysis of cheese for histamine, tyramine, tryptamine, histidine, tyrosine, and tryptophane, J. Dairy Sci. 68(11):2840–2846, 1985.

Chin, K. D. H. and P. E. Koehler, Identification and estimation of histamine, tryptamine, phenethylamine and tyramine in soy sauce by thin-layer chromatography of dansyl derivatives, J. Food Sci. 48(6):1826–1828, 1983.

Hui, J. Y. and S. L. Taylor, High-pressure liquid chromatographic determination of putrefactive amines in foods, J. Assoc. Off. Anal. Chem. 66(4):853–857, 1983.

Jalon, M., C. Santos-Buelga, J. C. Rivas-Gonzalo and A. Marine-Font, Tyramine in cocoa and derivatives, J. Food Sci. 48:545–547, 1983.

Joosten, H. M. L. and C. Olieman, Determination of biogenic amines in cheese and some other food products by high-performance liquid chromatography in combination with thermo-sensitized reaction detection, J. Chromatogr. 356(2):311–319, 1986.

McCabe, B. J., Dietary tyramine and other pressor amines in MAOI regimens: a review, J. Am. Diet. Assoc. 86(8):1059–1064, 1986.

Reuvers, T. B. A., M. M. de Pozuelo, M. Ramos and R. Jimenez, A rapid ion-pair HPLC procedure for the determination of tyramine in dairy products, J. Food Sci. 51(1):84–86, 1986.

Santos-Buelga, C., M. J. Pena-Egido and J. C. Rivas-Gonzalo, Changes in tyramine during chorizo sausage ripening, J. Food Sci. 51(2):518–513, 1986.

Sayem-El-Daher, N., R. E. Simard and L. L'Heureux, Determination of mono-, di- and polyamines in foods using a single-column amino acid auto-analyzer, J. Chromatogr. 256(2):313–321, 1983.

Sem, N. P., Analysis and significance of tyramine in foods, J. Food Sci. 34:22–26, 1969.

Sullivan, E. A. and K. I. Shulman, Diet and monoamine oxidase inhibitors: A re-examination, Can J. Psychiatry 29(8):707–711, 1984.

Voigt, M. N., R. R. Eitenmiller, P. E. Koehler and M. K. Hamdy, Tyramine, histamine, and tryptamine content of cheese, J. Milk Food Technol. 37(7):377–381, 1974.

Zee, J. A., R. E. Simard and L. L'Heureux, Evaluation of analytical methods for determination of biogenic amines in fresh and processed meat, J. Food Prot. 46(12):1044–1049, 1054, 1983.

Vitamin D

Berg, H. van den, P. G. Boshuis and W. H. P. Schreurs, Determination of vitamin D in fortified and nonfortified milk powder and infant formula using a specific radioassay after purification by high-performance liquid chromatography, J. Agr. Food Chem. 34(2):264–268, 1986.

Kunz, C., M. Niesen, H. von Lilienfeld-Toal and W. Burmeister, Vitamin D, 25-hydroxy-vitamin D and 1,25-dihydroxy-vitamin D in cow's milk, infant formulas and breast milk during different stages of lactation, Int. J. Vitam. Nutr. Res. 54(2-3):141–148, 1984.

Okano, T, K. Yokoshima and T. Kobayashi, High-performance liquid chromatographic determination of vitamin D-3 in bovine colostrum, early and later milk, J. Nutr. Sci. Vitaminol. (Tokyo) 30(5):431–439, 1984.

Sertl, D. C. and B. E. Molitor, Liquid chromatographic determination of vitamin D in milk and infant formula, J. Assoc. Off. Anal. Chem. 68(2):177–182, 1985.

Takeuchi, A., T. Okano, M. Ayame, H. Yoshikawa, S. Teraoka, Y. Murakami and T. Kobayashi, High-performance liquid chromatographic determination of vitamin D-3 in fish liver oils and eel body oils, J. Nutr. Sci. Vitaminol. (Tokyo) 30(5):421–430, 1984.

Vitamin E (Tocopherols)

Atuma, S. S., Electrochemical determination of vitamin E in margarine, butter and palm oil, J. Sci. Food Agric. 26(4):393–399, 1975.

Bunnell, R. H., J. Keating, A. Quaresimo and G. K. Parman, Alpha-tocopherol content of foods, Am. J. Clin. Nutr. 17(1):1–10, 1965.

Candlish, J. K., Tocopherol content of some Southeast Asian foods, J. Agric. Food Chem. 31(1):166–168, 1983.

Herting, D. C. and E. E. Drury, Alpha-tocopherol content of cereal grains and processed cereals, Agr. Food Chem. 17(4):785–790, 1969.

Koehler, H. H., H. C. Lee and M. Jacobson, Tocopherols in canned entrees and vended sandwiches, J. Am. Diet. Assoc. 70(6):616–620, 1977.

Lambertsen, G., H. Myklestad and O. R. Braekkan, Tocopherols in nuts, J. Sci. Food Agric. 13:617–620, 1962.

Landen, W. O. Jr., D. M. Hines, T. W. Hamill, J. I. Martin, E. R. Young, R. R. Eitenmiller and A. G. Soliman, Vitamin A and vitamin E content of infant formulas produced in the United States, J. Assoc. Off. Anal. Chem. 68(3):509–511, 1985.

Lehmann, J., H. L. Martin, E. L. Lashley, M. W. Marshall and

J. T. Judd, Vitamin E in foods from high and low linoleic acid diets, J. Am. Diet. Assoc. 86(9):1208–1216, 1986.

McLaughlin, P. J. and J. L. Weihrauch, Vitamin E content of foods, J. Am. Diet. Assoc. 75:647–665, 1979.

Syvaoja, E. L., V. Piironen, P. Varo, P. Koivistoinen and K. Salminen, Tocopherols and tocotrienols in Finnish foods: human milk and infant formulas, Int. J. Vitam. Nutr. Res. 55(2):159–166, 1985.

Vitamin K

Doisy, E. A., Availability of vitamin K in human diets (abstract), Fed. Proc. 31(2):713, 1972.

Fletcher, D. C., Do clotting factors in vitamin K-rich vegetables hinder anticoagulant therapy? (questions and answers), JAMA 237(17):1871, 1977.

Geigy Scientific Tables, Vol. 1, Units of Measurement, Body Fluids, Composition of the Body, Nutrition, 8th ed., Ciba-Geigy, 1981.

Haroon, Y., M. J. Shearer, G. McEnery, V. E. Allan and P. Barkhan, Assay of vitamin K-1 (phylloquinone) by high-performance liquid chromatography: values for human and cow's milk, Proc. Nutr. Soc. 39(2):49A, 1980.

Haroon, Y., M. J. Shearer, S. Rahim, W. G. Gunn, G. McEnery and P. Barkhan, The content of phylloquinone (vitamin K-1) in human milk, cows' milk and infant formula foods determined by high-performance liquid chromatography, J. Nutr. 112(6):1105–1117, 1982.

Howard, P. A. and K. N. Hannaman, Warfarin resistance linked to enteral nutrition products, J. Am. Diet. Assoc. 85(6):713–715, 1985.

Hwang, S. M., Liquid chromatographic determination of vitamin K-1 trans- and cis-isomers in infant formula, J. Assoc. Off. Anal. Chem. 68(4):684–689, 1985.

Kempin, S. J., Warfarin resistance caused by broccoli (letter), N. Engl. J. Med. 308(20):1229–1230, 1983.

Kutsop, J. J., Vitamin K-1 content of enteral products (letter), Am. J. Hosp. Pharm. 40(12):2120, 2123, 1983.

Matschiner, J. T. and E. A. Doisy, Vitamin K content of ground beef, J. Nutr. 90:331–334, 1966.

O'Reilly, R. A. and D. A. Rytand, "Resistance" to warfarin due to unrecognized vitamin K supplementation (letter), N. Engl. J. Med. 303(3):160–161, 1980.

Parrish D. B., Determination of vitamin K in foods: a review, Crit. Rev. Food Sci. Nutr. 13(4):337–352, 1980.

Schneider, D. L., H. B. Fluckiger and J. D. Manes, Vitamin K-1 content of infant formula products, Pediatrics 53(2):273–275, 1974.

Seifert, R. M., Analysis of vitamin K-1 in some green leafy vegetables by gas chromatography, J. Agric. Food Chem. 27(6):1301–1304, 1979.

REFERENCES BY FOOD GROUP

Beverages

Meranger, J. C., K. S. Subramanian and C. Chalifoux, Survey for cadmium, cobalt, chromium, copper, nickel, lead, zinc, calcium, and magnesium in Canadian drinking water supplies, J. Assoc. Off. Anal. Chem. 64(1):44–53, 1981.

Ough, C. S., E. A. Crowell and J. Benz, Metal content of California wines, J. Food Sci. 47:825–828, 1982. (Ca, Cu, Fe, K, Mg, Na, Pb, Zn)

Chips and Other Snack Foods

Khan, M. A. and J. A. Martin, Salt content of selected snack foods, J. Food Sci. 48:656–657, 1983.

Eggs

Naber, E. C., The effect of nutrition on the composition of eggs, Poult. Sci. 58(3):518–528, 1979.

Robel, E. J., The effect of age of breeder hen on the levels of vitamins and minerals in turkey eggs, Poult. Sci. 62(9):1751–1756, 1983.

Sivell, L. M., R. W. Wenlock and P. A. Jackson, Determination of vitamin D and retinoid activity in eggs by HPLC, Hum. Nutr. Appl. Nutr. 36A:430–437, 1982.

Entrees and Meals

De Ritter, E., M. Osadca, J. Scheiner and J. Keating, Vitamins in frozen convenience dinners and pot pies, J. Am. Diet. Assoc. 63(4):391–397, 1974. (vitamin A, carotene, vitamin E, thiamin, riboflavin, vitamin B-6, vitamin B-12, niacin, ascorbic acid)

Porrini, M., S. Ciappellano, P. Simonetti and G. Testolin, Chemical composition of Italian cooked dishes, Int. J. Vitam. Nutr. Res. 56(3):263–268, 1986.

Wiles, S. J., P. A. Nettleton, A. E. Black and A. A. Paul, The nutrient composition of some cooked dishes eaten in Britain: A supplementary food composition table, J. Hum. Nutr. 34:189–223, 1980.

Fast Foods

Appledorf, H. and L. S. Kelly, Proximate and mineral content of fast foods, J. Am. Diet. Assoc. 74(1):35–40, 1978. (Ca, P, Na, K, Mg, Fe, Zn, Mn, Cu)

Bowers, J. A., J. A. Craig, T. J. Tucker, J. M. Holden and L. P. Posati, Vitamin and proximate composition of fast-food chicken, J. Am. Diet. Assoc. 87:736–739, 1987.

Li, B. W., J. M. Holden, S. G. Brownlee and S. G. Korth, A nationwide sampling of fast-food fried chicken: Starch and moisture content, J. Am. Diet. Assoc. 87:740–743, 1987.

Young, E. A., E. H. Brennan and G. L. Irving, Update: Nutritional analysis of fast foods, Public Health Currents (Ross Labs) 21(3), May–June 1981. (proximate nutrients, 8 vitamins, 8 minerals)

Fats and Oils

Buss, D. H., P. A. Jackson and D. Scuffam, Composition of butters on sale in Britain, J. Dairy Res. 51:637–641, 1984.

Elson, C. M., R. G. Ackman and A. Chatt, Determination of selenium, arsenic, iodine and bromine in fish, plant and mammalian oils by cyclic instrumental neutron activation analysis, J. Am. Oil Chem. Soc. 60(4):829–838, 1983.

Peart, J., A comprehensive study on the composition of margarines on the market, J. Can. Diet. Assoc. 39(1):11–15, 1978.

Fish, Shellfish and Crustacea

Ahamad, I. H., R. M. Rao, J. A. Liuzzo and M. A. Khan, Comparison of nutrients in raw, commercially breaded and hand-breaded shrimp, J. Food Sci. 48:307–308, 1983. (proximates, thiamin, riboflavin, niacin, Ca, Fe)

Anthony, J. E., P. N. Hadgis, R. S. Milam, G. A. Herzfeld, L. J. Taper and S. J. Ritchey, Yields, proximate composition and mineral content of finfish and shellfish, J. Food Sci. 48:313–314, 316, 1983. (Ca, P, Mg, Na, K, Cu, Fe, Zn, Mn)

Berg, C. J. Jr., J. Krzynowek, P. Alatalo and K. Wiggin, Sterol and fatty acid composition of the clam, *Codakia orbicularis,* with chemoautotrophic symbionts, Lipids 20(2):116–120, 1985.

Crawford, D. L., D. K. Law and J. K. Babbitt, Nutritional characteristics of marine food. Fish carcass waste and machine-separated flesh, J. Agr. Food Chem. 20(5):1048–1051, 1972. (proximates, 14 minerals)

Grieg, R. A., Comparison of atomic absorption and neutron activation analyses for the determination of silver, chromium, and zinc in various marine organisms, Anal. Chem. 47(9):1682–1684, 1975.

Iwasaki, M. and R. Harada, Proximate and amino acid composition of the roe and muscle of selected marine species, J. Food Sci. 50:1585–1587, 1985.

Jhaveri, S. N., P. A. Karakoltsidis, J. Montecalvo Jr. and S. M. Constantinides, Chemical composition and protein quality of some southern New England marine species, J. Food Sci. 49:110–113, 1984. (protein, lipid, moisture, amino acids, Ca, Mg, Na, Zn, Fe, Cu, Mn, Ni, Cd)

Jhaveri, S. N., P. A. Karakoltsidis, S. Y. K. Shenouda and S. M. Constantinides, Ocean pout (*Macrozoarces americanus*): Nutrient analysis and utilization, J. Food Sci. 50:719–722, 1985. (amino acids, fatty acids, Ca, Mg, Na, Zn, Fe, Cu, Mn, Ni, Cd)

Lopez, A., D. R. Ward and H. L. Williams, Essential elements in oysters (*Crassostrea virginica*) as affected by processing method, J. Food Sci. 48:1680–1691, 1983. (Ca, Cl, Cr, Co, Cu, Fe, Mg, Mn, Mo, Ni, P, K, Si, Sn, Zn)

Lucas, H. F. Jr., D. N. Edgington and P. J. Colby, Concentrations of trace elements in Great Lakes fishes, J. Fisheries Res. Bd. Can. 27(4):677–684, 1970. (15 elements)

Mustafa, F. A. and D. M. Medeiros, Proximate composition, mineral content, and fatty acids of catfish (*Ictalurus punctatus, Rafinesque*) for different seasons and cooking methods, J. Food Sci. 50:585–588, 1985. (Na, K, P, Mg, Ca, Zn, Fe, Cu)

Seet, S. T. and W. D. Brown, Nutritional quality of raw, precooked and canned albacore tuna (*Thunnus alalunga*), J. Food Sci. 48:288–289, 1983. (amino acids, thiamin, riboflavin, niacin, Na, K, Ca, Fe, Cu)

Sidwell, V. D., P. R. Foncannon, N. S. Moore and J. C. Bonnet, Composition of the edible portion of raw (fresh or frozen) crustaceans, finfish, and mollusks. I. Protein, fat, moisture, ash, carbohydrate, energy value and cholesterol, Marine Fisheries Review 36(3):21–35, 1974.

Fruits

Bushway, R. J., D. F. McGann, W. P. Cook and A. A. Bushway, Mineral and vitamin content of lowbush blueberries (*Vaccinium angustifolium Ait.*), J. Food Sci. 48:1878–1880, 1983. (11 minerals, 5 vitamins)

Edem, D. O., O. U. Eka and E. T. Ifon, Chemical evaluation of nutritive value of the fruit of African starapple (*Chrysophyllum albidum*), Food Chem. 14(4):303–311, 1984.

Edem, D. O., O. U. Eka and E. T. Ifon, Chemical evaluation of the nutritive value of the raffia palm fruit (*Raphia hook-*

eri), Food Chem. 15(11):9–17, 1984. (proximates, 10 minerals, phytate, oxalate)

Johnson, C. D., R. R. Eitenmiller, J. B. Jones Jr., V. N. M. Rao and S. E. Gebhardt, Composition of red delicious apples, J. Food Sci. 49:952–953, 1984. (7 vitamins, 9 minerals)

Grains and Grain Products

Albrecht, J. A., E. H. Asp and I. M. Buzzard, Content and retention of sodium and other minerals in pasta cooked in unsalted or salted water, Cereal Chem. 64:106–109, 1987.

Burk, R. F. and N. W. Solomons, Trace elements and vitamins and bioavailability as related to wheat and wheat foods, Am. J. Clin. Nutr. 41(5 suppl.):1091–1102, 1985.

Douglass, J. S. and R. H. Matthews, Nutrient content of pasta products, Cereal Foods World 27(11):558–561, 1982. (proximate nutrients, 10 minerals, 5 vitamins)

Englyst, H. N., V. Anderson and J. H. Cummings, Starch and non-starch polysaccharides in some cereal foods, J. Sci. Food Agric. 34:1434–1440, 1983.

Morgan, D. E., Note on variations in the mineral composition of oat and barley grain grown in Wales, J. Sci. Food Agr. 19(7):393–395, 1968. (Ca, Mg, K, Na, P, Cl, Cu, Co, Mn, Zn)

Schweizer, T. F., W. Frolich, S. del Vedovo and R. Besson, Minerals and phytate in the analysis of dietary fiber from cereals. I. Cereal Chem. 61(2):116–119, 1984.

Simwemba, C. G., R. C. Hoseney, E. Varriano-Marston and K. Zeleznak, Certain B vitamin and phytic acid contents of pearl millet (*Pennisetum americanum (L.) Leeke*), J. Agric. Food Chem. 32(1):31–34, 1984.

Singh, B. and L. M. Dodda, Studies on the preparation and nutrient composition of bulgur from triticale, J. Food Sci. 44(2):449–452, 1979.

Sivell, L. M. and R. W. Wenlock, The nutritional composition of British bread: London area study, Hum. Nutr. Appl. Nutr. 37A:459–469, 1983.

Tarone, C. M. and R. H. Matthews, Proximate and mineral content of selected baked products, Cereal Foods World 27(7):308–313, 1982. (Ca, Fe, Mg, P, K, Na, Zn, Cu)

Wenlock, R. W., The nutrient content of UK wheat flours between 1957 and 1980, J. Sci. Food Agric. 33:1310–1318, 1982.

Wenlock, R. W., L. M. Sivell, R. T. King, D. Scuffam and R. A. Wiggins, The nutritional composition of British bread—a nationwide study, J. Sci. Food Agric. 34:1302–1318, 1983.

Zook, E. G., F. E. Greene and E. R. Morris, Nutrient composition of selected wheats and wheat products, VI. Distribution of manganese, copper, nickel, zinc, magnesium, lead, tin, cadmium, chromium, and selenium as determined by atomic absorption spectroscopy and colorimetry, Cereal Chem. 47:720–731, 1970.

Meat

Bentley, R., A fresh look at lamb, J. New Zealand Diet. Assoc. 40:81–88, 1986. (proximates, 10 vitamins, 9 minerals, amino acids, fatty acids)

Marchello, M. J., D. B. Milne and W. D. Slanger, Selected macro and micro minerals in ground beef and longissimus muscle, J. Food Sci. 49:105–106, 1984. (K, Na, P, Ca, Cu, Fe, Mg, Zn, Mn)

Ono, K., B. W. Berry, H. K. Johnson, E. Russek, C. F. Parker, V. R. Cahill and P. G. Althouse, Nutrient composition of lamb of two age groups, J. Food Sci. 49:1233–1257, 1984. (proximates, cholesterol, 8 vitamins, 8 minerals, fatty acids)

Paul, A. A. and D. A. T. Southgate, A study on the composition of retail meat: dissection into lean, separable fat and inedible portion, J. Hum. Nutr. 31:259–272, 1977.

Riss, T. L., P. J. Bechtel, R. M. Forbes, B. P. Klein and F. K. McKeith, Nutrient content of special fed veal ribeyes, J. Food Sci. 48:1868–1871, 1983. (proximates, cholesterol, 8 minerals, 8 vitamins, amino acids)

Young, L. L., G. K. Searcy, L. C. Blankenship, J. Salinsky and D. Hamm, Selected nutrients in ground and mechanically separated veal, J. Food Sci. 48:1576–1578, 1983. (proximates, cholesterol, purines, nucleic acids)

Milk, Cream and Cheese

Ahmed, A. A., Y. L. Awad and F. Fahmy, Studies on some minor constituents of camel's milk, Vet. Med. J. 25:51–56, 1977.

Dang, H. S., D. D. Jaiswal, C. N. Wadhwani, S. Somasunderam and H. Dacosta, Breast feeding: Mo, As, Mn, Zn, Cu concentrations in milk of economically poor Indian tribal and urban women, Sci. Total Environ. 44(2):177–182, 1985.

Department of Health and Social Security, The composition of mature human milk, Report on Health and Social Subjects No. 12, London, HMSO, 1977.

Florence, E., D. J. Knight, J. A. Owen, D. F. Milner and W. M. Harris, Nutrient content of liquid milk as retailed in the United Kingdom, J. Soc. Dairy Technol. 38:121–127, 1985.

Florence, E., D. F. Milner and W. M. Harris, Nutrient composition of dairy products. I. Cheeses, J. Soc. Dairy Technol. 37:13–16, 1984.

Florence, E., D. F. Milner and W. M. Harris, Nutrient composition of dairy products. II. Creams, J. Soc. Dairy Technol. 37:16–18, 1984.

Goldsmith, S. J., R. R. Eitenmiller, R. T. Toledo and H. M. Barnhart, Effects of processing and storage on the water-soluble vitamin content of human milk, J. Food Sci. 48:994–997, 1983. (9 vitamins)

Larson, L. L., S. E. Wallen, F. G. Owen and S. R. Lowry, Relation of age, season, production, and health indices to iodine and beta-carotene concentrations in cow's milk, J. Dairy Sci. 66(12):2557–2562, 1983.

Pennington, J. A. T., D. B. Wilson, B. E. Young, R. D. Johnson and J. E. Vanderveen, Mineral content of market samples of fluid whole milk, J. Am. Diet. Assoc. 87:1036–1042, 1987.

Sawaya, W. N., J. K. Khalil and A. F. Al-Shalhat, Mineral and vitamin content of goat's milk, J. Am. Diet. Assoc. 84(4):433–435, 1984. (Na, K, Ca, Mg, P, Fe, Cu, Zn, Mn, vitamin A, thiamin, riboflavin, vitamin B-6, vitamin B-12, folacin, niacin, pantothenic acid, ascorbic acid)

Sawaya, W. N., J. K. Khalil, A. Al-Shalhat and H. Al-Mohammad, Chemical composition and nutritional quality of camel milk, J. Food Sci. 49:744–747, 1984. (proximates, Ca, Mg, P, Na, K, Fe, Cu, Zn, Mn, 9 vitamins, fatty acids, amino acids)

Scott, K. J. and D. R. Bishop, Nutrient content of milk and milk products: water soluble vitamins in baby milk formulae, J. Dairy Res. 52:521–528, 1985.

Scott, K. J. and D. R. Bishop, The nutrient content of milk

and milk products. Vitamins of the B complex and vitamin C in retail market milk and milk products, J. Soc. Dairy Technol. 39:32–35, 1986.

Scott, K. J., D. R. Bishop, A. Zechalko and J. D. Edwards-Webb, Nutrient content of liquid milk. II. Content of vitamin C, riboflavin, folic acid, thiamin, vitamins B-12 and B-6 in pasteurized milk as delivered to the home and after storage in the domestic refrigerator, J. Dairy Res. 51:51–57, 1984.

Scott, K. J., D. R. Bishop, A. Zechalko, J. D. Edwards-Webb, P. A. Jackson and D. Scuffam, Nutrient content of liquid milk. I. Vitamins A, D-3, C, and of the B complex in pasteurized bulk liquid milk, J. Dairy Res. 51:37–50, 1984.

Nuts and Nut Products

Galvao, L. C. A., A. Lopez and H. L. Williams, Essential mineral elements in peanuts and peanut butter, J. Food Sci. 41: 1305–1307, 1976. (Ca, Cl, K, Mg, Na, P, Co, Cr, Cu, Fe, Mn, Mo, Zn)

Poultry

Mast, M. G., R. M. Leach and J. H. MacNeil, Performance, composition, and quality of broiler chickens fed dried whole eggs, Poultry Sci. 63(10):1940–1945, 1984.

Vegetables

Abdullah, A. and R. E. Baldwin, Mineral and vitamin contents of seeds and sprouts of newly available small-seeded soybeans and market samples of mungbeans, J. Food Sci. 49: 656–657, 1984. (Cu, Fe, Mg, Mn, Mo, P, K, Na, Zn, ascorbic acid, thiamin, riboflavin, niacin)

Bradbury, J. H. and U. Singh, Thiamin, riboflavin, and nicotinic acid contents of tropical root crops from the South Pacific, J. Food Sci. 51:1563–1564, 1986.

Finglas, P. M. and R. M. Faulks, Nutritional composition of UK retail potatoes, both raw and cooked, J. Sci. Food Agric. 35:1437–1356, 1984.

Kadam, S. S. and D. K. Salunkhe, Nutritional composition, processing, and utilization of horse gram and moth bean, CRC Crit. Rev. Food Sci. Nutr. 22(1):1–26, 1985.

Lopez, A., H. L. Williams and F. W. Cooler, Essential elements, cadmium and lead in fresh and canned corn (*Zea mays* L.). J. Food Sci. 50:1760–1761, 1985. (Ca, Cl, Cr, Co, Cu, Fe, Mg, Mn, Mo, Ni, P, K, Si, Na, Sn, Zn)

Sawaya, W. N., A. Al-Shalhat, A. Al-Sogair and M. Al-Mohammad, Chemical composition and nutritive value of truffles of Saudi Arabia, J. Food Sci. 50:450–453, 1985. (moisture, protein, fat, amino acids, ascorbic acid, Ca, Mg, P, Na, K, Fe, Mn, Cu, Zn)

Singh, U., K. C. Jain, R. Jambunathan and D. G. Faris, Nutritional quality of vegetable pigeonpeas (*Cajanus cajan (L.) Mill sp.*): Dry matter accumulation, carbohydrates and proteins, J. Food Sci. 49:799–802, 1984. (dietary fiber, starch, sugars, amino acids)

Singh, U., K. C. Jain, R. Jambunathan and D. G. Faris, Nutritional quality of vegetable pigeonpeas (*Cajanus cajan (L) Mill sp.*): Mineral and trace elements, J. Food Sci. 49: 645–646, 1984. (P, K, Ca, Mg, Zn, Cu, Fe, Mn)

Singh, U., M. S. Kherdekar, D. Sharma and K. B. Saxena,

Cooking quality and chemical composition of some early, medium and late maturing cultivars of pigeon pea (*Cajanus cajan (L.) Mill*)., J. Food Sci. Tech., India 21(6):367–373, 1984.

Slater, G. G., S. Shankman, J. S. Shepherd and R. B. Alfin-Slater, Seasonal variation in the composition of California avocados, J. Agr. Food Chem. 23(2):468–474, 1975. (proximates, fatty acids, 11 vitamins, 17 minerals)

Thomas, B., J. A. Roughan and E. D. Watters, Cobalt, chromium and nickel content of some vegetable foodstuffs, J. Sci. Food Agric. 25:771–776, 1974.

Yao, J. J., L. S. Wei and M. P. Steinberg, Effect of maturity on chemical composition and storage stability of soybeans, J. Am. Oil Chem. Soc. 60(7):1245–1249, 1983.

Miscellaneous Foods

Glass, L. and T. I. Hedrick, Nutritional composition of sweet- and acid-type wheys, II. Vitamin, mineral, and calorie contents, J. Dairy Sci. 60(2):190–196, 1977.

Sarwar, G., B. G. Shah, R. Mongeau and K. Hoppner, Nucleic acid, fiber and nutrient composition of inactive dried food yeast products, J. Food Sci. 50:353–357, 1985. (purine, pyrimidine, nucleic acids, amino acids, dietary fiber, Ca, Mg, P, Na, K, Fe, Zn, Cu, Mn, folacin, pantothenic acid, biotin)

REFERENCES FOR VARIOUS SUBSTANCES IN VARIOUS FOODS

Davies, N. T. and S. Warrington, The phytic acid, mineral, trace element, protein and moisture content of UK Asian immigrant foods, Hum. Nutr. Appl. Nutr. 40(1):49–59, 1986.

Gormican, A., Inorganic elements in foods used in hospital menus, J. Am. Diet. Assoc. 56:397–403, 1970. (P, K, Ca, Mg, Na, Al, Ba, Fe, Sr, B, Cu, Zn, Mn, Cr)

Holak, W., Analysis of foods for lead, cadmium, copper, zinc, arsenic, and selenium using closed-system sample digestion: Collaborative study, J. Assoc. Off. Anal. Chem. 63(3):485–495, 1980. (strained chicken, strained applesauce, reference materials)

Koehler, H. H. and M. M. Hard, Protein, fat, and amino acid content and protein quality of selected pre-prepared foods, J. Am. Diet. Assoc. 82(3):241–245, 1983.

Monro, J. A., W. D. Holloway and J. Lee, Elemental analysis of fruit and vegetables from Tonga, J. Food Sci. 51(2): 522–523, 1986. (Ca, K, Mg, P, S, B, Cu, Fe, Mn, Na, Zn)

Pennington, J. A. T., B. E. Young, D. B. Wilson, R. D. Johnson and J. E. Vanderveen, Mineral content of foods and total diets: The selected minerals in foods survey, 1982–84, J. Am. Diet. Assoc. 86(7):876–891, 1986. (Na, K, Ca, P, Mg, Fe, Zn, Cu, Mn, Se, I)

Tan, S. P., R. W. Wenlock and D. H. Buss, Second supplement to McCance and Widdowson's 'The composition of foods': immigrant foods. London, HMSO, 1985.

Wolnik, K. A., F. L. Fricke, S. G. Capar, G. L. Braude, M. W. Meyer, R. D. Satzger and R. W. Kuennen, Elements in major raw agricultural crops in the United States. 2. Other elements in lettuce, peanuts, potatoes, soybeans, sweet corn, and wheat, J. Agric. Food Chem. 31(6):1244–1249, 1983. (Ca, Cu, Fe, K, Mg, Mn, Mo, Na, Ni, P, Se, Zn)

Index